Advances in Experimental Medicine and Biology

For further volumes:
http://www.springer.com/series/5584

Eric C. Schirmer • Jose I. de las Heras
Editors

Cancer Biology and the Nuclear Envelope

Recent Advances May Elucidate Past Paradoxes

Springer

Editors
Eric C. Schirmer
University of Edinburgh
Wellcome Trust Centre for Cell Biology
Edinburgh, UK

Jose I. de las Heras
University of Edinburgh
Wellcome Trust Centre for Cell Biology
Edinburgh, UK

ISSN 0065-2598 ISSN 2214-8019 (electronic)
ISBN 978-1-4899-8031-1 ISBN 978-1-4899-8032-8 (eBook)
DOI 10.1007/978-1-4899-8032-8
Springer New York Heidelberg Dordrecht London

Library of Congress Control Number: 2014931492

Printed on acid-free paper

Springer is part of Springer Science+Business Media (www.springer.com)

This book is dedicated to the memory of Sir Kenneth and Noreen Murray whose devotion to science has been a major support for research at the University of Edinburgh.

Preface

It used to be said that "All roads lead to Damascus," and this was subsequently changed to "Rome." Today, it might be more appropriate to say "All roads lead to cancer." Half a century of focused modern research efforts have failed to find a "cure" for cancer because of the plethora of causes and mechanisms that can instigate tumorigenesis. Despite these many roads, the resultant tumor cells nonetheless share a handful of characteristics. To proliferate, cancer cells must have reactivated the cell cycle and often cell cycle regulators and signaling pathways that maintain a differentiated state are altered in tumors. Loss of genome integrity may or may not be causative in the progenitor cell, but it clearly becomes a characteristic within the tumor with chromosome translocations, DNA damage, and significant changes in transcriptional profiles all characteristic of pretty much all tumors. Moreover, the degree of metastasis is often correlated with the extent of DNA damage and chromosome translocations. Component cells of metastatic tumors migrate to spread and so cytoskeletal changes that enable cell migration are highly characteristic of more malignant tumors.

Even before any of the above-mentioned characteristics of tumors were identified, it was noted that most tumor cells exhibited changes in the shape and size of the nuclear envelope. Thus in the modern era as soon as the first nuclear envelope proteins were discovered—the nuclear lamins—they became a focus of research. Many correlations between lamin levels and increasing cancer grade were observed, and so lamin levels were added to nuclear size and shape changes in tumor diagnostics and prognostics. However, in some tumor types increased metastasis correlated with increases in certain lamins, while in other tumor types it correlated with decreases in the lamins. Therefore, the nuclear envelope was dropped as a major focus of cancer research.

In recent years, the nuclear envelope has been found to play important roles in cell cycle regulation and signaling, genome organization, the regulation of gene expression, DNA damage repair pathways and genome stability, and cytoskeletal organization, cell mechanical stability, and cell migration—all of the above noted general characteristics of cancer cells. Many recent studies revisiting the nuclear envelope as a player in tumorigenesis and cancer metastasis have found cancer associations through the above-mentioned central mechanisms/characteristics as well as several unexpected links. On this basis alone it is clearly time to make the nuclear envelope a major focus of cancer research. However, there may be an even more compelling reason in recent findings that nuclear envelope protein composition is highly tissue specific. Indeed, with the many general cancer functions already linked to the nuclear envelope this finding could be the Rosetta Stone that explains much of the tissue/tumor type-specific aspects of cancer and the reason that in the early studies certain nuclear envelope characteristics correlated with increased metastasis in one direction or another based on the tumor type.

This volume brings together many different researchers and perspectives covering the historical and current use of the nuclear envelope in cancer diagnosis and grading, clear and potentially relevant functions of the nuclear envelope in cell cycle regulation and signaling, chromatin organization and gene expression, genome stability, nucleocytoplasmic transport, cell mechanical stability and migration, as well as unexpected links between the nuclear envelope and tumorigenesis. We have tried to collect some divergent viewpoints as well as representing both clinical and basic research and both facts and conceptual ideas. Our hope is that this collection will inspire new directions in cancer research as well as a new focus on the nuclear envelope. We now know that the nuclear envelope is as complex a signaling node as the plasma membrane and perhaps the next phrasing of that old quote will be "all roads lead to the nuclear envelope."

Edinburgh, UK

Eric C. Schirmer
Jose I. de las Heras

Acknowledgements

We would like to thank all the authors for their well thought-out and written chapters and particularly thank Fiona Sarne at Springer for inspiring us with the vision for this volume and making the process of editing it relatively pleasant.

Contents

Part I
History and Use of the Nuclear Envelope in Cancer Prognosis

Introduction

Reports of differences in cell morphology in tumor cells go back to at least the mid-1800s, and many consider Sir Lionel Beale at this time to be the true father of cytology, when he described the aberrant morphology of cancer cells in various tumor types and ascribed a diagnostic and prognostic value to nuclear size and shape differences [1, 2]. Over 150 years later, nuclear size and shape are still used extensively in the clinic with clear statistical correlations having been observed in particular tumor types between nuclear size and shape defects and worse clinical outcomes. Eukaryotic cells tend to maintain a roughly constant ratio of nuclear to cell volume, the karyoplasmic ratio [3, 4], and changes in this nuclear size ratio are used as a prognostic indicator for the clinical outcome of various tumor types (e.g., [5, 6]). However, increased malignancy is linked to increased nuclear size for some tumor types, while it is linked to decreased nuclear size for other tumor types [7]. For example, increased nuclear volume is linked to malignancy for invasive meningiomas and bladder carcinoma [8, 9], while smaller nuclear volumes correlated with malignancy for squamous cell carcinoma of the lung [10]. In contrast, greater nuclear shape changes tend to always correlate with increased metastasis.

It would seem intuitive that the nuclear envelope is a nexus for such changes in nuclear size and shape, but this could not even begin to be tested until over 100 years later when the first nuclear envelope proteins were discovered. These were the lamins, among the most abundant proteins in the nucleus besides histones, at ~3 million copies per average mammalian nucleus [11]. There are three lamin genes, A, B1, and B2, and, of these lamin A was strongly reduced in certain cancers (e.g., [12, 13]). The subsequent finding that lamin A only appeared at later stages in differentiation [14] birthed the hypothesis that loss of lamin A reflected a dedifferentiation event in tumorigenesis [15]. However, it was soon noted that, in other tumor types, increases in lamin A expression, instead of decreases, correlated with worse clinical outcomes [16]. Other lamins have also been observed to change levels or phosphorylation state in particular tumor types. For example lamin B1 is reduced in colon carcinomas, colon adenomas, and gastric cancers [17], while lamin B2 is hyperphosphorylated in leukemia [18].

Other nuclear envelope proteins besides lamins may play roles in nuclear size and shape changes in tumors, and these are covered in later sections of this book. Other sections also address the molecular mechanisms behind these changes and other cellular functions influenced by the nuclear envelope that when perturbed can lead to pathogenesis. This first section focuses on the historical and current clinical use of lamin levels, nuclear shape and size changes, and nuclear envelope markers to better detect nuclear shape and size changes in cancer diagnostics and prognostics. The first chapter is more of a short introduction, starting with the work of Professor Müller, Professor Bennett, and Dr Beale in the 1800s, focusing on the long history of using nuclear characteristics in cancer diagnosis and the technological developments that made this possible, and providing an overview of the nuclear envelope as a hub of connections to cancer biology. In the second chapter Jos Broers and Frans Ramaekers of Maastricht University, who have truly led the way for understanding differences in the individual lamin subtypes in different cancer types and tissues, present a beautifully detailed history of the use of expression levels of different lamin subtypes in cancer diagnosis and prognosis, starting right at the time that lamins themselves were discovered. Regulation of apoptosis is also critical in cancer pathology and, as a lamin A mutant intransigent to cleavage delayed apoptosis [19], the role of lamins in apoptosis and its relation to cancer are also discussed. The remaining chapters in this section are contributed by three clinical world leaders who are studying and perfecting the use of the nuclear envelope in cancer diagnosis and prognosis. In the third chapter Andrew (Andy) Fisher from the University of Massachusetts Memorial Medical Center gives a delightful discussion of the value of different characteristics of the nuclear envelope including size, chromothripsis, and various types of shape changes in cancer prognostics. He presents a very insightful view on the appropriate grouping and weighting of these parameters as well as theories on how they reflect the processes of tumorigenesis and malignancy. In the fourth chapter Robert (Bob) Veltri and Christhunesa Christudass of Johns Hopkins Hospital delve into the history of the modern fusion between microscope and computer in developing methods to evaluate nuclear morphometry and applying this to clinical grading of prostate tumors for optimizing treatment. Their chapter brilliantly conveys the practical aspects of quantifying nuclear envelope differences in cancer pathology. The fifth and last chapter in this section by Gianni Bussolati and colleagues from the University of Turin pushes for changes in the methods used for assessing nuclear shape differences. They clearly demonstrate that enormous improvements in resolution are obtained when staining for nuclear envelope markers by immunofluorescence compared to standard approaches of hematoxylin and eosin staining [20]. This new approach enables different thyroid cancers and diseases to be distinguished based on biopsy that could not be before and increases the confidence of clinical grading for breast cancer. Together these chapters provide a solid overview and discussion of the existing methods and future directions in using the nuclear envelope for cancer grading and prognostics.

1. Beale LS (1860) Examination of sputum from a case of cancer of the pharynx and the adjacent parts. Arch Med (Lond) 2:44
2. Beale LS (1854) The microscope in its application to practical medicine. Highley, London
3. Cavalier-Smith T (2005) Economy, speed and size matter: evolutionary forces driving nuclear genome miniaturization and expansion. Ann Bot 95(1):147–175. doi:95/1/147 [pii] 10.1093/aob/mci010
4. Wilson EB (1925) The karyoplasmic ratio. In: The cell in development and heredity, 3rd edn. The Macmillan Company, New York, pp 727–733
5. Ishizuka Y, Oota K, Masubachi K (1972) Practical cytodiagnosis. Lippincott, Philadelphia
6. Weibel ER (1972) The value of stereology in analysing structure and function of cells and organs. J Microsc 95(1):3–13
7. Sorensen FB (1992) Quantitative analysis of nuclear size for objective malignancy grading: a review with emphasis on new, unbiased stereologic methods. Lab Invest 66(1):4–23
8. Fukuzawa S, Hashimura T, Sasaki M, Yamabe H, Yoshida O (1995) Nuclear morphometry for improved prediction of the prognosis of human bladder carcinoma. Cancer 76(10):1790–1796
9. Madsen C, Schroder HD (1996) Stereological estimation of nuclear mean volume in invasive meningiomas. APMIS 104(2):103–107
10. Ladekarl M, Boek-Hansen T, Henrik-Nielsen R, Mouritzen C, Henriques U, Sorensen FB (1995) Objective malignancy grading of squamous cell carcinoma of the lung. Stereologic estimates of mean nuclear size are of prognostic value, independent of clinical stage of disease. Cancer 76(5):797–802
11. Gerace L, Burke B (1988) Functional organization of the nuclear envelope. Annu Rev Cell Biol 4:335–374
12. Kuzmina SN, Buldyaeva TV, Akopov SB, Zbarsky IB (1984) Protein patterns of the nuclear matrix in differently proliferating and malignant cells. Mol Cell Biochem 58(1–2):183–186
13. Prokocimer M, Margalit A, Gruenbaum Y (2006) The nuclear lamina and its proposed roles in tumorigenesis: projection on the hematologic malignancies and future targeted therapy. J Struct Biol 155(2):351–360. doi:S1047-8477(06)00105-5 [pii] 10.1016/j.jsb.2006.02.016
14. Rober RA, Weber K, Osborn M (1989) Differential timing of nuclear lamin A/C expression in the various organs of the mouse embryo and the young animal: a developmental study. Development 105(2):365–378
15. Hass R, Giese G, Meyer G, Hartmann A, Dork T, Kohler L, Resch K, Traub P, Goppelt-Strube M (1990) Differentiation and retrodifferentiation of U937 cells: reversible induction and suppression of intermediate filament protein synthesis. Eur J Cell Biol 51(2):265–271

16. Kaufmann SH, Mabry M, Jasti R, Shaper JH (1991) Differential expression of nuclear envelope lamins A and C in human lung cancer cell lines. Cancer Res 51(2):581–586
17. Moss SF, Krivosheyev V, de Souza A, Chin K, Gaetz HP, Chaudhary N, Worman HJ, Holt PR (1999) Decreased and aberrant nuclear lamin expression in gastrointestinal tract neoplasms. Gut 45(5):723–729
18. Meier R, Muller PR, Hirt A, Leibundgut K, Ridolfi-Luthy A, Wagner HP (1997) Differential phosphorylation of lamin B2 in normal and leukemic cells. Leuk Res 21(9):841–847
19. Rao L, Perez D, White E (1996) Lamin proteolysis facilitates nuclear events during apoptosis. J Cell Biol 135(6 Pt 1):1441–1455
20. Asioli S, Bussolati G (2009) Emerin immunohistochemistry reveals diagnostic features of nuclear membrane arrangement in thyroid lesions. Histopathology 54(5):571–579. doi:HIS3259 [pii] 10.1111/j.1365-2559.2009.03259.x

The Nuclear Envelope and Cancer: A Diagnostic Perspective and Historical Overview

Jose I. de las Heras and Eric C. Schirmer

Abstract Cancer has been diagnosed for millennia, but its cellular nature only began to be understood in the mid-nineteenth century when advances in microscopy allowed detailed specimen observations. It was soon noted that cancer cells often possessed nuclei that were altered in size and/or shape. This became an important criterion for cancer diagnosis that continues to be used today. The mechanisms linking nuclear abnormalities and cancer only started to be understood in the second half of the twentieth century, with the discovery of nuclear lamina composition differences in cancer cells compared to normal cells. The nuclear envelope, rather than providing a mere physical barrier between the genetic material in the nucleus and the cytoplasm, is a very important functional hub for many cellular processes. In this review we give an overview of the links between cancer biology and nuclear envelope, from the early days of microscopy until the present day's understanding of some of the molecular mechanisms behind those links.

Keywords Cytology • Diagnostics • Karyoplasmic ratio • NETs • NPC • Nuclear lamina • Nuclear size

J.I. de las Heras (✉)
Wellcome Trust Centre for Cell Biology, University of Edinburgh, Kings Buildings, Michael Swann Building, Room 5.21, Edinburgh EH9 3JR, UK
e-mail: j.delasheras@ed.ac.uk

E.C. Schirmer
Wellcome Trust Centre for Cell Biology, University of Edinburgh, Kings Buildings, Michael Swann Building, Room 5.22, Edinburgh EH9 3JR, UK
e-mail: e.schirmer@ed.ac.uk

E.C. Schirmer and J.I. de las Heras (eds.), *Cancer Biology and the Nuclear Envelope*, Advances in Experimental Medicine and Biology 773, DOI 10.1007/978-1-4899-8032-8_1,

Abbreviations

H&E Hematoxylin and eosin
NET Nuclear envelope transmembrane protein
NPC Nuclear pore complex

The Nature of Cancer: From Ancient Egypt Until the Early Twentieth Century

We often talk about efforts to cure cancer as if they had only been going on for the past 60 years or so, but several papyri dating from roughly 2000 to 1500 BC indicate that the ancient Egyptians were able to distinguish between benign and malignant tumors and described the surgical removal of tumors, cauterization, and pharmacological as well as magical treatments for the disease [1]. Hippocrates (460–370 BC), father of Western medicine, used the word karkinos (crab) to name the disease that he described as producing hard swellings that were of a noninflammatory nature and had a tendency to spread through the body, causing death. At the time, all diseases were attributed to an imbalance in the body's four elemental humors: blood, phlegm, yellow bile, and black bile. The humoralist theory remained popular until the mid-1800s, when the cellular nature of cancer was identified. The reason for this change in attitude is simply a technical one: the improvement in the microscope's optics allowed much more detailed examination of specimens.

Microscopy was well established and used in biology for nearly 200 years before it became of assistance to cancer biology [2]. However, early microscopes suffered from chromatic and spherical aberrations that made detailed observations difficult. The modern microscope was born when the English physicist Joseph Lister (1786–1869) showed that spherical aberration could be minimized by a careful combination of lenses. He published his work in 1830 [3], and by the 1840s his microscope was used widely around the world. This microscope represented a significant improvement over previous models, bringing down the resolution to about 1 μm. Improved optics and development of differential staining techniques facilitated the examination of cancer cells (as well as from other pathologies) with a degree of detail unimaginable merely decades earlier. It was soon recognized that microscopic study of pathological specimens provided a very useful tool for the diagnosis of diseases, including cancer.

In the early 1890s the German zoologist Theodor Boveri recognized the genetic basis of cancer [4]. Boveri is principally credited with the discovery of chromosome territories, but he made some of the biggest and most significant leaps in cancer theory in history. He postulated that chromosomes were distinct from each other and transmitted heritable traits. He suggested that chromosome mutations could give rise to a cell with the ability to grow without limits and that this cell could pass on this ability to its descendants. He also proposed that there could be checkpoints, tumor-suppressor genes and oncogenes, and that cancers could be caused by radiation, physical or chemical insults, or pathogenic microorganisms.

The Early Observations of Cancer Cells

Although cancer had been diagnosed as a disease and studied for at least four millennia, its diagnosis remained relatively basic, with no significant advancement in understanding until the mid-1800s. Suddenly, improvements in microscopy led to a flurry of activity between the late 1830s and the 1860s that completely changed modern medicine and its attitude to cancer.

The German scientist Johannes Peter Müller (1801–1858) is considered to be the father of medical microscopy and pioneer of clinical cytology. In his 1838 "Über den feineren Bau und die Formen der krankhaften Geschwülste" (which translates as "On the Nature and Structural Characteristics of Cancer, and of Those Morbid Growths Which May Be Confounded with It") he was the first to describe cancer cells in detail and to note how they lose adherence when compared to normal cells [5]. Based on the physical characteristics he observed, such as altered cell morphology, reduced cell adherence, and altered tumor mass rigidity compared to the surrounding tissue, Müller developed criteria to diagnose benign and malignant neoplasms as well as to distinguish between sarcomas (tumors with abundant connective tissue) and carcinomas (tumors with little or no connective tissue). He ran a state-of-the-art laboratory at the Humboldt University in Berlin, with the best microscopes of the day that could resolve down to 1 μm. Many of his assistants became prominent microscopists themselves: these included Friedrich Henle who developed the early germ theory of disease, Robert Koch who founded the field of bacteriology and received the Nobel Prize in Physiology or Medicine in 1905 for his work on bacterial pathogens, Theodor Schwann who developed the cell theory, and Rudolf Virchow who built on Schwann's work and became the father of modern pathology, rejecting the notion of spontaneous generation with his "omnis cellula e cellula" (which can be translated as "every cell comes from another cell") and bringing an end to the humoralist theory of human disease that had been prevalent for the previous 2,000 years.

Müller's monograph in 1838 appears to have had the effect of turning the attention of physicians and scientists sharply on to cancer. In the next few years, several very important scientific articles were published that marked the path for pathological cytology.

Illustrations in scientific journals during most of the nineteenth century consisted generally of drawings carved in wood blocks that were subsequently stained and used to print the illustrations. In the 1840s the French physician Alfred Francois Donne (1801–1878) was the first person to apply photography to microscopy. He invented the photoelectric microscope, which enabled the projection of microscopy images onto a wall. These projections could then be captured as a daguerreotype, an early form of photography. In 1844 he published his "Cours de Microscopie Complementaire des Etudes Medicales," the first atlas of microscopic anatomy, illustrated with numerous photographs [6]. Donne was the first to describe leukemia and show photographs of blood cells from both autopsy specimens and living patients. The following year, in 1845, leukemia was recognized as a blood disorder by the English physician John Hughes Bennett (who had been a student of Donne's), in Edinburgh [7]. Microphotography did not become popular until nearly 50 years later, and despite Bennett's relationship with Donne, his publications only contained relatively basic drawings.

The first detailed and comprehensive description of the altered morphologies of cancer cells, as well as tumor anatomy and the different behavior of cancers in a variety of organs, came from the Irish physician and Edinburgh University graduate Walter Hayle Walshe (1812–1892) in 1846 [8]. His work is also one of the earliest examples of statistical analysis of cancer frequency according to age and gender, looking at lung cancer, which was already by then recognized as one of the most common forms of cancer. Unfortunately, despite the great detail of description, Walshe included no illustrations in his work, thus limiting its impact and utility to train other physicians.

A year later, in 1847, the physician Julius Vogel, a disciple of Müller, published his pioneering book on pathological anatomy [9]. He was one of the first to diagnose cancer using a method that later became known as exfoliative cytology (the microscopic examination of cells that are shed with a gentle scrape from various surfaces of the body, such as the inside of the mouth), rediscovered and brought to the fore by George Papanicolaou 80 years later, in the early twentieth century.

Then, in 1849, Professor Bennett published "On cancerous and cancroid growths" where he described cancers of a variety of organs [10]. In this work Bennett experimented using acetic acid treatment to aid the visualization of specimens, in which he noted cancer cell polymorphism and presence of multinucleated cells as well as cells with an increased number of nucleoli, which we now know to be a reflection of the increased ploidy level that is frequently observed in cancer cells (Fig. 1). This work was published with the publishers advertizing "190 illustrations, copied from nature, and drawn on wood by the author."

In 1851, Hermann Lebert (1813–1878) published a treatise [11] where he described the characteristics of malignant cells, their variation of sizes, and noted the commonly increased size of the nucleus compared to the cytoplasm (later known as the "karyoplasmic ratio" [12]). This is the first description of altered karyoplasmic ratios in cancer cells. Alteration of karyoplasmic ratios is a morphometric criterion still used today in diagnostics, well over 100 years later, and is only now beginning to be understood.

By the early 1850s, barely over a decade after Müller's monograph, the literature on cancer anatomy and pathology had multiplied and commonly included very useful—if still a bit crude—drawings of cancer cells. This was in great part due to the rapid advances in light microscopy that took place in those days. However, the microscopes were not easy to use and without stains to aid visualization, diagnosis remained a difficult and time-consuming task, as Lebert had noted in 1845 [13].

Sir Lionel Smith Beale (1828–1906) was an English physician and microscopist at King's College in London and is now considered the true father of cytology. He learnt from Professor Bennett that some acid or alkali treatments of specimens resulted in differential staining of cells. He further developed the differential staining technique to improve microscopic observations, noticing that active nuclei stain intensely using basic dyes whereas dead cells could be stained with acid dyes. In 1854 Beale published "The microscope and its application to practical medicine" [14].

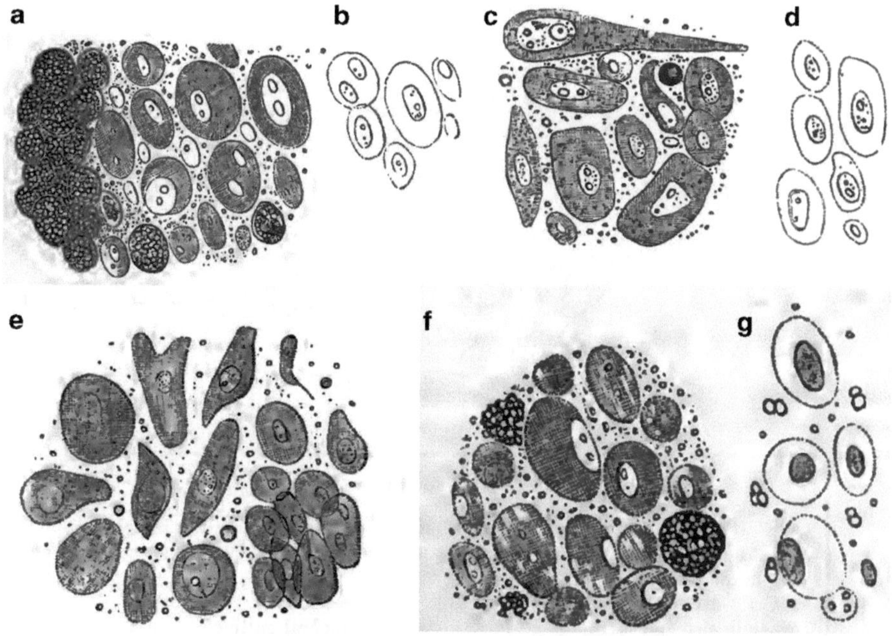

Fig. 1 Cell and nucleus size polymorphism in cancer cells. Adapted from Bennett [10]. (**a**) Cancer cells from a breast tumor, showing cellular and nuclear size polymorphism. (**b**) Same sample as (**a**), after treatment with acetic acid, which renders cytoplasm partially transparent. (**c**) Cells from a recurrent breast tumor, from a different individual than (**a**). (**d**) Same as (**c**), after treatment with acetic acid. (**e**) Uterine cancer cells, with cell and nuclear size and shape polymorphism. (**f**) Cancer cells from a liver tumor. (**g**) Same sample as (**f**), after treatment with acetic acid

In the first part of this volume Beale describes various types of microscopes available at the time and staining techniques that can be used to improve the visualization of clinical specimens. In the second part of the volume he describes a wide range of pathologies, diagnosis, and treatments and includes many illustrations of microscopic observations. In particular, he goes on to describe cancer cells of a variety of tumors, noting as diagnostic features the differences in their cell sizes and shapes, number and sizes of nuclei, and loss of adherence to adjacent cells in the biopsies. He discussed in detail ways in which cancerous cells could be distinguished from benign growths that may have a similar clinical appearance in a variety of tissues (Fig. 2). On the surface, these observations are not very different from those that Müller had noted and published 16 years earlier. What made Beale's work stand out was the quality of his illustrations and descriptions. His drawing abilities coupled to the use of basic specimen preparation and staining techniques meant that he was able to demonstrate with clarity what he saw under the microscope. In 1860, Beale published his now classic illustration of cells from sputum from a patient with pharyngeal cancer [15] (Fig. 3). His drawings were of such quality that a diagnosis can be derived from them today: the prominent cytologist Bernard Naylor stated

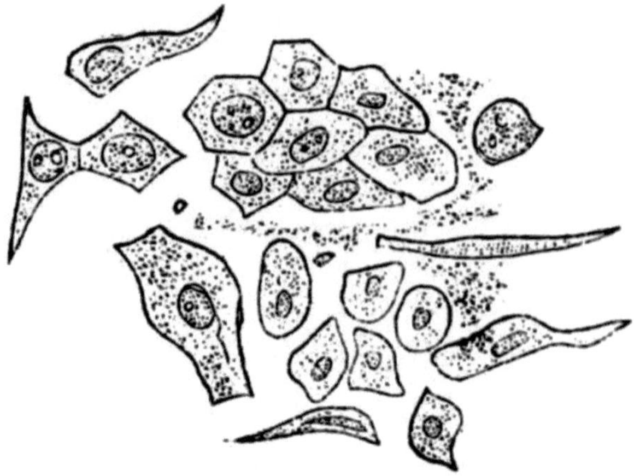

Cancerous.	*Cancroid.*
Cells not connected with the matrix in a regular manner, or forming laminæ.	Cells connected with the matrix, often forming distinct laminæ.
Cells differing much from each other in size and form.	Cells resembling each other in size and general outline.
Cells readily separable from each other.	
Cells not connected together at their margins; their edges seldom forming straight lines.	Cells often cohering by their edges, which generally form straight lines; three or four cells being frequently found united together.
Cells containing several smaller cells in their interior often met with.	Cells usually containing one nucleus.
Nuclei varying much in size and number in different cells.	Nuclei not varying much in size in different cells.
Juice scraped from the cut surface containing many cells floating freely in the fluid, and not connected with each other.	Juice scraped from the cut surface containing small collections of cells, which are often connected with each other.

Fig. 2 Epithelial cancer cells, and diagnostic criteria to distinguish between malignant (cancerous) and benign (cancroid) growths. Adapted from Beale [14]

about this illustration: "It is obvious to us today that the patient died of keratinizing squamous cell carcinoma" [16]. Although Lionel Beale's work was perhaps not the most important in volume, he clarified the importance of cytological diagnosis and effectively communicated this to the rest of the scientific community. One of his most prominent supporters was Rudolf Virchow, whose greatest achievements were in microscopic pathology. Virchow published several major pathology textbooks, including "Cellular Pathology" in 1858 and a three-part series on tumors in 1863–1865 [17–20].

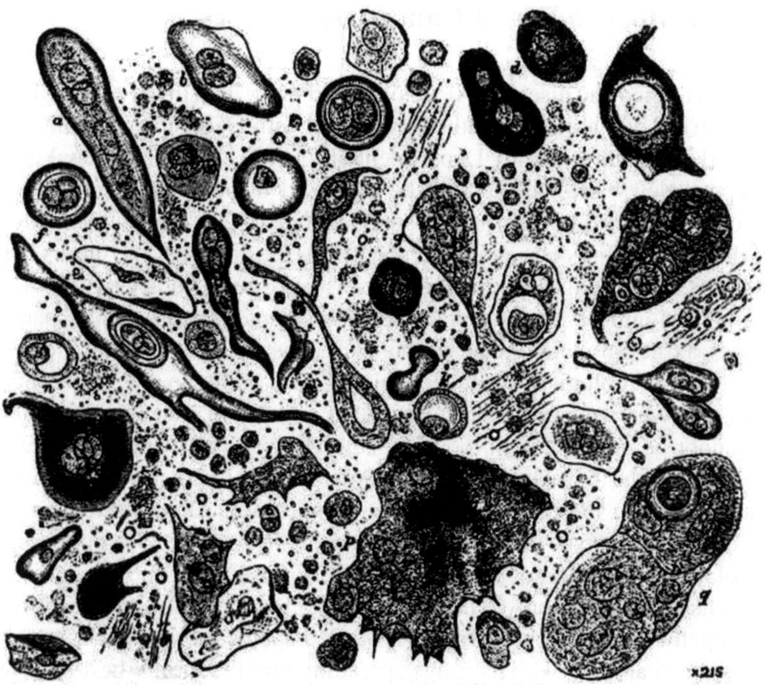

Fig. 3 Cancer cells in a sputum sample from a patient with cancer of the larynx. From Beale [15]

During the rest of the nineteenth century and early twentieth century, the advances in cancer diagnostics were mostly due to the development of specimen treatment techniques, such as formaldehyde fixation of tissues, and of novel stains, which helped physicians all over the world to publish their observations, as well as the development of microphotography. One of the most notable advances in staining was the development of the hematoxylin and eosin (H&E) stain in 1876 by A. Wissowzky [21], which is still in wide use today. With this method the nuclei are overstained dark blue in alum mordanted hematoxylin, followed by destain in dilute acid alcohol and blue color developing in slightly alkaline water. The cytoplasm is then stained orange-pink with eosin. H&E staining remains the gold standard for diagnosis of many cancer types.

The advances in cancer diagnosis developed in the mid-1800s resulted in the general public becoming more aware of cancer as a disease. Moreover, the increasing number of cancer diagnoses resulted in the perception of cancer as a rapidly growing disease and some degree of public fear. That the advances in diagnosis were not coupled with advances in treatment also gave the term cancer and its diagnosis the appearance of a death shroud, as can clearly be observed in the literature of the period. In response to this rising public fear and ignorance concerning cancer special research agencies dedicated to the investigation, education, care, and eradication of cancer were instigated in both the UK and the USA in the early 1900s.

The Early Modern Era of Cancer Diagnostics

Cytology as a scientific discipline developed and flourished in the twentieth century. The modern era of cytological pathology started with George Papanicolaou (1883–1962), working at the Anatomy Department of Cornell University, New York. In his 1928 paper "New cancer diagnosis" he proposed using vaginal smears to detect uterine cancer, using a polychromatic stain technique [22]. Papanicolaou described cancer diagnosis using cells gently scraped from the cervix of the uterus, based on a combination of changes in staining, size, shape, and characteristics of nuclear chromatin, assigning a numeric grade to each sample based on these parameters. This paper, a true hallmark of cancer diagnosis, was not received with much interest initially. Many pathologists were sceptical about the ability to diagnose cancer from scraped cells, when one of the most important features of cancer is tissue invasion, which cannot be inferred from loosened cells. Eleven years later, in 1939, Joseph Hinsey became the new director of the Anatomy Department and together with Henricus Stander, the director of the Gynecology Department, encouraged Papanicolaou to pursue his cancer research full time. The importance of Papanicolaou's work did not go unnoticed the second time, publishing mostly the same results in his commonly referenced 1942 Science article and two more papers written together with Herbert Traut [23–25]. Papanicolaou's smear test became known as the "Pap test" with its usage spreading rapidly during the 1940s, arriving in Europe after the end of World War II and becoming established as a routine check for uterine and cervical cancer. As a result of the establishment of such routine checks, cervical cancer mortality has greatly decreased from being the leading cause to the eighth most common cause of death from cancer in women [26].

Pap staining is not only used for uterine and cervical cytology. It was quickly discovered that it could be used for oral specimens [27, 28], and today it is used for a wide range of specimens, such as urine samples, cerebrospinal fluid, abdominal fluid, synovial and pleural fluid, fine needle aspiration biopsies, and many others.

The reason the Pap staining was such a success is that it retains nuclear detail and definition and cytoplasmic transparency and can indicate cellular differentiation of squamous epithelium. It is a polychromatic staining method that depends on the degree of cell maturity and metabolism, resulting in very detailed and distinct cellular staining. The basic Pap stain is derived from the classic H&E but contains several other ingredients:

1. Hematoxylin: Stains cell nuclei and allows a coarse observation of chromatin compaction.
2. Orange G: Stains keratin effectively. It stains small cells of keratinizing squamous cell carcinoma that may be present in sputum and other samples. The counterstain Orange G is high in alcohol and provides cytoplasmic transparency, enabling clear visualization of overlapping cells.
3. Eosin Y: Stains in pink superficial epithelial squamous cells, nucleoli, cilia, and red blood cells.
4. Acid Green: Stains cytoplasm.
5. Bismarck Brown Y: Stains cartilage and is nowadays often omitted.

The Late Modern Era: Automation and Computer-Assisted Image Analysis

The proper recognition of normal and cancerous cells is fundamental to diagnostic cytopathology, but the morphology of normal cells can vary greatly, depending on the tissue, and this can overlap with features of cancer cells. There is normally a continuum in the tissue variability. Diagnosis becomes critically dependent on both the availability of a marker for "abnormality" and the recognition of what is normal, typically by the eye of a well-trained pathologist.

The cytopathologist Stanley Patten (1924–1997) was one of the pioneers in the field of automation of diagnostic methods using a slit-scan cytofluorometer. Patten's initial interest centered around standardizing morphometric measurements of diagnostic potential to better define pathology and establish reliable and reproducible diagnostic criteria [29, 30]. George Wied (1921–2004), a disciple of Papanicolaou, also worked towards a standardization of cytologic terminology and morphological measurements, using acridine orange-stained material to obtain fluorescence intensity measurements that could be used to objectively calculate sample metrics [31–34]. With Wied and Patten the field of quantitative cytology was born. The morphometry parameters used include nuclear size, karyoplasmic ratio, and nuclear contour shape. Because microscope-based diagnosis is a demanding yet tedious task, the idea of automating screening of cervical smears and other samples soon arose. Wied was very interested in the possibility of automating sample analysis, but in the 1950s and 1960s computers were not yet widely used and were of minimal computing power. Despite that, by the late 1960s Wied had established a program to acquire and process cytological data. In 1970, his TICAS-MLD device was able to analyze cytological samples and produce an output with various cellular parameters that used clinical probability data for diagnosis [35]. As computing power and robotics rapidly increased in subsequent decades, full automation became possible, allowing the analysis of much larger samples for increased statistical power.

Wied and Patten are the pioneers in the field of automated diagnostics.

Today the work they started continues in the exciting research of clinicians such as Dr. Bob Veltri at the Johns Hopkins Hospital in Baltimore, Professor Gianni Bussolati at the University of Turin, and Professor Andy Fischer at the University of Massachusetts. Bob Veltri's team patented and commercialized, in 1996, the first statistical based algorithm to predict prostate cancer postoperative stage based on pretreatment biopsy data and quantitative digital image analysis. Professor Bussolati's laboratory has developed a cell nucleus 3D-reconstruction image analysis system, using the nuclear envelope protein emerin, to greatly aid the diagnosis of papillary thyroid carcinoma and breast cancer. Besides his interest in the molecular aspects of cancer diagnosis, Andy Fischer has invented the Cellient Automated Cell Block System, which automatically recovers small tissue fragments from a specimen container, using an improved microbiopsy needle, and delivers them rapidly to an indexable plane in paraffin for histologic sectioning.

These automated and/or computer-assisted diagnostic protocols outperform standard diagnostic procedures by pathologists in certain situations. It is interesting that the diagnostic parameters employed are still largely morphological and nucleus centric, essentially the same type of features that cytologists have been looking at for the past 160 years.

The Use of Nuclear Morphometry in Cancer Diagnosis

Cytopathologists have long been using nuclear morphology alterations in cancer cells for diagnostic and prognostic purposes. Nuclear size changes, in particular, have a great diagnostic value for many cancer types. Tumor cells were often observed to have enlarged nuclei, although in a few cases the opposite is true and a reduction of nuclear size correlates with a worse prognosis (Table 1).

However, nuclear size observations alone are not enough for a reliable diagnostic. For example, in osteosarcoma a reduction in nuclear size is an indicator for poor prognosis, but only if accompanied by a reduction of the round appearance of the nucleus [53]. In general, cancer is diagnosed by a pathologist using a combination of morphological features. Nuclear size is only one of the nuclear metrics used in cancer diagnosis. There are other visible nuclear changes that the trained eye of the cytopathologist can use to diagnose, classify, and even differentiate between tumor types with different prognoses. Principal among these are the karyoplasmic ratio, nuclear roundness, nuclear envelope smoothness, chromatin distribution as

Table 1 Nuclear size alteration correlates with grade and poorer prognosis in many cancer types

Cancer type	Nuclear size change	References
Breast cancer	+	[36–38]
Male breast cancer	+	[39]
Cervical cancer	+	[40, 41]
Small-cell cervical carcinoma	+	[42]
Colorectal cancer	+	[43]
Epidermal squamous carcinoma	+	[44]
Cutaneous soft tissue sarcoma	+	[45]
Gastric carcinoma	+	[46]
Lung squamous cell carcinoma	–	[47]
Liver cancer	+	[48]
Melanoma	+	[49, 50]
Invasive meningioma	+	[51]
Oral squamous carcinoma	+	[52]
Osteosarcoma	–	[53]
Ovarian cancer	+	[54]
Pancreatic cancer	+	[55]
Prostate adenocarcinoma	+	[56]
Papillary thyroid carcinoma	+	[57]
Urinary bladder carcinoma	+	[58–60]

In most cases, an enlargement of the nucleus is associated with worse prognosis. The "+" symbol denotes nuclear enlargement in cancer, and conversely, the "–" symbol denotes nuclear size reduction visualized with hematoxylin and other stains, and presence of nuclear envelope invaginations and grooves.

Though it is often difficult to pinpoint the original cause of a tumor because of the myriad of changes that occur, one general feature is that faulty control of cellular growth allows a particular "rogue" cell to proliferate in situations where it should not normally proliferate and which often develops the ability to invade surrounding tissue and ultimately migrate—metastasize—to other tissues. The genetics of cancer have been the focus of intense research for the past several decades. Tumor-suppressor genes, a class of genes that restrict cell proliferation, are often mutated or epigenetically silenced in cancer. Oncogenes can be abnormally activated, promoting cellular division. Mutations in checkpoint genes can allow a damaged cell to escape apoptosis and to continue to proliferate. DNA repair pathways can be impaired and promote further mutations and genome instability. However, despite all we have learned about the many mechanisms behind cancer, invariably a cytopathologist still makes the official diagnosis based on microscopic observations of biopsy material that are principally focused on nuclear morphological features.

Why is the nuclear envelope so good at diagnosis and predicting clinical outcomes for cancer? Francis Crick is alleged to have said: "If you can't study function, study structure." There are many structural ways that nuclear shape and size could provide tumor cells with an advantage in cancer.

The fact that very different cancers can arise by a variety of mechanisms and originate in different tissues, yet they tend to share a substantial number of the nuclear abnormalities mentioned earlier, suggests that these structural alterations have a significant functional consequence. The structure of the nuclear envelope is that of a double-membrane system with two completely separate lipid bilayers separated by a relatively uniform luminal space of ~50 nm in human cells. The two membranes are connected at sites where nuclear pore complexes (NPCs) are inserted, which direct the regulated transport of macromolecules in and out of the nucleus. The outer nuclear membrane contains integral proteins that connect it to the cytoskeleton, and in the luminal space these connect to the luminal parts of inner nuclear membrane proteins that in turn connect to the nucleoskeleton and chromatin. The primary structural support to the nucleus comes from the specific lamin nucleoskeleton that underlies the inner nuclear membrane and should be considered distinct from the nuclear matrix that supports chromatin inside the nucleus. Over the past decade or so it has become apparent that cancer cells have reduced stiffness and are strongly influenced by their biomechanical environment (reviewed in [61]). We now know that the nucleoskeleton is interconnected with the cytoskeleton. Thus, these biophysical/structural properties could also be involved in signaling to the nucleus through mechanotransduction, which could be very important in the unique microenvironment of a tumor that is very distinct from that of the surrounding normal tissue. It is also possible that an altered, less rigid, nuclear envelope could confer a significant advantage to metastasizing cells so that they can more easily migrate and invade surrounding tissue. The nucleus is the largest and most rigid of subcellular organelles, so a smaller or a less rigid nucleus would allow cells to squeeze through constrictions smaller than the diameter of their nucleus such as between adjacent cells to escape from the vasculature endothelium or the epithelium surrounding a tissue. Disruption of nucleoskeletal–cytoskeletal connections has profound effects on nuclear positioning, nuclear migration, and cell migration [62, 63]. An advantage of increased nuclear size could be to provide a greater surface area for sequestration of regulatory factors. The lamins and several NETs have been shown to sequester proteins such as the tumor-suppressor retinoblastoma protein [64] and transcriptional regulators involved in tissue differentiation (e.g., Smads [65, 66]). Thus in theory a larger nucleus could sequester more of the tumor-suppressor or other transcription factors important for both cell cycle regulation and differentiation state of a cell.

From Microscopy to Biochemistry

The question "what is different in the nuclear envelope between a normal and a cancer cell?" was addressed initially by means of microscopy observations, but what is different between the nuclear envelopes of cancer and normal cells at a biochemical level?

Professor Ilya B. Zbarsky began to address this question in his laboratory by electrophoretic analysis of the proteins fractionated and extracted in different ways from crude nuclear preparations. In 1964 Zbarsky and co-workers identified a number of differences between the electrophoretic patterns obtained with normal and cancer cells [67]. Over the following decades his laboratory improved extraction procedures, using various nonionic detergents and nucleases to aid the extraction of proteins tightly bound to the nuclear membrane. In the meantime other laboratories specifically studying the nuclear envelope, particularly that of Nobel Laureate Günter Blobel at the Rockefeller University, developed procedures to specifically isolate nuclear envelopes [68]. It had been observed that there was a thick protein layer resistant to most chemical extractions used in biology that underlay the nuclear envelope and had been referred to as the fibrous layer or the nuclear lamina. From these studies with isolated nuclear envelopes they found that the most abundant proteins by far, almost certainly those of this lamina layer, were three polypeptides of around 65–70 kDa that were named lamins and corresponded to lamin A, lamin B1, and lamin B2 [68]. This enabled the Zbarsky laboratory in 1984 to identify lamins as the most prominent bands changing when comparing electrophoretic profiles of rat hepatoma against quiescent and regenerating normal liver cells. Furthermore, they found that proliferating cells showed an increase in lamin B and reduction of lamins A/C compared to non-proliferating cells [69].

Despite the biochemical identification of lamins in the mid-1970s, they were not known to be relatives of cytoskeletal proteins until a decade later. In 1984 Bob Goldman's laboratory isolated lamins from cultured cells and characterized them as keratin-like proteins, but did not himself realize that they were from the protein polymer underlying the inner nuclear membrane [70]. Finally in 1986 Frank McKeon, Marc Kirschner, and Daniel Caput [71] and Daniel Fisher, Nilabh Chaudhary, and Gunter Blobel [72] separately identified the lamins as intermediate filament proteins. As such, the lamins have short N-terminal head domains (~33 amino acids) followed by a long rod domain (~350 amino acids) that homodimerizes to form four separate coiled coils separated by linkers for a linear length of ~52 nm followed by a large globular and variable C-terminal domain. The homodimers assemble into strands by head-to-tail interactions, and these strands then layer in an antiparallel fashion until there are 32 molecules in cross section to generate 10 nm wide filaments [73]. This assembly gives the lamins and other intermediate filaments unique properties compared to the other cytoskeletal proteins. Microtubules and actin filaments are built like stacked cinder blocks in a wall, whereas intermediate filaments are more like the entwined fibers of a rope, yet they are more tensile as the fibers can potentially move relative to one another—thus, it is not surprising that intermediate filaments are the primary components of spider's webs. Accordingly, actin filaments and microtubules will break under compression or stretching forces that leave intermediate filaments undamaged [74]. These characteristics are more important as the lamins are the only one of the three major cytoskeletal proteins giving structure to the nuclear envelope. However, even among the different lamin subtypes there are large differences in their contributions to mechanical stability. Lamin A was found to exhibit stronger binding in assembly assays compared to lamin B1, and lamin B2 was much weaker than both [75]. Correspondingly, lamin A has been found to be the most critical for mechanical stability [76]. Thus, though it provides the primary structural support for the nucleus, the nuclear envelope can nonetheless bend considerably in a migrating cell invading tissues and more so if lamin A is absent. This observation is more prescient in light of the fact that the most common observation with lamin levels in tumors is that lamin A is reduced, linking lamin abnormalities to the morphometric parameters used by cytologists.

While lamins initially received a great deal of attention, there are many other proteins in the nuclear envelope. The NPCs are large structures of >60 MDa in mammals containing around 30 different proteins in multiple copies (reviewed in [77]), and an average mammalian nucleus contains 2,000–3,000 NPCs. In addition to the NPCs, both outer and inner nuclear membranes contain a host of integral transmembrane proteins called NETs (for nuclear envelope transmembrane proteins). Just a decade ago, only a handful of NETs were known; however, in 2003, 67 novel NETs were identified in the laboratory of Larry Gerace by Eric Schirmer and colleagues [78]. A large proportion of the NETs were largely uncharacterized proteins of unknown function, with many of them exhibiting a marked tissue specificity in their expression. Today, close to 1,000 NETs have been identified [79–81], and a recent study comparing the nuclear envelope proteome of liver, muscle, and white blood cells showed that up to 60 % of the NETs may be preferentially expressed in a subset of tissues [81].

The tissue specificity of many NETs may contribute to the tissue-specific pathologies that occur with a set of nuclear envelope-linked diseases termed laminopathies. Many of these disorders manifest in a restricted number of tissues. For example, defects in the NETs emerin and the nesprins SYNE1 and SYNE2 as well as in lamins may result in Emery–Dreifuss muscular dystrophy. Intriguingly, different mutations in the *LMNA* gene (which encodes lamins A and C) can result in a variety of completely distinct diseases, each with different tissue-specific pathologies that can affect heart (dilated cardiomyopathy), motor and sensory nerves (Charcot–Marie–Tooth disease), skeletal muscle (Emery–Dreifuss muscular dystrophy), fat (familial partial lipodystrophy), or skin (restrictive dermopathy). Lamin A mutations can also be associated with various forms of premature ageing, such as Werner's syndrome and Hutchinson–Gilford Progeria syndrome. How can mutations in a single ubiquitously expressed protein give rise to disease in some tissues and not others? The simplest answer would be through interaction with other factors that are tissue specific, a role for which many of these newly identified NETs stand out as good candidates.

These tissue-specific NETs could also contribute to the tissue-specific nuclear characteristics of many tumor types. In addition to the unexpected degree of tissue specificity present in the nuclear envelope proteome, NETs and lamins are being found to have functions in a variety of cellular processes, many of which can be linked to tumorigenesis (Fig. 4). Proteins of the nuclear envelope participate in cell cycle regulation, mitosis, apoptosis, DNA repair, ageing, nuclear architecture, signaling, chromatin organization, gene expression regulation, and cell migration. All these various functions are critical for processes of tumorigenesis, tumor growth, and metastasis (reviewed in [82, 83]).

We have recently investigated the gene expression profiles of nuclear envelope proteins in a microarray of tumor and normal samples from nine tissues available at the BioGPS database [82]. The microarrays contained probes for lamins A, B1, and B2 and for 29 NETs that had been verified by our lab and others [78–80, 84–94]. Most of the genes showed small and/or inconsistent levels of misregulation between and within tissues, but other genes showed some general tendencies. For instance, *LMNB1*, *LMNB2*, and *NUP210* were generally upregulated, and *METTL7A*, *SYNE1*, and *SYNE2* were generally downregulated (Fig. 5a). These tendencies were not absolute. *LMNB1* and *LMNB2* were not upregulated in prostate tumors, and in kidney tumors only *LMNB2* was upregulated. Additionally, we observed that in most gastrointestinal tumors *METTL7A* was upregulated rather than downregulated (de las Heras and Schirmer, unpublished results). Different tissues express lamins with subtype ratios that are characteristic of each tissue [95]. This coupled with the marked tissue-restricted patterns of NET expression may account for the tissue variability in the lamin and NET misregulation observed between tumors and suggests that some of these expression patterns may be exploited for diagnostic purposes. Some NETs show a particular potential to be used as markers for particular tumor types, such as *LPCAT3/MBOAT5* among a few others. *LPCAT3* does not show significantly consistent misregulation in eight of the nine tumor types studied but appears to be strongly upregulated in all of the ovarian cancer samples studied (Fig. 5b). We have also observed that some NETs were only strongly misregulated in a subset of tumors of only one type of cancer, such as *SLC22A24*, *NCLN*, and *FAM105A*, which were all upregulated in a subset of breast tumors (de las Heras and Schirmer, unpublished results). These differences may additionally reflect differences in tumor subtype or grade, but the BioGPS data did not contain enough information about the tumor samples to explore this possibility.

One area of study that is already showing translational promise is the targeting of nuclear import/export of proteins and RNAs through the NPC. Nucleocytoplasmic

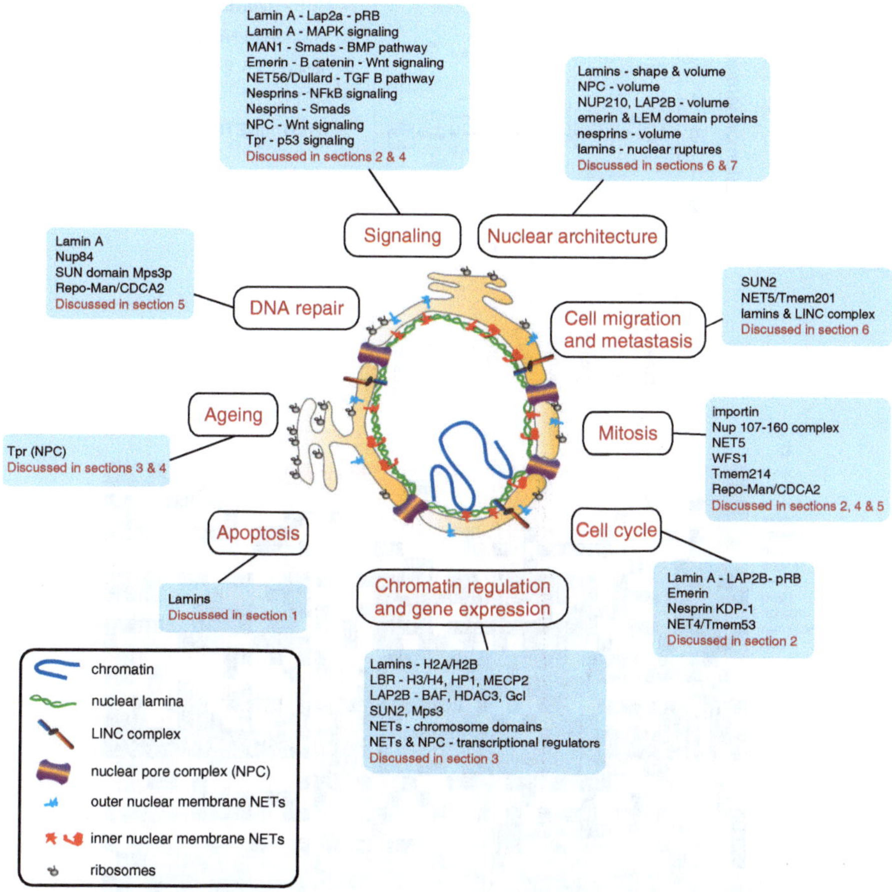

Fig. 4 Nuclear envelope functions with cancer links. The nuclear envelope comprises a double-membrane system studded with nuclear pore complexes and an underlying layer of intermediate filaments: the nuclear lamina. The nuclear envelope is connected to the cytoskeleton on the one side and chromatin on the other and acts as a powerful signaling node including pathways that are very relevant to cancer, such as Wnt and MAPK signaling. In addition, the nuclear envelope has been shown to play a role in many other functions that are relevant to cancer, such as control of nuclear architecture, cell migration, DNA repair, ageing, apoptosis, mitosis, and cell cycle regulation as well as genome organization and regulation of gene expression

transport is essential for cell growth and is often upregulated in tumors. Accordingly, the key nuclear export protein exportin 1 (XPO1/CRM1) has been found to be expressed at abnormally high levels in a number of cancers, and its inhibition promoted apoptosis and cell cycle arrest in cancer cells in vitro [96–99]. Clinical trials with initially promising results are currently under way using XPO1 inhibition in Philadelphia chromosome-positive (Ph+) leukemias, which are refractory to tyrosine kinase inhibitor therapy but appear to respond to an XPO1 inhibitor by triggering apoptosis of leukemic but not normal CD34+ progenitors [99].

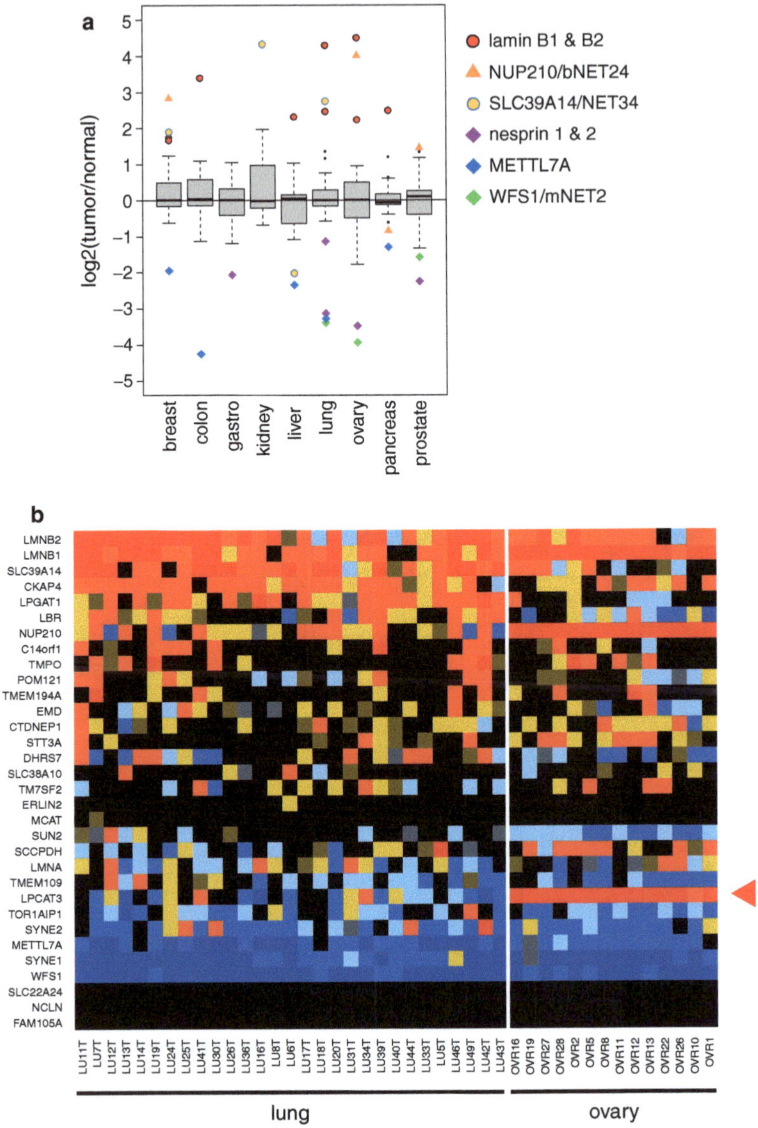

Fig. 5 Many nuclear envelope proteins are misregulated in tumors. (**a**) *Boxplot* showing the distribution of log2(tumor/normal) microarray signals for 29 nuclear envelope genes in nine tissues. The majority of the genes do not show a clear general misregulation in most tumors, but the genes that are most strongly misregulated are generally the same. Lamins B1 and B2 (*LMNB1* and *LMNB2*) and the nucleoporin *NUP210* are usually upregulated in tumors, while the protein methyltransferase *METTL7A* and nesprins *SYNE1* and *SYNE2* are almost always downregulated. However, some NETs, such as *WFS1*, are strongly downregulated in some tumors but not others, while *SLC39A14/NET34* is strongly upregulated in lung, kidney, and breast tumors and downregulated in liver cancer. (**b**) Heatmap illustrating the expression of 29 nuclear envelope genes in individual lung and ovary cancer patients, compared to their normal counterparts. A gradient of *reds* and *blues* indicate relative levels of up- and downregulation, respectively. The overall gene expression pattern is reasonably similar in lung and ovary patients; however, the tissue-specific NET *LPCAT3* (*red arrowhead*), which is normally expressed in the majority of normal tissues but not in ovary, is strongly upregulated in all ovarian tumors and downregulated in most lung tumors. Reproduced with permission from [82]

Concluding Remark

The more we learn about the nuclear envelope and its component proteins, the more it becomes apparent that the nuclear envelope, rather than representing an inert barrier between the cytoplasm and the nucleus, is at the center of many central cellular functions and processes, many of which have direct relevance to cancer biology. Over the past few years, the nuclear envelope has been shown to contain hundreds of NETs that are poorly characterized and of unknown function, many of which are altered in expression in various cancers. Many NETs showed altered expression patterns in cancer that suggest correlations with tissue and tumor grade. Together with the many clear links between lamins and NPC proteins and various cancers, this indicates that the nuclear envelope represents a novel, largely untapped, and potentially huge source for diagnostic and prognostic markers as well as for therapeutic intervention.

Acknowledgments Work in the Schirmer lab is supported by Wellcome Trust Senior Research Fellowship 095209.

References

1. David AR, Zimmerman MR (2010) Cancer: an old disease, a new disease or something in between? Nat Rev Cancer 10(10):728–733. doi:10.1038/Nrc2914
2. Jay V (2002) Antony van Leeuwenhoek. Arch Pathol Lab Med 126(6):658–659
3. Lister JJ (1830) On some properties in achromatic object-glasses applicable to the improvement of the microscope. Phil Trans R Soc Lond 120:187–200
4. Boveri T (2008) Concerning the origin of malignant tumours by Theodor Boveri. Translated and annotated by Henry Harris. J Cell Sci 121(Suppl 1):1–84. doi:10.1242/jcs.025742
5. Müller J (1838) Über den feineren Bau und die Formen der krankhaften Geschwülste. (On the nature and structural characteristics of cancer, and of those morbid growths which may be confounded with it) (trans: West C). G. Reimer, Berlin
6. Donne A (1844) Cours de Microscopie Complementaire des Etudes Medicales. Chez J.B Bailliere, Paris
7. Bennett JH (1845) Case 2, case of hypertrophy of the spleen and liver, in which death took place from suppuration of the blood. Edinb Med Surg J 64:413–423
8. Walshe WH (1846) The nature and treatment of cancer. Taylor and Walton, London
9. Vogel J (1847) The pathological anatomy of the human body. (The pathological anatomy of the human body) (trans: Day GEC, L.M.). H. Bailliere, London
10. Bennett JH (1849) On cancerous and cancroid growths. Sutherland & Knox, Edinburgh
11. Lebert H (1851) Traité Pratique des Maladies Cancéreuses et des Affections Curables Confondues aver le Cancer. Bailliere, Paris
12. Wilson EB (1925) The karyoplasmic ratio. In: Wilson EB (ed) The cell in development and heredity, 3rd edn. The Macmillan Company, New York, pp 727–733
13. Lebert H (1845) Physiologie pathologique ou récherches clinique, experimentales et microscopiques. Bailliere, Paris
14. Beale LS (1854) The microscope and its application to clinical medicine. Highley, London
15. Beale LS (1860) Examination of sputum from a case of cancer of the pharynx and the adjacent parts. Arch Med Lond 2:44

16. Naylor B (2000) The century for cytopathology. Acta Cytol 44(5):709–725
17. Virchow R (1863) Die Krankhaften Geschwülste, vol 1. Hirschwald, Berlin
18. Virchow R (1865) Die Krankhaften Geschwülste, vol 2. Hirschwald, Berlin
19. Virchow R (1865) Die Krankhaften Geschwülste, vol 3. Hirschwald, Berlin
20. Virchow R (1858) Die Cellularpathologie in ihrer Begrundung auf physiologische und pathologische Gewebelehre. (Cellular pathology as based upon physiological and pathological histology) (trans: Chance F). Hirschwald, Berlin
21. Wissowzky A (1876) Ueber das Eosin als reagenz auf Hämoglobin und die Bildung von Blutgefässen und Blutkörperchen bei Säugetier und Hühnerembryonen. Arch Mikrosk Anat 13:479–496
22. Papanicolaou GN (1928) New cancer diagnosis. Proceedings of the 3rd race betterment conference. RBF, Battle Creek, MI
23. Papanicolaou GN (1942) A New Procedure for Staining Vaginal Smears. Science 95(2469):438–439. doi:10.1126/science.95.2469.438
24. Papanicolaou GN, Traut HF (1941) The diagnostic value of vaginal smears in carcinoma of the uterus. Am J Obstet Gynecol 42:193–206
25. Traut HF, Papanicolaou GN (1943) Cancer of the uterus: the vaginal smear in its diagnosis. Cal West Med 59(2):121–122
26. Fu YS, Reagan JW, Benington JL (1989) Pathology of the uterine cervix, vagina, and vulva. W.B. Saunders, Philadelphia
27. Morrison LF, Hopp ES, Wu R (1949) Diagnosis of malignancy of the nasopharynx; cytological studies by the smear technic. Ann Otol Rhinol Laryngol 58(1):18–32
28. Montgomery PW, Vonhaam E (1951) A study of the exfoliative cytology in patients with carcinoma of the oral mucosa. J Dent Res 30(3):308–313. doi:10.1177/00220345510300030201
29. Patten SF, Reagan JW, Obenauf M (1959) An analytical approach to the diagnosis of endometrial adenocarcinoma in cytologic material. Trans Intersoc Cytol Council 7:153–156
30. Reagan JW, Patten SF (1961) An analytical study of the cellular changes in carcinoma in situ, squamous cell cancer and adenocarcinoma of the uterine cervix. Clin Obstet Gynecol 4: 1047–1127
31. Wied GL, Meier P, Clark LM (1964) An electronic data processing program for cytologic screening projects for uterine carcinoma. Acta Cytol 8(6):385–397
32. Wied GL, Messina AM, Manglano JI, Meier P, Blough RR (1964) Fluorospectrophotometric analyses of endometrial and endocervical epithelial cells. Acta Cytol 8(6):408–415
33. Wied GL, Bahr GF, Oldfield DG, Bartels PH (1968) Computer-assisted identification of cells from uterine adenocarcinoma. A clinical feasibility study with Ticas. 1. Measurements at wavelength 530 Nm. Acta Cytol 12(5):357–370
34. Bartels PH, Wied GL (1997) Automated screening for cervical cancer: Diagnostic decision procedures. Acta Cytol 41(1):6–10
35. Bartels PH, Bahr GF, Calhoun DW, Wied GL (1970) Cell recognition by neighborhood grouping techniques in Ticas. Acta Cytol 14(6):313–324
36. Abdalla FB, Markus R, Buhmeida A, Boder J, Syrjanen K, Collan Y (2012) Estrogen receptor, progesterone receptor, and nuclear size features in female breast cancer in Libya: correlation with clinical features and survival. Anticancer Res 32(8):3485–3493
37. Nandakumar V, Kelbauskas L, Hernandez KF, Lintecum KM, Senechal P, Bussey KJ, Davies PC, Johnson RH, Meldrum DR (2012) Isotropic 3D nuclear morphometry of normal, fibrocystic and malignant breast epithelial cells reveals new structural alterations. PLoS One 7(1):e29230. doi:10.1371/journal.pone.0029230
38. Tan PH, Goh BB, Chiang G, Bay BH (2001) Correlation of nuclear morphometry with pathologic parameters in ductal carcinoma in situ of the breast. Mod Pathol 14(10):937–941. doi:10.1038/modpathol.3880415
39. Veta M, Kornegoor R, Huisman A, Verschuur-Maes AH, Viergever MA, Pluim JP, van Diest PJ (2012) Prognostic value of automatically extracted nuclear morphometric features in whole slide images of male breast cancer. Mod Pathol 25(12):1559–1565. doi:10.1038/modpathol.2012.126

40. Saad RS, Kanbour-Shakir A, Lu E, Modery J, Kanbour A (2006) Cytomorphologic analysis and histological correlation of high-grade squamous intraepithelial lesions in postmenopausal women. Diagn Cytopathol 34(7):467–471. doi:10.1002/dc.20475
41. Slater DN, Rice S, Stewart R, Melling SE, Hewer EM, Smith JH (2005) Proposed Sheffield quantitative criteria in cervical cytology to assist the grading of squamous cell dyskaryosis, as the British Society for Clinical Cytology definitions require amendment. Cytopathology 16(4):179–192. doi:10.1111/j.1365-2303.2005.00271.x
42. Giorgadze T, Kanhere R, Pang C, Ganote C, Miller LE, Tabaczka P, Brown E, Husain M (2012) Small cell carcinoma of the cervix in liquid-based Pap test: utilization of split-sample immunocytochemical and molecular analysis. Diagn Cytopathol 40(3):214–219. doi:10.1002/dc.21542
43. Eynard HG, Soria EA, Cuestas E, Rovasio RA, Eynard AR (2009) Assessment of colorectal cancer prognosis through nuclear morphometry. J Surg Res 154(2):345–348. doi:10.1016/j.jss.2008.06.022
44. Malhotra S, Kazlouskaya V, Andres C, Gui J, Elston D (2013) Diagnostic cellular abnormalities in neoplastic and non-neoplastic lesions of the epidermis: a morphological and statistical study. J Cutan Pathol 40(4):371–378. doi:10.1111/cup.12090
45. Meachem MD, Burgess HJ, Davies JL, Kidney BA (2012) Utility of nuclear morphometry in the cytologic evaluation of canine cutaneous soft tissue sarcomas. J Vet Diagn Invest 24(3):525–530. doi:10.1177/1040638712440988
46. Ikeguchi M, Oka S, Saito H, Kondo A, Tsujitani S, Maeta M, Kaibara N (1999) Computerized nuclear morphometry: a new morphologic assessment for advanced gastric adenocarcinoma. Ann Surg 229(1):55–61
47. Ladekarl M, Baek-Hansen T, Henrik-Nielsen R, Mouritzen C, Henriques U, Sorensen FB (1995) Objective malignancy grading of squamous cell carcinoma of the lung. Stereologic estimates of mean nuclear size are of prognostic value, independent of clinical stage of disease. Cancer 76(5):797–802
48. Yan J, Zhuo S, Chen G, Wu X, Zhou D, Xie S, Jiang J, Ying M, Jia F, Chen J, Zhou J (2012) Preclinical study of using multiphoton microscopy to diagnose liver cancer and differentiate benign and malignant liver lesions. J Biomed Opt 17(2):026004. doi:10.1117/1.JBO.17.2.026004
49. Na YR, Seok SH, Kim DJ, Han JH, Kim TH, Jung H, Lee BH, Park JH (2009) Isolation and characterization of spheroid cells from human malignant melanoma cell line WM-266-4. Tumour Biol 30(5–6):300–309. doi:10.1159/000261073
50. Mossbacher U, Knollmayer S, Binder M, Steiner A, Wolff K, Pehamberger H (1996) Increased nuclear volume in metastasizing "thick" melanomas. J Invest Dermatol 106(3):437–440
51. Madsen C, Schroder HD (1996) Stereological estimation of nuclear mean volume in invasive meningiomas. APMIS 104(2):103–107
52. Natarajan S, Mahajan S, Boaz K, George T (2010) Prediction of lymph node metastases by preoperative nuclear morphometry in oral squamous cell carcinoma: a comparative image analysis study. Indian J Cancer 47(4):406–411. doi:10.4103/0019-509X.73580
53. de Andrea CE, Petrilli AS, Jesus-Garcia R, Bleggi-Torres LF, Alves MT (2011) Large and round tumor nuclei in osteosarcoma: good clinical outcome. Int J Clin Exp Pathol 4(2):169–174
54. Zeimet AG, Fiegl H, Goebel G, Kopp F, Allasia C, Reimer D, Steppan I, Mueller-Holzner E, Ehrlich M, Marth C (2011) DNA ploidy, nuclear size, proliferation index and DNA-hypomethylation in ovarian cancer. Gynecol Oncol 121(1):24–31. doi:10.1016/j.ygyno.2010.12.332
55. Taira T, Kawahara A, Yamaguchi T, Abe H, Ishida Y, Okabe Y, Naito Y, Yano H, Kage M (2012) Morphometric image analysis of pancreatic disease by ThinPrep liquid-based cytology. Diagn Cytopathol 40(11):970–975. doi:10.1002/dc.21704
56. Rashid F, Ul Haque A (2011) Frequencies of different nuclear morphological features in prostate adenocarcinoma. Ann Diagn Pathol 15(6):414–421. doi:10.1016/j.anndiagpath.2011.06.002
57. Shih SR, Chang YC, Li HY, Liau JY, Lee CY, Chen CM, Chang TC (2013) Preoperative prediction of papillary thyroid carcinoma prognosis with the assistance of computerized

morphometry of cytology samples obtained by fine-needle aspiration: preliminary report. Head Neck 35(1):28–34. doi:10.1002/hed.22909
58. Helander K, Hofer PA, Holmberg G (1984) Karyometric investigations on urinary bladder carcinoma, correlated to histopathological grading. Virchows Arch A Pathol Anat Histopathol 403(2):117–125
59. van Velthoven R, Petein M, Oosterlinck WJ, Zandona C, Zlotta A, Van der Meijden AP, Pasteels JL, Roels H, Schulman C, Kiss R (1995) Image cytometry determination of ploidy level, proliferative activity, and nuclear size in a series of 314 transitional bladder cell carcinomas. Hum Pathol 26(1):3–11
60. Fukuzawa S, Hashimura T, Sasaki M, Yamabe H, Yoshida O (1995) Nuclear morphometry for improved prediction of the prognosis of human bladder carcinoma. Cancer 76(10):1790–1796
61. Suresh S (2007) Biomechanics and biophysics of cancer cells. Acta Biomater 3(4):413–438. doi:10.1016/j.actbio.2007.04.002
62. Friedl P, Wolf K, Lammerding J (2011) Nuclear mechanics during cell migration. Curr Opin Cell Biol 23(1):55–64. doi:10.1016/j.ceb.2010.10.015
63. Wolf K, Te Lindert M, Krause M, Alexander S, Te Riet J, Willis AL, Hoffman RM, Figdor CG, Weiss SJ, Friedl P (2013) Physical limits of cell migration: control by ECM space and nuclear deformation and tuning by proteolysis and traction force. J Cell Biol 201(7):1069–1084. doi:10.1083/jcb.201210152
64. Ozaki T, Saijo M, Murakami K, Enomoto H, Taya Y, Sakiyama S (1994) Complex formation between lamin A and the retinoblastoma gene product: identification of the domain on lamin A required for its interaction. Oncogene 9(9):2649–2653
65. Osada S, Ohmori SY, Taira M (2003) XMAN1, an inner nuclear membrane protein, antagonizes BMP signaling by interacting with Smad1 in Xenopus embryos. Development 130(9):1783–1794
66. Pan D, Estevez-Salmeron LD, Stroschein SL, Zhu X, He J, Zhou S, Luo K (2005) The integral inner nuclear membrane protein MAN1 physically interacts with the R-Smad proteins to repress signaling by the transforming growth factor-{beta} superfamily of cytokines. J Biol Chem 280(16):15992–16001. doi:10.1074/jbc.M411234200
67. Zbarsky IB, Samarina OP, Yermolayeva LP (1964) Nuclear Proteins and Their Biosynthesis in the Tumour Cells. Acta Unio Int Contra Cancrum 20:937–940
68. Aaronson RP, Blobel G (1975) Isolation of nuclear pore complexes in association with a lamina. Proc Natl Acad Sci U S A 72(3):1007–1011
69. Kuzmina SN, Buldyaeva TV, Akopov SB, Zbarsky IB (1984) Protein patterns of the nuclear matrix in differently proliferating and malignant cells. Mol Cell Biochem 58(1–2):183–186
70. Zackroff RV, Goldman AE, Jones JC, Steinert PM, Goldman RD (1984) Isolation and characterization of keratin-like proteins from cultured cells with fibroblastic morphology. J Cell Biol 98(4):1231–1237
71. McKeon FD, Kirschner MW, Caput D (1986) Homologies in both primary and secondary structure between nuclear envelope and intermediate filament proteins. Nature 319(6053):463–468. doi:10.1038/319463a0
72. Fisher DZ, Chaudhary N, Blobel G (1986) cDNA sequencing of nuclear lamins A and C reveals primary and secondary structural homology to intermediate filament proteins. Proc Natl Acad Sci U S A 83(17):6450–6454
73. Stuurman N, Heins S, Aebi U (1998) Nuclear lamins: their structure, assembly, and interactions. J Struct Biol 122(1–2):42–66. doi:10.1006/jsbi.1998.3987
74. Janmey PA, Euteneuer U, Traub P, Schliwa M (1991) Viscoelastic properties of vimentin compared with other filamentous biopolymer networks. J Cell Biol 113(1):155–160
75. Schirmer EC, Gerace L (2004) The stability of the nuclear lamina polymer changes with the composition of lamin subtypes according to their individual binding strengths. J Biol Chem 279(41):42811–42817. doi:10.1074/jbc.M407705200
76. Lammerding J, Fong LG, Ji JY, Reue K, Stewart CL, Young SG, Lee RT (2006) Lamins A and C but not lamin B1 regulate nuclear mechanics. J Biol Chem 281(35):25768–25780. doi:10.1074/jbc.M513511200

77. Suntharalingam M, Wente SR (2003) Peering through the pore: nuclear pore complex structure, assembly, and function. Dev Cell 4(6):775–789
78. Schirmer EC, Florens L, Guan T, Yates JR 3rd, Gerace L (2003) Nuclear membrane proteins with potential disease links found by subtractive proteomics. Science 301(5638):1380–1382. doi:10.1126/science.1088176
79. Wilkie GS, Korfali N, Swanson SK, Malik P, Srsen V, Batrakou DG, Heras J, Zuleger N, Kerr AR, Florens L, Schirmer EC (2011) Several novel nuclear envelope transmembrane proteins identified in skeletal muscle have cytoskeletal associations. Mol Cell Proteomics 10(1):M110 003129. doi:10.1074/mcp.M110.003129
80. Korfali N, Wilkie GS, Swanson SK, Srsen V, Batrakou DG, Fairley EA, Malik P, Zuleger N, Goncharevich A, de Las HJ, Kelly DA, Kerr AR, Florens L, Schirmer EC (2010) The leukocyte nuclear envelope proteome varies with cell activation and contains novel transmembrane proteins that affect genome architecture. Mol Cell Proteomics 9(12):2571–2585. doi:10.1074/mcp.M110.002915
81. Korfali N, Wilkie GS, Swanson SK, Srsen V, de Las HJ, Batrakou DG, Malik P, Zuleger N, Kerr AR, Florens L, Schirmer EC (2012) The nuclear envelope proteome differs notably between tissues. Nucleus 3(6):552–564. doi:10.4161/nucl.22257
82. de Las Heras JI, Batrakou DG, Schirmer EC (2013) Cancer biology and the nuclear envelope: a convoluted relationship. Semin Cancer Biol 23(2):125–137. doi:10.1016/j.semcancer.2012.01.008
83. Chow KH, Factor RE, Ullman KS (2012) The nuclear envelope environment and its cancer connections. Nat Rev Cancer 12(3):196–209. doi:10.1038/nrc3219
84. Malik P, Korfali N, Srsen V, Lazou V, Batrakou DG, Zuleger N, Kavanagh DM, Wilkie GS, Goldberg MW, Schirmer EC (2010) Cell-specific and lamin-dependent targeting of novel transmembrane proteins in the nuclear envelope. Cell Mol Life Sci 67(8):1353–1369. doi:10.1007/s00018-010-0257-2
85. Brachner A, Reipert S, Foisner R, Gotzmann J (2005) LEM2 is a novel MAN1-related inner nuclear membrane protein associated with A-type lamins. J Cell Sci 118(Pt 24):5797–5810. doi:10.1242/jcs.02701
86. Chen IH, Huber M, Guan T, Bubeck A, Gerace L (2006) Nuclear envelope transmembrane proteins (NETs) that are up-regulated during myogenesis. BMC Cell Biol 7:38. doi:10.1186/1471-2121-7-38
87. Rolls MM, Stein PA, Taylor SS, Ha E, McKeon F, Rapoport TA (1999) A visual screen of a GFP-fusion library identifies a new type of nuclear envelope membrane protein. J Cell Biol 146(1):29–44
88. Lin F, Blake DL, Callebaut I, Skerjanc IS, Holmer L, McBurney MW, Paulin-Levasseur M, Worman HJ (2000) MAN1, an inner nuclear membrane protein that shares the LEM domain with lamina-associated polypeptide 2 and emerin. J Biol Chem 275(7):4840–4847
89. Worman HJ, Yuan J, Blobel G, Georgatos SD (1988) A lamin B receptor in the nuclear envelope. Proc Natl Acad Sci U S A 85(22):8531–8534
90. Senior A, Gerace L (1988) Integral membrane proteins specific to the inner nuclear membrane and associated with the nuclear lamina. J Cell Biol 107(6 Pt 1):2029–2036
91. Foisner R, Gerace L (1993) Integral membrane proteins of the nuclear envelope interact with lamins and chromosomes, and binding is modulated by mitotic phosphorylation. Cell 73(7):1267–1279
92. Hodzic DM, Yeater DB, Bengtsson L, Otto H, Stahl PD (2004) Sun2 is a novel mammalian inner nuclear membrane protein. J Biol Chem 279(24):25805–25812. doi:10.1074/jbc.M313157200
93. Apel ED, Lewis RM, Grady RM, Sanes JR (2000) Syne-1, a dystrophin- and Klarsicht-related protein associated with synaptic nuclei at the neuromuscular junction. J Biol Chem 275(41):31986–31995. doi:10.1074/jbc.M004775200
94. Clements L, Manilal S, Love DR, Morris GE (2000) Direct interaction between emerin and lamin A. Biochem Biophys Res Commun 267(3):709–714. doi:10.1006/bbrc.1999.2023
95. Broers JL, Machiels BM, Kuijpers HJ, Smedts F, van den Kieboom R, Raymond Y, Ramaekers FC (1997) A- and B-type lamins are differentially expressed in normal human tissues. Histochem Cell Biol 107(6):505–517

96. Inoue H, Kauffman M, Shacham S, Landesman Y, Yang J, Evans CP, Weiss RH (2013) CRM1 blockade by selective inhibitors of nuclear export attenuates kidney cancer growth. J Urol 189(6):2317–2326. doi:10.1016/j.juro.2012.10.018
97. Tai YT, Landesman Y, Acharya C, Calle Y, Zhong MY, Cea M, Tannenbaum D, Cagnetta A, Reagan M, Munshi AA, Senapedis W, Saint-Martin JR, Kashyap T, Shacham S, Kauffman M, Gu Y, Wu L, Ghobrial I, Zhan F, Kung AL, Schey SA, Richardson P, Munshi NC, Anderson KC (2013) CRM1 inhibition induces tumor cell cytotoxicity and impairs osteoclastogenesis in multiple myeloma: molecular mechanisms and therapeutic implications. Leukemia. doi:10.1038/leu.2013.115
98. Salas Fragomeni RA, Chung HW, Landesman Y, Senapedis W, Saint-Martin JR, Tsao H, Flaherty KT, Shacham S, Kauffman M, Cusack JC (2013) CRM1 and BRAF inhibition synergize and induce tumor regression in BRAF-mutant melanoma. Mol Cancer Ther 12(7):1171–1179. doi:10.1158/1535-7163.MCT-12-1171
99. Walker CJ, Oaks JJ, Santhanam R, Neviani P, Harb JG, Ferenchak G, Ellis JJ, Landesman Y, Eisfeld AK, Gabrail NY, Smith CL, Caligiuri MA, Hokland P, Roy DC, Reid A, Milojkovic D, Goldman JM, Apperley J, Garzon R, Marcucci G, Shacham S, Kauffman MG, Perrotti D (2013) Preclinical and clinical efficacy of XPO1/CRM1 inhibition by the karyopherin inhibitor KPT-330 in Ph+leukemias. Blood 122(17):3034–3044. doi:10.1182/blood-2013-04-495374

The Role of the Nuclear Lamina in Cancer and Apoptosis

Jos L.V. Broers and Frans C.S. Ramaekers

Abstract Not long after the discovery of lamin proteins, it became clear that not all lamin subtypes are ubiquitously expressed in cells and tissues. Especially, A-type lamins showed an inverse correlation with proliferation and were thus initially called statins. Here we compare the findings of both A- and B-type lamin expression in various normal tissues and their neoplastic counterparts. Based on immunocytochemistry it becomes clear that lamin expression patterns are much more complicated than initially assumed: while normally proliferative cells are devoid of A-type lamin expression, many neoplastic tissues do show prominent A-type lamin expression. Conversely, cells that do not proliferate can be devoid of lamin expression. Yet, within the different types of tissues and tumors, lamins can be used to distinguish between tumor subtypes. The link between the appearance of A-type lamins in differentiation and the appearance of A-type lamins in a tumor likely relates the proliferative capacity of the tumor to its differentiation state.

While lamins are targets for degradation in the apoptotic process, and accordingly are often used as markers for apoptosis, intriguing studies on an active role of lamins in the initiation or the prevention of apoptosis have been published recently and give rise to a renewed interest in the role of lamins in cancer.

J.L.V. Broers, Ph.D. (✉)
Department of Molecular Cell Biology,
CARIM School for Cardiovascular Diseases, Maastricht University,
P.O. box 616, 6200 MD Maastricht, The Netherlands
e-mail: jos.broers@maastrichtuniversity.nl

F.C.S. Ramaekers, Ph.D.
Department of Molecular Cell Biology, GROW—School for Oncology and Developmental Biology, Maastricht University, P.O. box 616, 6200 MD Maastricht, The Netherlands
e-mail: f.ramaekers@maastrichtuniversity.nl

E.C. Schirmer and J.I. de las Heras (eds.), *Cancer Biology and the Nuclear Envelope*, Advances in Experimental Medicine and Biology 773, DOI 10.1007/978-1-4899-8032-8_2,

Keywords Lamins • A-type lamins • B-type lamins • Immunocytochemistry • Epidermis • Skin cancer • Normal blood • Hematologic malignancies • Male germ line tumors • Cervical tissues • Lung cancer • Apoptosis

Abbreviations

SCLC	Small cell lung cancer
CIS	Carcinoma in situ
CIN	Cervical intra-epithelial neoplasia
EBV	Epstein–Barr Virus
HPV	Human papillomavirus
HSV	Herpes simplex virus
HIV	Human immunodeficiency virus
GFP	Green fluorescent protein
FTI	Farnesyl transferase inhibitor

Introduction

The number of detailed studies on lamin subtype expression in normal tissues is remarkably low. The first studies on differential A-type lamin expression came from the group of E. Wang [1, 2], who used an antibody against a protein initially called statin. They stated that statin in general is absent in proliferating cells, while non-proliferating cells, induced to senescence, showed a pronounced statin expression [1]. Later studies confirmed that statin was in fact lamin A [3]. The differential expression of A-type lamins in cancer is remarkable. In a variety of epithelia and corresponding carcinomas several attempts have been made to correlate proliferation with the absence of A-type lamins. Also a positive correlation between the degree of differentiation and the presence of A-type lamins has been suggested. A refined insight into differences in protein expression has been obtained by generating antibodies that specifically recognize the main splice variants of the LMNA gene, lamin A or lamin C, as well as antibodies that differentiate between the products of the two different B-type lamin genes, lamin B1 and lamin B2. Older studies did not differentiate between these B-type lamins and just mentioned the expression of lamin B protein. In addition, antibodies that recognize different phosphorylation states of lamins have enabled studying altered lamina associations in the nucleus at the cellular level [4].

In contrast to A-type lamins, there is a general consensus that B-type lamins are ubiquitously expressed in epithelial tissues and carcinomas. Yet, also B-type lamins, and especially lamin B1, are often downregulated in a subset of tumor cells within the same tumor.

In this review we discuss (mainly immunocytochemical) lamin expression studies performed in different normal epithelia and their corresponding carcinomas and discuss the impact of these findings for disease diagnosis.

Programmed cell death or apoptosis is a key mechanism in maintaining a balance between tissue growth and shrinkage. Consequently, suppression of apoptosis is a major process enabling tumor development. However most tumors display extensive apoptosis; especially tumors with high proliferation rates show prominent levels of apoptosis. A main challenge in cancer treatment is shifting the balance in favor of the apoptotic process. Since long it is known that cleavage of lamin proteins by caspases is a necessary step in apoptosis allowing for nuclear membrane degradation to proceed, followed by chromatin condensation [5]. Now, there is mounting evidence that, in turn, abnormal lamina organization can lead to apoptosis. However, knowledge about the exact mechanisms supporting the relationship between the lamina and apoptosis is as yet merely speculative.

Normal Epidermis and Skin Cancer

Epidermis

Comparing the expression of A-type lamins as reported by different groups, one can immediately observe that there is a discrepancy between findings of different research groups. Initial studies by Röber et al. [6] showed a gradual increase in A-type lamin expression in all cell layers of the epidermis of mice starting at the later stages of embryonic development and continuing to increase after birth. Expression of these proteins in all epidermal layers was confirmed in paraffin sections of human epidermis [7]. In contrast, several groups noticed the reduced expression of A-type lamins in a large number of (but not all) basal cells and an increase of expression in suprabasal cells [8–10]. Upon ageing, A-type lamin expression becomes more heterogeneous, when comparing lamin A/C expression in the skin of a young child (1 year) versus old people (>60 years [11]). Using A-type lamin subtype-specific antibodies the absence of lamin A from basal cells was striking. Nuclei were completely devoid of lamin staining. In contrast, lamin C expression was still present in basal cells, however often not forming a clear lamina but rather giving a diffuse intranuclear staining pattern [9]. Basal cells that did not express lamin C appeared to be resting basal cells [10]. These findings are not in line with a recent study [12], showing an increase of A-type lamin expression in basal cells and a decrease in the suprabasal cells. How can these discrepancies be explained? First of all, the research was performed on different species (mouse vs. human). Secondly, different fixation and permeabilization methods were applied (formalin fixation, paraffin embedding, and antigen retrieval vs. unfixed frozen sections or methanol-fixed frozen sections). Thirdly, different A-type lamin antibodies were used. Apparently, these factors can influence the recognition of lamin A/C epitopes.

To further complicate this issue, several studies have stressed the importance of different phosphorylation states of lamins. Several lamin A/C antibodies only recognize certain phosphorylation states of lamins, while others will give a more general labeling [4, 13]. Moreover, epitope masking not due to changes in phosphorylation state can occur, as has been shown for lamin B1 [14]. This problem could be overcome by applying a large panel of antibodies on tissues that were fixed in different ways. Unfortunately, these studies were only performed with a limited type of tissues and few antibodies (e.g., see [4, 14]).

Several groups have investigated the expression of B-type lamins in human epidermal tissues. Most reports mention the presence of lamin B1 throughout the epidermis [7, 9, 10], with a prevalence for decoration of the lower, proliferative layers of the epidermis and the absence of lamin B1 in a subset of the basal cells [8]. Most groups demonstrate the uniform staining of all epidermal layers with lamin B2 antibodies [8–10], whereas sometimes a decrease of lamin B2 expression was found in granular cells [7].

Skin Cancer

In basal cell carcinomas most studies found a reduction of A-type lamins. For instance Oguchi et al. [7] showed a reduction in most of the basal carcinomas examined, using an antibody that did not differentiate between lamin A and lamin C.

Using A-type lamin subtype-specific antibodies, Venables et al. [10] showed a reduced expression of lamin A in the majority of tumors. These tumors appeared to be hyperproliferative based on the expression of the proliferation marker Ki67. Downregulation of lamin C was less common in these carcinomas (5/16 tumors [10]). Using similar antibodies, Tilli et al. [9] found more lamin A-expressing cells than lamin C expressing cells in basal cell carcinomas. Strikingly, both studies showed a nucleolar rather than a nuclear lamina staining in some basal cell carcinomas using a lamin C antibody. Whether this staining corresponds to the intranuclear foci seen in early embryonal development [15] remains to be examined.

In squamous cell carcinomas of the skin, a study by Oguchi et al. [7] showed that most tumors were strongly positive for A-type lamins, with only a minority of cancers showing a reduction. Another study confirmed the expression of both lamin A and lamin C in squamous cell carcinomas of the skin [9]. Most of the tumors with reduced lamin A/C expression were poorly differentiated, confirming the general notion that A-type lamin expression is decreased with loss of differentiation.

Germ-Line Cells and Germ Cell Tumors

Since in general the largest differences in lamin expression can be found upon changes in differentiation, one would expect a large number of studies on lamin expression in development of germ-line cells and in developing embryos. Noticeably,

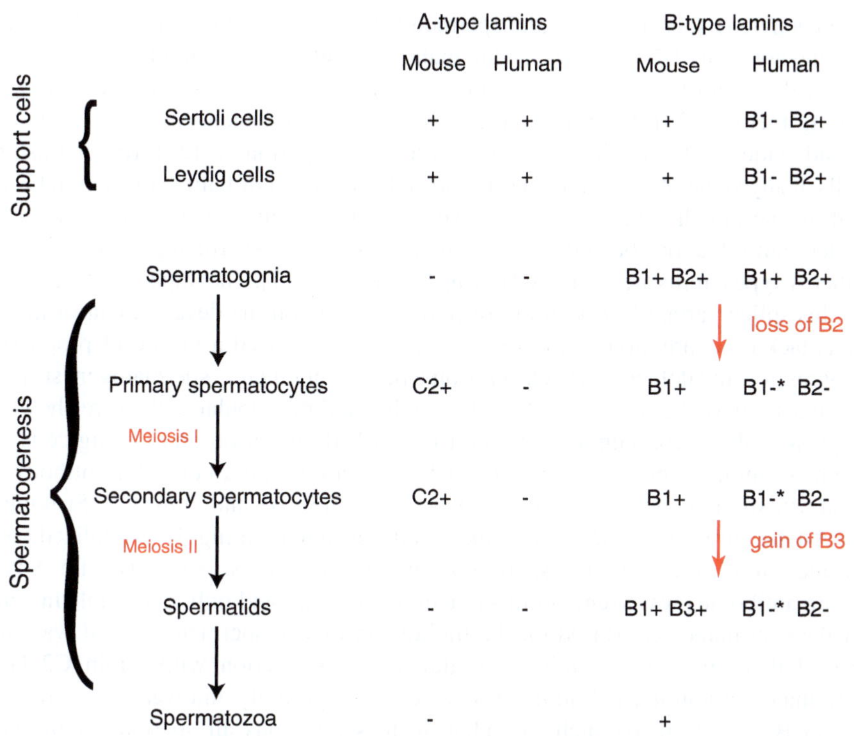

Fig. 1 Overview of reactivity of lamin antibodies in testicular germ cells. Note the prominent changes in lamin expression upon spermatogenesis. *epitope masking? Based on the more extensive studies in mouse tissues, it is likely that lamin B1 is also present in human sperm cell development

this is not the case. To our knowledge only few studies have investigated lamin expression in germ-line tissues and tumors. In fact, only one study has been done on lamin expression in testis and testis tumors using human material [16]. A few more studies have compared lamin expression patterns in normal male and female germ cell types of other species [17–21]. Figure 1 shows an overview of lamin reactivity in male germ-line cells in mouse and human tissue.

Normal Male Germ-Line Cells

Initial studies claimed that no lamins were present during spermatogenesis in chicken [21]; however, subsequent studies in mouse tissues showed that lamins are indeed present in several cells during sperm development [18, 20]. In mouse male gonads, lamin A/C as well as lamin B antibodies reacted with isolated prepuberal Sertoli cells. In addition, anti-lamin B stained the nuclear lamina of all germ-line cell types examined, including primitive spermatogonia, preleptotene, leptotene–zygotene, and pachytene spermatocytes and spermatids [20]. Later studies showed that lamin B1 and not lamin B2 was expressed in male germ cells [22]. Lamin B3, a splice variant of the lamin B2 gene, was initially suggested to be expressed in spermatocytes during meiosis [23], but a more recent study has challenged this observation and found evidence that lamin B3 is only expressed in spermatids [24]. By contrast, no cells at any stage of spermatogenesis showed expression of lamin A or lamin C [19, 20]. In sperm cell development, however, an alternatively spliced form of lamin C, called lamin C2, has been detected in meiotic stages of spermatogenesis, while no other A-type lamins are expressed during this process [17]. The impact of the absence of this splice variant has been discovered in a study on the development of mouse cells lacking expression of the *Lmna* gene, which showed a failure of prophase I progression and defective sex chromosome pairing in *Lmna*$^{-/-}$ spermatogenesis [25].

In a study on human testes, Sertoli, Leydig, and peritubular cells were shown to express both A-type lamins and lamin B2 [16]. Both Sertoli and Leydig cells did express lamin B2 but in general showed no reaction with a lamin B1 antibody. In contrast, spermatogonia were positive for both lamin B1 and lamin B2. Strikingly, no A-type lamins were detected in these cells, despite their highly specialized commitment to differentiate into spermatocytes. In some cases, reactivity with A-type lamin antibodies was seen, but this reaction was weak and only detectable in some of the spermatocytes. Based on the findings in mouse spermatocytes, it was suggested that this weak staining was due to cross-reaction with lamin C2 [16]. Spermatogonia in normal human testis were only partially and weakly positive for lamin B2, while in parenchyma adjacent to seminomas all spermatogonia were clearly positive. Lamin B2 expression in spermatogonia adjacent to seminomas seems, therefore, slightly increased [16].

Using B-type lamin antibodies, human spermatocytes showed no reactivity. While the absence of lamin B2 was in accordance with findings in mouse spermatocytes [22], the absence of lamin B1 in human spermatocytes was unexpected and may be due to epitope masking that has been shown to occur also with this particular antibody in heart tissues [14].

Male Germ Cell Tumors

A study by Machiels et al. [16] showed that the seminomas examined could be divided into two groups: one group contained a mutation in K- or N-RAS (RAS positive), and the second group of seminomas had no detectable mutation in the RAS genes (RAS negative). RAS is one of a family of small GTPases, many of which have been linked to cancers. Using an antibody to A-type lamins, striking differences were observed between these groups: the RAS-negative seminomas were negative for lamin A, and the RAS-positive seminomas were positive, although sometimes weakly. Interestingly, one case was known to contain a heterogeneous population of tumor cells with and without RAS mutation, and this tumor showed a heterogeneous staining pattern with the lamin A antibody. Most seminomas were negative with another lamin A/C antibody, with only one RAS-positive seminoma case being positive. The lamin A antibody 133A2 showed partial reactivity with only two RAS-positive seminomas. None of the RAS-negative seminomas gave a staining reaction with the A-type lamin antibodies.

Most non-seminomas were positive with lamin A/C antibodies. Strikingly, embryonal carcinomas were found to be negative for lamin A, but positive for lamin A/C using two different lamin A/C antibodies. Low expression of lamin A together with normal or high expression of lamin C may explain this reactivity pattern, although epitope masking for the lamin A antibody cannot be excluded. Normally lamin A and C proteins are expressed to comparable degrees, but an imbalance in the expression ratio of lamin A over C may occur. To examine this phenomenon further, samples of three embryonal carcinomas were used for immunoblotting. When these blots were stained with the lamin A/C antibodies, it was obvious that the reactivity level of the lamin C bands was much stronger than that of lamin A, which confirmed the immunohistochemical observations. The presence of lamin A in the blots was confirmed by a weak reactivity with the lamin A antibody 133A2, which may even be overrepresented as a result of non-tumor components that are present in the tumor tissue, such as small blood vessels, and express A-type lamins. The very low expression level of lamin A and the imbalanced expression of lamin A and lamin C using two different antibodies argued in favor of the interpretation that embryonal carcinomas indeed did not express lamin A and were not negative due to epitope masking.

Yolk sac tumors, choriocarcinoma, and teratoma could not be distinguished from each other by studying A-type lamin expression. All three histologically distinct tumor types gave similar perinuclear staining with the lamin A antibody as well as the lamin A/C antibodies.

When a carcinoma in situ (CIS) adjacent to non-seminomas (Table 1) was negative for lamin B1, spermatogonia were also negative. When a CIS was positive for lamin B1, spermatogonia were also positive. In addition, the reaction of the Sertoli cells in these sections was always opposite to that of spermatogonia and CIS. In normal testis, Sertoli cells were negative and spermatogonia were positive for lamin B1, being the physiological expression pattern.

Uterine Cervical Tissues and Premalignant Cervical Lesions

An extensive study was performed on expression patterns of lamins in normal cervical epithelium and premalignant epithelium lesions, known as CIN (for cervical intra-epithelial neoplasia) [8]. In normal ectocervical stratified epithelium, lamin B2 is expressed in all cell layers. In contrast to other non-keratinizing stratified squamous cell epithelia, lamin B1 is most strongly expressed both in the basal and in the parabasal epithelial cells, with a reduction of staining in the upper cell layers. Also, in contrast to other stratified epithelia, lamin A/C antibodies as well as a specific lamin A antibody showed prominent staining of the entire epithelium.

Table 1 Comparison of expression of lamin subtypes in normal tissues and their corresponding tumors

Tissue	Lamin A	Lamin C	Lamin A/C	Lamin B1[a]	Lamin B2	References
Skin and skin cancer						
Basal cell	–	±	±	++	++	[8–10]
Suprabasal cells	++	++	++	+	– to ++	[7–10]
Squam. cell carcinomas	++	++	++	++	++	[7, 9]
Basal cell carcinomas	– to +	± to +	– to +	++	++	[7, 9, 10]
Germ cells and tumors (male)						
Sertoli cells	+	+	+	–	+	[16, 18–20]
Leydig cells	+	+	+	–	+	[16, 18–20]
Spermatogonia	–	–	–	+	+	[16, 18–20]
Spermatocytes	–	– (C2 +)	–	–/+[a]	–	[16, 18–20]
Spermatids	–	–	–	–/+[a]	– (B3+[b])	[24]
Seminomas						
RAS positive	– to +	– to +	+	+	++	[16]
RAS negative	–	–	–	+	++	[16]
Non-seminomas						
Embryonal cell ca.	–	+	+	+	++	[16]
Other	+	+	+	+	++	[16]
Uterine Ectocervix and CIN						
Basal cell	+		+	++	++	[8]
Suprabasal cell	++		++	± to +	++	[8]
CIN (*I*→*III*)	++ to ±		+ to ±	±	+	[8]
Lymphoid cells and tumors						
Hematopoietic cells			–	+	+	[35]
Granulocytes			–	+	+	[35, 38]
Early lymphoid cells			–	+	+	[37, 39]
T-cells			–	+	+(react –)	[35, 39]
B-cells			– to+ (CD30pos)	+	+(react –)	[35, 39]
Mononuclear cells			+	+	+	[40]
PMN cells			–	+	+	[40]
Malignancies						
Lymphoid cells			–	+	+	[34]
Myeloid cells			– to+(diff)	+	+	[34]
Hodgkin's lymphoma	– to ++		– to ++	+	+	[39]
Lung tissues and cancer						
Bronchial basal cells	– to +		– to +	+	++	[8, 44]
Bronchial columnar cells	++		++	– to +	++	[8, 44]
Alveolar cells	++		++	– to +	++	[8, 44]
SCLC			– to ±	– to ++	++	[44]
Adenocarcinomas			– to ++ (cyt!)	– to ++	++	[44]
Squam. cell carcinomas			++	– to ++	++	[44]

Squam. cell squamous cell, *react* reactive lymph nodes, *PMN cells* polymorphonuclear leukocytes, *diff* differentiated tumors, *cyt!* cytoplasmic reaction

[a]Possibly negative due to epitope masking

[b]Mouse only

In endocervical columnar epithelium, most cells are positive with the antibodies examined except for the reserve cells, which are largely negative for lamin B1. In connective tissue cells of the endocervix all antibodies are positive except for the lamin B1 antibody, which is negative in the fibroblasts but positive with most lymphoid cells.

Metaplastic cervical epithelium showed expression of A-type lamins in all epithelial layers, albeit that not all cells were labeled, similar to normal cervical epithelia. B-type lamins showed a homogenous staining in basal and intermediate layers but a striking decrease in the superficial layers of the tissue. Especially the loss of lamin B1 from these superficial metaplastic cells was striking.

More than 30 cases of CIN were examined for their lamin expression patterns. Strikingly, metaplastic epithelium next to CIN lesions showed an aberrant lamin expression pattern. Expression of A-type lamins was increased in the basal layer and decreased in the suprabasal cell, an inversion of the "normal" A-type lamin expression pattern in stratified squamous cell epithelia. Also, lamin B1 expression was largely decreased in these tissues, while lamin B2 expression was more heterogeneous than in comparable regions of normal epithelia.

From low-grade (CIN I) to high-grade (CIN III) CIN lesions the normal differentiation of the squamous epithelium is increasingly lost. Expression of A-type lamins was highly variable within each layer with strong, weak, and even absence of labeling of nuclei at very close distance from each other. Lamin staining patterns were similar in different layers of either grade/lesion type (CIN I–CIN III) and lamin B2 was present in most epithelial cells, while lamin B1 expression seemed to remain confined to the lower layers of this epithelium. A-type lamins were expressed heterogeneously throughout all layers of the epithelium. However, an overall decrease in the number of cells with lamin staining, as well as an average decrease in intensity of staining, was noticeable in high-grade CIN lesions.

The heterogeneous lamin A/C staining patterns in CIN lesions invite speculation about the correlation between the integration and/or episomal presence of the human papillomavirus (HPV), occurring in nearly all of these tumors [26], and the concurrent loss of A-type lamin expression. Other viruses such as Epstein–Barr virus (EBV), herpes simplex virus (HSV), cytomegalovirus (CMV), and human immunodeficiency virus (HIV) have been shown to disrupt the nuclear lamina after infection, enabling the release of virion particles from the nucleus [27]. This disruption can be due to conformational changes of the lamina, as shown for HSV-1 [28], resulting in increased solubility, but also physical disruption of the nuclear lamina can occur, as seen in lamin-GFP-transfected cells, infected with HSV-1 [29]. In HIV infections the viral protein Vpr induces perforations in the nuclear lamina, leading to dynamic disruptions in the nuclear envelope [30].

Until now, only quantitative tissue studies were performed on the impact of the HPV viral proteins on lamin expression. These studies showed that HPV 16 E5 does cause a downregulation of lamins A/C [31], while HPV 16 E1–4 and HPV 16 E6 do not seem to have a prominent effect on lamin A/C expression [32]. Possibly, HPV integration benefits from a weakened lamina in cases of coinfection with HSV-2 or HIV, since epidemiologic studies indicate an increased risk of developing cervical cancer in these cases [33]. However, direct effects of HPV infections on the lamina structure have not yet been studied at the cellular level.

Lymphoid Cells and Tumors

Normal Blood Cells

Cells, very suitable to study the correlation between lamina expression and proliferation versus differentiation, can be found in the lymphoid system and tumor cells derived from these cells, since the different stages of blood cell development have been very well defined. An extensive review shows an overview of the most important findings in normal blood cells and hematologic malignancies [34].

Initial studies indicated that hematopoietic cells were devoid of A-type lamins, showing that in mouse cells both T and B lymphocytes as well as granulocytes and monocytic cells directly isolated from spleen, thymus, blood, or bone marrow did not express lamin A/C but only lamin B [35]. Comparable studies in rat showed that thymocytes and human pre-B lymphoblasts do not express A-type lamins, while in purified T and B lymphocytes isolated from blood samples A-type lamins could be detected [36, 37]. Human peripheral blood granulocytes express little if any lamin A or lamin C [38]. In a study on human lymph node tissue in patients with Hodgkin's lymphoma it was shown that in most of the reactive lymph nodes investigated A-type lamins were absent [39]. These last findings, combined with those of other animal studies, led us to the suggestion that both B- and T-cells express very low amounts of A-type lamins or no lamins at all. The positive findings by Guilly et al. [37] on immunoblots of isolated B- and T-cells could easily have been caused by contamination with other blood cells that do contain A-type lamins, such as mononuclear neutrophils [40].

Cells expressing CD20 (a marker for B-cell differentiation and present on B-cells but not on plasma cells) do not express A-type lamins, while CD30-positive cells (a marker for activated B-cells) in the paracortex as well as in the medulla in general did show A-type lamin expression. These findings suggest that cells with a higher degree of lymphocyte differentiation do express A-type lamins. Another intriguing observation was the absence of lamin B2 in both centrocytes and centroblasts of the follicle center of the lymph nodes, while the paracortex showed a high expression of lamin B2. A lamin B1 antibody showed reactivity with all cells in all regions of the lymph nodes.

Hematologic Malignancies

Cell lines derived from different lineages of hematologic malignancies in general showed variable A-type lamin expression. A T-lymphoblast cell line (KE 37) was negative for A-type lamin expression [36], while fully differentiated stages in the B-cell lineage such as represented by the RPMI-6666 line (an EBV-transformed lymphoblastoid B-cell line) and the U266 plasmacytoma line (an IgE-producing human myeloma) showed a strong lamina A/C labeling. In contrast, a previous study showed that Ig-secreting mouse myeloma cells lack lamin A and C expression [41]. To explain these apparently conflicting findings a detailed study was undertaken by Kaufmann [38], who showed that human myeloid leukaemia cell lines and marrow samples from patients with acute non-lymphocytic leukaemia do express low but detectable levels of A-type lamins. Also, lamins A and C were detected in cell lines of myeloid (KGla), erythroid (HEL), and megakaryocytic (Mo-7e) lineages. Strikingly expression of A-type lamins can be strongly enhanced in the HL-60 human progranulocytic leukaemia cells by inducing differentiation into monocytes using *O*-tetra-decanoylphorbol-13-myristate acetate (TPA) [38]. Samples from patients with chronic myelogenous leukaemia did not show expression of A-type lamins [38]. A study on patients with acute lymphoblastic leukaemia and non-Hodgkin's lymphomas revealed that these tumors do not contain A-type lamins [42].

Jansen et al. [39] have studied lamin expression in nodular sclerosing Hodgkin's disease and noticed a prominent labeling of both Reed–Sternberg cells and Hodgkin cells with A-type lamin antibodies. They also tried to find a correlation between proliferation (using the Ki67 antigen marker) and presence of A-type lamins. While in normal cells Ki67 expression was limited to cells with a reduced A-type lamin expression, this was not the case in neoplastic cells. Both lamin A-positive and lamin A-negative cells expressed Ki67.

Lung Epithelium and Lung Cancer

In normal lung as well as lung cancers the expression of lamins has been investigated using different techniques, including Northern blotting, Western blotting, and immunocytochemistry [8, 43]. From these studies it became clear that both in normal and neoplastic lung tissue a dramatic reduction of A-type lamins can occur. In normal lung lamin A/C expression is only observed in a subset of cells, relating to the differentiation stage of individual cells. A wide range of lamin A/C expression levels is also observed in lung cancers, which may reflect a change in the lamin levels or the differentiation stage of the cell that initiated the tumor or changes in differentiation within these tumors.

Normal Lung Epithelium

Studies on lamin expression in normal bronchial and alveolar cells [8] revealed that A-type lamins were expressed in bronchial columnar cells but showed large differences in expression levels between patient samples in bronchial basal cells. In some samples, no or few basal cells were stained with these antibodies, whereas other samples revealed a uniform positive staining in all bronchial cells. Since these normal lung samples were obtained from (ex-)smokers who developed lung cancers in other regions of the lung, it is tempting to speculate that differences in A-type lamin expression between samples reflect premalignant stages of the diseases, which are not yet visible upon histological examination of these samples. All alveolar pneumocytes were positive for A-type lamin antibodies.

Lamin B2 was present in virtually all epithelial cells. In contrast, expression of lamin B1 was quite heterogeneous, showing a strong decoration/staining of basal cells of bronchi but not in suprabasal, columnar cells. A similar heterogeneity was revealed in alveolar cells, with only a limited number of cells stained.

Lung Cancer Cell Lines

Kaufmann et al. [43] were the first to describe the prominent decrease of A-type lamins in small cell lung cancer (SCLC) cell lines. They showed that lamin A/C levels were more than 80 % lower in SCLC cell lines compared to non-SCLC lines, as detected by Western and Northern blotting. These findings were confirmed in another study using a different panel of cell lines [44]. In this latter study immunofluorescence confirmed the absence or very weak expression of A-type lamins in SCLC cell lines. From Fig. 2 it becomes clear that at the individual cell level a pronounced variation in lamin C expression can be seen, ranging from complete absence of labeling, via weak diffuse intranuclear labeling, towards cells with a clearly visible nuclear membrane labeling. In contrast, lamin B1 expression appears to be quite homogenous in these cells. In general, nuclei containing a lamina with lamin A and/or lamin C are larger in size than those without A-type lamins (Fig. 2, NCI-H82), and indeed cell lines that differentiate from the classic via the variant SCLC phenotype towards non-SCLC show increased nuclear size along with the appearance of pronounced lamin A/C staining (unpublished). Kaufmann et al. [38] performed an additional and very interesting study on SCLC cell lines. They compared A-type lamin expression in the SCLC cell line NCI-H249 before and after transfection with v-Ha-RAS and found a dramatic increase in lamin A/C expression after transfection. How this v-Ha-RAS transfection, which alters the phenotype of this cell line from SCLC to non-SCLC, impacts on the expression of A-type lamins is not entirely clear yet. A correlation between RAS activation and increased lamin A/C expression has also been found in other studies. As mentioned above, a positive correlation between RAS expression and A-type lamin expression was found in human seminomas [16]. In parallel to these results, a recent study showed that in osteoblast differentiation by FGF3 activation, leading to RAS and ERK activation, expression of lamin A/C is increased [45]. As for SCLC, the mechanism by which increased LMNA expression is achieved upon RAS activation is unclear so far. The recent finding that phospho-ERK, a prominent downstream target of the RAS signaling pathway, is increased in laminopathy cells is intriguing and suggests that not only A-type lamin expression can be induced by RAS signaling but also, conversely, the RAS signaling route can be triggered by defective A-type lamin expression in a feedback loop [46]. Interestingly, both RAS and lamins undergo the same posttranslational modifications, including farnesylation, so a common expression regulation

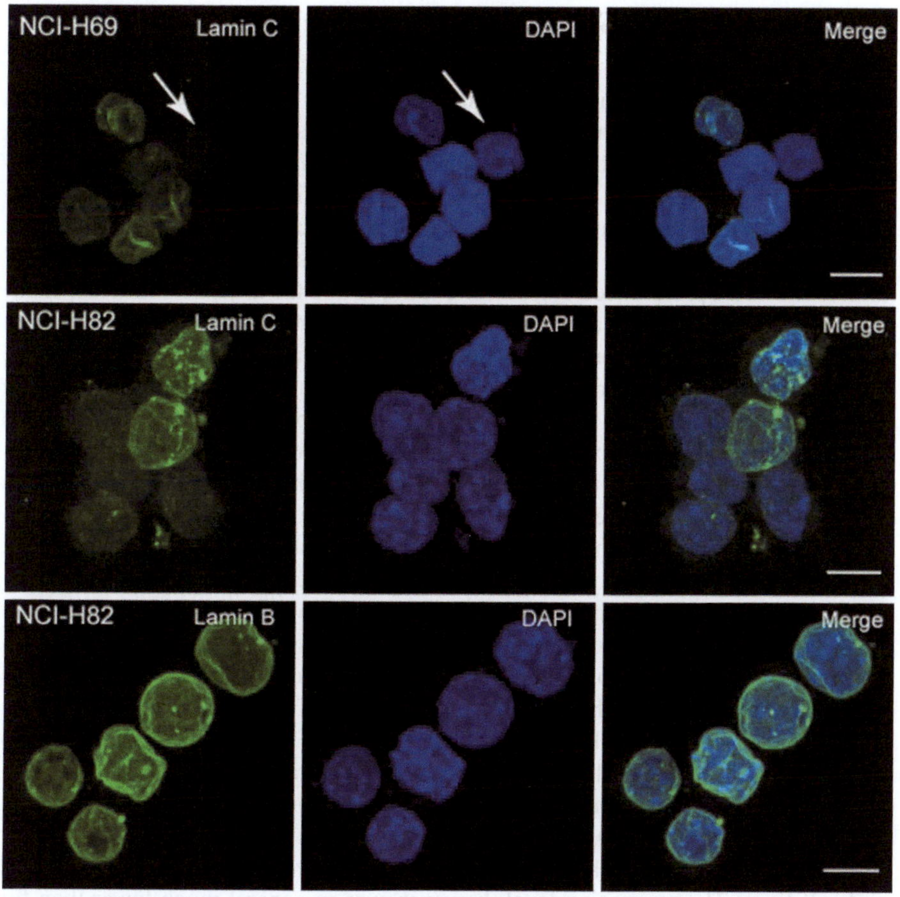

Fig. 2 Confocal z-projections of SCLC cultures immunostained with lamin C or lamin B1 antibodies (*green*). Note the large variation of lamin C immunostaining within a single clump of tumor cells, ranging from invisible (*arrow*) to a clear decoration of the nuclear rim. Note also that nuclei with lamin C in their lamina appear larger than in the neighboring lamin C-negative cells (NCI-H82). Lamin B2 is ubiquitously expressed in all tumor cells. Nuclei were counterstained with DAPI. Scale bars represent 10 μm

pathway should not be excluded. The correlation between RAS activation and increased lamin A/C expression could explain why lamin A/C seems to be a marker for differentiation in some cell systems and a marker for increased proliferation in tumors. It is known that v-Ha-RAS can cause proliferation in some cells and senescence induction in other cells. In the SCLC cell line NCI-H249 it seems that v-Ha-RAS expression leads to a cell phenotype with both increased growth and increased levels of differentiation [47], accompanied by an increase in lamin A/C expression [43], while, as mentioned above, in osteoblasts, enhanced normal RAS expression causes differentiation and senescence [45].

Lung Cancer Tissue Specimens

A detailed study on the differential expression of A- and B-type lamins in both SCLC and non-SCLC revealed that A-type lamin expression is strikingly reduced in most SCLC. In fact, 6 out of 15 cases were scored as having no lamin A/C expression at all [44]. Cells with a higher degree of differentiation, including lung carcinoids (4 out of 6 cases) and non-SCLC (23 out 25 cases), showed a prominent labeling using the A-type lamin antibodies. To our surprise, several non-SCLC tumors showed a pronounced cytoplasmic staining and the absence of nuclear staining. While at the histological section level no cellular or nuclear abnormalities were detected, lamin A/C staining was found in the cytoplasm and not decorating the nuclear rim of these tumors (Fig. 3). These findings urged us to search for lamin abnormalities both at RNA and protein level as well as by mRNA sequencing. In the limited number of tumors examined, no A-type lamin mutations were found (Broers et al., unpublished). Thus, the cause of this cytoplasmic labeling of A-type lamins in lung cancer remains unresolved. Possibly, a disturbed nuclear import mechanism of lamins gives rise to cytoplasmic accumulation of these proteins. A study by Mical and Monteiro [48] showed that the presence of a correct nuclear localization signal is not sufficient for nuclear translocation but that also the CAAX domain and an extra 42 AA central rod domain are needed. Possibly these motifs are not recognized for nuclear import, or alternatively the nuclear import machinery for lamins itself is not functioning in these cells. A similar study in colon adenomas and gastric cancer also revealed cytoplasmic lamin staining [49]. Also, in this latter study no follow-up studies were performed explaining this aberrant lamin localization; however, in our search for lamin mutations in these tumors, we did discover a novel splice variant, lamin AΔ10, that appeared to be present not only in lung cancer cell lines and solid lung cancer but also in other tumors as well as in most normal tissues and cell lines. The expression level, however, was in general much lower than for full-length lamins A and C in the tissues examined, and in these older studies only a nested PCR allowed a reliable detection of the transcript [50]. However, in a recent study in neonatal ductal arterial tissue this transcript could be detected in a single RT-PCR run (35 cycles) [51]. Until now, the number of studies on this splice variant has been very limited. Initial attempts to generate lamin AΔ10-specific antibodies have failed. Lamin AΔ10-tagged GFP localizes normally to the nuclear membrane, forming a network with lamins A and C (Broers et al. unpublished). It is unknown whether lamin AΔ10 needs full-length lamins A and C for incorporation into the nuclear lamina. Since the protein is processed like lamin A, including farnesylation and cleavage of the C-terminus [50], independent incorporation seems likely.

In most lung cancer specimens B-type lamins were expressed in all tumor cells, but in a minority of cases a reduced lamin B2 staining was found [36]. Also, lamin B1 was expressed in all lung cancers examined, albeit with a larger variation of staining intensity in general compared to lamin B2 within tumors (Broers et al. unpublished).

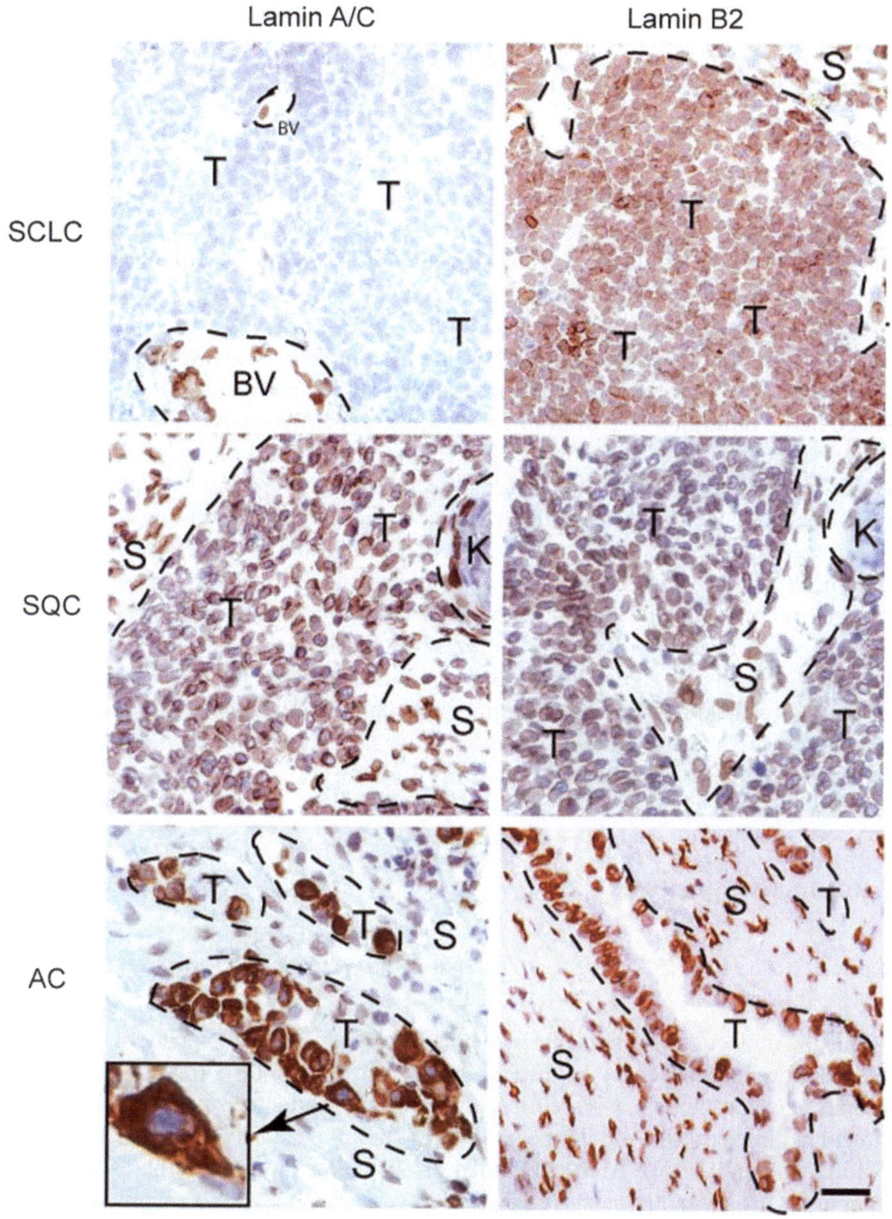

Fig. 3 Immunocytochemical staining of different lung cancer subtypes using antibodies to lamin A/C and lamin B2. Specific immunostaining can be appreciated as a red-brown deposition of aminoethylcarbazole. Nuclei are counterstained with hematoxylin. Tumor areas (T) are indicated, next to surrounding (reactive) stromal areas (S) and blood vessels (BV). Small cell lung cancers (SCLC) in general do not express A-type lamins with a specific decoration of vascular endothelial cells (BV) only. In contrast, a lamin B2 antibody stains all (tumor and non-tumor) cells of SCLC. Squamous cell carcinomas (SQC) show nuclear membrane staining in virtually all tumor cells with a lamin A/C antibody as well as with a lamin B2 antibody. Note the absence of both lamins A/C and B2 in keratinizing areas (K) of the tumor. An adenocarcinoma (AC) shows next to nuclear staining a pronounced cytoplasmic staining in part of the tumor cells (*insert*) using an A-type lamin antibody. Lamin B2 is present in all tumor cells of this adenocarcinoma. Scale bar is 25 μm

Lamins as Markers for Differentiation, Proliferation, and Tumor Progression

As stated in the previous sections, it has become clear that especially A-type lamin expression can be used as a marker for the developmental stage of a tumor. However, in general this only holds true within a certain group of tumors with otherwise similar characteristics. Table 1 summarizes the relationship between normal cells and their derived tumors as well as within tumors. The staining results of all these tissues give rise to a lot of questions regarding their usefulness. How can any conclusions be drawn on general expression patterns, if so many exceptions occur? As stated in the introductory part of this review, the general idea that differential lamin expression can be used for marking differentiation or proliferation is too simple. It is not possible to distinguish whether a decrease in A-type lamin expression in a tumor cell is due to local dedifferentiation of this cell or due to the fact that this tumor cell is derived from a particular normal cell with a lower degree of differentiation. While many factors such as epitope masking and selective recognitions of phosphorylation state of the lamina have obscured the results of these stainings [4, 13, 14], the impact of other factors on lamin expression has been largely overlooked. For instance, to our knowledge, no correlation has been examined in vivo between RAS expression and staining with lamin antibodies. Also, the relative expression of the different A-type lamin isoforms (lamin A, lamin AΔ10, lamin C) has not been addressed thoroughly in most studies.

Lamins in Apoptosis

Lamins as Target Molecules in the Execution of Apoptosis

While the molecular mechanisms by which the unique expression patterns of lamins in tumors remain obscure, the critical importance of apoptosis to the regulation of tumors is well established as are roles of lamins in the apoptotic process. Thus lamin functions in apoptosis could be a link to lamin changes observed in tumors. Numerous studies have been performed on the role of lamins in the execution of apoptosis. In cancer, apoptosis is a common event. Strikingly, most highly proliferative tumors show increased levels of apoptosis, as compared to tumors with a low proliferative capacity. In these former tumors even a small alteration in the percentage of apoptotic cells can lead to a dramatic expansion or shrinkage of tumor size. Lamins appear to be specifically targeted by caspases 3 and 6 that become activated both via the intrinsic and extrinsic pathway of apoptosis. Upon induction of apoptosis cytochrome c release activates procaspases, which cleave target molecules in an amino acid sequence-specific manner. A-type lamins are cleaved at their conserved VEID site, which is located in the non-helical linker region L12 at position 230. Cleavage is mediated specifically by caspase 6 and not by other caspases [52, 53].

It was initially assumed that caspase 6 is also responsible for B-type lamin cleavage at their conserved VEVD site [54]; however, for complete cleavage of lamin B in vivo the presence of caspase 3 seems to be indispensable [55], possibly by forming essential components of the apoptosome complex [56, 57]. Several studies have indicated that an intact lamina can prevent or at least delay the execution of apoptosis, preventing chromatin condensation and fragmentation. For instance, a study by Rao et al. showed that an intact lamina, rendered uncleavable by caspases after mutating the VEVD/VEID lamin cleavage site, could delay the onset of apoptosis for 12 h [58]. In this way the lamina composition could prevent or delay apoptosis in certain tumors, depending on the amount of lamins present and the accessibility of lamins for caspases. Until now, however, no studies have confirmed an altered sensitivity of lamins to degradation in any cancer subtypes. Absence of appropriate (pro-)caspases could be a mechanism to prevent apoptosis indirectly. Indeed, a recent study showed that a caspase-3 gene product (caspase-3s) could counteract on caspase 3 activity, preventing a proper apoptosome assembly [57].

Lamin Mutants Promote the Execution of Apoptosis

From the literature it is not clear whether lamins play an active role in preventing or promoting apoptosis in cancer. As mentioned, an intact lamina is capable of preventing chromatin condensation and fragmentation. Yet, such an intact lamina does not warrant prevention from apoptosis. Lamin phosphorylation can cause depolymerization of the complete lamina network within minutes, as seen in vital imaging studies during mitosis [59, 60]. Also, in the apoptotic process the lamina becomes solubilized very rapidly, even before A-type lamin cleavage has been completed, as seen in CHO cells, transfected with lamin A [54]. In laminopathies, several studies speculate on the direct effects of A-type lamin mutations on the occurrence of apoptosis. Indeed, in cell cultures from laminopathy patients with different lamin mutations increased apoptosis can be found [61]. More pronounced effects could be achieved by exposing patient cells to mechanical strain that elicited an increase in apoptosis in these cells [62]. Yet, studies in laminopathy animal models resulted in conflicting results. Heterozygous *lmna*$^{+/-}$ mice subjected to 6 weeks of moderate or strenuous exercise training did not show induction of apoptosis and even seemed to protect these mice from developing symptoms reflective of laminopathy diseases [63]. In contrast, Lu et al. [64] found a dramatic increase in frequency of apoptosis in the heart of transgenic mice with a human LMNA E82K mutation. They showed that in the heart tissue of these mice both FAS and mitochondrial pathways of apoptosis were activated, leading to increased expression and activation of caspases 8, 9, and 3. Next to mutant A-type lamins (which are not or only rarely found in cancer as far as is currently known), unprocessed lamins, especially progerin, could be responsible for the induction of apoptosis. While a study by McClintock et al. [65] showed that progerin can be expressed by normal cells, and can be associated with the normal ageing process, a recent study, investigating A-type lamin expression during closure of the ductus arteriosus revealed that progerin is expressed during this process, at which a prominent induction of cell death via apoptosis can be seen in this tissue [51].

How the presence of lamin mutations and/or unprocessed lamins can induce apoptosis is not yet clear. One route to apoptosis could be through mechanical weakening of cells with aberrant lamin expression. Lamin mutations lead to mechanical nuclear weakness [66, 67], which can lead to nuclear and cellular damage and even nuclear ruptures [68]. These events can lead to excessive chromatin damage, which in turn will lead to apoptosis. Indeed, several studies have shown the increase in DNA repair and increased apoptosis and senescence in cell cultures of laminopathy cells [61, 69, 70]. Even in normal cells, overexpression of normal lamin A leads to increased senescence and apoptosis [71]. In cancer, the impact of lamin overexpression may be different since a recent study showed that in prostate cancer cells, transfection of A-type lamins leading to overexpression causes enhanced growth, invasion, and migration by activation of the PI3K/AKT/PTEN pathway [72], while knockdown of lamins in the same cell culture has opposite effects. On the other hand, knockdown of lamin B1 leads to apoptosis rather than necrosis after induction of cell death in the mouse mammary tumor FM3A cell line [73]. In this respect, it is tempting to speculate about the mechanism by which statins, known to block, amongst others, farnesylation of A- and B-type lamins, have a beneficial effect in cancer treatment. Several studies showed that statins can induce apoptosis (e.g., see [74]), while also the more specific lamin-processing inhibitors (farnesyl transferase inhibitors, FTIs) can induce apoptosis in lymphomas [75]. Since these FTIs are also known to inhibit farnesylation of the RAS protein [76], a possible synergistic effect is evoked in these cancers.

Taken together, it can be stated that an active role of lamins in the induction but also the prevention of apoptosis is beginning to emerge, indicating the vital role of these proteins in cell survival. Clearly, the research on the role of lamins in cancer has only just begun.

References

1. Wang E (1985) Rapid disappearance of statin, a nonproliferating and senescent cell-specific protein, upon reentering the process of cell cycling. J Cell Biol 101(5 Pt 1):1695–1701
2. Wang E, Krueger JG (1985) Application of a unique monoclonal antibody as a marker for nonproliferating subpopulations of cells of some tissue. J Histochem Cytochem 33(6): 587–594
3. Coates PJ, Hobbs RC, Crocker J, Rowlands DC, Murray P, Quinlan R, Hall PA (1996) Identification of the antigen recognized by the monoclonal antibody BU31 as lamins A and C. J Pathol 178(1):21–29. doi:10.1002/(SICI)1096-9896(199601)178:1<21::AID-PATH439>3.0.CO;2-Q [pii] 10.1002/(SICI)1096-9896(199601)178:1<21::AID-PATH439>3.0.CO;2-Q
4. Dyer JA, Kill IR, Pugh G, Quinlan RA, Lane EB, Hutchison CJ (1997) Cell cycle changes in A-type lamin associations detected in human dermal fibroblasts using monoclonal antibodies. Chromosome Res 5(6):383–394

5. Oberhammer FA, Hochegger K, Froschl G, Tiefenbacher R, Pavelka M (1994) Chromatin condensation during apoptosis is accompanied by degradation of lamin A+B, without enhanced activation of cdc2 kinase. J Cell Biol 126(4):827–837
6. Rober RA, Weber K, Osborn M (1989) Differential timing of nuclear lamin A/C expression in the various organs of the mouse embryo and the young animal: a developmental study. Development 105(2):365–378
7. Oguchi M, Sagara J, Matsumoto K, Saida T, Taniguchi S (2002) Expression of lamins depends on epidermal differentiation and transformation. Br J Dermatol 147(5):853–858. doi:4948 [pii]
8. Broers JL, Machiels BM, Kuijpers HJ, Smedts F, van den Kieboom R, Raymond Y, Ramaekers FC (1997) A- and B-type lamins are differentially expressed in normal human tissues. Histochem Cell Biol 107(6):505–517
9. Tilli CM, Ramaekers FC, Broers JL, Hutchison CJ, Neumann HA (2003) Lamin expression in normal human skin, actinic keratosis, squamous cell carcinoma and basal cell carcinoma. Br J Dermatol 148(1):102–109. doi:5026 [pii]
10. Venables RS, McLean S, Luny D, Moteleb E, Morley S, Quinlan RA, Lane EB, Hutchison CJ (2001) Expression of individual lamins in basal cell carcinomas of the skin. Br J Cancer 84(4):512–519. doi:10.1054/bjoc.2000.1632 S000709200091632X [pii]
11. Dreesen O, Chojnowski A, Ong PF, Zhao TY, Common JE, Lunny D, Lane EB, Lee SJ, Vardy LA, Stewart CL, Colman A (2013) Lamin B1 fluctuations have differential effects on cellular proliferation and senescence. J Cell Biol. doi:jcb.201206121 [pii] 10.1083/jcb.201206121
12. Hanif M, Rosengardten Y, Sagelius H, Rozell B, Eriksson M (2009) Differential expression of A-type and B-type lamins during hair cycling. PLoS One 4(1):e4114. doi:10.1371/journal.pone.0004114
13. Kuga T, Nozaki N, Matsushita K, Nomura F, Tomonaga T (2010) Phosphorylation statuses at different residues of lamin B2, B1, and A/C dynamically and independently change throughout the cell cycle. Exp Cell Res 316(14):2301–2312. doi:S0014-4827(10)00280-6 [pii] 10.1016/j.yexcr.2010.05.017
14. Tunnah D, Sewry CA, Vaux D, Schirmer EC, Morris GE (2005) The apparent absence of lamin B1 and emerin in many tissue nuclei is due to epitope masking. J Mol Histol 36(5):337–344. doi:10.1007/s10735-005-9004-7
15. Foster HA, Stokes P, Forsey K, Leese HJ, Bridger JM (2007) Lamins A and C are present in the nuclei of early porcine embryos, with lamin A being distributed in large intranuclear foci. Chromosome Res 15(2):163–174. doi:10.1007/s10577-006-1088-8
16. Machiels BM, Ramaekers FC, Kuijpers HJ, Groenewoud JS, Oosterhuis JW, Looijenga LH (1997) Nuclear lamin expression in normal testis and testicular germ cell tumours of adolescents and adults. J Pathol 182(2):197–204. doi:10.1002/(SICI)1096-9896(199706)182:2<197::AID-PATH823>3.0.CO;2-P [pii] 10.1002/(SICI)1096-9896(199706)182:2<197::AID-PATH823>3.0.CO;2-P
17. Alsheimer M, Benavente R (1996) Change of karyoskeleton during mammalian spermatogenesis: expression pattern of nuclear lamin C2 and its regulation. Exp Cell Res 228(2):181–188. doi:10.1006/excr.1996.0315
18. Maul GG, French BT, Bechtol KB (1986) Identification and redistribution of lamins during nuclear differentiation in mouse spermatogenesis. Dev Biol 115(1):68–77
19. Moss SB, Burnham BL, Bellve AR (1993) The differential expression of lamin epitopes during mouse spermatogenesis. Mol Reprod Dev 34(2):164–174. doi:10.1002/mrd.1080340208
20. Moss SB, Donovan MJ, Bellve AR (1987) The occurrence and distribution of lamin proteins during mammalian spermatogenesis and early embryonic development. Ann N Y Acad Sci 513:74–89
21. Stick R, Schwarz H (1982) The disappearance of the nuclear lamina during spermatogenesis: an electron microscopic and immunofluorescence study. Cell Differ 11(4):235–243
22. Vester B, Smith A, Krohne G, Benavente R (1993) Presence of a nuclear lamina in pachytene spermatocytes of the rat. J Cell Sci 104(Pt 2):557–563
23. Furukawa K, Hotta Y (1993) cDNA cloning of a germ cell specific lamin B3 from mouse spermatocytes and analysis of its function by ectopic expression in somatic cells. EMBO J 12(1):97–106

24. Schutz W, Alsheimer M, Ollinger R, Benavente R (2005) Nuclear envelope remodeling during mouse spermiogenesis: postmeiotic expression and redistribution of germline lamin B3. Exp Cell Res 307(2):285–291. doi:10.1016/j.yexcr.2005.03.023
25. Alsheimer M, Liebe B, Sewell L, Stewart CL, Scherthan H, Benavente R (2004) Disruption of spermatogenesis in mice lacking A-type lamins. J Cell Sci 117(Pt 7):1173–1178. doi:10.1242/jcs.00975
26. Schiffman M, Castle PE, Jeronimo J, Rodriguez AC, Wacholder S (2007) Human papillomavirus and cervical cancer. Lancet 370(9590):890–907. doi:10.1016/S0140-6736(07)61416-0
27. Cibulka J, Fraiberk M, Forstova J (2012) Nuclear actin and lamins in viral infections. Viruses 4(3):325–347. doi:10.3390/v4030325 viruses-04-00325 [pii]
28. Reynolds AE, Liang L, Baines JD (2004) Conformational changes in the nuclear lamina induced by herpes simplex virus type 1 require genes U(L)31 and U(L)34. J Virol 78(11): 5564–5575. doi:10.1128/JVI.78.11.5564-5575.2004
29. Simpson-Holley M, Colgrove RC, Nalepa G, Harper JW, Knipe DM (2005) Identification and functional evaluation of cellular and viral factors involved in the alteration of nuclear architecture during herpes simplex virus 1 infection. J Virol 79(20):12840–12851. doi:10.1128/JVI.79.20.12840-12851.2005
30. de Noronha CM, Sherman MP, Lin HW, Cavrois MV, Moir RD, Goldman RD, Greene WC (2001) Dynamic disruptions in nuclear envelope architecture and integrity induced by HIV-1 Vpr. Science 294(5544):1105–1108. doi:10.1126/science.1063957
31. Kivi N, Greco D, Auvinen P, Auvinen E (2008) Genes involved in cell adhesion, cell motility and mitogenic signaling are altered due to HPV 16 E5 protein expression. Oncogene 27(18):2532–2541. doi:10.1038/sj.onc.1210916
32. Niu XY, Peng ZL, Duan WQ, Wang H, Wang P (2006) Inhibition of HPV 16 E6 oncogene expression by RNA interference in vitro and in vivo. Int J Gynecol Cancer 16(2):743–751. doi:10.1111/j.1525-1438.2006.00384.x
33. Bosch FX, de Sanjose S (2007) The epidemiology of human papillomavirus infection and cervical cancer. Dis Markers 23(4):213–227
34. Prokocimer M, Margalit A, Gruenbaum Y (2006) The nuclear lamina and its proposed roles in tumorigenesis: projection on the hematologic malignancies and future targeted therapy. J Struct Biol 155(2):351–360. doi:10.1016/j.jsb.2006.02.016
35. Rober RA, Sauter H, Weber K, Osborn M (1990) Cells of the cellular immune and hemopoietic system of the mouse lack lamins A/C: distinction versus other somatic cells. J Cell Sci 95(Pt 4):587–598
36. Guilly MN, Bensussan A, Bourge JF, Bornens M, Courvalin JC (1987) A human T lymphoblastic cell line lacks lamins A and C. EMBO J 6(12):3795–3799
37. Guilly MN, Kolb JP, Gosti F, Godeau F, Courvalin JC (1990) Lamins A and C are not expressed at early stages of human lymphocyte differentiation. Exp Cell Res 189(1):145–147
38. Kaufmann SH (1992) Expression of nuclear envelope lamins A and C in human myeloid leukemias. Cancer Res 52(10):2847–2853
39. Jansen MP, Machiels BM, Hopman AH, Broers JL, Bot FJ, Arends JW, Ramaekers FC, Schouten HC (1997) Comparison of A and B-type lamin expression in reactive lymph nodes and nodular sclerosing Hodgkin's disease. Histopathology 31(4):304–312
40. Yabuki M, Miyake T, Doi Y, Fujiwara T, Hamazaki K, Yoshioka T, Horton AA, Utsumi K (1999) Role of nuclear lamins in nuclear segmentation of human neutrophils. Physiol Chem Phys Med NMR 31(2):77–84
41. Paulin-Levasseur M, Scherbarth A, Traub U, Traub P (1988) Lack of lamins A and C in mammalian hemopoietic cell lines devoid of intermediate filament proteins. Eur J Cell Biol 47(1):121–131
42. Stadelmann B, Khandjian E, Hirt A, Luthy A, Weil R, Wagner HP (1990) Repression of nuclear lamin A and C gene expression in human acute lymphoblastic leukemia and non-Hodgkin's lymphoma cells. Leuk Res 14(9):815–821
43. Kaufmann SH, Mabry M, Jasti R, Shaper JH (1991) Differential expression of nuclear envelope lamins A and C in human lung cancer cell lines. Cancer Res 51(2):581–586

44. Broers JLV, Raymond Y, Klein Rot M, Kuijpers H, Wagenaar SS, Ramaekers FCS (1993) Nuclear A-type lamins are differentially expressed in human lung cancer subtypes. Am J Pathol 143(1):211–220
45. Krejci P, Prochazkova J, Smutny J, Chlebova K, Lin P, Aklian A, Bryja V, Kozubik A, Wilcox WR (2010) FGFR3 signaling induces a reversible senescence phenotype in chondrocytes similar to oncogene-induced premature senescence. Bone 47(1):102–110. doi:10.1016/j.bone.2010.03.021
46. Muchir A, Wu W, Worman HJ (2009) Reduced expression of A-type lamins and emerin activates extracellular signal-regulated kinase in cultured cells. Biochim Biophys Acta 1792(1): 75–81. doi:10.1016/j.bbadis.2008.10.012
47. Falco JP, Baylin SB, Lupu R, Borges M, Nelkin BD, Jasti RK, Davidson NE, Mabry M (1990) v-rasH induces non-small cell phenotype, with associated growth factors and receptors, in a small cell lung cancer cell line. J Clin Invest 85(6):1740–1745. doi:10.1172/JCI114630
48. Mical TI, Monteiro MJ (1998) The role of sequences unique to nuclear intermediate filaments in the targeting and assembly of human lamin B: evidence for lack of interaction of lamin B with its putative receptor. J Cell Sci 111(Pt 23):3471–3485
49. Moss SF, Krivosheyev V, de Souza A, Chin K, Gaetz HP, Chaudhary N, Worman HJ, Holt PR (1999) Decreased and aberrant nuclear lamin expression in gastrointestinal tract neoplasms. Gut 45(5):723–729
50. Machiels BM, Zorenc AH, Endert JM, Kuijpers HJ, van Eys GJ, Ramaekers FC, Broers JL (1996) An alternative splicing product of the lamin A/C gene lacks exon 10. J Biol Chem 271(16):9249–9253
51. Bokenkamp R, Raz V, Venema A, DeRuiter MC, van Munsteren C, Olive M, Nabel EG, Gittenberger-de Groot AC (2011) Differential temporal and spatial progerin expression during closure of the ductus arteriosus in neonates. PLoS One 6(9):e23975. doi:10.1371/journal.pone.0023975
52. Okinaga T, Kasai H, Tsujisawa T, Nishihara T (2007) Role of caspases in cleavage of lamin A/C and PARP during apoptosis in macrophages infected with a periodontopathic bacterium. J Med Microbiol 56(Pt 10):1399–1404. doi:10.1099/jmm.0.47193-0
53. Ruchaud S, Korfali N, Villa P, Kottke TJ, Dingwall C, Kaufmann SH, Earnshaw WC (2002) Caspase-6 gene disruption reveals a requirement for lamin A cleavage in apoptotic chromatin condensation. EMBO J 21(8):1967–1977. doi:10.1093/emboj/21.8.1967
54. Broers JL, Bronnenberg NM, Kuijpers HJ, Schutte B, Hutchison CJ, Ramaekers FC (2002) Partial cleavage of A-type lamins concurs with their total disintegration from the nuclear lamina during apoptosis. Eur J Cell Biol 81(12):677–691. doi:10.1078/0171-9335-00282
55. Slee EA, Adrain C, Martin SJ (2001) Executioner caspase-3, -6, and -7 perform distinct, nonredundant roles during the demolition phase of apoptosis. J Biol Chem 276(10):7320–7326. doi:10.1074/jbc.M008363200
56. Kivinen K, Kallajoki M, Taimen P (2005) Caspase-3 is required in the apoptotic disintegration of the nuclear matrix. Exp Cell Res 311(1):62–73. doi:10.1016/j.yexcr.2005.08.006
57. Vegran F, Boidot R, Solary E, Lizard-Nacol S (2011) A short caspase-3 isoform inhibits chemotherapy-induced apoptosis by blocking apoptosome assembly. PLoS One 6(12):e29058. doi:10.1371/journal.pone.0029058
58. Rao L, Perez D, White E (1996) Lamin proteolysis facilitates nuclear events during apoptosis. J Cell Biol 135(6 Pt 1):1441–1455
59. Moir RD, Yoon M, Khuon S, Goldman RD (2000) Nuclear lamins A and B1: different pathways of assembly during nuclear envelope formation in living cells. J Cell Biol 151(6):1155–1168
60. Broers JL, Machiels BM, van Eys GJ, Kuijpers HJ, Manders EM, van Driel R, Ramaekers FC (1999) Dynamics of the nuclear lamina as monitored by GFP-tagged A-type lamins. J Cell Sci 112(Pt 20):3463–3475
61. Meaburn KJ, Cabuy E, Bonne G, Levy N, Morris GE, Novelli G, Kill IR, Bridger JM (2007) Primary laminopathy fibroblasts display altered genome organization and apoptosis. Aging Cell 6(2):139–153. doi:10.1111/j.1474-9726.2007.00270.x

62. Verstraeten VL, Ji JY, Cummings KS, Lee RT, Lammerding J (2008) Increased mechanosensitivity and nuclear stiffness in Hutchinson-Gilford progeria cells: effects of farnesyltransferase inhibitors. Aging Cell 7(3):383–393. doi:10.1111/j.1474-9726.2008.00382.x
63. Chandar S, Yeo LS, Leimena C, Tan JC, Xiao XH, Nikolova-Krstevski V, Yasuoka Y, Gardiner-Garden M, Wu J, Kesteven S, Karlsdotter L, Natarajan S, Carlton A, Rainer S, Feneley MP, Fatkin D (2010) Effects of mechanical stress and carvedilol in lamin A/C-deficient dilated cardiomyopathy. Circ Res 106(3):573–582. doi:10.1161/CIRCRESAHA.109.204388
64. Lu D, Lian H, Zhang X, Shao H, Huang L, Qin C, Zhang L (2010) LMNA E82K mutation activates FAS and mitochondrial pathways of apoptosis in heart tissue specific transgenic mice. PLoS One 5(12):e15167. doi:10.1371/journal.pone.0015167
65. McClintock D, Ratner D, Lokuge M, Owens DM, Gordon LB, Collins FS, Djabali K (2007) The mutant form of lamin A that causes Hutchinson-Gilford progeria is a biomarker of cellular aging in human skin. PLoS One 2(12):e1269. doi:10.1371/journal.pone.0001269
66. Broers JL, Peeters EA, Kuijpers HJ, Endert J, Bouten CV, Oomens CW, Baaijens FP, Ramaekers FC (2004) Decreased mechanical stiffness in LMNA-/- cells is caused by defective nucleo-cytoskeletal integrity: implications for the development of laminopathies. Hum Mol Genet 13(21):2567–2580. doi:10.1093/hmg/ddh295
67. Lammerding J, Schulze PC, Takahashi T, Kozlov S, Sullivan T, Kamm RD, Stewart CL, Lee RT (2004) Lamin A/C deficiency causes defective nuclear mechanics and mechanotransduction. J Clin Invest 113(3):370–378. doi:10.1172/JCI19670
68. De Vos WH, Houben F, Kamps M, Malhas A, Verheyen F, Cox J, Manders EM, Verstraeten VL, van Steensel MA, Marcelis CL, van den Wijngaard A, Vaux DJ, Ramaekers FC, Broers JL (2011) Repetitive disruptions of the nuclear envelope invoke temporary loss of cellular compartmentalization in laminopathies. Hum Mol Genet 20(21):4175–4186. doi:10.1093/hmg/ddr344
69. Liu B, Wang J, Chan KM, Tjia WM, Deng W, Guan X, Huang JD, Li KM, Chau PY, Chen DJ, Pei D, Pendas AM, Cadinanos J, Lopez-Otin C, Tse HF, Hutchison C, Chen J, Cao Y, Cheah KS, Tryggvason K, Zhou Z (2005) Genomic instability in laminopathy-based premature aging. Nat Med 11(7):780–785. doi:10.1038/nm1266
70. Richards SA, Muter J, Ritchie P, Lattanzi G, Hutchison CJ (2011) The accumulation of un-repairable DNA damage in laminopathy progeria fibroblasts is caused by ROS generation and is prevented by treatment with N-acetyl cysteine. Hum Mol Genet 20(20):3997–4004. doi:10.1093/hmg/ddr327
71. Candelario J, Sudhakar S, Navarro S, Reddy S, Comai L (2008) Perturbation of wild-type lamin A metabolism results in a progeroid phenotype. Aging Cell 7(3):355–367. doi:10.1111/j.1474-9726.2008.00393.x
72. Kong L, Schafer G, Bu H, Zhang Y, Zhang Y, Klocker H (2012) Lamin A/C protein is overexpressed in tissue-invading prostate cancer and promotes prostate cancer cell growth, migration and invasion through the PI3K/AKT/PTEN pathway. Carcinogenesis 33(4):751–759. doi:10.1093/carcin/bgs022
73. Sato A, Hiramoto A, Satake A, Miyazaki E, Naito T, Wataya Y, Kim HS (2008) Association of nuclear membrane protein lamin B1 with necrosis and apoptosis in cell death induced by 5-fluoro-2′-deoxyuridine. Nucleosides Nucleotides Nucleic Acids 27(5):433–438. doi:10.1080/15257770802086864
74. Goc A, Kochuparambil ST, Al-Husein B, Al-Azayzih A, Mohammad S, Somanath PR (2012) Simultaneous modulation of the intrinsic and extrinsic pathways by simvastatin in mediating prostate cancer cell apoptosis. BMC Cancer 12:409. doi:10.1186/1471-2407-12-409
75. Rolland D, Camara-Clayette V, Barbarat A, Salles G, Coiffier B, Ribrag V, Thieblemont C (2008) Farnesyltransferase inhibitor R115777 inhibits cell growth and induces apoptosis in mantle cell lymphoma. Cancer Chemother Pharmacol 61(5):855–863. doi:10.1007/s00280-007-0543-3
76. Sousa SF, Fernandes PA, Ramos MJ (2008) Farnesyltransferase inhibitors: a detailed chemical view on an elusive biological problem. Curr Med Chem 15(15):1478–1492

The Diagnostic Pathology of the Nuclear Envelope in Human Cancers

Andrew H. Fischer

Abstract Cancer is still diagnosed on the basis of altered tissue and cellular morphology. The criteria that pathologists use for diagnosis include many morphologically distinctive alterations in the nuclear envelope (NE). With the expectation that diagnostic NE changes will have biological relevance to cancer, a classification of the various types of NE structural changes into three groups is proposed. The first group predicts chromosomal instability. The changes in this group include pleomorphism of lamina size and shape, as if constraints to maintain a spherical shape were lost. Also characteristic of chromosomal instability are the presence of micronuclei, a specific structural feature likely related to the newly described physiology of chromothripsis. The second group is predicted to be functionally important during clonal evolution, because the NE changes in this group are conserved during the clonal evolution of genetically unstable tumors. Two examples of this group include increased ratio of nuclear volume to cytoplasmic volume and the relatively fragile nuclei of small-cell carcinomas. The third and most interesting group develops in a near-diploid, genetically stable background. Many of these (perhaps ultimately all) are directly related to the activation of particular oncogenes. The changes in this group so far include long inward folds of the NE and spherical invaginations of cytoplasm projecting partially into the nucleus ("intranuclear cytoplasmic inclusions"). This group is exemplified by papillary thyroid carcinoma in which RET and TRK tyrosine kinases, and probably B-Raf mutations, directly lead to diagnostic longitudinal folds of the lamina ("nuclear grooves") and intranuclear cytoplasmic inclusions. B-Raf activation may also be linked to intranuclear cytoplasmic inclusions in melanoma and to nuclear grooves in Langerhans cell histiocytosis. Nuclear grooves in granulosa cell tumor may be related to mutations in the *FOXL2* oncogene. Uncovering the precise mechanistic basis for any of these lamina alterations would

A.H. Fischer, M.D. (✉)
Department of Pathology, University of Massachusetts Memorial Medical Center,
RM 213, Biotech 3, 1 Innovation Dr, Worcester, MA 01605, USA
e-mail: Fischa01@ummhc.org

E.C. Schirmer and J.I. de las Heras (eds.), *Cancer Biology and the Nuclear Envelope*, Advances in Experimental Medicine and Biology 773, DOI 10.1007/978-1-4899-8032-8_3,

provide a valuable objective means for improving diagnosis, and will likely reflect new types of functional changes, relevant to particular forms of cancer.

Keywords Criteria of malignancy • Histogenetic classification • Hallmarks of cancer • Relation between oncogenes • Nuclear envelope morphology

Abbreviations

H&E	Hematoxylin and eosin
N:C ratio	Nuclear volume compared to cytoplasmic volume ratio
NE	Nuclear envelope
PTC	Papillary thyroid carcinoma

Introduction

In spite of major progress in cancer research, the actual diagnosis of cancer is still made by pathologists who visually inspect cells and tissues for alterations at the light microscope level. The criteria that pathologists use to diagnose cancer are called the "criteria of malignancy." The large-scale organization of the nuclear envelope (NE) defines nuclear shape, and alterations of nuclear shape are a very important subset of the criteria of malignancy. The underlying thesis of this review is that diagnostic alterations in nuclear shape will ultimately be found to be related to specific cancer cell physiologies. With the increasing evidence for associations between NE proteins and physiologies that appear relevant to cancer, and the increasing number of NE proteins that can be manipulated to alter shape, it is clear that nuclear shape changes are relevant to cancer [1].

It is not a simple problem, though. There are many forms of cancer, with many different types of NE changes, different genetic underpinnings, and different clinical manifestations (different prognoses, doubling times, risk of metastasis, patterns of spread, response to therapies, etc.). It is clearly not useful to make generalizations about "the" cancer cell or "the" NE change in cancer cells. The relevance of the NE to cancer is likely to be manifest in many different ways in different forms of cancer. Tumors have been historically named according to the cell of origin—the histogenetic classification. However, the histogenetic classification provides only a very limited representation of the biologic features of a tumor, and various NE abnormalities in cancer do not segregate neatly with histogenesis. Tumors are subclassified by the "grade" or the "differentiation," yet these two terms also do not directly relate to NE alterations. Thus, researchers are typically unable to find information in pathology reports or tumor bank databases that may correlate with a specific change in the NE. Another problem is that a spherical (normal) nuclear contour is actually typical of some forms of cancer, and a wide variety of nuclear shape changes can be seen in some normal cells, sometimes overlapping with the appearance in some cancer cells. Finally, when carefully studied, at least some, if not many, diagnostic changes in the NE do not appear to reflect any existing Hallmarks of Cancer physiologies [2, 3]. The Hallmarks of Cancer are the expected changes that can theoretically account for the accumulation of more cells over time in a tumor. For example, the hallmarks include more rapid cell division, resistance to apoptosis, ability to induce a blood vascular supply, ability to escape the immune system, autonomy from the need for exogenous growth factors, and several other hypothetical changes. Since diagnostic changes in the NE may not be able to be explained by existing Hallmarks, it appears that researchers interested in NE changes in cancer should expect to uncover new cancer physiologies [4]. The goal of this review is to sort through these difficulties and construct a framework to help dissect specific NE-based cancer-associated physiologies. A three-tiered classification of the various diagnostic NE abnormalities in cancer is proposed as an important start. Ultimately, it is argued that attention will have to be restricted to particular measurable NE alterations within particular cell types and their specific microenvironments, probably with dynamic live cell imaging in order to uncover what will likely be unpredicted cell physiologies involving the NE in various cancers.

It Should Be Expected That Diagnostic Changes in the NE in Cancer Reflect Important Physiologies

At all levels of biology—from the molecular level to the level of the whole organism and its ecology—there is a fundamental reciprocal relation between structure and function. We recognize that altering the function of a protein often requires that its structure be modified. Reciprocally, an alteration in the structure of a protein—particularly an alteration conserved in evolution—is readily accepted to reflect the existence of an important functional attribute. This essential accommodation between structure and function is also obvious at the level of the whole organism. For example, it is obvious that changes in the wing structures of different insect species enable differences in flight physiologies. At the cellular level, the reciprocal relation of structure and function should be viewed in the same manner as any other level of biology, but our understanding of this level remains very limited. Many of the diagnostic features of cancers are found at the cellular level, involving the NE (Fig. 1). Cancer is a cellular-level biological process whereby Darwinian natural selection acts on heritable variation (whether genetic or epigenetic) to result in expansion of cells into new microniches [4]. The unit of natural selection in cancer is the whole cancer cell, and therefore it should be expected that abnormal physiologies of cancer cells will be reflected in altered cellular-level morphology. The criteria of malignancy span the tissue level to the subcellular level evident by light microscopy: electron microscopy has not generally been able to expose more specific, fine-scale structural features that define malignancy [5], supporting the notion that key physiologies operate at the cellular–subcellular level.

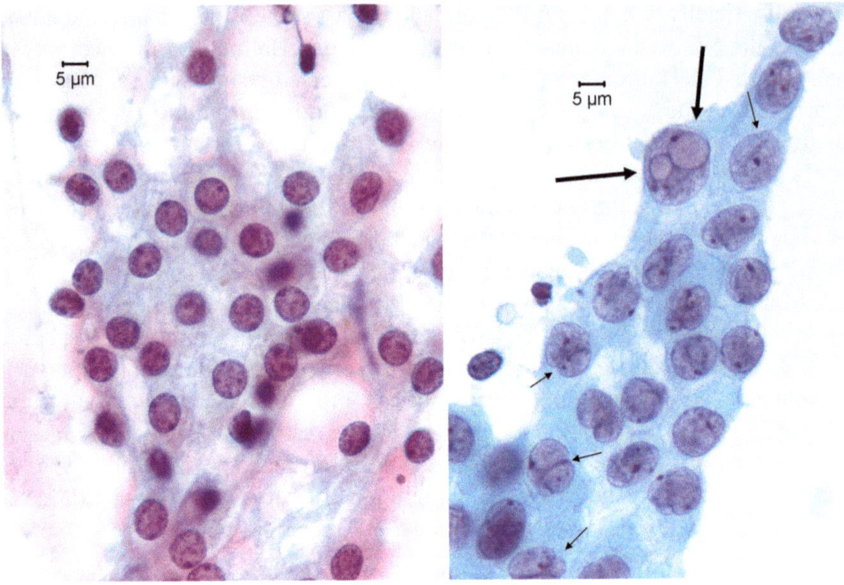

Fig. 1 NE structural changes are key diagnostic traits of some cancers. On the *left* is an alcohol-fixed, Papanicolaou-stained fine-needle aspirate of normal thyroid epithelial cells. The Papanicolaou stain is a modified hematoxylin and eosin stain that is still used for cancer diagnosis by cytopathologists. Note the rigidly *round*-to-*ovoid shape* of the normal thyroid nuclei, and the presence of small aggregates of heterochromatin, many of which are positioned against the nuclear lamina. On the *right* is a fine-needle aspiration biopsy with diagnostic features of papillary thyroid carcinoma (PTC), fixed and stained in the same manner as normal thyroid epithelium. Diagnostic features include the long linear infoldings of the NE (referred to in the literature as "nuclear grooves" (e.g., *short thin arrows*)). Also very important diagnostically are the spherical invaginations of cytoplasm into the nucleus termed "intranuclear cytoplasmic inclusions" (*long thick arrows*). Two intranuclear inclusions are present in one nucleus in this case. In addition to the lamina changes, PTC shows a relative dispersal of heterochromatin. PTC such as this is relatively genetically stable and commonly bears only a single detectable mutation in B-RAF, RET, or TRK (see text). A prediction is that B-RAF, RET, or TRK function by altering NE (and chromatin) organization to enable a new physiology (see text)

The reciprocal relation between structure and function is not exactly cause and effect, but generally structure must first be altered to alter function rather than the other way around. In accordance with our view at any other level of biology, cell structural features that distinguish normal from cancer cells should not be viewed as being merely a consequence of a preconceived notion of a cancer "hallmark." In general, new physiologies are only able to be characterized when structure is accounted for. For example, without studying human anatomy, it was hard to disprove the theory of the four humors of the ancient Greeks. Since some of the best diagnostic traits of cancers involve alterations in the NE, it seems imperative to fund research on the structural basis of diagnostic NE changes in order to gain insight into the functional changes of cancer cells.

Relation of the Histogenetic Classification of Cancer to the Criteria of Malignancy

Neoplasms are still mostly classified according to the cell from which the neoplasm arises. For example, pathology reports and cancer registries list "lung cancer" and distinguish it from "pancreatic cancer" because lung cancers arise from one or another normal cell within the lung, and pancreatic cancers arise from a pancreatic cell. The histogenetic classification has been important for predicting the existence of carcinogens (e.g., cigarette smoke in lung cancers) and for envisioning that cancer is a multistep process and predicting that premalignant phases should exist in the development of some cancers. In fact there remains uncertainty about whether there is one particular cell in any organ that gives rise to cancers within that organ or whether multiple cell types in one organ each have the potential to give rise to a cancer [6]. What is clear, however, is the same histogenetic type of cancer can have widely different morphologic features. This is true even if one restricts attention to tumors that share evidence of a common lineage or show evidence for a similar pattern of differentiation. For example, small-cell lung carcinomas (carcinomas are tumors that show evidence of epithelial differentiation, possibly because they arise from an epithelial cell) are defined in part by the presence of a fragile-appearing NE, whereas non-small-cell lung carcinomas show a relatively more rigid-appearing though often irregularly shaped NE (Fig. 2).

On the other hand, morphologically similar features are sometimes seen in cancers that arise from different cell types. For example, morphologically distinctive and diagnostically important "intranuclear cytoplasmic inclusions" are rarely exhibited by normal cells but are highly characteristic of papillary thyroid carcinoma (PTC) (Fig. 1), early stages in the development of some pancreatic adenocarcinomas (Lee PJ, Owens CL, Hutchinson L, Fischer AH, manuscript in press, Journal American Society of Cytopathology) (Fig. 3), some lung adenocarcinomas, melanomas and benign melanocytic nevi, and in some other tumors.

Thus the actual morphologic features that are diagnostic of a cancer are not necessarily related to the name given the tumor, and sometimes different histogenetic "types" of cancer seem to have more in common than two different cancers that can be given the same name.

The histogenetic classification is increasingly recognized to be incongruous with a characterization of tumors based on their genetic alterations: Similar genetic changes can be shared by tumors of diverse histogenesis, and different genetic changes can be seen in different forms of cancer that arise from the same cell type. In general, when the morphologic features are distinctly different for tumors arising from the same cell type, the underlying genetic abnormalities and clinical features are found to differ. In sharply demarcated examples, the distinctive tumors are given different names. For example, thyroid epithelial cancers include two principal types: follicular and papillary. These two types have different morphologic features (the former has a round nucleus similar to normal thyroid—see Fig. 1), different non-overlapping sets of mutations [7], and different clinical behaviors (the former metastasize by the bloodstream, whereas the latter metastasize early via

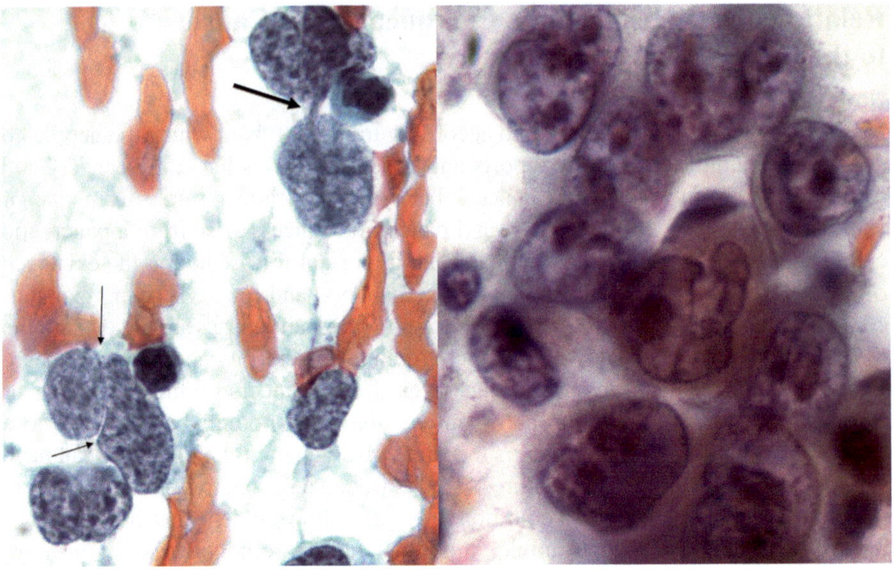

Fig. 2 Lung cancer NE changes. Fine-needle aspirations of small-cell lung cancer (*left*) and non-small-cell lung cancer (*right*, at slightly higher magnification) were prepared and stained as in Fig. 1. Small-cell carcinoma and non-small-cell carcinoma have a similar or an identical histogenesis, yet their morphology is distinctly different. Note the relatively fragile-appearing NE of small-cell carcinoma with rupture of one nucleus (*long thick arrow*) and nuclear molding in which the shape of one nucleus conforms passively with the shape of an adjacent nucleus (*short thin arrows*). In comparison, non-small-cell carcinoma shows more rigid-appearing nuclei that do not crush easily in biopsy preparations, but show stochastic nuclear shape abnormalities. Other important features that distinguish small-cell carcinoma include the scant cytoplasm, tendency for small-cell carcinomas to have internal foci of heterochromatin that seem to lack an affinity for the lamina, and inconspicuous nucleoli. These two morphologically distinguishable tumors, both highly genetically unstable, have different clinical features and different underlying patterns of cancer gene activations (see text)

lymphatics) [8]. Thus, the behavior of a cancer cell is reflected at least partly by its morphology which in turn is related to its genetic alterations, and the behavior is not always related to the histogenesis.

Nuclear Grading, the Degree of Differentiation, and the Relation to NE Changes

The "grade" of a tumor is based on the pathologist's qualitative assessment of how aggressive the tumor will be. Grading evolved based on the correlation of morphologic impressions with autopsy or other follow-up data. A higher grade denotes a more aggressive tumor. The criteria for grading differ for different histogenetic types of cancers. For renal cell carcinomas, grade is largely based on nucleolar

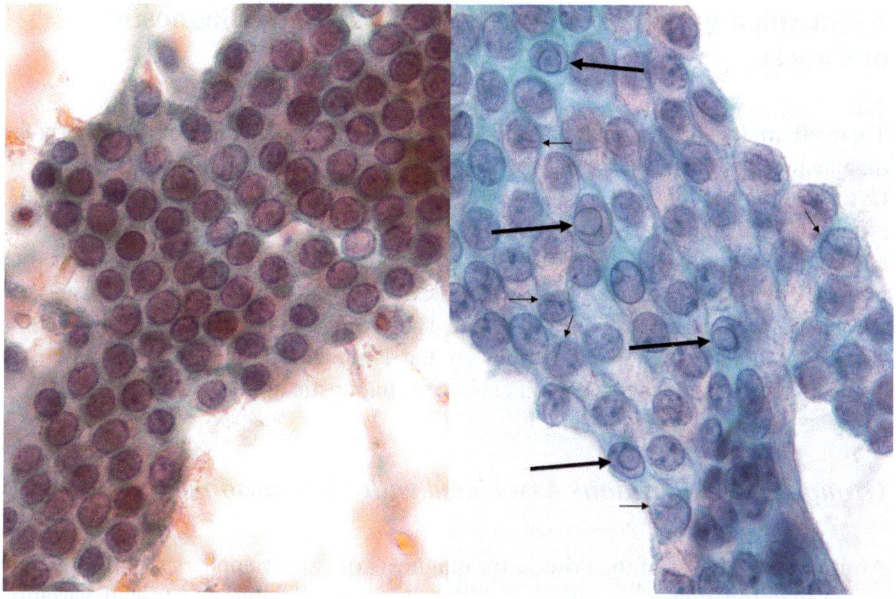

Fig. 3 Similar diagnostic structural features are often seen in tumors of different histogenetic origins. Fine-needle aspirations, fixed and stained as in Fig. 1, are shown with normal pancreatic ductal cells on the *left*, and the earliest known stage of a pancreatic adenocarcinoma is shown on the *right* (in this case an "intraductal papillary mucinous neoplasm"). Note the intranuclear cytoplasmic inclusions (*long thick arrows*) and nuclear grooves (*short thin arrows*) similar to the findings in PTC. By Papanicolaou staining, this pancreatic neoplasm is predicted to be chromosomally stable and diploid (see text). Both images are at identical magnification

prominence, with a smaller contribution from a subjective assessment of the irregularity of the NE [9]. For breast cancer, the grade is a composite of three features [10]: architectural features (how well the tumor cells produce glandular structures), the mitotic rate, and a "nuclear grade." The nuclear grade is the most subjective of the three, and it includes primarily a consideration of overall nuclear pleomorphism (which correlates very closely with measures of chromosomal instability and may or may not include loss of round-to-oval shape of the nuclear contour [11, 12]). The three features are given a score of 1–3, and the sum of all three is used to determine grade, where 3–5 points are Grade 1 (good prognosis), 6–7 points are Grade 2, and 8–9 points are Grade 3 (worst prognosis). Some Grade 1 tumors may have nuclear shape abnormalities (for example, some near-diploid lobular carcinomas that can show deep infoldings of the NE), and some Grade 3 tumors may have relatively round–oval nuclei. Thus, nuclear shape change is only a component of breast cancer grading, and the grade does not specify an exact phenotype. Pathology reports that include a histogenetic classification and a grade may not be able to disclose important associations between morphologic changes in the NE and particular NE proteins or functional measures. Therefore, it will be useful to engage pathologists in studies of NE proteins or physiologies to identify correlations with NE morphology.

A Classification of the NE Structural Changes Diagnostic of Cancer

There are many varied interpretations of the biologic significance of the criteria for diagnosing cancer (the "criteria of malignancy") [13, 14]. The American Society of Cytopathology assembled a Cell Biology Liaison Working Group to interpret the criteria from the perspective of the relevant cell biology literature [15]. The Working Group derived a classification of all of the criteria in a manner that was hoped to provide a common perspective between the cell biologists studying cancer and the cytologists who actually diagnose cancer. The present review is an extension of this concept, focusing on just alterations of the NE. The classification is shown in Table 1, and it includes three main classes of diagnostic abnormalities.

Group 1: NE Alterations Associated with Chromosomal Instability

An important subset of the criteria for diagnosis of cancer appears to simply relate to the identification of genetic instability. An unpredictable cell-to-cell variation within a population ("pleomorphism") in any morphologic feature is an indication for instability in the phenotype of the cells, and an unstable phenotype predicts some form of genetic or epigenetic instability. This group of cytologic criteria is the most familiar to non-pathologists, and it usually includes prominent pleomorphism in the size and shape of nuclei (Fig. 4). The type and degree of nuclear shape change can differ in different chromosomally unstable tumors (for example, see Fig. 2).

Table 1 Classification of NE abnormalities in cancer cells

Group 1. Changes associated with chromosomal instability.
• Cell-to-cell variation in NE size and shape. Deep infoldings, aneurismal outpouchings, and polylobulation that vary unpredictably from cell to cell, accompanied by unpredictable variation from cell to cell in total DNA content and variation in other cellular features.
• Presence of micronuclei. Probably functionally related to chromothripsis.
Group 2. Conserved NE structural features within a genetically unstable population, of unknown functional significance.
• Increased nuclear lamina surface area or nuclear volume compared to cytoplasmic volume ("increased N/C ratio"). Frequently associated with nucleolar prominence.
• Fragile nuclear lamina of small-cell carcinoma.
Group 3. NE alterations that occur in the absence of chromosomal instability, directly linked or possibly linked to activations of specific cancer genes, all of unknown functional significance.
• Long nuclear infoldings ("grooves") induced by RET/PTC, TRK/PTC, and probably activated B-RAF in papillary thyroid carcinoma.
• Intranuclear cytoplasmic inclusions induced by RET/PTC and possibly TRK/PTC and activated B-RAF in papillary thyroid carcinoma.
• Nuclear grooves in granulosa cell tumor, possibly mediated by FOXL2.
• Nuclear grooves in Langerhans cell histiocytosis, possibly mediated by activated B-RAF.

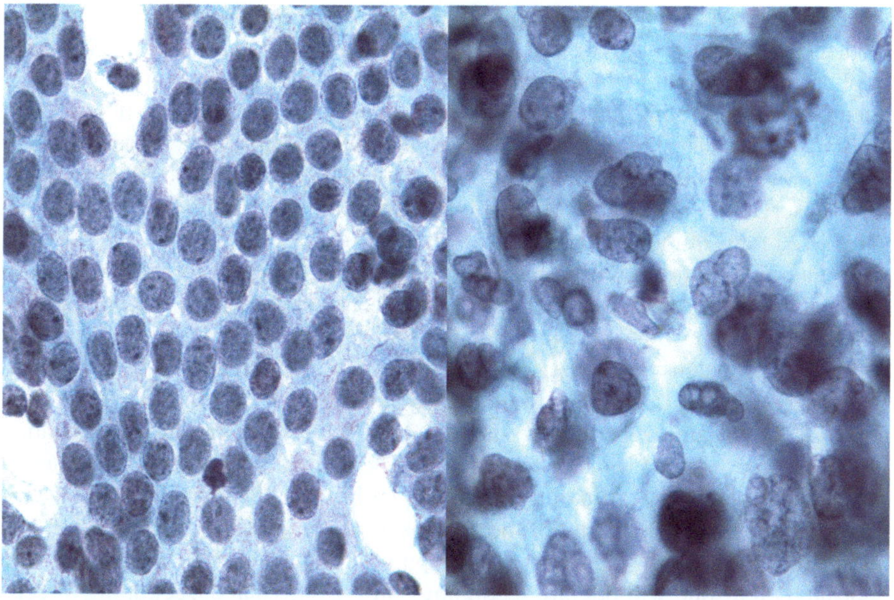

Fig. 4 Normal breast ductal cells and ductal carcinoma. Fine-needle aspiration samples prepared and stained as in Fig. 1 to show a comparison of normal ductal cells (*left*) with a "high-grade," chromosomally unstable, and aneuploid ductal carcinoma (*right*). Note the wide variety of nuclear shapes of the aneuploid carcinoma. Some nuclei are relatively spherical, while others show various lobulations, aneurismal like outpouchings, and some longitudinal infoldings. In addition to unpredictable variation in nuclear shape and variation in total amount of hematoxylin (reflecting chromosomal instability) there is also unpredictable variation in heterochromatin patterns, possibly representing an "epigenetic instability." Both images are at identical magnification

The major form of genetic instability directly recognized by pathologists corresponds to the cell biology concept of "chromosomal instability," present in about 90 % of solid tumors [16]. Chromosomal instability was recognized in the nineteenth century using stains and microscopy techniques that are still in use today, before the chromosomal basis of inheritance was fully articulated [17]. In studying the morphology of cancer cells, von Hansemann noted that the variation in the degree of chromasia and the size and shape of interphase nuclei correlated with variation in the numbers of chromosomes segregated per daughter cell through mitosis [18]. Within the decade, Theodor Boveri demonstrated that abnormal numbers of centrosomes could induce abnormal segregation of chromosomes in the mitotic progeny of early-developing sea urchin embryos, and he later postulated the link between centrosome abnormalities and chromosomal instability [19]. Centrosome abnormalities and asymmetries in the mitotic spindle apparatus still provide strong evidence for diagnosing a chromosomally unstable cancer. Chromosomal instability is a complex heterogeneous phenotype involving potentially many specific defects [16, 20]. Micronuclei are one particular manifestation of chromosomal instability, commonly representing mis-segregated chromosomes that acquire an independent NE from the bulk of the chromosomes [16]. They can be detected with hematoxylin and eosin (H&E) staining [21]. Recent studies have implicated an aberrant DNA damage response within micronuclei that results in the phenotype of chromothripsis [22]. Thus, distinctive structural changes in the NE of chromosomally unstable cells have distinct functional significance. Other subtleties of chromosomal instability have not been obvious from just classical histopathologic observation, and in the following discussion the phenotype is simplified and considered to be homogeneous.

In interphase, chromosomal instability is manifest as a cell-to-cell variation in the total amount of hematoxylin (variable degree of chromasia, reflecting variable numbers of chromosomes) per nucleus. The total amount of hematoxylin that stains an interphase nucleus bears a reasonably close approximation to total DNA content, as evidenced by parallel Feulgen staining (a quantitative DNA stain for light microscopy) or by restaining pathology samples with fluorescent in situ hybridization probes to enumerate chromosome numbers. The estimation of total DNA content per cell is difficult in paraffin-embedded samples because the sections do not necessarily include the full diameter of the nucleus. Cytopathology preparations are better suited for disclosing this particular diagnostic feature because whole cells are deposited on a slide without any sectioning. On the other hand, sectioning of cells and tissues is helpful for disclosing other types of diagnostic changes, particularly changes in the architecture of tissues. There are stable forms of aneuploidy that can probably be predicted just on H&E staining. For example, a distinct subset of breast cancers have an aneuploid but stable chromosome content and lack centrosome abnormalities [23]. Other measures of chromosomal instability besides FISH or measurement of total DNA content are being developed for diagnosis and prognostication [24]. Such new techniques may be easier to apply than FISH and would likely be more sensitive and reproducible than the estimates that are obtained with the unaided human eye.

Polyploidization is biologically distinct from aneuploidy [25], and it is also manifest as a variation from cell to cell in total hematoxylin staining. However the variation from polyploidization results in discrete stepwise, geometric doublings of total hematoxylin or DNA content. Polyploidization is also accompanied by a predictable approximate doubling of cytoplasmic volume such that the ratio of nuclear to cytoplasmic volume does not change. For most cell types, polyploidization is not associated with an alteration in the texture of the cytoplasm or in the pattern of heterochromatin formation in the nucleus compared to corresponding diploid cells: Polyploid cells look like big variants of normal cells. Importantly, polyploid cells usually have the same overall nuclear shape as their diploid counterparts (generally round to oval). Megakaryocytes are an important exception to this rule, since these cells characteristically develop irregularly shaped nuclei during polyploidization, along with maturation of the cytoplasm to enable formation of platelets. Polyploidization is physiological in megakaryocytes and trophoblastic cells, it is common in cardiac myocytes that have been mechanically stressed, it is well documented in aging hepatocytes, it is common in several endocrine tissues such as adrenal and thyroid, it is common in umbrella cells of the bladder [26], and it appears to be an occasional finding in cervical squamous cells in Pap tests [15, 25]. Polyploidization should predispose cells to the development of aneuploidy [25]; however, it is paradoxical that polyploidy appears very commonly in clinical samples, and polyploidization is documented to lack prediction of a risk for subsequent tumor development [26].

Microsatellite instability is another form of genetic instability, and microsatellite unstable carcinomas commonly have a grossly diploid DNA content [27]. Nevertheless, microsatellite unstable carcinomas can have highly irregular nuclear shapes [28].

Dissecting the significance of NE irregularity that accompanies chromosomal instability—Most chromosomally unstable malignancies can show extensive variation in the size and shape of the nucleus (Figs. 2 and 4). The degree of variation in nuclear size and shape persists at a grossly similar level during the natural history of chromosomally unstable cancers. Nuclear shape does not appear to become progressively more irregular or less irregular over time, or in different metastatic sites, at least not to a degree that can be discerned with the unaided human eye. Thus, nuclear shape pleomorphism is not used by pathologists to distinguish the degree of invasiveness of a sample or to determine whether a sample is from a primary site or a metastasis. The diagnosis of invasion is based generally on large-scale tissue architectural features (reviewed in [15]), whereas the diagnosis of a metastatic focus is based on knowing where the biopsy comes from rather than on any specific cellular-level structural change. Conservation of structure through evolution predicts that the particular structure is functionally significant. In the evolution of metazoans, a rounded to ovoid nuclear shape seems to be selected for. Yet, in the clonal evolution associated with chromosome instability, nuclear shape is apparently NOT conserved, as if there were no Darwinian selective pressure at all for a particular shape! Genetically unstable cancer cells seem to lose control over the retention of a normal round-to-ovoid shape. At the same time, many cancer cells may also remain relatively rounded, and there is no indication from classical pathologic examination that cells with rounded or irregular shapes have different properties. Extracting biological significance from the NE pleomorphism associated with chromosomal instability seems like a difficult prospect, and the evolutionary argument can easily be raised that there may be NO biological significance to NE pleomorphism accompanying chromosomal instability.

On the other hand, classical pathologic examination is subjective and qualitative. Improved objective measures of NE alterations in cancer cells could help disclose structural features that have physiologic significance (e.g., objective measurements that may be able to predict a higher risk of metastasis). Classical pathologic examination has been limited to snapshot images of cells, hindering identification of any associations between morphology and subsequent cellular behavior. Techniques for visualizing the lamina in living explants of human cancer cells (i.e., in their native tissue environment) have only become available within the past 10–15 years, but these techniques have not yet been widely applied to studying human tumor dynamics in their native microenvironment. Experience at all other levels of biology suggests that observation of the dynamic features of a system provides much greater insight than static observations. Dynamic studies of the NE in transient cultures of human microbiopsy samples would seem likely to expose biologically relevant associations that cannot be predicted or identified in static images. The tremendous biological variation and pleomorphism of the nuclear lamina in chromosomally unstable tumors provides an easy opportunity (a veritable biological library of phenotypes) to scan for correlations between NE morphology and cell behaviors.

In vitro studies of the NE have shown unexpected dynamics, including transient rupturing of the NE in interphase [29]. We performed live cell imaging of chromosomally unstable DU-145 prostate cancer cells, growing in monolayer cultures and expressing GFP lamin A. Unexpectedly, some cells had statically irregular nuclei whereas other cells showed dynamic interphase deflections of the lamina [30]. While we did not observe an obvious correlation between dynamic interphase deflections of the NE and cell migration, it seems possible that within tissue, nuclear shape pleomorphism could be relevant to cell migration. The extremely lobulated nuclear contour of a normal mature neutrophil is likely to have functional significance by allowing the nucleus to fit through tight spaces between cells [31]. Lamin B receptor (LBR) is required for complete lobulation of the neutrophil nucleus, and LBR mutations in humans give rise to the Pelger–Huet anomaly characterized by hypolobulated neutrophil nuclei and potential defects in neutrophil migration [32, 33]. We were able to show that neither cytochalasin nor nocodazole (drugs that block polymerization of actin and microtubules, respectively) arrested the dynamic deflections of the lamina suggesting that biomechanical forces that deform the NE are not based on actin microfilaments or microtubules. Further, the appearance of cells with a collapsed cytoskeleton suggested the unanticipated presence of an intranuclear, possibly chromatin-based, force that actively deforms the interphase NE in a subset of chromosomally unstable DU145 cells. Taking a dynamic approach can permit many questions to be asked: Are dynamic NE shape changes restricted to particular phases of the cell cycle? Are the dynamics of NE reassembly or smoothening altered in cells with particular forms of interphase NE irregularity? Do cells with dynamic interphase NE irregularities have a greater or a lesser degree of sensitivity to chemotherapeutic drugs? Do transient NE disruptions [29] occur with different frequencies in NEs with statically irregular, dynamically irregular, or statically round shapes?

In interpreting studies that attempt to identify prognostic implications of NE changes, it will be important to be aware that chromosomal instability and NE pleomorphism are tightly correlated. As noted above, low-grade carcinomas tend to be genetically stable whereas high-grade carcinomas tend to show evidence of chromosomal instability. Many low-grade tumors also tend to have relatively more rounded nuclear shapes than high-grade tumors (see below in Group 3 for the important exceptions). It is well established that the "grade" of a tumor is closely related to the clinical or the biological aggressiveness of the tumor. Thus, a priori, one expects that cancers with high degrees of NE irregularity will yield an excess of high-grade chromosomally unstable tumors that have a worse prognosis. Separating any effects of NE irregularity per se from other consequences of the various types of chromosomal instability is a very difficult problem.

Certain changes in NE configuration are sometimes conserved in spite of the accelerated evolution of tumor cells conferred by genetic instability. The conservation of a structural feature in evolution is usually an indication that the structural feature has functional significance. Two such conserved NE structural features in chromosomally unstable tumor cells are included in the next group of diagnostic NE changes in cancer.

Group 2: Diagnostic NE Alterations That Are Conserved in Chromosomally Unstable Tumors, as if They Reflect an Important Physiology

(a) *Nucleocytoplasmic ratio*—An increased nuclear volume compared to the cytoplasmic volume (the so-called N:C ratio, often called karyoplasmic ratio) is commonly observed in cancer cells compared to normal cells [34–36]. Increased N:C ratio is particularly prominent in tumors with morphologic features suggestive of chromosomal instability, but its conservation within the genetically unstable population is a compelling evidence that it has an important functional relation.

The ratio of the size of the nucleus to the amount of cytoplasm for particular normal cell types is highly conserved through evolution (reviewed in [37, 38]), as if the N:C ratio has functional significance in normal cells. Nuclear size can be thought of as being dependent on DNA content and average DNA density [39, 40]. In addition to these two factors (which can both vary within a population of tumor cells), nuclear size bears an uncharacterized relation to cytoplasmic volume. Experimental manipulations that alter overall cell size in yeast, *Drosophila*, and *Caenorhabditis elegans* are associated with commensurate changes in nuclear volume (apparently through changes in DNA compaction) to preserve a relatively constant ratio of nuclear volume to cytoplasmic volume [41] (reviewed in [37, 39, 42]). It is important to point out that different types of normal human cells have different N:C ratios, and the same cell type can have widely different N:C ratios depending on physiologic changes. For example, during activation of a fibroblast in a healing wound, there is enlargement of the nucleus associated with conversion of their abundant heterochromatin to euchromatin, but there is an even greater increase in the amount of cytoplasm.

The increased nuclear volume is clearly related to the fact that tumor cells with chromosomal instability generally have a mean DNA content that is higher than diploid. If average DNA density is otherwise similar between hyperdiploid cancer cells and a normal diploid cell, then necessarily the nuclear volume would be increased. Nuclear enlargement per se could therefore be a functionless, passive consequence of a selection for increased DNA content in chromosomally unstable tumors. Several explanations for the apparent selection for a hyperdiploid DNA content of tumor cells have been offered [16]: (1) Increasing the chromosome number decreases the chance that all copies of a vital chromosome would be lost in the asymmetric cell divisions of chromosomal instability.

(2) Larger chromosome numbers also tend to stabilize relative transcript and protein abundances when cells lose or gain an extra copy or two of a chromosome. Thus, hyperdiploidy reduces "proteotoxic stresses" such as overactivity of proteasome degradation pathways and unfolded protein responses [16].

N:C ratio by itself is less important for cancer diagnosis than the combination of an inappropriately scant cytoplasm compared to the signs in the nucleus of activation [36]. During physiological activation, cells acquire a euchromatic appearance accompanied by overall nuclear enlargement including enlargement of nucleoli and an increase in the total cytoplasm and basophilia of the cytoplasm (the basophilia or the weak hematoxylin staining of the cytoplasm reflects increased numbers of ribosomes as evidenced in electron micrographs). In contrast, the phenotype of cancer cells with an increased N:C ratio includes the increased euchromatinization and enlargement of nuclear and nucleolar size but without signs of increased protein production in the cytoplasm. This phenotype implies that nucleolar or ribosomal function is somehow closely related to the abnormal N:C ratio of cancer cells [30].

(b) *Nuclear fragility and rigidity*—Nuclear fragility is a particularly prominent diagnostic trait of "small-cell carcinoma" (Fig. 2). Small-cell carcinoma is a morphologically distinctive tumor that can arise from nearly any site in the body. The most common site of origin is lung. Inactivating mutations in both p53 and Rb are found in about 90 % of small-cell carcinomas of lung, and amplification of Myc is seen in about 20 % of cases [43]. Small-cell carcinoma is among the most aggressive of human malignancies, essentially always exhibiting metastases when first diagnosed, and it is rapidly fatal if untreated. The tumor is highly genetically unstable, as shown by whole-genome sequencing [44]. In spite of the obvious genetic instability, there is a striking conservation of a series of morphologic characteristics (Fig. 2) [15, 45]: (1) The NE of small-cell carcinoma appears to lack strength, as estimated by observing how easily nuclei crush during the biopsy or the brushing procedure or crush during preparation of a cytologic smear of the sample. While the adjacent normal nuclei withstand the biopsy and smearing, small-cell carcinoma nuclei selectively are ruptured or crushed. (2) The nuclei seem to have so little rigidity that they easily "mold" or conform to each other or to other objects in their vicinity. (3) Possibly mechanistically related to the fragility is a peculiar chromatin organization of small-cell carcinoma. The NE of small-cell carcinomas appear to lack the usual affinity for heterochromatin of other cell types. It is as if heterochromatin aggregates are as likely to be present in a central location as they are in a peripheral location. (4) The chromatin of small-cell carcinomas also generally lacks confluent areas of euchromatin. (5) Another conserved feature in small-cell carcinoma is a very high N:C ratio with scant cytoplasm. (6) Moreover, the cytoplasm lacks circumferential arrangement of cytoplasmic intermediate filaments (keratins and vimentin). The scant cytoplasmic intermediate filaments commonly condense into a single dot-like focus. The scant cytoplasm with scant disorganized intermediate filaments may not explain the fragility of the nuclei of small-cell carcinoma since the cytoplasm of many tumor types can be seen to easily strip away from the nuclei during preparation of a smear, yet these stripped naked nuclei appear relatively rigid compared to small-cell carcinoma. (7) A final diagnostic trait of small-cell carcinomas is that they lack large nucleoli, a paradoxical finding given their extremely high turnover.

"Small" is relative. Most small-cell carcinomas are larger than the normal cells that they arise from, but other carcinomas tend to be much larger. Small-cell carcinomas lack lamin A/C compared to non-small-cell carcinomas of lung [46] (e.g., right side of Fig. 2). However, carcinoid tumors (a low-grade genetically stable tumor believed to be histogenetically related to small-cell carcinoma) also lack lamin A/C [46] and yet have relatively rigid round-to-ovoid nuclei.

Group 3: Diagnostic Changes in the NE in Near-Diploid Tumors

The background of genetic instability in the first two classes makes it difficult to dissect particular physiologies associated with specific cancer genes or specific NE proteins. In contrast, the diagnostic traits in this third group appear far more approachable, and these changes deserve more attention.

(a) *Papillary thyroid carcinoma*—PTC is described in detail elsewhere [15, 47–49]. Briefly, PTC is diagnosed on the basis of nuclear structural changes involving the NE as well as chromatin. Two principal types of tumors arise from normal thyroid epithelial cells: PTC and follicular neoplasms [50]. Both normal thyroid epithelium and thyroid follicular neoplasms show a round-to-oval nucleus with distinct aggregates of heterochromatin (Fig. 1), whereas PTC nuclei show various NE irregularities [51] and dispersal of heterochromatin into fine aggregates. Nuclear size is modestly increased. Two distinctive types of irregularity of the NE in PTC are present. "Nuclear grooves" or long longitudinal inward folds of the lamina traversing half or more of the nuclear diameter are common. Often several folds may be present in one nucleus, intersecting at relatively random angles to each other. The folds in the nucleus can be very shallow or extend half way or deeper into the nucleus. In addition to these relatively linear infoldings of the NE, a highly distinctive and very diagnostically important "intranuclear cytoplasmic inclusion" is present in a generally small proportion of the nuclei of PTCs. These inclusions are rarely present in more than about 5 % of tumor cells [51]. The inclusions appear as a very spherical shaped invagination of otherwise unremarkable cytoplasm part way into the nucleus. The diameter of the inclusions can vary from near the limit of light microscopic resolution to diameter of the whole nucleus. Sometimes more than one inclusion can be found in one nucleus, and the inclusions sometimes merge with a relatively straight line demarcating the two. Electron micrograph studies have not disclosed a specific cytoplasmic feature in the invaginated segment [52, 53]. One immunohistochemical study showed an accumulation of beta catenin in the center on the cytoplasmic side of some of the inclusions [54]. Another study has reported that centrosomes may be located within the center of some of the inclusions [55]. The nuclear lamina underlying these inclusions contains lamins A/C, B1, emerin, and Lap2 epitopes at the same relative intensity of staining as the rest of the lamina [48, 56].

In addition to the lamina changes, chromatin changes are useful for diagnosing PTC. Compared to normal thyroid epithelial cells, there is a marked dispersion of heterochromatin into smaller aggregates (Fig. 1); however, the total amount of heterochromatin is unaltered [57].

Importantly, different clinical features and different sets of mutations are found in PTC compared to thyroid follicular neoplasms [7]. Translocations in RET or TRK tyrosine kinases and point mutations in B-RAF are restricted to PTC. While many other epithelial cancers progress through obvious dysplastic or in situ changes, there is no other histologic intermediate between normal thyroid and PTC. Both PTC and follicular neoplasms tend to be relatively genetically stable, and translocations in RET were shown to be present within all of the cells in some small tumors. These observations raised the question of whether RET translocations were sufficient to cause PTC and led to an experiment to test whether the RET oncogene would be sufficient to induce the nuclear features of PTC if the tumor indeed arose in one step. Normal thyroid epithelium is easy to culture, and direct demonstration of the induction of chromatin dispersal and NE irregularity was shown after introducing the RET tyrosine kinase [58]. Subsequent studies showed that analogous translocations of the TRK tyrosine kinase found in some PTCs were sufficient to induce the same nuclear findings, and a phosphotyrosine docking site for SHC/FRS2 shared by RET and TRK was shown to be essential for nuclear restructuring [59].

The time course for the development of nuclear irregularity following expression of RET was shown to be within 6 h, without a need for an intervening mitosis and post-mitotic nuclear reassembly [60]. Thus RET induces NE irregularity within interphase.

Point mutations in B-RAF were later found to be present, mutually exclusive with RET translocations, at an early time in the development of many PTCs, and mouse transgenic models of B-RAF activation in the thyroid suggest that B-RAF functions identically to RET or TRK in altering the NE. Activating mutations of H-RAS do not alter nuclear shape in normal thyroid epithelium [58]. This is paradoxical because RAS may be required to transduce a signal between tyrosine kinases RET or TRK and B-RAF [7]. H-RAS is an uncommon mutation in the thyroid, but it is specifically associated with follicular neoplasms that do not show nuclear contour irregularities of PTC.

B-RAF mutations are also found as an early event in melanocytic neoplasia, present in over half of benign moles ("melanocytic nevi") and a similar proportion of melanomas. Benign nevi and melanomas both show intranuclear cytoplasmic inclusions identical to those of PTC. Compared to the clean genetic background of PTCs, melanomas can have many complex abnormalities and, according to the classification of NE in cancers, intranuclear inclusions in melanocytic neoplasms therefore would appear to belong in the second group of diagnostic NE changes in cancers. Intranuclear cytoplasmic inclusions are uncommon in any nonneoplastic cells, but they can be seen in ostensibly normal hepatocytes. They are common in hepatocellular carcinoma, some lung adenocarcinomas, and particularly pulmonary adenocarcinomas with EGFR mutations [61]. Recently, intranuclear inclusions were shown to be present in about one-third of cases of the earliest known stage of development of pancreatic neoplasia (Fig. 3). Understanding the precise structural basis of intranuclear cytoplasmic inclusions would improve the diagnosis of thyroid and other tumors.

The actual targets of RET or TRK that effect the alteration in chromatin and NE are completely unknown. Regarding the functional significance of the changes induced by RET (and probably B-RAF), there are some important observations: (1) Transcription patterns change relatively little compared to either follicular-type neoplasms or normal thyroid tissue [62, 63], yet the change in the chromatin is quite dramatic compared to differences between other cell types. (2) The distribution of active RNA Pol II, distribution of splicing factors, and intranuclear distribution of nuclease hypersensitive sites are not grossly different between PTC and follicular neoplasms (reviewed in [15]). There is therefore essentially no evidence that the nuclear restructuring associated with oncogene activations functions to alter transcription. (3) Measurements of cell cycle kinetics and apoptotic rates do not distinguish PTC from the histogenetically related follicular neoplasms, suggesting that the functional changes are independent of these central "hallmark" cancer traits [64, 65] (reviewed in [15]).

(b) *Langerhans cell histiocytosis*—Langerhans cell histiocytosis is a relatively genetically stable tumor with a variable but usually indolent clinical course. B-RAF mutations are found in a majority of cases (Fig. 5) [66]. A key diagnostic feature is the presence of long nuclear grooves very similar to PTC: intranuclear cytoplasmic inclusions are not a common feature of this tumor.

(c) *Adult granulosa cell tumor*—Adult granulosa cell tumor is a rare type of ovarian cancer with a mutation-free background except for a point mutation in *FOXL2* gene [67]. This tumor is clinically unpredictable and can metastasize. "Coffee bean"-shaped nuclei in which there is a modestly deep and long nuclear groove is a key diagnostic trait (Fig. 6). Normal granulosa cells have relatively round nuclei. A direct test of whether FOXL2 induces nuclear grooves in granulosa cells has not been done.

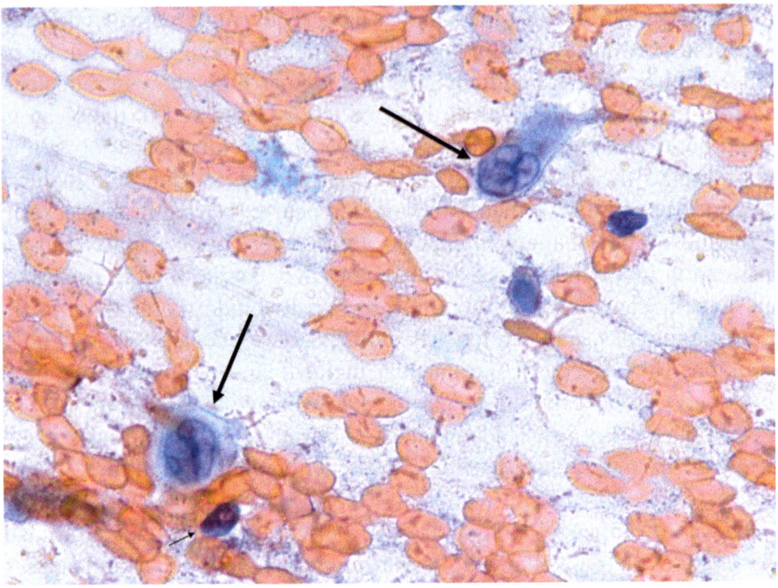

Fig. 5 Fine-needle aspiration sample of Langerhans cell histiocytosis. The neoplastic cells (*long thick arrows*) are characterized by the presence of long nuclear grooves similar to papillary thyroid carcinoma. One lymphocyte is present (*short thin arrow*) along with a couple of degenerated cells. Red blood cells are stained *orange* in this and the other images. B-RAF mutations are present in a majority of cases of Langerhans cell histiocytosis (and were present in this case)

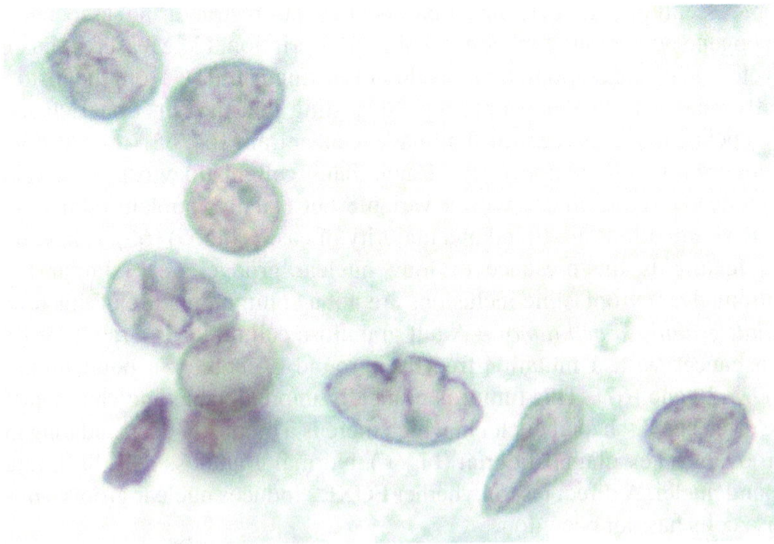

Fig. 6 Fine-needle aspiration sample of a metastatic granulosa cell tumor. This diploid tumor is characterized by "coffee bean"-shaped nuclei bearing long nuclear grooves similar to Langerhans cell histiocytosis and papillary thyroid carcinoma. Mutations are restricted to a single gene (*FOXL2*) in this tumor. The hypothesis that *FOXL2* mutations directly induce NE irregularity has not been tested

Dissecting the Functional Significance of the NE Changes in Group 3

Nuclear grooves and intranuclear cytoplasmic inclusions are not seen in every single cell in a given tumor. It is not as if a cancer gene like RET was acting like Gregor Mendel's famous wrinkled pea gene. Pathologist's observations suggest that cancer genes loosen the otherwise tight requirement for a rounded or an ovoid NE shape, releasing the constraint in a particular and reproducible manner in at least a portion of cells. In spite of the indirect nature of cancer genes' effects on the NE, the analogy with conventional Darwinian evolution is compelling. NE structural changes in this third group are akin to the "speciating" characteristics that a field biologist would use to identify differences between related species [4]. Evolutionary biologists widely recognize that the structural differences that distinguish related species are due to the genetic differences that were functionally relevant to their speciation. From this same perspective, the NE structural changes in Group 3 that distinguish normal cells from neoplastic cells should be expected to be induced by cancer genes, and the structural changes should, in some manner (possibly very indirectly), reflect the functional effects of the cancer genes.

Of all these models, PTC is the most manipulable. Relatively abundant normal human thyroid tissue is frequently excised as part of the treatment of thyroid nodules, and cultures of normal human thyroid epithelium are easy to establish [60]. Normal melanocytes, Langerhans cells, or granulosa cells are relatively scarce, and protocols for culturing these cells are not well established.

Caution needs to be taken in adapting studies to in vitro models. While RET induces nuclear shape/size changes that are essentially identical to PTC when expressed transiently in normal thyroid cells growing in monolayers, we and others [68] have consistently observed that cultures become dominated within a few passages by cells that lose the characteristic nuclear morphology of PTC (in spite of selecting for expression of the transgene). It is as if the artificial environment of a plastic dish selects against the phenotype of an irregular NE of PTC. Likewise, PTC cell lines established from patients do not show nuclear features of PTC in our experience. This author has little experience with other cell lines such as small-cell carcinomas, but based on the differences observed between patient samples of PTC and in vitro models of PTC, researchers should be careful about using monolayer culture systems to model the NE changes of in vivo neoplasms. This caveat is understandable from an ecological or an evolutionary perspective: changes in evolutionary fitness are always closely dependent on the particular environment in which the fitness change is selected.

Sparse Association of NE Lamina Features with Tissue-Level Diagnostic Changes

The criteria of malignancy include important tissue-level architectural changes. Most fatal cancers in humans are carcinomas (derived from epithelial cells), and normal epithelial cells bear a strict dependence on a basal lamina (not to be confused with nuclear lamina) connection for their survival. Even if mitotic rates went through the roof, cell growth is constrained by this requirement and cells quickly run out of room for clonal expansion. The tissue-level criteria of malignancy expose constraints on cell growth, and these criteria appear to relate directly to the mechanisms by which normal cells overcome the various growth constraints. One change that takes place at an early stage is that epithelial cells become more crowded on the available basal lamina. Such crowded cells develop a more columnar shape, and the nucleus becomes elongated. This is a prominent feature of the earliest stage of colon cancer development ("cigar-shaped nuclei"). The nuclear elongation could easily be considered to be a consequence rather than a cause of the crowding. Still, the nucleus must be able to accommodate the change in shape during crowding, and such an accommodation may place a selection pressure on a wide variety of NE physiologies. For example, the spacing between genes and the NE would have to alter, and either the lamina surface area or the nuclear volume must adapt to crowding. Further, the stage of crowding seems to put a selection pressure for the acquisition of the NE-associated pathophysiology of "pseudostratification" or the loosening of the otherwise strict basal positioning of nuclei.

Another early mechanism for clonal expansion is reflected in the tissue-level criterion of "true stratification," probably a manifestation of the cell biology concept of resistance to anoikis. Anoikis refers to the death of cells that detach from a basal lamina connection (from the Greek, meaning "without a home"). The main diagnostic feature of true stratification is identification of epithelial cells that grow freely whether or not they are attached to a basal lamina. A lesser diagnostic feature of true stratification is that elongation of nuclear shape tends to diminish (nuclei tend to become more rounded) and nuclear size tends to increase along with some increase in the amount of euchromatin. Pathologists describe this diagnostic change as the acquisition of "room to breathe" for the nucleus.

The ability to grow independent of a basal lamina connection is not sufficient to allow a clone of epithelial neoplastic cells to "invade" or metastasize. The diagnostic feature of "invasion" is very specifically the loss of responsiveness of epithelial cells to normal stromal tissue landmarks, as described in detail in [15]. Importantly, there are not any NE shape changes that reliably distinguish an in situ from an invasive cancer, and there are not any NE shape changes that reliably distinguish an early invasive cancer from a metastasis, at least not with conventional morphologic techniques.

Nuclear Shape Changes That May Be Seen in Normal Cells

Poorly preserved cells often have a shriveled appearance, as though the nucleus has lost volume, and the NE of such cells is irregular but with a characteristic "raisin-like" appearance [36]. Pathologists know to ignore such shriveled-appearing nuclei. During differentiation, the spherical nuclear shape of a promyelocyte becomes converted over a period of days into the multiple-lobed heterochromatinized nucleus of the terminally differentiated neutrophil. The shape of the neutrophil nucleus apparently permits migration through tight intracellular junctions. During lymphocyte

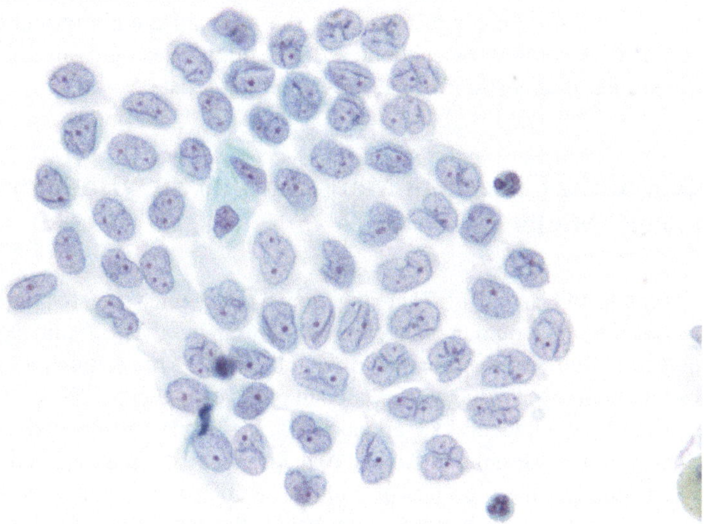

Fig. 7 Nuclear shape changes in normal cells. Normal mesothelial cells usually have a *round*-to-*ovoid nuclear* shape, but on occasion (with unclear clinical correlation), mesothelial cells can show long nuclear grooves and polylobulations essentially identical to papillary thyroid carcinoma or other malignant tumors. Any hypothesis for a functional significance of NE changes in cancer needs to account for observations such as these

activation, the nucleus can become irregular, accompanied by massive chromatin reorganization. The exact timing of the NE irregularity during lymphocyte activation is not well characterized but is presumed to coincide with precise stages of differentiation. The functional significance has not been well explored, but lymphocytes share with neutrophils the requirement to migrate between cell junctions. Despite abundant research on lymphocyte activation, the morphologic features of this process are surprisingly obscure.

For reasons that are not at all apparent, various epithelial cells can uncommonly show dramatic NE irregularity. For example, normal and physiologically activated mesothelial cells generally have a very round nuclear shape, but occasionally highly lobulated nuclear contours may be encountered with no obvious clinical significance (Fig. 7). Irregularity such as this can lead to diagnostic errors. The pathologist must develop familiarity with the occurrence of such anomalies and be able to use more than one diagnostic feature to justify a diagnosis of malignancy.

Any hypothesis for a functional significance of NE irregularity in cancer needs to account for observations such as these. One reasonable hypothesis is that normal cells sometimes require an irregular NE shape for a transient physiology; the physiology in normal cells can then switch off, whereas in malignant cells the physiology is locked in. Other explanations can invoke peculiarities of the particular cell of origin (e.g., NE irregularity is only associated with an altered physiology in some cell types, such as nuclear shape changes in neutrophils to accommodate cell migration). Until there is a better explanation for the precise structural basis of NE irregularities, it also remains possible that there are different physical bases for the NE irregularities sometimes encountered in benign cells compared with morphologically similar diagnostic irregularities in tumor cells.

Relation of the Criteria of Malignancy to the "Hallmarks of Cancer" Model

The "Hallmarks of Cancer" [2, 3] are the known or expected (i.e., hypothetical) physiological changes that could explain clonal evolution, but they offer no explanation for why there are criteria of malignancy or what the criteria may signify. Objective measures of mitotic rate have diagnostic value in only very specific cellular contexts (e.g., for distinguishing endocervical adenocarcinoma from benign endocervical cells [69]). It may at first seem paradoxical, but the presence of apoptotic cells is actually a feature that predicts the presence of a clonal neoplastic expanding population [35]. Also seemingly paradoxical, mitotic rates bear little relation to tumor progression. For example, the percentage of mitotic figures, or the percent labeling of cells by Ki67 (a label for cells that are not in G0), is not useful for distinguishing invasive carcinomas from clones that are still growing in situ. The Hallmarks of Cancer concept has shifted interest away from studying what we know are real diagnostic structural changes in cancer cells toward trying to find evidence in support of a hypothesis.

Limitations of the Hallmarks of Cancer model are evident from a classical Darwinian perspective: cancer is a microevolutionary process in which mutations (mutations can be epigenetic or genetic) that confer a "growth advantage" are naturally selected for. What is clear from classic Darwinian evolution is that the mechanisms for altering fitness are essentially infinite in scope and cannot be classified into a handful of "hallmarks of evolution." For example, increased fitness in the Darwinian sense has little predictable relation to alterations in reproductive kinetics or longevity per se. We should fully expect a wide range of different mechanisms for increasing "cellular fitness," and morphology remains an essential guide for pointing us toward such new physiologies. Experimental evidence is showing that NE proteins can affect cell physiologies in completely unexpected ways, for example, several nuclear pore complex proteins can alter gene expression independent of their roles in transport of mRNAs [70].

Future Prospects

At this point, there are tantalizing clues that diagnostic NE alterations have a functional significance, but the full evidence for the proof of this hypothesis is lacking. A handful of cancer genes seem to "signal" for loss of nuclear regularity, but virtually nothing is known about the intermediates in the signaling pathway or the targets in the NE. There are more than enough candidate nuclear lamina proteins that can produce NE irregularity that could be relevant to cancer [1], but there are still no precisely characterized changes or specific probes (e.g., an antibody against a phospho-epitope of an NE protein) that can be applied to a human tumor sample to predict whether a cell will show a diagnostic NE irregularity. Even more difficult will be the demonstration that any particular NE protein or protein complex is responsible for a selectable phenotype that relates in some manner to the diagnostic large-scale alteration of the NE. It would not be a proof to show that a particular NE protein or complex is relevant to cancer just because its alteration affects a "Hallmarks of Cancer" physiology, since the hallmark physiologies themselves remain hypothetical and may not be relevant to a particular cell in a particular environment.

A problem in trying to establish links between particular NE proteins and functional changes has been the lack of documentation of model systems in which particular NE proteins are manipulated. The expectation that cancer genes will sometimes function by impacting NE shape is difficult to test because documentation of the cellular level morphologic effects of expression of various cancer genes is frequently lacking in the literature. It is not actually difficult to have a cytologist document the morphologic features of most model systems [15]. The criteria of malignancy are based on transmitted light microscopy, using H&E, or modified H&E, staining. Since cell biologists are accustomed to using immunofluorescence, it can be useful to create a surrogate H&E image using fluorescence in order to create a common platform that can be shared with pathologists. Fluorescence can be used to re-create an H&E image if DAPI is used to stain the DNA (like hematoxylin) and Sypro ruby is used to stain proteins (like eosin); the emissions from these dyes can be digitally inverted and pseudocolored to re-create an H&E stain. H&E staining is crucial for diagnosis because it discloses such a broad range of cellular level structural changes [71].

What would it take to prove that a diagnostic NE morphologic change is functionally significant and evolutionarily selected for in a cancer? This difficult question needs to be broken into several parts (Table 2). An alteration in an NE protein/complex would need to be found in association with spontaneous human tumors in which a diagnostic NE morphologic change is seen. The alteration could be a direct mutation of a component of the NE protein/complex, or it could be the end result of a signaling pathway to the NE protein/complex from a cancer gene active in the tumor, among other possibilities. Identifying such correlations may be facilitated by collaborations between pathologists and basic scientists because pathology reports will not be able to convey what specific NE changes may be present in a particular sample. There will then be a need to demonstrate that the alteration in the NE protein/complex leads to the putative NE diagnostic structural change. Such manipulations may need to occur within a particular cell type or even a particular tissue microenvironment. Arriving at a clear functional significance is most difficult. It may require direct time-lapse observation of cells with an inducible alteration of the NE protein/complex to study cellular behavior within the exact native microenvironment. Fortunately, we are at a point in time in which such techniques are feasible, using human tissue samples and gene transfer techniques in transient ex vivo microbiopsy/tissue cultures.

Table 2 How can diagnostic NE shape abnormalities be tied to specific NE components, and how can these specific lamina pathologies be definitively linked to particular cell physiologies relevant to cancer?

A putative NE-associated protein or complex needs to be shown to have an alteration (e.g., a mutation, an absence, a specific posttranslational modification perhaps mediated by signaling downstream of a cancer gene, a mislocalization, or a specific new association with another protein) in a human cancer that has NE shape abnormalities. Candidates are emerging [1], but documentation of the morphologic associations with the presence of these candidates in human tumors is generally lacking. Since there are so many different types of cancer, one should not expect that every cancer of the same histogenesis will have this alteration (e.g., an NE protein alteration may be present in only a minority of pancreatic cancers).
Inducing the alteration of the putative NE protein should lead to an abnormal NE shape in the cell type and microenvironment in which it can be identified. Alternatively, the occurrence of the alteration should be able to be documented on a per-cell basis to correlate with the NE shape change. Transduction of NE proteins or cancer genes in short-term cultures of human microbiopsy samples appears feasible and may provide the best model to demonstrate the inducibility of an altered NE; monolayer cultures of normal cells may not allow expression of the NE phenotype (see text).
Direct visualization of cells with the alteration in the cells' native tissue microenvironment in which the NE alteration occurs may be required to develop hypotheses for the functional significance of the alteration. It is not sufficient to prove functionality by showing that the altered NE protein or complex can affect a Hallmark of Cancer trait (see text). Time-lapse study of short-term cultures of human microbiopsy samples or transgenic animal models may be needed to prove that the NE alteration permits clonal expansion.

References

1. Chow K, Factor RE, Ullman KS (2012) The nuclear envelope environment and its cancer connections. Nat Rev Cancer 12(3):196–209
2. Hanahan D, Weinberg RA (2000) The hallmarks of cancer. Cell 100(1):57–70
3. Hanahan D, Weinberg RA (2011) Hallmarks of cancer: the next generation. Cell 144(5):646–674
4. Fischer AH, Young KA, DeLellis RA (2004) Incorporating the criteria of malignancy into the evolutionary framework for cancer. J Cell Biochem 93:28–36
5. Ghadially FN (1985) Diagnostic electron microscopy of tumours, 2nd edn. Butterworths, London
6. Sutherland KD, Berns A (2010) Cell of origin of lung cancer. Mol Oncol 4(5):397–403
7. Nikiforov YE, Nikiforova MN (2011) Molecular genetics and diagnosis of thyroid cancer. Nat Rev Endocrinol 7(10):569–580
8. Tallini G (2002) Molecular pathobiology of thyroid neoplasms. Endocr Pathol 13(4):271–288
9. Novara G, Martignoni G, Artibani W, Ficarra V (2007) Grading systems in renal cell carcinoma. J Urol 177(2):430–436
10. Elston CW, Ellis IO (1991) Pathological prognostic factors in breast cancer. I. The value of histological grade in breast cancer: experience from a large study with long-term follow-up. Histopathology 19(5):403–410 [Reprint in Histopathology. 2002: 41(3A):154–161; PMID: 12405947]
11. Davey DD, Banks ER, Jennings D, Powell DE (1993) Comparison of nuclear grade and DNA cytometry in breast carcinoma aspirates to histologic grade in excised cancers. Am J Clin Pathol 99(6):708–713

12. Bozzetti C, Nizzoli R, Naldi N, Guazzi A, Camisa R, Manotti L, Pilato FP, Mazzini G, Cocconi G (1996) Nuclear grading and flow cytometric DNA pattern in fine-needle aspirates of primary breast cancer. Diagn Cytopathol 15(2):116–120
13. Dey P (2010) Cancer nucleus: morphology and beyond. Diagn Cytopathol 38:382–390
14. Bignold LP (ed) (2006) Cancer: cell structures, carcinogens and genomic instability. Birkhauser Verlag, Basel
15. Fischer AH, Zhao C, Li QK, Gustafson KS, Eltoum IE, Tambouret R, Benstein B, Savaloja LC, Kulesza P (2010) The cytologic criteria of malignancy. J Cell Biochem 110:795–811
16. Holland AJ, Cleveland DW (2012) Losing balance: the origin and impact of aneuploidy in cancer. EMBO Rep 13(6):501–514
17. Voeller BR (1968) The chromosomal theory of inheritance: classic papers in development and heredity. Appleton-Centry-Crofts, New York
18. Hansemann D (1890) Ueber asymmetrische Zelltheilung in epithelkrebsen und deren biologische bedeutung. Arch Pathol Anat Physiol Klin Medicin 119:299–326
19. Boveri T (2007) Concerning the origin of malignant tumors (trans: Harris H). The Company of Biologists Limited and Cold Spring Harbor Laboratory Press (copublishers), New York
20. Shimizu N (2011) Molecular mechanisms of the origin of micronuclei from extrachromosomal elements. Mutagenesis 26:119. doi:10.1093/mutage/geq053
21. Samanta S, Dey P (2012) Micronucleus and its applications. Diagn Cytopathol 40(1):84–90
22. Crasta K, Ganem NJ, Dagher R, Lantermann AB, Ivanova EV, Pan Y, Nezi L, Protopopov A, Chowdhury D, Pellman D (2012) DNA breaks and chromosome pulverization from errors in mitosis. Nature 482(7383):53–58
23. Kronenwett U, Huwendiek S, Castro J, Ried T, Auer G (2004) Characterisation of breast fine-needle aspiration biopsies by centrosome aberrations and genomic instability. Br J Cancer 92(2):389–395
24. McGranahan N, Burrell RA, Endesfelder D, Novelli MR, Swanton C (2012) Cancer chromosomal instability: therapeutic and diagnostic challenges. EMBO Rep 13(6):528–538
25. Davoli T, de Lange T (2011) The causes and consequences of polyploidy in normal development and cancer. Annu Rev Cell Dev Biol 27:585. doi:10.1146/annurev-cellbio-092910-154234
26. Bubendorf L, Grilli B, Sauter G, Mihatsch MJ, Gasser TC, Dalquen P (2001) Multiprobe FISH for enhanced detection of bladder cancer in voided urine specimens and bladder washings. Am J Clin Pathol 116(1):79–86
27. Sinicrope FA, Rego RL, Halling KC, Foster N, Sargent DJ, La Plant B, French AJ, Laurie JA, Goldberg RM, Thibodeau SN, Witzig TE (2006) Prognostic impact of microsatellite instability and DNA ploidy in human colon carcinoma patients. Gastroenterology 131(3):729–737
28. Shia J, Black D, Hummer AJ, Boyd J, Soslow RA (2008) Routinely assessed morphological features correlate with microsatellite instability status in endometrial cancer. Hum Pathol 39(1):116–125
29. Vargas JD, Hatch EM, Anderson DJ, Hetzer MW (2012) Transient nuclear envelope rupturing during interphase in human cancer cells. Nucleus 3(1):88–100
30. Fischer AH, Bardarov S, Jiang Z (2004) Molecular aspects of diagnostic nucleolar and nuclear envelope changes in prostate cancer. J Cell Biochem 91:170–184
31. Sanchez JA, Wangh LJ (1999) New insights into the mechanisms of nuclear segmentation in human neutrophils. J Cell Biochem 73(1):1–10
32. Hoffmann K, Dreger CK, Olins AL, Olins DE, Shultz LD, Lucke B, Karl H, Kaps R, Muller D, Vaya A, Aznar J, Ware RE, Cruz NS, Lindner TH, Herrmann H, Reis A, Sperling K (2002) Mutations in the gene encoding the lamin B receptor produce an altered nuclear morphology in granulocytes (Pelger-Huet anomaly). Nat Genet 31(4):410–414
33. Gaines P, Tien CW, Olins AL, Olins DE, Shultz LD, Carney L, Berliner N (2008) Mouse neutrophils lacking lamin B-receptor expression exhibit aberrant development and lack critical functional responses. Exp Hematol 36(8):965–976
34. Cibas ES, Ducatman BS (2009) Cytology: diagnostic principles and clinical correlates. Saunders Elsevier, Philadelphia

35. DeMay RM (1996) The art and science of cytopathology, 1st edn. American Society of Clinical Pathologists Press, Chicago
36. Frost JK (1986) The cell in health and disease: an evaluation of cellular morphologic expression of biologic behavior, 2nd edn. Karger, New York, Revised edition
37. Edens LJ, White KH, Jevtic P, Li X, Levy DL (2012) Nuclear size regulation: from single cells to development and disease. Trends Cell Biol 23(4):151–159
38. Levy DL, Heald R (2012) Mechanisms of intracellular scaling. Annu Rev Cell Dev Biol 28:113–135
39. Webster M, Witkin KL, Cohen-Fix O (2009) Sizing up the nucleus: nuclear shape, size and nuclear-envelope assembly. J Cell Sci 122(10):1477–1486. doi:10.1242/jcs.037333
40. Fujimoto S, Ito M, Matsunaga S, Fukui K, Fujimoto S, Ito M, Matsunaga S, Fukui K (2005) An upper limit of the ratio of DNA volume to nuclear volume exists in plants. Genes Genet Syst 80(5):345–350
41. Neumann FR, Nurse P (2007) Nuclear size control in fission yeast. J Cell Biol 179(4):593–600
42. Huber MD, Gerace L, Huber MD, Gerace L (2007) The size-wise nucleus: nuclear volume control in eukaryotes. J Cell Biol 179(4):583–584
43. Wistuba II, Gazdar AF, Minna JD (2001) Molecular genetics of small cell lung carcinoma. Semin Oncol 28(2 Suppl 4):3–13
44. Pleasance ED, Stephens PJ, O'Meara S, McBride DJ, Meynert A, Jones D, Lin ML, Beare D, Lau KW, Greenman C, Varela I, Nik-Zainal S, Davies HR, Ordonez GR, Mudie LJ, Latimer C, Edkins S, Stebbings L, Chen L, Jia M, Leroy C, Marshall J, Menzies A, Butler A, Teague JW, Mangion J, Sun YA, McLaughlin SF, Peckham HE, Tsung EF, Costa GL, Lee CC, Minna JD, Gazdar A, Birney E, Rhodes MD, McKernan KJ, Stratton MR, Futreal PA, Campbell PJ (2010) A small-cell lung cancer genome with complex signatures of tobacco exposure. Nature 463(7278):184–190
45. Zink D, Fischer AH, Nickerson JA (2004) Nuclear structure in cancer cells. Nat Rev Cancer 4(9):677–687
46. Broers JL, Raymond Y, Rot MK, Kuijpers H, Wagenaar SS, Ramaekers FC (1993) Nuclear A-type lamins are differentially expressed in human lung cancer subtypes. Am J Pathol 143(1):211–220
47. Papotti M, Manazza AD, Chiarle R, Bussolati G (2004) Confocal microscope analysis and tridimensional reconstruction of papillary thyroid carcinoma nuclei. Virchows Arch 444(4):350–355
48. Asioli S, Bussolati G (2009) Emerin immunohistochemistry reveals diagnostic features of nuclear membrane arrangement in thyroid lesions. Histopathology 54(5):571–579. doi:10.1111/j.1365-2559.2009.03259.x
49. Asioli S, Maletta F, Pacchioni D, Lupo R, Bussolati G (2010) Cytological detection of papillary thyroid carcinomas by nuclear membrane decoration with emerin staining. Virchows Arch 457(1):43–51. doi:10.1007/s00428-010-0910-z
50. Kondo T, Ezzat S, Asa SL (2006) Pathogenetic mechanisms in thyroid follicular-cell neoplasia. Nat Rev Cancer 6(4):292–306
51. Das DK (2005) Intranuclear cytoplasmic inclusions in fine-needle aspiration smears of papillary thyroid carcinoma: a study of its morphological forms, association with nuclear grooves, and mode of formation. Diagn Cytopathol 32(5):264–268
52. Johannessen JV, Sobrinho-Simoes M, Finseth I, Pilstrom L (1983) Ultrastructural morphometry of thyroid neoplasms. Am J Clin Pathol 79(2):166–171
53. Oyama T (1989) A histopathological, immunohistochemical and ultrastructural study of intranuclear cytoplasmic inclusions in thyroid papillary carcinoma. Virchows Arch A Pathol Anat Histopathol 414(2):91–104
54. Rezk S, Brynes RK, Nelson V, Thein M, Patwardhan N, Fischer A, Khan A (2004) beta-Catenin expression in thyroid follicular lesions: potential role in nuclear envelope changes in papillary carcinomas. Endocr Pathol 15(4):329–337

55. Niu DF, Murata S, Kondo T, Nakazawa T, Kawasaki T, Ma DF, Yamane T, Nakamura N, Katoh R (2009) Involvement of centrosomes in nuclear irregularity of thyroid carcinoma cells. Virchows Arch 455(2):149–157. doi:10.1007/s00428-009-0802-2
56. Fischer AH, Taysavang P, Weber C, Wilson K (2001) Nuclear envelope organization in papillary thyroid carcinoma. Histol Histopathol 16(1):1–14
57. Johannessen JV, Gould VE, Jao W, Johannessen JV, Gould VE, Jao W (1978) The fine structure of human thyroid cancer. Hum Pathol 9(4):385–400
58. Fischer AH, Bond JA, Taysavang P, Battles OE, Wynford-Thomas D (1998) Papillary thyroid carcinoma oncogene (RET/PTC) alters the nuclear envelope and chromatin structure. Am J Pathol 153:1443–1450
59. Fischer AH, Greco A, Jhiang SM, Pierotti M, Taysavang P, Shepherd D, Roccato E, Ranzi V, Weber C (2000) RET/PTC1 and TRK-T3 both induce large scale nuclear envelope and chromatin changes, and their nuclear restructuring activity requires an SHC docking site (platform presentation). In: Cold Spring Harbor symposium: dynamic organization of nuclear function. Cold Spring Harbor, New York
60. Fischer AH, Taysavang P, Jhiang SM (2003) Nuclear envelope irregularity is induced by RET/PTC during interphase. Am J Pathol 163(3):1091–1100
61. Zakowski MF, Hussain S, Pao W, Ladanyi M, Ginsberg MS, Heelan R, Miller VA, Rusch VW, Kris MG (2009) Morphologic features of adenocarcinoma of the lung predictive of response to the epidermal growth factor receptor kinase inhibitors Erlotinib and Gefitinib. Arch Pathol Lab Med 133(3):470–477
62. Huang Y, Prasad M, Lemon WJ, Hampel H, Wright FA, Kornacker K, LiVolsi V, Frankel W, Kloos RT, Eng C, Pellegata NS, de la Chapelle A (2001) Gene expression in papillary thyroid carcinoma reveals highly consistent profiles. Proc Natl Acad Sci U S A 98(26):15044–15049
63. Borrello MG, Alberti L, Fischer A, Degl'innocenti D, Ferrario C, Gariboldi M, Marchesi F, Allavena P, Greco A, Collini P, Pilotti S, Cassinelli G, Bressan P, Fugazzola L, Mantovani A, Pierotti MA (2005) Induction of a proinflammatory program in normal human thyrocytes by the RET/PTC1 oncogene. Proc Natl Acad Sci U S A 102(41):14825–14830
64. Yoshida A, Nakamura Y, Imada T, Asaga T, Shimizu A, Harada M (1999) Apoptosis and proliferative activity in thyroid tumors. Surg Today 29(3):204–208
65. Biesterfeld S, Rickert D, Eichler S, Furste K, Mrusek S, Alfer J (2003) TV-image analysis based quantification of the proliferative activity and the apoptotic rate in thyroid tumors and thyroiditis. Anticancer Res 23(5b):4269–4275
66. Badalian-Very G, Vergilio J-A, Degar BA, MacConaill LE, Brandner B, Calicchio ML, Kuo FC, Ligon AH, Stevenson KE, Kehoe SM, Garraway LA, Hahn WC, Meyerson M, Fleming MD, Rollins BJ (2010) Recurrent BRAF mutations in Langerhans cell histiocytosis. Blood 116:1919. doi:10.1182/blood-2010-04-279083
67. Shah SP, Kobel M, Senz J, Morin RD, Clarke BA, Wiegand KC, Leung G, Zayed A, Mehl E, Kalloger SE, Sun M, Giuliany R, Yorida E, Jones S, Varhol R, Swenerton KD, Miller D, Clement PB, Crane C, Madore J, Provencher D, Leung P, DeFazio A, Khattra J, Turashvili G, Zhao Y, Zeng T, Glover JN, Vanderhyden B, Zhao C, Parkinson CA, Jimenez-Linan M, Bowtell DD, Mes-Masson AM, Brenton JD, Aparicio SA, Boyd N, Hirst M, Gilks CB, Marra M, Huntsman DG (2009) Mutation of FOXL2 in granulosa-cell tumors of the ovary. N Engl J Med 360(26):2719–2729
68. Bond JA, Wyllie FS, Rowson J, Radulescu A, Wynford-Thomas D (1994) In vitro reconstruction of tumour initiation in a human epithelium. Oncogene 9(1):281–290
69. Pirog EC, Isacson C, Szabolcs MJ, Kleter B, Quint W, Richart RM (2002) Proliferative activity of benign and neoplastic endocervical epithelium and correlation with HPV DNA detection. Int J Gynecol Pathol 21(1):22–26
70. D'Angelo MA, Gomez-Cavazos JS, Mei A, Lackner DH, Hetzer MW (2012) A change in nuclear pore complex composition regulates cell differentiation. Dev Cell 22(2):446–458
71. Schmolze DB, Standley C, Fogarty KE, Fischer AH (2011) Advances in microscopy techniques. Arch Pathol Lab Med 135(2):255–263 [Review]

Nuclear Morphometry, Epigenetic Changes, and Clinical Relevance in Prostate Cancer

Robert W. Veltri and Christhunesa S. Christudass

Abstract Nuclear structure alterations in cancer involve global genetic (mutations, amplifications, copy number variations, translocations, etc.) and epigenetic (DNA methylation and histone modifications) events that dramatically and dynamically spatially change chromatin, nuclear body, and chromosome organization. In prostate cancer (CaP) there appears to be early (<50 years) versus late (>60 years) onset clinically significant cancers, and we have yet to clearly understand the hereditary and somatic-based molecular pathways involved. We do know that once cancer is initiated, dedifferentiation of the prostate gland occurs with significant changes in nuclear structure driven by numerous genetic and epigenetic processes. This review focuses upon the nuclear architecture and epigenetic dynamics with potential translational clinically relevant applications to CaP. Further, the review correlates changes in the cancer-driven epigenetic process at the molecular level and correlates these alterations to nuclear morphological quantitative measurements. Finally, we address how we can best utilize this knowledge to improve the efficacy of personalized treatment of cancer.

Keywords Prostate cancer • Epigenetics • Nuclear morphology • Nuclear roundness

Abbreviations

AR	Androgen receptor
CaP	Prostate cancer
CT	Chromosome territory

R.W. Veltri, Ph.D. (✉) • C.S. Christudass
The Brady Urological Research Institute, Park 213, Johns Hopkins Hospital,
600 N Wolfe Street, Baltimore, MD 21287, USA
e-mail: rveltri1@jhmi.edu; cchris38@jhmi.edu

E.C. Schirmer and J.I. de las Heras (eds.), *Cancer Biology and the Nuclear Envelope*, Advances in Experimental Medicine and Biology 773, DOI 10.1007/978-1-4899-8032-8_4,

CCD	Charge coupled device
HGPIN	High-grade prostate intraepithelial neoplasia
IHC	Immunohistochemistry
NE	Nuclear envelope
NET	Nuclear envelope transmembrane protein
NMD	Nuclear morphometric descriptor
NRF	Nuclear roundness factor
NRV	Nuclear roundness variance
PRC2	Polycomb repressive complex 2
PSA	Prostate-specific antigen
QNG	Quantitative nuclear grade
RP	Radical prostatectomy
TSA	Trichostatin A

Introduction

Rudolf Virchow [1] published his famous aphorism "omnis cellula e cellula" ("every cell stems from another cell"), and he launched the field of cellular pathology and stated that all diseases involve changes in normal cells, that is, all pathology ultimately is cellular pathology. Further, for over 140 years it has been shown that nuclear morphology is often disrupted in cancer. In the 1860s, Lionel S. Beale [2, 3] of King's College Hospital examined unstained sputum from a patient with cancer of the pharynx and observed nuclear morphology variations in the cancerous cells. Lionel Beale also established a private laboratory near the King's College Hospital and gave a course of lectures on "The Microscope in Medicine" which included practical demonstrations in clinical pathology. He also wrote books on infectious disease theory and the practical value of the microscope in medicine to exam urine, blood, tumor tissue, and infectious agents. Subsequently, with many advances in microscopy, cytologic and anatomic pathologists recognized the importance of cell as well as nuclear structure in cancer diagnosis and prognosis.

In terms of early advances in cell biology, Theodor Boveri (1862–1915) was the first to use the term "chromosome territory" (CT). Although Boveri was able to observe nuclear dynamics, he was reliant solely on fixed materials and inferior microscopic instrumentation, whereas many decades later the efforts of Cremer et al. [4, 5] gave additional meaning to CT. In Boveri's 1909 publication, he described chromatin movements and organization in three observational hypotheses [6]. First, CT arrangements are stably maintained during interphase. Second, that chromosome stability is lost during prometaphase and there are greater movements of CTs. Finally, the daughter nuclei exhibit symmetry with each other and the general radial CT positioning between mother/daughter nuclei is maintained. Chromatin is organized into specific structural domains, likely by association with distinct nuclear compartments that are enriched in regulatory or nuclear structural proteins such as the nuclear matrix and associated attachment proteins as well as nuclear envelope transmembrane protein (NET)/lamina proteins, etc. [4, 5, 7]. Importantly, gene activity is modulated by interactions with several of these subnuclear compartments and specific protein elements of the nuclear envelope (NE). The organization of the chromosomes is based on CT positioning and allows late replicating genes and gene-poor chromosomes to be located at the nuclear periphery, while early replicating genes and gene-rich chromosomes are more centrally disposed, suggesting that many inactive genes are located at the periphery of the normal cell nucleus [8]. In spite of our increased understanding of how genomes are organized into CTs and where genes tend to be spatially expressed in normal cells; once cancer is initiated and progresses the chromosomes often become disorganized with either approximately the same amount of chromosomal material observed after the genetic alteration (balanced) or a major loss and/or gain of chromosomal material involved after the alteration (unbalanced) [4, 5, 7].

The "gold standard" for detection of cancer remains the pathologist's detection of gross changes in cellular (nucleus and cytoplasm) and tissue structure and organization.

Today, nuclear morphology measures include nuclear size, shape, DNA content (ploidy), and chromatin organization. The microscope and several improvements in the microscope lens, lighting, charge-coupled device (CCD) digital cameras, and novel software for analyzing images over the years have allowed for the detailed observation and study of nuclear size, shape and chromatin texture in cells, which clearly indicated abnormalities in cancer cells [7, 9]. Also, the development of histochemical stains provided significant improvements to study cancer cell and tissue morphology [10]. Hematoxylin was demonstrated to form a dye–metal complex with arginine-rich basic (cationic) nucleoproteins such as histones. Eosin dye is acidic in nature and tends to bind to more eosinophilic cellular structures (cytoplasm, collagen and muscle fibers) producing various shades of pink. Combining hematoxylin and eosin (H and E) enabled study of nuclear structure and its internal organization. George Papanicolaou developed a stain that enables visualization of many cytoplasmic and nuclear structural features of cells in the 1930s, and applied the stain to cervical cells to test for cancer—the so-called "Pap test" [11]. The Pap stain for cytology combines hematoxylin stain for tissues with phosphotungstic acid-Orange G solution and two sulfonic groups (SO_3Na) and the eosin with two auxochromic groups (COONa and NaO). The latter are acid dyes that demonstrate an attraction to basic proteins, such as prekeratin. H and E staining is usually performed on paraffin-embedded formalin fixed tissues and is read and interpreted by an anatomic pathologist, while the Pap stained slides are fixed in alcohol preparations and read and interpreted by a cytopathologist.

Also, the Feulgen staining reagent was developed for nuclei because it specifically and quantitatively stoichiometrically binds to DNA. The Feulgen reagent binds to DNA by uncovering the free aldehyde groups in DNA during the acid hydrolysis process, which then reacts with the reagent via a Schiff-Base interaction to form a stable, bluish/purple colored compound that absorbs light at 560 nm [12, 13]. In order to best interpret the Feulgen stained nuclei, a microspectrophotometer microscope fitted with a 3CCD color camera is employed to capture the information based on equations that calculate nuclear size, shape, texture and DNA content with DNA ploidy based on a single step pixel map of each nucleus [13]. Our laboratory employs the AutoCyte Pathology Workstation (APW, TriPath Inc., Burlington, NC, USA) with QUIC-DNA V1.201 software that is capable of measuring several nuclear morphometric descriptors (NMDs) to calculate a quantitative nuclear grade (QNG) from the NMDs in. An example of the Feulgen stain and an artificially colored 3D single nucleus is shown for a normal, high grade prostate intraepithelial neoplasia (HGPIN) and prostate cancer (CaP). The information collected on about 150 cancer epithelial cells can be used to predict grade, stage, biochemical recurrence, metastasis, and survival for CaP [7, 9].

Clinical Translational Relevance of Nuclear Structure in Prostate Cancer

It is imperative to be aware that the Gleason System for CaP histopathological grading is not based at all on nuclear grading; rather it is based on the assessment of dedifferentiation of glandular tissue architecture in CaP area (their altered size, shape, and distribution) when viewed under a low power microscope (final magnification 100–200×) by an expert pathologist. Donald Gleason devised the original scheme that established five patterns (Grade 1–5) to describe well differentiated to moderately and poorly differentiated cancer glands that has held up for the most part [14, 15]. Prognosis is based upon the fact that less aggressive prostate tumors have more of an appearance of normal glandular tissue, whereas more aggressive tumors that are more likely to invade and metastasize differ significantly from normal tissue owing to a loss of benign glandular architecture in terms of their size, shape, and distribution (poorly differentiated), as well as other histological features of tissue architecture including changes in the nuclear chromatin structure seen with H and E staining. To assign a Gleason score, the pathologist first looks for a dominant (primary) pattern of tumor cell growth or grade (the area where the cancer is most prominent) and then looks for a less widespread pattern or grade (secondary), and gives each one a grade number. The Gleason score is the sum of the dominant, or primary, tissue pattern grade (representing the majority of tumor) and the less dominant, or secondary, tissue pattern grade (assigned to the minority of the tumor). Today, pathologists tend to describe a Gleason score of 5 or 6 as a low-grade cancer, 7 (3+4 or 4+3) as medium-grade, and 8, 9, or 10 as high-grade cancer and then interpret a prognosis that includes the Gleason score as well as additional clinical information [16]. Occasionally, a pathologist may note a small area of a higher grade pattern in a biopsy or radical prostatectomy (RP) specimen known as a "tertiary pattern" and may record this result, because it may be prognostically relevant with time [16, 17]. A lower-grade cancer tends to grow more slowly and is less likely to invade and spread than a cancer with a higher grade pattern. Some limitations for the Gleason score system involve interpretations when comparing a biopsy to RP specimens, reproducibility of Gleason grading due to subjective interpretation amongst multiple pathologists and difficulty in diagnosing small acinar atypical lesions [17]. Our research has been focused on extracting information from the cancer and the benign adjacent nucleus, which can exceed the subjective evaluation of the CaP patient glandular architecture (Gleason grade patterns) as a variable to predict CaP outcomes and be used for intervention decisions [7, 9, 16, 17].

Alternative approaches to assess cancer involve characterization of nuclear structure through several approaches including manual, semi-automated, or automated machine vision techniques to assess architecture from H and E formalin-fixed paraffin-embedded tissue preparations. Diamond et al. [18] utilized a manual Graphpad software with a microscope to trace up to 300 malignant and benign nuclei from each CaP patient. Next, they compared nuclear size and shape in a set of prostate organ-confined CaP cases that had long-term follow-up and determined that they could distinguish those with a good prognosis from those with a poor prognosis (metastasis) with high accuracy ($p<0.005$). Defining a circle as 1.0, they calculated the nuclear roundness factor (NRF) as follows: NRF = $(C/2\pi)/(A/\pi)1/2$ (C = circumference and A = area), whereas the circularity form factor = $4\pi A/C^2$. The text below illustrates several applications of this technology; however, it has not been commercialized for practical use by pathologists. Dr Donald Coffey's laboratory and Dr. Mitchell Benson compared the use of flow cytometry (where the nuclei were labeled with acridine orange) to measure light scatter (forward and perpendicular) with the nuclear roundness factor performed on the same nuclei to assess tumor aggressiveness and heterogeneity of several well to poorly differentiated rat Dunning prostate tumor cell lines [19, 20]. The correlation between flow cytometry and nuclear roundness factor variance (NRV) using nuclear tracing was exceptional. Later, others using commercially available hardware and software validated the clinical value of NRV measurements using a microscope. The images were analyzed with the DynaCell Motility Morphometry Measurement workstation (JAW Associates, Inc., Annapolis, MD, USA). With this method, measurements varied by less than 5 % among examiners, and the authors confirmed that this NRV shape variable readily predicts progressive disease and mortality of CaP [21–23]. Finally, Veltri et al. [24] showed that the accuracy of NRV assessed by DynaCell technology is significantly higher than the Gleason score to predict metastasis and CaP-specific death in men with long-term follow-up (median follow-up of 17 years). Therefore, nuclear architecture (irregularity of nuclear shape) when accurately quantified is a significant variable to predict aggressive CaP outcomes and NRV exceeds the prognostic value of Gleason grade patterns or score to predict the long-term survival in this patient sample.

Another alternative digital imaging approach described by Veltri et al. [7] used the APW and Feulgen stained prostatic nuclei to study the CaP in biopsy and RP specimens [7, 9]. Our laboratory uses these nuclear images and the ~40 NMDs captured by the APW using DNA QUIC DNA V1.201 software to process the nuclear images and then calculate a QNG illustrated in Fig. 1 determined from the NMDs to make predictions of grade, stage, metastasis, and survival [7, 9]. The technology was also used by Badalament et al. [25] to create a nuclear morphometric QNG signature combined with serum prostate-specific antigen (PSA) to predict stage using ROC analysis with an AUC = 86 % (sensitivity = 85.7 %; specificity = 71.3 %). This was at a time when the staging of CaP based on biopsy informatics was about 50 % accurate. A limitation of this early algorithm was the number of nuclear features

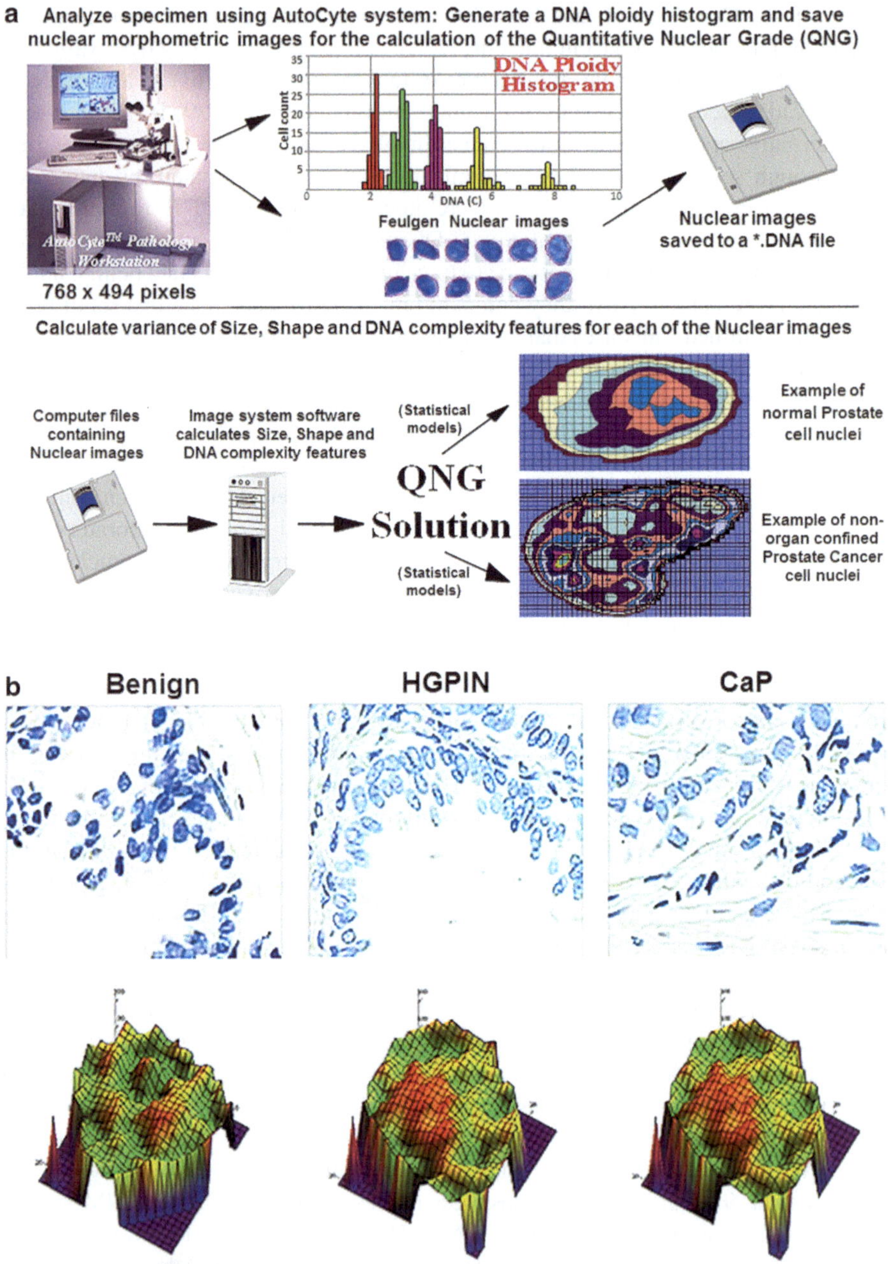

Fig. 1 Automated analysis of nuclear pathology in prostate cancer (**a**) General Description of the AutoCyte Pathology Workstation's operation. (**b**) Images of single 2D Feulgen stained prostate benign, high grade prostate intraepithelial neoplasia (HGPIN), and prostate cancer nuclei (*left to right, upper panels*). These *blue* colored epithelial nuclei are captured by the APW software (QUIV DNA) and 40 nuclear morphometric descriptors (NMDs) are used to calculated image-based solutions for CaP outcomes. In the *bottom panel* is a 3D construction of the nuclear pixel grey level map (made using Mathcad) shows variations in nuclear chromatin labeled with the Feulgen DNA stain

available and the stringency for the Multivariate Logistic Regression (MLR) modeling. However, when the model was applied to incoming biopsy specimens at a urology pathology company, the algorithm performed within 5 % of specifications. Veltri et al. [26] also studied the biopsies of 557 consecutive men that underwent RP at Johns Hopkins Hospital from October 1998 to January 2000. Combining QNG, the Gleason score and complexed PSA density (complete model) yielded a ROC AUC = 82.4 % (sensitivity = 73.5 %; specificity = 83 %) to predict non-organ-confined CaP from a biopsy. Next, Veltri et al. [27] used the APW system and Feulgen stained nuclei to capture 38 nuclear morphometric descriptors to predict CaP biochemical progression. The patient cohort included 115 patients with clinically localized CaP, and the mean follow-up period in 70/115 patients without disease progression was 10.4 ± 1.7 years. Using backward stepwise MLR and the variances of 11/38 of the nuclear morphometric descriptors to calculate QNG were found to be significant for predicting biochemical progression ($p = 0.00001$; ROC AUC = 86 %; sensitivity = 78 %; specificity = 83 %). Furthermore, the QNG and the postoperative Gleason score, when combined, created a MLR model for the prediction of biochemical progression, yielding a ROC AUC = 92 % and having a sensitivity of 89 % and specificity of 84 %. These two parameters (QNG and Gleason score) separated the 115 patients into three statistically significant risk groups based upon Kaplan–Meier plot analysis. Predicting aggressive CaP effectively depends on having a sufficient sample size and long-term follow-up data for the successful application of nuclear morphometry as a variable in addition to routine pathological and clinical variables. In order to assess aggressive CaP using QNG Khan et al. [28] successfully predicted progression to metastasis and/or CaP mortality in 227 RP surgical specimens by employing the APW imaging system and applying the QNG analysis. The combined pathology-QNG model retained lymph node status, prostatectomy Gleason score, and QNG, yielding a ROC AUC = 86 % with an accuracy of 76 % at 90 % sensitivity. Next, Veltri et al. [29] employed the same digital imaging technology and the APW to calculate a QNG solution using a tissue microarray made from 0.6 mm tissue cores of 182 patients (cancer and adjacent benign areas) to evaluate the use of QNG alone and with pathological and clinical variables to predict metastasis and death due to CaP. The pathology model yielded a ROC AUC = 72.5 %. We assessed the QNG solution determined by MLR statistical models for the adjacent benign and cancer areas and yielded a ROC AUC = 81.6 % and 79.9 %, respectively. Hence, semi-automated digital image analysis can use nuclear NMDs to make clinical outcome predictions; however, the technology requires time and expertise to perform reproducibly whether or not it is a manual or semi-automated NRV single variable or a QNG signature methodology. Hence, commercialization continues to be a challenge unless automation can be readily accomplished.

Other applications for quantitative nuclear morphometry based on the APW system permit studies that correlate alterations in nuclear structure with biological and clinical aspects of CaP. Using a NCI Cooperative Prostate Cancer Tissue Resource tissue microarray of 92 cases with long-term follow-up (56 non-recurrences and 36 recurrences), our laboratory [24] demonstrated that the histone acetyltransferase p300 protein (p300, HAT) alters CaP cancer cell nuclear structure and predicts biochemical progression. In this study we also demonstrated that specific nuclear

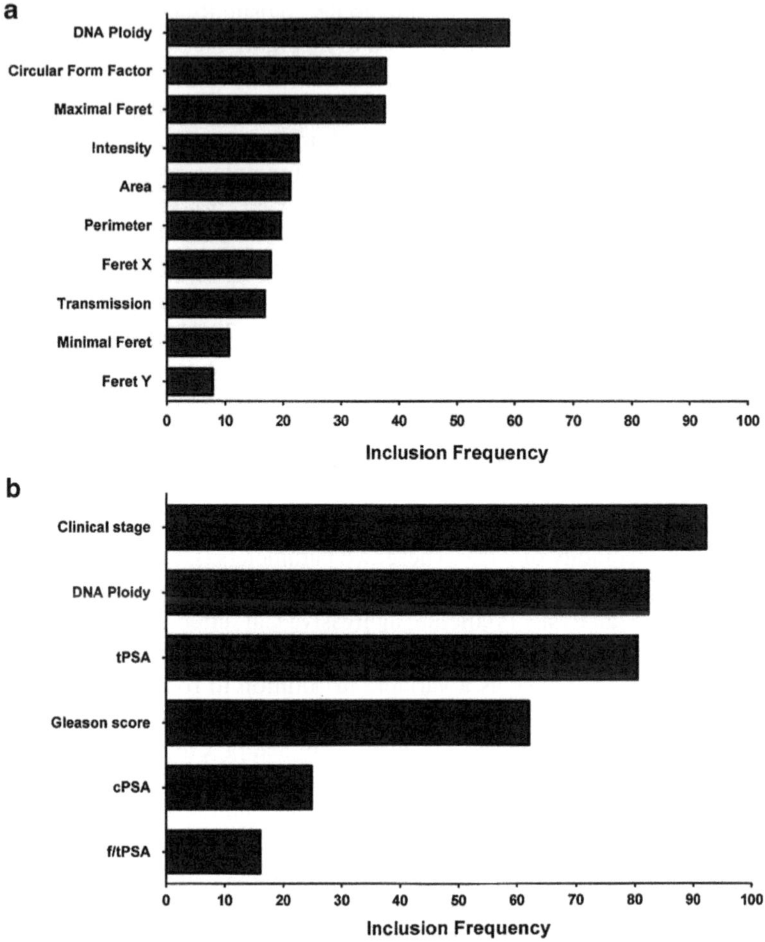

Fig. 2 Statistical contribution of nuclear morphometry in predicting prostate cancer. Bar graph of the statistical contribution of nuclear morphometry (**a**) and clinical pathological features combined with morphometry (**b**) based on boot strapping (200×) a cox proportional hazards model analysis to predict organ-confined prostate cancer. Notably DNA Ploidy is retained in a multivariate prediction model for organ-confined PCa

features, i.e., circular form factor (rho = −0.26; $p = 0.012$) and minimum Feret (rho = −0.21; $p = 0.048$) exhibited significant correlations with p300 protein expression. The quantitative immunohistochemistry (qIHC) of the p300 protein expression in high grade tumors (Gleason score ≥ 7) was significantly higher compared to low grade tumors (17.7 % versus 13.7 %, respectively, $p = 0.03$). Further, p300 expression remained significant in the Cox multivariate model independent of Gleason score ($p = 0.03$). Also, CaP patients with a Gleason score ≥ 7 and p300 IHC expression >24 % showed the highest risk for CaP biochemical recurrence ($p = 0.002$) in a Kaplan–Meier plot. Using the same imaging technology we showed that nuclear features predict non-organ-confined CaP [26]. In Fig. 2a we show the nuclear morphometric features correlate with organ-confined disease status in CaP. Note that DNA ploidy was the most frequently included feature in a MLR bootstrap model and that several nuclear shape factors were also useful. In Fig. 2b using the same MLR method, we compared the contribution of clinical and pathological features to make the same decision and of note is that DNA ploidy was very comparable to clinical stage in this patient cohort ($n=370$) and when combined in a clinicopathological model discriminates organ confined from non-organ-confined CaP [26]. Another application is the correlation of nuclear morphometry changes to demonstrate the response of CaP cells to histone deacetylase Inhibitors (e.g., Valproic acid; VPA) [30]. In vitro tissue microarrays consisted of CaP cell lines that were treated for 3, 7 or 14 days with 0, 0.6 or 1.2 mM VPA. In vivo the tissue microarrays consisted of cores from CaP xenografts from nude mice treated for 30 days with similar concentrations of VPA achieved in drinking water. Digital images of at least 200 Feulgen stained nuclei were captured and nuclear alterations were measured. Both in vitro and in vivo VPA treatment of CaP cells resulted in significant dose- and time-dependent changes in nuclear structure. Hence, quantitative nuclear morphometry may be useful as a biomarker to assess pathological status of men with CaP, and pave the way for therapeutics based on the proteins or genes that alter chromatin structure and nuclear morphometry [7, 9].

Today, the emergence of the rapid scanning microscope image analysis and the development of novel machine vision imaging techniques is aiding pathologists to analyze histologic tissue images and distinguish cancer grades. Automated image applications have been the recent focus for CaP and other cancers [31, 32]. The development of machine vision techniques has been applied to H and E stained tissue sections, aiding pathologists to analyze CaP tissue images and evaluate the grade patterns of CaP, which has made steady progress during the past decade. As the CaP malignancy is manifested by the loss of the normal glandular architecture (i.e., shape, size, and differentiation of the glands, i.e., Gleason grade patterns) [16, 17], applications of image analysis to improve segmentation and texture analysis to assess different Gleason grading patterns based on H and E and Feulgen stained tissue images have been reported [7, 13, 33, 34]. Numerous machine vision approaches to nuclear size, shape and texture analysis of these images have been applied. Wavelet and multiwavelet transforms, fractal analysis, texton forest/random tree, and cell network cycles have been utilized for texture feature extraction and classification in studies of the automated Gleason grading [33–43]. Collaborating with Dr. Anant Madabhushi at Case Western Reserve University, we codeveloped an image computational method to assess nuclei in Gleason graded CaP. Dr. Madabhushi applied a novel adaptive active contour scheme (AdACM) machine vision method that combines nuclear segmentation boundary and a solid geometry graphic term that includes shape etc. (Fig. 3) [35]. The technique reduces the computational time required in half (250 s for 120 nuclei), measured in seconds; the approach uses the nuclear shape "prior term" in the variational formula and is only invoked for those instances in the image where nuclear overlaps between objects are identified. By not having to invoke all three nuclear feature terms (shape, boundary, and region) for segmenting every nuclear object in the image, the computational expense of the

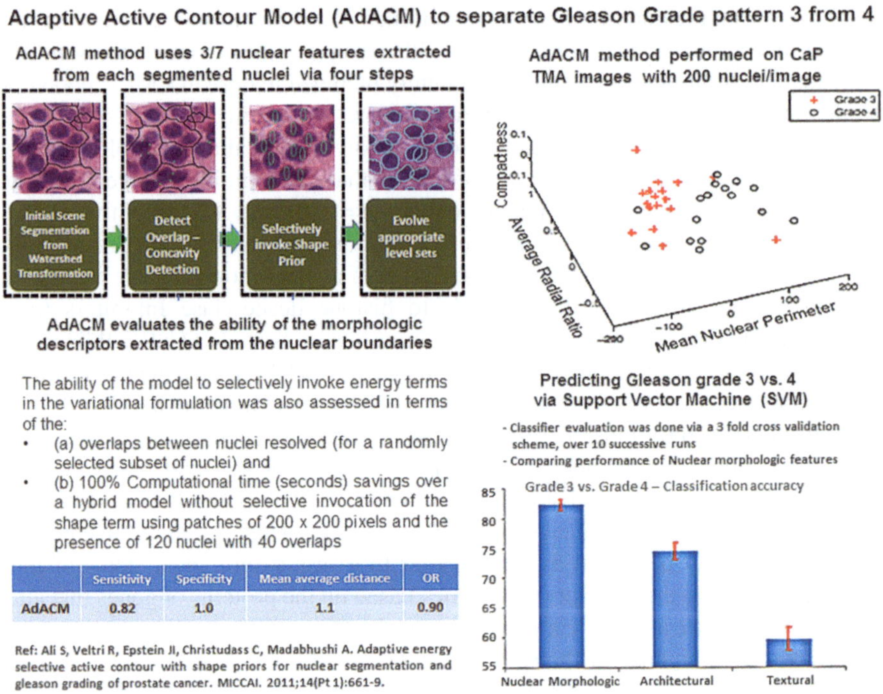

Fig. 3 This figure demonstrates how AdACM computer-assisted image analysis can separate Gleason grade pattern 3 from 4. The graph in the *upper right* panel shows how three features can accurately separate 3 from 4 (Odds Ratio = 0.90). In the *top left hand* panel of the figure, the segmentation method is described. In the *bottom right* panel of the figure, the plot depicts the contribution of nuclear morphology, architecture, and texture to the computational solution plot in the *upper right* hand space

integrated active contour model is dramatically reduced. The AdACM [35] method was employed for the task of segmenting nuclei on CaP tissue microarray core images. Morphological, architectural and textural features extracted from these segmented nuclei were found to able to discriminate different Gleason grade patterns 3 (indolent) and 4 (aggressive) with a ROC AUC = 86 % via a mathematically derived classifier and using only three nuclear features. The "nuclear morphologic features" proved to be the best predictor of the three features captured for the study (Fig. 3). Additional collaborative machine vision computational techniques should help to determine if our approaches can predict time-dependent CaP outcomes such as biochemical recurrence, metastasis, and survival.

Using the same CaP tissue microarray in collaboration with Dr. Li with Yoon at the University of Pittsburgh Electrical Engineering department we applied wavelet machine vision technology called cardinal multiridgelet transform (CMRT) [44] to analyze CaP histological H and E images and extract nuclear texture features in the transform domain. CMRT provides cardinality, orthogonally, approximate translation invariance and rotation invariance of the transform. With 48 tissue microarray images of Gleason grade 3 and grade 4 as a training set and using nuclear texture features extracted there from, a support vector machine with Gaussian kernel was trained to classify grade 3 and grade 4. The leave-one-out cross-validation assessment showed the model accuracy was 93.75 % and a ROC AUC = 0.96 to make this critical pathological separation. Please note this wavelet approach produced similar results to AdACM in terms of time and accuracy. At this point we realize the value of an automated approach to nuclear morphometry in a clinical setting, but yet we do not clearly understand why and how the nuclear shape may be altered in normal differentiation versus cancer dedifferentiation to a malignant state. Hence, in the future we can apply automated computer machine vision technology to process tissue images and extract pathologically relevant prognostic features such as a new cancer grade concept and combine this data with molecular biomarkers.

Why Does Nuclear Architecture Change in CaP?

Epigenetics involves alterations in gene expression or cellular phenotype that are caused by other mechanisms beyond changes in the DNA sequence through mutations, amplifications, deletions, copy number variations, etc. Examples of epigenetic change include chemical modifications of the histone tails as well as DNA methylation, which over time have often been mired in controversy regarding the heritability of such changes. It is difficult to sort out the concept of heritability in this review. However, we need to accept the premise that environment may be playing an important role in "phenotype plasticity" through transcription of genes that alter cellular and tissue phenotype. Hence, rather than argue this point I have chosen the option to accept the concept in order to address the question of epigenetic events that play key roles in altering the cancer phenotype during initiation and progression [7, 9, 24].

Since the nucleus is a major focus in this review, the anatomy of the NE and its interactions with the key nuclear components of chromatin and DNA will be highlighted. Under normal conditions the NE separates nuclear and cytoplasmic functions and at its inner surface it provides a docking site for chromatin via several NETs and the intermediate filament lamins [45, 46]. The major structural elements of the NE are the inner nuclear membrane, the outer nuclear membrane, the nuclear pore complexes, and the nuclear lamins. Notable, is the importance of alterations in nuclear structure in cancer and the role of the NE and its NETs and associated inner and outer membrane parts [45, 46] (i.e., lumen/perinuclear space [45], ribosomes [45–47], nuclear pores [45, 47–49], nuclear lamina (A, B, and C) [48, 49], nuclear matrix [50, 51], etc.) and their functional interplay during normal cell proliferation, cell differentiation, and carcinogenesis [46]. All of these NE components can impact nuclear architecture (size, shape, and integrity), genome stability (chromosome spatial topology, chromatin regulation, nuclear matrix organization, and gene expression) as well as cell functions (e.g., DNA repair, cell signaling, cell cycle, and mitosis) during carcinogenic progression [45, 46, 52–58]. Additionally, histone modifications such as acetylation, methylation, ubiquitination, and phosphorylation are extremely critical to regulation of gene transcription and chromatin organization in normal, differentiating stem and cancer cells [57, 59, 60]. Further, critical environmentally driven factors such as occupational or behavioral exposure to carcinogens, diet and metabolism, inflammation and infection, etc., can produce dramatic epigenetic changes that drive alterations in gene activation and suppression causing multiple structural changes in nuclear shape, size and chromatin organization that may generate valuable early diagnostic and prognostic information regarding the pathology and pathogenesis of malignancy [53, 59–63].

One facet of the epigenetic molecular machinery that could drive cancer events involves chromatin remodeling by proteins in the Polycomb group (PcG) and their interaction with nucleosomes (linked by histone H1). Nucleosomes are composed of 140–145 bp of DNA wrapped around the histone octamer that consists of two copies each of H2A, H2B, H3, and H4 (Fig. 4). The enzyme-catalyzed chemical modification of selected amino acids of histones is a mechanism used throughout the living world to increase and regulate the functional plasticity of gene expression. Such molecular plasticity involves several histone modifications at the N-terminal tails that methylate lysine or arginine, acetylate lysine, phosphorylate serine, threonine, or tyrosine, and ubiquitinate lysine, each of which can influence specific gene expression to alter phenotypic changes via modifications to chromatin structure and architecture [54, 57, 59, 60]. Several residues on the tails of histone H3 (e.g., H3K4, H3K9, H3K27, H3K36), as well as in the core of histone H3 (e.g., H3K79) have been shown to be sites for such modifications that are involved in transcriptional regulation and alterations in chromatin organization. Additionally, such histone modification–demodification cycles can directly or indirectly influence DNA methylation. For example, high levels of H3K4 methylation correlates with low levels cytosine methylation at CpG dimers; levels of H3K4 methylation are influenced by other H3 modifications, including acetylation, which can exert an indirect effect on DNA methylation; and methylation of H3 at K9 or K36 can influence levels or positioning of DNA methylation [54, 57, 59, 60]. In mammals DNA methylation occurs at the cytosines of CpG dimers in DNA. The deamination of 5-methyl cytosine (meC) forms thymidine, resulting in a G-T base mismatch, the repair of which could result in the replacement of either base. Replacement of the G with an A results in a mutated DNA sequence, in which the original meC is replaced with T. Hence, epigenetic changes inevitably weave together chemical modifications of histones with DNA methylation events causing phenotypic changes through the influence of environmental agents, which can also produce genetic changes (i.e., mutations, deletions, amplifications, etc.) that promote cancer [54]. Clearly, in cancer, histone modifications lie at the heart of mechanisms by which a variety of functionally significant nuclear proteins activate (oncogenes) or silence specific regions (i.e., tumor suppressor genes) of the human genome. These alterations involve transcription factors, chromatin modifying enzymes, the complexes that methylate DNA, or the chromatin remodelers that reposition nucleosomes along the DNA strand [58, 59, 64]. Recently, in a breast cancer model (MCF-7), Tropberger et al. [63] have functionally characterized acetylation of H3K122 and revealed that H3K122 acetylation

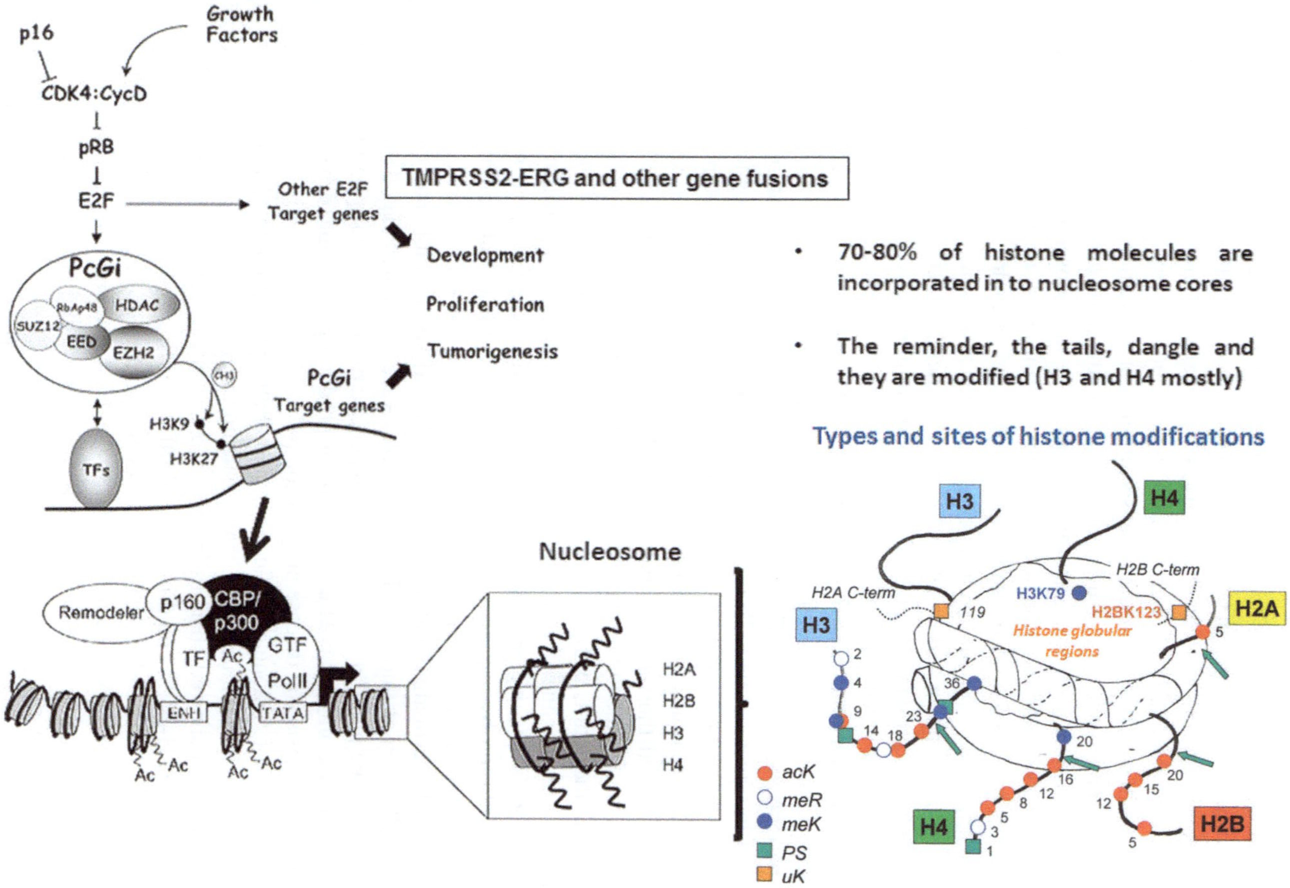

Fig. 4 Epigenetic marks involved in tumorigenesis. The illustration shows the interaction of Polycomb 2 transcription complex with the nucleosome transcription complex as well as sites for modification of the histones

is catalyzed by p300/CBP and can be sufficient to stimulate transcription in vitro. They showed that H3K122 acetylation is specifically enriched at active transcription start sites and enhancers as well as on H3.3- and H2A.Z-containing nucleosomes. H3K122 is modified by acetylation at estrogen-regulated genes and marks enhancers actively engaged in transcriptional regulation. Finally, the authors showed that mutation of H3K122 can impair transcriptional activation in vivo and have proposed a model for H3K122 acetylation on the lateral nucleosome surface changing chromatin structure to promote transcription in breast tumors.

Aberrant epigenetic events such as DNA hypomethylation and hypermethylation and altered histone acetylation and methylation have been observed in CaP affecting the expression and function of a large array of genes that can lead to tumorigenesis, tumor progression, and metastasis. Initially CaP is androgen dependent, but can eventually become androgen independent after androgen deprivation therapy. Androgen-independent CaP is characterized by a heterogeneous loss of androgen receptor (AR) expression [61, 62, 65, 66]. AR promoter methylation is more prevalent in androgen-independent CaP than in primary androgen-dependent CaP, suggesting that epigenetic silencing of AR by DNA hypermethylation could be an alternative mechanism leading to androgen independence in a subset of advanced CaP patients. Similarly, in CaP the importance of histone modifications and progression has been studied. To be clinically applicable, an ideal prognostic tumor biomarker must be readily detectable in noninvasive clinical specimens. DNA hypermethylation and histone modifications alter nuclear architecture, fulfilling this requirement, and thus are promising biomarkers [67]. Jarrard et al. [68] reported aberrant promoter methylation in AR-negative CaP cell lines. These results are consistent with the results of Izbicka et al. [69] that showed 5,6-dihydro-5-azacytidine, an inhibitor of cytosine DNA methyltransferase, could restore androgen sensitivity in androgen insensitive human CaP cell lines, which then become sensitive to growth inhibition by anti-androgens. Human cancers almost ubiquitously harbor epigenetic alterations. There is strong evidence that some epigenetic alterations (e.g., DNA hypermethylation and hypomethylation) are heritable and can also be dynamically altered during CaP progression. Recent research has demonstrated using "cityscape plots" a wide range of epigenetic plasticity and support that DNA methylation alterations have the potential for producing selectable driver events in CaP carcinogenesis and disease progression [67].

In the area of histone modifications and their application to CaP prognosis, Seligson et al. [70] conducted IHC on a tissue microarray of 226 CaP cases of which 183/226 (81 %) showed changes in IHC expression for histones: acetylated (Ac) H3K9, H3K18, H4K12, and dimethylated (diMe) H4R3 and H3K4. The objective was to predict biochemical recurrence, defined as a postoperative serum PSA of 0.2 ng/ml or greater and was seen in 61 (34 %) of all study patients, and 20 (19 %) of patients with low grade tumors. The median follow-up time within the recurring and non-recurring patient groups was 22.0 (range 1.0–115.0) and 65.5 months (range 2.0–163.0). In a multivariate Cox Proportional Hazards Ratio model the histone modification panel had a value of 3.86 (95 % CI = 1.18–12.62), $p = 0.025$. The two groups are identified on the basis of the "simple clustering rule" involving only

H3K18Ac and H3K4diMe modifications. The study also included a validation set of 39 cases with low grade CaP that were analyzed according to the above simple rule involving H3K18Ac and H3K4diMe and the IHC staining distinguishes between two groups of patients with risks of tumor recurrence: 4 % in group A versus 31 % in group B (log-rank $p=0.016$; hazard ratio=9.2; 95 % CI 1.02–82.2). Recent studies by Bianco-Miotto et al. [71] on global patterns of specific histone modifications revealed an epigenetic signature for CaP involving H3K18Ac and H3K4diMe. The authors studied histone modifications in 279 cases of CaP and they showed that H3K18Ac and H3K4diMe when combined are predictors of relapse-free survival, with high global levels associated with a 1.71-fold ($p<0.0001$) and 1.80-fold ($p=0.006$) increased risk of tumor recurrence, respectively. These high levels of both histone modifications were associated with a threefold increased risk of relapse ($p<0.0001$). Further, the study revealed an epigenetic gene expression candidate gene signature for CaP that included several interesting epigenetic genes (DNMT3A, MBD4, MLL2, MLL3, NSD1, and SRCAP), which significantly discriminated non-malignant from CaP tumor tissue ($p=0.0063$). Notably, of those six genes altered between primary and metastatic CaP, DNMT3A, MLL2, NSD1, and MLL3 were significantly downregulated and MBD4 and SRCAP upregulated tumor in the primary prostate cancer samples with biochemical recurrence when compared with the primary samples without recurrence. In the metastatic samples, these same genes were also significantly altered, with DNMT3A, MLL2, NSD1, MBD4, and MLL3 upregulated and SRCAP downregulated when compared with the primary prostate tumors. The prognostic classification on the validation set therefore confirmed the predictive power of histone modifications as markers of CaP prognosis.

In another study Watson et al. [72] used digital texture analysis to assess global chromatin patterns following treatment of normal (PNT1A) and CaP (LNCaP) cell lines with trichostatin-A (TSA) and observed significant alterations in the TSA induced H3K9 hyperacetylation resulting in decondensation of heterochromatin, which was associated with altered gene expression profiles in both the immortalized normal PNT1A prostate cell line and a malignant androgen-dependent CaP cell line LNCaP. Though some changes were TSA dose dependent and cell cycle dependent, flow cytometric analysis enabled the observation of clear differences in chromatin decondensation and H3K9 acetylation between the normal and tumor lines.

Our laboratory studied the protein expression profiling of the Dunning rat CaP cell lines of varying metastatic potential [G (0 %), AT-1 (>20 %), and MLL (100 %)] using SELDI-TOF-MS [73]. We identified a 17.5 km/z SELDI-TOF-MS peak that was found to retain discriminatory value in each of two separate study sets that was verified as histone H2B. The increases in the histone H2B peak correlate with the metastatic potential of the Dunning cell lines, going up slightly in the AT-1 subline and consistently increasing more strongly in the MLL subline. Clearly, the above results obtained to date support that signatures of global histone modifications and histone levels are associated with prognostic features of CaP. Also, other publications demonstrate that alterations in the expression of histone remodeling enzymes may represent novel diagnostic and prognostic markers of CaP and potentially new targets for therapeutics [74, 75]. Therefore, global epigenetic modifications in androgen sensitive and resistant CaP can activate or repress multiple genes that impact nuclear chromatin architecture as well as CaP progression to metastasis [61, 65–75].

PcG, which is best known for its role in silencing the HOX gene cluster during embryonic development [76, 77], acts by forming multiprotein complexes that, through modification of chromatin structure, repress target gene expression (Fig. 4). Three potential E2F regulated PcG genes, Enhancer of Zeste Homolog 2 (EZH2), Embryonic Ectoderm Development (EED), and Suppressor of Zeste 12 (SUZ12), constitute the Polycomb repressive complex 2 (PRC2) [78, 79] and it requires an intact SET domain (for methylation of histone tails) and endogenous histone deacetylase activity for its function [80, 81]. EZH2 and EED are also essential for the proliferation of both transformed and non-transformed cells and are under the regulation of the pRB-E2F pathway. EZH2 overexpression is associated with poor prognosis in patients with metastatic disease [78–80]. EZH2 promotes a reduction in the pool of insoluble F-actin and regulates cell adhesion and migration in invasive CaP cells [82, 83] and may control gene function via regulation of nuclear actin that is associated with the chromatin remodeling complex. Su et al. [84] demonstrated the existence of a cytosolic EZH2-containing methyltransferase complex that controls cellular signaling via ligand induced actin polymerization. Pharmacologic interference of EZH2 function selectively induces apoptosis in cancer, but not in normal cells and accessibility is dictated broadly by the degree of chromatin compaction, which is influenced in part by polycomb group proteins [85]. PRC2 catalyzes trimethylation of histone H3 lysine 27 (H3K27me3) [86]. H3K27me3 may also recruit DNA methyltransferases, and histone deacetylases, resulting in additional transcriptional repressive marks and heterochromatin compaction. Hence, overexpression of EZH2 is a marker of advanced and metastatic disease in many solid tumors, including prostate and breast cancer [86].

As another related clinically translational event, Laitinen et al. [87] suggested that demonstration by IHC of low Ki-67 (0–1 %) (a measure of cell proliferation) and EZH2 (<50 %) identifies a subgroup of patients with a very low risk of CaP, and could be candidates for active surveillance instead of immediate prostatectomy. Jhavar et al. [88] showed that Ki-67 expression is an independent determinant of very high risk among men enrolled in an active surveillance cohort. Hence, the degrees of expression of EZH2 combined with a measure of cell proliferation are potential prognostic biomarkers of the severity of CaP and other solid tumors.

Because our group has characterized the clinical relevance of nuclear features that can predict biochemical recurrence, metastasis, and CaP-specific survival we also have been studying what molecular mechanisms may cause such changes [7, 9, 24–29, 89]. The literature supports that changes in nuclear morphology are associated with deregulation of nuclear matrix proteins [50, 63] and abnormal expression of lamins [45, 47–49] and PcG [76, 90] genes and such changes have been found in undifferentiated neoplastic cells. Nuclear size and shape factors, especially mean nuclear area, have been shown to correlate with the Gleason score tissue architecture [91]. Debes et al. [92] demonstrated that the p300 histone acetyltransferase (HAT), a transcriptional regulator, is overexpressed in CaP and correlates specifically to nuclear alterations in terms of DNA content, size, and shape. These nuclear alterations were seen in prostate biopsies and in CaP cell lines transfected with p300. Subsequently, Isharwal et al. [24] confirmed that p300 protein expression measured by IHC significantly correlated with nuclear alterations seen in tumor cells; specifically with DNA content ($p=0.016$), circular form factor ($p=0.012$) and minimum feret ($p=0.048$). Nuclear size and shape factors, especially mean nuclear area, were concordant with the Gleason score. Activation of the PcG proteins through p300 and perhaps EZH2 may regulate, in part, nuclear size and shape via the histone modifications and may provide a tool for evaluation of the pathological status of CaP [93]. Recently, Imbalzano et al. [94] have demonstrated that the SWI/SNF chromatin remodeling enzyme ATPase and the Brahma-related gene 1 (BRG1) contributes to the regulation of overall nuclear size and shape of immortalized mammary epithelial cells. Notably, they observed in BRG1 knockdown cells the formation of grooves at the nuclear periphery; however, there were no changes in levels of the nuclear structure markers lamin A/C, lamin B, emerin, nesprin, nurim, and the splicing speckle component SRm160. In addition, no changes in immunostaining for H1 or the modified histones phospho-H3Ser10 and H3triMeK4 were observed. This recent finding suggests that BRG1 can also mediate cancer nuclear shape by internal nuclear mechanisms that likely control chromatin dynamics. Hence, BRG1, p300, and other epigenetic histone-mediated processes of the PcG complexes noted above can alter nuclear structure in cancer.

Another central aspect of nuclear architecture is the nuclear matrix [95–97], which is composed in large part of the ribonucleoprotein (RNP) network, packaged amongst a multitude of proteins (<400) that form a non-chromatin structure throughout the nucleoplasm. Infrequent and specific matrix attachment regions (MARs) and scaffold-associated regions (SARs) of chromatin fibers bind the nucleoskeleton and support the chromatin loop domains and high mobility group nucleosome-non-histone binding (HMGN) [98] proteins that play an intricate role in chromatin structure and function [46, 50, 91]. The nuclear matrix also interconnects with the nuclear lamina (a fibrous meshwork of intermediate filament lamins and associated proteins underlying the inner nuclear membrane) and intranuclear lamin subassemblies, which interact with chromatin [35, 38, 39, 41, 44, 97]. Since the nuclear lamins are attached directly to NETs in the inner nuclear membrane and are bound to the heterochromatin structure, they provide a scaffold for organization of numerous nuclear functions tied to a variety of proteins [45, 46, 98–101]. The nuclear lamins have roles in epigenetics, chromatin organization, DNA replication, transcription, and DNA repair, normal cellular aging, stem cell renewal, virus infections, and cancer [101]. Mutations in the lamin genes are linked to a variety of degenerative laminopathies, whereas changes in the expression of lamins are associated with tumorigenesis and also telomere structure, length, and function, and in the stabilization of the DNA damage repair response pathway [102]. The NE and its NET proteins are involved in maintaining and/or disrupting chromatin organization and nuclear architecture during cell division, human embryonic stem cell differentiation, and tumorigenesis dedifferentiation, and therefore, understanding these processes has potential clinical translational value [53, 65, 95–104].

Lamins play key roles in preserving several genome functions (e.g., higher-order genome organization and stability, chromatin regulation, transcription, DNA replication, and maintenance of telomeres) [45–48, 95–104] as well as being critical for maintaining nuclear architecture [95–97]. The importance of the NE and the lamins are well known in tumor development and progression [35]. Lamins are associated with proliferation and cell motility and they can serve as prognostic biomarkers in solid tumors [104]. Coradeghini et al. demonstrated differential expression of lamins A/C and B in CaP, with lamin B expression correlating with increasing Gleason grade [105]. Skvortsov et al. [106] showed that lamin-A/C expression correlated with the different Gleason groups. Compared to paired benign samples, lower Gleason score tumors showed down-regulation of lamin A/C in 60 % of CaP cases while higher Gleason score tumors revealed upregulation in 70 % of cases. To confirm lamin A/C regulation the authors used IHC to successfully confirm the differences between benign tissue, lower and higher Gleason score tumors using tissue microarrays of an independent set of some 90 tumor cases (ROC AUC = 0.88). Kong et al. [107] demonstrated that lamin A/C is overexpressed in invasive CaP. Their data showed that lamin A/C proteins are positively involved in malignant behavior of CaP cells in vitro and confirmed their data using IHC with a tissue microarray made up of 376 tissue cores of 94 CaP cases. Also, their data support that the mechanism goes through the PI3K/AKT/PTEN pathway and lamin A/C may represent a new and a novel therapeutic target for CaP. Though the lamins appear to be closely involved in the tumor biology events such as motility, proliferation, and invasiveness; their role in altering nuclear morphology occasionally has become controversial [94, 108, 109].

Conclusions

In summary, the nuclear envelope and its numerous associated proteins (lamins A/C and B, emerin, LAP2, BRG1, nesprin 1 and 2, the nucleoporins NUP88, 98, 133, 214, etc.) and key nuclear structural elements (i.e., nuclear matrix, actin, and lamins) play significant roles in chromatin spatial organization. Additionally, these elements maintain internal nuclear architecture, genome stability, and normal cellular processes (e.g., DNA repair, signaling, cell cycle, and mitosis). The human cell has evolved into a highly ordered biological machine driven by energy and the need to sustain spatial geometry of DNA and chromatin and the protein-related functions associated with maintenance of the nuclear apparatus. However, in disease this well-engineered cellular machine fails and often the built-in repair mechanisms also fail or may in fact accelerate the disease (e.g., autoimmune and malignant disease). Given all that we have discussed; where are the best molecular pathways or targets in either the primary cancer biopsy specimen or a benign area that identifies a lethal cancer and does so early? In both androgen-dependent and -independent CaP, we have noted the importance of critical targets including the PcG, enzyme-driven histone modifications, lamins A/C and B, BRG1, and p300. Also, there exists strong evidence that the environment (androgens, infection, diet, metabolism, temperature, etc.) can produce several genetic and epigenetic changes that disrupt normal cellular functions related to nuclear architecture, genome stability, DNA repair mechanisms, and some have been noted above [52, 53, 55, 56, 58]. One certainty is that nuclear morphology is often disrupted early in cancer with respect to nuclear size, shape, DNA content (ploidy), and chromatin organization. Does the entire target organ possess molecular and/or structural changes (field effects) that may differentiate a lethal and nonlethal cancer? Given the importance of nuclear shape to prognosis of cancer phenotypes, it is surprising and frustrating that we currently lack a detailed understanding to explain these changes and how they might arise and relate to specific molecular pathways in the cancer cell. This review offers an attempt to explain parts of this dilemma, at least in CaP. Finally, what are some of the NETs and their multiple attachments (at the periphery and internally) to chromatin, DNA, telomeres, etc. Additionally, how do these interactions play a role in modification of nuclear morphometry, chromosome organization, and molecular regulatory events that are clinically more useful in early prognosis and identifying new potential therapeutic targets of hormone-dependent and independent tumors?

References

1. Virchow R (1863) Cellular pathology as based upon physiological and pathological histology. J B Lippincott, Philadelphia, PA
2. Beale L (1860) Examination of sputum from a case of cancer of the pharynx and the adjacent parts. Arch Med (Lond) 2:44–46
3. Long SR, Cohen MB (1993) Classics in cytology. VI: the early cytologic discoveries of Lionel S. Beale. Diagn Cytopathol 9(5):595–598
4. Cremer T, Cremer C (2006) Rise, fall and resurrection of chromosome territories: a historical perspective. Part II. Fall and resurrection of chromosome territories during the 1950s to 1980s. Part III. Chromosome territories and the functional nuclear architecture: experiments and models from the 1990s to the present. Eur J Histochem 50(4):223–272
5. Cremer T, Cremer C (2006) Rise, fall and resurrection of chromosome territories: a historical perspective. Part I. The rise of chromosome territories. Eur J Histochem 50(3):161–176
6. Boveri T (2008) Concerning the origin of malignant tumours by Theodor Boveri. Translated and annotated by Henry Harris. J Cell Sci 121(Suppl 1):1–84
7. Veltri RW, Christudass CS, Isharwal S (2012) Nuclear morphometry, nucleomics and prostate cancer progression. Asian J Androl 14(3):375–384
8. Cremer T, Cremer M (2010) Chromosome territories. Cold Spring Harb Perspect Biol 2(3):a003889
9. Veltri RW, Partin AW, Miller CM (2005) Quantitative nuclear grade (QNG): the clinical applications of the quantitative measurement of nuclear structure using image analysis. In: Kelloff GJ, Hawk ET, Sigman CC (eds) Cancer chemoprevention. Humana Press, Totowa, NJ, pp 97–108
10. Sheehan DC, Hrapchak BB (1980) Theory and practice of histotechnology, 2nd edn. Battelle, Columbus, OH
11. Papanicolaou GN, Traut HF (1997) The diagnostic value of vaginal smears in carcinoma of the uterus. 1941. Arch Pathol Lab Med 121(3):211–224
12. Gill JE, Jotz MM (1976) Further observations on the chemistry of pararosaniline-Feulgen staining. Histochemistry 46(2):147–160

13. Schulte E, Wittekind D (1989) Standardization of the Feulgen-Schiff technique. Staining characteristics of pure fuchsin dyes; a cytophotometric investigation. Histochemistry 91(4):321–331
14. Gleason DF (1966) Classification of prostatic carcinomas. Cancer Chemother Rep 50(3):125–128
15. Gleason DF (1992) Histologic grading of prostate cancer: a perspective. Hum Pathol 23(3):273–279
16. Epstein JI (2010) An update of the Gleason grading system. J Urol 183(2):433–440
17. Brimo F et al (2013) Contemporary grading for prostate cancer: implications for patient care. Eur Urol 63(5):892–901
18. Diamond DA et al (1982) Computerized image analysis of nuclear shape as a prognostic factor for prostatic cancer. Prostate 3(4):321–332
19. Benson MC, McDougal DC, Coffey DS (1984) The application of perpendicular and forward light scatter to assess nuclear and cellular morphology. Cytometry 5(5):515–522
20. Benson MC, McDougal DC, Coffey DS (1984) The use of multiparameter flow cytometry to assess tumor cell heterogeneity and grade prostate cancer. Prostate 5(1):27–45
21. Mohler JL et al (1988) Nuclear roundness factor measurement for assessment of prognosis of patients with prostatic carcinoma. II. Standardization of methodology for histologic sections. J Urol 139(5):1085–1090
22. Mohler JL et al (1988) Nuclear roundness factor measurement for assessment of prognosis of patients with prostatic carcinoma. I. Testing of a digitization system. J Urol 139(5): 1080–1084
23. Partin AW et al (1989) A comparison of nuclear morphometry and Gleason grade as a predictor of prognosis in stage A2 prostate cancer: a critical analysis. J Urol 142(5):1254–1258
24. Isharwal S et al (2008) p300 (histone acetyltransferase) biomarker predicts prostate cancer biochemical recurrence and correlates with changes in epithelia nuclear size and shape. Prostate 68(10):1097–1104
25. Badalament RA et al (1996) An algorithm for predicting nonorgan confined prostate cancer using the results obtained from sextant core biopsies with prostate specific antigen level. J Urol 156(4):1375–1380
26. Veltri RW et al (2002) Prediction of pathological stage in patients with clinical stage T1c prostate cancer: the new challenge. J Urol 168(1):100–104
27. Veltri RW et al (1996) Ability to predict biochemical progression using Gleason score and a computer-generated quantitative nuclear grade derived from cancer cell nuclei. Urology 48(5):685–691
28. Khan MA et al (2003) Quantitative alterations in nuclear structure predict prostate carcinoma distant metastasis and death in men with biochemical recurrence after radical prostatectomy. Cancer 98(12):2583–2591
29. Veltri RW et al (2004) Ability to predict metastasis based on pathology findings and alterations in nuclear structure of normal-appearing and cancer peripheral zone epithelium in the prostate. Clin Cancer Res 10(10):3465–3473
30. Kortenhorst MS et al (2009) Valproic acid causes dose- and time-dependent changes in nuclear structure in prostate cancer cells in vitro and in vivo. Mol Cancer Ther 8(4):802–808
31. Doyle S et al (2010) A boosted Bayesian multiresolution classifier for prostate cancer detection from digitized needle biopsies. IEEE Trans Biomed Eng 59(5):1205–1218
32. Madabhushi A et al (2011) Computer-aided prognosis: predicting patient and disease outcome via quantitative fusion of multi-scale, multi-modal data. Comput Med Imaging Graph 35(7–8):506–514
33. Gao M, Bridgman P, Kumar S (2003) Computer aided prostate cancer diagnosis using image enhancement and JPEG2000. In: Proceedings of the SPIE Annual Meeting, 2003, San Diego, CA
34. Tiwari P et al (2011) Multi-modal data fusion schemes for integrated classification of imaging and non-imaging biomedical data. In: ISBI, 2011, Chicago, IL, pp 165–168

35. Ali S et al (2011) Adaptive energy selective active contour with shape priors for nuclear segmentation and Gleason grading of prostate cancer. Med Image Comput Comput Assist Interv 14(Pt 1):661–669
36. Doyle S et al (2007) Automated grading of prostate cancer using architectural and textural image features. In: ISBI, 2007, Arlington, VA, pp 1284–1287
37. Huang PW, Lee CH (2009) Automatic classification for pathological prostate images based on fractal analysis. IEEE Trans Med Imaging 28(7):1037–1050
38. Khurd P et al (2010) Computer-aided Gleason grading of prostate cancer histopathological images using texton forests. In: ISBI, 2010, Rotterdam, pp 636–639
39. Khurd P et al (2011) Network cycle features: application to computer-aided Gleason grading of prostate cancer histopathological images. In: ISBI, 2011, Chicago, IL, pp 1632–1636
40. Naik S et al (2008) Automated gland and nuclei segmentation for grading of prostate and breast cancer histopathology. In: ISBI, Paris, 2008, pp 284–287
41. Nguyen K, Jain AK, Allen RL (2010) Automated gland segmentation and classification for gleason grading of prostate tissue images. In: 20th international conference on pattern recognition (ICPR), 2010, Istanbul, pp 1497–1500
42. Ou Y et al (2009) Sampling the spatial patterns of cancer: optimized biopsy procedures for estimating prostate cancer volume and Gleason Score. Med Image Anal 13(4):609–620
43. Tabesh A et al (2007) Multifeature prostate cancer diagnosis and Gleason grading of histological images. IEEE Trans Med Imaging 26(10):1366–1378
44. Yoon HY et al (2011) Cardinal multiridgelet-based prostate cancer histological image classification for Gleason grading. In: BIBM, 2011, Atlanta, GA, pp 315–320
45. Chow KH, Factor RE, Ullman KS (2012) The nuclear envelope environment and its cancer connections. Nat Rev Cancer 12(3):196–209
46. de Las Heras JI, Batrakou DG, Schirmer EC (2013) Cancer biology and the nuclear envelope: a convoluted relationship. Semin Cancer Biol 23(2):125–137
47. Verstraeten VL et al (2007) The nuclear envelope, a key structure in cellular integrity and gene expression. Curr Med Chem 14(11):1231–1248
48. Hozák P et al (1995) Lamin proteins form an internal nucleoskeleton as well as a peripheral lamina in human cells. J Cell Sci 108(Pt 2):635–644
49. Naetar N, Foisner R (2009) Lamin complexes in the nuclear interior control progenitor cell proliferation and tissue homeostasis. Cell Cycle 8(10):1488–1493
50. Berezney R, Coffey DS (1974) Identification of a nuclear protein matrix. Biochem Biophys Res Commun 60(4):1410–1417
51. Coffey DS (2002) Nuclear matrix proteins as proteomic markers of preneoplastic and cancer lesions: commentary re: G. Brunagel et al., nuclear matrix protein alterations associated with colon cancer metastasis to the liver. Clin. Cancer Res., 8: 3039–3045, 2002. Clin Cancer Res 8(10):3031–3033
52. Getzenberg RH, Coffey DS (2011) Changing the energy habitat of the cancer cell in order to impact therapeutic resistance. Mol Pharm 8(6):2089–2093
53. Hanahan D, Weinberg RA (2011) Hallmarks of cancer: the next generation. Cell 144(5):646–674
54. He S et al (2008) Chromatin organization and nuclear microenvironments in cancer cells. J Cell Biochem 104(6):2004–2015
55. Podlaha O et al (2012) Evolution of the cancer genome. Trends Genet 28(4):155–163
56. Schetter AJ, Heegaard NH, Harris CC (2010) Inflammation and cancer: interweaving microRNA, free radical, cytokine and p53 pathways. Carcinogenesis 31(1):37–49
57. Strahl BD, Allis CD (2000) The language of covalent histone modifications. Nature 403(6765):41–45
58. Turner BM (2011) Environmental sensing by chromatin: an epigenetic contribution to evolutionary change. FEBS Lett 585(13):2032–2040
59. Wu JI, Lessard J, Crabtree GR (2009) Understanding the words of chromatin regulation. Cell 136(2):200–206

60. Zhang Y, Reinberg D (2001) Transcription regulation by histone methylation: interplay between different covalent modifications of the core histone tails. Genes Dev 15(18): 2343–2360
61. Sasaki M et al (2002) Methylation and inactivation of estrogen, progesterone, and androgen receptors in prostate cancer. J Natl Cancer Inst 94(5):384–390
62. Suzuki H, Ito H (1999) Role of androgen receptor in prostate cancer. Asian J Androl 1(3):81–85
63. Tropberger P et al (2013) Regulation of transcription through acetylation of H3K122 on the lateral surface of the histone octamer. Cell 152(4):859–872
64. Jones PA, Baylin SB (2002) The fundamental role of epigenetic events in cancer. Nat Rev Genet 3(6):415–428
65. Cai C et al (2011) Androgen receptor gene expression in prostate cancer is directly suppressed by the androgen receptor through recruitment of lysine-specific demethylase 1. Cancer Cell 20(4):457–471
66. Chlenski A et al (2001) Androgen receptor expression in androgen-independent prostate cancer cell lines. Prostate 47(1):66–75
67. Aryee MJ et al (2013) DNA methylation alterations exhibit intraindividual stability and interindividual heterogeneity in prostate cancer metastases. Sci Transl Med 5(169):169ra10
68. Jarrard DF et al (1998) Methylation of the androgen receptor promoter CpG island is associated with loss of androgen receptor expression in prostate cancer cells. Cancer Res 58(23):5310–5314
69. Izbicka E et al (1999) 5,6 Dihydro-5′-azacytidine (DHAC) restores androgen responsiveness in androgen-insensitive prostate cancer cells. Anticancer Res 19(2A):1285–1291
70. Seligson DB et al (2005) Global histone modification patterns predict risk of prostate cancer recurrence. Nature 435(7046):1262–1266
71. Bianco-Miotto T et al (2010) Global levels of specific histone modifications and an epigenetic gene signature predict prostate cancer progression and development. Cancer Epidemiol Biomarkers Prev 19(10):2611–2622
72. Watson JA et al (2010) Hyperacetylation in prostate cancer induces cell cycle aberrations, chromatin reorganization and altered gene expression profiles. J Cell Mol Med 14(6B): 1668–1682
73. Malik G et al (2007) SELDI protein profiling of dunning R-3327 derived cell lines: identification of molecular markers of prostate cancer progression. Prostate 67(14):1565–1575
74. Huisman A et al (2007) Discrimination between benign and malignant prostate tissue using chromatin texture analysis in 3-D by confocal laser scanning microscopy. Prostate 67(3):248–254
75. Li LC, Carroll PR, Dahiya R (2005) Epigenetic changes in prostate cancer: implication for diagnosis and treatment. J Natl Cancer Inst 97(2):103–115
76. Brock HW, van Lohuizen M (2001) The Polycomb group–no longer an exclusive club? Curr Opin Genet Dev 11(2):175–181
77. Jacobs JJ, van Lohuizen M (2002) Polycomb repression: from cellular memory to cellular proliferation and cancer. Biochim Biophys Acta 1602(2):151–161
78. Czermin B et al (2002) Drosophila enhancer of Zeste/ESC complexes have a histone H3 methyltransferase activity that marks chromosomal Polycomb sites. Cell 111(2):185–196
79. Varambally S et al (2002) The polycomb group protein EZH2 is involved in progression of prostate cancer. Nature 419(6907):624–629
80. van der Vlag J, Otte AP (1999) Transcriptional repression mediated by the human polycomb-group protein EED involves histone deacetylation. Nat Genet 23(4):474–478
81. Yu J et al (2007) A polycomb repression signature in metastatic prostate cancer predicts cancer outcome. Cancer Res 67(22):10657–10663
82. Bryant RJ et al (2008) The Polycomb Group protein EZH2 regulates actin polymerization in human prostate cancer cells. Prostate 68(3):255–263
83. Bettinger BT, Gilbert DM, Amberg DC (2004) Actin up in the nucleus. Nat Rev Mol Cell Biol 5(5):410–415

84. Su IH et al (2005) Polycomb group protein ezh2 controls actin polymerization and cell signaling. Cell 121(3):425–436
85. Tan J et al (2007) Pharmacologic disruption of Polycomb-repressive complex 2-mediated gene repression selectively induces apoptosis in cancer cells. Genes Dev 21(9):1050–1063
86. Chase A, Cross NC (2011) Aberrations of EZH2 in cancer. Clin Cancer Res 17(9): 2613–2618
87. Laitinen S et al (2008) EZH2, Ki-67 and MCM7 are prognostic markers in prostatectomy treated patients. Int J Cancer 122(3):595–602
88. Jhavar S et al (2009) Biopsy tissue microarray study of Ki-67 expression in untreated, localized prostate cancer managed by active surveillance. Prostate Cancer Prostatic Dis 12(2):143–147
89. Veltri RW et al (2010) Nuclear roundness variance predicts prostate cancer progression, metastasis, and death: a prospective evaluation with up to 25 years of follow-up after radical prostatectomy. Prostate 70(12):1333–1339
90. Cao R et al (2002) Role of histone H3 lysine 27 methylation in Polycomb-group silencing. Science 298(5595):1039–1043
91. Bektas S et al (2009) The relation between Gleason score, and nuclear size and shape factors in prostatic adenocarcinoma. Turk J Med Sci 39(3):381–387
92. Debes JD et al (2005) p300 modulates nuclear morphology in prostate cancer. Cancer Res 65(3):708–712
93. Wang Z et al (2008) Combinatorial patterns of histone acetylations and methylations in the human genome. Nat Genet 40(7):897–903
94. Imbalzano KM et al (2013) Nuclear shape changes are induced by knockdown of the SWI/SNF ATPase BRG1 and are independent of cytoskeletal connections. PLoS One 8(2):e55628
95. Chernov IP, Akopov SB, Nikolaev LG (2004) Structure and functions of nuclear matrix associated regions (S/MARs). Russ J Bioorg Chem 30(1):1–11
96. Elcock LS, Bridger JM (2008) Exploring the effects of a dysfunctional nuclear matrix. Biochem Soc Trans 36(Pt 6):1378–1383
97. Nickerson J (2001) Experimental observations of a nuclear matrix. J Cell Sci 114(Pt 3): 463–474
98. Postnikov Y, Bustin M (2010) Regulation of chromatin structure and function by HMGN proteins. Biochim Biophys Acta 1799(1–2):62–68
99. Dechat T et al (2008) Nuclear lamins: major factors in the structural organization and function of the nucleus and chromatin. Genes Dev 22(7):832–853
100. Prokocimer M et al (2009) Nuclear lamins: key regulators of nuclear structure and activities. J Cell Mol Med 13(6):1059–1085
101. Stuurman N, Heins S, Aebi U (1998) Nuclear lamins: their structure, assembly, and interactions. J Struct Biol 122(1–2):42–66
102. Gonzalez-Suarez I et al (2009) Novel roles for A-type lamins in telomere biology and the DNA damage response pathway. EMBO J 28(16):2414–2427
103. Butler JT et al (2009) Changing nuclear landscape and unique PML structures during early epigenetic transitions of human embryonic stem cells. J Cell Biochem 107(4):609–621
104. Goldman RD et al (2002) Nuclear lamins: building blocks of nuclear architecture. Genes Dev 16(5):533–547
105. Coradeghini R et al (2006) Differential expression of nuclear lamins in normal and cancerous prostate tissues. Oncol Rep 15(3):609–613
106. Skvortsov S et al (2011) Proteomics profiling of microdissected low- and high-grade prostate tumors identifies Lamin A as a discriminatory biomarker. J Proteome Res 10(1):259–268
107. Kong L et al (2012) Lamin A/C protein is overexpressed in tissue-invading prostate cancer and promotes prostate cancer cell growth, migration and invasion through the PI3K/AKT/PTEN pathway. Carcinogenesis 33(4):751–759
108. Foster CR et al (2010) Lamins as cancer biomarkers. Biochem Soc Trans 38(Pt 1):297–300
109. Reddy KL, Feinberg AP (2013) Higher order chromatin organization in cancer. Semin Cancer Biol 23(2):109–115

"To Be or Not to Be in a Good Shape": Diagnostic and Clinical Value of Nuclear Shape Irregularities in Thyroid and Breast Cancer

Gianni Bussolati, Francesca Maletta, Sofia Asioli, Laura Annaratone, Anna Sapino, and Caterina Marchiò

Abstract Variation in both nuclear shape and size ("pleomorphism"), coupled with changes in chromatin amount and distribution, remains the basic criteria for microscopy in a cytologic diagnosis of cancer. The biological determinants of nuclear shape irregularities are not clarified, so, rather than on the genesis of nuclear irregularities, we here focus our attention on a descriptive analysis of nuclear pleomorphism. We keep in mind that evaluation of nuclear shape as currently practiced in routine preparations is improper because it is indirectly based on the distribution of DNA as revealed by the affinity for basic dyes. Therefore, over the last years we have been using as criteria morphological features of nuclei of thyroid and breast carcinomas as determined by immunofluorescence, *in situ* hybridization, and 3D reconstruction. We have translated this approach to routine diagnostic pathology on tissue sections by employing immunoperoxidase staining for emerin. Direct detection of nuclear envelope irregularities by tagging nuclear membrane proteins such as lamin B and emerin has resulted in a more objective definition of the shape of the nucleus. In this review we discuss in detail methodological issues as well as diagnostic and prognostic implications provided by decoration/staining of the nuclear envelope in both thyroid and breast cancer, thus demonstrating how much it matters "to be in the right shape" when dealing with pathological diagnosis of cancer.

Keywords Nuclei • Pleomorphism • Papillary carcinoma • Breast cancer • 3D reconstruction

G. Bussolati (✉)
Department of Medical Sciences, University of Turin, via Santena 7, 10126 Turin, Italy

Institut Victor Babes, Bucharest, Romania
e-mail: gianni.bussolati@unito.it

F. Maletta • S. Asioli • L. Annaratone • A. Sapino • C. Marchiò
Department of Medical Sciences, University of Turin, via Santena 7, 10126 Turin, Italy
e-mail: francesca.maletta@unito.it; sofia.asioli@unito.it; laura.annaratone@unito.it; anna.sapino@unito.it; caterina.marchio@unito.it

E.C. Schirmer and J.I. de las Heras (eds.), *Cancer Biology and the Nuclear Envelope*, Advances in Experimental Medicine and Biology 773, DOI 10.1007/978-1-4899-8032-8_5,

Abbreviations

FISH	Fluorescence *in situ* hybridization
H&E	Hematoxylin and eosin
NE	Nuclear envelope
PTC	Papillary thyroid carcinoma
PDC	Poorly differentiated carcinoma
NEP	Nuclear envelope pleomorphism

Introduction

Irregularity of nuclear shape and an increased nuclear-cytoplasmic ratio (also called karyoplasmic ratio) characterize most, though not all, neoplastic conditions. In fact, we can roughly consider, as far as nuclear shape in cancer is concerned, three types of events. In some tumors, nuclei are roundish, with a smooth nuclear membrane not dissimilar from the corresponding normal epithelium. However, in the vast majority of cancers most nuclei are pleomorphic, as defined by the presence of irregularities in both nuclear shape and size coupled with changes in chromatin amount and distribution within the nucleus [1]. Such features remain the basic microscopy criteria for a cytologic diagnosis of cancer: indeed, indentations, undulations, and folds of the nuclear membrane, as originally reported by ultrastructural observations [2], occur early in neoplastic processes [3].

Finally, in some types of cancer, and notably in thyroid cancer, nuclear shape irregularity presents a typical and reproducible pattern, acquiring clear diagnostic significance. Typically, papillary thyroid carcinoma (PTC) is characterized by the presence of indentations, grooves (the so-called coffee-bean nuclei), pseudo-inclusions (or "Orphan-Annie-eyed" nuclei), and nuclear clearing. These characteristics derive from finely dispersed chromatin or deep and complex cytoplasmic longitudinal invaginations into the double-membrane nuclear envelope (NE), as demonstrated by electron microscopy [4–7]. The presence of these features is the only clue to the diagnosis of PTC, which alone represents almost 80 % of all thyroid carcinomas [8].

The Nuclear Envelope and Rationale for Its Use in Pathology

Light microscopy appreciation of nuclear pleomorphism in cytopathology and histopathology is indirect, as it is currently based on staining of nucleic acids with basic dyes such as hematoxylin. Since peripheral chromatin is bound to the nuclear membrane, this provides crude evidence of nuclear shape. However, a method to decorate the NE could provide direct detection of the NE and its components. Indeed, by highlighting NE-associated proteins we could provide a more objective and direct appreciation of nuclear shape and definitively reconstruct nuclear shape based on the distribution of NE proteins [1].

Detection, appreciation, and rendering of nuclear shape are, for intrinsic reasons, different in cytopathology and in histopathology. In fact, while in the cytological approach whole nuclei are available for investigations, in histological sections only nuclear segments are available, which makes images partial and seldom conclusive. Moreover, in diagnostic cytology, the preservation of nuclear shape is heavily influenced by the technical procedure for preparations, since in liquid cytology, the shape of the nucleus is fully preserved. By contrast, smearing followed by cell drying is bound to produce a collapse of the nuclear shape leading to misdiagnosis.

With these caveats in mind, the following approaches have been followed by our group:

A. Tagging of components of the nuclear membrane by immunofluorescence and immunoperoxidase staining (Figs. 1, 2, and 3)
B. Immunofluorescence decoration of the nuclear membrane, associated with gene labeling by fluorescence *in situ* hybridization (FISH) (Fig. 4)
C. Confocal microscopy and image capture, followed by 3D reconstruction using specific software (e.g. Amira 3D Analysis Software for Life Sciences—http://www.vsg3d.com/) (Figs. 4, 5, and 6)
D. Image analysis using specific softwares (e.g. Image-Pro Plus, MediaCybernetics, http://www.mediacy.com, which is an image analysis software package for fluorescence imaging and for recording sequential images)

Following these procedures, we have been able to trace the distribution of the NE with immunofluorescence and immunoperoxidase procedures by using antibodies targeting lamin B and another NE marker, namely, emerin [9] (approaches A, B). These two proteins label different structural components, since lamin B is located in the proteinaceous layer at the interface between the chromatin and the membrane, while emerin is a transmembrane protein of the inner nuclear membrane [9]. We also obtained a proper 3D reconstruction of the nuclear shape (approaches B, C). Confocal microscopy analysis allows the creation of a stack of images along the z-axis that can be uploaded into dedicated software for advanced 3D visualization and volume modeling. The nuclear outline is obtained after segmentation of sequential images of nuclear sections [10]. The segmented areas are then employed to generate 3D polygonal surface models using macros in the dedicated software (approaches A, B, C). An alternative procedure to confocal microscopy for generating sharp images from tissue specimens is provided by deconvolution technology (approach D). Briefly, immunofluorescent preparations are examined typically with a wide-field fluorescence microscope, equipped with either a motorized stage or a piezo focus lens positioner, a camera, and a dedicated software that allows 3D image stacks to be recorded. Subsequent deconvolution of the image stacks improves the clarity of images by applying an algorithm that uses pixel information in the adjacent sections to remove out-of-focus light [11, 12].

With confocal microscopy a single nucleus can be observed at different and sequential cutting planes, and a 3D reconstruction can be obtained by adding each section to build up the entire nuclear volume. Similarly, after removal of the out-of-focus light by deconvolution, the individual sections can be reconstructed to obtain 3D models.

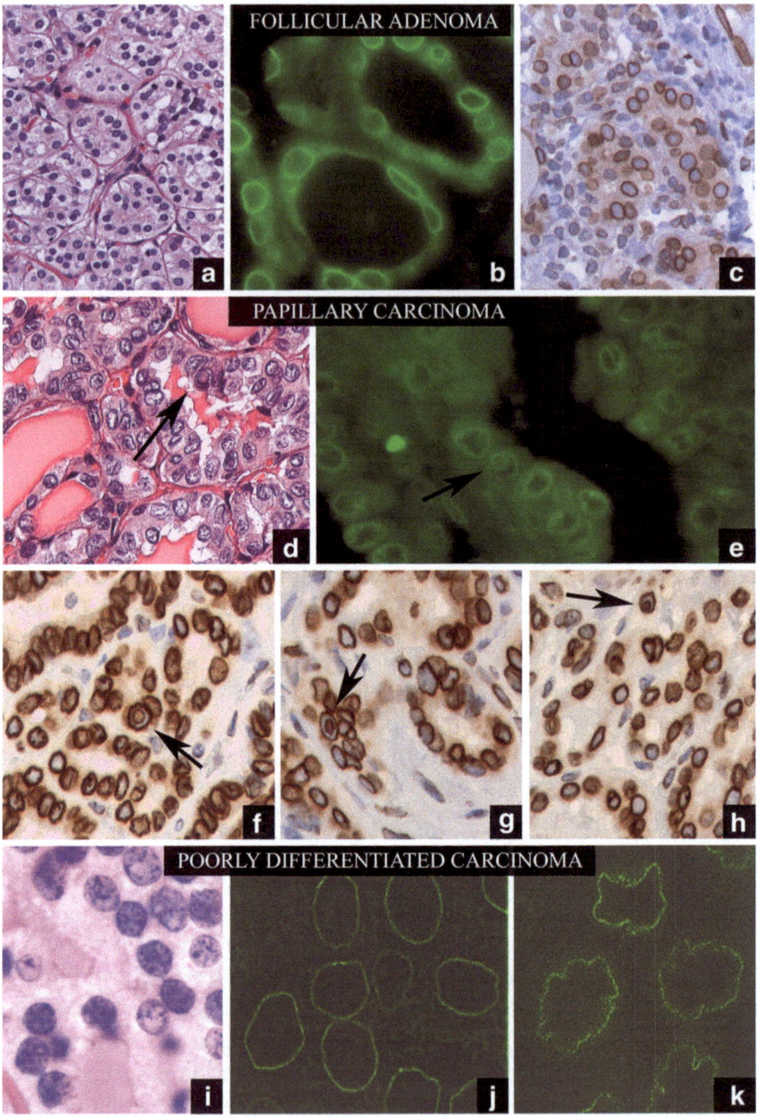

Fig. 1 How emerin staining in both immunofluorescence and immunoperoxidase highlights nuclear shape in a spectrum of thyroid lesions. Panels (**a–c**) are a follicular adenoma stained in various ways. The presence of round and regular nuclei is evident. Parallel sections were stained with (**a**) hematoxylin and eosin (H&E), (**b**) immunofluorescence for emerin, and (**c**) immunoperoxidase for emerin. Panels (**d–h**) are an example of papillary thyroid carcinoma showing NE irregularities. (**d**) An H&E-stained section of a thyroid proliferation with irregular nuclei and scarce pseudo-inclusions (*arrow*). (**e**) The irregularities become extremely evident with immunofluorescence for emerin. (**f–h**) Similarly immunoperoxidase staining for emerin reveals the presence of several pseudo-inclusions by marking nuclear shape and highlighting its foldings (*arrows* in **e–h**: evident and widespread nuclear pseudo-inclusions). Panels (**i–k**) are a case of poorly differentiated carcinoma. (**i**) H&E staining reveals nuclei to look quite regular with only scarce irregularities of the nuclear contours. (**j**) Immunofluorescence staining for emerin reveals nuclei that can look quite regular (as in follicular lesions). (**k**) However, in some cells this staining reveals the so-called star-shaped or raisin-like features

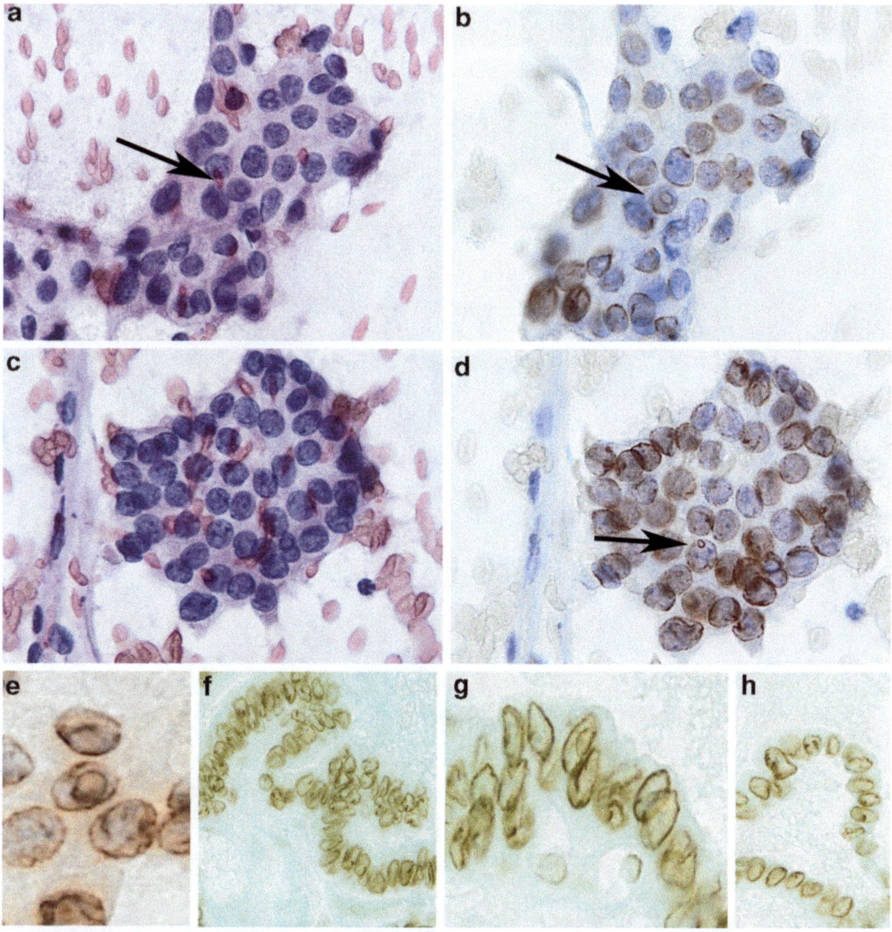

Fig. 2 Emerin staining in cytological specimens of papillary thyroid carcinoma. (**a–d**) Cases of papillary thyroid carcinoma (PTC) in alcohol-fixed smears, (**e**) with Thin Prep, and panels (**f–h**) are cell blocks obtained from fine needle aspirations on thyroid nodules. In smears (**a**, **c**) cells are stained with hematoxylin and eosin (H&E), while (**b**, **d**) are the same fields and nuclei stained with immunoperoxidase for emerin. Direct comparison of the same fields reveals the superior ability of emerin staining to highlight diagnostic nuclear features, such as nuclear pseudo-inclusions (*arrows* in **a**, **b**), even of very small size (*arrow* in **d**), grooves, and crescent-like figures. (**e**) On Thin Prep preparations, pseudo-inclusions are evident. (**f–h**) The emerin-stained sections obtained from cell block highlight other features typical of PTC, such as the garland-like appearance (**f**) and deep irregularities of nuclear shape (**g**, **h**)

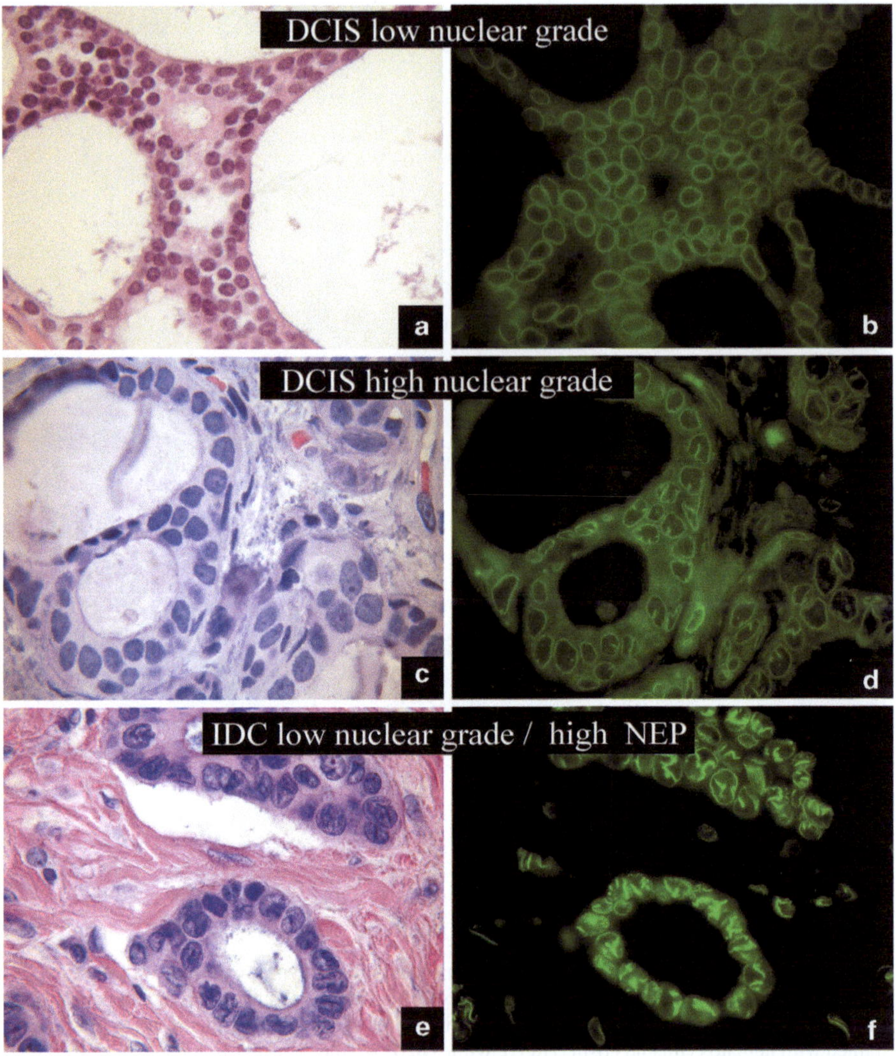

Fig. 3 Micrographs depicting different scenarios in the evaluation of nuclear pleomorphism in breast cancer pathology. Ductal carcinoma *in situ* (DCIS) of low nuclear grade (**a**, **b**) shows a regular lining of the nuclear envelope by immunofluorescence for emerin (**b**). (**c**, **d**) Immunofluorescence for emerin best shows fine irregularities of the nuclei in ductal carcinoma *in situ* of high nuclear grade. (**e**, **f**) Finally, an example of infiltrating ductal carcinoma (IDC) of low histological grade shows high-grade nuclear envelope pleomorphism (NEP), as best highlighted by immunofluorescence for emerin

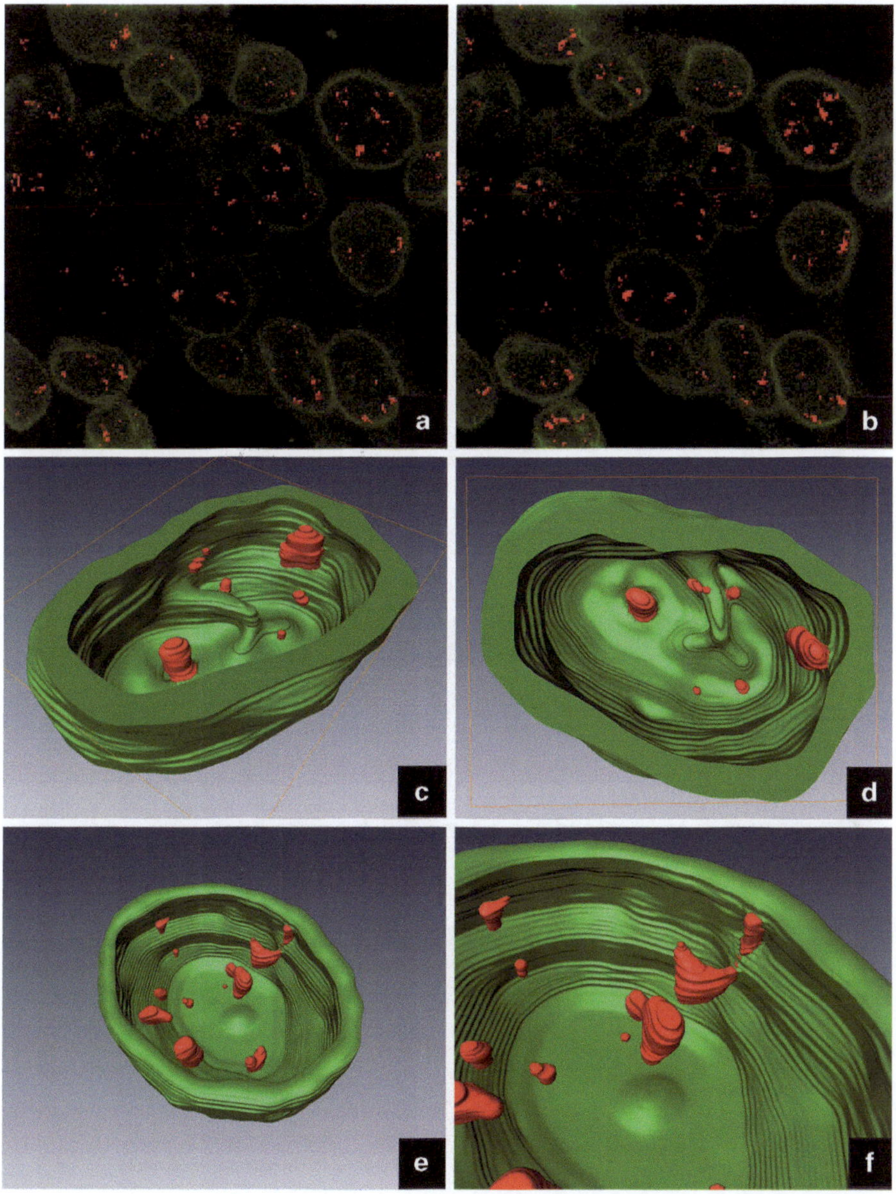

Fig. 4 3D reconstruction of nuclei with visualization of *HER2* gene. (**a, b**) Immunofluorescence for lamin B (*green*) is performed together with fluorescence *in situ* hybridization (FISH) for the *HER2* gene (*red* signals) in BT-474 cells (*HER2* amplified, as exemplified by the gene clusters). (**c–f**) 3D reconstruction of these nuclei shows the relationship between *HER2* gene clusters and nuclear envelope

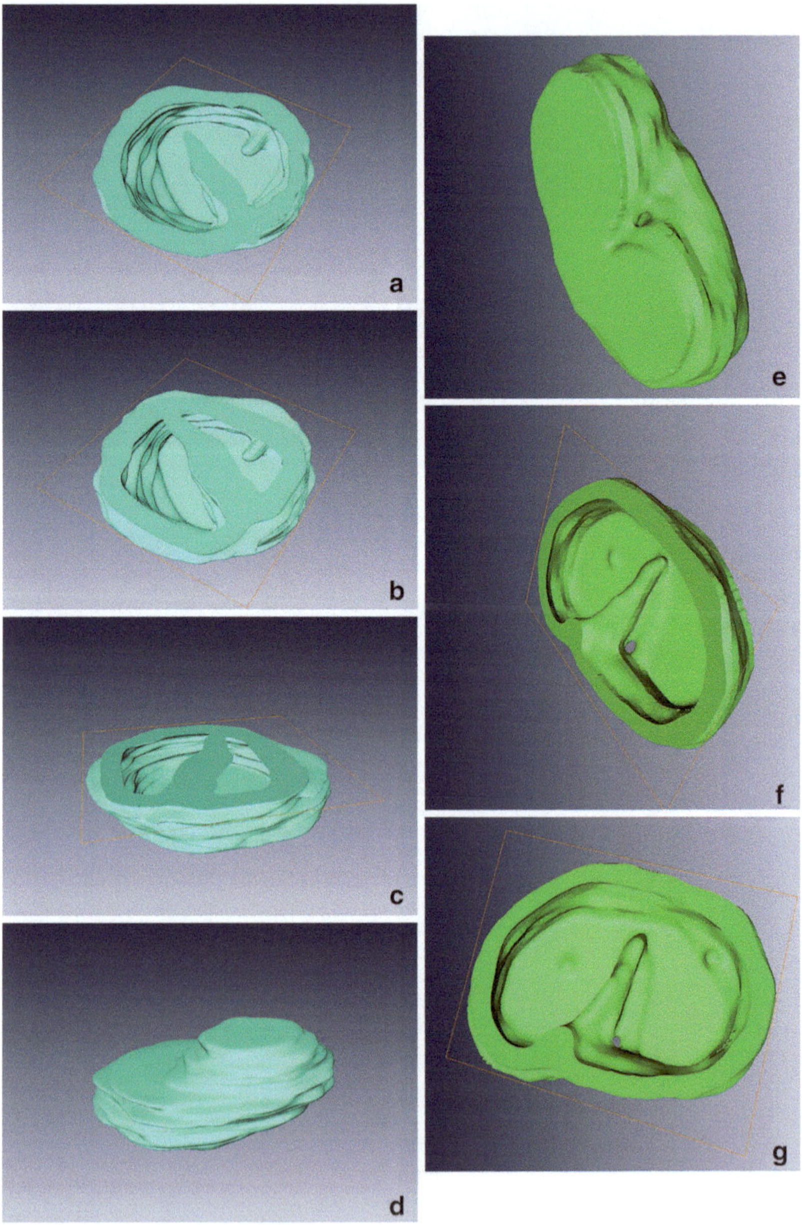

Fig. 5 3D reconstruction of PTC nuclei. Panels (**a–d**) are images obtained from sequential cutting planes of a single papillary thyroid carcinoma (PTC) nucleus, while images from (**e–g**) are different perspectives of a 3D reconstruction of another example of PTC nucleus. Immunofluorescence for emerin was performed on sequential sections of nuclei from PTC cell lines, and a 3D reconstruction was obtained using software Amira (Amira 3D Analysis Software for Life Sciences—http://www.vsg3d.com). The models of nuclear shape here shown revealed the presence of irregularities of the nuclear membrane with foldings and invaginations, which corresponds to the so-called coffee bean (or grooves) on traditional H&E-stained nuclei

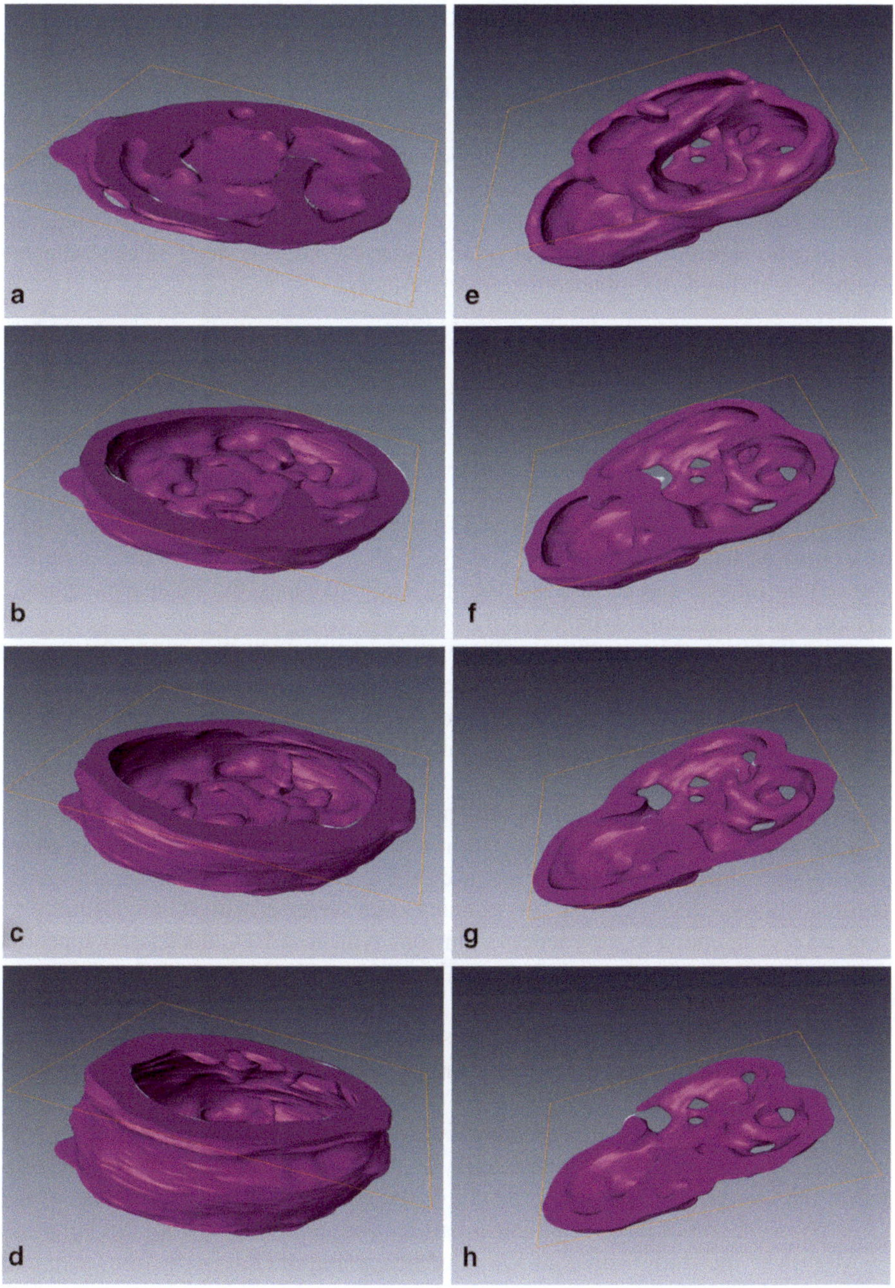

Fig. 6 3D reconstruction of breast cancer nuclei. Panels (**a–d**) and (**e–h**) are images obtained from sequential cutting planes of two different nuclei of breast cancer. Immunofluorescence for lamin B was performed on sequential sections of nuclei from breast cancer cells (BT-474). The software Amira (Amira 3D Analysis Software for Life Sciences—http://www.vsg3d.com) was used to obtain a 3D reconstruction. The 3D models highlight irregularities of nuclear contour and several intranuclear tubules

Diagnostic and Clinical Impact of Nuclear Shape in PTC

Over the years, we have focused our attention on nuclear pleomorphisms and alterations in shape (dysmorphisms) that occur in cancers and specifically in thyroid and breast carcinomas, two areas in which nuclear pleomorphism holds high biological significance and great diagnostic and prognostic impact. In breast carcinoma the nuclear shape varies according to the histological subtype and grade, involving also a prognostic significance. In thyroid carcinomas the nuclear shape is instead paradigmatic and diagnostic of specific types of cancer.

Nuclear Shape in PTC and PDC Versus Other Thyroid Pathologies

A study conducted on cell lines derived from PTC and from follicular carcinomas, as well as on histological sections and cytological fine needle aspiration samples, showed an intense and diffuse staining for lamin B along the nuclear membrane irrespective of the tumor type [10].

Remarkable nuclear deformities, infolding, and "tubelike" invaginations were evident in the vast majority of PTC nuclei, and the typical intranuclear pseudo-inclusions were also lined by lamin B. Moreover, PTC nuclei were larger and much more irregular than the corresponding control cases of follicular tumors (Fig. 1). The invaginations and indentations of PTC nuclei, as revealed by the sequential reconstructions obtained with the use of the confocal microscope, appeared to penetrate into the nucleus to a variable degree from a minimal fraction up to reach an entire penetration, which, as a consequence, acquires a "donut-like" configuration. Serial sections showed that "pouches" or "tunnels" that were seen in 3D reconstructions (Fig. 5) corresponded to the pseudo-inclusions typical of PTC nuclei and appeared to be always connected to the cell cytoplasm and lined by intact nuclear membrane.

In control cases of follicular tumors, nuclei were smaller than in PTC, with a round or an oval shape and a regular and smooth contour. Confocal microscope and 3D-reconstruction images highlighted the presence of only slight and occasional deformities.

It can be thus concluded that the typical irregularities of PTC nuclei may appear at the light microscopy level alternatively as grooves or pseudo-inclusions according to the viewpoint from which the cell is explored, but they are all facets of the same phenomenon of large-scale invaginations with reciprocal cytoplasm bulging.

Moreover, the study with confocal microscopy and 3D reconstructions acquired diagnostic usefulness, since it opened the possibility to apply knowledge on nuclear shape and volume to the so-called grey area of thyroid pathology, which comprises follicular patterned lesions with optically clear nuclei but without clear-cut features of PTC. Irregularly shaped nuclei in fact can also be found in thyroiditis, hyperplastic lesions, goiter with degenerative changes, oxyphilic tumors or be the consequence of the fine needle aspiration procedure or the fixation artefacts [13]. These benign lesions may have nuclear irregularities that mimic those of PTC: besides clear nuclei, occasional grooves can be appreciated and widespread alterations of nuclear contour are often present. What distinguishes these lesions from PTC is a combination of several factors, both histological (architecture, presence of vascular and/or capsular invasion) and cytological (extent, frequency, and intensity of nuclear irregularities, presence of nuclear pseudo-inclusions), but traditional staining may not be sufficient to fully appreciate these differences and distinction may be challenging. For this reason, the use of immunohistochemical staining to improve detection of nuclear shape might be of help in this differential diagnosis.

In order to apply this method of nuclear stain to routine histological and cytological diagnosis, immunohistochemical staining with anti-emerin antibodies was evaluated [9, 14, 15]. Emerin is a protein of the inner nuclear membrane which appears to interact with the lamina and chromatin; it is a serine-rich nuclear membrane protein involved in mediating membrane anchorage to the cytoskeleton [16]. Fischer et al. [17] demonstrated that its expression is not reduced or abolished in cytoplasmic pseudo-inclusions or grooves of PTC, but it simply conforms to nuclear irregularities and foldings. Thus, cases of PTC, follicular adenoma, follicular carcinoma, Hashimoto's thyroiditis, goiter, Graves disease, and normal thyroid tissues were stained with anti-emerin antibodies.

In PTC, emerin staining allowed an easy identification of all previously described nuclear irregularities (invaginations, pseudo-inclusions, grooves, crescent-like nuclei, and deep-stellate nuclear shape) but also a peculiar pattern never described before, which is the presence of minute curls along the periphery of the nucleus, leading to a garland-like pattern (Figs. 1 and 2); moreover, when directly comparing the same nuclei stained with immunofluorescence for emerin and subsequently restained with hematoxylin and eosin (H&E), it was evident that only some of the grooves seen with immunofluorescence were appreciable with H&E as well. For this reason, emerin staining was tested on cases of follicular variant of PTC (FVPTC). This controversial variant is in fact characterized by follicles lined by cells that lack the typical features of PTC: nuclei are dark, and irregularities of shape are often borderline. Grooves are scarce and pseudo-inclusions rare or totally absent [6, 18, 19]. For this reason, the diagnosis of FVPTC is traditionally affected by a high rate of inter-observer discordance, even among the so-called expert thyroid pathologists [20–22]. The distinction between FVPTC on one side and benign lesions on the other (follicular adenoma, goiter, nodule in the context of thyroiditis) is based on the shape of the nucleus, presenting grooves and invaginations in the former while roundish in the latter. The differential diagnosis is important since it carries a profound therapeutic and prognostic impact, but it is sometimes difficult and problematic, because of improper preservation of the nuclear shape in histological sections.

After emerin staining of cases of FVPTC, invaginations of the nuclear membrane were more evident than on H&E slides, and emerin tracing of the envelope allowed the recognition of some pseudo-inclusions that were "hidden" by the presence of dark nuclei in H&E preparations.

Staining for emerin shows distinct and different patterns between PTC nuclei and other conditions, because it reveals more clearly nuclear irregularities in cases of PTC, while it confirms a regular nuclear profile in normal thyroid gland and other lesions (follicular lesions, goiter, thyroiditis). In fact, we have demonstrated that in thyroid lesions other than PTC, the vast majority of cells have smooth and round nuclei and only occasional cells may have irregularities of shape and invaginations similar to PTC nuclei [9]. Of note, such irregularities are only occasional and never reach the degree so typical and diagnostic of PTC.

The diagnosis of poorly differentiated carcinoma (PDC) is based on a diagnostic algorithm involving the presence of a solid, trabecular, or insular histological pattern as well as of necrosis and increased rate of mitoses [23]. However, a role in the diagnosis is played by nuclei as well. In PDC, nuclei are small (if compared with PTC nuclei), round, and hyperchromatic and lack typical clear-cut features of PTC (pseudo-inclusions, grooves, crescent-like features). Nuclei in PDC appear as "convoluted" because of the presence of an irregular ("convoluted" or "raisin-like") contour membrane. Only occasional grooves are observed, and no ground-glass appearance or pseudo-inclusions. By decorating/staining the NE with anti-emerin antibodies, PDC-convoluted or raisin-like nuclei showed humps and plicae, thus giving the appearance of a star-shaped structure (Fig. 1).

Thyroid Cytology

Tracing the nuclear membrane by emerin decoration/staining could improve the preoperative cytological diagnosis of thyroid carcinomas.

In particular, one of the main issues in thyroid cytopathology is the so-called indeterminate category, which includes cases where the lesion cannot be clearly defined as benign or malignant based on morphology alone; these cases are collectively grouped into the III and IV categories according to the Bethesda System for reporting Thyroid Cytopathology [24]. The categories III and IV (see Table 1a) are considered a sort of "grey zone" of thyroid cytology, and several authors have discussed the issue of "indeterminate" thyroid fine needle aspiration diagnosis. Efforts to detect cytological features or ancillary procedures that could distinguish between benign and malignant follicular patterned lesions (in need of surgical removal) have been the subject of several studies, but none was found to have absolute value or reproducibility [25–31].

Our results showed that emerin correctly traced the nuclear membrane in all types of cytological specimens (smear, cell block, Thin Prep) (see Table 1b). Smears and Thin Preps from cases with a definite cytological diagnosis of malignancy (category VI according to Bethesda System) [24] showed evident nuclear irregularities with foldings, grooves, and pseudo-inclusions. Comparison on the same nuclei of the H&E and immunoperoxidase slides (by recording H&E cytological images, demounting, and then restaining for emerin) (Fig. 2) clearly demonstrates the increased ability to define the nuclear membrane and its irregularities.

Table 1 Categories for reporting thyroid cytopathologies (according to Bethesda System) (a) and cytology processing tissue methods (b)

Categories for reporting thyroid cytopathology (Bethesda System) (a)		
	Category	*Risk of malignancy*
I	Nondiagnostic or Unsatisfactory	1-4%
II	Benign	0-3%
III	Atypia of Undetermined Significance or Follicular Lesion of Undetermined Significance	5-15%
IV	Follicular Neoplasm or Suspicious for a Follicular Neoplasm	15-30%
V	Suspicious for Malignancy	60-75%
VI	Malignant	97-99%

Cytology processing-tissue methods (b)	
Smear	Specimens from FNA are immediately spread thinly on a microscope slide, air-dried or alcohol- fixed and stained for examination.
Thin Prep	Specimens from FNA are put in a special fluid collection system and the slides for cytologic examination are filtered out in one-cell-thick layers on a slide.
Cell-block	Specimens from FNA are directly fixed in alcohol, centrifugated, paraffin-embedded, thus obtaining cell-blocks from which 3-5 μm sections can be cut.

FNA = fine needle aspiration

This approach proved particularly useful in the definition of unclear and problematic cases classified as III/IV categories: by highlighting and amplifying nuclear irregularities (e.g., invaginations, true inclusions, grooves), it helped in identifying, among all the indeterminate cases, the malignant lesions, which, after surgery, proved to be PTC or FVPTC. Those nuclear irregularities which were barely perceivable or borderline on H&E preparations proved instead more evident with emerin staining, and this helped in raising the suspicion of a malignant lesion.

In conclusion, emerin staining proved a useful tool to correctly identify PTC nuclei and to discriminate FVPTC cases among lesions classified as III/IV categories according to the Bethesda System [24]. It can be performed on smears, even after H&E staining, thus allowing for the accurate and straightforward identification of nuclear changes characteristic of PTC even in fine needle aspiration samples with very scant cellularity (number of cells obtained by FNA).

Diagnostic and Clinical Impact of Nuclear Shape in Breast Cancer

In breast cancer diagnostic pathology it is well known that irregularities in nuclear shape as observed by H&E staining play a crucial role in the diagnosis of both *in situ* and infiltrative lesions. Indeed, *in situ* carcinomas are classified using a three-tier system (Table 2) into low-, intermediate-, and high-grade lesions based on the degree of nuclear pleomorphism. Nuclear pleomorphism represents one of just

Table 2 Schematic representation of how histological grade is performed in breast cancer diagnostic pathology

Growth pattern	***Mitotic count**** *(applied to HPF diameter of 0.46 mm)*	***Nuclear pleomorphism***	
>75% of tubule formation **SCORE 1**	1-4 mitoses **SCORE 1**	Small and roundish nuclei with uniform chromatin **SCORE 1**	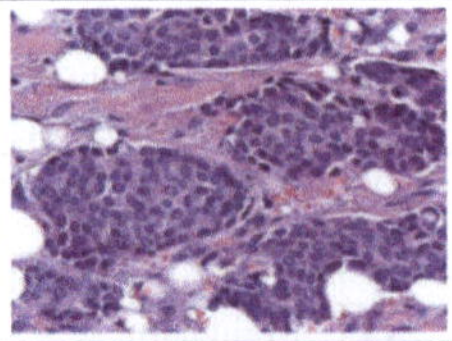
10-75% of tubule formation **SCORE 2**	5-11 mitoses **SCORE 2**	Variable shape and size, vescicolous chromatin, nucleoli present **SCORE 2**	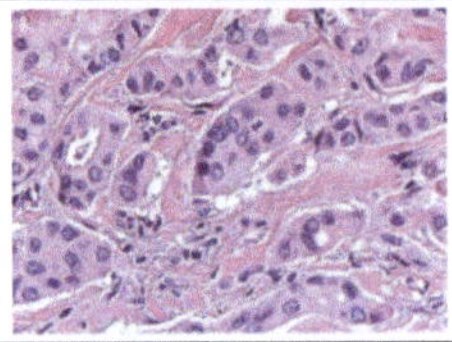
<10% of tubule formation **SCORE 3**	>=12 mitoses **SCORE 3**	High variability in shape and size, prominent nucleoli **SCORE 3**	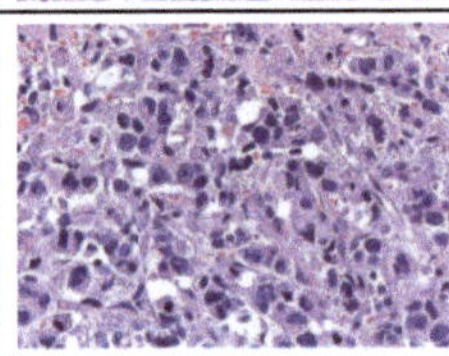

TOTAL SCORE

Score 3, 4, 5: **G1**
Score 6, 7: **G2**
Score 8, 9: **G3**

Three parameters are assessed: evaluation of tubule formation, number of mitosis, and nuclear pleomorphism (the latter corresponding to nuclear grade)

Scores attributed to single parameters are summed up, and the final score labels the lesion as G1 (low grade), G2 (intermediate grade), or G3 (high grade)

*Mitotic count depends on the diameter of the microscopic field of the microscope used to analyze the tissue specimen, in the figure we reported values corresponding to 0.46 mm

three components to be evaluated in the grading system of invasive breast carcinomas, the others being the number of mitoses and architectural growth pattern [32]. Histological grade (Table 2) holds a universally acknowledged robust prognostic value [32]; however, regrettably intra- and inter-pathologist agreement on grading in breast cancer is reported between poor and moderate [33, 34]. Indeed, the inter-observer agreement ranges between 50 and 85 %, and about 40–50 % of breast cancers are diagnosed as grade 2 cancers [32, 33].

With respect to nuclear grade, the seminal work by Elston and Ellis [32] grades nuclear pleomorphism by using three score values. These score values are given by comparing tumor nuclei with nuclei of normal breast, and at least four features are considered: size, shape, uniformity of nuclear chromatin, and nucleoli. Score 1 nuclei are little larger but very similar to normal cell nuclei, while score 3 nuclei show marked variation in size and a "bizarre" morphology. Yet, as noted above with PTC, light microscopy appreciation of foldings and indentations of the nuclear membrane is rough and indirect, being based on the staining of membrane-bound chromatin. Based on these premises it is not surprising that systematic differences between pathologists in scoring nuclear pleomorphism in breast cancer potentially contribute to differences in allocating cases to the correct grade, and the observed discrepancies confirm the need for improved nuclear grading criteria [35, 36].

Despite their considerable biological interest, the intranuclear tubular extensions of the NE have not gained much attention in pathology. We have therefore endeavoured to investigate whether direct observation of the NE could provide a more objective and direct appreciation of nuclear pleomorphism of breast cancer cells with the final aim to ameliorate definition of prognosis in breast cancer diagnostic pathology.

First, we have carried out a project in which various cell lines (primary cultures of normal mammary epithelium and established breast cancer cell lines) in addition to isolated cells and tissue sections from primary human breast cancer of different grades and stages were examined. Finally, the degree of pleomorphism of the NE was extended to other pathological parameters (histological grade, number of metastatic lymph nodes, vascular invasion, staging) in a series of 273 breast cancers. Results with *in vitro*-immortalized cultures showed that nuclei of "normal" breast epithelium when put into 2D cultures displayed a uniformly smooth *silhouette*, while lamin B and emerin patterns in most breast cancer cell lines resulted to build up, upon 3D reconstruction, a complex scaffold of intranuclear tubular structures (Fig. 6). As for tumor cells in human surgical samples, we showed that high nuclear pleomorphism, as defined by staining of the NE proteins emerin and lamin (Fig. 3), may potentially recognize within the histologically low-grade cancer group (G1) and in tumors with low proliferation activity, those more prone to metastasize [37].

Basically, from a practical standpoint, decoration/staining of the NE may be regarded as a novel diagnostic and prognostic parameter that may complement information obtained by conventional cytohistological techniques, and it can be postulated that fine detection of the nuclear shape and pleomorphism of the NE represents a novel parameter of interest in pathological grading, holding also a potential impact for planning therapy in breast cancer. Although the significance of this complex scaffold of intranuclear tubular structures is presently unknown it can be hypothesized that irregularities and intranuclear tubules might be involved in or reactive to defects in the nuclear-cytoplasmic transport, reportedly a feature typical of cancer cells [38].

As an additional remark, we have also investigated the possibility to visualize the spatial organization of gene signals with respect to the NE. This can be achieved by coupling immunofluorescence for lamins (or other NE proteins) and FISH for target genes. In particular, for breast cancer we have investigated *HER2* gene amplification in BT-474 breast carcinoma cells (Fig. 4).

HER2 gene amplification is found in about 15–20 % of all breast carcinomas and represents the main mechanism driving HER2 activation in breast cancer, which has a negative prognostic impact [33, 39]. Proper documentation of the presence of *HER2* gene amplification represents the crucial step to deliver a specific target therapy in breast carcinoma patients (the humanized antibody trastuzumab, i.e., Herceptin®) [33]. This is performed routinely by using an *in situ* technique, i.e., FISH, with specific probes directed against the target gene, on sections of human tissue samples. Usually a dual-color probe (one for the gene, the other for the centromere of the chromosome 17 (CEP17), the chromosome where *HER2* maps to) is employed, and results can be scored either based on *HER2*/CEP17 ratio (*HER2/CEP17*$\geq$*2= amplification*) or on the basis of the absolute numbers of the *HER2* gene (amplification whenever *HER2*>6) [39, 40]. For the sake of detection of amplification, only numerical count of signals is performed and no attention is currently paid to the spatial organization of signals.

With our immune-FISH followed by 3D reconstruction we showed in *HER2*-amplified breast cancer cells a range of patterns in the spatial distribution of gene signals (both single and clustered) with respect to the NE, some being anchored to the NE and others haphazardly spaced within the nucleus. Implications of the relationship between amplified regions of the genome and anchorage to the NE are unknown at present, but it is generally thought that interactions demonstrated between NE proteins and epigenetic heterochromatin marks would correlate peripheral localization with silencing. Nonetheless, further experimental studies would be warranted to properly investigate the implications in terms of activation or inactivation of genes. Indeed, the spatial localization of chromatin within the mammalian nucleus has been shown to be important for several genomic processes [41], including transcription [42], RNA processing [43], as well as DNA repair and recombination [44]. In addition, studies based on 3D-immuno-FISH suggest a key function for the inner nuclear membrane–lamina compartment in transcriptional silencing of large segments of the genome [41]. Finally, very recently it has been demonstrated that the yeast nuclear pore complex protein Nup170p interacts with regions of the genome that contain ribosomal protein and subtelomeric genes, where it functions as a repressor of transcription [45]. These results suggest that nuclear pore proteins are active participants in silencing and the formation of peripheral heterochromatin [45].

Conclusions

Variation in both nuclear shape and size ("pleomorphism"), coupled with changes in chromatin amount and distribution, remains the basic microscopy criteria for a cytologic diagnosis of cancer. The biological determinants of nuclear shape irregularities are not clarified. It has been suggested that alterations in nuclear shape might be related to genetic imbalances in cancer [46], and Fischer [17] gave experimental evidence using *in vitro* models of PTC that induced gene mutations are associated with the structural features typical of this type of thyroid carcinoma that involve rearrangement of the NE and chromatin distribution [16, 17, 47]. On the other hand, some diseases characterized by genetically determined abnormalities in lamin proteins [48, 49] suggest that irregularities in nuclear shape are due to the abnormal farnesylation of lamin proteins, perhaps through interaction of the farnesylated lamins with the phospholipid bilayer [50]. This raises the possibility that both genetic and posttranslational events might be involved in the origin of nuclear shape abnormalities.

Indeed, small GTPases appear to represent a candidate for playing a central role in this process, since they are important in the nuclear envelope assembly [51] and are notoriously a key player in oncogenesis [52, 53]. Moreover, recent evidence has been presented [54] suggesting that prenylation of small GTPases is impaired in cancer cells.

Other reviews focus on possible mechanisms to generate nuclear shape abnormalities, but here we focus on using these diagnostically. Standard H&E staining cannot adequately distinguish fine abnormalities of the nuclear shape, as it is indirectly based on the distribution of DNA as revealed by the affinity for basic dyes.

A more objective definition of the shape of the nucleus can be provided by decoration/staining of the NE, followed by image capture and 3D reconstruction. We applied this approach to two areas of tumor pathology: thyroid and breast cancer. In the papillary type of thyroid cancer, most nuclei show a variation in shape so typical as to be paradigmatic and diagnostic, while in breast cancer nuclear irregularities vary according to the subtype and the aggressiveness of cancer. For instance it is minimal in tubular carcinoma while marked in grade 3 cancers.

The technical approach presented here proved feasible on both isolated cells and tissue sections and ultimately provides a reproducible approach of diagnostic and clinical interest.

The pathological diagnosis of PTC is usually straightforward since the majority of cases of PTC are easy to diagnose on routine-stained preparations, with overt irregularities, such as grooves, pseudo-inclusions, and ground-glass appearance. Although these nuclear changes help to define PTC, these features are only diagnostic when widespread and in combination. However, in some cases both in histology and cytology diagnosis of PTC can be challenging, and the classical microscope observation of PTC nuclei (based on nucleic acid staining with basic dyes, such as hematoxylin) is clearly insufficient to appreciate the complete spectrum of PTC nuclear irregularities.

By tracing in immunofluorescence and immunoperoxidase proteins of the NE (e.g., lamins, emerin), it is possible to obtain a clear, evident, and direct representation of nuclear shape, thus highlighting those microscopical features barely visible with H&E.

When shifting the attention from the "content" (chromatin) to the "container" (nuclear membrane), confocal microscopy and 3D reconstructions provided us models of nuclear structure in PTC cells, and emerin immunostaining on cytological and histological samples proved a feasible tool to improve diagnosis in "difficult" PTC cases.

In breast cancer, the presence of an extensive network of invaginated projections of the NE inside the nucleus, as revealed by tagging lamin B and emerin in immunofluorescence preparations, opens prospects of biological and diagnostic interest.

Intranuclear tubules are an interesting and intriguing phenomenon, possibly involved in or reactive to defects in the nuclear-cytoplasmic transport, reportedly a feature typical of cancer cells. In addition, this scaffold might also be a drug target since Lee et al. [55] already demonstrated a selective binding of doxorubicin to intranuclear tubules. Moreover, the combined 3D detection of the spatial distribution of genes and intranuclear invaginations, as exemplified in the present study, might provide a novel interpretation on active versus inactive genes. Indeed, other studies based on 3D-immuno-FISH seem to suggest the inner nuclear membrane–lamina compartment as a key player in transcriptional silencing of large segments of the genome [41].

Finally, we gave evidence that immunofluorescence decoration/staining of the NE provides a reproducible and objective evaluation of nuclear shape irregularities associated with pleomorphism and provides prognostic information to parallel and enhance that provided by routine histological procedures.

Acknowledgements GB is funded by Persother (SMIS-CSRN:549/12.024); AS is funded by AIRC (IG10787) and Ricerca Sanitaria Finalizzata (RF-2010-2310674); and CM is funded by AIRC (MFAG 13310).

References

1. Bussolati G (2008) Proper detection of the nuclear shape: ways and significance. Rom J Morphol Embryol 49(4):435–439, 490408435439 [pii]
2. Ghadially F (1988) Ultrastructural pathology of the cell and matrix: a text and atlas of physiological and pathological alterations in the fine structure of cellular and extracellular components, vol 1, 3rd edn. Butterworths, London
3. Frost J (1986) The cell in health and disease: an evaluation of cellular morphologic expression of biologic behavior . Karger, Basel, 2nd revised edn
4. Johannessen JV, Gould VE, Jao W (1978) The fine structure of human thyroid cancer. Hum Pathol 9(4):385–400
5. Batistatou A, Scopa CD (2009) Pathogenesis and diagnostic significance of nuclear grooves in thyroid and other sites. Int J Surg Pathol 17(2):107–110. doi:10.1177/1066896908316071, 1066896908316071 [pii]
6. LiVolsi VA (2011) Papillary thyroid carcinoma: an update. Mod Pathol 24(Suppl 2):S1–S9. doi:10.1038/modpathol.2010.129, modpathol2010129 [pii]
7. Arora SK, Dey P (2012) Intranuclear pseudoinclusions: morphology, pathogenesis, and significance. Diagn Cytopathol 40(8):741–744. doi:10.1002/dc.21714
8. Cancer TIAfRo (2004) Pathology and genetics of tumours of endocrine organs (IARC WHO classification of tumours), 1st edn. IARC Press, Lyon
9. Asioli S, Bussolati G (2009) Emerin immunohistochemistry reveals diagnostic features of nuclear membrane arrangement in thyroid lesions. Histopathology 54(5):571–579. doi:10.1111/j.1365-2559.2009.03259.x, HIS3259 [pii]
10. Papotti M, Manazza AD, Chiarle R, Bussolati G (2004) Confocal microscope analysis and tridimensional reconstruction of papillary thyroid carcinoma nuclei. Virchows Arch 444(4):350–355. doi:10.1007/s00428-003-0962-4
11. Holmes TJ (1992) Blind deconvolution of quantum-limited incoherent imagery: maximum-likelihood approach. J Opt Soc Am A 9(7):1052–1061
12. Holmes TJ, O'Connor NJ (2000) Blind deconvolution of 3D transmitted light brightfield micrographs. J Microsc 200(Pt 2):114–127, jmi751 [pii]

13. Baloch ZW, LiVolsi VA (2002) Etiology and significance of the optically clear nucleus. Endocr Pathol 13(4):289–299, EP:13:4:289 [pii]
14. Asioli S, Maletta F, Pacchioni D, Lupo R, Bussolati G (2010) Cytological detection of papillary thyroid carcinomas by nuclear membrane decoration with emerin staining. Virchows Arch 457(1):43–51. doi:10.1007/s00428-010-0910-z
15. Kinsella MD, Hinrichs B, Cohen C, Siddiqui MT (2012) Highlighting nuclear membrane staining in thyroid neoplasms with emerin: review and diagnostic utility. Diagn Cytopathol 41(6):497–504. doi:10.1002/dc.22870
16. Fischer AH, Taysavang P, Weber CJ, Wilson KL (2001) Nuclear envelope organization in papillary thyroid carcinoma. Histol Histopathol 16(1):1–14
17. Fischer AH, Taysavang P, Jhiang SM (2003) Nuclear envelope irregularity is induced by RET/PTC during interphase. Am J Pathol 163(3):1091–1100
18. Bell CD, Coire C, Treger T, Volpe R, Baumal R, Fornasier VL (2001) The 'dark nucleus' and disruptions of follicular architecture: possible new histological aids for the diagnosis of the follicular variant of papillary carcinoma of the thyroid. Histopathology 39(1):33–42, his1137 [pii]
19. Liu J, Singh B, Tallini G, Carlson DL, Katabi N, Shaha A, Tuttle RM, Ghossein RA (2006) Follicular variant of papillary thyroid carcinoma: a clinicopathologic study of a problematic entity. Cancer 107(6):1255–1264. doi:10.1002/cncr.22138
20. Chan J (2002) Strict criteria should be applied in the diagnosis of encapsulated follicular variant of papillary thyroid carcinoma. Am J Clin Pathol 117(1):16–18. doi:10.1309/P7QL-16KQ-QLF4-XW0M
21. Lloyd RV, Erickson LA, Casey MB, Lam KY, Lohse CM, Asa SL, Chan JK, DeLellis RA, Harach HR, Kakudo K, LiVolsi VA, Rosai J, Sebo TJ, Sobrinho-Simoes M, Wenig BM, Lae ME (2004) Observer variation in the diagnosis of follicular variant of papillary thyroid carcinoma. Am J Surg Pathol 28(10):1336–1340, 00000478-200410000-00009 [pii]
22. Suster S (2006) Thyroid tumors with a follicular growth pattern: problems in differential diagnosis. Arch Pathol Lab Med 130(7):984–988. doi:10.1043/1543-2165(2006)130[984:TTWAFG]2.0.CO;2, 2006-0069-RAR [pii]
23. Volante M, Collini P, Nikiforov YE, Sakamoto A, Kakudo K, Katoh R, Lloyd RV, LiVolsi VA, Papotti M, Sobrinho-Simoes M, Bussolati G, Rosai J (2007) Poorly differentiated thyroid carcinoma: the Turin proposal for the use of uniform diagnostic criteria and an algorithmic diagnostic approach. Am J Surg Pathol 31(8):1256–1264. doi:10.1097/PAS.0b013e3180309e6a, 00000478-200708000-00017 [pii]
24. Cibas ES, Ali SZ (2009) The Bethesda System For Reporting Thyroid Cytopathology. Am J Clin Pathol 132(5):658–665. doi:10.1309/AJCPPHLWMI3JV4LA, 132/5/658 [pii]
25. Baloch ZW, Fleisher S, LiVolsi VA, Gupta PK (2002) Diagnosis of "follicular neoplasm": a gray zone in thyroid fine-needle aspiration cytology. Diagn Cytopathol 26(1):41–44, 10.1002/dc.10043 [pii]
26. Baloch ZW, Sack MJ, Yu GH, Livolsi VA, Gupta PK (1998) Fine-needle aspiration of thyroid: an institutional experience. Thyroid 8(7):565–569
27. Gharib H (1994) Fine-needle aspiration biopsy of thyroid nodules: advantages, limitations, and effect. Mayo Clin Proc 69(1):44–49
28. Asa SL (2005) The role of immunohistochemical markers in the diagnosis of follicular-patterned lesions of the thyroid. Endocr Pathol 16(4):295–309, EP:16:4:295 [pii]
29. Papotti M, Rodriguez J, De Pompa R, Bartolazzi A, Rosai J (2005) Galectin-3 and HBME-1 expression in well-differentiated thyroid tumors with follicular architecture of uncertain malignant potential. Mod Pathol 18(4):541–546. doi:10.1038/modpathol.3800321, 3800321 [pii]
30. Bartolazzi A, Orlandi F, Saggiorato E, Volante M, Arecco F, Rossetto R, Palestini N, Ghigo E, Papotti M, Bussolati G, Martegani MP, Pantellini F, Carpi A, Giovagnoli MR, Monti S, Toscano V, Sciacchitano S, Pennelli GM, Mian C, Pelizzo MR, Rugge M, Troncone G, Palombini L, Chiappetta G, Botti G, Vecchione A, Bellocco R (2008) Galectin-3-expression analysis in the surgical selection of follicular thyroid nodules with indeterminate fine-needle aspiration cytology: a prospective multicentre study. Lancet Oncol 9(6):543–549. doi:10.1016/S1470-2045(08)70132-3, S1470-2045(08)70132-3 [pii]

31. Saggiorato E, De Pompa R, Volante M, Cappia S, Arecco F, Dei Tos AP, Orlandi F, Papotti M (2005) Characterization of thyroid 'follicular neoplasms' in fine-needle aspiration cytological specimens using a panel of immunohistochemical markers: a proposal for clinical application. Endocr Relat Cancer 12(2):305–317. doi:10.1677/erc.1.00944, 12/2/305 [pii]
32. Elston CW, Ellis IO (1991) Pathological prognostic factors in breast cancer. I. The value of histological grade in breast cancer: experience from a large study with long-term follow-up. Histopathology 19(5):403–410
33. Marchio C, Reis-Filho JS (2008) Molecular diagnosis in breast cancer. Diagn Histopatho 14:5
34. Robbins P, Pinder S, de Klerk N, Dawkins H, Harvey J, Sterrett G, Ellis I, Elston C (1995) Histological grading of breast carcinomas: a study of interobserver agreement. Hum Pathol 26(8):873–879
35. Meyer JS, Alvarez C, Milikowski C, Olson N, Russo I, Russo J, Glass A, Zehnbauer BA, Lister K, Parwaresch R (2005) Breast carcinoma malignancy grading by Bloom-Richardson system vs proliferation index: reproducibility of grade and advantages of proliferation index. Mod Pathol 18(8):1067–1078. doi:10.1038/modpathol.3800388, 3800388 [pii]
36. Sloane JP, Amendoeira I, Apostolikas N, Bellocq JP, Bianchi S, Boecker W, Bussolati G, Coleman D, Connolly CE, Eusebi V, De Miguel C, Dervan P, Drijkoningen R, Elston CW, Faverly D, Gad A, Jacquemier J, Lacerda M, Martinez-Penuela J, Munt C, Peterse JL, Rank F, Sylvan M, Tsakraklides V, Zafrani B (1999) Consistency achieved by 23 European pathologists from 12 countries in diagnosing breast disease and reporting prognostic features of carcinomas. European Commission Working Group on Breast Screening Pathology. Virchows Arch 434(1):3–10
37. Bussolati G, Marchio C, Gaetano L, Lupo R, Sapino A (2008) Pleomorphism of the nuclear envelope in breast cancer: a new approach to an old problem. J Cell Mol Med 12(1):209–218. doi:10.1111/j.1582-4934.2007.00176.x, JCMM176 [pii]
38. Kau TR, Way JC, Silver PA (2004) Nuclear transport and cancer: from mechanism to intervention. Nat Rev Cancer 4(2):106–117. doi:10.1038/nrc1274, nrc1274 [pii]
39. Sapino A, Goia M, Recupero D, Marchio C (2013) Current challenges for HER2 testing in diagnostic pathology: state of the art and controversial issues. Front Oncol 3:129. doi:10.3389/fonc.2013.00129
40. Wolff AC, Hammond ME, Hicks DG, Dowsett M, McShane LM, Allison KH, Allred DC, Bartlett JM, Bilous M, Fitzgibbons P, Hanna W, Jenkins RB, Mangu PB, Paik S, Perez EA, Press MF, Spears PA, Vance GH, Viale G, Hayes DF; American Society of Clinical Oncology; College of American Pathologists (2013) Recommendations for human epidermal growth factor receptor 2 testing in breast cancer: American Society of Clinical Oncology/College of American Pathologists clinical practice guideline update. J Clin Oncol. 2013;31(31):3997–4013. doi: 10.1200/JCO.2013.50.9984. Epub 2013 Oct 7
41. Zullo JM, Demarco IA, Pique-Regi R, Gaffney DJ, Epstein CB, Spooner CJ, Luperchio TR, Bernstein BE, Pritchard JK, Reddy KL, Singh H (2012) DNA sequence-dependent compartmentalization and silencing of chromatin at the nuclear lamina. Cell 149(7):1474–1487. doi:10.1016/j.cell.2012.04.035, S0092-8674(12)00591-0 [pii]
42. Sutherland H, Bickmore WA (2009) Transcription factories: gene expression in unions? Nat Rev Genet 10(7):457–466. doi:10.1038/nrg2592, nrg2592 [pii]
43. Lamond AI, Spector DL (2003) Nuclear speckles: a model for nuclear organelles. Nat Rev Mol Cell Biol 4(8):605–612. doi:10.1038/nrm1172, nrm1172 [pii]
44. Misteli T, Soutoglou E (2009) The emerging role of nuclear architecture in DNA repair and genome maintenance. Nat Rev Mol Cell Biol 10(4):243–254. doi:10.1038/nrm2651, nrm2651 [pii]
45. Van de Vosse DW, Wan Y, Lapetina DL, Chen WM, Chiang JH, Aitchison JD, Wozniak RW (2013) A role for the nucleoporin Nup170p in chromatin structure and gene silencing. Cell 152(5):969–983. doi:10.1016/j.cell.2013.01.049, S0092-8674(13)00141-4 [pii]
46. Fischer AH, Zhao C, Li QK, Gustafson KS, Eltoum IE, Tambouret R, Benstein B, Savaloja LC, Kulesza P (2010) The cytologic criteria of malignancy. J Cell Biochem 110(4):795–811. doi:10.1002/jcb.22585
47. Fischer AH, Bond JA, Taysavang P, Battles OE, Wynford-Thomas D (1998) Papillary thyroid carcinoma oncogene (RET/PTC) alters the nuclear envelope and chromatin structure. Am J Pathol 153(5):1443–1450

48. Eriksson M, Brown WT, Gordon LB, Glynn MW, Singer J, Scott L, Erdos MR, Robbins CM, Moses TY, Berglund P, Dutra A, Pak E, Durkin S, Csoka AB, Boehnke M, Glover TW, Collins FS (2003) Recurrent de novo point mutations in lamin A cause Hutchinson-Gilford progeria syndrome. Nature 423(6937):293–298. doi:10.1038/nature01629, nature01629 [pii]
49. Maraldi NM, Lattanzi G, Capanni C, Columbaro M, Merlini L, Mattioli E, Sabatelli P, Squarzoni S, Manzoli FA (2006) Nuclear envelope proteins and chromatin arrangement: a pathogenic mechanism for laminopathies. Eur J Histochem 50(1):1–8
50. Polychronidou M, Grobhans J (2011) Determining nuclear shape: the role of farnesylated nuclear membrane proteins. Nucleus 2(1):17–23. doi:10.4161/nucl.2.1.13992, 1949-1034-2-1-4 [pii]
51. Hetzer M, Gruss OJ, Mattaj IW (2002) The Ran GTPase as a marker of chromosome position in spindle formation and nuclear envelope assembly. Nat Cell Biol 4(7):E177–E184. doi:10.1038/ncb0702-e177, ncb0702-e177 [pii]
52. Bendris N, Arsic N, Lemmers B, Blanchard JM (2012) Cyclin A2, Rho GTPases and EMT. Small GTPases 3(4):225–228. doi:10.4161/sgtp.20791, 20791 [pii]
53. Recchi C, Seabra MC (2012) Novel functions for Rab GTPases in multiple aspects of tumour progression. Biochem Soc Trans 40(6):1398–1403. doi:10.1042/BST20120199, BST20120199 [pii]
54. Ntantie E, Gonyo P, Lorimer EL, Hauser AD, Schuld N, McAllister D, Kalyanaraman B, Dwinell MB, Auchampach JA, Williams CL (2013) An adenosine-mediated signaling pathway suppresses prenylation of the GTPase Rap1B and promotes cell scattering. Sci Signal 6(277):ra39. doi:10.1126/scisignal.2003374, 6/277/ra39 [pii]
55. Lee RK, Lui PP, Ngan EK, Lui JC, Suen YK, Chan F, Kong SK (2006) The nuclear tubular invaginations are dynamic structures inside the nucleus of HeLa cells. Can J Physiol Pharmacol 84(3–4):477–486. doi:10.1139/y05-110, y05-110 [pii]

Part II
The Nuclear Envelope in Cell Cycle Regulation and Signaling

Introduction

A central mechanism by which the nuclear envelope could influence cancer progression is through influencing the cell cycle, which typically loses critical controls in tumorigenesis. The first link between the nuclear envelope and the cell cycle was the observation by Larry Gerace and Gunter Blobel that the intermediate filament lamin polymer was depolymerized by hyperphosphorylation at the beginning of mitosis [1]. Ten years later it was found by the McKeon, Nigg, and Kirschner labs that mitotic kinases direct this hyperphosphorylation [2–4]. Moreover, mutating critical residues in lamin A to prevent their phosphorylation blocked both lamin disassembly and mitotic progression [2]. Thus, just the inability to disassemble the nuclear envelope yields a physical barrier to progression through mitosis. It follows logically that even intermediate defects in nuclear envelope disassembly could have negative consequences for successful mitosis—for example partial maintained connections between nuclear envelope fragments and chromatin could block proper chromosome segregation, resulting in micronuclei and aneuploidy. Defects in nuclear envelope proteins also adversely affect nuclear envelope reassembly at the end of mitosis as this process is thought to be driven by binding of certain nuclear envelope proteins to mitotic chromosomes. In this light it is perhaps not surprising that disruption of both the nuclear envelope transmembrane proteins MAN1 and emerin or their chromatin-binding partner BAF in *Caenorhabditis elegans* resulted in defects such as anaphase chromatin bridges [5, 6].

Although one might think that nuclear envelope proteins would become irrelevant during mitosis since the nuclear envelope is gone, it turns out that many nuclear envelope proteins have separate functions during mitosis. Major functions described thus far are supporting the mitotic spindle [7–10] and associating with kinetochores [11], presumably to strengthen the complex binding to spindle microtubules.

Interphase functions of the nuclear envelope can also affect the ability of cells to enter the cell cycle, and evidence exists for interactions of both lamins and nuclear envelope transmembrane proteins with signaling pathways controlling cell cycle entry.

The most investigated nexus for this is the interaction between lamin A and master cell cycle regulator and tumor-suppressor protein pRb, first observed in 1994 [12]. In the first chapter in this section, Juniper Pennypacker and Brian Kennedy, President of the Buck Institute for Research on Aging and among the first to characterize lamin-pRB interactions, address what is known about pRb interactions with lamins in regulation of the cell cycle. They also address aspects of the lamin–pRb nexus in aging. They present a body of data that strongly argues that accumulation of the progerin form of lamin A with aging could impair p53 networks and enable tumorigenesis. Next Roland Foisner, discoverer of the LAP2 family of NETs and Deputy Director at the Max F. Perutz Laboratories, and Andreas Brachner from the Medical University of Vienna address the function of a variant of the lamin–pRb nexus that also includes the alpha soluble splice variant of the nuclear envelope transmembrane protein LAP2β. This particular complex sequesters pRb from its target genes in the nuclear interior, but at the same time it stabilizes pRb so that in some cell types levels can build up. This means that when the cell cycle is activated it can proceed in these cell types in a much more robust manner. Thus depending on cell type and circumstances this lamin–LAP2α–pRb complex can either inhibit cell cycle progression or enhance it. Next, Eric Schirmer and colleagues discuss the role of nuclear envelope transmembrane proteins and NPC proteins during mitosis and also interphase roles of nuclear envelope transmembrane proteins in regulating the cell cycle. Interestingly, roughly 20 % of nuclear envelope transmembrane proteins tested have some roles in the cell cycle [13]. Finally, signaling pathways are critical for achieving and maintaining tissue differentiation and often additionally exert controls on the cell cycle, and so in the last chapter Jason Choi and Howard Worman from Columbia University, discoverer of one of the first NETs -LBR- and of many aspects of lamin and NET functions in disease, discuss this function of the nuclear envelope. In particular, NF-κB, Wnt, and TGFβ signaling cascades from the plasma membrane get several additional layers of regulation from the nuclear envelope before they can activate transcription: first getting into the nucleus through the NPCs and second having cascade proteins β-catenin and smads sequestered at the nuclear envelope by multiple nuclear envelope transmembrane proteins.

1. Gerace L, Blobel G (1980) The nuclear envelope lamina is reversibly depolymerized during mitosis. Cell 19(1):277–287
2. Heald R, McKeon F (1990) Mutations of phosphorylation sites in lamin A that prevent nuclear lamina disassembly in mitosis. Cell 61(4):579–589
3. Peter M, Nakagawa J, Doree M, Labbe JC, Nigg EA (1990) In vitro disassembly of the nuclear lamina and M phase-specific phosphorylation of lamins by cdc2 kinase. Cell 61(4):591–602
4. Ward GE, Kirschner MW (1990) Identification of cell cycle-regulated phosphorylation sites on nuclear lamin C. Cell 61(4):561–577
5. Liu J, Lee K, Segura-Totten M, Neufeld E, Wilson K, Gruenbaum Y (2003) MAN1 and emerin have overlapping function(s) essential for chromosome segregation and cell division in Caenorhabditis elegans. Proc Natl Acad Sci U S A 100(8):4598–4603

6. Margalit A, Segura-Totten M, Gruenbaum Y, Wilson KL (2005) Barrier-to-autointegration factor is required to segregate and enclose chromosomes within the nuclear envelope and assemble the nuclear lamina. Proc Natl Acad Sci U S A 102(9):3290–3295. doi:0408364102 [pii] 10.1073/pnas.0408364102
7. Suntharalingam M, Wente SR (2003) Peering through the pore: nuclear pore complex structure, assembly, and function. Dev Cell 4(6):775–789
8. Ohba T, Nakamura M, Nishitani H, Nishimoto T (1999) Self-organization of microtubule asters induced in Xenopus egg extracts by GTP-bound Ran [see comment]. Science 284(5418):1356–1358
9. Wiese C, Wilde A, Moore MS, Adam SA, Merdes A, Zheng Y (2001) Role of importin-beta in coupling Ran to downstream targets in microtubule assembly. Science 291(5504):653–656
10. Tsai MY, Wang S, Heidinger JM, Shumaker DK, Adam SA, Goldman RD, Zheng Y (2006) A mitotic lamin B matrix induced by RanGTP required for spindle assembly. Science 311(5769):1887–1893
11. Belgareh N, Rabut G, Bai SW, van Overbeek M, Beaudouin J, Daigle N, Zatsepina OV, Pasteau F, Labas V, Fromont-Racine M, Ellenberg J, Doye V (2001) An evolutionarily conserved NPC subcomplex, which redistributes in part to kinetochores in mammalian cells. J Cell Biol 154(6):1147–1160.
12. Ozaki T, Saijo M, Murakami K, Enomoto H, Taya Y, Sakiyama S (1994) Complex formation between lamin A and the retinoblastoma gene product: identification of the domain on lamin A required for its interaction. Oncogene 9(9):2649–2653
13. Korfali N, Srsen V, Waterfall M, Batrakou DG, Pekovic V, Hutchison CJ, Schirmer EC (2011) A flow cytometry-based screen of nuclear envelope transmembrane proteins identifies NET4/Tmem53 as involved in stress-dependent cell cycle withdrawal. PLoS One 6(4):e18762. doi:10.1371/journal.pone.0018762

RB and Lamins in Cell Cycle Regulation and Aging

Brian K. Kennedy and Juniper K. Pennypacker

Abstract While speculation has centered on a role for nuclear lamins in tumor progression for many years, most of the diseases that have been linked to lamin mutation are dystrophic in nature, often limiting the proliferation potential of affected cells in vivo and in vitro. Nevertheless, these lamin mutations, particularly in the *LMNA* gene that encodes A-type lamins, have provided an interesting tool set to understand functions of nuclear intermediate filament proteins in cell cycle progress and various means of exit, including quiescence, senescence, and differentiation down various lineages. The picture that has emerged is complex with lamins controlling the activity of key cell cycle factors such as the retinoblastoma protein (RB) and interacting with several important signal transduction pathways. Here we describe the current state of knowledge and speculate that lamins may be intimately involved in the regulation of cell proliferation, acting at the interface between cancer and aging.

Keywords A-type lamins • B-type lamins • Retinoblastoma protein • *LMNA* gene • Lamins • Aging • Progerin • Senescence • Cancer • Cell cycle progression • Telomeres • p53

Abbreviations

ATR	Ataxia telangiectasia and Rad3-related protein
ADLD	Autosomal dominant leukodystrophy
CDK	Cyclin-dependent kinase

B.K. Kennedy, Ph.D. (✉) • J.K. Pennypacker
The Buck Institute for Research on Aging, 8001 Redwood Boulevard, Novato, CA 94945, USA
e-mail: bkennedy@buckinstitute.org; JPennypacker@buckinstitute.org

E.C. Schirmer and J.I. de las Heras (eds.), *Cancer Biology and the Nuclear Envelope*, Advances in Experimental Medicine and Biology 773, DOI 10.1007/978-1-4899-8032-8_6,

DCM1A	Dilated cardiomyopathy type 1A
ERK	Extracellular signal-regulated kinases
HGPS	Hutchinson–Gilford Progeria syndrome
LAP2α	Lamina-associated polypeptide 2α
MDM2	Mouse double minute 2 homolog
ROS	Reactive oxygen species
RB	Retinoblastoma protein
SASP	Senescence-associated secretory phenotype
SIRT1	Silent mating-type information regulation 2 homolog 1
VHL	von Hippel–Lindau gene

Introduction

Since the discovery of A-type and B-type lamins as components of the nuclear lamina [1], they have been the subject of intense scrutiny regarding possible roles in a range of nuclear functions. The finding that they are targets for mutation in degenerative and progeroid diseases has further driven research in the area [2]. Partially overlapping research threads for nearly three decades have implicated lamins in the control of cell proliferation and differentiation, leading to speculation that lamins may have roles in cancer progression. One obvious connection between lamin function and cell cycle progression comes from the fact that the nuclear envelope breaks down during mitosis in mammalian cells, leading to a dissociation of the lamin intermediate filament structure that exists between the chromatin and the envelope [3]. Upon reformation of the nucleus after mitosis, the lamina also reassembles, and numerous studies have been performed to define a role for A- and B-type lamins in this process.

A full description of A- and B-type lamins is provided in other reviews. Here we provide basic facts relevant to lamin roles in cell cycle regulation and aging. All A-type lamins (lamins A and C in most settings) are encoded by the *LMNA* gene, which is targeted for mutation in a wide range of pathologies [2]. Among these, forms of dilated cardiomyopathy and muscular dystrophy are generally associated with reduced A-type lamin function and can be phenocopied by knockout of the *LMNA* locus in the mouse [4]. In contrast, dominant gain-of-function or neomorphic mutations in *LMNA* can lead to progeroid syndromes [5–7]. The most common of these is Hutchinson–Gilford Progeria syndrome (HGPS), which is most often associated with the *LMNA* G608G mutation, which is silent with respect to coding sequence but activates a cryptic splice site leading to the production of progerin, a variant of lamin A that lacks 50 amino acids in the C-terminus [6, 7]. Whether progerias are mechanistically linked to normal aging has been an ongoing debate in the aging research field for decades with no consensus yet emerging [8, 9]. Interestingly, however, alternative splicing of *LMNA* can occur in normal cells leading to low-level production of progerin, and recent studies have demonstrated progerin accumulation with organismal age or with increasing passage in cell culture [10–13].

These findings at least raise the possibility that progerin may in part promote the normal aging process.

B-type lamins, encoded by *LMNB1* and *LMNB2*, are also linked to disease states and have been implicated in cell cycle progression. Interestingly, recent studies indicate that lamin B expression may be altered as cells approach senescence [14], a topic discussed in Sects. 3 and 4. One major difference between A- and B-type lamins is their expression patterns. While B-type lamins are expressed in all cell types, A-type lamins are regulated during development and differentiation. In a mouse, for instance, A-type lamin expression is not evident until mid-gestation, when it can be detected in cells committing to different lineages [15]. This finding has led to the possibility that A-type lamins are cell commitment factors, being expressed when cells adopt certain fates and perhaps ensuring gene expression programs that define those fates. A-type lamins are also not expressed in stem cells (or at least at very low levels), a fact made particularly evident in studies aimed to generate induced pluripotent stem cells from fibroblasts of HGPS patients [16–18]. These fibroblasts have proliferation defects and altered nuclear shape but can be induced to become stem cells. Upon this transition, the cells lose A-type lamin expression (including that of progerin) and no longer display proliferation or nuclear shape abnormalities. When induced to differentiate again, they resume progerin expression and regain abnormal behavior.

A deeper role for lamins in cell cycle regulation was proposed when the discovery was made that lamin tethering of many cell cycle regulators, including c-Myc and retinoblastoma protein (RB), was important for their function [19–23]. This has stimulated investigation by many laboratories into the role of lamins in coordinating the transit through the G1 phase of the cell cycle. Lamins have also been linked to the control of DNA replication (below) and checkpoint pathways, including those involved in repair of DNA lesions [24].

If lamins control cell cycle progression and exit, then their function might be impaired during cancer progression, either through direct mutation or through other events during cancer expression that affect their activity. This theory is augmented by findings in several tumors that nuclear shape and organization are commonly altered in cancer cell lines [25, 26]. Many investigators have examined the possibility of lamin impairment in cancer progression, and, particularly with regard to A-type lamins, expression patterns often change in cancer although few mutations have been identified. This finding has led to the possibility that lamins may be used as biomarkers for cancer progression, and in many different tumor types the expression of A-type lamins or lamin B1 changes during different stages of tumor development. However, there are no clear generalizations to be made, with expression pattern changes often complex and specific to tumor type [27].

In this review, we focus on the role of A-type lamins in cell cycle progression and exit, discussing how these specific functions may relate to both cancer and aging. The answers to these questions remain unresolved, but a number of tantalizing findings have been reported in the last few years that may finally lead to the primary functions of lamins in the nucleus, as well as how they relate to disease progression, both those of a dystrophic and hyper-proliferative nature.

Cell Proliferation

The G1-to-S phase transition of the mammalian cell cycle is tightly regulated in normal cells and a primary target for dysregulation in cancer. Many of the regulatory factors have been associated with the nuclear matrix. Rather than cover a compendium of these factors, which in many cases interact with nuclear lamins, the focus of this section will be to define the phenotypic roles of nuclear lamins in cell proliferation and, wherever possible, to connect these roles to regulation of proliferative factors. In general, proliferative effects of *LMNA* mutations will be separated into those associated with reduced A-type lamin function and those associated with the expression of progeria alleles, which are either hypermorphs or neomorphs.

Lmna$^{-/-}$ mice develop skeletal muscle dystrophy and dilated cardiomyopathy [4], succumbing between 6 and 8 weeks of age likely due to cardiac conduction defects [28]. To clarify, recent findings suggest that *Lmna*$^{-/-}$ mice may not be true nulls for the *LMNA* locus, as a truncated allele of lamin A appears to be expressed in these mice [29]. The most likely scenario is that this mouse is actually a hypomorph, a theory supported by the phenotype of another *Lmna* disruption in a mouse that results in lethality before weaning [30] and the one known case of a homozygous nonsense mutation of *LMNA* identified in human patient who died shortly after birth [31]. For purposes of clarity and consistency, the term *Lmna*$^{-/-}$ will still be applied to the original mouse generated by Sullivan et al. [4].

While speculation about reduced A-type lamin expression and cancer has a long history, none of the mouse models or human patients with *LMNA* mutations linked to striated and cardiac muscle have been associated with oncogenesis. However, there is strong evidence that A-type lamins regulate key factors involved in controlling the G1-to-S transition, and the most evidence exists for lamin A effects on the RB [32]. The connection between A-type lamins and RB has been examined in a variety of settings, and, while there may be differences, the general consensus is that A-type lamins are required for normal RB function.

One of the first tumor suppressors identified, loss of both copies of the *RB* gene leads to a range of different cancers including retinoblastomas and osteosarcomas [33]. More broadly, RB activity is deregulated in a wide range of tumors, generally through unchecked activity of cyclin-dependent kinases (CDKs) [34]. RB has myriad binding partners and has been ascribed to a number of functions in the nucleus [35]. Most notably, RB acts as a repressor of the transcriptional factor E2F, which controls a range of genes important for entry into S phase of the cell cycle. When hypophosphorylated and active, RB binds to E2F complexes and acts as a repressor of S-phase genes, retaining cells in G1. CDK-dependent phosphorylation promotes release of RB from E2F and cell cycle progression. Along with p53 (discussed below) and telomere regulation, the RB pathway is a major determinant of cell senescence and also plays a role in cell differentiation in multiple lineages. Finally, control of G1-dependent gene expression is but one function ascribed to RB, which is also linked to DNA replication, mitosis, and checkpoint pathways, including those initiated by DNA damage where its roles are still being fully defined [34].

Both A-type lamins and their binding partner, lamina-associated polypeptide 2α (LAP2α), have been reported to interact with RB [21, 36], repress E2F-dependent transcription, and promote cell cycle arrest [37]. Consistently, loss of LAP2α or lamin A/C impairs normal cell cycle regulation leading to inappropriate S-phase entry. Mouse cells lacking A-type lamins or LAP2α have altered cell cycle profiles with premature S-phase entry and in some contexts enhanced proliferation [38–40]. In contrast, one study in human primary fibroblasts indicated that reduced lamin A/C or *LAP2α* expression led to cell cycle arrest [41]. The reason(s) for these different observations remains unknown.

In addition to promoting RB-dependent transcriptional repression of E2F target genes, A-type lamins regulate RB by at least three other mechanisms by coordination of RB phosphorylation, localization, and protein stability [39, 42]. Some aspects of control of RB protein stability are beginning to be understood. For instance, in cells lacking lamin A/C, enhanced levels of RB degradation occur through a proteasome-dependent mechanism [39]. Reduced RB levels make $Lmna^{-/-}$ fibroblasts insensitive to $p16^{INK4A}$-mediated cell cycle arrest [38]. However, the E3 ligase responsible for RB degradation remains to be identified and appears to be independent of the MDM2 and gankyrin pathways that have been linked to RB turnover in other contexts [43]. A number of other proteins are destabilized by loss of RB, including the RB-related protein p107 [39], emerin [44], and ATR kinase [45]. These findings raise the possibility that A-type lamins might coordinate nuclear proteasome function, and altered activity of ubiquitin ligase components has been detected in cells expressing mutant forms of lamin A [45].

Other A-type lamin functions may promote G1 maintenance. Serum stimulation of G1 arrested cells promotes ERK1/2 mitogen-activated protein kinase-dependent phosphorylation of c-Fos, leading to its association with c-Jun- and AP-1-dependent transcription [46]. Prior to stimulation, c-Fos and ERK1/2 were found to be in a complex with lamin A/C at the nuclear periphery preventing premature activation of AP-1, which occurred in $Lmna^{-/-}$ fibroblasts [47, 48]. Independent studies have reported enhanced ERK1/2 activity in cells with reduced A-type lamin expression, and this has been linked to cardiac pathology in dilated cardiomyopathy type 1A (DCM1A) laminopathy patients [49]. Interestingly, a more recent study indicates that ERK1/2–lamin A/C and RB–lamin A/C complexes are mutually exclusive and finds that ERK1/2-dependent lamin A/C binding upon serum stimulation displaces RB, thereby promoting cell cycle progression [50]. If ERK1/2 levels are elevated in cycling cells, this may lead to RB dysregulation and underlie some of the altered cell cycle parameters evident in $Lmna^{-/-}$ cells.

A number of reports have implicated lamins in regulation of DNA replication. For instance, early studies showed that disruption of the lamin structure impaired initiation of DNA synthesis [51–53]. In immortalized cells, lamin B was localized to intranuclear sites of late S-phase replication [54], whereas in primary fibroblasts, intranuclear A-type lamins associate with initial sites of DNA synthesis upon S-phase entry [55]. The impact of lamins on S-phase progression in mammalian cells is less clear. S phase is elongated in fibroblasts lacking A-type lamins, although this could be an indirect effect of premature S-phase entry due to defective RB function [39].

A recent study has shed light on a different aspect of replication by comparing the response of *Lmna*$^{-/-}$ cells to forms of DNA damage-induced cell cycle arrest, finding that lamin A/C was required for restart of stalled replication forks and genome maintenance after hydroxyurea-induced replication stress [56]. While the roles of A-type and B-type lamins in DNA replication remain to be fully elaborated, this is clearly an important area for further studies.

Mitotic defects have not been reported for cells with reduced A-type lamin function. However, a recently identified novel allele of LMNA, *Lmna*^{DHE}, was identified as a spontaneous mouse mutation with a subset of progeroid phenotypes [57]. Fibroblasts heterozygous for this *Lmna* allele exhibit, in addition to reduced levels of hypophosphorylated RB, a reduction in a mitosis-specific centromere condensing subunit that depends on RB activity [58]. These alterations result in a range of chromosome segregation defects. It will be of interest to determine whether other *Lmna* disease-associated alleles lead to similar defects.

Cell Senescence: A-Type Lamins

Both A-type and B-type lamins have been linked to cell senescence, the process by which primary cells withdraw from the cell cycle in response to extended passaging, irreparable damage, or unbalanced proliferative signals. Cell senescence has typically been viewed as an impedance to cancer progression and not a driving force in aging, but recent findings paint a more complex picture [59]. Senescent cells do accumulate with age, and while they generally never reach a large percentage of the population of a tissue, recent findings indicate that they adopt an altered secretory profile, the senescence-associated secretory phenotype (SASP), that leads to paracrine release of a number of inflammatory cytokines. These cytokines may promote tissue aging and stimulate tumor development in neighboring cells. In this section, we cover links between lamins and senescence, discussing the still tenuous connections between lamins and normal aging process.

Several studies have implicated A-type lamins in cell senescence, although the phenotype is most clearly associated with the expression of progeria-associated *LMNA* alleles such as progerin [9]. Progerin expression, in addition to delaying cell cycle progression, brings about premature senescence in a variety of contexts [60, 61]. Lamin A is normally farnesylated at its C-terminus but only for a short time because the last 18 amino acid residues are removed in two cleavage steps. The protease that removes the farnesyl group is Zmpste24 in mice (FACE-1 in humans). In HGPS the deleted exon also removes this cleavage site so that the progerin form of lamin A is permanently farnesylated. *Zmpste24*$^{-/-}$ cells with defective lamin A processing also exhibit enhanced levels of senescence [62, 63]. The mechanisms behind these effects remain unclear. As cells approach senescence the p16INK4A/RB and p53 pathways both become engaged, leading cells to stop proliferation and enter a permanently arrested state [59]. In addition, telomere attrition during passaging in culture drives senescence, particularly as telomere ends shorten beyond critical thresholds, invoking DNA damage checkpoint response pathways. All three of these networks (p16INK4A/RB, p53, and telomeres) are interrelated, and progerin has been linked functionally to each.

The connection between progerin expression and RB pathway is not entirely understood presently. In one study, where *LMNA–progerin* alleles were expressed in a lamin A/C-deficient background, it was found that progerin restored RB stability [38]. Moreover, inactivation of the RB pathway by expression of HPV E7 failed to suppress the proliferation defects of human fibroblasts stably expressing progerin [64]. However, an analysis of global gene expression profiles in fibroblasts from HGPS patients identified the RB-E2F pathway as dysregulated [65] due in part to RB gene expression. The mechanisms underlying this effect were unknown. Interestingly, exposure of cells to farnesyl transferase inhibitors mostly restored the normal gene expression profile.

It is generally thought that *LMNA* mutations are not associated with tumors; however, two instances have been reported in progeria models. In one case, an osteosarcoma was identified in an HGPS patient [66, 67]. This is intriguing since osteosarcomas are commonly associated with *RB* mutations [68]. Interestingly, this patient expressed a smaller 35 amino acid C-terminal deletion in the C-terminus of *LMNA* and not progerin [66, 67]. It would be interesting to determine the levels of RB and activity of the RB pathway in this context. One issue possibly limiting cancer progression in progeria patients is the early progression of the disease leading to mortality for patients usually in their teens. A recent study has identified a novel late-onset progeria syndrome, *LMNA*-associated cardiocutaneous progeria, that is associated with possible cancer susceptibility [69]. This syndrome is associated with a heterozygous novel mutation in the lamin A/C coiled-coil domain that is largely uncharacterized.

The RB pathway is not the only cell cycle regulatory network that is influenced by A-type lamins. Whereas inactivation of RB did not rescue proliferation defects and premature senescence in human fibroblasts stably expressing progerin, either inactivation of p53 (by HPV E7) or expression of telomerase did [64]. Several studies have connected *LMNA* mutation to p53 engagement due to enhanced DNA damage [24], but a recent study has elaborated this connection further. In this case, depletion of lamin A/C in primary human fibroblasts led to dramatic destabilization of RB as expected; however, the cells also had proliferation defects and a senescent phenotype instead of the expected short G1 phase due to enhanced specific activation of the p53–p21 axis [70]. p53 did not display enhanced levels or activating phosphorylation, and many targets were not upregulated. Instead, a subset of targets including p21 were upregulated leading to repression of E2F targets even in the absence of normal levels of RB. Cross talk between the RB and p53 pathways is not unprecedented. Moreover, these findings suggest that the A-type lamins interact with both pathways in a nuanced manner and whether *LMNA* mutations lead to altered proliferation with early G1 cell cycle exit or reduced proliferation leading to senescence may depend on the specific nature of the *LMNA* mutation.

Another interesting recent finding has connected lamins to the p53 pathway in a different manner. In renal carcinoma cells, where genetic inactivation of the von Hippel–Lindau (VHL) gene E3 ligase is a frequent event, progerin expression may play a role in controlling the p53 pathway [71]. p53 is not generally mutated in these tumors but is inactivated functionally. Jung et al. found that progerin was a target of VHL-mediated proteasomal degradation. Loss of VHL led to stabilization of progerin, which was otherwise bound to p14/ARF, sequestering it from MDM2 and leading in turn to p53 degradation. Thus, progerin was required for p53 inactivation in the absence of VHL. If progerin has this activity in a wide array of cell types, this finding may provide a potential link between aging and cancer:

> Accumulation of progerin with aging would lead to inactivation of the p53 network, impairing its tumor suppressive and checkpoint activities. More research is needed to test this intriguing hypothesis.

Recent studies have also pointed to a role for lamins in the maintenance of telomere metabolism, another activity that could be closely linked to cell senescence [72]. For instance, HGPS fibroblasts are reported to have faster rates of telomere attrition [73]. This finding does not on its own suggest a direct role for A-type lamins at telomeres; however, a number of studies have reported that telomeres associate with the nuclear matrix and more specifically with A-type lamins [74–77]. *Lmna*$^{-/-}$ fibroblasts have shorter telomeres but no differences in telomerase activity [78]. Instead, the answer may be related to altered chromatin structure at telomeres and trace back to the reduced function of RB and its related proteins p107 and p130 [72]. Cells lacking A-type lamins exhibit a decrease in histone H4K20me3 [78], a known feature of cells lacking RB family members [79, 80]. However, the latter cells have increased telomere length, leading the authors of the lamin study to suggest that A-type lamins might be required for telomere elongation in the absence of RB family members [81, 82]. A relocalization of telomeres from peripheral to central regions of the nucleus has also been reported in *Lmna*$^{-/-}$ cells, through at present unknown mechanisms [83].

HGPS fibroblasts have altered telomere chromatin as well, although the changes are distinct from those in *Lmna*$^{-/-}$ cells [83, 84]. In this case, decreased H4K20me3 and increased H4K20me were found. This finding suggests that, not unexpectedly, progerin influences telomere metabolism in a manner distinct from hypomorphic mutation of *LMNA*. Recently, it was reported that progerin-induced DNA damage is localized specifically to telomeres [85]. Expression of telomerase resolves this DNA damage, and, once repaired, HGPS fibroblasts regain full potential to proliferate. It will be intriguing to see how DNA damage is restricted in the genome by progerin.

Finally, the proliferative defects leading to senescence with expression of progerin and/or unprocessed lamin A appear to extend to adult stem cell populations. Studies in mesenchymal stem cells have indicated that progerin expression leads to elevated Notch signaling, causing perturbations in stem cell differentiation and maintenance of stem cell identity [13]. In addition, adult bone marrow-derived stem cells from *Zmpste24*$^{-/-}$ mice have decreased *SIRT1* function due to its dissociation from the nuclear matrix and leading to reduced proliferation and premature senescence [86]. This phenotype may be relevant for aging as restoration of *SIRT1* function was associated with improved stem cell function and enhanced survival. Studies in fibroblasts and other cell culture models have provided important insights into lamin function; however, more emphasis needs to be placed on the role of lamins in adult stem cell populations, which could be central to a subset of the pathologies associated with laminopathies.

Cell Senescence: B-Type Lamins

Lamin B1 has also been tightly associated with cell senescence, with altered expression in either direction possibly having deleterious consequences [87]. Initial suggestions of altered lamin B expression came from studies of HGPS cells, where lamin B1 was found to be reduced [87]. Two virtually contemporaneous recent reports have shown that loss of lamin B1 expression is a marker for cell senescence induced by a variety of causes, including replicative exhaustion. Loss of lamin B1 expression was not dependent on many molecular inducers of senescence but was driven by activation of either the p53 or the RB pathway [88]. Changes in lamin B1 could be traced back to reduced mRNA stability, although reduced protein stability could be detected as well in mouse liver induced to senescence by irradiation. In the second study, Shimi et al. also reported loss of lamin B1 as a much needed biomarker of cell senescence [14]. Shimi et al. also examined the consequences of RNAi-mediated knockdown of lamin B1 expression, finding that this was sufficient to both slow proliferation and reduce senescence. The proliferative delay was dependent on p53, and the senescent phenotype was dependent on both p53 and RB. Yet a third very recent study has confirmed and extended the observed reduction in lamin B1 expression to senescent keratinocytes and to chronologically aged human skin tissue [89]. However, in this study enforced reduction in lamin B1 expression failed to lead to senescence. The disparities between the two studies are not known [14, 89]. Together these findings (1) indicate that loss of lamin B1 may serve as an effective biomarker of in vivo senescence and (2) suggest the existence of a complex regulatory loop connecting B-type lamins to the RB and p53 pathways.

Whereas mutations affecting the *LMNB1* coding sequence have not been reported, overexpression is linked to at least two diseases, suggesting that too much lamin B1 may be as deleterious as too little. Duplication of the *LMNB1* locus results in adult-onset autosomal dominant leukodystrophy (ADLD) [90], and lamin B1 overexpression has been detected in lymphoblasts and fibroblasts from ataxia telangiectasia patients [91]. Interestingly, lamin B1 overexpression also drives cell senescence, a phenomenon also observed in ataxia telangiectasia cells, which are rescued by restoration of normal lamin B1 expression. Induction of senescence by overexpression of lamin B1 has been repeated in a second study [89]. In this case, the senescent phenotype could be rescued by expression of telomerase or inactivation of p53, paralleling observations for progerin-induced senescence [64]. Finally, senescence induced by overexpression of lamins is not restricted to B1. Increased levels of lamin A also reduce the replicative life-span of primary human fibroblasts [92].

Several studies indicate a connection between A-type and B-type lamins in the formation of intermediate filament networks, and there appears to be an interplay between the two nuclear intermediate filament families with respect to cell senescence. For instance, reduced expression of A-type lamins exacerbates the senescent phenotype of cells overexpressing lamin B1 [89]. In the context of senescence in normal cells, one thing that needs to be resolved is whether increased progerin levels and reduced lamin B1 expression are related. Does one family of nuclear intermediate filaments regulate the other during aging and senescence?

Whether reactive oxygen species (ROS) drive aspects of the aging process remains highly debated [93]. Both A- and B-type lamins have been linked to reactive oxygen production and sensing in recent years, and while the details remain murky, these findings represent another promising set of leads as the relationship between lamins and aging is elucidated [87]. There may be a direct connection as conserved cysteines in the C-terminal tail of lamin A have been found to be oxidized in senescent cells [94]. This led to the formation of intra- and intermolecular disulfide bonds and perturbation of the lamina. These cysteine residues may serve as a reservoir or a sensor for oxidation, as mutating the cysteines to alanine led to oxidative stress sensitivity and premature senescence. In the case of B-type lamins, a number of conflicting results have been reported. In some contexts, increased ROS has been reported to lead to elevated and reduced lamin B1 levels [14, 88, 91]. Similarly, both higher and lower levels of lamin B1 lead to reduced ROS levels [14]. Further studies with ROS and lamins will likely clarify this complex and potentially mechanistically rewarding relationship.

Conclusions

While nuclear lamins have been speculated to control cell cycle progression for decades, this area of research has exploded in recent years and many labs have investigated the effect of laminopathy-associated mutations in *LMNA* and diseases associated with *LMNB1* overexpression. These studies have linked nuclear lamins to virtually every major aspect of cell cycle progression and, more recently, cell senescence. This latter connection may be particularly interesting given that *LMNA* mutations are associated with progeria as well as the findings that both progerin and lamin B1 have altered expression with normal aging.

However, several big questions remain to be resolved: Does altered lamin expression promote tumor progression? Are lamins important regulators of the aging process? With respect to cancer, an increasing number of studies have linked altered expression of both A-type and B-type to different tumors, but the relationships are complex and causal links are generally lacking. It is critical to resolve these issues in more detail to determine for which tumors lamins might be effective biomarkers and perhaps more importantly to understand why changes in lamin expression may promote tumorigenesis.

With respect to aging, the findings are certainly becoming more intriguing. That *LMNA* mutations cause HGPS is not sufficient to ascribe A-type lamins a role in normal aging. That progerin is expressed in aging and/or senescent cells is also not sufficient, but together the data certainly justify further analysis of lamin roles in the normal aging process. To provide convincing evidence, it will be necessary to manipulate A-type (or B-type) lamins in a manner that leads to enhanced organismal longevity. For instance, it would be informative to apply technologies developed to reduce progerin expression in HGPS models to wild-type mice to determine whether suppression of progerin in this context leads to longer life-span. Experiments such as these will begin to answer the critical questions surrounding aging and lamins.

Progress in understanding disease-relevant functions of lamins has escalated dramatically in recent years, and the next few years will without a doubt provide exciting new findings relevant to cancer and aging. Perhaps the most exciting aspect of the field is that researchers are beginning to understand the roles of lamins at the mechanistic level. Further progress on this front will likely yield effective therapeutic approaches for treatment of laminopathies and, importantly, an increasingly elegant understanding of the organization of the mammalian nucleus.

Acknowledgements Lamin-related research in the lab of B.K.K. is supported by a grant from the National Institute of Aging (R01 AG024287).

References

1. Gerace L, Blum A, Blobel G (1978) Immunocytochemical localization of the major polypeptides of the nuclear pore complex-lamina fraction. Interphase and mitotic distribution. J Cell Biol 79(2 Pt 1):546–566
2. Schreiber KH, Kennedy BK (2013) When lamins go bad: nuclear structure and disease. Cell 152(6):1365–1375
3. Gerace L, Blobel G (1980) the nuclear envelope lamina is reversibly depolymerized during mitosis. Cell 19:277–287
4. Sullivan T, Escalante-Alcalde D, Bhatt H, Anver M, Bhat N, Nagashima K, Stewart CL, Burke B (1999) Loss of A-type lamin expression compromises nuclear envelope integrity leading to muscular dystrophy. J Cell Biol 147:913–920
5. Cao H, Hegele RA (2003) LMNA is mutated in Hutchinson-Gilford progeria (MIM 176670) but not in Wiedemann-Rautenstrauch progeroid syndrome (MIM 264090). J Hum Genet 48:271–274
6. De Sandre-Giovannoli A, Bernard R, Cau P, Navarro C, Amiel J, Boccaccio I, Lyonnet S, Stewart CL, Munnich A, Le Merrer M, Levy N (2003) Lamin A truncation in Hutchison-Gilford progeria. Science 300:2055
7. Eriksson M, Brown WT, Gordon LB, Glynn MW, Singer J, Scott L, Erdos MR, Robbins CM, Moses TY, Berglund P, Dutra A, Pak E, Durkin S, Csoka AB, Boehnke M, Glover TW, Collins FS (2003) Recurrent de novo point mutations in lamin A cause Hutchison-Gilford progeria syndrome. Nature 423:293–298
8. Burtner CR, Kennedy BK (2010) Progeria syndromes and ageing: what is the connection? Nat Rev Mol Cell Biol 11(8):567–578
9. Reddy S, Comai L (2012) Lamin A, farnesylation and aging. Exp Cell Res 318(1):1–7
10. Verdy C, Branka JE, Mekideche N (2011) Quantitative assessment of lactate and progerin production in normal human cutaneous cells during normal ageing: effect of an Alaria esculenta extract. Int J Cosmet Sci 33(5):462–466
11. McClintock D, Ratner D, Lokuge M, Owens DM, Gordon LB, Collins FS, Djabali K (2007) The mutant form of lamin A that causes Hutchinson-Gilford progeria is a biomarker of cellular aging in human skin. PLoS One 2(12):e1269

12. Scaffidi P, Misteli T (2006) Lamin A-dependent nuclear defects in human aging. Science 312(5776):1059–1063
13. Scaffidi P, Misteli T (2008) Lamin A-dependent misregulation of adult stem cells associated with accelerated ageing. Nat Cell Biol 10(4):452–459
14. Shimi T, Butin-Israeli V, Adam SA, Hamanaka RB, Goldman AE, Lucas CA, Shumaker DK, Kosak ST, Chandel NS, Goldman RD (2011) The role of nuclear lamin B1 in cell proliferation and senescence. Genes Dev 25(24):2579–2593
15. Rober RA, Weber K, Osborn M (1989) Differential timing of nuclear lamin A/C expression in the various organs of the mouse embryo and the young animal: a developmental study. Development 105:365–378
16. Ho JC, Zhou T, Lai WH, Huang Y, Chan YC, Li X, Wong NL, Li Y, Au KW, Guo D, Xu J, Siu CW, Pei D, Tse HF, Esteban MA (2011) Generation of induced pluripotent stem cell lines from 3 distinct laminopathies bearing heterogeneous mutations in lamin A/C. Aging (Albany NY) 3(4):380–390
17. Zhang J, Lian Q, Zhu G, Zhou F, Sui L, Tan C, Mutalif RA, Navasankari R, Zhang Y, Tse HF, Stewart CL, Colman A (2011) A human iPSC model of Hutchinson Gilford progeria reveals vascular smooth muscle and mesenchymal stem cell defects. Cell Stem Cell 8(1):31–45
18. Liu GH, Barkho BZ, Ruiz S, Diep D, Qu J, Yang SL, Panopoulos AD, Suzuki K, Kurian L, Walsh C, Thompson J, Boue S, Fung HL, Sancho-Martinez I, Zhang K, Yates J 3rd, Izpisua Belmonte JC (2011) Recapitulation of premature ageing with iPSCs from Hutchinson-Gilford progeria syndrome. Nature 472(7342):221–225
19. Eisenman RN, Tachibana CY, Abrams HD, Hann SR (1985) V-myc- and c-myc-encoded proteins are associated with the nuclear matrix. Mol Cell Biol 5(1):114–126
20. Mancini MA, Shan B, Nickerson JA, Penman S, Lee WH (1994) The retinoblastoma gene product is a cell cycle-dependent, nuclear matrix-associated protein. Proc Natl Acad Sci U S A 91(1):418–422
21. Ozaki T, Saijo M, Murakami K, Enomoto H, Taya Y, Sakiyama S (1994) Complex formation between Lamin A and the retinoblastoma gene product: identification of the domain on Lamin A required for its interaction. Oncogene 9:2649–2653
22. Templeton DJ, Park SH, Lanier L, Weinberg RA (1991) Nonfunctional mutants of the retinoblastoma protein are characterized by defects in phosphorylation, viral oncoprotein association, and nuclear tethering. Proc Natl Acad Sci U S A 88:3033–3037
23. Mittnacht S, Weinberg RA (1991) G1/S phosphorylation of the retinoblastoma protein is associated with altered affinity for the nuclear compartment. Cell 65:381–393
24. Redwood AB, Gonzalez-Suarez I, Gonzalo S (2011) Regulating the levels of key factors in cell cycle and DNA repair: new pathways revealed by lamins. Cell Cycle 10(21):3652–3657
25. Zink D, Fischer AH, Nickerson JA (2004) Nuclear structure in cancer cells. Nat Rev Cancer 4(9):677–687
26. Foster CR, Przyborski SA, Wilson RG, Hutchison CJ (2010) Lamins as cancer biomarkers. Biochem Soc Trans 38(Pt 1):297–300
27. Butin-Israeli V, Adam SA, Goldman AE, Goldman RD (2012) Nuclear lamin functions and disease. Trends Genet 28(9):464–471
28. Nikolova V, Leimena C, McMahon AC, Tan JC, Chandar S, Jogia D, Kesteven SH, Michalicek J, Otway R, Verheyen F, Rainer S, Stewart CL, Martin D, Feneley MP, Fatkin D (2004) Defects in nuclear structure and function promote dilated cardiomyopathy in lamin A/C-deficient mice. J Clin Invest 113(3):357–369
29. Jahn D, Schramm S, Schnolzer M, Heilmann CJ, de Koster CG, Schutz W, Benavente R, Alsheimer M (2012) A truncated lamin A in the $Lmna^{-/-}$ mouse line: implications for the understanding of laminopathies. Nucleus 3(5):463–474
30. Kubben N, Voncken JW, Konings G, van Weeghel M, van den Hoogenhof MM, Gijbels M, van Erk A, Schoonderwoerd K, van den Bosch B, Dahlmans V, Calis C, Houten SM, Misteli T, Pinto YM (2011) Post-natal myogenic and adipogenic developmental: defects and metabolic impairment upon loss of A-type lamins. Nucleus 2(3):195–207

31. van Engelen BG, Muchir A, Hutchison CJ, van der Kooi AJ, Bonne G, Lammens M (2005) The lethal phenotype of a homozygous nonsense mutation in the lamin A/C gene. Neurology 64(2):374–376
32. Boban M, Braun J, Foisner R (2010) Lamins: 'structure goes cycling'. Biochem Soc Trans 38(Pt 1):301–306
33. Friend SH, Bernards R, Rogelj S, Weinberg RA, Rapaport JM, Albert DM, Dryja TP (1986) A human DNA segment with properties of the gene that predisposes to retinoblastoma and osteosarcoma. Nature 323(6089):643–646
34. Dick FA, Rubin SM (2013) Molecular mechanisms underlying RB protein function. Nat Rev Mol Cell Biol 14(5):297–306
35. Morris EJ, Dyson N (2001) Retinoblastoma protein partners. Adv Cancer Res 82:1–54
36. Markiewicz E, Dechat T, Foisner R, Quinlan RA, Hutchison CJ (2002) Lamin A/C binding protein LAP2a is required for nuclear anchorage of retinoblastoma protein. Mol Biol Cell 13:4401–4413
37. Dorner D, Vlcek S, Foeger N, Gajewski A, Makolm C, Gotzmann J, Hutchison CJ, Foisner R (2006) Lamina-associated polypeptide 2alpha regulates cell cycle progression and differentiation via the retinoblastoma-E2F pathway. J Cell Biol 173(1):83–93
38. Nitta RT, Jameson SA, Kudlow BA, Conlan LA, Kennedy BK (2006) Stabilization of the retinoblastoma protein by A-type nuclear lamins is required for INK4A-mediated cell cycle arrest. Mol Cell Biol 26:5360–5372
39. Johnson BR, Nitta RT, Frock RL, Mounkes L, Barbie DA, Stewart CL, Harlow E, Kennedy BK (2004) A-type lamins regulate retinoblastoma protein function by promoting subnuclear localization and preventing proteasomal degradation. Proc Natl Acad Sci U S A 101:9677–9682
40. Naeger LK, Goodwin EC, Hwang ES, DeFilippis RA, Zhang H, DiMaio D (1999) Bovine papillomavirus E2 protein activates a complex growth-inhibitory program in p53 negative HT-3 cervical carcinoma cells that includes repression of cyclin A and cdc25A phosphatase genes and accumulation of hypophosphorylated retinoblastoma protein. Cell Growth Differ 10(6):413–422
41. Pekovic V, Harborth J, Broers JL, Ramaekers FC, van Engelen B, Lammens M, von Zglinicki T, Foisner R, Hutchison C, Markiewicz E (2007) Nucleoplasmic LAP2{alpha}-lamin A complexes are required to maintain a proliferative state in human fibroblasts. J Cell Biol 176(2):163–172
42. Van Berlo JH, Voncken JW, Kubben N, Broers JL, Duisters R, van Leeuwen RE, Crijns HJ, Ramaekers FC, Hutchison CJ, Pinto YM (2005) A-type lamins are essential for TGF-beta1 induced PP2A to dephosphorylate transcription factors. Hum Mol Genet 14(19):2839–2849
43. Nitta RT, Smith CL, Kennedy BK (2007) Evidence that proteasome-dependent degradation of the retinoblastoma protein in cells lacking A-type lamins occurs independently of gankyrin and MDM2. PLoS One 2(9):e963
44. Muchir A, Massart C, van Engelen BG, Lammens M, Bonne G, Worman HJ (2006) Proteasome-mediated degradation of integral inner nuclear membrane protein emerin in fibroblasts lacking A-type lamins. Biochem Biophys Res Commun 351(4):1011–1017
45. Muralikrishna B, Chaturvedi P, Sinha K, Parnaik VK (2012) Lamin misexpression upregulates three distinct ubiquitin ligase systems that degrade ATR kinase in HeLa cells. Mol Cell Biochem 365(1–2):323–332
46. Verde P, Casalino L, Talotta F, Yaniv M, Weitzman JB (2007) Deciphering AP-1 function in tumorigenesis: fra-ternizing on target promoters. Cell Cycle 6(21):2633–2639
47. Ivorra C, Kubicek M, Gonzalez JM, Sanz-Gonzalez SM, Alvarez-Barrientos A, O'Connor JE, Burke B, Andres V (2006) A mechanism of AP-1 suppression through interaction of c-Fos with lamin A/C. Genes Dev 20(3):307–320
48. Gonzalez JM, Navarro-Puche A, Casar B, Crespo P, Andres V (2008) Fast regulation of AP-1 activity through interaction of lamin A/C, ERK1/2, and c-Fos at the nuclear envelope. J Cell Biol 183(4):653–666

49. Muchir A, Wu W, Worman HJ (2010) Mitogen-activated protein kinase inhibitor regulation of heart function and fibrosis in cardiomyopathy caused by lamin A/C gene mutation. Trends Cardiovasc Med 20(7):217–221
50. Rodriguez J, Calvo F, Gonzalez JM, Casar B, Andres V, Crespo P (2010) ERK1/2 MAP kinases promote cell cycle entry by rapid, kinase-independent disruption of retinoblastoma-lamin A complexes. J Cell Biol 191(5):967–979
51. Ellis DJ, Jenkins H, Whitfield WG, Hutchison CJ (1997) GST-lamin fusion proteins act as dominant negative mutants in Xenopus egg extract and reveal the function of the lamina in DNA replication. J Cell Sci 110:2507–2518
52. Spann TP, Moir RD, Goldman AE, Stick R, Goldman RD (1997) Disruption of nuclear lamin organization alters the distribution of replication factors and inhibits DNA synthesis. J Cell Biol 136:1201–1212
53. Moir RD, Spann TP, Herrmann H, Goldman RD (2000) Disruption of nuclear lamin organization blocks the elongation phase of DNA replication. J Cell Biol 149:1179–1192
54. Moir RD, Spann TP, Goldman RD (1994) Dynamic properties of nuclear lamins: lamin B is associated with sites of DNA replication. J Cell Biol 125:1201–1212
55. Kennedy BK, Barbie DA, Classon M, Dyson N, Harlow E (2000) Nuclear organization of DNA replication in primary mammalian cells. Genes Dev 14:2855–2868
56. Singh M, Hunt CR, Pandita RK, Kumar R, Yang CR, Horikoshi N, Bachoo R, Serag S, Story MD, Shay JW, Powell SN, Gupta A, Jeffery J, Pandita S, Chen BP, Deckbar D, Lobrich M, Yang Q, Khanna KK, Worman HJ, Pandita TK (2013) Lamin A/C depletion enhances DNA damage-induced stalled replication fork arrest. Mol Cell Biol 33(6):1210–1222
57. Odgren PR, Pratt CH, Mackay CA, Mason-Savas A, Curtain M, Shopland L, Ichicki T, Sundberg JP, Donahue LR (2010) Disheveled hair and ear (Dhe), a spontaneous mouse Lmna mutation modeling human laminopathies. PLoS One 5(4):e9959
58. Pratt CH, Curtain M, Donahue LR, Shopland LS (2011) Mitotic defects lead to pervasive aneuploidy and accompany loss of RB1 activity in mouse LmnaDhe dermal fibroblasts. PLoS One 6(3):e18065
59. Campisi J (2013) Aging, cellular senescence, and cancer. Annu Rev Physiol 75: 685–705
60. Dechat T, Shimi T, Adam SA, Rusinol AE, Andres DA, Spielmann HP, Sinensky MS, Goldman RD (2007) Alterations in mitosis and cell cycle progression caused by a mutant lamin A known to accelerate human aging. Proc Natl Acad Sci U S A 104(12):4955–4960
61. Cao K, Capell BC, Erdos MR, Djabali K, Collins FS (2007) A lamin A protein isoform overexpressed in Hutchinson-Gilford progeria syndrome interferes with mitosis in progeria and normal cells. Proc Natl Acad Sci U S A 104(12):4949–4954
62. Liu B, Wang J, Chan KM, Tjia WM, Deng W, Guan X, Huang JD, Li KM, Chau PY, Chen DJ, Pei D, Pendas AM, Cadinanos J, Lopez-Otin C, Tse HF, Hutchison C, Chen J, Cao Y, Cheah KS, Tryggvason K, Zhou Z (2005) Genomic instability in laminopathy-based premature aging. Nat Med 11(7):780–785
63. Varela I, Cadinanos J, Pendas AM, Gutierrez-Fernandez A, Folgueras AR, Sanchez LM, Zhou Z, Rodriguez FJ, Stewart CL, Vega JA, Tryggvason K, Freije JM, Lopez-Otin C (2005) Accelerated ageing in mice deficient in Zmpste24 protease is linked to p53 signalling activation. Nature 437(7058):564–568
64. Kudlow BA, Stanfel MN, Burtner CR, Johnston ED, Kennedy BK (2008) Suppression of proliferative defects associated with processing-defective lamin A mutants by hTERT or inactivation of p53. Mol Biol Cell 19(12):5238–5248
65. Marji J, O'Donoghue SI, McClintock D, Satagopam VP, Schneider R, Ratner D, Worman HJ, Gordon LB, Djabali K (2010) Defective lamin A-Rb signaling in Hutchinson-Gilford progeria syndrome and reversal by farnesyltransferase inhibition. PLoS One 5(6):e11132
66. King CR, Lemmer J, Campbell JR, Atkins AR (1978) Osteosarcoma in a patient with Hutchinson-Gilford progeria. J Med Genet 15(6):481–484
67. Shalev SA, De Sandre-Giovannoli A, Shani AA, Levy N (2007) An association of Hutchinson-Gilford progeria and malignancy. Am J Med Genet A 143A(16):1821–1826

68. Wadayama B, Toguchida J, Shimizu T, Ishizaki K, Sasaki MS, Kotoura Y, Yamamuro T (1994) Mutation spectrum of the retinoblastoma gene in osteosarcomas. Cancer Res 54(11):3042–3048
69. Kane MS, Lindsay ME, Judge DP, Barrowman J, Ap Rhys C, Simonson L, Dietz HC, Michaelis S (2013) LMNA-associated cardiocutaneous progeria: an inherited autosomal dominant premature aging syndrome with late onset. Am J Med Genet A 161(7):1599–1611
70. Moiseeva O, Bourdeau V, Vernier M, Dabauvalle MC, Ferbeyre G (2011) Retinoblastoma-independent regulation of cell proliferation and senescence by the p53-p21 axis in lamin A/C-depleted cells. Aging Cell 10(5):789–797
71. Jung YS, Lee SJ, Lee SH, Chung JY, Jung YJ, Hwang SH, Ha NC, Park BJ (2013) Loss of VHL promotes progerin expression, leading to impaired p14/ARF function and suppression of p53 activity. Cell Cycle 12(14)
72. Gonzalez-Suarez I, Gonzalo S (2010) Nurturing the genome: A-type lamins preserve genomic stability. Nucleus 1(2):129–135
73. Allsopp RC, Vaziri H, Patterson C, Goldstein S, Younglai EV, Futcher AB, Greider CW, Harley CB (1992) Telomere length predicts replicative capacity of human fibroblasts. Proc Natl Acad Sci U S A 89(21):10114–10118
74. Shoeman RL, Wadle S, Scherbarth A, Traub P (1988) The binding in vitro of the intermediate filament protein vimentin to synthetic oligonucleotides containing telomere sequences. J Biol Chem 263(35):18744–18749
75. de Lange T (1992) Human telomeres are attached to the nuclear matrix. EMBO J 11(2):717–724
76. Luderus ME, van Steensel B, Chong L, Sibon OC, Cremers FF, de Lange T (1996) Structure, subnuclear distribution, and nuclear matrix association of the mammalian telomeric complex. J Cell Biol 135(4):867–881
77. Raz V, Vermolen BJ, Garini Y, Onderwater JJ, Mommaas-Kienhuis MA, Koster AJ, Young IT, Tanke H, Dirks RW (2008) The nuclear lamina promotes telomere aggregation and centromere peripheral localization during senescence of human mesenchymal stem cells. J Cell Sci 121(Pt 24):4018–4028
78. Gonzalez-Suarez I, Redwood AB, Perkins SM, Vermolen B, Lichtensztejin D, Grotsky DA, Morgado-Palacin L, Gapud EJ, Sleckman BP, Sullivan T, Sage J, Stewart CL, Mai S, Gonzalo S (2009) Novel roles for A-type lamins in telomere biology and the DNA damage response pathway. EMBO J 28(16):2414–2427
79. Gonzalo S, Garcia-Cao M, Fraga MF, Schotta G, Peters AH, Cotter SE, Eguia R, Dean DC, Esteller M, Jenuwein T, Blasco MA (2005) Role of the RB1 family in stabilizing histone methylation at constitutive heterochromatin. Nat Cell Biol 7(4):420–428
80. Gonzalo S, Blasco MA (2005) Role of Rb family in the epigenetic definition of chromatin. Cell Cycle 4(6):752–755
81. Benetti R, Gonzalo S, Jaco I, Schotta G, Klatt P, Jenuwein T, Blasco MA (2007) Suv4-20h deficiency results in telomere elongation and derepression of telomere recombination. J Cell Biol 178(6):925–936
82. Garcia-Cao M, O'Sullivan R, Peters AH, Jenuwein T, Blasco MA (2004) Epigenetic regulation of telomere length in mammalian cells by the Suv39h1 and Suv39h2 histone methyltransferases. Nat Genet 36(1):94–99
83. Gonzalez-Suarez I, Redwood AB, Gonzalo S (2009) Loss of A-type lamins and genomic instability. Cell Cycle 8(23):3860–3865
84. Shumaker DK, Dechat T, Kohlmaier A, Adam SA, Bozovsky MR, Erdos MR, Eriksson M, Goldman AE, Khuon S, Collins FS, Jenuwein T, Goldman RD (2006) Mutant nuclear lamin A leads to progressive alterations of epigenetic control in premature aging. Proc Natl Acad Sci U S A 103(23):8703–8708
85. Benson EK, Lee SW, Aaronson SA (2010) Role of progerin-induced telomere dysfunction in HGPS premature cellular senescence. J Cell Sci 123(Pt 15):2605–2612
86. Liu B, Ghosh S, Yang X, Zheng H, Liu X, Wang Z, Jin G, Zheng B, Kennedy BK, Suh Y, Kaeberlein M, Tryggvason K, Zhou Z (2012) Resveratrol rescues SIRT1-dependent adult stem cell decline and alleviates progeroid features in laminopathy-based progeria. Cell Metab 16(6):738–750

87. Barascu A, Le Chalony C, Pennarun G, Genet D, Zaarour N, Bertrand P (2012) Oxidative stress alters nuclear shape through lamins dysregulation: a route to senescence. Nucleus 3(5):411–417
88. Freund A, Laberge RM, Demaria M, Campisi J (2012) Lamin B1 loss is a senescence-associated biomarker. Mol Biol Cell 23(11):2066–2075
89. Dreesen O, Chojnowski A, Ong PF, Zhao TY, Common JE, Lunny D, Lane EB, Lee SJ, Vardy LA, Stewart CL, Colman A (2013) Lamin B1 fluctuations have differential effects on cellular proliferation and senescence. J Cell Biol 200(5):605–617
90. Padiath QS, Saigoh K, Schiffmann R, Asahara H, Yamada T, Koeppen A, Hogan K, Ptacek LJ, Fu YH (2006) Lamin B1 duplications cause autosomal dominant leukodystrophy. Nat Genet 38(10):1114–1123
91. Barascu A, Le Chalony C, Pennarun G, Genet D, Imam N, Lopez B, Bertrand P (2012) Oxidative stress induces an ATM-independent senescence pathway through p38 MAPK-mediated lamin B1 accumulation. EMBO J 31(5):1080–1094
92. Candelario J, Sudhakar S, Navarro S, Reddy S, Comai L (2008) Perturbation of wild-type lamin A metabolism results in a progeroid phenotype. Aging Cell 7(3):355–367
93. Harman D (1956) Aging: a theory based on free radical and radiation chemistry. J Gerontol 11:298–300
94. Pekovic V, Gibbs-Seymour I, Markiewicz E, Alzoghaibi F, Benham AM, Edwards R, Wenhert M, von Zglinicki T, Hutchison CJ (2011) Conserved cysteine residues in the mammalian lamin A tail are essential for cellular responses to ROS generation. Aging Cell 10(6):1067–1079

Lamina-Associated Polypeptide (LAP)2α and Other LEM Proteins in Cancer Biology

Andreas Brachner and Roland Foisner

Abstract The LEM proteins comprise a heterogeneous family of chromatin-associated proteins that share the LEM domain, a structural motif mediating interaction with the DNA associated protein, Barrier-to-Autointegration Factor (BAF). Most of the LEM proteins are integral proteins of the inner nuclear membrane and associate with the nuclear lamina, a structural scaffold of lamin intermediate filament proteins at the nuclear periphery, which is involved in nuclear mechanical functions and (hetero-)chromatin organization. A few LEM proteins, such as Lamina-associated polypeptide (LAP)2α and Ankyrin and LEM domain-containing protein (Ankle)1 lack transmembrane domains and localize throughout the nucleoplasm and cytoplasm, respectively. LAP2α has been reported to regulate cell proliferation by affecting the activity of retinoblastoma protein in tissue progenitor cells and numerous studies showed upregulation of LAP2α in cancer. Ankle1 is a nuclease likely involved in DNA damage repair pathways and single nucleotide polymorphisms in the *Ankle1* gene have been linked to increased breast and ovarian cancer risk. In this review we describe potential mechanisms of the involvement of LEM proteins, particularly of LAP2α and Ankle1 in tumorigenesis and we provide evidence that LAP2α expression may be a valuable diagnostic and prognostic marker for tumor analyses.

Keywords LEM-domain • Lamina associated polypeptide • Retinoblastoma protein • Lamin A • Cell cycle • Ankyrin and LEM domain containing • DNA repair • Telomere • E2F • TMPO • Thymopoietin

A. Brachner • R. Foisner (✉)
Max F. Perutz Laboratories, Medical University Vienna, Dr. Bohr-Gasse 9, 1030 Vienna, Austria
e-mail: andreas.brachner@univie.ac.at; roland.foisner@meduniwien.ac.at

E.C. Schirmer and J.I. de las Heras (eds.), *Cancer Biology and the Nuclear Envelope*, Advances in Experimental Medicine and Biology 773, DOI 10.1007/978-1-4899-8032-8_7,

Abbreviations

CMV	Cytomegalovirus
HPV	Human papilloma virus
INM	Inner nuclear membrane
ONM	Outer nuclear membrane
PARP	Poly(ADP-ribose) polymerase
SNP	Single nucleotide polymorphism

Introduction

The nucleus of eukaryotic cells is surrounded by a specialized internal membrane system, the nuclear envelope [1], which is composed of two membrane sheets, the inner (INM) and outer (ONM) nuclear membranes (Fig. 1). INM and ONM merge at the sites where nuclear pore complexes are inserted into the nuclear envelope, and the ONM is continuous with the endoplasmic reticulum [2]. In metazoan organisms, the nuclear envelope also includes the nuclear lamina that underlies the INM and serves as a structural scaffold for the nucleus [3, 4]. It is formed by the type V intermediate filament proteins, the A- and B-type lamins, and by a number of integral and associated proteins of the INM. The nuclear envelope and in particular the nuclear lamina are involved in nuclear architecture and nuclear mechanical functions [3, 5], in chromatin organization and gene regulation through tethering and silencing heterochromatic regions [6–8], and in signaling through recruiting

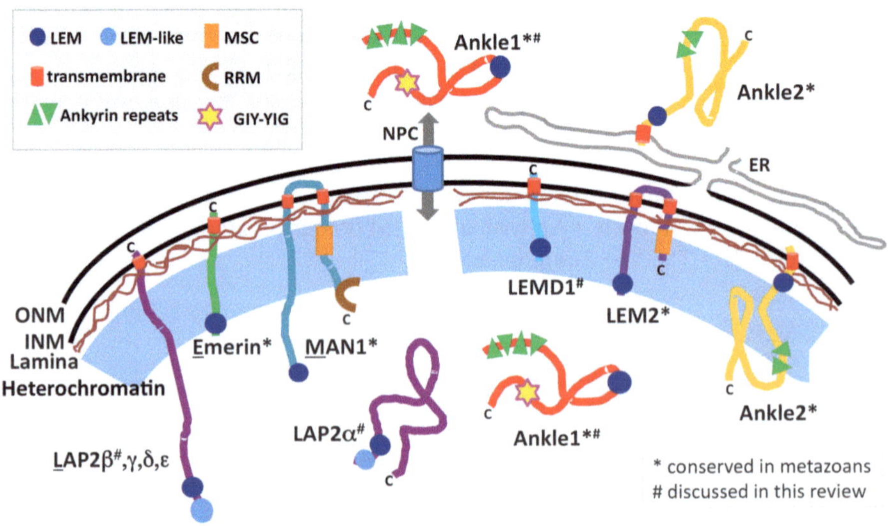

Fig. 1 Schematic overview of the mammalian LEM-protein family. Protein localization and domains are indicated

transcription regulators, epigenetic modifier enzymes and signaling molecules to the nuclear periphery [9, 10]. In view of the multitude of functions of the nuclear envelope, it is not surprising that several components of the nuclear lamina have been linked to various diseases ranging from muscular dystrophies and cardiomyopathies over lipodystrophies to systemic diseases like the premature aging syndrome Hutchinson–Gilford progeria [11].

LEM-Domain Containing Proteins: A Prominent Family of Nuclear (Envelope) Proteins

Mass spectrometric approaches revealed that the INM of mammalian cells contains over 80 integral proteins that are expressed in a tissue specific manner [12, 13]. The *L*AP2-*E*merin-*M*AN1 (LEM)-domain containing proteins (Fig. 1) represent one of the best studied family of INM proteins [14]. These proteins share the LEM domain, a 40 aa long bi-helical structure motif, which mediates binding to an abundant and essential chromatin protein in metazoan species termed Barrier-to-Autointegration-Factor (BAF) [15–18]. Therefore, all LEM-proteins can associate with chromatin via BAF and it is generally assumed that they are involved in tethering chromatin to the nuclear periphery during interphase [19]. Most characterized INM LEM proteins bind lamins in the lamina and link the membrane to the lamina scaffold. In addition, several LEM proteins were shown to recruit and regulate signaling molecules [20] such as Smads involved in transforming growth factor beta (TGFβ) and bone morphogenetic protein (BMP) signaling [21, 22], β-catenin, a transcriptional co-activator of the Wnt signaling pathway [23], the Lmo7 transcription factor [24], the germ cell less (GCL) transcriptional repressor [25, 26], and histone deacetylase 3 (HDAC3) [27].

Mammalian genomes contain seven individual genes that encode LEM-domain proteins (Fig. 1): Lamina-associated polypeptide 2 (LAP2), Emerin, MAN1, LEM2, LEMD1, Ankle1, and Ankle2 [14, 28, 29]. In addition, Ankle1, LAP2 and LEMD1 generate various isoforms by alternative splicing. Most LEM proteins contain either one or two transmembrane domains and are integral components of the INM. However, two isoforms of the LAP2 gene (LAP2α and ζ) and Ankle1 lack a membrane-spanning domain and localize to the nucleoplasm and cytoplasm [14]. In this review we describe and discuss evidence that the LAP2 isoform LAP2α and potentially other LEM proteins may be involved in cancer development or may serve as useful diagnostic and prognostic markers for some types of cancers.

LAP2 Proteins in Cancer

The mammalian *LAP2* (*TMPO*) gene, *also known as thymopoietin*, encodes six splice isoforms (α, β, γ, δ, ε, ζ), all sharing a common ~180 aa long N-terminal domain including the LEM-motif (interacting with the chromatin protein BAF) and an additional LEM-like motif in the very N-terminus, which interacts with DNA

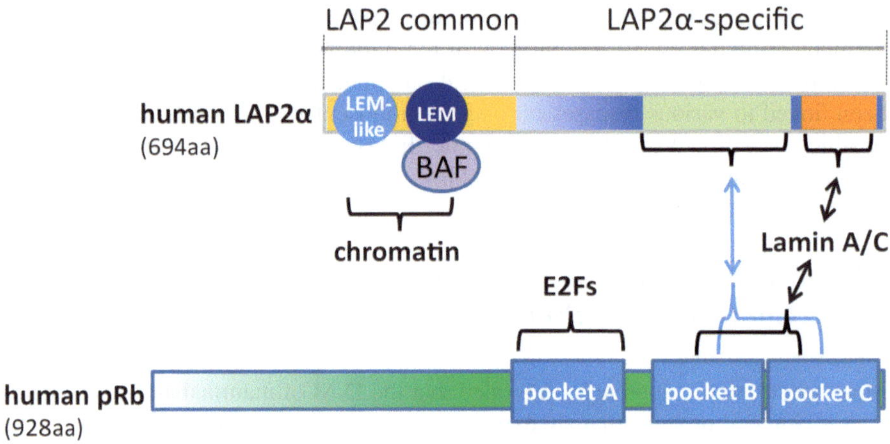

Fig. 2 LAP2α, lamin A/C, and pRb interact directly and may form a trimeric complex. Domain organization of human LAP2α and pRb and interaction domains are shown

directly [15] (Fig. 2). While most LAP2 isoforms differ in their up to ~300 aa long C-terminus only due to the inclusion/exclusion of small, alternatively spliced domains, the ~500 aa long LAP2α C-terminus is encoded by a single exon unique for LAP2α [30]. Unlike the other major LAP2 isoforms, LAP2α's C-terminus lacks a C-terminal transmembrane domain and folds as an extensive four-stranded anti-parallel coiled coil dimer [31] that can also form higher oligomers [32]. In addition, while the membrane bound LAP2 isoforms localize at the INM and interact primarily with B-type lamins of the peripheral lamina [33], LAP2α specifically binds A-type lamins via its unique C-terminus [34] in the nucleoplasm [35]. Furthermore, LAP2α's C-terminus mediates interaction with the cell cycle regulator and tumor suppressor, retinoblastoma protein (pRb) [35–37]. Recent observations that LAP2α may be involved in the regulation of pRb localization and repressor activity led to the hypothesis that LAP2α may play a role in tumorigenesis.

Potential Role of LAP2α and A-Type Lamins in pRb-Mediated Cell Cycle Control

The retinoblastoma protein (pRb) is one of the three pocket proteins (pRb, p107, p130) [38], which regulate cell cycle transition from G1 to S-phase, cell cycle exit, and differentiation in multicellular eukaryotes [39–42]. A plethora of data has shown that pRb is one of the major tumor suppressors by preventing cell proliferation in the absence of strong mitogenic signals, and concordantly, in the majority of tumors the pRb pathway was found deregulated [43]. However, impaired pRb functions in cancer cells are rarely linked to mutations in the *RB1* gene (except in the hereditary form of the childhood retinoblastoma disease), but are caused by defects in the expression

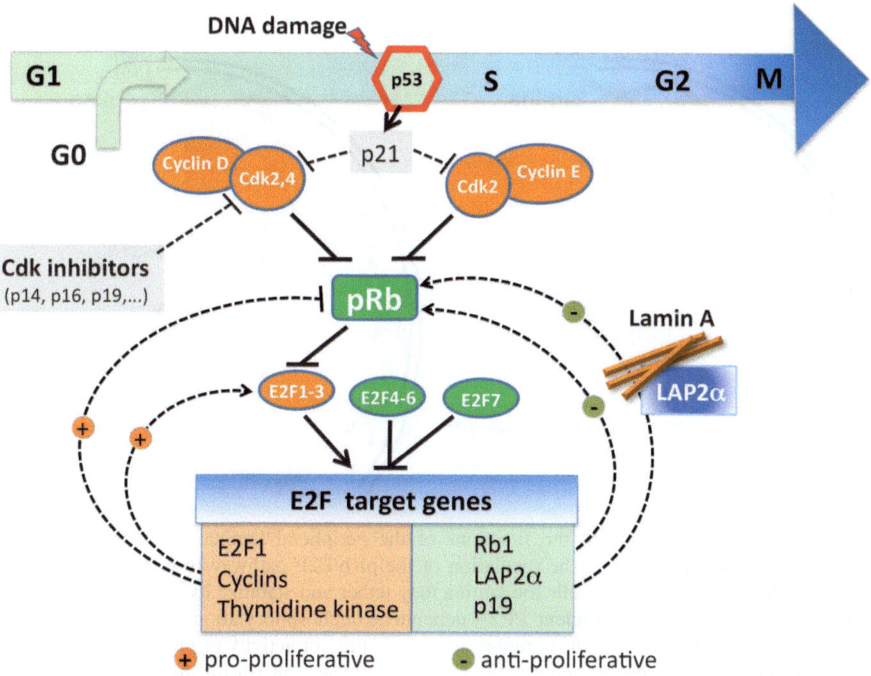

Fig. 3 The pRb/E2F pathway regulates cell cycle progression. The scheme depicts major upstream regulators of pRb and downstream effectors, as well as important pro-proliferative and anti-proliferative feedback mechanisms upon transition from a resting (G0) to a proliferating state and during G1 to S-phase progression. The potential role of nucleoplasmic LAP2α-lamin A/C complexes in the pRb regulatory network is indicated

or activity of upstream regulators or downstream effectors of pRb. Also certain human viruses, e.g., human papilloma virus (HPV) or cytomegalovirus (CMV), impair pRb function by expressing proteins (HPV-E7 and CMV-IE86) that bind pRb with high affinity and affect binding and activity of normal cellular pRb regulators.

In non-tumor cells pRb is regulated by posttranslational modifications (Fig. 3), among which phosphorylation by cyclin D/cyclin-dependent kinase (cdk) 4 at the G1/S phase transition, and cyclin E/cdk2 during S-phase, are the best studied ones. Only hypophosphorylated pRb binds to and represses the cell cycle-activating E2F transcription factors (E2F1, E2F2, E2F3), to allow cells to efficiently exit the cell cycle. Hyperphosphorylation of pRb by mitogen-activated cyclin-dependent kinases inhibits its repressor function, since E2Fs are released from the complex, leading to transcriptional activation of E2F target genes required for S-phase progression (e.g., cyclin E, PCNA, thymidine kinase) [44]. This basic cell cycle-dependent regulation of pRb and E2Fs is fine-tuned by a complex network of proteins with pro-and anti-proliferative activities, which feed into reinforcing and attenuating signaling loops (Fig. 3). Cell cycle entry initiated by cyclin D/cdk4-dependent pRb phosphorylation and activation of E2F1 is reinforced by a positive feedback loop through

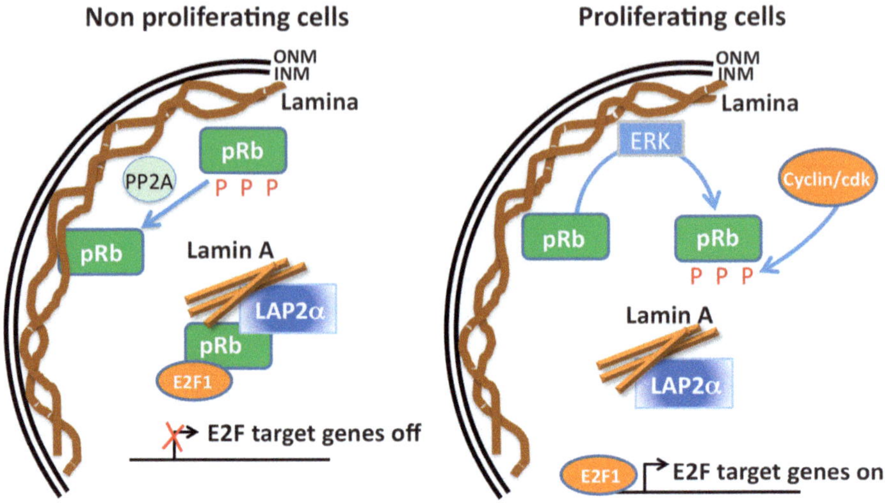

Fig. 4 Hypothetical model of the functions of the peripheral lamina and the nucleoplasmic LAP2α-lamin A/C complex in the regulation of the pRb/E2F pathway in non-proliferating and proliferating cells. In arrested cells the lamina may tether and stabilize hypophosphorylated pRb and serve as platform for efficient PP2A-dependent dephosphorylation of phospho-pRb. The nucleoplasmic LAP2α-lamin A/C complex activates repressor activity of pRb leading to E2F target gene repression. In proliferating cells, ERK may release pRb from the lamina, favoring its cdk-mediated phosphorylation. LAP2α-lamin A/C complexes dissociate from phospho-Rb, allowing E2F activation and E2F target gene expression

E2F-dependent upregulation of E2Fs themselves and cyclins, maintaining hyperphosphorylated pRb during S-phase [45]. On the other hand, E2Fs activate also negative cell cycle regulators [46], which provide a negative feedback loop to prevent uncontrolled proliferation. Among others, E2F1 activates transcription of *RB1*, anti-proliferative factors like p19^{ink4d} (an inhibitor of cdk4) [47], various checkpoint and DNA repair genes (e.g., p73 and ATM) and pro-apoptotic genes (e.g., APAF1, caspases) [44], as well as the transcriptional repressor E2F7 [48] which together with E2F4-6 silence promoters of cell cycle promoting genes independently of pRb [49].

How does LAP2α fit into this complex regulatory network modulating pRb function? Both LAP2α [36, 37] and its nucleoplasmic binding partners lamins A and C (A-type lamins) [50] bind pRb directly. Several studies have revealed different mechanisms by which these proteins may affect pRb regulation (Fig. 4). LAP2α was found to preferentially interact with hypophosphorylated pRb and is required for pRb anchorage in the nucleus [37]. LAP2α overexpression in cells caused downregulation of E2F-dependent reporter gene activity and repressed endogenous E2F target genes [36], suggesting that LAP2α promotes pRb-mediated repression of target genes. Accordingly, overexpression of LAP2α in pre-adipocytes promoted cell cycle exit and initiation of differentiation to adipocytes in vitro [36]. In contrast, fibroblasts derived from LAP2α-deficient mice showed impaired pRb repressor activity, upregulated E2F/pRb target gene expression, and delayed cell cycle exit upon contact inhibition. Also at the tissue and organismal level, progenitor cells in proliferative tissues showed impaired cell cycle regulation in the mouse model. Proliferating progenitors in the paw epidermis, in colon crypts, and in skeletal muscle were significantly upregulated, leading to epidermal hyperplasia, increased crypt length, and an increased number of fiber-associated satellite cells, respectively [35]. Overall, LAP2α seems to be important for an additional level of pRb regulation on top of the basic regulatory machinery, allowing an efficient activation of pRb repressor activity and pRb target gene repression to promote cell cycle exit (Fig. 3). Upon loss of LAP2α, pRb activity is not lost but impaired, which may explain the observation that LAP2α-deficient mice did not show a clearly increased incidence of cancer during their life time [35]. However, these mice have not been challenged yet with cancer promoting agents or treatments.

A-type lamins have been suggested to be required for pRb protein stabilization (Fig. 4) by preventing its proteasomal degradation [51, 52]. Another study has suggested that A-type lamins form a platform allowing efficient PP2A-dependent dephosphorylation of pRb, which is required for efficient TGFβ-induced cell cycle arrest of fibroblasts [53]. Yet another study reported a complex of pRb-lamin A in non-proliferating cells, keeping pRb in a hypophosphorylated state [54]. Mitogenic signal-dependent activation of MAPK signaling leads to translocation of ERK to the nucleus, where it displaces pRb from lamin A, which in turn becomes hyperphosphorylated by cyclin/cdks. Interestingly, displacement of pRb from lamins did not require ERK's kinase activity.

Overall, these results indicate that LAP2α and A-type lamins can activate pRb repressor activity and thereby act in an anti-proliferative manner (Figs. 3 and 4). The molecular mechanisms how these proteins affect pRb activity are not yet clear. One could imagine various ways, such as stabilization of pRb protein, stable tethering of pRb to chromatin and/or promoters, preventing pRb phosphorylation or mediating efficient pRb dephosphorylation. Interestingly, lamin A has been shown to have repressive activity when artificially tethered to reporter gene promoters [55]. Furthermore, a recent study in *Caenorhabditis elegans* showed that a muscular dystrophy-linked lamin mutant impaired tissue-specific gene regulation during worm development [7]. Thus, lamin A may not only increase pRb-dependent gene repression, but may have more general roles maybe through recruiting epigenetic modifiers or changing overall chromatin state at promoters.

Another open question concerns the functional relationship between LAP2α and A-type lamins in pRb-mediated gene regulation. A-type lamins are assumed to exist in two different sub-compartments in the nucleus: An estimated 90 % of total A-type lamins localize to the nuclear envelope as a component of the nuclear lamina scaffold in a LAP2α-independent pool, while ~10 % localize throughout the nucleoplasm in a mobile and dynamic pool, most likely in association with LAP2α [35, 56, 57]. It is unclear whether pRb binds preferentially to a LAP2α–lamin complex in the nucleoplasm or to the peripheral lamina network (Fig. 4). Most studies on the role of A-type lamins in pRb-mediated cell cycle control do not discriminate between these two lamin pools, or assume, without clear experimental evidence that the peripheral nuclear lamina is involved. We favor a predominant role of

Table 1 Tumors and tumor derived cell lines, which show upregulated LAP2 expression

Cancer samples and cell lines with upregulated LAP2 expression	References
Cervix carcinoma; human	[78, 111]
Colorectal cancer cell lines; human	[112]
Colon cancer; human	[70]
Hepatocellular carcinoma; TKO mice	[71]
Breast cancer cell line (MCF7); human	[73]
Myeloma; human	[69]
Pancreatic cancer; human	[113]
Gastric cancer; human	[114]
Medulloblastoma; human	[115]
Lymphoma; human	[81, 97]
Larynx, stomach, colon, lymphoma, sarcoma; human	[116]

nucleoplasmic lamins in the regulation of pRb-dependent gene expression for the following reasons: (1) Peripheral localization of pRb, as predicted by a preferential docking of pRb to the lamina, has not been observed [58]. (2) Both knockout of LAP2α in mice, which leads to specific loss of the nucleoplasmic lamin pool only, and total lamin A/C knockout, which affects both the peripheral and nucleoplasmic lamin A, showed the same misregulation of pRb and hyper-proliferation phenotype in epidermal progenitor cells [35].

LAP2α Expression During the Cell Cycle and in the Context of Tumorigenesis

A number of studies have shown that LAP2α is highest expressed in proliferating cells and is down-regulated upon cell cycle exit and differentiation [35, 37, 59, 60] or is differentially expressed during the cell cycle in proliferating cells [61]. LAP2α transcripts were also upregulated during liver [62] and muscle regeneration in vivo [63, 64], processes that involve transient controlled proliferation of progenitor cells. Independent studies on the LAP2α promoter, based on chromatin immunoprecipitation and microarray techniques identified E2F1 and c-Myc [65], E2F1 and E2F4 [66], E2F3b [67], and E2F7 [68] on the LAP2 promoter, suggesting that the expression of the LAP2 gene is under direct control of major cell cycle regulators, such as E2Fs. Interestingly both, cell cycle driving (i.e., c-Myc, E2F1) and repressing transcription factors (i.e., E2F3b, E2F4, E2F7) seem to regulate the LAP2 promoter, which points to a complex feedback mechanism. From these studies it remains unclear, though, whether all LAP2 isoforms are similarly regulated. In line with its high expression in proliferating cells, LAP2α was found to be overexpressed in various human tumor samples and cancer-derived cell lines at transcript and protein levels (examples summarized in Table 1).

Is Cancer-Related LAP2α Overexpression Cause or Consequence of Proliferation?

Numerous studies found LAP2 or LAP2α overexpressed in tumor cells (Table 1), frequently correlated with the upregulation of tumor-relevant signaling pathways such as the NEK2 [69], Gli [70], Notch [71], and Estrogen [72, 73] pathways. Thus, many studies have shown a consistent correlation between active proliferation and upregulation of LAP2 transcription during physiological processes and in a variety of cancers. These results are inconsistent with an anti-proliferative activity of LAP2α, predicted from the studies in mice and several cell lines mentioned above. On the one hand, LAP2α seems to be upregulated in most if not all proliferating, particularly cancer cells: on the other hand, LAP2α expression is predicted to have an anti-proliferative function. How can one solve this apparent discrepancy? Is LAP2α causally involved in promoting proliferation of cancer cells or is its upregulation a consequence of the high proliferation activity of these cells?

A potential causal positive relation between LAP2α levels and proliferation is only supported by a handful of reports consistent with a pro-proliferative effect of the LAP2α-lamin A complex. Both RNA interference-mediated knockdown of lamin A and LAP2α caused cell cycle arrest in primary human dermal fibroblasts [58]. Furthermore muscular dystrophy-linked homozygous mutations in the *LMNA* gene in patient-derived fibroblasts, leading to loss of lamin protein expression, severely impaired cell proliferation [74]. Additionally, fibroblasts derived from Hutchinson–Gilford progeria patients expressing the very different lamin A variant progerin, lacking 50 amino acids and permanently farnesylated, caused passage-dependent proliferation defects in culture [75]. Similarly postnatal fibroblasts, but not embryonic fibroblasts, from a progeria mouse model showed proliferative arrest and cell death [76]. However, in all these cases, it seems likely that the anti-proliferative effect is caused indirectly by DNA damage and/or deregulation of signaling pathways, which lead to the activation of cell cycle checkpoints. For example, progeria cells have been shown to accumulate DNA damage during in vitro passage [75], and mouse fibroblasts from a progeria model have a defective Wnt signaling causing misregulation of extracellular matrix components [76].

Overall, it seems very likely that LAP2α and lamin A have an anti-proliferative activity through promoting pRb repressor activity as described above. The increased LAP2α protein levels in proliferating versus non-proliferating cells and in tumor versus normal cells may simply be a consequence of the mitogen-induced increase in E2F1 activity. We propose that in normal cells the E2F1-dependent upregulation of LAP2α provides a negative feedback loop that ensures efficient LAP2α-mediated activation of pRb repressor activity causing inhibition of E2F-dependent transcription and cell cycle exit. Cancer cells acquire multiple changes impacting on the pRb and p53 checkpoint pathways. If these changes impair the pRb pathway, LAP2α overexpression would be unable to activate the pRb-mediated negative cell cycle feedback loop. In line with this hypothesis, several studies reported that viral oncoproteins affect LAP2 expression levels via inactivation of pRb [77, 78] and p53 pathways [78]. For example, expression of HPV proteins E6 and E7, which are known to inhibit p53 and pRb by direct binding [79], is linked to an upregulation of LAP2α in cervical cancer cells [78]. Knockdown of E6 or E7 in these cells restored p53 and pRb activity, respectively, and caused down-regulation of LAP2α. Conversely, knockdown of p53 in normal human fibroblasts increased LAP2α levels [78]. Similarly, expression of CMV IE86, an inhibitor of pRb led to upregulation of LAP2 levels [77], and the LAP2 promoter was shown to be regulated by the pRb-p16ink4a pathway [80] and by E2F7, which is a target of p53 [68]. Unfortunately, these studies did not look at the relationship between p53/pRb-mediated LAP2 expression control and tumorigenic behavior of cells.

Overall we conclude that LAP2α overexpression in cancer cells is possible only upon additional changes that promote cell cycle progression, like amplification of c-myc, hyperactivity of mitogenic signaling, or loss of repressive factors. This may also explain why overexpression of LAP2α was linked to a worse patient prognosis [81, 82]. LAP2α may thus be a useful diagnostic and prognostic gene in tumorigenesis.

LAP2α at Telomeres: A Link to DNA Damage Repair?

Several microscopy studies revealed that LAP2α is highly dynamic and changes its localization during the cell cycle. While it localizes throughout the nucleoplasm in interphase and disperses in the cytoplasm upon nuclear envelope breakdown in prophase, it associates with (sub-)telomeric regions on chromosomes during late anaphase and telophase [56] (Fig. 5). In addition, a proteomic approach identified LAP2α in a complex with the telomere repeat binding factor 1 (TRF1) [83], a component of the telomeric shelterin complex that protects and regulates telomeres. The physiological relevance of these findings is still unclear, but several recent observations are consistent with a potential role of LAP2α in a DNA damage-response pathway at telomeres, which is known to be required for functional telomeres [84].

Several studies showed that lamin A is involved in DNA repair and telomere maintenance [85–87]. LAP2α was found to associate with Werner helicase, WRN [88], a protein well known for its role in telomere maintenance and DNA repair [89]. Both Werner helicase and LAP2α were found in two independent studies in a complex with Ku86 [90], a key protein involved in non-homologous end joining DNA repair pathways and in telomere protection [91–93]. LAP2α also appeared among the top hits in an interaction screen of proteins modified with Poly(ADP-ribose) scaffolds [94], which are generated by poly(ADP-ribose) polymerases (PARPs) at sites of DNA damage and serve as docking site for DNA damage signaling and repair proteins. Interestingly, the PARP tankyrase localizes to telomeres and targets TRF1 in S/G2 phase, thereby triggering the release of TRF1 from telomeres and allowing access for telomerase and telomere-processing enzymes.

Overall, LAP2α may be involved in telomere maintenance or protection pathways. However, as LAP2α deficient mice did not show any of the phenotypes associated

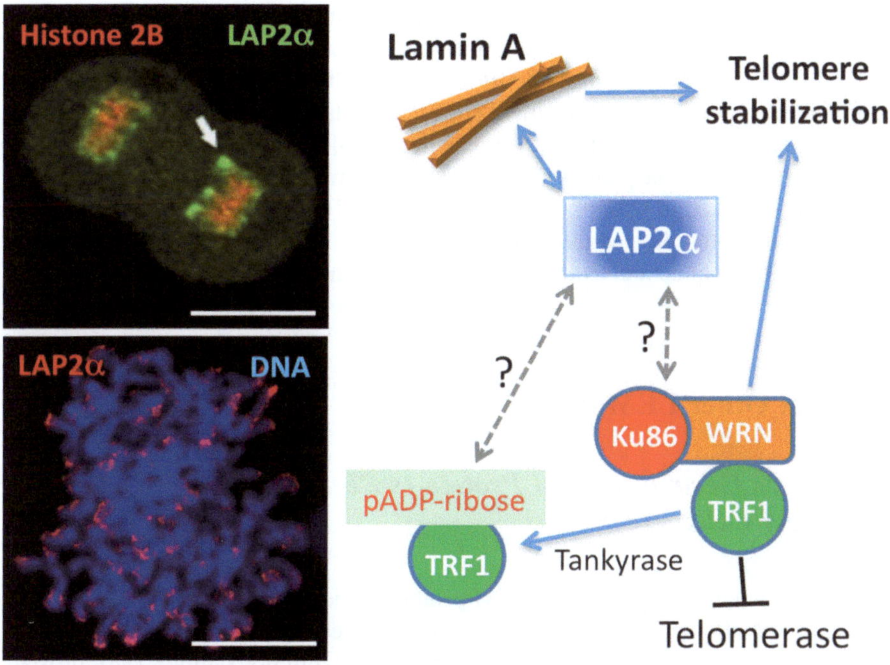

Fig. 5 LAP2α transiently localizes to telomeric regions during late anaphase-telophase (*top*) and is detected on the chromosome tips in metaphase spreads (*bottom*). Confocal fluorescence images showing the localization of ectopic, fluorescently tagged LAP2α and Histone 2B (*upper panel*) or stained for LAP2α and DNA (courtesy of T. Dechat and A. Gajewski, MFPL). Bars, 5 μm. *Arrow* indicates localization of YFP-LAP2α at chromosome tips in late anaphase. *Right panel* shows a hypothetical involvement of LAP2α in telomere stability based on recently reported protein interactions. See text for details

with dysfunctional telomeres (e.g., premature aging, infertility, increased occurrence of tumors), it seems unlikely that LAP2α is essential for telomere maintenance. Having said this, one has to take into account that telomere biology is substantially different in rodents and humans, including the broader expression of telomerase in mouse versus human cells and tissues and the about five to ten times longer telomeres in murine versus human cells. Therefore, it is still possible that aberrant LAP2α function may have an impact on telomere maintenance in human cancer.

The Potential Relevance of Other LAP2 Isoforms in Cancer

Two studies have reported a potential link between cancer and LAP2β, the largest membrane-bound LAP2 isoform. LAP2β was upregulated in various digestive tract cancers (stomach, liver, pancreas, and bile duct) [95]. Knockdown of LAP2β in cancer cells reduced—whereas ectopic expression of LAP2β increased—cell motility, but had no effect on cell proliferation. LAP2β knockdown also induced significant changes in gene expression [95], which is likely linked to the previously reported interaction of LAP2β with histone deacetylase 3 (HDAC3) and its involvement in gene regulation at the nuclear envelope [96]. LAP2β upregulation was also reported in lymphoma patient samples and in normal human lymphocytes upon mitogenic stimulation with phytohemagglutinin [97]. Since also LAP2α was found upregulated in phytohemagglutinin stimulated cells, this is likely a consequence of proliferation-dependent activation of the LAP2 promoter.

Links of Other LEM Proteins to Cancer

LEMD1, a Germ line LEM Protein Is Reexpressed in Cancer Cells

LEMD1 is a mammalian-specific LEM-protein of the INM expressed exclusively in testis as six alternatively spliced isoforms. Interestingly, LEMD1 was initially described as a component overexpressed in colorectal tumor samples [98], prostate cancer [99], and lymphoma cells [100]. Hence, LEMD1 was postulated to belong to the cancer/testis antigens [98], a group of about 40 germ line-specific genes that are reexpressed in tumors originating from unrelated cell types [101]. However, neither the biological functions of LEMD1 nor its contribution to tumorigenesis is known.

Polymorphisms in the Ankle1 Gene Are Linked to Breast and Ovarian Cancer Risk

In breast and ovarian cancer, female and male carriers of mutant alleles of the breast cancer associated genes BRCA1 and BRCA2 bear a dramatically increased risk to develop cancer compared to the reference population. Inherited BRCA1 and BRCA2 mutations account for approximately 5 % of all cases of breast and 14 % of ovarian cancers [102]. In families with inherited predisposition for breast cancer or with a combined risk for breast and ovarian cancers, the frequency of BRCA mutations is 40 % and >80 %, respectively [103]. Some other cases of families with an inherited predisposition for breast cancer show no mutations in the coding sequences of BRCA1 and BRCA2. Therefore, it is assumed that other high risk breast cancer genes exist, or that a combination of genetic variants of low penetrance genes exerts additive effects manifesting in a significantly increased risk to develop breast or ovarian cancer [102]. Several recently published studies employed high-throughput genomic analyses of breast and ovarian cancer patient samples aiming at the identification of loci containing such low penetrance genes that may modulate cancer risk [104–108].

Intriguingly, two single nucleotide polymorphisms (SNPs) locating at chromosomal locus 19p13.11 consistently appeared in all of these studies showing a statistically significant link to breast cancer susceptibility. These SNPs are localized within the coding sequence of the gene encoding Ankyrin and LEM-domain containing protein 1 (Ankle1) and lead to amino acid changes within the polypeptide. Ankle1 is a conserved gene in metazoan species encoding an unusual LEM protein described in two recent studies in *C. elegans* and mammalian cells [109, 110]. Ankle1 lacks a transmembrane domain and shuttles between the nucleoplasm and cytoplasm in human cells [109]. Intriguingly, Ankle1 contains a GIY-YIG-type endonuclease domain, which was shown to cleave DNA in vitro and in vivo [109, 110]. While Ankle1 overexpression in mammalian cells activates DNA damage response pathways, a *C. elegans* strain carrying a point mutation in the *lem-3* gene—the *C. elegans* ortholog of Ankle1—was hypersensitive towards DNA damaging agents. Altogether, these studies suggest that Ankle1 may be an enzyme involved in DNA repair pathways.

Considering that the vast majority of mutations predisposing carriers to breast cancer were identified in genes involved in DNA damage signaling and repair (e.g., BRCA1, BRCA2, Rad51, Chek2, ATM, p53), the endonuclease Ankle1 may indeed be a relevant factor for tumorigenesis.

Acknowledgements We thank Andreas Gajewski and Thomas Dechat, MFPL, Vienna, for generously providing immunofluorescence images shown in Fig. 5. Work in the authors' laboratory was supported by grants from the Austrian Science Fund (FWF P22569-B09 and P23805-B20) to R.F. and from the "Herzfelder'sche Familienstiftung" to A.B.

References

1. Wilson KL, Dawson SC (2011) Evolution: functional evolution of nuclear structure. J Cell Biol 195(2):171–181. doi:10.1083/jcb.201103171
2. Mekhail K, Moazed D (2010) The nuclear envelope in genome organization, expression and stability. Nat Rev Mol Cell Biol 11(5):317–328. doi:10.1038/nrm2894
3. Dechat T, Pfleghaar K, Sengupta K, Shimi T, Shumaker DK, Solimando L, Goldman RD (2008) Nuclear lamins: major factors in the structural organization and function of the nucleus and chromatin. Genes Dev 22(7):832–853. doi:10.1101/gad.1652708
4. Parnaik VK (2008) Role of nuclear lamins in nuclear organization, cellular signaling, and inherited diseases. Int Rev Cell Mol Biol 266:157–206. doi:10.1016/S1937-6448(07)66004-3
5. Dahl KN, Ribeiro AJ, Lammerding J (2008) Nuclear shape, mechanics, and mechanotransduction. Circ Res 102(11):1307–1318. doi:10.1161/CIRCRESAHA.108.173989
6. Kind J, van Steensel B (2010) Genome-nuclear lamina interactions and gene regulation. Curr Opin Cell Biol 22(3):320–325. doi:10.1016/j.ceb.2010.04.002
7. Mattout A, Pike BL, Towbin BD, Bank EM, Gonzalez-Sandoval A, Stadler MB, Meister P, Gruenbaum Y, Gasser SM (2011) An EDMD mutation in C. elegans lamin blocks muscle-specific gene relocation and compromises muscle integrity. Curr Biol 21(19):1603–1614. doi:10.1016/j.cub.2011.08.030
8. Solovei I, Wang AS, Thanisch K, Schmidt CS, Krebs S, Zwerger M, Cohen TV, Devys D, Foisner R, Peichl L, Herrmann H, Blum H, Engelkamp D, Stewart CL, Leonhardt H, Joffe B (2013) LBR and lamin A/C sequentially tether peripheral heterochromatin and inversely regulate differentiation. Cell 152(3):584–598. doi:10.1016/j.cell.2013.01.009

9. Andres V, Gonzalez JM (2009) Role of A-type lamins in signaling, transcription, and chromatin organization. J Cell Biol 187(7):945–957. doi:10.1083/jcb.200904124
10. Heessen S, Fornerod M (2007) The inner nuclear envelope as a transcription factor resting place. EMBO Rep 8(10):914–919. doi:10.1038/sj.embor.7401075
11. Worman HJ (2012) Nuclear lamins and laminopathies. J Pathol 226(2):316–325. doi:10.1002/path.2999
12. Korfali N, Wilkie GS, Swanson SK, Srsen V, de las Heras J, Batrakou DG, Malik P, Zuleger N, Kerr AR, Florens L, Schirmer EC (2012) The nuclear envelope proteome differs notably between tissues. Nucleus 3(6):552–564. doi:10.4161/nucl.22257
13. Schirmer EC, Florens L, Guan T, Yates JR 3rd, Gerace L (2003) Nuclear membrane proteins with potential disease links found by subtractive proteomics. Science 301(5638):1380–1382. doi:10.1126/science.1088176
14. Brachner A, Foisner R (2011) Evolvement of LEM proteins as chromatin tethers at the nuclear periphery. Biochem Soc Trans 39(6):1735–1741. doi:10.1042/BST20110724
15. Cai M, Huang Y, Ghirlando R, Wilson KL, Craigie R, Clore GM (2001) Solution structure of the constant region of nuclear envelope protein LAP2 reveals two LEM-domain structures: one binds BAF and the other binds DNA. EMBO J 20(16):4399–4407. doi:10.1093/emboj/20.16.4399
16. Shumaker DK, Lee KK, Tanhehco YC, Craigie R, Wilson KL (2001) LAP2 binds to BAF. DNA complexes: requirement for the LEM domain and modulation by variable regions. EMBO J 20(7):1754–1764. doi:10.1093/emboj/20.7.1754
17. Lin F, Blake DL, Callebaut I, Skerjanc IS, Holmer L, McBurney MW, Paulin-Levasseur M, Worman HJ (2000) MAN1, an inner nuclear membrane protein that shares the LEM domain with lamina-associated polypeptide 2 and emerin. J Biol Chem 275(7):4840–4847
18. Cai M, Huang Y, Suh JY, Louis JM, Ghirlando R, Craigie R, Clore GM (2007) Solution NMR structure of the barrier-to-autointegration factor-Emerin complex. J Biol Chem 282(19):14525–14535. doi:10.1074/jbc.M700576200
19. Wilson KL, Foisner R (2010) Lamin-binding proteins. Cold Spring Harb Perspect Biol 2(4):a000554. doi:10.1101/cshperspect.a000554
20. Wilson KL, Berk JM (2010) The nuclear envelope at a glance. J Cell Sci 123(Pt 12): 1973–1978. doi:10.1242/jcs.019042
21. Pan D, Estevez-Salmeron LD, Stroschein SL, Zhu X, He J, Zhou S, Luo K (2005) The integral inner nuclear membrane protein MAN1 physically interacts with the R-Smad proteins to repress signaling by the transforming growth factor-{beta} superfamily of cytokines. J Biol Chem 280(16):15992–16001. doi:10.1074/jbc.M411234200
22. Lin F, Morrison JM, Wu W, Worman HJ (2005) MAN1, an integral protein of the inner nuclear membrane, binds Smad2 and Smad3 and antagonizes transforming growth factor-beta signaling. Hum Mol Genet 14(3):437–445. doi:10.1093/hmg/ddi040
23. Markiewicz E, Tilgner K, Barker N, van de Wetering M, Clevers H, Dorobek M, Hausmanowa-Petrusewicz I, Ramaekers FC, Broers JL, Blankesteijn WM, Salpingidou G, Wilson RG, Ellis JA, Hutchison CJ (2006) The inner nuclear membrane protein emerin regulates beta-catenin activity by restricting its accumulation in the nucleus. EMBO J 25(14):3275–3285. doi:10.1038/sj.emboj.7601230
24. Dedeic Z, Cetera M, Cohen TV, Holaska JM (2011) Emerin inhibits Lmo7 binding to the Pax3 and MyoD promoters and expression of myoblast proliferation genes. J Cell Sci 124(Pt 10):1691–1702. doi:10.1242/jcs.080259
25. Holaska JM, Lee KK, Kowalski AK, Wilson KL (2003) Transcriptional repressor germ cell-less (GCL) and barrier to autointegration factor (BAF) compete for binding to emerin in vitro. J Biol Chem 278(9):6969–6975. doi:10.1074/jbc.M208811200M208811200
26. Nili E, Cojocaru GS, Kalma Y, Ginsberg D, Copeland NG, Gilbert DJ, Jenkins NA, Berger R, Shaklai S, Amariglio N, Brok-Simoni F, Simon AJ, Rechavi G (2001) Nuclear membrane protein LAP2beta mediates transcriptional repression alone and together with its binding partner GCL (germ-cell-less). J Cell Sci 114(Pt 18):3297–3307

27. Demmerle J, Koch AJ, Holaska JM (2012) The nuclear envelope protein emerin binds directly to histone deacetylase 3 (HDAC3) and activates HDAC3 activity. J Biol Chem 287(26):22080–22088. doi:10.1074/jbc.M111.325308
28. Wagner N, Krohne G (2007) LEM-domain proteins: new insights into lamin-interacting proteins. Int Rev Cytol 261:1–46. doi:10.1016/S0074-7696(07)61001-8
29. Lee KK, Wilson KL (2004) All in the family: evidence for four new LEM-domain proteins Lem2 (NET-25), Lem3, Lem4 and Lem5 in the human genome. Symp Soc Exp Biol 56:329–339
30. Harris CA, Andryuk PJ, Cline SW, Mathew S, Siekierka JJ, Goldstein G (1995) Structure and mapping of the human thymopoietin (TMPO) gene and relationship of human TMPO beta to rat lamin-associated polypeptide 2. Genomics 28(2):198–205. doi:10.1006/geno.1995.1131
31. Bradley CM, Jones S, Huang Y, Suzuki Y, Kvaratskhelia M, Hickman AB, Craigie R, Dyda F (2007) Structural basis for dimerization of LAP2alpha, a component of the nuclear lamina. Structure 15(6):643–653. doi:10.1016/j.str.2007.04.007
32. Snyers L, Vlcek S, Dechat T, Skegro D, Korbei B, Gajewski A, Mayans O, Schofer C, Foisner R (2007) Lamina-associated polypeptide 2-alpha forms homo-trimers via its C terminus, and oligomerization is unaffected by a disease-causing mutation. J Biol Chem 282(9):6308–6315. doi:10.1074/jbc.M605782200
33. Foisner R, Gerace L (1993) Integral membrane proteins of the nuclear envelope interact with lamins and chromosomes, and binding is modulated by mitotic phosphorylation. Cell 73(7):1267–1279. doi:10.1016/0092-8674(93)90355-T
34. Dechat T, Korbei B, Vaughan OA, Vlcek S, Hutchison CJ, Foisner R (2000) Lamina-associated polypeptide 2alpha binds intranuclear A-type lamins. J Cell Sci 113(Pt 19):3473–3484
35. Naetar N, Korbei B, Kozlov S, Kerenyi MA, Dorner D, Kral R, Gotic I, Fuchs P, Cohen TV, Bittner R, Stewart CL, Foisner R (2008) Loss of nucleoplasmic LAP2alpha-lamin A complexes causes erythroid and epidermal progenitor hyperproliferation. Nat Cell Biol 10(11):1341–1348. doi:10.1038/ncb1793
36. Dorner D, Vlcek S, Foeger N, Gajewski A, Makolm C, Gotzmann J, Hutchison CJ, Foisner R (2006) Lamina-associated polypeptide 2alpha regulates cell cycle progression and differentiation via the retinoblastoma-E2F pathway. J Cell Biol 173(1):83–93. doi:10.1083/jcb.200511149
37. Markiewicz E, Dechat T, Foisner R, Quinlan RA, Hutchison CJ (2002) Lamin A/C binding protein LAP2alpha is required for nuclear anchorage of retinoblastoma protein. Mol Biol Cell 13(12):4401–4413. doi:10.1091/mbc.E02-07-0450
38. Grana X, Garriga J, Mayol X (1998) Role of the retinoblastoma protein family, pRB, p107 and p130 in the negative control of cell growth. Oncogene 17(25):3365–3383. doi:10.1038/sj.onc.1202575
39. Poznic M (2009) Retinoblastoma protein: a central processing unit. J Biosci 34(2):305–312
40. Zhu L (2005) Tumour suppressor retinoblastoma protein Rb: a transcriptional regulator. Eur J Cancer 41(16):2415–2427. doi:10.1016/j.ejca.2005.08.009
41. Bartek J, Bartkova J, Lukas J (1996) The retinoblastoma protein pathway and the restriction point. Curr Opin Cell Biol 8(6):805–814
42. Weinberg RA (1995) The retinoblastoma protein and cell cycle control. Cell 81(3):323–330
43. Malumbres M, Barbacid M (2001) To cycle or not to cycle: a critical decision in cancer. Nat Rev Cancer 1(3):222–231. doi:10.1038/35106065
44. Bracken AP, Ciro M, Cocito A, Helin K (2004) E2F target genes: unraveling the biology. Trends Biochem Sci 29(8):409–417. doi:10.1016/j.tibs.2004.06.006
45. Wong JV, Dong P, Nevins JR, Mathey-Prevot B, You L (2011) Network calisthenics: control of E2F dynamics in cell cycle entry. Cell Cycle 10(18):3086–3094
46. Polager S, Ginsberg D (2008) E2F—at the crossroads of life and death. Trends Cell Biol 18(11):528–535. doi:10.1016/j.tcb.2008.08.003
47. Carcagno AL, Marazita MC, Ogara MF, Ceruti JM, Sonzogni SV, Scassa ME, Giono LE, Canepa ET (2011) E2F1-mediated upregulation of p19INK4d determines its periodic expression during cell cycle and regulates cellular proliferation. PLoS One 6(7):e21938. doi:10.1371/journal.pone.0021938

48. Lammens T, Li J, Leone G, De Veylder L (2009) Atypical E2Fs: new players in the E2F transcription factor family. Trends Cell Biol 19(3):111–118. doi:10.1016/j.tcb.2009.01.002
49. Chen HZ, Tsai SY, Leone G (2009) Emerging roles of E2Fs in cancer: an exit from cell cycle control. Nat Rev Cancer 9(11):785–797. doi:10.1038/nrc2696
50. Ozaki T, Saijo M, Murakami K, Enomoto H, Taya Y, Sakiyama S (1994) Complex formation between lamin A and the retinoblastoma gene product: identification of the domain on lamin A required for its interaction. Oncogene 9(9):2649–2653
51. Johnson BR, Nitta RT, Frock RL, Mounkes L, Barbie DA, Stewart CL, Harlow E, Kennedy BK (2004) A-type lamins regulate retinoblastoma protein function by promoting subnuclear localization and preventing proteasomal degradation. Proc Natl Acad Sci U S A 101(26): 9677–9682. doi:10.1073/pnas.0403250101
52. Nitta RT, Jameson SA, Kudlow BA, Conlan LA, Kennedy BK (2006) Stabilization of the retinoblastoma protein by A-type nuclear lamins is required for INK4A-mediated cell cycle arrest. Mol Cell Biol 26(14):5360–5372. doi:10.1128/MCB.02464-05
53. Van Berlo JH, Voncken JW, Kubben N, Broers JL, Duisters R, van Leeuwen RE, Crijns HJ, Ramaekers FC, Hutchison CJ, Pinto YM (2005) A-type lamins are essential for TGF-beta1 induced PP2A to dephosphorylate transcription factors. Hum Mol Genet 14(19):2839–2849. doi:10.1093/hmg/ddi316
54. Rodriguez J, Calvo F, Gonzalez JM, Casar B, Andres V, Crespo P (2010) ERK1/2 MAP kinases promote cell cycle entry by rapid, kinase-independent disruption of retinoblastoma-lamin A complexes. J Cell Biol 191(5):967–979. doi:10.1083/jcb.201004067
55. Lee DC, Welton KL, Smith ED, Kennedy BK (2009) A-type nuclear lamins act as transcriptional repressors when targeted to promoters. Exp Cell Res 315(6):996–1007. doi:10.1016/j.yexcr.2009.01.003
56. Dechat T, Gajewski A, Korbei B, Gerlich D, Daigle N, Haraguchi T, Furukawa K, Ellenberg J, Foisner R (2004) LAP2alpha and BAF transiently localize to telomeres and specific regions on chromatin during nuclear assembly. J Cell Sci 117(Pt 25):6117–6128. doi:10.1242/jcs.01529
57. Moir RD, Yoon M, Khuon S, Goldman RD (2000) Nuclear lamins A and B1: different pathways of assembly during nuclear envelope formation in living cells. J Cell Biol 151(6):1155–1168
58. Pekovic V, Harborth J, Broers JL, Ramaekers FC, van Engelen B, Lammens M, von Zglinicki T, Foisner R, Hutchison C, Markiewicz E (2007) Nucleoplasmic LAP2alpha-lamin A complexes are required to maintain a proliferative state in human fibroblasts. J Cell Biol 176(2):163–172. doi:10.1083/jcb.200606139
59. Gotic I, Schmidt WM, Biadasiewicz K, Leschnik M, Spilka R, Braun J, Stewart CL, Foisner R (2010) Loss of LAP2 alpha delays satellite cell differentiation and affects postnatal fiber-type determination. Stem Cells 28(3):480–488. doi:10.1002/stem.292
60. Markiewicz E, Ledran M, Hutchison CJ (2005) Remodelling of the nuclear lamina and nucleoskeleton is required for skeletal muscle differentiation in vitro. J Cell Sci 118(Pt 2): 409–420. doi:10.1242/jcs.01630
61. Whitfield ML, Sherlock G, Saldanha AJ, Murray JI, Ball CA, Alexander KE, Matese JC, Perou CM, Hurt MM, Brown PO, Botstein D (2002) Identification of genes periodically expressed in the human cell cycle and their expression in tumors. Mol Biol Cell 13(6): 1977–2000. doi:10.1091/mbc.02-02-0030
62. Dong HM, Xu CS (2008) Analysis of the relevance of e2fs and their target genes with rat liver regeneration. Indian J Gastroenterol 27(1):31–32
63. Melcon G, Kozlov S, Cutler DA, Sullivan T, Hernandez L, Zhao P, Mitchell S, Nader G, Bakay M, Rottman JN, Hoffman EP, Stewart CL (2006) Loss of emerin at the nuclear envelope disrupts the Rb1/E2F and MyoD pathways during muscle regeneration. Hum Mol Genet 15(4):637–651. doi:10.1093/hmg/ddi479
64. Bakay M, Wang Z, Melcon G, Schiltz L, Xuan J, Zhao P, Sartorelli V, Seo J, Pegoraro E, Angelini C, Shneiderman B, Escolar D, Chen YW, Winokur ST, Pachman LM, Fan C, Mandler R, Nevo Y, Gordon E, Zhu Y, Dong Y, Wang Y, Hoffman EP (2006) Nuclear envelope dystrophies show a transcriptional fingerprint suggesting disruption of Rb-MyoD pathways in muscle regeneration. Brain 129(Pt 4):996–1013. doi:10.1093/brain/awl023

65. Li Z, Van Calcar S, Qu C, Cavenee WK, Zhang MQ, Ren B (2003) A global transcriptional regulatory role for c-Myc in Burkitt's lymphoma cells. Proc Natl Acad Sci U S A 100(14):8164–8169. doi:10.1073/pnas.1332764100
66. Ren B, Cam H, Takahashi Y, Volkert T, Terragni J, Young RA, Dynlacht BD (2002) E2F integrates cell cycle progression with DNA repair, replication, and G(2)/M checkpoints. Genes Dev 16(2):245–256. doi:10.1101/gad.949802
67. Asp P, Acosta-Alvear D, Tsikitis M, van Oevelen C, Dynlacht BD (2009) E2f3b plays an essential role in myogenic differentiation through isoform-specific gene regulation. Genes Dev 23(1):37–53. doi:10.1101/gad.1727309
68. Westendorp B, Mokry M, Groot Koerkamp MJ, Holstege FC, Cuppen E, de Bruin A (2012) E2F7 represses a network of oscillating cell cycle genes to control S-phase progression. Nucleic Acids Res 40(8):3511–3523. doi:10.1093/nar/gkr1203
69. Zhou W, Yang Y, Xia J, Wang H, Salama ME, Xiong W, Xu H, Shetty S, Chen T, Zeng Z, Shi L, Zangari M, Miles R, Bearss D, Tricot G, Zhan F (2013) NEK2 induces drug resistance mainly through activation of efflux drug pumps and is associated with poor prognosis in myeloma and other cancers. Cancer Cell 23(1):48–62. doi:10.1016/j.ccr.2012.12.001
70. Shi T, Mazumdar T, Devecchio J, Duan ZH, Agyeman A, Aziz M, Houghton JA (2010) cDNA microarray gene expression profiling of hedgehog signaling pathway inhibition in human colon cancer cells. PLoS One 5(10). doi:10.1371/journal.pone.0013054
71. Viatour P, Ehmer U, Saddic LA, Dorrell C, Andersen JB, Lin C, Zmoos AF, Mazur PK, Schaffer BE, Ostermeier A, Vogel H, Sylvester KG, Thorgeirsson SS, Grompe M, Sage J (2011) Notch signaling inhibits hepatocellular carcinoma following inactivation of the RB pathway. J Exp Med 208(10):1963–1976. doi:10.1084/jem.20110198
72. Moudgil VK, Dinda S, Khattree N, Jhanwar S, Alban P, Hurd C (2001) Hormonal regulation of tumor suppressor proteins in breast cancer cells. J Steroid Biochem Mol Biol 76(1–5):105–117
73. Buterin T, Koch C, Naegeli H (2006) Convergent transcriptional profiles induced by endogenous estrogen and distinct xenoestrogens in breast cancer cells. Carcinogenesis 27(8): 1567–1578. doi:10.1093/carcin/bgi339
74. Muchir A, van Engelen BG, Lammens M, Mislow JM, McNally E, Schwartz K, Bonne G (2003) Nuclear envelope alterations in fibroblasts from LGMD1B patients carrying nonsense Y259X heterozygous or homozygous mutation in lamin A/C gene. Exp Cell Res 291(2):352–362
75. Goldman RD, Shumaker DK, Erdos MR, Eriksson M, Goldman AE, Gordon LB, Gruenbaum Y, Khuon S, Mendez M, Varga R, Collins FS (2004) Accumulation of mutant lamin A causes progressive changes in nuclear architecture in Hutchinson-Gilford progeria syndrome. Proc Natl Acad Sci U S A 101(24):8963–8968. doi:10.1073/pnas.0402943101
76. Hernandez L, Roux KJ, Wong ES, Mounkes LC, Mutalif R, Navasankari R, Rai B, Cool S, Jeong JW, Wang H, Lee HS, Kozlov S, Grunert M, Keeble T, Jones CM, Meta MD, Young SG, Daar IO, Burke B, Perantoni AO, Stewart CL (2010) Functional coupling between the extracellular matrix and nuclear lamina by Wnt signaling in progeria. Dev Cell 19(3): 413–425. doi:10.1016/j.devcel.2010.08.013
77. Song YJ, Stinski MF (2002) Effect of the human cytomegalovirus IE86 protein on expression of E2F-responsive genes: a DNA microarray analysis. Proc Natl Acad Sci U S A 99(5): 2836–2841. doi:10.1073/pnas.052010099
78. Ward MC, van der Watt PJ, Tzoneva G, Leaner VD (2011) Deregulated LAP2alpha expression in cervical cancer associates with aberrant E2F and p53 activities. IUBMB Life 63(11):1018–1026. doi:10.1002/iub.528
79. Munger K, Scheffner M, Huibregtse JM, Howley PM (1992) Interactions of HPV E6 and E7 oncoproteins with tumour suppressor gene products. Cancer Surv 12:197–217
80. Vernell R, Helin K, Muller H (2003) Identification of target genes of the p16INK4A-pRB-E2F pathway. J Biol Chem 278(46):46124–46137. doi:10.1074/jbc.M304930200
81. Gentles AJ, Alizadeh AA, Lee SI, Myklebust JH, Shachaf CM, Shahbaba B, Levy R, Koller D, Plevritis SK (2009) A pluripotency signature predicts histologic transformation and influences survival in follicular lymphoma patients. Blood 114(15):3158–3166. doi:10.1182/blood-2009-02-202465

82. Agnelli L, Forcato M, Ferrari F, Tuana G, Todoerti K, Walker BA, Morgan GJ, Lombardi L, Bicciato S, Neri A (2011) The reconstruction of transcriptional networks reveals critical genes with implications for clinical outcome of multiple myeloma. Clin Cancer Res 17(23):7402–7412. doi:10.1158/1078-0432.CCR-11-0596
83. Crabbe L, Cesare AJ, Kasuboski JM, Fitzpatrick JA, Karlseder J (2012) Human telomeres are tethered to the nuclear envelope during postmitotic nuclear assembly. Cell Rep 2(6): 1521–1529. doi:10.1016/j.celrep.2012.11.019
84. Verdun RE, Crabbe L, Haggblom C, Karlseder J (2005) Functional human telomeres are recognized as DNA damage in G2 of the cell cycle. Mol Cell 20(4):551–561. doi:10.1016/j.molcel.2005.09.024
85. Redwood AB, Perkins SM, Vanderwaal RP, Feng Z, Biehl KJ, Gonzalez-Suarez I, Morgado-Palacin L, Shi W, Sage J, Roti-Roti JL, Stewart CL, Zhang J, Gonzalo S (2011) A dual role for A-type lamins in DNA double-strand break repair. Cell Cycle 10(15):2549–2560
86. Gonzalez-Suarez I, Redwood AB, Perkins SM, Vermolen B, Lichtensztejin D, Grotsky DA, Morgado-Palacin L, Gapud EJ, Sleckman BP, Sullivan T, Sage J, Stewart CL, Mai S, Gonzalo S (2009) Novel roles for A-type lamins in telomere biology and the DNA damage response pathway. EMBO J 28(16):2414–2427. doi:10.1038/emboj.2009.196
87. Huang S, Risques RA, Martin GM, Rabinovitch PS, Oshima J (2008) Accelerated telomere shortening and replicative senescence in human fibroblasts overexpressing mutant and wild-type lamin A. Exp Cell Res 314(1):82–91. doi:10.1016/j.yexcr.2007.08.004
88. Lachapelle S, Gagne JP, Garand C, Desbiens M, Coulombe Y, Bohr VA, Hendzel MJ, Masson JY, Poirier GG, Lebel M (2011) Proteome-wide identification of WRN-interacting proteins in untreated and nuclease-treated samples. J Proteome Res 10(3):1216–1227. doi:10.1021/pr100990s
89. Kudlow BA, Kennedy BK, Monnat RJ Jr (2007) Werner and Hutchinson-Gilford progeria syndromes: mechanistic basis of human progeroid diseases. Nat Rev Mol Cell Biol 8(5): 394–404. doi:10.1038/nrm2161
90. Sand-Dejmek J, Adelmant G, Sobhian B, Calkins AS, Marto J, Iglehart DJ, Lazaro JB (2011) Concordant and opposite roles of DNA-PK and the "facilitator of chromatin transcription" (FACT) in DNA repair, apoptosis and necrosis after cisplatin. Mol Cancer 10:74. doi:10.1186/1476-4598-10-74
91. Jaco I, Munoz P, Blasco MA (2004) Role of human Ku86 in telomere length maintenance and telomere capping. Cancer Res 64(20):7271–7278. doi:10.1158/0008-5472.CAN-04-1381
92. Wang Y, Ghosh G, Hendrickson EA (2009) Ku86 represses lethal telomere deletion events in human somatic cells. Proc Natl Acad Sci U S A 106(30):12430–12435. doi:10.1073/pnas.0903362106
93. Myung K, Ghosh G, Fattah FJ, Li G, Kim H, Dutia A, Pak E, Smith S, Hendrickson EA (2004) Regulation of telomere length and suppression of genomic instability in human somatic cells by Ku86. Mol Cell Biol 24(11):5050–5059. doi:10.1128/MCB.24.11.5050-5059.2004
94. Gagne JP, Pic E, Isabelle M, Krietsch J, Ethier C, Paquet E, Kelly I, Boutin M, Moon KM, Foster LJ, Poirier GG (2012) Quantitative proteomics profiling of the poly(ADP-ribose)-related response to genotoxic stress. Nucleic Acids Res 40(16):7788–7805. doi:10.1093/nar/gks486
95. Kim HJ, Hwang SH, Han ME, Baek S, Sim HE, Yoon S, Baek SY, Kim BS, Kim JH, Kim SY, Oh SO (2012) LAP2 is widely overexpressed in diverse digestive tract cancers and regulates motility of cancer cells. PLoS One 7(6):e39482. doi:10.1371/journal.pone.0039482
96. Somech R, Shaklai S, Geller O, Amariglio N, Simon AJ, Rechavi G, Gal-Yam EN (2005) The nuclear-envelope protein and transcriptional repressor LAP2beta interacts with HDAC3 at the nuclear periphery, and induces histone H4 deacetylation. J Cell Sci 118(Pt 17):4017–4025. doi:10.1242/jcs.02521
97. Somech R, Gal-Yam EN, Shaklai S, Geller O, Amariglio N, Rechavi G, Simon AJ (2007) Enhanced expression of the nuclear envelope LAP2 transcriptional repressors in normal and malignant activated lymphocytes. Ann Hematol 86(6):393–401. doi:10.1007/s00277-007-0275-9

98. Yuki D, Lin YM, Fujii Y, Nakamura Y, Furukawa Y (2004) Isolation of LEM domain-containing 1, a novel testis-specific gene expressed in colorectal cancers. Oncol Rep 12(2):275–280
99. Ghafouri-Fard S, Ousati Ashtiani Z, Sabah Golian B, Hasheminasab SM, Modarressi MH (2010) Expression of two testis-specific genes, SPATA19 and LEMD1, in prostate cancer. Arch Med Res 41(3):195–200. doi:10.1016/j.arcmed.2010.04.003
100. Matsuyama H, Suzuki HI, Nishimori H, Noguchi M, Yao T, Komatsu N, Mano H, Sugimoto K, Miyazono K (2011) miR-135b mediates NPM-ALK-driven oncogenicity and renders IL-17-producing immunophenotype to anaplastic large cell lymphoma. Blood 118(26): 6881–6892. doi:10.1182/blood-2011-05-354654
101. Simpson AJ, Caballero OL, Jungbluth A, Chen YT, Old LJ (2005) Cancer/testis antigens, gametogenesis and cancer. Nat Rev Cancer 5(8):615–625. doi:10.1038/nrc1669
102. Berliner JL, Fay AM (2007) Risk assessment and genetic counseling for hereditary breast and ovarian cancer: recommendations of the National Society of Genetic Counselors. J Genet Couns 16(3):241–260. doi:10.1007/s10897-007-9090-7
103. Ford D, Easton DF, Stratton M, Narod S, Goldgar D, Devilee P, Bishop DT, Weber B, Lenoir G, Chang-Claude J, Sobol H, Teare MD, Struewing J, Arason A, Scherneck S, Peto J, Rebbeck TR, Tonin P, Neuhausen S, Barkardottir R, Eyfjord J, Lynch H, Ponder BA, Gayther SA, Zelada-Hedman M et al (1998) Genetic heterogeneity and penetrance analysis of the BRCA1 and BRCA2 genes in breast cancer families. The Breast Cancer Linkage Consortium. Am J Hum Genet 62(3):676–689
104. Stevens KN, Vachon CM, Lee AM, Slager S, Lesnick T, Olswold C, Fasching PA, Miron P, Eccles D, Carpenter JE, Godwin AK, Ambrosone C, Winqvist R, Brauch H, Schmidt MK, Cox A, Cross SS, Sawyer E, Hartmann A, Beckmann MW, Schulz-Wendtland R, Ekici AB, Tapper WJ, Gerty SM, Durcan L, Graham N, Hein R, Nickels S, Flesch-Janys D, Heinz J, Sinn HP, Konstantopoulou I, Fostira F, Pectasides D, Dimopoulos AM, Fountzilas G, Clarke CL, Balleine R, Olson JE, Fredericksen Z, Diasio RB, Pathak H, Ross E, Weaver J, Rudiger T, Forsti A, Dunnebier T, Ademuyiwa F, Kulkarni S, Pylkas K, Jukkola-Vuorinen A, Ko YD, Van Limbergen E, Janssen H, Peto J, Fletcher O, Giles GG, Baglietto L, Verhoef S, Tomlinson I, Kosma VM, Beesley J, Greco D, Blomqvist C, Irwanto A, Liu J, Blows FM, Dawson SJ, Margolin S, Mannermaa A, Martin NG, Montgomery GW, Lambrechts D, dos Santos Silva I, Severi G, Hamann U, Pharoah P, Easton DF, Chang-Claude J, Yannoukakos D, Nevanlinna H, Wang X, Couch FJ (2011) Common breast cancer susceptibility loci are associated with triple-negative breast cancer. Cancer Res 71(19):6240–6249. doi:10.1158/0008-5472.CAN-11-1266
105. Chen F, Chen GK, Millikan RC, John EM, Ambrosone CB, Bernstein L, Zheng W, Hu JJ, Ziegler RG, Deming SL, Bandera EV, Nyante S, Palmer JR, Rebbeck TR, Ingles SA, Press MF, Rodriguez-Gil JL, Chanock SJ, Le Marchand L, Kolonel LN, Henderson BE, Stram DO, Haiman CA (2011) Fine-mapping of breast cancer susceptibility loci characterizes genetic risk in African Americans. Hum Mol Genet 20(22):4491–4503. doi:10.1093/hmg/ddr367
106. Zhang B, Beeghly-Fadiel A, Long J, Zheng W (2011) Genetic variants associated with breast-cancer risk: comprehensive research synopsis, meta-analysis, and epidemiological evidence. Lancet Oncol 12(5):477–488. doi:10.1016/S1470-2045(11)70076-6
107. Antoniou AC, Wang X, Fredericksen ZS, McGuffog L, Tarrell R, Sinilnikova OM, Healey S, Morrison J, Kartsonaki C, Lesnick T, Ghoussaini M, Barrowdale D, Peock S, Cook M, Oliver C, Frost D, Eccles D, Evans DG, Eeles R, Izatt L, Chu C, Douglas F, Paterson J, Stoppa-Lyonnet D, Houdayer C, Mazoyer S, Giraud S, Lasset C, Remenieras A, Caron O, Hardouin A, Berthet P, Hogervorst FB, Rookus MA, Jager A, van den Ouweland A, Hoogerbrugge N, van der Luijt RB, Meijers-Heijboer H, Gomez Garcia EB, Devilee P, Vreeswijk MP, Lubinski J, Jakubowska A, Gronwald J, Huzarski T, Byrski T, Gorski B, Cybulski C, Spurdle AB, Holland H, Goldgar DE, John EM, Hopper JL, Southey M, Buys SS, Daly MB, Terry MB, Schmutzler RK, Wappenschmidt B, Engel C, Meindl A, Preisler-Adams S, Arnold N, Niederacher D, Sutter C, Domchek SM, Nathanson KL, Rebbeck T, Blum JL, Piedmonte M, Rodriguez GC, Wakeley K, Boggess JF, Basil J, Blank SV, Friedman E, Kaufman B, Laitman Y, Milgrom R, Andrulis IL, Glendon G, Ozcelik H, Kirchhoff T, Vijai J, Gaudet MM,

Altshuler D, Guiducci C, Loman N, Harbst K, Rantala J, Ehrencrona H, Gerdes AM, Thomassen M, Sunde L, Peterlongo P, Manoukian S, Bonanni B, Viel A, Radice P, Caldes T, de la Hoya M, Singer CF, Fink-Retter A, Greene MH, Mai PL, Loud JT, Guidugli L, Lindor NM, Hansen TV, Nielsen FC, Blanco I, Lazaro C, Garber J, Ramus SJ, Gayther SA, Phelan C, Narod S, Szabo CI, Benitez J, Osorio A, Nevanlinna H, Heikkinen T, Caligo MA, Beattie MS, Hamann U, Godwin AK, Montagna M, Casella C, Neuhausen SL, Karlan BY, Tung N, Toland AE, Weitzel J, Olopade O, Simard J, Soucy P, Rubinstein WS, Arason A, Rennert G, Martin NG, Montgomery GW, Chang-Claude J, Flesch-Janys D, Brauch H, Severi G, Baglietto L, Cox A, Cross SS, Miron P, Gerty SM, Tapper W, Yannoukakos D, Fountzilas G, Fasching PA, Beckmann MW, Dos Santos Silva I, Peto J, Lambrechts D, Paridaens R, Rudiger T, Forsti A, Winqvist R, Pylkas K, Diasio RB, Lee AM, Eckel-Passow J, Vachon C, Blows F, Driver K, Dunning A, Pharoah PP, Offit K, Pankratz VS, Hakonarson H, Chenevix-Trench G, Easton DF, Couch FJ (2010) A locus on 19p13 modifies risk of breast cancer in BRCA1 mutation carriers and is associated with hormone receptor-negative breast cancer in the general population. Nat Genet 42(10):885–892. doi:10.1038/ng.669

108. Bolton KL, Tyrer J, Song H, Ramus SJ, Notaridou M, Jones C, Sher T, Gentry-Maharaj A, Wozniak E, Tsai YY, Weidhaas J, Paik D, Van Den Berg DJ, Stram DO, Pearce CL, Wu AH, Brewster W, Anton-Culver H, Ziogas A, Narod SA, Levine DA, Kaye SB, Brown R, Paul J, Flanagan J, Sieh W, McGuire V, Whittemore AS, Campbell I, Gore ME, Lissowska J, Yang HP, Medrek K, Gronwald J, Lubinski J, Jakubowska A, Le ND, Cook LS, Kelemen LE, Brook-Wilson A, Massuger LF, Kiemeney LA, Aben KK, van Altena AM, Houlston R, Tomlinson I, Palmieri RT, Moorman PG, Schildkraut J, Iversen ES, Phelan C, Vierkant RA, Cunningham JM, Goode EL, Fridley BL, Kruger-Kjaer S, Blaeker J, Hogdall E, Hogdall C, Gross J, Karlan BY, Ness RB, Edwards RP, Odunsi K, Moyisch KB, Baker JA, Modugno F, Heikkinenen T, Butzow R, Nevanlinna H, Leminen A, Bogdanova N, Antonenkova N, Doerk T, Hillemanns P, Durst M, Runnebaum I, Thompson PJ, Carney ME, Goodman MT, Lurie G, Wang-Gohrke S, Hein R, Chang-Claude J, Rossing MA, Cushing-Haugen KL, Doherty J, Chen C, Rafnar T, Besenbacher S, Sulem P, Stefansson K, Birrer MJ, Terry KL, Hernandez D, Cramer DW, Vergote I, Amant F, Lambrechts D, Despierre E, Fasching PA, Beckmann MW, Thiel FC, Ekici AB, Chen X, Johnatty SE, Webb PM, Beesley J, Chanock S, Garcia-Closas M, Sellers T, Easton DF, Berchuck A, Chenevix-Trench G, Pharoah PD, Gayther SA (2010) Common variants at 19p13 are associated with susceptibility to ovarian cancer. Nat Genet 42(10):880–884. doi:10.1038/ng.666

109. Brachner A, Braun J, Ghodgaonkar M, Castor D, Zlopasa L, Ehrlich V, Jiricny J, Gotzmann J, Knasmuller S, Foisner R (2012) The endonuclease Ankle1 requires its LEM and GIY-YIG motifs for DNA cleavage in vivo. J Cell Sci 125(Pt 4):1048–1057. doi:10.1242/jcs.098392

110. Dittrich CM, Kratz K, Sendoel A, Gruenbaum Y, Jiricny J, Hengartner MO (2012) LEM-3—a LEM domain containing nuclease involved in the DNA damage response in C. elegans. PLoS One 7(2):e24555

111. Rosty C, Sheffer M, Tsafrir D, Stransky N, Tsafrir I, Peter M, de Cremoux P, de la Rochefordiere A, Salmon R, Dorval T, Thiery JP, Couturier J, Radvanyi F, Domany E, Sastre-Garau X (2005) Identification of a proliferation gene cluster associated with HPV E6/E7 expression level and viral DNA load in invasive cervical carcinoma. Oncogene 24(47): 7094–7104. doi:10.1038/sj.onc.1208854

112. Futschik M (2002) Gene expression profiling of metastatic and nonmetastatic colorectal cancer cell lines. Genome Lett 1:26–34

113. Ohuchida K, Mizumoto K, Ishikawa N, Fujii K, Konomi H, Nagai E, Yamaguchi K, Tsuneyoshi M, Tanaka M (2005) The role of S100A6 in pancreatic cancer development and its clinical implication as a diagnostic marker and therapeutic target. Clin Cancer Res 11(21):7785–7793. doi:10.1158/1078-0432.CCR-05-0714

114. Kang W, Tong JH, Chan AW, Lee TL, Lung RW, Leung PP, So KK, Wu K, Fan D, Yu J, Sung JJ, To KF (2011) Yes-associated protein 1 exhibits oncogenic property in gastric cancer and its nuclear accumulation associates with poor prognosis. Clin Cancer Res 17(8):2130–2139. doi:10.1158/1078-0432.CCR-10-2467

115. Yokota N, Mainprize TG, Taylor MD, Kohata T, Loreto M, Ueda S, Dura W, Grajkowska W, Kuo JS, Rutka JT (2004) Identification of differentially expressed and developmentally regulated genes in medulloblastoma using suppression subtraction hybridization. Oncogene 23(19):3444–3453. doi:10.1038/sj.onc.1207475
116. Parise P, Finocchiaro G, Masciadri B, Quarto M, Francois S, Mancuso F, Muller H (2006) Lap2alpha expression is controlled by E2F and deregulated in various human tumors. Cell Cycle 5(12):1331–1341

NETs and Cell Cycle Regulation

Michael I. Robson, Phu Le Thanh, and Eric C. Schirmer

Abstract There are many ways that the nuclear envelope can influence the cell cycle. In addition to roles of lamins in regulating the master cell cycle regulator pRb and nuclear envelope breakdown in mitosis, many other nuclear envelope proteins influence the cell cycle through regulatory or structural functions. Of particular note among these are the nuclear envelope transmembrane proteins (NETs) that appear to influence cell cycle regulation through multiple separate mechanisms. Some NETs and other nuclear envelope proteins accumulate on the mitotic spindle, suggesting functional or structural roles in the cell cycle. In interphase exogenous overexpression of some NETs promotes an increase in G1 populations, while others promote an increase in G2/M populations, sometimes associated with the induction of senescence. Intriguingly, most of the NETs linked to the cell cycle are highly restricted in their tissue expression; thus, their misregulation in cancer could contribute to the many tissue-specific types of cancer.

Keywords Cell cycle • Lamin • Mitosis • Nuclear envelope transmembrane protein • p53 • pRb

M.I. Robson • P. Le Thanh
Wellcome Trust Centre for Cell Biology, University of Edinburgh,
Kings Buildings, Michael Swann Building, Edinburgh EH9 3JR, UK
e-mail: m.robson@sms.ed.ac.uk; phult91@gmail.com

E.C. Schirmer (✉)
Wellcome Trust Centre for Cell Biology, University of Edinburgh, Kings Buildings,
Michael Swann Building, Room 5.22, Edinburgh EH9 3JR, UK
e-mail: e.schirmer@ed.ac.uk

E.C. Schirmer and J.I. de las Heras (eds.), *Cancer Biology and the Nuclear Envelope*,
Advances in Experimental Medicine and Biology 773, DOI 10.1007/978-1-4899-8032-8_8,

Abbreviations

EDMD	Emery–Dreifuss muscular dystrophy
FACS	Fluorescence activated cell sorting
NE	Nuclear envelope
NET	Nuclear envelope transmembrane protein
NPC	Nuclear pore complex
Nup	Nucleoporin

Introduction

Cell cycle and mitotic misregulation are major factors in both the cause and progression of many cancers. While tumor suppressor proteins like p53, pRb, p16, and p21 are well-known proteins associated with both the cell cycle and cancer progression, a plethora of other kinds of cell cycle proteins from kinases and phosphatases to microtubules in the mitotic spindle have been linked to cancer [1–3]. While one might not think of the nuclear envelope (NE) as being involved in these processes—especially since it is disassembled during mitosis—there are also NE links to all these aspects of the cell cycle.

To begin with, the absence of the NE during mitosis does not mean the absence of NE effects on mitosis, as failure to properly disassemble in prophase or reassemble in telophase would have profound effects on the success of mitosis. For example, maintained interactions with chromatin could lead to aneuploidy. Moreover, proteins of the disassembled NE must go somewhere during mitosis and indeed several have been found to interact with key cell cycle regulators such as the protein phosphatase 1 regulator RepoMan [4] and mitotic structures such as the mitotic spindle [5] and the centrosome [6]. In some of these cases the mitotic associations are clearly functional, while others may be a storage form to prevent aberrant functions during mitosis. NE proteins also influence cell cycle in interphase by forming interactions with key cell cycle regulators, which can determine whether or not a cell will begin cycling. For example, the NE protein lamin A sequesters the cell cycle master regulator/tumor suppressor pRb away from its target genes, preventing entrance into S-phase [7–9]. Lamin A similarly affects the apoptosis regulator E1B 19K protein [10]; thus, linking its functions to apoptosis, another critical cellular mechanism in cancer biology. Thus, the consequences of these NE functions can be diverse, ranging from controlling entry into S-phase to initiating cell proliferation or cell cycle withdrawal to disrupting various stages of mitosis.

Structurally, the NE is a complex double membrane system that surrounds the genetic material (Fig. 1). Historically it was thought to function simply as a barrier isolating and protecting the genetic material from potentially damaging cytoplasmic enzymatic activities such as oxidative metabolism. While it certainly serves this function, it also acts as the gatekeeper for the necessary translocation of proteins into the nucleus to regulate nuclear activities such as transcription, replication, and

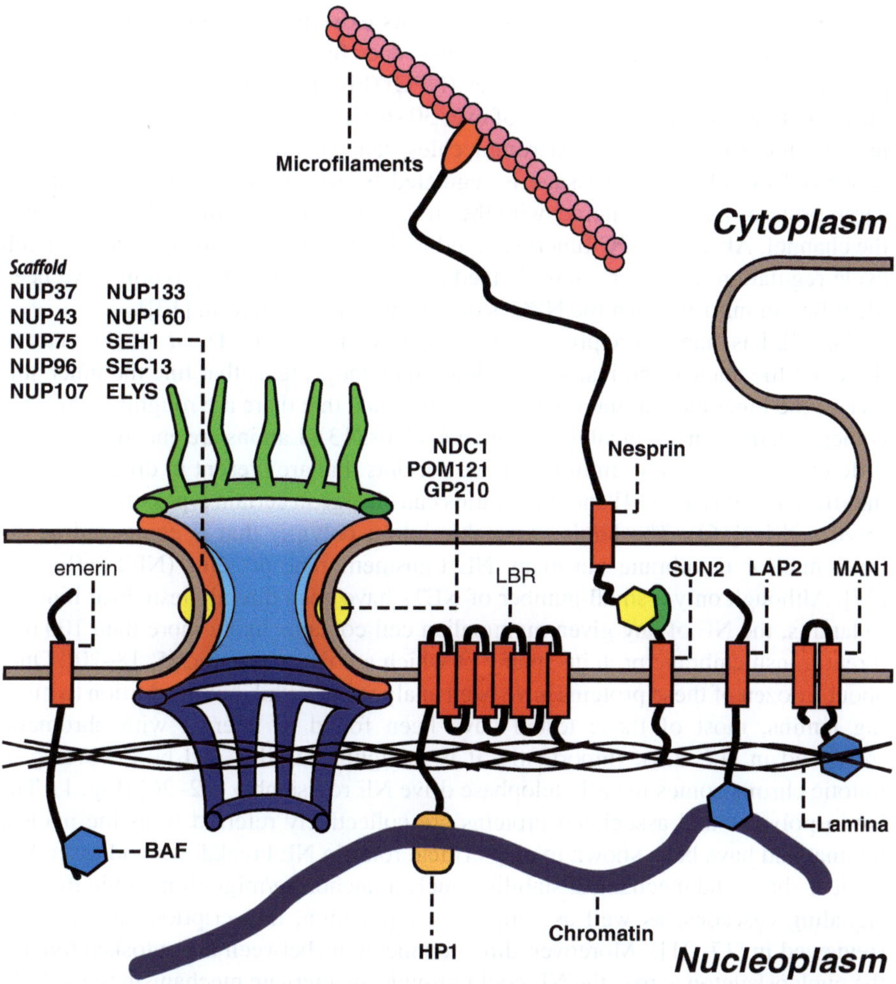

Fig. 1 Nuclear envelope transmembrane proteins (NETs) interconnect the NE. The nuclear envelope (NE) consists of the inner and outer nuclear membranes, the latter of which is contiguous with the endoplasmic reticulum (ER). A plethora of interactions interconnect NE structures. Inner nuclear membrane (NETs) connects the NE to chromatin. The lamin B receptor (LBR) and the LEM domain proteins *L*ap2, *E*merin and *M*an1 interact with chromatin via the chromatin-associated heterochromatin protein 1 (HP1) and barrier-to-autoinegration factor (BAF), respectively. However, it is believed many more NETs also contribute to NE–chromatin interactions. LBR, and probably many more NETs, also interact with the nuclear lamina intermediate filament network, which lines the nucleoplasmic side of the inner nuclear membrane. By contrast, inner nuclear membrane SUN NETs interact with the KASH domains of outer nuclear membrane Nesprin NETs to form the LINC complex, connecting the NE and nucleoskeleton to the cytoskeleton. Collectively, the interconnectivity of these interactions, in addition to the elasticity of the lamina, provides the NE with both mechanical stability and flexibility. Three NETs, NDC1, Pom121, and gp210, also form the core of the nuclear pore complex (NPC), a >60 MDa NE-penetrating channel which regulates the transport of macromolecules and out of the nucleus. Of the many subcomplexes that form the NPC core scaffold, the Nup107-160 complex is one of the most prominent and binds on both side of the NPC

entry into mitosis. Thus, the NE is just by its existence and physical structure a major regulator of the cell cycle. The gatekeeper function of the NE is controlled principally by the nuclear pore complex (NPC) (Fig. 1). In mammalian cells, NPCs are >60 MDa structures comprised of over 30 structural proteins which form a channel through which soluble macromolecules can pass in and out of the nucleus (reviewed in [11]). This passage is regulated by transport receptors that bind to cargo proteins and then interact with the structural proteins of the NPC to negotiate the channel. Aberrant NPC function could logically impact on many aspects of cell cycle regulation; however, more defined roles for several NPC proteins have been identified in mitosis when the NPC is disassembled (reviewed in [12]).

The NE has many more proteins besides those of the NPC. The most abundant of these are the nuclear lamins, intermediate filament proteins that line the inner surface of the inner nuclear membrane. It is estimated that there are roughly 3,000,000 copies of lamins in a typical mammalian nucleus [13]. Lamins are encoded by three different genes and have multiple splice variants that are present in different ratios in different cell types [14] and these ratios can change in certain types of cancer [15] (reviewed in [16]). The lamins assemble into a polymer that is connected to the inner nuclear membrane via many NE transmembrane proteins (NETs) (Fig. 1) [17]. Although only a small number of NETs have been directly tested for binding to lamins, the NE of any given mammalian cell contains likely more than 100 different transmembrane proteins, many of which are tissue specific [5, 18–20]. Only about a dozen of these proteins have been analyzed in detail, but in addition to binding lamins, most of those tested have been found to interact with chromatin (reviewed in [17, 21]). Importantly, it is thought that NET and NPC binding to mitotic chromosomes in early telophase drive NE reassembly [22–26] (Fig. 1). The lamin polymer and associated proteins are collectively referred to as the nuclear lamina, and have been shown to play critical roles in NE breakdown and assembly, nuclear shape and mechanical stability, nuclear anchoring/migration within the cell, signaling cascades, as well as support of replication, transcription, and splicing (reviewed in [17, 21]). Moreover, direct connections between the cytoskeleton and the nucleoskeleton across the NE could provide an alternate mechanism to the NPC for transducing signals between the cytoplasm and nucleus (reviewed in [27]).

Links Between Cell Cycle Misregulation in Nuclear Envelopathies and Cancer

Several NE proteins have been linked to a wide spectrum of inherited diseases collectively known as laminopathies or nuclear envelopathies (reviewed in [28, 29]). Those caused by mutations in NETs range from muscular dystrophy to bone and blood disorders. One proposed mechanism for how mutations in NETs could cause pathology is a disruption in cell cycle regulation.

The first indication that the cell cycle might be altered in a nuclear envelopathy was found by expressing mutations in the NET emerin in tissue culture cells. Two emerin mutations linked to Emery–Dreifuss muscular dystrophy (EDMD) caused a near doubling in the length of the cell cycle, prolonging S-phase from 12 to 22 h when overexpressed in COS-7 cells [30]. While this points to a possible role of the cell cycle in EDMD, four other EDMD mutations tested had no effect indicating that there must be multiple pathways to disease pathology. The function of emerin in the cell cycle was also investigated in *Caenorhabditis elegans*, where its knockdown alone did not have any notable effect on cell division but a combined knockdown of emerin and the NET MAN1 strongly blocked cell division [31].

Another potential link between EDMD and cell cycle misregulation comes from microarray studies in EDMD patient samples that revealed defects in the Rb pathway [32]. The tumor suppressor pRb regulates the cell cycle at the G1/S transition by binding the E2F family of transcription factors (reviewed in [33]). Most emerin-linked EDMD patients exhibit reduced emerin levels, and in an emerin knockout mouse pRb pathway genes were aberrantly regulated. Genes affected included regulators of protein acetylation, including histone acetyltransferases, which could potentially lead to hypo/hyperacetylation and altered gene expression patterns [34]. Separate work on lamin–LAP2 interactions with pRb revealed that misregulation of pRb through NE defects causes hyperproliferation of erythroid and epidermal progenitor cells [35], suggesting that EDMD pathology could result from an early depletion of satellite cells. This is consistent with the timing of disease onset in late childhood and has resulted in misregulation of the cell cycle becoming one of the three favored hypotheses for how NE proteins can cause disease.

NET31/TMEM209 has also been shown to affect cell growth. This NET is upregulated in lung cancer, and its ectopic overexpression in Cos7 cells increases the rate of cell proliferation, while depletion blocks growth. NET31 also interacts with the nucleoporin Nup205. This interaction was shown to stabilize Nup205 with a corresponding increase in the level of nuclear c-Myc, suggesting that the NET31 upregulation in lung cancer reflects a function as a driving force in the cell proliferation due to its effects on c-Myc rather than a downstream consequence of the cancer [36].

Overexpression of NETs does not necessarily lead to cell cycle defects. For instance, LAP2β is increased in expression in a diverse range of digestive tract cancers and appears to interact with HA95 to mediate chromatin–NE interactions implicated in initiation of DNA replication [37]. However, though LAP2β was overexpressed in several cancers, its knockdown only affected cell proliferation in pancreatic cancer. Separately, however, knockdown of LAP2β decreased cell motility in all tested cancer cell types [38]. Thus, NETs may have multiple roles besides affecting the cell cycle in cancers. With many more NETs to be characterized, there is a good possibility that many will have various effects on cancers.

NE Breakdown and Reassembly

NE disassembly at the onset of mitosis/meiosis—NE proteins can affect the cell cycle most centrally in higher eukaryotes when the NE disassembles during mitosis. This releases chromatin from its NE anchors and enables tubulin, normally completely excluded from the nucleus, to gain access to chromatin in order to assemble the mitotic spindle. Failure to properly disassemble the NE not only could block spindle assembly, but once chromosome segregation is complete an only partially disassembled NE could increase the incidence of lagging chromosomes and aneuploidy, a common feature of cancer cells, and could prevent successful cytokinesis resulting in polyploidy. The process of NE disassembly is fundamentally driven by the mitotic phosphorylation of NE proteins and their binding partners resulting in the coordinated disruption of NE interactions and structures (Fig. 2). Indeed, depolymerization of the NE's structural lamina scaffold is induced by phosphorylation of lamins [39] which, if inhibited by mutation of certain critical residues that are phosphorylated at the onset of mitosis, prevents NE disassembly and blocks cell cycle progression into mitosis [40]. Like lamins, a number of NETs are phosphorylated by mitotic kinases to facilitate NE breakdown. For instance, phosphorylation of gp210, a transmembrane protein component of the NPC, prevents its association with the NPC and is suggested to promote NPC disassembly [41–43]. Similarly, both the NETs LAP2β and LBR are phosphorylated during mitosis in a manner which causes their dissociation from the lamina and/or chromatin, thereby breaking the link between the NE and chromatin and allowing disassembly [24, 44–46]. On the chromatin side, a chromatin binding partner of several NETs, barrier-to-autointegration-factor (BAF), is also phosphorylated in a manner that further breaks the chromatin interactions with NETs to support NE breakdown in mitosis [47, 48]. The phosphorylation of BAF is also critical for Karyosome formation during meiosis in *Drosophila* where the NHK-1 (Vrk-1) kinase must phosphorylate BAF in order for chromatin to lose its physical association with the NE much earlier than when NE breakdown occurs [49]. Thus, phosphorylation of lamins, NETs, and chromatin are all required to break the interrelated protein interaction network that makes up the NE and allow its controlled disassembly.

NE reassembly—By contrast to NE disassembly, NE reassembly is driven by the dephosphorylation-induced binding of NETs and their associated membranes to chromatin during late anaphase/telophase which subsequently permits the reformation of the lamina and reassembly of NPCs (Fig. 3). Dephosphorylation of NETs restores their affinity for chromatin and is tightly controlled to prevent NE reassembly occurring too early, as happens when LBR is prematurely dephosphorylated by Cdk1 inhibition [46]. While this dephosphorylation is ultimately controlled by levels of mitotic kinases and phosphatases, NETs can contribute to it by serving as a coordinating scaffold. For example, during reassembly the NET Lem4 (ANKLE2) serves as a signaling scaffold which promotes the dephosphorylation of BAF by inhibiting BAF's mitotic kinase NHK-1/Vrk-1 and simultaneously recruiting its phosphatase PP2A [50].

Many NETs are capable of binding chromatin and therefore driving NE reassembly (Fig. 1). The LEM domain proteins MAN1 [31], emerin [51], and Lap2β [44, 52] bind chromatin via their soluble chromatin binding partner BAF [53–55], while LBR binds heterochromatin protein 1 (HP1) [56, 57] and histone H3 [58]. Some of these interactions may be more significant than others in driving reassembly, however, since depletion of LBR from reconstituted NE-derived vesicles prevented vesicle binding to chromatin [24].

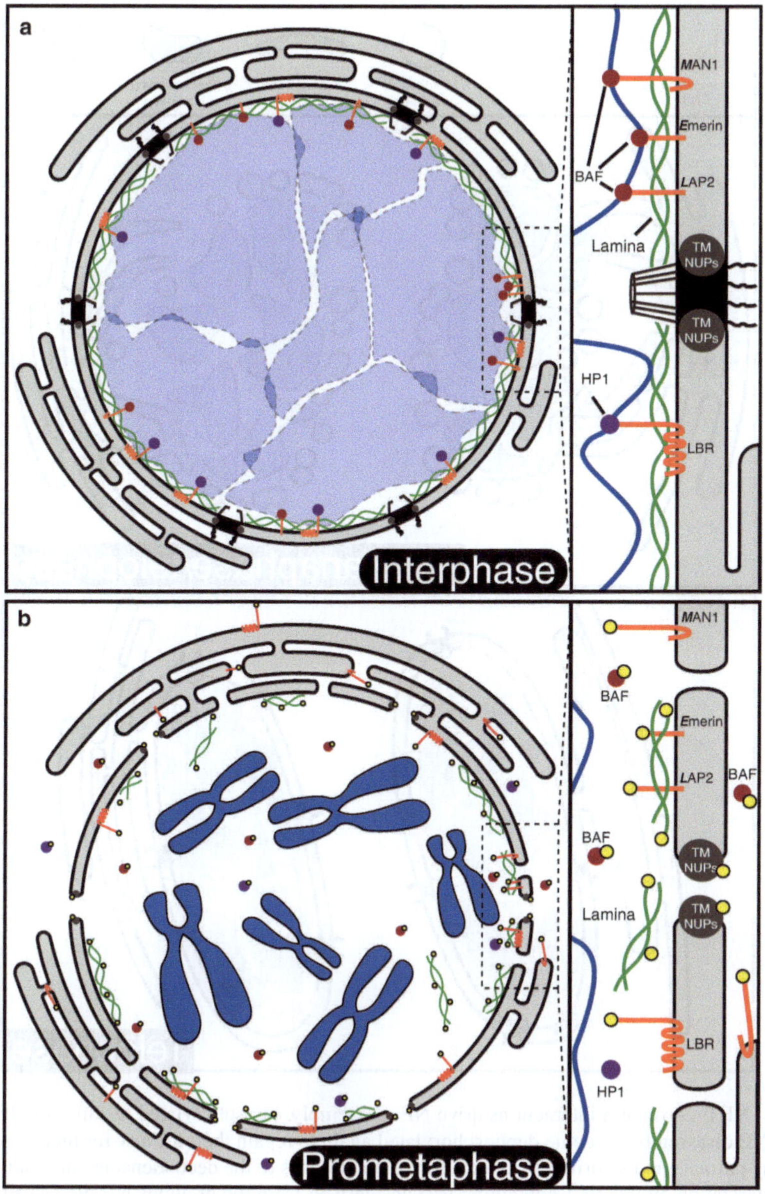

Fig. 2 Phosphorylation of NE proteins drives disassembly. During disassembly the many interactions at the NE are broken in a coordinated manner by mitotic phosphorylation (*yellow circles*) of proteins. Lamins are phosphorylated at many residues, resulting in their depolymerization. Phosphorylation of LBR, and probably many more NETs, breaks their interaction with lamins and chromatin proteins such as HP1 and BAF, while phosphorylation of HP1 and BAF reduces their affinity for both their NET partner proteins and histones/DNA. Collectively, the loss of these interactions disperses NETs away from the NE, Finally, phosphorylation of gp210 and other NPC components promotes NPC disassembly. Dispersal of these structures then allows for disassembly of the NE, either by its vesiculation into mitotic vesicles or through its tubularization and absorption into the ER. These phosphorylations are maintained until the proper timing for NE reassembly because premature dephosphorylation causes inappropriate reassembly during mitosis

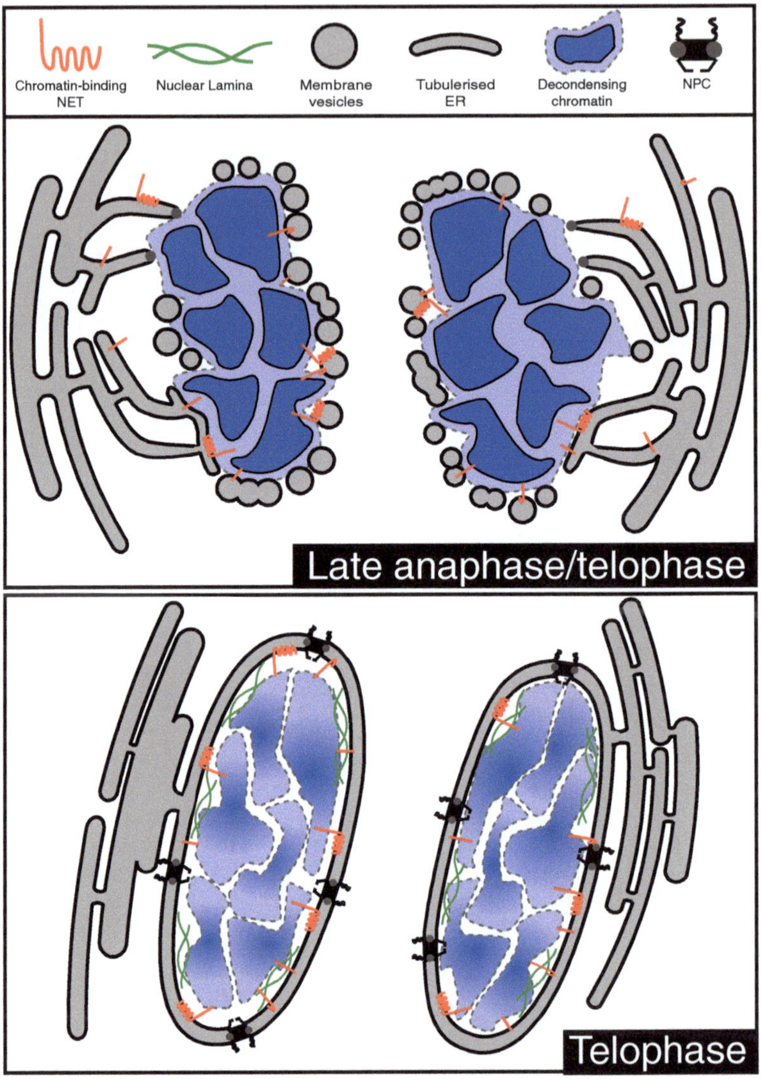

Fig. 3 NET–chromatin interactions drive NE reassembly. Phosphorylated residues of NETs that drive NE disassembly become dephosphorylated again to regain their affinity for their chromatin-binding partners and so drive the association of membranes to the decondensing chromatin. In the vesicularization model NET-containing vesicles derived from the previous NE bind to chromatin and subsequently fuse to form a new continuous NE containing NPCs. By contrast, in the tubularization model, tubules from the mitotic ER (where NETs are dispersed during NE disassembly) diffuse until they bind chromatin. Subsequently once a connection has been made they steadily accumulate by diffusion–retention at the tubule-chromatin interface. At the same time non-NE proteins diffuse away to accumulate in the general ER. In either case, NET interactions with chromatin drive membrane association and NE assembly. Separately, NPC reformation is initiated by the recruitment of the Nup107-160 complex by chromatin-bound ELYS which then recruits Pom121 and begins NPC assembly. The assembled NPCs can then import additional lamins for lamina assembly and growth

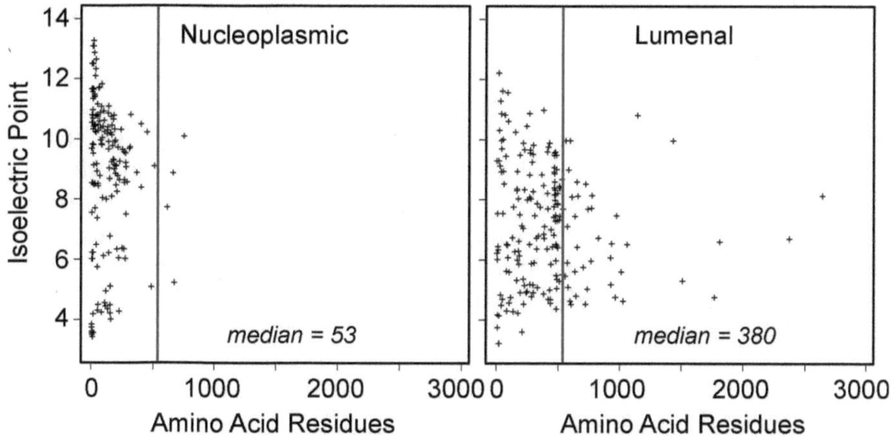

Fig. 4 Many NETs likely interact with DNA directly as they tend to have nucleoplasmic domains with high isoelectric points. The analysis of predicted nucleoplasmic and lumenal domains of 199 NETs identified in a proteomics study of rat liver NEs indicates that NET nucleoplasmic domains are shorter and significantly more basic than their luminal domains. The *line* represents the predicted size limit of the NPC peripheral channels through which nucleoplasmic domains must navigate for NETs to gain entrance to the inner nuclear membrane. Taken with permission from Zuleger et al., 2011 *Journal of Cell Biology*, published by the Rockefeller Press [59]

In addition to binding BAF, Lap2β can also bind core histones, indicating NETs may also bind to chromatin more directly [44]. Indeed, many more NETs may exhibit a more direct chromatin binding capability as a recent bioinformatic analysis of larger sets of NETs identified by proteomics revealed that the nucleoplasmic regions of hundreds of NETs tend to have similarly high isoelectric points (Fig. 4) [59]. Indeed, trypsinization of the vesicles used for in vitro NE assembly abolishes their binding to chromatin and similarly pre-clearing of the vesicle populations by binding the vesicles to DNA-coated beads blocks their function in NE reassembly [22]. Thus, at the end of mitosis, the NE reassembles through a process driven by the interaction of NETs with chromatin.

NETs are also key to NE reassembly and cell cycle progression through their role in NPC formation. The NPC contains three known NETs, namely, Pom121, NDC1, and gp210, the latter of which segregates onto a vesicle population distinct from that shared by Pom121 and NDC1 in the *Xenopus* in vitro assembly system and is absent in a number of cell types indicating it is not absolutely required for NPC function [60–63]. Immunodepletion of both Pom121 and NDC1, but not gp210, prevents in vitro NE reassembly by inhibiting the fusion of vesicles surrounding chromatin, although it does not inhibit the binding of these vesicles to chromatin indicating a link between NPC formation and NE reassembly [60, 64]. However, while Pom121 and NDC1 depletion results in similar NE fusion defects, the introduction of additional Pom121 to NDC1-depleted extracts or vice versa fails to rescue the formation of closed NEs indicating each serves a distinct function in NE fusion [60].

NPC NETs are targeted to mitotic chromosomes by the DNA binding protein ELYS/Mel-28. ELYS/Mel-28, present on chromatin during anaphase, recruits the Nup107-160 complex, and subsequently Pom121 associated vesicles to chromatin [65–67]. This ELYS/Nup107-160 complex dependent recruitment of NET NPC components to chromatin appears critical for successful NPC formation and NE reassembly since depletion of ELYS or Nup107-160 results in the formation of a chromatin-associated closed NE devoid of assembled NPCs [65, 68, 69]. Hence, NETs collectively serve to drive NE reassembly primarily by forming interactions with chromatin and aiding in the formation of NPCs. Defects in any of these mitotic functions could affect the quality of cell division and lead to aneuploidy, a common feature of tumors [70].

NET Functions During Mitosis

Once the NE breaks down in mitosis the NE proteins could either go into a storage form or they could engage in separate functions. Indeed, several soluble proteins associated with the NPC have critical functions during mitosis. For example, Ran, critical for cargo release and receptor shuttling in interphase nucleocytoplasmic transport (reviewed in [11]), and the transport receptor importin/karyopherinβ are required for aster formation in the mitotic spindle [71, 72]. Importinβ has a second mitotic function through an interaction with RepoMan [4]. As a transport receptor, importinβ might utilize its ability to interact with its cargos that it transports into the nucleus during interphase to bind them in mitosis until needed or bring different proteins together for assembling various complexes—for example, bringing protein phosphatase 1 together with its substrates. Some NPC structural components likely have mitotic functions as well as evidenced by the involvement of both RAE1 and the Nup107-160 complex in spindle formation and the additional co-localization of the latter with kinetochores during mitosis [73–75]. However, the specific function of these interactions remains unclear.

Notably, for the over 100 different NETs in any given cell, there would obviously be some restrictions on potential mitotic functions because these proteins presumably must remain membrane bound. What happens to the membranes during mitosis remains unclear with some studies supporting NE vesiculation [5, 76–82] and others supporting tubular structures merged with the ER (Fig. 5) [83–86]. Consistent with at least some vesiculation, the distribution of certain NETs appears to be distinct from others in mitosis in vivo, while in an artificial *Xenopus* NE assembly system it was observed that two different vesicle types were required for reassembly with some NETs differentially segregating between the two populations [78, 80]. In tissue culture cells expressing NETs fused to GFP, distinct distribution patterns have been observed in mitosis for different NETs (Fig. 5). Most NETs appear to be excluded from the mitotic spindle while being distributed evenly throughout the rest of the mitotic cell, but some NETs (NET5/Samp1/Tmem201, WFS1, Tmem214, and otefin) have been observed to partially concentrate on or around the mitotic spindle and,

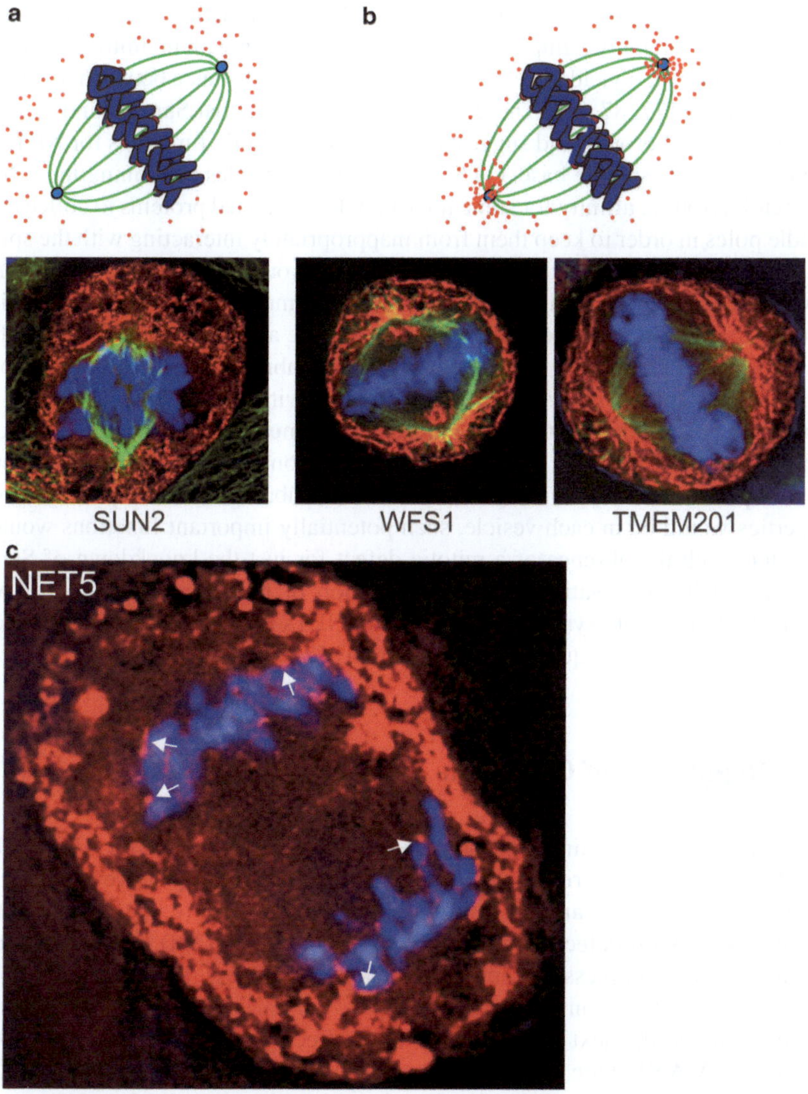

Fig. 5 Different NETs display distinct localization patterns during mitosis. During mitosis most NETs are excluded from the mitotic spindle while others accumulate on mitotic structures, suggesting they may have separate mitotic functions. (**a** and **b**) SUN2 concentrates at the NE during interphase and during mitosis it is excluded from the spindle (**a**) while both WFS1 and NET5 (also called Tmem201 and Samp1) are also at the NE during interphase but localize in part to spindle poles during mitosis (**b**). NET-RFP fusions are stained in red whilst the DAPI stained chromatin and mitotic spindle are coloured blue and green, respectively. (**c**) Additionally, at least some of NET5 appears to bind directly to a subset of chromosomes during anaphase as the membranes begin to bind chromatin. Together, these data indicate the likelihood of mitotic functions for some NETs, suggest that at least some of the NE vesiculates during mitosis, and strongly supports the idea that NET-containing vesicle binding to mitotic chromosomes drives NE reformation at the end of mitosis. Some images taken with permission from Wilkie et al., 2011 *Molecular and Cellular Proteomics*, published by the American Society for Biochemistry and Molecular Biology (ASBMB) [5]

in the case of the latter, the centrosome [5, 6, 77]. The function of this localization, if any, is not clear. Only Samp1 was tested by knockdown with no mitotic phenotype reported; however, it yielded an interphase defect in the distance between the NE and the centrosome [77, 87, 88]. These observations indicate that Samp1 binds a microtubule associated protein and suggest three possible mitotic functions for NETs that accumulate at the spindle base. (1) Storage: mitotic vesicles containing these NETs use their interphase affinity for some microtubule associated proteins to collect at the spindle poles in order to keep them from inappropriately interacting with the spindle itself. (2) Sequestration: if these NETs, particularly outer nuclear membrane NETs, serve as platforms to mediate interactions between multiple microtubule associated proteins they could sequester particular microtubule associated proteins away from the spindle that could alter its functionality. (3) Stabilization: the accumulation of vesicles at the spindle carrying NETs that interact with microtubule associated proteins could potentially function like glue, acting as multifaceted adaptors to hold the minus ends of the multiple microtubules coming from the spindle loosely together. As multiple NETs with probably redundant microtubule associated protein binding properties would be in each vesicle, such potentially important functions would be consistent with the absence of a mitotic defect for just the knockdown of Samp1. Consistent with observations of the segregation of NETs in mitosis, an in vitro *Xenopus* NE assembly system also segregated NETs into two distinct vesicle populations that were both required for NE assembly [78, 80].

NET Regulation of Cell Cycle Progression and Withdrawal

A complex between lamin A, the soluble splice variant of the NET LAP2, LAP2α, and pRb is well characterized in cell cycle regulation [8, 9, 89]. However, the mechanism by which emerin and other NETs might regulate pRb (e.g., the disruption of pRb pathways with defects in emerin in EDMD patients [32]) or other cell cycle regulators/tumor suppressors remains unclear. Nonetheless, it is certain that many NETs can have effects on the cell cycle.

After emerin, the next NET found to have effects on the cell cycle was MAN1. Emerin and MAN1 share a common weakly conserved domain known as the LEM domain because it was first identified from homology between the three NE proteins, *L*AP2, *e*merin, and *M*AN1 [90, 91]. The LEM domain has since been found in additional NETs and some soluble cytoplasmic proteins [92–95]. Whereas emerin knockdown alone in *C. elegans* had no notable phenotype, the combined RNAi knockdown of emerin and MAN1 resulted in no embryos reaching the 100-cell stage [31]. The cellular phenotype was mostly complete cytokinesis failure due to unresolvable anaphase chromatin bridges. The NET LAP2β has separately been implicated in promoting cell cycle progression through an interaction with HA95, a homolog of the nuclear-A kinase anchoring protein AKAP95. Disruption of the interaction between LAP2β and HA95 abolished the initiation of DNA replication through the proteasome-mediated degradation of a component of the pre-replication complex, namely cdc6, thereby preventing DNA replication and cell cycle progression [37].

The next NET to be linked to cell cycle regulation was a member of the KASH domain-containing nesprin family. Nesprins are proteins involved in linking the NE to the cytoskeleton through direct interactions with actin and potentially with intermediate filaments and microtubules through indirect interactions [27]. Nesprins also bind SUN family NETs and through these interactions connect the cytoskeleton across the NE lumen to the nucleoskeleton [27]. A *C. elegans* protein, KDP-1, was identified in a 2-hybrid screen for SUN-binding proteins and found to contain a KASH domain and thus appears to be a novel member of the nesprin family. KDP-1 targets to the NE and its knockdown in *C. elegans* embryos caused delayed cell entry into mitosis after replication, resulting in a notable reduction of germ line cells in the mitotic zone [96]. The mechanism of action for KDP-1 remains unknown. As work from the Noegel, Karakesisoglou, and Burke laboratories have shown a function of nesprins in centrosome positioning [97, 98], it is possible that KDP-1 could function through a similar mechanism in interphase to affect cell cycle progression.

The finding of cell cycle functions for these NETs inspired another study to screen for additional NETs functioning in cell cycle regulation [99]. Thirty-nine novel NETs identified from proteomic studies of liver and blood cells were fused to monomeric RFP and expressed in HEK293T cells. The DNA was stained and the cells analyzed by FACS (fluorescence activated cell sorting), which measures DNA content and could distinguish NET expressing cells by the monomeric RFP from untransfected cells in the same population. Thus, the 4/2 N ratios, representing the G2/M and G1/S-phase populations respectively, were separately measured within the NET transfected and untransfected populations and compared. Eight of the thirty-nine NETs tested—over 20 %—yielded notable changes with seven increasing and one decreasing the 4:2 N ratio [99]. These NETs were NET4/Tmem53, NET11/Sccpdh, Tmub1, Fam3c, Magt1, Tmem126a, NET59/Ncln, and NET31/Tmem209, which was separately found to be upregulated and affect cell growth in lung cancer [36].

To determine if these cell cycle effects were mediated by classical pathways such as those of the p53 or pRb master cell cycle regulator/tumor suppressor proteins, the NETs that were positive in the screen were also tested in cell lines deficient in these cell cycle regulators. For two NETs, NET4/Tmem53 and NET59/Ncln, the effects were significantly diminished or lost when tested in cells lacking the p53 master regulator. Tmem53/NET4 also lost its effects in the pRb deficient cells [99].

NET59/Ncln has previously been found to indirectly affect Smad transcriptional regulators to modulate TGFβ signaling pathways [100]. This is particularly interesting in that Smads also interact with MAN1 [101, 102], a NET previously linked to cell cycle regulation. TGFβ signaling is also affected by another LEM domain NET in *Drosophila*, Otefin [103]. Several other NETs have also been shown to affect signaling cascades: emerin can regulate β-catenin/wnt signaling [104]; NET25/LEM2 influences Erk1/2 pathways [105]; and NET39/PPAPDC3 inhibits the mTOR signaling pathway [106]. Of these, NET25/LEM2 was not tested in the screen, and NET39/PPAPDC3 had no effect with the FACS profiles of the transfected and untransfected populations exactly overlaid [99], but these findings indicate that NETs can yield effects on the cell cycle through interactions with likely multiple well-characterized signaling pathways.

The effect of expression of NET4/Tmem53 on the cell cycle was unique in being reversed by both p53 and pRb reduction and thus being dependent on components of both pathways. Correspondingly, knockdown of NET4/Tmem53 yielded effects on both proteins. Levels of p53 were doubled with NET4/Tmem53 knockdown and p21 levels increased sevenfold. For pRb, overall levels remained unaffected, but its regulation was strongly affected with a dramatic reduction in pRb phosphorylation [99]. This shows that NET4/Tmem53 functions on pRb pathways in a different manner from lamins and LAP2α as lamin or LAP2α knockdown affected the actual levels of the pRb protein [89]. All of these effects depended on p38 MAP kinase, which is activated upon loss of NET4/Tmem53 and has been previously shown to be associated with stress pathways [107]. The effect on the cell cycle profiles observed in the FACS-based screen appears to be largely caused by cells withdrawing from the cell cycle and entering senescence [99].

The other NETs that affected cell cycle progression have not been further analyzed for mechanism. However, as their effects were all independent of pRb and p53 (with the exception of NET59/Ncln) they must affect the cell cycle by distinct mechanisms.

NET Spatial and Tissue Distribution and the Cell Cycle

The different NETs exerted their effects on the cell cycle from discrete spatial positions. NET4/Tmem53 and NET31/Tmem209 appear to be restricted to the outer nuclear membrane based on data obtained from super resolution microscopy using the structured illumination (OMX) platform [19, 108]. In contrast, NET59/Ncln and Magt1 are in the inner nuclear membrane [18, 19]. Thus, NETs on both sides of the NE can influence the regulation of the cell cycle by distinct mechanisms.

Strikingly, most of the NETs that altered cell cycle profiles were highly restricted in the tissues where they were expressed. The BioGPS transcriptome database compares the expression of thousands of genes over 84 different human tissues [109]. According to this resource, Tmub1, Fam3c, and Magt1 are all expressed more than 5× higher in blood compared to the median value for the 84 tissues examined while Tmem126A is expressed in blood at roughly 50× higher than the median. This is not completely surprising as these four NETs were all identified in a proteomic study of NEs isolated from blood [18]; however, each NET was also restricted in expression to specific blood cell types. Magt1 was expressed highest in natural killer and dendritic cells but poorly expressed in lymphocytes, endothelial cells, and CD34+ cells. In contrast Tmem126A was expressed highest in lymphoblasts, CD34+ cells and CD56+ natural killer cells. Among the NETs identified in the liver proteomic study, NET4/Tmem53, as expected, was preferentially expressed in liver, but NET11/Sccpdh turned out to be expressed much higher in brain and testis than in liver [109]. Only NET59/Ncln was widely expressed and this is more interesting as it was the only of the new NETs that has been shown to intersect with the TGFβ family, like MAN1, which is also widely expressed and exhibits cell cycle defects when levels are reduced.

Concluding Remarks

Prior to the explosion of NETs identified in proteomic studies 10 years ago, only roughly a dozen NETs had been identified. Among these well-characterized NETs, emerin, MAN1, and nesprins have been linked to the cell cycle. Screening a large set of NETs identified by proteomics has roughly quadrupled this number, netting over 1/5th of those tested. While emerin, MAN1, NET4/Tmem53 and NET59/Ncln effects appear to be mediated through known signaling pathways, the mechanism by which KDP-1 affects cell cycle regulation remains unclear, and those of NET11/Sccpdh, NET31/Tmem209, Tmub1, Fam3c, Magt1, and Tmem126a are independent of p53 and pRb pathways. Thus, the majority of the novel NETs identified recently by proteomics that exhibited cell cycle effects regulate the cell cycle through novel or less-characterized pathways. These novel NETs are generally uncharacterized proteins with no known functions.

The tissue specificity of many of the NETs affecting the cell cycle is striking and also suggestive. Nuclear size (the karyoplasmic ratio) is linked to cancer prognosis in a tissue or tumor specific manner. For example, large nuclear size is linked to increased risk of metastasis in bladder tumors but smaller nuclear size is linked to increased risk of metastasis in lung cell carcinomas [110, 111]. Thus, tissue-specific NETs that influence the cell cycle could influence the accumulation of larger or smaller nuclei by increasing the population of cells in G2 as opposed to G1. Thus, by functions in structural aspects of mitosis as well as direct effects on pathways regulating the cell cycle, NETs have a clear effect on control of the cell cycle—an essential aspect of a tumor's ability to proliferate. That some of these NETs can act directly on tumor suppressor proteins and that some are tissue-specific suggests that NETs may play roles—either through their own mutation or through mutated partners losing the ability to interact—in tissue specific tumor formation, one of the least understood aspects of cancer biology.

Acknowledgements Work in the Schirmer lab is supported by Wellcome Trust Senior Research Fellowship 095209. Michael Robson was supported by a Wellcome Trust Studentship and Phu Le Thanh by an MRC Studentship.

References

1. Sherr CJ, McCormick F (2002) The RB and p53 pathways in cancer. Cancer Cell 2(2):103–112
2. Malumbres M, Barbacid M (2009) Cell cycle, CDKs and cancer: a changing paradigm. Nat Rev Cancer 9(3):153–166. doi:10.1038/nrc2602
3. Pease JC, Tirnauer JS (2011) Mitotic spindle misorientation in cancer—out of alignment and into the fire. J Cell Sci 124(Pt 7):1007–1016. doi:10.1242/jcs.081406
4. Vagnarelli P, Earnshaw WC (2012) Repo-Man-PP1: a link between chromatin remodelling and nuclear envelope reassembly. Nucleus 3(2):138–142. doi:10.4161/nucl.19267

5. Wilkie GS, Korfali N, Swanson SK, Malik P, Srsen V, Batrakou DG, de las Heras J, Zuleger N, Kerr AR, Florens L, Schirmer EC (2011) Several novel nuclear envelope transmembrane proteins identified in skeletal muscle have cytoskeletal associations. Mol Cell Proteomics 10(1):M110.003129. doi:10.1074/mcp.M110.003129
6. Habermann K, Mirgorodskaya E, Gobom J, Lehmann V, Muller H, Blumlein K, Deery MJ, Czogiel I, Erdmann C, Ralser M, von Kries JP, Lange BM (2012) Functional analysis of centrosomal kinase substrates in *Drosophila melanogaster* reveals a new function of the nuclear envelope component otefin in cell cycle progression. Mol Cell Biol 32(17):3554–3569. doi:10.1128/MCB.00814-12
7. Ozaki T, Saijo M, Murakami K, Enomoto H, Taya Y, Sakiyama S (1994) Complex formation between lamin A and the retinoblastoma gene product: identification of the domain on lamin A required for its interaction. Oncogene 9(9):2649–2653
8. Markiewicz E, Dechat T, Foisner R, Quinlan R, Hutchison C (2002) Lamin A/C binding protein LAP2alpha is required for nuclear anchorage of retinoblastoma protein. Mol Biol Cell 13(12):4401–4413
9. Johnson BR, Nitta RT, Frock RL, Mounkes L, Barbie DA, Stewart CL, Harlow E, Kennedy BK (2004) A-type lamins regulate retinoblastoma protein function by promoting subnuclear localization and preventing proteasomal degradation. Proc Natl Acad Sci U S A 101(26):9677–9682
10. Rao L, Modha D, White E (1997) The E1B 19K protein associates with lamins in vivo and its proper localization is required for inhibition of apoptosis. Oncogene 15(13):1587–1597. doi:10.1038/sj.onc.1201323
11. Suntharalingam M, Wente SR (2003) Peering through the pore: nuclear pore complex structure, assembly, and function. Dev Cell 4(6):775–789
12. Guttinger S, Laurell E, Kutay U (2009) Orchestrating nuclear envelope disassembly and reassembly during mitosis. Nat Rev Mol Cell Biol 10(3):178–191. doi:10.1038/nrm2641
13. Gerace L, Burke B (1988) Functional organization of the nuclear envelope. Annu Rev Cell Biol 4:335–374
14. Broers JL, Machiels BM, Kuijpers HJ, Smedts F, van den Kieboom R, Raymond Y, Ramaekers FC (1997) A- and B-type lamins are differentially expressed in normal human tissues. Histochem Cell Biol 107(6):505–517
15. Jansen M, Machiels B, Hopman A, Broers J, Bot F, Arends J, Ramaekers F, Schouten H (1997) Comparison of A and B-type lamin expression in reactive lymph nodes and nodular sclerosing Hodgkin's disease. Histopathology 31(4):304–312
16. de Las Heras JI, Batrakou DG, Schirmer EC (2012) Cancer biology and the nuclear envelope: a convoluted relationship. Semin Cancer Biol 23:125. doi:10.1016/j.semcancer.2012.01.008
17. Schirmer EC, Foisner R (2007) Proteins that associate with lamins: many faces, many functions. Exp Cell Res 313(10):2167–2179
18. Korfali N, Wilkie GS, Swanson SK, Srsen V, Batrakou DG, Fairley EA, Malik P, Zuleger N, Goncharevich A, de Las Heras J, Kelly DA, Kerr AR, Florens L, Schirmer EC (2010) The leukocyte nuclear envelope proteome varies with cell activation and contains novel transmembrane proteins that affect genome architecture. Mol Cell Proteomics 9(12):2571–2585. doi:10.1074/mcp.M110.002915
19. Malik P, Korfali N, Srsen V, Lazou V, Batrakou DG, Zuleger N, Kavanagh DM, Wilkie GS, Goldberg MW, Schirmer EC (2010) Cell-specific and lamin-dependent targeting of novel transmembrane proteins in the nuclear envelope. Cell Mol Life Sci 67(8):1353–1369. doi:10.1007/s00018-010-0257-2
20. Korfali N, Wilkie GS, Swanson SK, Srsen V, de Las Heras J, Batrakou DG, Malik P, Zuleger N, Kerr AR, Florens L, Schirmer EC (2012) The nuclear envelope proteome differs notably between tissues. Nucleus 3(6):552–564. doi:10.4161/nucl.22257
21. Mattout-Drubezki A, Gruenbaum Y (2003) Dynamic interactions of nuclear lamina proteins with chromatin and transcriptional machinery. Cell Mol Life Sci 60(10):2053–2063
22. Ulbert S, Platani M, Boue S, Mattaj IW (2006) Direct membrane protein-DNA interactions required early in nuclear envelope assembly. J Cell Biol 173(4):469–476

23. Anderson DJ, Vargas JD, Hsiao JP, Hetzer MW (2009) Recruitment of functionally distinct membrane proteins to chromatin mediates nuclear envelope formation in vivo. J Cell Biol 186(2):183–191. doi:10.1083/jcb.200901106
24. Pyrpasopoulou A, Meier J, Maison C, Simos G, Georgatos SD (1996) The lamin B receptor (LBR) provides essential chromatin docking sites at the nuclear envelope. EMBO J 15(24):7108–7119
25. Ellenberg J, Lippincott-Schwartz J (1999) Dynamics and mobility of nuclear envelope proteins in interphase and mitotic cells revealed by green fluorescent protein chimeras. Methods 19(3):362–372. doi:10.1006/meth.1999.0872
26. Haraguchi T, Koujin T, Hayakawa T, Kaneda T, Tsutsumi C, Imamoto N, Akazawa C, Sukegawa J, Yoneda Y, Hiraoka Y (2000) Live fluorescence imaging reveals early recruitment of emerin, LBR, RanBP2, and Nup153 to reforming functional nuclear envelopes. J Cell Sci 113(Pt 5):779–794
27. Stewart CL, Roux KJ, Burke B (2007) Blurring the boundary: the nuclear envelope extends its reach. Science 318(5855):1408–1412
28. Foisner R, Aebi U, Bonne G, Gruenbaum Y, Novelli G (2007) 141st ENMC international workshop inaugural meeting of the EURO-Laminopathies project "nuclear envelope-linked rare human diseases: from molecular pathophysiology towards clinical applications", Naarden, The Netherlands, 10–12 March 2006. Neuromuscul Disord 17(8):655–660
29. Worman HJ, Bonne G (2007) "Laminopathies": a wide spectrum of human diseases. Exp Cell Res 313(10):2121–2133
30. Fairley E, Riddell A, Ellis J, Kendrick-Jones J (2002) The cell cycle dependent mislocalisation of emerin may contribute to the Emery-Dreifuss muscular dystrophy phenotype. J Cell Sci 115(2):341–354
31. Liu J, Lee K, Segura-Totten M, Neufeld E, Wilson K, Gruenbaum Y (2003) MAN1 and emerin have overlapping function(s) essential for chromosome segregation and cell division in *Caenorhabditis elegans*. Proc Natl Acad Sci U S A 100(8):4598–4603
32. Bakay M, Wang Z, Melcon G, Schiltz L, Xuan J, Zhao P, Sartorelli V, Seo J, Pegoraro E, Angelini C, Shneiderman B, Escolar D, Chen YW, Winokur ST, Pachman LM, Fan C, Mandler R, Nevo Y, Gordon E, Zhu Y, Dong Y, Wang Y, Hoffman EP (2006) Nuclear envelope dystrophies show a transcriptional fingerprint suggesting disruption of Rb-MyoD pathways in muscle regeneration. Brain 129(Pt 4):996–1013
33. Genovese C, Trani D, Caputi M, Claudio PP (2006) Cell cycle control and beyond: emerging roles for the retinoblastoma gene family. Oncogene 25(38):5201–5209
34. Melcon G, Kozlov S, Cutler DA, Sullivan T, Hernandez L, Zhao P, Mitchell S, Nader G, Bakay M, Rottman JN, Hoffman EP, Stewart CL (2006) Loss of emerin at the nuclear envelope disrupts the Rb1/E2F and MyoD pathways during muscle regeneration. Hum Mol Genet 15(4):637–651
35. Naetar N, Korbei B, Kozlov S, Kerenyi MA, Dorner D, Kral R, Gotic I, Fuchs P, Cohen TV, Bittner R, Stewart CL, Foisner R (2008) Loss of nucleoplasmic LAP2alpha-lamin A complexes causes erythroid and epidermal progenitor hyperproliferation. Nat Cell Biol 10(11):1341–1348. doi:10.1038/ncb1793
36. Fujitomo T, Daigo Y, Matsuda K, Ueda K, Nakamura Y (2012) Critical function for nuclear envelope protein TMEM209 in human pulmonary carcinogenesis. Cancer Res 72(16):4110–4118. doi:10.1158/0008-5472.CAN-12-0159
37. Martins S, Eikvar S, Furukawa K, Collas P (2003) HA95 and LAP2 beta mediate a novel chromatin-nuclear envelope interaction implicated in initiation of DNA replication. J Cell Biol 160(2):177–188. doi:10.1083/jcb.200210026
38. Kim HJ, Hwang SH, Han ME, Baek S, Sim HE, Yoon S, Baek SY, Kim BS, Kim JH, Kim SY, Oh SO (2012) LAP2 is widely overexpressed in diverse digestive tract cancers and regulates motility of cancer cells. PLoS One 7(6):e39482. doi:10.1371/journal.pone.0039482
39. Gerace L, Blobel G (1980) The nuclear envelope lamina is reversibly depolymerized during mitosis. Cell 19(1):277–287
40. Heald R, McKeon F (1990) Mutations of phosphorylation sites in lamin A that prevent nuclear lamina disassembly in mitosis. Cell 61(4):579–589

41. Favreau C, Worman HJ, Wozniak RW, Frappier T, Courvalin JC (1996) Cell cycle-dependent phosphorylation of nucleoporins and nuclear pore membrane protein Gp210. Biochemistry 35(24):8035–8044. doi:10.1021/bi9600660
42. Galy V, Antonin W, Jaedicke A, Sachse M, Santarella R, Haselmann U, Mattaj I (2008) A role for gp210 in mitotic nuclear-envelope breakdown. J Cell Sci 121(Pt 3):317–328. doi:10.1242/jcs.022525
43. Onischenko EA, Crafoord E, Hallberg E (2007) Phosphomimetic mutation of the mitotically phosphorylated serine 1880 compromises the interaction of the transmembrane nucleoporin gp210 with the nuclear pore complex. Exp Cell Res 313(12):2744–2751. doi:10.1016/j.yexcr.2007.05.011
44. Foisner R, Gerace L (1993) Integral membrane proteins of the nuclear envelope interact with lamins and chromosomes, and binding is modulated by mitotic phosphorylation. Cell 73(7):1267–1279
45. Courvalin JC, Segil N, Blobel G, Worman HJ (1992) The lamin B receptor of the inner nuclear membrane undergoes mitosis- specific phosphorylation and is a substrate for p34cdc2-type protein kinase. J Biol Chem 267(27):19035–19038
46. Tseng LC, Chen RH (2011) Temporal control of nuclear envelope assembly by phosphorylation of lamin B receptor. Mol Biol Cell 22(18):3306–3317. doi:10.1091/mbc.E11-03-0199
47. Nichols RJ, Wiebe MS, Traktman P (2006) The vaccinia-related kinases phosphorylate the N′ terminus of BAF, regulating its interaction with DNA and its retention in the nucleus. Mol Biol Cell 17(5):2451–2464. doi:10.1091/mbc.E05-12-1179
48. Gorjanacz M, Klerkx EP, Galy V, Santarella R, Lopez-Iglesias C, Askjaer P, Mattaj IW (2007) *Caenorhabditis elegans* BAF-1 and its kinase VRK-1 participate directly in post-mitotic nuclear envelope assembly. EMBO J 26(1):132–143. doi:10.1038/sj.emboj.7601470
49. Lancaster OM, Cullen CF, Ohkura H (2007) NHK-1 phosphorylates BAF to allow karyosome formation in the *Drosophila* oocyte nucleus. J Cell Biol 179(5):817–824. doi:10.1083/jcb.200706067
50. Asencio C, Davidson IF, Santarella-Mellwig R, Ly-Hartig TB, Mall M, Wallenfang MR, Mattaj IW, Gorjanacz M (2012) Coordination of kinase and phosphatase activities by Lem4 enables nuclear envelope reassembly during mitosis. Cell 150(1):122–135. doi:10.1016/j.cell.2012.04.043
51. Hirano Y, Segawa M, Ouchi FS, Yamakawa Y, Furukawa K, Takeyasu K, Horigome T (2005) Dissociation of emerin from barrier-to-autointegration factor is regulated through mitotic phosphorylation of emerin in a *Xenopus* egg cell-free system. J Biol Chem 280(48):39925–39933. doi:10.1074/jbc.M503214200
52. Furukawa K, Glass C, Kondo T (1997) Characterization of the chromatin binding activity of lamina-associated polypeptide (LAP) 2. Biochem Biophys Res Commun 238(1):240–246
53. Furukawa K (1999) LAP2 binding protein 1 (L2BP1/BAF) is a candidate mediator of LAP2-chromatin interaction. J Cell Sci 112(Pt 15):2485–2492
54. Cai M, Huang Y, Ghirlando R, Wilson KL, Craigie R, Clore GM (2001) Solution structure of the constant region of nuclear envelope protein LAP2 reveals two LEM-domain structures: one binds BAF and the other binds DNA. EMBO J 20(16):4399–4407
55. Shumaker DK, Lee KK, Tanhehco YC, Craigie R, Wilson KL (2001) LAP2 binds to BAF. DNA complexes: requirement for the LEM domain and modulation by variable regions. EMBO J 20(7):1754–1764. doi:10.1093/emboj/20.7.1754
56. Ye Q, Worman HJ (1996) Interaction between an integral protein of the nuclear envelope inner membrane and human chromodomain proteins homologous to *Drosophila* HP1. J Biol Chem 271(25):14653–14656
57. Ye Q, Callebaut I, Pezhman A, Courvalin JC, Worman HJ (1997) Domain-specific interactions of human HP1-type chromodomain proteins and inner nuclear membrane protein LBR. J Biol Chem 272(23):14983–14989
58. Polioudaki H, Kourmouli N, Drosou V, Bakou A, Theodoropoulos PA, Singh PB, Giannakouros T, Georgatos SD (2001) Histones H3/H4 form a tight complex with the inner nuclear membrane protein LBR and heterochromatin protein 1. EMBO Rep 2(10):920–925. doi:10.1093/embo-reports/kve199

59. Zuleger N, Kelly DA, Richardson AC, Kerr AR, Goldberg MW, Goryachev AB, Schirmer EC (2011) System analysis shows distinct mechanisms and common principles of nuclear envelope protein dynamics. J Cell Biol 193(1):109–123. doi:10.1083/jcb.201009068
60. Mansfeld J, Guttinger S, Hawryluk-Gara LA, Pante N, Mall M, Galy V, Haselmann U, Muhlhausser P, Wozniak RW, Mattaj IW, Kutay U, Antonin W (2006) The conserved transmembrane nucleoporin NDC1 is required for nuclear pore complex assembly in vertebrate cells. Mol Cell 22(1):93–103
61. Eriksson C, Rustum C, Hallberg E (2004) Dynamic properties of nuclear pore complex proteins in gp210 deficient cells. FEBS Lett 572(1–3):261–265. doi:10.1016/j.febslet.2004.07.044
62. Olsson M, Scheele S, Ekblom P (2004) Limited expression of nuclear pore membrane glycoprotein 210 in cell lines and tissues suggests cell-type specific nuclear pores in metazoans. Exp Cell Res 292(2):359–370
63. Stavru F, Nautrup-Pedersen G, Cordes VC, Gorlich D (2006) Nuclear pore complex assembly and maintenance in POM121- and gp210-deficient cells. J Cell Biol 173(4):477–483 [comment]
64. Antonin W, Franz C, Haselmann U, Antony C, Mattaj IW (2005) The integral membrane nucleoporin pom121 functionally links nuclear pore complex assembly and nuclear envelope formation. Mol Cell 17(1):83–92. doi:10.1016/j.molcel.2004.12.010
65. Franz C, Walczak R, Yavuz S, Santarella R, Gentzel M, Askjaer P, Galy V, Hetzer M, Mattaj IW, Antonin W (2007) MEL-28/ELYS is required for the recruitment of nucleoporins to chromatin and postmitotic nuclear pore complex assembly. EMBO Rep 8(2):165–172. doi:10.1038/sj.embor.7400889
66. Rasala BA, Ramos C, Harel A, Forbes DJ (2008) Capture of AT-rich chromatin by ELYS recruits POM121 and NDC1 to initiate nuclear pore assembly. Mol Biol Cell 19(9):3982–3996. doi:10.1091/mbc.E08-01-0012
67. Gillespie PJ, Khoudoli GA, Stewart G, Swedlow JR, Blow JJ (2007) ELYS/MEL-28 chromatin association coordinates nuclear pore complex assembly and replication licensing. Curr Biol 17(19):1657–1662. doi:10.1016/j.cub.2007.08.041
68. Walther TC, Alves A, Pickersgill H, Loiodice I, Hetzer M, Galy V, Hulsmann BB, Kocher T, Wilm M, Allen T, Mattaj IW, Doye V (2003) The conserved Nup107-160 complex is critical for nuclear pore complex assembly. Cell 113(2):195–206
69. Harel A, Orjalo AV, Vincent T, Lachish-Zalait A, Vasu S, Shah S, Zimmerman E, Elbaum M, Forbes DJ (2003) Removal of a single pore subcomplex results in vertebrate nuclei devoid of nuclear pores. Mol Cell 11(4):853–864
70. Qi W, Yu H (2006) The spindle checkpoint and chromosomal stability. Genome Dyn 1:116–130. doi:10.1159/000092504
71. Ohba T, Nakamura M, Nishitani H, Nishimoto T (1999) Self-organization of microtubule asters induced in *Xenopus* egg extracts by GTP-bound Ran.[see comment]. Science 284(5418):1356–1358
72. Wiese C, Wilde A, Moore MS, Adam SA, Merdes A, Zheng Y (2001) Role of importin-beta in coupling Ran to downstream targets in microtubule assembly. Science 291(5504):653–656
73. Blower MD, Nachury M, Heald R, Weis K (2005) A Rae1-containing ribonucleoprotein complex is required for mitotic spindle assembly. Cell 121(2):223–234. doi:10.1016/j.cell.2005.02.016
74. Wong RW, Blobel G, Coutavas E (2006) Rae1 interaction with NuMA is required for bipolar spindle formation. Proc Natl Acad Sci U S A 103(52):19783–19787. doi:10.1073/pnas.0609582104
75. Belgareh N, Rabut G, Bai SW, van Overbeek M, Beaudouin J, Daigle N, Zatsepina OV, Pasteau F, Labas V, Fromont-Racine M, Ellenberg J, Doye V (2001) An evolutionarily conserved NPC subcomplex, which redistributes in part to kinetochores in mammalian cells. J Cell Biol 154(6):1147–1160
76. Audhya A, Desai A, Oegema K (2007) A role for Rab5 in structuring the endoplasmic reticulum. J Cell Biol 178(1):43–56
77. Buch C, Lindberg R, Figueroa R, Gudise S, Onischenko E, Hallberg E (2009) An integral protein of the inner nuclear membrane localizes to the mitotic spindle in mammalian cells. J Cell Sci 122(Pt 12):2100–2107. doi:10.1242/jcs.047373

78. Drummond S, Ferrigno P, Lyon C, Murphy J, Goldberg M, Allen T, Smythe C, Hutchison CJ (1999) Temporal differences in the appearance of NEP-B78 and an LBR-like protein during *Xenopus* nuclear envelope reassembly reflect the ordered recruitment of functionally discrete vesicle types. J Cell Biol 144(2):225–240
79. Hetzer M, Meyer HH, Walther TC, Bilbao-Cortes D, Warren G, Mattaj IW (2001) Distinct AAA-ATPase p97 complexes function in discrete steps of nuclear assembly. Nat Cell Biol 3(12):1086–1091 [see comment]
80. Salpingidou G, Rzepecki R, Kiseleva E, Lyon C, Lane B, Fusiek K, Golebiewska A, Drummond S, Allen TD, Ellis JA, Smythe C, Goldberg MW, Hutchison CJ (2008) NEP-A and NEP-B both contribute to nuclear pore formation in *Xenopus* eggs and oocytes. J Cell Sci 121(5):706–716
81. Vigers GP, Lohka MJ (1991) A distinct vesicle population targets membranes and pore complexes to the nuclear envelope in *Xenopus* eggs. J Cell Biol 112(4):545–556
82. Vigers GP, Lohka MJ (1992) Regulation of nuclear envelope precursor functions during cell division. J Cell Sci 102(Pt 2):273–284
83. Anderson DJ, Hetzer MW (2007) Nuclear envelope formation by chromatin-mediated reorganization of the endoplasmic reticulum. Nat Cell Biol 9(10):1160–1166. doi:10.1038/ncb1636
84. Anderson DJ, Hetzer MW (2008) Reshaping of the endoplasmic reticulum limits the rate for nuclear envelope formation. J Cell Biol 182(5):911–924. doi:10.1083/jcb.200805140
85. Ellenberg J, Siggia ED, Moreira JE, Smith CL, Presley JF, Worman HJ, Lippincott-Schwartz J (1997) Nuclear membrane dynamics and reassembly in living cells: targeting of an inner nuclear membrane protein in interphase and mitosis. J Cell Biol 138(6):1193–1206
86. Yang L, Guan T, Gerace L (1997) Integral membrane proteins of the nuclear envelope are dispersed throughout the endoplasmic reticulum during mitosis. J Cell Biol 137(6):1199–1210
87. Figueroa R, Gudise S, Larsson V, Hallberg E (2010) A transmembrane inner nuclear membrane protein in the mitotic spindle. Nucleus 1(3):249–253. doi:10.4161/nucl.1.3.11740
88. Sonnichsen B, Koski LB, Walsh A, Marschall P, Neumann B, Brehm M, Alleaume AM, Artelt J, Bettencourt P, Cassin E, Hewitson M, Holz C, Khan M, Lazik S, Martin C, Nitzsche B, Ruer M, Stamford J, Winzi M, Heinkel R, Roder M, Finell J, Hantsch H, Jones SJ, Jones M, Piano F, Gunsalus KC, Oegema K, Gonczy P, Coulson A, Hyman AA, Echeverri CJ (2005) Full-genome RNAi profiling of early embryogenesis in *Caenorhabditis elegans*. Nature 434(7032):462–469. doi:10.1038/nature03353
89. Pekovic V, Harborth J, Broers JL, Ramaekers FC, van Engelen B, Lammens M, von Zglinicki T, Foisner R, Hutchison C, Markiewicz E (2007) Nucleoplasmic LAP2alpha-lamin A complexes are required to maintain a proliferative state in human fibroblasts. J Cell Biol 176(2):163–172
90. Laguri C, Gilquin B, Wolff N, Romi-Lebrun R, Courchay K, Callebaut I, Worman HJ, Zinn-Justin S (2001) Structural characterization of the LEM motif common to three human inner nuclear membrane proteins. Structure 9(6):503–511
91. Lin F, Blake DL, Callebaut I, Skerjanc IS, Holmer L, McBurney MW, Paulin-Levasseur M, Worman HJ (2000) MAN1, an inner nuclear membrane protein that shares the LEM domain with lamina-associated polypeptide 2 and emerin. J Biol Chem 275(7):4840–4847
92. Brachner A, Reipert S, Foisner R, Gotzmann J (2005) LEM2 is a novel MAN1-related inner nuclear membrane protein associated with A-type lamins. J Cell Sci 118(Pt 24):5797–5810
93. Lee KK, Wilson KL (2004) All in the family: evidence for four new LEM-domain proteins Lem2 (NET-25), Lem3, Lem4 and Lem5 in the human genome. Symp Soc Exp Biol 56:329–339
94. Schirmer EC, Florens L, Guan T, Yates JR 3rd, Gerace L (2003) Nuclear membrane proteins with potential disease links found by subtractive proteomics. Science 301(5638):1380–1382
95. Wagner N, Schmitt J, Krohne G (2004) Two novel LEM-domain proteins are splice products of the annotated *Drosophila melanogaster* gene CG9424 (Bocksbeutel). Eur J Cell Biol 82(12):605–616

96. McGee MD, Stagljar I, Starr DA (2009) KDP-1 is a nuclear envelope KASH protein required for cell-cycle progression. J Cell Sci 122(Pt 16):2895–2905. doi:10.1242/jcs.051607
97. Schneider M, Lu WS, Neumann S, Brachner A, Gotzmann J, Noegel AA, Karakesisoglou I (2011) Molecular mechanisms of centrosome and cytoskeleton anchorage at the nuclear envelope. Cell Mol Life Sci 68(9):1593–1610. doi:10.1007/s00018-010-0535-z
98. Roux KJ, Crisp ML, Liu Q, Kim D, Kozlov S, Stewart CL, Burke B (2009) Nesprin 4 is an outer nuclear membrane protein that can induce kinesin-mediated cell polarization. Proc Natl Acad Sci U S A 106(7):2194–2199. doi:10.1073/pnas.0808602106
99. Korfali N, Srsen V, Waterfall M, Batrakou DG, Pekovic V, Hutchison CJ, Schirmer EC (2011) A flow cytometry-based screen of nuclear envelope transmembrane proteins identifies NET4/Tmem53 as involved in stress-dependent cell cycle withdrawal. PLoS One 6(4):e18762. doi:10.1371/journal.pone.0018762
100. Haffner C, Dettmer U, Weiler T, Haass C (2007) The Nicastrin-like protein Nicalin regulates assembly and stability of the Nicalin-nodal modulator (NOMO) membrane protein complex. J Biol Chem 282(14):10632–10638
101. Osada S, Ohmori SY, Taira M (2003) XMAN1, an inner nuclear membrane protein, antagonizes BMP signaling by interacting with Smad1 in *Xenopus* embryos. Development 130(9):1783–1794
102. Pan D, Estevez-Salmeron LD, Stroschein SL, Zhu X, He J, Zhou S, Luo K (2005) The integral inner nuclear membrane protein MAN1 physically interacts with the R-Smad proteins to repress signaling by the transforming growth factor-{beta} superfamily of cytokines. J Biol Chem 280(16):15992–16001
103. Jiang X, Xia L, Chen D, Yang Y, Huang H, Yang L, Zhao Q, Shen L, Wang J (2008) Otefin, a nuclear membrane protein, determines the fate of germline stem cells in *Drosophila* via interaction with Smad complexes. Dev Cell 14(4):494–506. doi:10.1016/j.devcel.2008.02.018
104. Markiewicz E, Tilgner K, Barker N, van de Wetering M, Clevers H, Dorobek M, Hausmanowa-Petrusewicz I, Ramaekers FC, Broers JL, Blankesteijn WM, Salpingidou G, Wilson RG, Ellis JA, Hutchison CJ (2006) The inner nuclear membrane protein emerin regulates beta-catenin activity by restricting its accumulation in the nucleus. EMBO J 25(14):3275–3285
105. Huber MD, Guan T, Gerace L (2009) Overlapping functions of nuclear envelope proteins NET25 (Lem2) and emerin in regulation of extracellular signal-regulated kinase signaling in myoblast differentiation. Mol Cell Biol 29(21):5718–5728
106. Liu GH, Guan T, Datta K, Coppinger J, Yates J 3rd, Gerace L (2009) Regulation of myoblast differentiation by the nuclear envelope protein NET39. Mol Cell Biol 29(21):5800–5812
107. Zarubin T, Han J (2005) Activation and signaling of the p38 MAP kinase pathway. Cell Res 15(1):11–18. doi:10.1038/sj.cr.7290257
108. Schermelleh L, Carlton PM, Haase S, Shao L, Winoto L, Kner P, Burke B, Cardoso MC, Agard DA, Gustafsson MG, Leonhardt H, Sedat JW (2008) Subdiffraction multicolor imaging of the nuclear periphery with 3D structured illumination microscopy. Science 320(5881):1332–1336
109. Wu C, Orozco C, Boyer J, Leglise M, Goodale J, Batalov S, Hodge CL, Haase J, Janes J, Huss JW 3rd, Su AI (2009) BioGPS: an extensible and customizable portal for querying and organizing gene annotation resources. Genome Biol 10(11):R130. doi:10.1186/gb-2009-10-11-r130
110. Fukuzawa S, Hashimura T, Sasaki M, Yamabe H, Yoshida O (1995) Nuclear morphometry for improved prediction of the prognosis of human bladder carcinoma. Cancer 76(10):1790–1796
111. Ladekarl M, Boek-Hansen T, Henrik-Nielsen R, Mouritzen C, Henriques U, Sorensen FB (1995) Objective malignancy grading of squamous cell carcinoma of the lung. Stereologic estimates of mean nuclear size are of prognostic value, independent of clinical stage of disease. Cancer 76(5):797–802

Nuclear Envelope Regulation of Signaling Cascades

Jason C. Choi and Howard J. Worman

Abstract The ultimate purpose of signal transduction is to transmit extracellular or cytoplasmic stimuli to the nuclear interior to elicit a cellular response, mediated primarily through changes in gene expression. The evolution of the nuclear envelope and the consequent compartmentalization of the genome, which is a defining feature of eukaryotes, introduced a physical barrier to the free access of genes. Initially regarded as nothing more than this, a physical barrier with selective permeability, recent findings have transformed our view of the nuclear envelope and its diverse roles in various aspects of cell biology and human diseases, much of which is only beginning to be understood. The realization that mutations in genes encoding nuclear envelope proteins cause a diverse array of tissue-selective diseases often referred to as "laminopathies" has provided new insight into structural and regulatory functions of the nuclear envelope. Genetic mutations causing abnormalities in the nuclear envelope can lead to dysregulated signaling that underlies pathogenesis of these diseases. The emerging picture indicates that the nuclear envelope is a node that fine-tunes signaling output and as such it may play a role in the biology of cancer.

Keywords Nucleus • Nuclear envelope • Nuclear lamina • Signal transduction • Lamins • Integral membrane proteins • Laminopathies • Inner nuclear membrane

---

J.C. Choi • H.J. Worman (✉)
Department of Medicine, College of Physicians and Surgeons, Columbia University, 630W 168th Street P&S 9-435, New York, NY 10032, USA

Department of Pathology and Cell Biology, College of Physicians and Surgeons, Columbia University, 630W 168th Street P&S 9-508, New York, NY 10032, USA
e-mail: jcc2187@columbia.edu; hjw14@columbia.edu

E.C. Schirmer and J.I. de las Heras (eds.), *Cancer Biology and the Nuclear Envelope*, Advances in Experimental Medicine and Biology 773, DOI 10.1007/978-1-4899-8032-8_9,

Abbreviations

ECM	Extracellular matrix
EDMD	Emery–Dreifuss muscular dystrophy
ER	Endoplasmic reticulum
HGPS	Hutchinson–Gilford progeria syndrome
INM	Inner nuclear membrane
MAPK	Mitogen activated protein kinase
MSC	Mesenchymal stem cells
ONM	Outer nuclear membrane

Introduction

The eukaryotic nucleus is a double membrane-bound organelle that encloses and regulates the genome by providing a unique molecular and biochemical environment distinct from the rest of the cell. This is primarily achieved by the nuclear envelope, which consists of several discrete elements, the most prominent of which are the nuclear membranes. The nuclear membranes consist of three morphologically distinguishable but interconnected domains: the inner (INM) and outer (ONM) nuclear membranes and the pore membranes. The INM and ONM are separated by a 30–50 nm lumen also termed the perinuclear space. They periodically fuse to form pore membranes adjacent to the nuclear pore complexes that occupy the channel the two adjacent pore membranes create [1–3]. Because the INM and ONM are connected at the nuclear pore complex by the pore membranes, they actually represent discrete domains of a single membrane system [1, 2]. This membrane system is further contiguous with the rough endoplasmic reticulum (ER), which is directly connected to the ONM and contains ribosomes on its outer surface [1, 2]. Despite originating from a common continuous structure, the INM, pore membranes, ONM, and ER maintain their identities largely through the enrichment of a unique profile of integral membrane proteins and other associated proteins, as well as specific composition of lipids and cholesterols that constitute the membranes themselves [4–6]. Data from subtractive proteomics analysis indicate that the INM of interphase mammalian hepatocytes contains approximately 80 unique transmembrane proteins; however, only a few have been characterized in great detail [7–10]. Furthermore, the protein composition of the nuclear envelope membranes varies in different mammalian tissues [11].

In metazoans, a prominent feature of the nuclear envelope is the nuclear lamina, a 20–50 nm thick proteinaceous meshwork lining the inner surface of the INM [12–15]. The nuclear lamina is widely accepted to provide tensile and structural integrity to the nucleus, as well as anchorage for chromatin and various integral proteins of the INM [16–18]. Previously regarded as a relatively inert structural scaffold providing mechanical support to the nucleus, recent findings indicate that the nuclear lamina plays a much more dynamic role in regulating various cellular and molecular processes including nuclear architecture and size, epigenetic regulation and transcription, DNA repair and stability, cellular senescence, and signal transduction [16, 17, 19]. The nuclear lamina is composed of A-type and B-type lamins, which are members of the intermediate filament protein family. In mammalian somatic cells, primarily four lamin proteins are expressed that are encoded by three separate genes [15, 20]. Lamin B1 and B2 (B-type lamins) are encoded by *LMNB1* and *LMNB2*, respectively, whereas lamins A and C (A-type lamins) are both encoded by *LMNA* and produced by alternative mRNA splicing [19]. Germ-cell-specific isoforms lamin C2 and lamin B3 produced by alternative splicing of transcripts from *LMNA* and *LMNB1*, respectively, have also been described [21, 22]. The expression of B-type lamins appears to be ubiquitous; it occurs early in embryonic development and persists through adult life in most cell types [23–25]. In contrast, the expression of A-type lamins is asynchronous, developmentally regulated and detected after tissue differentiation, consistent with the notion that A-type lamins are a marker of differentiation [23–26]. In support of this idea, during murine gestation, the very first differentiated cells to emerge from the dividing zygote called trophoblast cells are also the first cell-type to express A-type lamins [26]. The expression of A-type lamins is virtually undetectable in the early phases of murine embryonic development and organogenesis; only after embryonic day 10 are appreciable levels of A-type lamins detected in the embryo proper [25, 26].

Reevaluation of A-type lamins as more than mere structural proteins began when a diverse array of human diseases, sometimes collectively called "laminopathies," was connected to mutations in *LMNA* [27, 28]. Currently, more than a dozen differentially named clinical diseases have been identified and can be grouped by the affected tissues. They are striated muscle (a spectrum of overlapping clinical entities with heart involvement including Emery–Dreifuss muscular dystrophy [EDMD], limb girdle muscular dystrophy 1B, and isolated dilated cardiomyopathy), adipose tissue (Dunnigan-type familial partial lipodystrophy and some unusual partial lipodystrophy syndromes), peripheral nerve (Charcot–Marie–Tooth disorder type 2B1), and those that involve multiple tissues systems such as mandibuloacral dysplasia and Hutchinson–Gilford progeria syndrome (HGPS). Mutations in genes encoding several integral proteins of the INM that bind to A-type lamins have also been shown to cause tissue-selective diseases also often referred to as laminopathies. While these rare inherited diseases do not include cancer as a phenotype, research on their pathogenesis has shed light on the role of the nuclear envelope in regulating signaling cascades that are often involved in carcinogenesis and metastasis. These discoveries are of potentially major significance to cancer biologists, as nuclear envelope structural alterations frequently observed in laminopathies also often occur in tumors with dysregulated signaling [29].

A-Type Lamins

The presence of numerous diseases arising from mutations in *LMNA*, as well as several mouse models that recapitulate the human diseases, have provided insights into and expanded on the cellular functions of A-type lamins. To understand the pathogenic mechanisms arising from *LMNA* mutations, it is necessary to understand

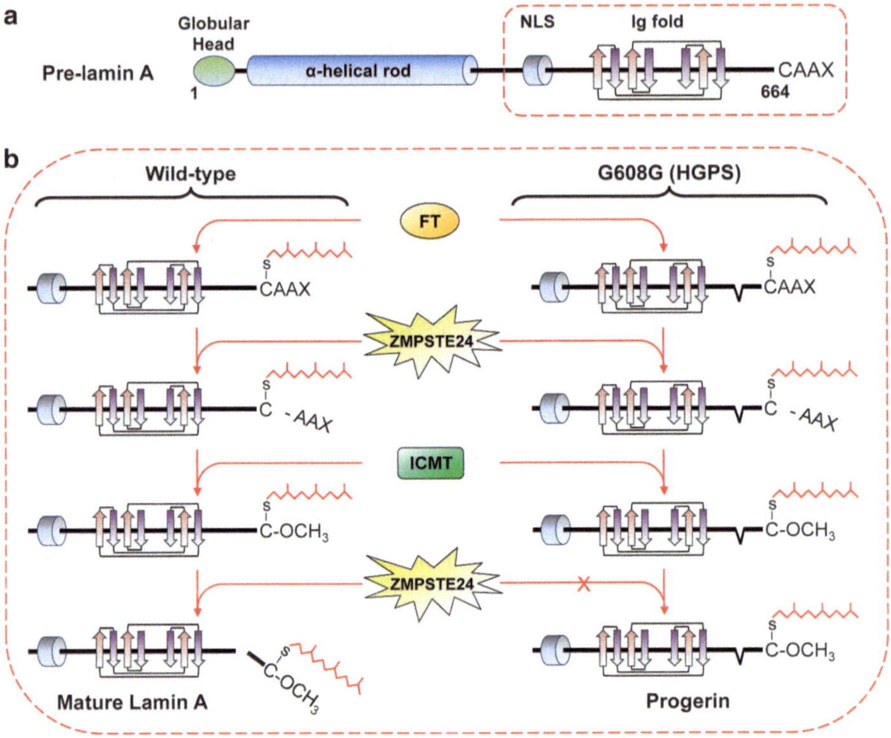

Fig. 1 Schematic diagram of prelamin A structure and processing. (**a**). Structural composition of prelamin A protein. C-terminus of prelamin A (indicated by *dashed red lines*) contains the NLS (nuclear localization signal), Ig (immunoglobulin-like) fold, and the CAAX box. (**b**). Catalytic processing of C-terminus of prelamin A to produce mature lamin A (*left*) or progerin (*right*). FT and ICMT denote protein farnesyltransferase and isoprenylcysteine carboxyl methyltransferase, respectively

the organization of the protein structures as well as the mechanisms involved in producing a mature protein. In humans, lamins A and C are identical for the first 566 amino acid residues and diverge only at their C-termini [13, 15, 30, 31]. To some extent, lamin A and C are functionally redundant, given that mice engineered to express only lamin C appear to be virtually normal [32]. However, unlike lamin C, lamin A undergoes complicated enzymatic processing to produce a mature protein (see below), the dysregulation of which directly contributes to the pathogenesis of a subset of laminopathies.

Similar to other members of the intermediate filament protein family, lamin A (as well as all lamins) contains a central α-helical rod domain that is flanked by globular N-terminal head and C-terminal tail domains [19, 33, 34] (Fig. 1a). The C-terminal domain contains an immunoglobulin-like β-fold and a nuclear localization signal between the α-helical rod domain and this fold [35, 36]. The central a-helical rod domains of lamins are highly conserved among other members of intermediate filament proteins and are thought to be essential for self-assembly into higher order structures [13, 15]. All the mammalian lamins except for lamin C and lamin C2 contain CAAX (where "C" is cysteine, "A" is an aliphatic amino acid, and "X" is a hydrophobic residue) motifs at their C-termini, which is a signal for farneslyation and other posttranslational modifications [37–39].

Lamin A is synthesized as a precursor molecule prelamin A that undergoes several rounds of catalytic processing to produce mature lamin A (Fig. 1b) [37, 39]. Prelamin A contains the C-terminal CAAX motif, which serves as a primer for a series of sequential enzymatic processing reactions [40–42]. The initial reaction involves farnesylation of the cysteine residue by protein farnesyltransferase. The farnesylation reaction is followed by cleavage of the AAX residues. The exposed farnesylcysteine is then carboxymethylated by isoprenylcysteine carboxyl methyltransferase. Lastly, mature lamin A is produced following a second cleavage reaction catalyzed by the metalloproteinase ZMPSTE24 that removes the last 15 C-terminal amino acid residues on prelamin A from the carboxymethylated farnesylcysteine. It is generally believed that the hydrophobic modification of the CAAX motif followed by its subsequent removal is required for proper localization and assembly of lamin A at the hydrophobic nuclear envelope environment [38, 43, 44].

Disease-Causing Mutations in *LMNA*

The first human disease shown to result from mutations in *LMNA* was autosomal dominant EDMD [45]. Disorders that selectively affect skeletal and heart muscle, which manifest clinically as a dilated cardiomyopathy with conduction system defects, are the most prevalent of the laminopathies [19]. Most *LMNA* mutations that cause striated muscle disease are autosomal dominant missense or small in-frame deletions with the encoded protein exhibiting a similar half-life as the wild-type proteins [46]. These mutations result in alterations scattered throughout the protein, suggesting that they cause disease by altering the global structure of the nuclear lamina. However, some *LMNA* mutations causing striated muscle disease are splice site or nonsense, leading to haploinsufficiency of A-type lamins. In contrast, dominantly inherited missense mutations that cause Dunnigan-type familial partial lipodystrophy, characterized by the selective loss of peripheral subcutaneous adipose tissue and the subsequent development of insulin resistance and diabetes mellitus, are centered around the immunoglobulin-like fold domain and predicted to alter its surface charge [35, 36, 47–49].

Perhaps the most unique *LMNA* mutations are those causing HGPS, which is a multiple system disorder with some features of accelerated aging. Although they appear grossly normal at birth, children with HGPS exhibit growth retardation, micrognathia, reduced subcutaneous fat, alopecia, osteoporosis, and skin mottling within the first year of life [19]. They invariably develop premature vascular occlusive disease, which is the most life threatening clinical manifestation of the disease. Most individuals with HGPS die in the second decade of life from coronary artery or cerebrovascular disease. HGPS is predominantly caused by a dominant de novo G608G (nucleotide 1824 C>T) mutation within exon 11 of *LMNA*. Although the amino acid sequence is unchanged, the mutation activates a cryptic splice donor site, causing a deletion of 150 bp within exon 11 [50, 51].

The G608G mutation causes an in-frame deletion of 50 amino acids near the C-terminus of prelamin A that includes the ZMPSTE24 cleavage site (see processing of prelamin A above). Therefore, ZMPSTE24 cannot catalyze cleavage of the C-terminal carboxymethylated farnesylcysteine and the 15 upstream amino acid residues, leading to the production of a truncated, permanently farnesylated variant of prelamin A, termed "progerin" (Fig. 1). Mice with genetic deletion of *Zmpste24* accumulate farnesylated lamin A in their tissue and exhibit progeroid symptoms similar to those caused by the *Lmna* G608G mutation [52, 53]. Additionally, the loss of *ZMPSTE24* in humans leads to the neonatal lethal progeria syndrome restrictive dermopathy [54, 55]. Thus, progerin and unprocessed prelamin A appear to be "toxic" proteins responsible for cellular alterations that lead to progeria phenotypes. Indeed, treatment of both *Zmpste24* null mice and $Lmna^{G608G/G608G}$ mice with inhibitors for protein farnesyltransferase ameliorate progeroid phenotypes, indicating that the farnesylated forms of these proteins are responsible for disease [56, 57]. Despite the promising results in mice, a clinical trial with the protein farnesyltransferase inhibitor lonafarnib only provided a modest benefit for patients with HGPS [58]. An emerging consensus from studying patients with laminopathies and mouse modes suggests that A-type lamins play a direct and dynamic role in regulating signal transduction and that the expression of mutant variants alters signaling pathways that underlie disease pathogenesis. The pleiotropic nature of laminopathies hints at a central and complex role of A-type lamins in altering the activity of signaling cascades, presumably acting in a cell type and tissue-specific fashion. Similarly, other proteins of the INM appear to regulate or influence the activity of cell signaling cascades (Fig. 2).

Mitogen Activated Protein Kinase (MAPK) Signaling

One well-characterized signaling pathway connected to lamin A is MAPK. The MAPK pathway consists of three main branches: extracellular signal-regulated kinase 1/2 (ERK1/2), c-Jun NH_2-terminal kinase (JNK), and p38 [59, 60]. The defining feature of the MAPK pathway is the three-tiered "core signaling module," which is a tri-layered dual-specificity kinase cascade system. Starting from upstream in the cascade, a diverse group of protein kinase families collectively termed MAPK kinase kinases (MAP3Ks) phosphorylate conserved serine/threonine residues within the MAPK kinases (MAP2Ks but also referred to as MEKs or MKKs). Once phosphorylated, MAP2Ks in turn phosphorylate conserved tyrosine/threonine residues within the MAPKs, which are the ERKs, JNKs, and p38. The phosphorylated MAPKs then transit to specific subcellular compartments such as the nucleus, the localization of which is predominantly mediated by binding interactions with

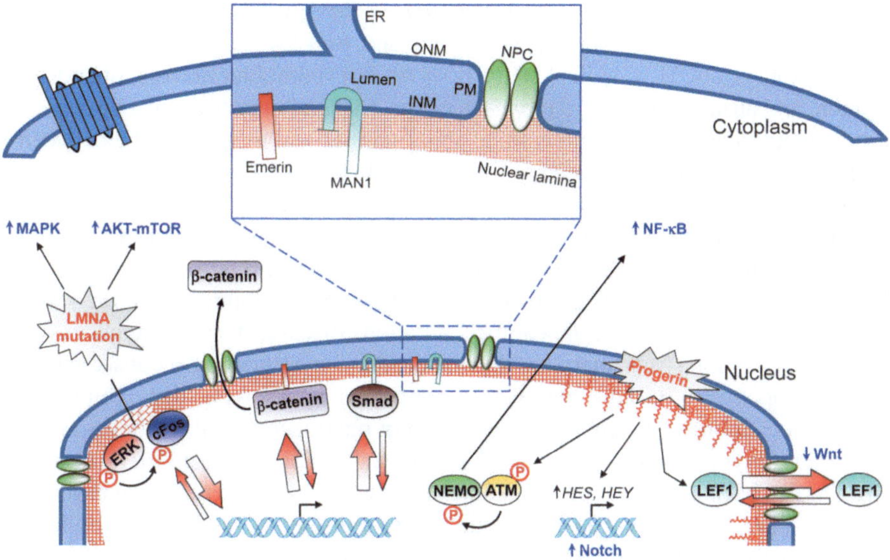

Fig. 2 Signaling pathways regulated by proteins of the nuclear envelope. The three domains of the nuclear membranes are shown in the *magnified box*: the inner (INM) and outer (ONM) nuclear membranes and the pore membranes (PM). The INM and ONM are separated by the lumen and the ONM is contiguous with the endoplasmic reticulum (ER). The signaling pathways altered by lamin A, emerin, and MAN1 as described in the current review are graphically illustrated

sequestering anchors and components of the nuclear transport machinery, where they regulate various cellular processes including growth, differentiation, metabolism, stress response, inflammation, apoptosis, and autophagy [59–61]. Dysregulation of MAPK pathways, leading to the disruption of these diverse cellular processes they regulate, has been shown to be both driving and contributing factors in oncogenesis [62–64].

The initial evidence potentially linking lamin A with MAPKs was the finding that c-Fos, which is an ERK1/2-activated transcription factor that positively regulates the cell cycle, is localized at the nuclear envelope and that this localization is dependent on its interaction with lamin A [65]. Under conditions of low mitogenic signaling, lamin A sequestered c-Fos at the nuclear envelope and this effect was reversed with increased mitogenic signaling. Moreover, enhanced cell proliferation was observed in cells lacking wild-type lamin A whereas a decreased proliferative capacity was noted with lamin A overexpression, demonstrating that it can function as a molecular switch that regulates c-Fos-dependent cell proliferation [65].

The linkage between MAPKs and lamin A was further revealed with the help of a mouse model of a laminopathy. It was discovered that the ERK1/2 and JNK branches of MAPK signaling, and later p38, are enhanced in hearts of a "knock in" mouse model of EDMD [66–68]. These mice (*Lmna*$^{\mathrm{H222P/H222P}}$) carry a point mutation in which a histidine at position 222 in lamin A and lamin C is changed to proline, mimicking a missense mutation that causes autosomal dominant EDMD in humans [66]. By 8 weeks of age, male *Lmna*$^{H222P/H222P}$ mice develop left ventricular dilatation and decreased fractional shortening and, by 12 weeks, abnormalities in the conduction system. Histological analysis shows left ventricular fibrosis, fiber degeneration, and atrial dilatation by 16 weeks of age. Skeletal muscle dystrophy develops with slower kinetics relative to cardiac abnormalities and these mice die between 4 and 8 months of age, as opposed to a ~2 years life span of a wild-type mice [66].

In 10-week-old male *Lmna*$^{H222P/H222P}$ mice, ERK1/2 and JNK are activated specifically in heart and skeletal muscle, resulting in their enhanced nuclear accumulation where they activate the expression of their downstream targets genes [67, 69]. The first sign of MAPK activation in hearts of *Lmna*$^{H222P/H222P}$ mice occurs at 4 weeks of age, which is prior to the appearance of cardiac disease, strongly suggesting that ERK1/2 activation is a primary pathogenic mechanism rather than a consequence of cardiomyopathy [67]. This notion was confirmed by showing that inhibition of MAPK activation by systemic administration of pharmacological inhibitors ameliorates cardiomyopathy in *Lmna*$^{H222P/H222P}$ mice [68, 70–72]. Increased ERK1/2 signaling has also been observed in human fibroblast as well as in hearts of human subjects carrying *LMNA* mutation causing striated muscle disease [73, 74]. These collective findings demonstrated a causative link between aberrant MAPK activation and the pathogenesis of striated muscle disease caused by *LMNA* mutations.

Demonstration that lamin A binding to c-Fos depends on the c-Fos phosphorylation status helped identify mechanistic insights into how lamin A regulates ERK1/2 activity [75]. Notably, the ERK1/2 interaction with c-Fos was mediated by lamin A, which may function as a scaffold that facilitates efficient binding. This is achieved by the ability of lamin A to bind to both ERK1/2 and c-Fos [65, 75]. The ERK1/2 binding site on lamin A was mapped to a span of amino acid sequence 247–355, which corresponds to the helical rod domain of lamin A [75]. Therefore, ERK1/2 co-localizes with c-Fos at the nuclear envelope, which is mediated by lamin A binding, leading to phosphorylation and release of c-Fos from the nuclear envelope.

Although the demonstration of a direct interaction of ERK1/2 with lamin A was an important step in elucidating the molecular mechanism of how the nuclear envelope regulates MAPK signaling, several unanswered questions remain. For example, do mutations leading to alterations in lamin A, including mutations altering the protein beyond the ERK1/2 binding domain, affect its interaction with ERK1/2? Also, does the phosphorylation status of ERK1/2 alter its binding capacity to lamin A? Lastly, what are the pathogenic mechanisms initiated by enhanced MAPK activation? Answering these questions will undoubtedly delineate a more complete picture of the complex interplay between lamin A and ERK1/2 at the nuclear envelope.

AKT-Mammalian Target of Rapamycin Signaling

Another signaling pathway, often dysregulated in cancer that has been linked to alterations in A-type lamins, is the AKT- mammalian target of rapamycin (mTOR) signaling pathway. Historically studied in the context of insulin and insulin-like growth factor signaling, AKT-mTOR signaling is frequently co-activated along with

ERK1/2 in response to growth factor signaling and in various forms of cancer [76, 77]. In response to insulin, AKT, which is a serine/threonine kinase, can activate mTOR by two separate mechanisms. AKT-mediated phosphorylation of PRAS40, an mTOR binding protein that acts as a repressor, disrupts its binding to mTOR [78, 79]. Alternatively, AKT can also provide an inhibitory phosphorylation on tuberous sclerosis complex 2 (TSC2) [80]. TSC2, together with its isoform TSC1, form a GTPase-activating protein for RHEB, a GTPase required for the protein kinase activity of the mTOR complex. As only GTP-bound RHEB can activate mTOR, AKT phosphorylation of TSC2 inhibits the GTPase activity of RHEB, leading to accumulation of GTP-bound RHEB and activation of mTOR. AKT-mTOR functions as the central signaling circuit that links diverse cellular metabolic processes such as glucose uptake, glycogen synthesis, and autophagy to properly coordinate cellular output in the form of protein synthesis, cell growth, and proliferation [80].

The initial indication that AKT-mTOR signaling may be activated in hearts of *Lmna*$^{H222P/H222P}$ mice that develop EDMD was evident in the same study that demonstrated enhanced ERK1/2 and JNK1/2 signaling [67]. Gene ontology term analyses on mRNA expression profiling of hearts from these mice revealed that the highest scoring signaling pathway was the insulin-like growth factor signaling [67]. A subsequent study confirmed that enhanced AKT-mTOR signaling occurred prior to the penetrance of disease phenotype, indicating that the enhanced signaling contributes to the disease pathogenesis [81]. This was confirmed with the demonstration that reducing mTOR signaling by systemic administration of temsirolimus, a rapamycin analog, improved heart function, confirming its contribution to disease pathogenesis [81]. Another group using a different mouse model of EDMD independently reported essentially identical findings [82]. Mice expressing a truncated form of lamin A (lamin AΔ8-11), which was originally believed to be *Lmna* null, exhibit severe postnatal growth retardation characterized by the appearance of muscular dystrophy and cardiomyopathy with none of the mice surviving by 8 weeks [83, 84]. Elevated mTOR signaling was observed in heart and skeletal muscle of these mice [82]. Pharmacologic reversal of elevated mTOR signaling by rapamycin improved both cardiac and skeletal muscle function. This treatment also increased survival in these mice, further confirming the causative role of mTOR signaling in striated muscle disease triggered by *LMNA* mutation [82]. Mechanistically, both groups identified that the enhanced mTOR signaling interfered with autophagy, an evolutionarily conserved self-degradative process that maintains cellular and energy homeostasis by recycling cytoplasmic or damaged/toxic proteins under periods of starvation or increased energy demand [85].

It is not known how alterations in A-type lamins trigger AKT-mTOR signaling. In addition to cardiac tissue from the two mouse models of EDMD described above, mouse embryonic fibroblasts from lamin AΔ8-11 expressing mice exhibit activated AKT signaling, suggesting that its activation is promiscuous and not specific to any mutation or tissue type [86]. Perhaps insulin-like growth factor, as identified in the original study demonstrating MAPK activation in hearts of *Lmna*$^{H222P/H222P}$ mice, may be involved given that enhanced AKT activity appears to be mediated in part by soluble factor(s) [87]. Adding an additional layer of complexity, lamin A itself can be phosphorylated by AKT, by which its expression can be regulated [88, 89].

Signaling Pathways Altered in HGPS

Physiological aging is associated with the progressive and irreversible deterioration of tissue maintenance and homeostasis. Given this simple fact, many have hypothesized that this homeostatic process is disrupted in the accelerated aging disorder HGPS. Aided by the availability of murine models as well as primary tissue from human patients, understanding of the pathogenic mechanisms involved in HGPS has expanded dramatically. Not surprisingly, given that HGPS is a multiple system disorder, multiple abnormalities in signaling pathways have been identified to contribute to the disease pathogenesis, many of which are indeed involved in tissue maintenance/homeostasis. Expression of the truncated prelamin A variant progerin appears to be the "culprit" in the dysregulation of these signaling pathways in HGPS [90]. The altered signaling pathways and their consequences to the disease pathogenesis will be described in the order they were discovered.

Notch signaling—HGPS primarily affects tissues that originate from the mesenchymal stem cell (MSC) lineage. Therefore, an attractive hypothesis is that progerin expression preferentially interferes with MSC function and/or viability that ultimately leads to differentiated tissue pathologies. This hypothesis gained support when progerin expression was shown to alter Notch signaling in MSCs [91]. The Notch signaling pathway is a highly conserved juxtacrine signaling system employed by adjacent cells to communicate with each other to synchronously regulate cell fate specification events and stem-cell differentiation [92]. Notch signaling is commonly dysregulated in many cancers and is thought to maintain cancer stem cells [93]. In human fibroblasts engineered to express progerin and in primary fibroblasts from patients with HGPS, several Notch effectors such as HES1, HES5, and HEY1 were all upregulated in expression [91]. Overexpression of progerin in immortalized MSCs also altered their differentiation capacity, leading to spontaneous preferential differentiation towards osteogenesis and away from adipogenesis [91]. This effect was reproduced when NICD, a key activator of Notch signaling, was overexpressed in immortalized MSCs, confirming that activated Notch causes differentiation defects.

Wnt signaling—The multiple system disorder HGPS has been attributed in part to deficient/defective deposition of extracellular matrix (ECM). Consisting of water, structural proteins and polysaccharides, the exact composition of which is specific to individual tissues, the ECM provides tissue architecture and binding surfaces for cellular constituents as well as biochemical and biomechanical signals required for tissue morphogenesis, differentiation, and response to injury [94]. A connection between Wnt signaling, the ECM, and HGPS was first suggested from a study using a progeria mouse model. *Lmna*$^{L530P/L530P}$ mice, which carry a mutation that causes EDMD in humans but a progeroid phenotype in mice, produce a truncated form of prelamin A that, like progerin, remains farnesylated [95, 96]. Postnatal fibroblasts isolated from *Lmna*$^{L530P/L530P}$ mice exhibit reduced expression of genes encoding ECM proteins but not fibroblasts isolated from their embryos, correlating with the fact that children with HGPS are apparently normal at birth and develop the first symptoms at 6–12 months of age [95]. This reduction in the expression of genes encoding ECM proteins was due to the reduced activity of the TCF4/LEF1 complex, a key downstream transcriptional effector of the Wnt signaling pathway. The reduced transcriptional activity was the result of decreased expression and nuclear accumulation of LEF1 [95]. Given the central role of Wnt signaling in ECM deposition, defective Wnt signaling is attractive as a pathogenic factor in HGPS, especially for the predominant abnormalities in bone. Wnt signaling is critical for cartilage development as well as osteoblast and chondrocyte differentiation during vertebrate skeletogenesis, and it is likely that its reduction as a consequence of progerin expression adversely affects these developmental processes [97, 98].

NF-κB signaling—Age related loss of tissue homeostasis is caused by various stress factors such as telomere shortening, reduced replenishment from depleted stem cells, and accumulation of damaged DNA and macromolecules [99, 100]. Cells from subjects with HGPS and *Zmpste24* null mice display premature senescence and accelerated accumulation of DNA damage, which may result from an abnormal nuclear lamina that interferes with overall nuclear structure, DNA integrity, and DNA replication [18, 101–103]. In addition, alterations in the capacity to sense cellular stress may exacerbate the loss of tissue homeostasis and this is widely believed to play an important role in progeroid syndromes [104]. Commonly associated with activated immune cells in response to pathogenic insult, NF-κB signaling functions as a sensor for stress signals emanating within the cell such as oxidative and genotoxic stress [105, 106]. Moreover, prolonged aberrant activation of NF-κB signaling has been linked to the aging process as well as age-related pathologies including atherosclerosis, diabetes, neurodegeneration, osteoporosis, and sarcopenia [107, 108]. Further supporting the involvement of aberrant NF-κB signaling and age-related diseases, transcriptional profiling of liver tissue isolated from *Zmpste24*$^{-/-}$ mice has revealed a signature indicative of enhanced NF-κB activation [109]. This activated NF-κB signaling was confirmed in both *Zmpste24*$^{-/-}$ and *Lmna*$^{G609G/G609G}$ (which is the murine equivalent of *LMNA*$^{G608G/G608G}$ mutation in human patients with HGPS) and occurred through ATM-NEMO-mediated mechanisms [109]. Reduction of NF-κB signaling by genetic and pharmacological means prevented progeroid features in the mice and extended longevity, confirming its contribution to pathogenesis [109].

Emerin

Integral proteins of the INM that bind to A-type lamins also affect signaling cascades and mutations in their genes cause human diseases. Emerin is a 29 kDa single pass transmembrane protein of the inner nuclear membrane that contains a LEM domain (*L*AP2, *E*merin, *M*AN1), a 45-residue motif that facilitates binding to a conserved metazoan chromatin protein termed barrier to autointegration factor (BAF) [110]. Encoded by *EMD* located on the X-chromosome, the first link between the nuclear envelope and inherited human disease was discovered based on the observation that X-linked EDMD is caused by mutations in emerin [111]. The localization of emerin at the INM is mediated by its interaction with A-type lamins [111–114].

The similar phenotypes of X-linked EDMD, caused by mutations in *EMD*, and the autosomal dominant form, caused by mutations in *LMNA*, suggests a functional coupling of emerin and A-type lamins at the nuclear lamina. Although a number of protein binding partners have been identified, no definitive function has emerged for emerin thus far.

One binding partner for emerin is β-catenin, a downstream effector of the canonical Wnt signaling pathway. In the absence of an activating signal, β-catenin is phosphorylated by the action of the Axin complex, composed of Axin, adenomatous polyposis coli, casein kinase 1, and glycogen synthase kinase 3 [98, 115]. This phosphorylation marks β-catenin for targeted degradation by the ubiquitin-proteasome pathway. Upon activation of Wnt signaling, β-catenin escapes proteasomal degradation and accumulates in the nucleus, where it activates Wnt target genes by acting as a transcriptional cofactor that interacts with transcription factors such as TCF/LEF [98, 115]. Emerin binding to β-catenin inhibits its activity by facilitating nuclear export, thereby preventing accumulation in the nucleus [116]. Emerin-deficient primary fibroblasts from patients with X-linked EDMD display increased β-catenin accumulation in the nucleus and a corresponding increase in their proliferative capacity [116]. These effects have been replicated in wild-type fibroblasts expressing constitutively active β-catenin, confirming that the abnormally increased growth phenotype in emerin null human fibroblasts is mediated by the enhanced β-catenin signaling [116]. There is also evidence that emerin-mediated β-catenin activity is co-regulated by a larger protein complex containing nesprin2, a 796 kDa ONM protein, and β-catenin [117].

MAN1

MAN1 (also known as LEMD3) is a 97 kDa double pass transmembrane protein with both N and C-termini facing the nucleoplasm [118–120]. Similar to emerin, MAN1 contains a LEM domain near its N-terminus [110, 119, 121]. The N-terminal portion of MAN1 also interacts with lamin A, which is necessary for its retention in the inner nuclear membrane [118, 122].

Several independent studies from different laboratories have demonstrated that MAN1 is a regulator of receptor-mediated Smads (rSmads). rSmads are intracellular mediators of the transforming growth factor-β (TGF-β) family of cytokines, which have been demonstrated to play an intimate role in cancer metastasis [123]. Two major subfamilies are the TGF-β and bone morphogenic protein (BMP) subfamilies, the signals of which are mediated by Smad1 and Smad2/3, respectively. Activation of cell surface receptors leads to phosphorylation and nuclear translocation of rSmads where they interact with Smad4 to form complexes that regulate transcription form various target genes [123]. The C-terminus of MAN1 binds to rSmads and short circuits TGF-β and BMP signaling [124–126]. This inhibitory effect has been confirmed in MAN1 knockout mice in which enhanced TGF-β signaling interferes with embryonic vasculogenesis [127, 128]. Consistent with these findings, heterozygous human subjects carrying a hypomorphic mutation in the gene encoding MAN1 display bone and skin abnormalities, likely due to the altered balance of TGF-β and BMP signaling in these tissues during postembryonic development [124]. Mechanistically, MAN1 suppression of TGF-β and BMP signaling is thought to be mediated through inhibitory sequestration of rSmads at the inner nuclear membrane, thereby preventing their ability to migrate to gene enhancer regions and activate transcription. Experimental results have also shown that MAN1 reduces rSmad phosphorylation and nuclear localization [126, 129, 130].

Concluding Remarks

Pathologists have historically used the detection of abnormal nuclear shape as one of the diagnostic markers of cancer. Alterations in nuclear morphology also occur in laminopathies [29, 131]. In both cancer and laminopathies, these nuclear structural abnormalities are frequently associated with corresponding defects in signaling cascades (Fig. 2). Hence, as in the laminopathies, alterations in the nuclear envelope may be involved in dysregulated signaling cascades in carcinogenesis. While there is a dearth of experimental results at this time, this hypothesis is extremely appealing. Future research directed at determining the nuclear envelope proteome in various cancers could be performed as a first step towards testing this hypothesis.

References

1. Gruenbaum Y, Margalit A, Goldman RD, Shumaker DK, Wilson KL (2005) The nuclear lamina comes of age. Nat Rev Mol Cell Biol 6(1):21–31. doi:10.1038/nrm1550
2. Stewart CL, Roux KJ, Burke B (2007) Blurring the boundary: the nuclear envelope extends its reach. Science 318(5855):1408–1412. doi:10.1126/science.1142034
3. Terry LJ, Shows EB, Wente SR (2007) Crossing the nuclear envelope: hierarchical regulation of nucleocytoplasmic transport. Science 318(5855):1412–1416. doi:10.1126/science.1142204
4. Schirmer EC, Gerace L (2005) The nuclear membrane proteome: extending the envelope. Trends Biochem Sci 30(10):551–558. doi:10.1016/j.tibs.2005.08.003
5. Cascianelli G, Villani M, Tosti M, Marini F, Bartoccini E, Magni MV, Albi E (2008) Lipid microdomains in cell nucleus. Mol Biol Cell 19(12):5289–5295. doi:10.1091/mbc.E08-05-0517
6. Ledeen RW, Wu G (2004) Nuclear lipids: key signaling effectors in the nervous system and other tissues. J Lipid Res 45(1):1–8. doi:10.1194/jlr.R300015-JLR200
7. Datta K, Guan T, Gerace L (2009) NET37, a nuclear envelope transmembrane protein with glycosidase homology, is involved in myoblast differentiation. J Biol Chem 284(43):29666–29676. doi:10.1074/jbc.M109.034041
8. Huber MD, Guan T, Gerace L (2009) Overlapping functions of nuclear envelope proteins NET25 (Lem2) and emerin in regulation of extracellular signal-regulated kinase signaling in myoblast differentiation. Mol Cell Biol 29(21):5718–5728. doi:10.1128/MCB.00270-09
9. Liu GH, Guan T, Datta K, Coppinger J, Yates J 3rd, Gerace L (2009) Regulation of myoblast differentiation by the nuclear envelope protein NET39. Mol Cell Biol 29(21):5800–5812. doi:10.1128/MCB.00684-09

10. Schirmer EC, Florens L, Guan T, Yates JR 3rd, Gerace L (2003) Nuclear membrane proteins with potential disease links found by subtractive proteomics. Science 301(5638):1380–1382. doi:10.1126/science.1088176
11. Korfali N, Wilkie GS, Swanson SK, Srsen V, de Las Heras J, Batrakou DG, Malik P, Zuleger N, Kerr AR, Florens L, Schirmer EC (2012) The nuclear envelope proteome differs notably between tissues. Nucleus 3(6):552–564. doi:10.4161/nucl.22257
12. Aebi U, Cohn J, Buhle L, Gerace L (1986) The nuclear lamina is a meshwork of intermediate-type filaments. Nature 323(6088):560–564. doi:10.1038/323560a0
13. Fisher DZ, Chaudhary N, Blobel G (1986) cDNA sequencing of nuclear lamins A and C reveals primary and secondary structural homology to intermediate filament proteins. Proc Natl Acad Sci U S A 83(17):6450–6454
14. Goldman AE, Maul G, Steinert PM, Yang HY, Goldman RD (1986) Keratin-like proteins that coisolate with intermediate filaments of BHK-21 cells are nuclear lamins. Proc Natl Acad Sci U S A 83(11):3839–3843
15. McKeon FD, Kirschner MW, Caput D (1986) Homologies in both primary and secondary structure between nuclear envelope and intermediate filament proteins. Nature 319(6053):463–468. doi:10.1038/319463a0
16. Burke B, Stewart CL (2013) The nuclear lamins: flexibility in function. Nat Rev Mol Cell Biol 14(1):13–24. doi:10.1038/nrm3488
17. Wilson KL, Foisner R (2010) Lamin-binding proteins. Cold Spring Harb Perspect Biol 2(4):a000554. doi:10.1101/cshperspect.a000554
18. Worman HJ, Foisner R (2010) The nuclear envelope from basic biology to therapy. Biochem Soc Trans 38(Pt 1):253–256. doi:10.1042/BST0380253
19. Dauer WT, Worman HJ (2009) The nuclear envelope as a signaling node in development and disease. Dev Cell 17(5):626–638. doi:10.1016/j.devcel.2009.10.016
20. Peter M, Kitten GT, Lehner CF, Vorburger K, Bailer SM, Maridor G, Nigg EA (1989) Cloning and sequencing of cDNA clones encoding chicken lamins A and B1 and comparison of the primary structures of vertebrate A- and B-type lamins. J Mol Biol 208(3):393–404
21. Furukawa K, Hotta Y (1993) cDNA cloning of a germ cell specific lamin B3 from mouse spermatocytes and analysis of its function by ectopic expression in somatic cells. EMBO J 12(1):97–106
22. Furukawa K, Inagaki H, Hotta Y (1994) Identification and cloning of an mRNA coding for a germ cell-specific A-type lamin in mice. Exp Cell Res 212(2):426–430. doi:10.1006/excr.1994.1164
23. Broers JL, Machiels BM, Kuijpers HJ, Smedts F, van den Kieboom R, Raymond Y, Ramaekers FC (1997) A- and B-type lamins are differentially expressed in normal human tissues. Histochem Cell Biol 107(6):505–517
24. Lehner CF, Stick R, Eppenberger HM, Nigg EA (1987) Differential expression of nuclear lamin proteins during chicken development. J Cell Biol 105(1):577–587
25. Stewart C, Burke B (1987) Teratocarcinoma stem cells and early mouse embryos contain only a single major lamin polypeptide closely resembling lamin B. Cell 51(3):383–392
26. Rober RA, Weber K, Osborn M (1989) Differential timing of nuclear lamin A/C expression in the various organs of the mouse embryo and the young animal: a developmental study. Development 105(2):365–378
27. Worman HJ (2012) Nuclear lamins and laminopathies. J Pathol 226(2):316–325. doi:10.1002/path.2999
28. Butin-Israeli V, Adam SA, Goldman AE, Goldman RD (2012) Nuclear lamin functions and disease. Trends Genet 28(9):464–471. doi:10.1016/j.tig.2012.06.001
29. Chow KH, Factor RE, Ullman KS (2012) The nuclear envelope environment and its cancer connections. Nat Rev Cancer 12(3):196–209. doi:10.1038/nrc3219
30. Lin F, Worman HJ (1993) Structural organization of the human gene encoding nuclear lamin A and nuclear lamin C. J Biol Chem 268(22):16321–16326
31. Machiels BM, Zorenc AH, Endert JM, Kuijpers HJ, van Eys GJ, Ramaekers FC, Broers JL (1996) An alternative splicing product of the lamin A/C gene lacks exon 10. J Biol Chem 271(16):9249–9253

32. Fong LG, Ng JK, Lammerding J, Vickers TA, Meta M, Cote N, Gavino B, Qiao X, Chang SY, Young SR, Yang SH, Stewart CL, Lee RT, Bennett CF, Bergo MO, Young SG (2006) Prelamin A and lamin A appear to be dispensable in the nuclear lamina. J Clin Invest 116(3):743–752. doi:10.1172/JCI27125
33. Dechat T, Pfleghaar K, Sengupta K, Shimi T, Shumaker DK, Solimando L, Goldman RD (2008) Nuclear lamins: major factors in the structural organization and function of the nucleus and chromatin. Genes Dev 22(7):832–853. doi:10.1101/gad.1652708
34. Stuurman N, Heins S, Aebi U (1998) Nuclear lamins: their structure, assembly, and interactions. J Struct Biol 122(1–2):42–66. doi:10.1006/jsbi.1998.3987
35. Dhe-Paganon S, Werner ED, Chi YI, Shoelson SE (2002) Structure of the globular tail of nuclear lamin. J Biol Chem 277(20):17381–17384. doi:10.1074/jbc.C200038200
36. Krimm I, Ostlund C, Gilquin B, Couprie J, Hossenlopp P, Mornon JP, Bonne G, Courvalin JC, Worman HJ, Zinn-Justin S (2002) The Ig-like structure of the C-terminal domain of lamin A/C, mutated in muscular dystrophies, cardiomyopathy, and partial lipodystrophy. Structure 10(6):811–823
37. Beck LA, Hosick TJ, Sinensky M (1990) Isoprenylation is required for the processing of the lamin A precursor. J Cell Biol 110(5):1489–1499
38. Kitten GT, Nigg EA (1991) The CaaX motif is required for isoprenylation, carboxyl methylation, and nuclear membrane association of lamin B2. J Cell Biol 113(1):13–23
39. Sinensky M, Fantle K, Trujillo M, McLain T, Kupfer A, Dalton M (1994) The processing pathway of prelamin A. J Cell Sci 107(Pt 1):61–67
40. Young SG, Fong LG, Michaelis S (2005) Prelamin A, Zmpste24, misshapen cell nuclei, and progeria—new evidence suggesting that protein farnesylation could be important for disease pathogenesis. J Lipid Res 46(12):2531–2558. doi:10.1194/jlr.R500011-JLR200
41. Zhang FL, Casey PJ (1996) Protein prenylation: molecular mechanisms and functional consequences. Annu Rev Biochem 65:241–269. doi:10.1146/annurev.bi.65.070196.001325
42. Clarke S (1992) Protein isoprenylation and methylation at carboxyl-terminal cysteine residues. Annu Rev Biochem 61:355–386. doi:10.1146/annurev.bi.61.070192.002035
43. Krohne G, Waizenegger I, Hoger TH (1989) The conserved carboxy-terminal cysteine of nuclear lamins is essential for lamin association with the nuclear envelope. J Cell Biol 109(5):2003–2011
44. Holtz D, Tanaka RA, Hartwig J, McKeon F (1989) The CaaX motif of lamin A functions in conjunction with the nuclear localization signal to target assembly to the nuclear envelope. Cell 59(6):969–977
45. Bonne G, Di Barletta MR, Varnous S, Becane HM, Hammouda EH, Merlini L, Muntoni F, Greenberg CR, Gary F, Urtizberea JA, Duboc D, Fardeau M, Toniolo D, Schwartz K (1999) Mutations in the gene encoding lamin A/C cause autosomal dominant Emery-Dreifuss muscular dystrophy. Nat Genet 21(3):285–288. doi:10.1038/6799
46. Ostlund C, Bonne G, Schwartz K, Worman HJ (2001) Properties of lamin A mutants found in Emery-Dreifuss muscular dystrophy, cardiomyopathy and Dunnigan-type partial lipodystrophy. J Cell Sci 114(Pt 24):4435–4445
47. Speckman RA, Garg A, Du F, Bennett L, Veile R, Arioglu E, Taylor SI, Lovett M, Bowcock AM (2000) Mutational and haplotype analyses of families with familial partial lipodystrophy (Dunnigan variety) reveal recurrent missense mutations in the globular C-terminal domain of lamin A/C. Am J Hum Genet 66(4):1192–1198. doi:10.1086/302836
48. Shackleton S, Lloyd DJ, Jackson SN, Evans R, Niermeijer MF, Singh BM, Schmidt H, Brabant G, Kumar S, Durrington PN, Gregory S, O'Rahilly S, Trembath RC (2000) LMNA, encoding lamin A/C, is mutated in partial lipodystrophy. Nat Genet 24(2):153–156. doi:10.1038/72807
49. Cao H, Hegele RA (2000) Nuclear lamin A/C R482Q mutation in Canadian kindreds with Dunnigan-type familial partial lipodystrophy. Hum Mol Genet 9(1):109–112
50. De Sandre-Giovannoli A, Bernard R, Cau P, Navarro C, Amiel J, Boccaccio I, Lyonnet S, Stewart CL, Munnich A, Le Merrer M, Levy N (2003) Lamin a truncation in Hutchinson-Gilford progeria. Science 300(5628):2055. doi:10.1126/science.1084125

51. Eriksson M, Brown WT, Gordon LB, Glynn MW, Singer J, Scott L, Erdos MR, Robbins CM, Moses TY, Berglund P, Dutra A, Pak E, Durkin S, Csoka AB, Boehnke M, Glover TW, Collins FS (2003) Recurrent de novo point mutations in lamin A cause Hutchinson-Gilford progeria syndrome. Nature 423(6937):293–298. doi:10.1038/nature01629
52. Pendas AM, Zhou Z, Cadinanos J, Freije JM, Wang J, Hultenby K, Astudillo A, Wernerson A, Rodriguez F, Tryggvason K, Lopez-Otin C (2002) Defective prelamin A processing and muscular and adipocyte alterations in Zmpste24 metalloproteinase-deficient mice. Nat Genet 31(1):94–99. doi:10.1038/ng871
53. Bergo MO, Gavino B, Ross J, Schmidt WK, Hong C, Kendall LV, Mohr A, Meta M, Genant H, Jiang Y, Wisner ER, Van Bruggen N, Carano RA, Michaelis S, Griffey SM, Young SG (2002) Zmpste24 deficiency in mice causes spontaneous bone fractures, muscle weakness, and a prelamin A processing defect. Proc Natl Acad Sci U S A 99(20):13049–13054. doi:10.1073/pnas.192460799
54. Navarro CL, Cadinanos J, De Sandre-Giovannoli A, Bernard R, Courrier S, Boccaccio I, Boyer A, Kleijer WJ, Wagner A, Giuliano F, Beemer FA, Freije JM, Cau P, Hennekam RC, Lopez-Otin C, Badens C, Levy N (2005) Loss of ZMPSTE24 (FACE-1) causes autosomal recessive restrictive dermopathy and accumulation of Lamin A precursors. Hum Mol Genet 14(11):1503–1513. doi:10.1093/hmg/ddi159
55. Navarro CL, De Sandre-Giovannoli A, Bernard R, Boccaccio I, Boyer A, Genevieve D, Hadj-Rabia S, Gaudy-Marqueste C, Smitt HS, Vabres P, Faivre L, Verloes A, Van Essen T, Flori E, Hennekam R, Beemer FA, Laurent N, Le Merrer M, Cau P, Levy N (2004) Lamin A and ZMPSTE24 (FACE-1) defects cause nuclear disorganization and identify restrictive dermopathy as a lethal neonatal laminopathy. Hum Mol Genet 13(20):2493–2503. doi:10.1093/hmg/ddh265
56. Fong LG, Frost D, Meta M, Qiao X, Yang SH, Coffinier C, Young SG (2006) A protein farnesyltransferase inhibitor ameliorates disease in a mouse model of progeria. Science 311(5767):1621–1623. doi:10.1126/science.1124875
57. Yang SH, Meta M, Qiao X, Frost D, Bauch J, Coffinier C, Majumdar S, Bergo MO, Young SG, Fong LG (2006) A farnesyltransferase inhibitor improves disease phenotypes in mice with a Hutchinson-Gilford progeria syndrome mutation. J Clin Invest 116(8):2115–2121. doi:10.1172/JCI28968
58. Gordon LB, Kleinman ME, Miller DT, Neuberg DS, Giobbie-Hurder A, Gerhard-Herman M, Smoot LB, Gordon CM, Cleveland R, Snyder BD, Fligor B, Bishop WR, Statkevich P, Regen A, Sonis A, Riley S, Ploski C, Correia A, Quinn N, Ullrich NJ, Nazarian A, Liang MG, Huh SY, Schwartzman A, Kieran MW (2012) Clinical trial of a farnesyltransferase inhibitor in children with Hutchinson-Gilford progeria syndrome. Proc Natl Acad Sci U S A 109(41):16666–16671. doi:10.1073/pnas.1202529109
59. Kyriakis JM, Avruch J (2012) Mammalian MAPK signal transduction pathways activated by stress and inflammation: a 10-year update. Physiol Rev 92(2):689–737. doi:10.1152/physrev.00028.2011
60. Kyriakis JM, Avruch J (2001) Mammalian mitogen-activated protein kinase signal transduction pathways activated by stress and inflammation. Physiol Rev 81(2):807–869
61. Plotnikov A, Zehorai E, Procaccia S, Seger R (2011) The MAPK cascades: signaling components, nuclear roles and mechanisms of nuclear translocation. Biochim Biophys Acta 1813(9):1619–1633. doi:10.1016/j.bbamcr.2010.12.012
62. Wagner EF, Nebreda AR (2009) Signal integration by JNK and p38 MAPK pathways in cancer development. Nat Rev Cancer 9(8):537–549. doi:10.1038/nrc2694
63. Raman M, Chen W, Cobb MH (2007) Differential regulation and properties of MAPKs. Oncogene 26(22):3100–3112. doi:10.1038/sj.onc.1210392
64. McKay MM, Morrison DK (2007) Integrating signals from RTKs to ERK/MAPK. Oncogene 26(22):3113–3121. doi:10.1038/sj.onc.1210394
65. Ivorra C, Kubicek M, Gonzalez JM, Sanz-Gonzalez SM, Alvarez-Barrientos A, O'Connor JE, Burke B, Andres V (2006) A mechanism of AP-1 suppression through interaction of c-Fos with lamin A/C. Genes Dev 20(3):307–320. doi:10.1101/gad.349506

66. Arimura T, Helbling-Leclerc A, Massart C, Varnous S, Niel F, Lacene E, Fromes Y, Toussaint M, Mura AM, Keller DI, Amthor H, Isnard R, Malissen M, Schwartz K, Bonne G (2005) Mouse model carrying H222P-Lmna mutation develops muscular dystrophy and dilated cardiomyopathy similar to human striated muscle laminopathies. Hum Mol Genet 14(1):155–169. doi:10.1093/hmg/ddi017
67. Muchir A, Pavlidis P, Decostre V, Herron AJ, Arimura T, Bonne G, Worman HJ (2007) Activation of MAPK pathways links LMNA mutations to cardiomyopathy in Emery-Dreifuss muscular dystrophy. J Clin Invest 117(5):1282–1293. doi:10.1172/JCI29042
68. Muchir A, Wu W, Choi JC, Iwata S, Morrow J, Homma S, Worman HJ (2012) Abnormal p38alpha mitogen-activated protein kinase signaling in dilated cardiomyopathy caused by lamin A/C gene mutation. Hum Mol Genet 21(19):4325–4333. doi:10.1093/hmg/dds265
69. Muchir A, Kim YJ, Reilly SA, Wu W, Choi JC, Worman HJ (2013) Inhibition of extracellular signal-regulated kinase 1/2 signaling has beneficial effects on skeletal muscle in a mouse model of Emery-Dreifuss muscular dystrophy caused by lamin A/C gene mutation. Skelet Muscle 3:17
70. Muchir A, Shan J, Bonne G, Lehnart SE, Worman HJ (2009) Inhibition of extracellular signal-regulated kinase signaling to prevent cardiomyopathy caused by mutation in the gene encoding A-type lamins. Hum Mol Genet 18(2):241–247. doi:10.1093/hmg/ddn343
71. Wu W, Muchir A, Shan J, Bonne G, Worman HJ (2011) Mitogen-activated protein kinase inhibitors improve heart function and prevent fibrosis in cardiomyopathy caused by mutation in lamin A/C gene. Circulation 123(1):53–61. doi:10.1161/CIRCULATIONAHA.110.970673
72. Wu W, Shan J, Bonne G, Worman HJ, Muchir A (2010) Pharmacological inhibition of c-Jun N-terminal kinase signaling prevents cardiomyopathy caused by mutation in LMNA gene. Biochim Biophys Acta 1802(7–8):632–638. doi:10.1016/j.bbadis.2010.04.001
73. Emerson LJ, Holt MR, Wheeler MA, Wehnert M, Parsons M, Ellis JA (2009) Defects in cell spreading and ERK1/2 activation in fibroblasts with lamin A/C mutations. Biochim Biophys Acta 1792(8):810–821. doi:10.1016/j.bbadis.2009.05.007
74. Muchir A, Reilly SA, Wu W, Iwata S, Homma S, Bonne G, Worman HJ (2012) Treatment with selumetinib preserves cardiac function and improves survival in cardiomyopathy caused by mutation in the lamin A/C gene. Cardiovasc Res 93(2):311–319. doi:10.1093/cvr/cvr301
75. Gonzalez JM, Navarro-Puche A, Casar B, Crespo P, Andres V (2008) Fast regulation of AP-1 activity through interaction of lamin A/C, ERK1/2, and c-Fos at the nuclear envelope. J Cell Biol 183(4):653–666. doi:10.1083/jcb.200805049
76. Shaw RJ, Cantley LC (2006) Ras, PI(3)K and mTOR signalling controls tumour cell growth. Nature 441(7092):424–430. doi:10.1038/nature04869
77. Jones JI, Clemmons DR (1995) Insulin-like growth factors and their binding proteins: biological actions. Endocr Rev 16(1):3–34
78. Sancak Y, Thoreen CC, Peterson TR, Lindquist RA, Kang SA, Spooner E, Carr SA, Sabatini DM (2007) PRAS40 is an insulin-regulated inhibitor of the mTORC1 protein kinase. Mol Cell 25(6):903–915. doi:10.1016/j.molcel.2007.03.003
79. Vander Haar E, Lee SI, Bandhakavi S, Griffin TJ, Kim DH (2007) Insulin signalling to mTOR mediated by the Akt/PKB substrate PRAS40. Nat Cell Biol 9(3):316–323. doi:10.1038/ncb1547
80. Zoncu R, Efeyan A, Sabatini DM (2011) mTOR: from growth signal integration to cancer, diabetes and ageing. Nat Rev Mol Cell Biol 12(1):21–35. doi:10.1038/nrm3025
81. Choi JC, Muchir A, Wu W, Iwata S, Homma S, Morrow JP, Worman HJ (2012) Temsirolimus activates autophagy and ameliorates cardiomyopathy caused by lamin A/C gene mutation. Sci Transl Med 4(144):144ra102. doi:10.1126/scitranslmed.3003875
82. Ramos FJ, Chen SC, Garelick MG, Dai DF, Liao CY, Schreiber KH, MacKay VL, An EH, Strong R, Ladiges WC, Rabinovitch PS, Kaeberlein M, Kennedy BK (2012) Rapamycin reverses elevated mTORC1 signaling in lamin A/C-deficient mice, rescues cardiac and skeletal muscle function, and extends survival. Sci Transl Med 4(144):144ra103. doi:10.1126/scitranslmed.3003802
83. Sullivan T, Escalante-Alcalde D, Bhatt H, Anver M, Bhat N, Nagashima K, Stewart CL, Burke B (1999) Loss of A-type lamin expression compromises nuclear envelope integrity leading to muscular dystrophy. J Cell Biol 147(5):913–920

84. Jahn D, Schramm S, Schnolzer M, Heilmann CJ, de Koster CG, Schutz W, Benavente R, Alsheimer M (2012) A truncated lamin A in the Lmna -/- mouse line: implications for the understanding of laminopathies. Nucleus 3(5):463–474. doi:10.4161/nucl.21676
85. Levine B, Kroemer G (2008) Autophagy in the pathogenesis of disease. Cell 132(1):27–42. doi:10.1016/j.cell.2007.12.018
86. Boguslavsky RL, Stewart CL, Worman HJ (2006) Nuclear lamin A inhibits adipocyte differentiation: implications for Dunnigan-type familial partial lipodystrophy. Hum Mol Genet 15(4):653–663. doi:10.1093/hmg/ddi480
87. Choi JC, Wu W, Muchir A, Iwata S, Homma S, Worman HJ (2012) Dual specificity phosphatase 4 mediates cardiomyopathy caused by lamin A/C (LMNA) gene mutation. J Biol Chem 287(48):40513–40524. doi:10.1074/jbc.M112.404541
88. Bertacchini J, Beretti F, Cenni V, Guida M, Gibellini F, Mediani L, Marin O, Maraldi NM, de Pol A, Lattanzi G, Cocco L, Marmiroli S (2013) The protein kinase Akt/PKB regulates both prelamin A degradation and Lmna gene expression. FASEB J 27:2145. doi:10.1096/fj.12-218214
89. Cenni V, Bertacchini J, Beretti F, Lattanzi G, Bavelloni A, Riccio M, Ruzzene M, Marin O, Arrigoni G, Parnaik V, Wehnert M, Maraldi NM, de Pol A, Cocco L, Marmiroli S (2008) Lamin A Ser404 is a nuclear target of Akt phosphorylation in C2C12 cells. J Proteome Res 7(11):4727–4735. doi:10.1021/pr800262g
90. Worman HJ, Fong LG, Muchir A, Young SG (2009) Laminopathies and the long strange trip from basic cell biology to therapy. J Clin Invest 119(7):1825–1836. doi:10.1172/JCI37679
91. Scaffidi P, Misteli T (2008) Lamin A-dependent misregulation of adult stem cells associated with accelerated ageing. Nat Cell Biol 10(4):452–459. doi:10.1038/ncb1708
92. Artavanis-Tsakonas S, Rand MD, Lake RJ (1999) Notch signaling: cell fate control and signal integration in development. Science 284(5415):770–776
93. Reya T, Morrison SJ, Clarke MF, Weissman IL (2001) Stem cells, cancer, and cancer stem cells. Nature 414(6859):105–111. doi:10.1038/35102167
94. Frantz C, Stewart KM, Weaver VM (2010) The extracellular matrix at a glance. J Cell Sci 123(Pt 24):4195–4200. doi:10.1242/jcs.023820
95. Hernandez L, Roux KJ, Wong ES, Mounkes LC, Mutalif R, Navasankari R, Rai B, Cool S, Jeong JW, Wang H, Lee HS, Kozlov S, Grunert M, Keeble T, Jones CM, Meta MD, Young SG, Daar IO, Burke B, Perantoni AO, Stewart CL (2010) Functional coupling between the extracellular matrix and nuclear lamina by Wnt signaling in progeria. Dev Cell 19(3):413–425. doi:10.1016/j.devcel.2010.08.013
96. Mounkes LC, Kozlov S, Hernandez L, Sullivan T, Stewart CL (2003) A progeroid syndrome in mice is caused by defects in A-type lamins. Nature 423(6937):298–301. doi:10.1038/nature01631
97. Miraoui H, Marie PJ (2010) Fibroblast growth factor receptor signaling crosstalk in skeletogenesis. Sci Signal 3(146):re9. doi:10.1126/scisignal.3146re9
98. MacDonald BT, Tamai K, He X (2009) Wnt/beta-catenin signaling: components, mechanisms, and diseases. Dev Cell 17(1):9–26. doi:10.1016/j.devcel.2009.06.016
99. Vijg J, Campisi J (2008) Puzzles, promises and a cure for ageing. Nature 454(7208):1065–1071. doi:10.1038/nature07216
100. Kirkwood TB (2005) Understanding the odd science of aging. Cell 120(4):437–447. doi:10.1016/j.cell.2005.01.027
101. Ramirez CL, Cadinanos J, Varela I, Freije JM, Lopez-Otin C (2007) Human progeroid syndromes, aging and cancer: new genetic and epigenetic insights into old questions. Cell Mol Life Sci 64(2):155–170. doi:10.1007/s00018-006-6349-3
102. Varela I, Cadinanos J, Pendas AM, Gutierrez-Fernandez A, Folgueras AR, Sanchez LM, Zhou Z, Rodriguez FJ, Stewart CL, Vega JA, Tryggvason K, Freije JM, Lopez-Otin C (2005) Accelerated ageing in mice deficient in Zmpste24 protease is linked to p53 signalling activation. Nature 437(7058):564–568. doi:10.1038/nature04019
103. Liu B, Wang J, Chan KM, Tjia WM, Deng W, Guan X, Huang JD, Li KM, Chau PY, Chen DJ, Pei D, Pendas AM, Cadinanos J, Lopez-Otin C, Tse HF, Hutchison C, Chen J, Cao Y, Cheah

KS, Tryggvason K, Zhou Z (2005) Genomic instability in laminopathy-based premature aging. Nat Med 11(7):780–785. doi:10.1038/nm1266
104. Burtner CR, Kennedy BK (2010) Progeria syndromes and ageing: what is the connection? Nat Rev Mol Cell Biol 11(8):567–578. doi:10.1038/nrm2944
105. Hayden MS, Ghosh S (2008) Shared principles in NF-kappaB signaling. Cell 132(3):344–362. doi:10.1016/j.cell.2008.01.020
106. Perkins ND (2007) Integrating cell-signalling pathways with NF-kappaB and IKK function. Nat Rev Mol Cell Biol 8(1):49–62. doi:10.1038/nrm2083
107. Adler AS, Sinha S, Kawahara TL, Zhang JY, Segal E, Chang HY (2007) Motif module map reveals enforcement of aging by continual NF-kappaB activity. Genes Dev 21(24):3244–3257. doi:10.1101/gad.1588507
108. Tak PP, Firestein GS (2001) NF-kappaB: a key role in inflammatory diseases. J Clin Invest 107(1):7–11. doi:10.1172/JCI11830
109. Osorio FG, Barcena C, Soria-Valles C, Ramsay AJ, de Carlos F, Cobo J, Fueyo A, Freije JM, Lopez-Otin C (2012) Nuclear lamina defects cause ATM-dependent NF-kappaB activation and link accelerated aging to a systemic inflammatory response. Genes Dev 26(20):2311–2324. doi:10.1101/gad.197954.112
110. Laguri C, Gilquin B, Wolff N, Romi-Lebrun R, Courchay K, Callebaut I, Worman HJ, Zinn-Justin S (2001) Structural characterization of the LEM motif common to three human inner nuclear membrane proteins. Structure 9(6):503–511
111. Ostlund C, Ellenberg J, Hallberg E, Lippincott-Schwartz J, Worman HJ (1999) Intracellular trafficking of emerin, the Emery-Dreifuss muscular dystrophy protein. J Cell Sci 112(Pt 11):1709–1719
112. Clements L, Manilal S, Love DR, Morris GE (2000) Direct interaction between emerin and lamin A. Biochem Biophys Res Commun 267(3):709–714. doi:10.1006/bbrc.1999.2023
113. Raharjo WH, Enarson P, Sullivan T, Stewart CL, Burke B (2001) Nuclear envelope defects associated with LMNA mutations cause dilated cardiomyopathy and Emery-Dreifuss muscular dystrophy. J Cell Sci 114(Pt 24):4447–4457
114. Vaughan A, Alvarez-Reyes M, Bridger JM, Broers JL, Ramaekers FC, Wehnert M, Morris GE, Whitfield WGF, Hutchison CJ (2001) Both emerin and lamin C depend on lamin A for localization at the nuclear envelope. J Cell Sci 114(Pt 14):2577–2590
115. Anastas JN, Moon RT (2013) WNT signalling pathways as therapeutic targets in cancer. Nat Rev Cancer 13(1):11–26. doi:10.1038/nrc3419
116. Markiewicz E, Tilgner K, Barker N, van de Wetering M, Clevers H, Dorobek M, Hausmanowa-Petrusewicz I, Ramaekers FC, Broers JL, Blankesteijn WM, Salpingidou G, Wilson RG, Ellis JA, Hutchison CJ (2006) The inner nuclear membrane protein emerin regulates beta-catenin activity by restricting its accumulation in the nucleus. EMBO J 25(14):3275–3285. doi:10.1038/sj.emboj.7601230
117. Neumann S, Schneider M, Daugherty RL, Gottardi CJ, Eming SA, Beijer A, Noegel AA, Karakesisoglou I (2010) Nesprin-2 interacts with {alpha}-catenin and regulates Wnt signaling at the nuclear envelope. J Biol Chem 285(45):34932–34938. doi:10.1074/jbc.M110.119651
118. Wu W, Lin F, Worman HJ (2002) Intracellular trafficking of MAN1, an integral protein of the nuclear envelope inner membrane. J Cell Sci 115(Pt 7):1361–1371
119. Lin F, Blake DL, Callebaut I, Skerjanc IS, Holmer L, McBurney MW, Paulin-Levasseur M, Worman HJ (2000) MAN1, an inner nuclear membrane protein that shares the LEM domain with lamina-associated polypeptide 2 and emerin. J Biol Chem 275(7):4840–4847
120. Paulin-Levasseur M, Blake DL, Julien M, Rouleau L (1996) The MAN antigens are non-lamin constituents of the nuclear lamina in vertebrate cells. Chromosoma 104(5):367–379
121. Lee KK, Gruenbaum Y, Spann P, Liu J, Wilson KL (2000) *C. elegans* nuclear envelope proteins emerin, MAN1, lamin, and nucleoporins reveal unique timing of nuclear envelope breakdown during mitosis. Mol Biol Cell 11(9):3089–3099
122. Mansharamani M, Wilson KL (2005) Direct binding of nuclear membrane protein MAN1 to emerin in vitro and two modes of binding to barrier-to-autointegration factor. J Biol Chem 280(14):13863–13870. doi:10.1074/jbc.M413020200

123. Massague J (2012) TGFbeta signalling in context. Nat Rev Mol Cell Biol 13(10):616–630. doi:10.1038/nrm3434
124. Hellemans J, Preobrazhenska O, Willaert A, Debeer P, Verdonk PC, Costa T, Janssens K, Menten B, Van Roy N, Vermeulen SJ, Savarirayan R, Van Hul W, Vanhoenacker F, Huylebroeck D, De Paepe A, Naeyaert JM, Vandesompele J, Speleman F, Verschueren K, Coucke PJ, Mortier GR (2004) Loss-of-function mutations in LEMD3 result in osteopoikilosis, Buschke-Ollendorff syndrome and melorheostosis. Nat Genet 36(11):1213–1218. doi:10.1038/ng1453
125. Lin F, Morrison JM, Wu W, Worman HJ (2005) MAN1, an integral protein of the inner nuclear membrane, binds Smad2 and Smad3 and antagonizes transforming growth factor-beta signaling. Hum Mol Genet 14(3):437–445. doi:10.1093/hmg/ddi040
126. Pan D, Estevez-Salmeron LD, Stroschein SL, Zhu X, He J, Zhou S, Luo K (2005) The integral inner nuclear membrane protein MAN1 physically interacts with the R-Smad proteins to repress signaling by the transforming growth factor-{beta} superfamily of cytokines. J Biol Chem 280(16):15992–16001. doi:10.1074/jbc.M411234200
127. Cohen TV, Kosti O, Stewart CL (2007) The nuclear envelope protein MAN1 regulates TGFbeta signaling and vasculogenesis in the embryonic yolk sac. Development 134(7):1385–1395. doi:10.1242/dev.02816
128. Ishimura A, Ng JK, Taira M, Young SG, Osada S (2006) Man1, an inner nuclear membrane protein, regulates vascular remodeling by modulating transforming growth factor beta signaling. Development 133(19):3919–3928. doi:10.1242/dev.02538
129. Bourgeois B, Gilquin B, Tellier-Lebègue C, Östlund C, Wu W, Pérez J, El Hage P, Lallemand F, Worman HJ, Zinn-Justin S (2013) Inhibition of transforming growth factor-b signaling at the nuclear envelope: characterization of MAN1-Smad2/3-PPM1A interactions. Sci Signal 6:ra49
130. Raju GP, Dimova N, Klein PS, Huang HC (2003) SANE, a novel LEM domain protein, regulates bone morphogenetic protein signaling through interaction with Smad1. J Biol Chem 278(1):428–437. doi:10.1074/jbc.M210505200
131. Diamond DA, Berry SJ, Umbricht C, Jewett HJ, Coffey DS (1982) Computerized image analysis of nuclear shape as a prognostic factor for prostatic cancer. Prostate 3(4):321–332

Part III
Nuclear Envelope Regulation of the Genome

Introduction

A ubiquitous characteristic of cancers is massive changes in the pattern of gene expression of tumor cells. Although the specific changes are quite variable, it would be hard to find a tumor type where this is not a feature. In many cancers, these gene changes correlate with changes in DNA methylation and other epigenetic marks, leading some to argue that cancer is an epigenetic disease. However, there are many other genome changes observed that correlate as well or even better and can also cause changes in gene expression. Two of the most critical of these are chromosome translocations and DNA damage/breaks, the latter of which will be discussed in detail later in Part V.

Chromosome translocations were first linked to tumorigenesis by Professor Janet Rowley at the University of Chicago in 1973 [1]. Checking karyotypes for translocations has since become a standard diagnostic tool for many cancer types. Physically, chromosome translocations need DNA damage and repair systems to occur, but the direction of chromosome translocations can depend on aspects of higher order chromosome structure in the nucleus. For example, certain tumor types have a propensity for particular translocations between specific chromosomes and these tend to be linked to tumors of particular tissues. Correspondingly, Tom Misteli's laboratory found that the chromosomes involved in these translocations are adjacent to one another in the interphase nuclei of the particular tissues where such tumors arise [2].

Much of chromosome positioning has now been linked to the nuclear envelope in work largely directed from the Bickmore and Bridger laboratories [3, 4] and normal chromosome positioning has been observed to be lost in many tumors [5]. Although the positioning of most chromosomes follows from issues of gene density and is likely related to nuclear envelope protein associations with silenced chromatin, recent work has identified several tissue-specific nuclear envelope proteins that can reposition chromosomes and are likely involved in the tissue-specific patterns of chromosome positioning [6]. Such proteins are likely also involved in specific gene positioning with corresponding effects on gene expression.

In addition to tissue-specific nuclear envelope tethering setting the stage for particular translocation events, one could easily postulate that the physical function of tethering also serves to stabilize the genome and its disruption could play a role in development of cancer. In fact, the loss of A-type lamins in many tumor types was postulated by some to result in less physical tethering and thus more mobile chromatin with a greater likelihood of genome damage, though this has never been properly tested. In this section we begin with Irina Stancheva, a leading researcher on DNA methylation, epigenetics, and cancer, and Eric Schirmer from the Wellcome Trust Centre for Cell Biology at the University of Edinburgh discussing the wider range of nuclear envelope interactions with chromatin that can influence gene expression and genome stability. They cover epigenetic changes in cancer and regulation of gene expression, spatial genome organization with respect to gene activation in development and in nuclear envelope linked diseases, and other mechanisms for transcriptional regulation from the nuclear envelope. Finally, they discuss the possibilities of genome stabilization from tethering of chromatin at the nuclear periphery, using the intermediate filament lamin polymer to stabilize the genome. Next, Roman Lyakhovetsky and Yosef (Yossi) Gruenbaum, one of the most longstanding and prolific lamin researchers, of Hebrew University discuss the range of physical properties and functions of lamins and lamin-like proteins in different model systems to give an overview of these important proteins, already strongly linked to cancer and used in cancer diagnosis and prognosis as discussed already in the chapter by Jos Broers and Frans Ramaekers in Section I. They also discuss some of their own work together with Susan Gasser's laboratory investigating the role of lamins in directing the specific positioning of genes during development. Finally in this section is an in-depth coverage of chromosome positioning by one of the world's leading experts in this area, Joanna Bridger, together with colleagues from Brunel University, the University of London, and George Washington University. They discuss chromosome-positioning changes in both nuclear envelope-linked diseases and cancer and also how such changes might reflect changes that occur during aging and senescence.

1. Rowley JD (1973) Identificaton of a translocation with quinacrine fluorescence in a patient with acute leukemia. Ann Genet 16(2):109–112
2. Parada LA, McQueen PG, Munson PJ, Misteli T (2002) Conservation of relative chromosome positioning in normal and cancer cells. Curr Biol 12(19):1692–1697
3. Bridger JM, Boyle S, Kill IR, Bickmore WA (2000) Re-modelling of nuclear architecture in quiescent and senescent human fibroblasts. Curr Biol 10(3): 149–152. doi:S0960-9822(00)00312-2 [pii]
4. Croft JA, Bridger JM, Boyle S, Perry P, Teague P, Bickmore WA (1999) Differences in the localization and morphology of chromosomes in the human nucleus. J Cell Biol 145(6):1119–1131
5. Cremer M, Kupper K, Wagler B, Wizelman L, von Hase J, Weiland Y, Kreja L, Diebold J, Speicher MR, Cremer T (2003) Inheritance of gene density-related higher order chromatin arrangements in normal and tumor cell nuclei. J Cell Biol 162(5):809–820. doi:10.1083/jcb.200304096 jcb.200304096 [pii]
6. Zuleger N, Boyle S, Kelly DA, de Las Heras JI, Lazou V, Korfali N, Batrakou DG, Randles KN, Morris GE, Harrison DJ, Bickmore WA, Schirmer EC (2013) Specific nuclear envelope transmembrane proteins can promote the location of chromosomes to and from the nuclear periphery. Genome Biol 14(2):R14. doi:gb-2013-14-2-r14 [pii] 10.1186/gb-2013-14-2-r14

Nuclear Envelope: Connecting Structural Genome Organization to Regulation of Gene Expression

Irina Stancheva and Eric C. Schirmer

Abstract For many years, the nuclear envelope was viewed as a passive barrier that separates the genetic material in the nucleus from the cytoplasm of the cell and permits regulated trafficking of various molecules through the nuclear pores. Research in the past two decades has shown that the nuclear envelope is a complex cellular compartment, which harbors tissue-specific resident proteins, extensively interacts with chromatin and contributes to spatial genome organization and regulation of gene expression. Chromatin at the nuclear periphery is organized into active and silenced domains punctuated by insulator elements. The nuclear envelope transmembrane proteins and the nuclear lamina serve as anchoring sites for heterochromatin. They recruit chromatin that has been modified with specific epigenetic marks, provide silencing factors that add new epigenetic modifications to genes located at the nuclear periphery, and sequester transcription factors away from the nuclear interior. On the other hand, proteins of the nuclear pores anchor as well as help generate active chromatin, promote transcription, and coordinate gene expression with mRNA export. The importance of these functions is underscored by aberrant distribution of peripheral chromatin and changes in gene expression that occur in cancer and heritable human diseases linked to mutations in nuclear envelope proteins. Although many mechanistic questions addressing the role of the nuclear envelope in genome organization and function have been answered in recent years, a great deal remains to be discovered in this exciting and rapidly moving field.

I. Stancheva (✉)
The Wellcome Trust Centre for Cell Biology, University of Edinburgh,
Edinburgh EH9 3JR, UK

Wellcome Trust Centre for Cell Biology, University of Edinburgh, Kings Buildings,
Michael Swann Building, Room 4.22, Edinburgh EH9 3JR, UK
e-mail: istancheva@ed.ac.uk

E.C. Schirmer
The Wellcome Trust Centre for Cell Biology, University of Edinburgh,
Edinburgh EH9 3JR, UK

E.C. Schirmer and J.I. de las Heras (eds.), *Cancer Biology and the Nuclear Envelope*,
Advances in Experimental Medicine and Biology 773, DOI 10.1007/978-1-4899-8032-8_10,

Keywords Nuclear envelope • Nuclear lamina • NPC • Chromatin • Gene expression • Insulator • Epigenetics • Cancer

Abbreviations

BAF	Barrier-to-autointegration factor
EDMD	Emery–Dreifuss muscular dystrophy
HP1	Heterochromatin protein 1
HDAC3	Histone deacetylase 3
HGPS	Hutchinson–Gilford progeria syndrome
INM	Inner nuclear membrane
IPTG	Isopropyl β-D-1-thiogalactopyranoside
lacO	Lac operator repeats
lacI	Bacterial lac repressor
NE	Nuclear envelope
NET	Nuclear envelope transmembrane protein
NPC	Nuclear pore complex

Introduction

The nuclear envelope (NE) forms extensive connections with the cytoplasm of the cell and chromatin in the nuclear space. Thus, the NE is uniquely positioned to integrate extracellular, cytoplasmic and nuclear signaling networks. There are multiple ways in which the NE contributes to the regulation of gene expression ranging from transmitting signals from the cytoplasm to the nucleus to spatial genome organization and fine-tuning of gene expression at the nuclear periphery. A wide range of cellular functions that are highly relevant to tissue homeostasis, human disease, and cancer biology are dependent on the composition and the organization of the NE. Therefore, there is an ever-growing interest in exploring the mechanisms of nuclear architecture and gene expression that are regulated by spatial cues at the nuclear periphery in conjunction with the NE and its resident proteins.

Structurally, the NE is a complex structure consisting of two separate concentric lipid bilayers, the outer (ONM) and the inner (INM) nuclear membranes, respectively (Fig. 1). The ONM is studded with ribosomes and continuous with the endoplasmic reticulum [1], but it has many unique proteins that connect to the cytoskeleton as well as to the proteins residing in the INM [2]. The connectivity of the NE to the endoplasmic reticulum also means that the NE lumen is continuous with the endoplasmic reticulum lumen. The ONM provides an impenetrable barrier for proteins and most small molecules except where the nuclear pore complexes (NPCs) are inserted. The NPCs regulate directional transport of soluble macromolecules in and

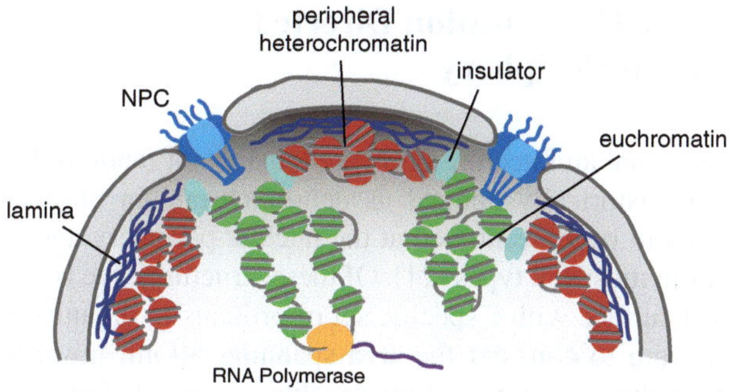

Fig. 1 Chromatin organization at the nuclear envelope. The nuclear envelope is a double membrane system perforated by nuclear pore complexes (NPCs) that regulate transport of molecules in and out of the nucleus. Between NPCs the inner nuclear membrane is lined by a filamentous polymer of the intermediate filament lamin proteins (lamina). At a simplistic level, peripheral heterochromatin tends to be associated with the lamina while peripheral euchromatin tends to be associated with the NPCs. Insulators on the DNA separate the peripheral euchromatin from peripheral heterochromatin

out of the nucleus and are extremely large (>40 MDa) structures made up of at least 30 distinct polypeptides present in multiple copies [3]. Where NPCs are inserted, the ONM curves around the outer face of the NPCs into the INM. The INM has its own unique set of transmembrane proteins, many of which interact with the intermediate filament lamins (in higher eukaryotes) that form a polymeric meshwork under the INM [4, 5]. The integral membrane proteins of both membranes are generally referred to as NETs for *n*uclear *e*nvelope *t*ransmembrane proteins [6].

Lamins, many NETs, and some NPC proteins bind directly to chromatin and/or transcriptional regulators and some have been reported to directly bind DNA (Fig. 1; reviewed in [7]). Moreover, some of the chromatin-binding partners of NETs specifically reflect transcriptionally silenced chromatin [8–10]. These interactions, accordingly, have been shown to direct the distribution of heterochromatin at the nuclear periphery [11, 12]. The affinity between lamins/NETs and chromatin provides a mechanism whereby binding interactions with lamina proteins can both sequester certain parts of the genome to the periphery and provide regulatory proteins to these genome regions.

Most of the NE proteins initially found to bind chromatin were widely expressed. However, subsequent proteomics studies found a much larger set of proteins at the NE than previously expected [13–16] and more recent studies further revealed that many of these NETs possess a high degree of tissue specificity [17–19]. Recent work indicates that tissue-specific NETs can influence spatial genome organization and gene expression [17, 20]. These findings suggest that the NE is a complex cellular compartment, which integrates both cytoplasmic and extracellular signals to regulate gene expression.

Spatial Genome Organization Directed from the Nuclear Periphery

The three-dimensional architecture of the genome is not random. For example, the heterochromatin, historically defined as denser chromatin observed by electron microscopy, tends to be concentrated at the nuclear periphery and around nucleoli and centromeres in most cell types [21]. Of these structures, the NE provides a large two-dimensional surface with a specific set of proteins to organize the genome. In general, if one were to consider the area spanning 50 nm inwards from the NE against the total volume of a typical mammalian nucleus (~5–10 μm in diameter), the NE could be considered to control roughly 1/30th of the nucleus. As each chromosome is one long folded strand of DNA, this NE tethering could physically propagate effects along the DNA polymer deep inside the nucleus. The idea that there appears to be a direct physical contact between chromatin and the NE has been supported by microscopy studies [22, 23] and biochemical experiments detecting retention of chromatin components after NE purification and extraction with high ionic strength buffers [24].

The notion that higher order chromatin structure might play regulatory roles was implied by observations that the distribution of heterochromatin is relatively uniform within a given cell type, yet it can vary widely between different cell types. For example, neurons tend to have very little peripheral heterochromatin while hepatocytes display a uniform and patchy heterochromatin distribution, epithelial cells have a less uniform patchy distribution, fibroblasts have a uniform distribution, and lymphoblasts tend to have an enormous amount of heterochromatin that extends several microns into the nucleus from the periphery [21]. In the case of lymphoblast cells, this dense chromatin largely dissipates upon immune activation, consistent with the idea that the strongly negatively stained material represents transcriptionally inert, but plastic facultative chromatin [25, 26].

Heterochromatin is now defined by specific chromatin epigenetic modifications such as histone H3 lysine 9 dimethylation and trimethylation (H3K9me2 and H3K9me3) and lack of histone acetylation and H3 K4 dimethylation and trimethylation (H3K4me2 and H3K4me3), which are normally present at genes that are actively transcribed or poised for transcription. Recent studies clearly demonstrate that antibodies against H3K9me2, a mark characteristic of facultative heterochromatin, detect chromatin positioned close to the NE, while antibodies against H3K9me3, a modification found at constitutive heterochromatin associated with telomeres and centromeres in higher eukaryotes, stains internal heterochromatic domains [27]. These data support earlier biochemical and microscopic observations indicating that the heterochromatin protein 1 alpha (HP1α) seems to have a distinct subpopulation at the NE [28]. Recently, many specific interactions have been reported between the NETs, lamins and silent chromatin as well as cross talk between NPCs and transcriptional regulatory mechanisms. However, the precise relationship between gene activity and higher order chromatin organization at the nuclear periphery is not yet entirely clear.

Centromere and Telomere Patterns of Spatial Genome Organization

While early observations made by clinicians suggested that changes in nuclear morphology in tumor cells might be functionally important, it was not until the late 1800s that cell biologists began to notice that genome organization is not completely random [29]. The first basic description of such nonrandom nuclear organization came from Carl Rabl who found that the centromeres in nuclei from salamander larvae were located at the nuclear periphery, concentrating at the side of the nucleus where the centrosome was located (though on the outside) [29]. This "Rabl configuration" has since been observed in many plants and higher mammals, though it is more often transiently occurring just before or during mitosis and particularly meiosis [30, 31]. It also has been observed for telomeres, with associations at one pole of the nucleus occurring in both mitotic cells and in some interphase cells. One possible role for the Rabl configuration in meiosis is to orient the synaptonemal complex and align chromosomes for homologous recombination [32].

In interphase mammalian cells centromeres are not typically located at the periphery, but they do accumulate at the NE in certain cell types such as human neutrophils [33]. Though NE tethering of telomeres is also usually transient, this connection appears to be permanent in budding and fission yeast [34, 35] and is maintained throughout sperm development in mammals [36]. The NET SUN proteins have been linked to both telomere and centromere tethering to the NE in certain systems [36–40]. Most recently, the interaction between the fission yeast Sad1, a SUN family member, and a novel kinetochore protein Csi has been found to direct the tethering of centromeres to the NE [40]. Older studies suggested that lamins are also able to bind to specific chromatin structures such as centromeres and telomeres [41, 42]; however, as lamins also bind SUN proteins, it is possible that these reports reflect the SUN interactions.

Centromere function can be very important for cancer biology both by being essential for proper chromosome segregation in mitosis and, through these NE connections, for the formation of the synaptonemal complex. Improper formation of synaptonemal complex could contribute to chromosome translocations and aneuploidy. Telomeres also have been shown to play important roles in aging, cancer, and cell immortalization as maintenance of telomere length is essential for the immortalization process. Although it is not known what effect telomere length has on spatial positioning, the NE tethering of telomeres and centromeres provides potential links between the NE, genomic instability and cancer, which merit further studies.

Large-Scale Patterns of Tissue-Specific Chromosome and Chromatin Organization

While the positioning of structurally important chromosome regions, such as centromeres and telomeres that are rich in repetitive sequences and present on all chromosomes, tends to be transient in higher organisms, the nonrandom distribution of individual chromosomes tends to be tissue-specific. Theodor Boveri first suggested roughly 100 years ago that chromosomes tend to occupy particular domains in the interphase nucleus while studying eggs of the worm *Ascaris* [43]. However, only recently has work indicated that specific chromosomes have higher than random probability to occupy characteristic positions within the three-dimensional framework of the nucleus with respect to the NE. Whole chromosome fluorescence in situ hybridization (FISH) revealed that in human fibroblasts chromosome 18 tends to be located at the nuclear periphery while chromosome 19 tends to be positioned internally [44]. Much of this positioning appears to correlate with gene density of individual chromosomes [45, 46]; however, gene density cannot fully account for differences in chromosome positioning. For example in hybrid nuclei containing mouse and human chromosomes, the human chromosomes adopt a spatial position in the mouse cell nucleus based on the synteny with the mouse chromosomes rather than their gene density [47]. Moreover, the positioning of some chromosomes can vary between cell types in a tissue-specific manner although the molecular mechanisms that determine this phenomenon are as yet very poorly understood. For example mouse chromosome 5 tends to be peripheral in lung cells while being internal in blood and liver cells; also chromosome 6 is peripheral in CD8+ T-cells but internal in CD4+ T-cells [48, 49].

The spatial organization of chromosomes is often altered in tumors. Determining the position of chromosomes 18 and 19 in both normal and tumor cell lines revealed that their respective radial positioning with respect to the NE was highly conserved in normal fibroblasts, but in seven of eight tumor cell lines this particular positioning was much less pronounced [46]. It is not clear whether these observations reflect tissue-specific differences in nuclear organization between fibroblasts and epithelial tumors or altered chromosome numbers that are characteristic of tumor cells: further studies using matched cancer and normal tissues should elucidate this question. Importantly, the maintenance of tissue-specific chromosome positioning patterns may explain why some chromosomal translocations occur with higher frequency in certain tumor types. Recently, a study aiming to test whether chromosomes most commonly involved in tissue-specific tumor translocations were positioned adjacent to one another during interphase has found a statistically significant correlation between adjacent positioning of chromosomes in normal tissues and translocation frequency observed in tumors [50].

This was not only one of the first indications that different cell types favor certain chromosome groupings [48, 49], but it also highlights that chromosome positioning may explain the tissue-specific patterns of genomic instability observed in cancer and other human disease conditions.

Large-scale differences in chromosome positioning may also reflect the above-mentioned tissue-specific patterns of heterochromatin distribution in different cell types or, as an extreme example, the evolutionary adaptation of similar cell types to function under diverse conditions. For example, a large comparative survey of heterochromatin organization in a variety of tissues and species detected profound tissue-specific differences in heterochromatin patterning as well as differences between similar cell types, such as the eye rod cell nuclei, in diurnal, nocturnal and aquatic mammals [11, 51]. Further experiments have suggested that heterochromatin positioning in some cases could be explained by tissue- and cell type-specific expression of lamin B receptor (LBR) and lamins A/C, which contribute to tethering of heterochromatin to the nuclear periphery [11]. Whether this is the most prominent mechanism to achieve tissue-specific genome architecture is yet to be determined. Notably, many tissue-specific NETs could also fulfill this function.

Specific Gene Positioning with Respect to the NE

The finding that centromeres, telomeres, and gene-poor chromosomes associate with the NE together with the knowledge that most of the genome is noncoding might suggest that this spatial positioning is directed by general repetitive elements in the genome. The first indication that specific coding gene loci could be preferentially positioned at the NE came from *Drosophila* where several individual gene loci were observed to be reproducibly proximal to the NE [52]. This conserved positioning would be expected to have functional consequences.

We now know that many individual genes have nonrandom positions in the nucleus and many genes change position under certain conditions, particularly those under strong regulation during development. For example, the immunoglobulin H (*IgH*) locus moves from the nuclear periphery to the nuclear interior during B lymphocyte development at a critical time when the locus undergoes *V(D)J* recombination. Specifically, the *IgH* locus is at the NE in early lymphocyte lineages such as Pro-B cells and T-cells but is in the nuclear interior in later stages such as Pre-B cells [53]. Also during neurogenesis the *Mash1* (*Ascl1*) locus moves away from the periphery [54] and during adipogenesis several genes involved in lipid biogenesis that are upregulated during differentiation have been observed to move from the nuclear periphery to the interior [55]. Finally, the cystic fibrosis transmembrane conductance receptor (*CFTR*) gene is positioned at the NE in some cell types and in the interior in other cell types in a reproducible fashion [56]. Notably, in most cases described genes associate with the nuclear periphery in their inactive state and reposition towards the nuclear interior when active. Some genes, however, have been observed to exhibit a spatial preference for the nuclear periphery that is maintained rather than changing with differentiation and activation. These include the proteolipid protein (*PLP*) gene locus [57], the interferon-γ locus [58], the breast cancer *ERBB2* locus [59], and the osteogenesis collagen type 1 alpha 1 (*COL1A1*) locus [60].

In most studies, when the repositioning of individual genes between the nuclear periphery and interior was observed the rest of the chromosome was not tested. In cases when this was tested, many genes were found to move between the periphery and interior without a corresponding change in the position of the whole chromosome territory [55, 61]. This is consistent with numerous observations that chromosome territories often exhibit "looping out" of small regions [62–64]. However, the *FABP4* gene and its host chromosome both strongly shift from the nuclear periphery to the nuclear interior during adipocyte differentiation [55]. Thus genes and chromosomes may not always exhibit synergistic behavior and it remains unclear if the spatial gene positioning regulates or simply reflects gene activity.

Tethering of Chromosomes and Loci to the NE

In attempts to investigate the consequences of gene and/or chromosome positioning within the nuclear space several laboratories developed artificial tethering systems allowing recruitment of a single locus to the NE [65–67]. Three independent studies used mammalian cell lines with bacterial lac operator repeats (lacO) inserted into a genomic region that tended to be in the nuclear interior. Separately, the bacterial lac repressor (lacI) that specifically binds these repeats was fused to either GFP or to a NE protein (lamin B1 or the NETs LAP2β and emerin) and expressed in the cells carrying the array. Expression of the lacI-reporter fusion to GFP had no effect on the position of the lacO locus within the three-dimensional organization of the nucleus, but when the lacI fused to a NE protein was expressed the lacO array repositioned from the nuclear interior to the NE [65–67]. Once at the NE, the tethering could be reversed because the binding of lacI to lacO sequences can be disrupted by addition of isopropyl β-d-1-thiogalactopyranoside (IPTG). Addition of IPTG to disrupt this interaction caused the locus to be released from the periphery. Several important conclusions stemmed from these studies. In all cases, tethering of the lacO array to the nuclear envelope led to repositioning of the entire chromosome carrying the lacO array from the nuclear interior to the nuclear periphery. However, the effect of this repositioning on gene expression was not consistent between individual studies (see below). It was also observed that when chromosome 11 containing the lacO repeats moved to the periphery, another chromosome, chromosome 4, moved away from the periphery suggesting that genes/chromosomes may compete for space based on the strength of affinity interactions. As the NE represents only ~1/30th of the nuclear volume it would not be likely to be able to accommodate all chromosomes in a typical nucleus. Lastly, the positioning of lacO repeats and their recipient chromosome was not heritable - the loss of lacO-lacI affinity interaction upon treatment of the cells with IPTG resulted in loss of peripheral localization of the lacO array in daughter cells. This argues that in the case of endogenous loci/chromosomes a particular pattern of affinity interactions must be restored at the end of each mitosis for a specific cell type to achieve a particular chromosome configuration. Such reestablishment would not be expected to be completely accurate. Thus, not surprisingly, a particular organizational pattern using the directed lac array system was never achieved in more than 89 % of cells and endogenous patterns tend to be achieved at frequencies of 60–80 %. A recent study using an elegant experimental set up to follow specifically the localization of peripheral chromatin through several rounds of mitosis also clearly demonstrates that sequences labeled as peripheral in mother cells may (with some degree of probability) end up localizing internally in daughter cells [27]. This is likely to reflect the randomness in chromosome movements when aligning at the metaphase plate that may result in some chromosomes not being accessible when NETs bind to reform the NE in telophase. Such lack of accurate heritable propagation places the spatial genome organization outside the classical definition of epigenetics.

The lacO-lacI system provides an extremely strong tether not just because of the high binding affinities between lacO and lacI, but also because the lacO sequence usually is amplified 128–256 times in the array. Though this is certain to be stronger than any individual chromatin–NE interactions in mammalian cells, interaction sites on human chromosomes responsible for spatial genome organization would likely be both abundant and widely distributed, thus providing many tether points that would in the end have the same effect as the amplified lacO array. Alternatively, large gene clusters such as at the *IgH* locus, *Hox* loci, and olfactory receptor gene clusters might provide unique binding sites that would create distinct microenvironments at the NE. Indeed, a recent study found that olfactory receptor gene clustering is associated with the NE [68].

So how are loci and/or chromosomes tethered to the NE? The NE tethers for chromatin should be strongly embedded core components of the NE, e.g., lamins and NETs. Lamins form an intermediate filament polymer resistant to most standard cell extraction conditions such as 1 M NaCl and 2 % detergent and some lamin-lamin interactions can withstand 6 M urea extraction [69]. Thus, the nuclear lamina provides a strong scaffold to tether chromosomes. Lamins have been shown to bind core histones, particularly the H2A/H2B subtypes [70–72]. These interactions would not be expected to discriminate any particular areas of the genome, however, these studies were performed before the identification of many histone modifications and it is possible that if revisited some specificity might be observed. Some specificity was also observed in lamin binding to DNA, in particular to repetitive AT-rich sequences and the minor groove of single-stranded DNA in matrix- and scaffold-attachment regions (MARs and SARs; [41, 73–75]).

Just as the NET SUN proteins are responsible for the tethering of centromeres and telomeres to the NE, other NETs are likely to contribute to specific chromosome attachments. As the NETs are embedded in the membrane, they provide a strong anchor to chromatin, but this strength is further increased when considering that most NETs tested thus far have been shown to bind the lamin polymer [6]. That the NETs could contribute specificity to genome organization was first supported by observations that LBR and emerin bind to distinct positions on chromosomes at the earliest stages in NE assembly in telophase [76]. LBR has, like lamins, been found to bind histones, though in this case histones H3/H4 [77], and also to heterochromatin protein 1 (HP1) [10], which binds with high affinity to methylated histone H3 tails (H3K9me2 and H3K9me3) [78, 79]. These interactions provide a mechanistic explanation for the enrichment of H3K9me2 modified heterochromatin at the nuclear periphery. Notably, either pharmacological inhibition or knockdown by RNA interference of the enzymes responsible for depositing H3K9me2 (G9a and GLP, also known as EHMT2 and EHMT1, respectively) led to dissociation of peripheral chromatin from its proximity to the nuclear lamina [27].

The NET LAP2β has also been shown to bind to core histones [80], but also binds directly to DNA and to the barrier-to-autointegration factor (BAF) [81, 82]. BAF is a soluble protein that binds both to histones and DNA and so can contribute to higher order chromatin structure [83]. BAF binding could bring some specificity to NE–chromatin interactions because it has particular affinity for selected linker histones including H1.1 [84]. The NET MAN1 can also bind BAF, but separately binds directly to DNA through a winged helix fold domain in its carboxyl-terminal domain [85].

The NETs mentioned above are all widely expressed and so it is relatively easy to imagine how they could contribute to general spatial genome organization patterns such as those related to gene density—and indeed LAP2 and LBR have been linked to such general organization [11, 12, 68]. However, it is not easy to explain how these ubiquitously expressed proteins, except perhaps LBR, could contribute to the tissue-specific patterns of genome organization observed for certain chromosomes and gene loci. It would seem more likely that tissue-specific NETs recently discovered in proteomic analyses of NEs isolated from several different tissues [17–19] might perform this function. Indeed, screening of novel blood-specific NETs identified one that promoted chromatin condensation and two others that repositioned a gene locus [17]. Moreover, a recent study indicates that several NETs expressed preferentially in either liver, fat, or muscle cells can reposition chromosomes to the NE when exogenously expressed in fibroblasts and, in the case of the liver-specific NETs 45 and 47, their knockdown resulted in release of certain chromosomes from the NE in liver cells [20]. The same study showed that different NETs are able to affect the positioning of different subsets of chromosomes, indicating that these are indeed the likely endogenous players that provide tissue specificity to spatial genome organization [20].

Relationships Between Nuclear Positioning and Gene Expression State

The idea that changing the position of a gene with respect to the nuclear periphery could lead to a change in gene expression received its first strong support in *Drosophila* when an insulator sequence called *gypsy* was found to be preferentially located at the nuclear periphery. When *gypsy* and a reporter gene were inserted into a more internal area of the genome, the locus was translocated to the periphery and the expression from the reporter was correspondingly reduced [86]. This came to be known as one of the many examples of position effect variegation in gene expression that have been observed in *Drosophila*, yeast and mammalian cells. The example above is somewhat anecdotal and may reflect caveats with experimental design as insulator sequences, such as *gypsy*, serve as boundary elements protecting active genes from spreading of nearby heterochromatin [87, 88]. Insulator elements bind specific proteins and are involved in long-range chromatin interactions (chromatin looping), blocking of enhancer activity, and notably, delineation of subnuclear localization of chromosomes [87, 89].

In mammalian cells, insulator sequences are bound by zinc-finger proteins CTCF (CC*CTC*-binding *f*actor) and its relative BORIS (*B*rother *O*f the *R*egulator of *I*mprinted *S*ites) [90, 91]. In recent years, the mapping of CTCF binding sites in the genome [92] and the lamina-associated chromatin domains [93, 94] in conjunction with maps of chromatin modifications and gene expression profiles have provided important insights into the relationship between nuclear positioning, chromatin architecture, chromatin modifications and gene expression state.

Gene Activity at the Nuclear Periphery

Apart from the *gypsy* insulator experiment mentioned above, multiple examples can be found in literature suggesting that genes located at the nuclear periphery tend to be transcriptionally inactive, late replicating and marked by chromatin modifications indicative of silenced chromatin [95]. When genes were found to migrate to the nuclear interior upon differentiation or under the influence of external stimuli, this in some cases was accompanied by upregulation of transcriptional activity, increase of histone acetylation and the presence of RNA polymerase II at gene promoters [53, 54]. However, as alluded to earlier, examples exist demonstrating that not all genes located at the nuclear periphery are transcriptionally silenced. Moreover, the cause and consequence relationship between gene location and gene activity has always been difficult to infer from studies on individual gene loci.

The experiments using lacO arrays to tether genomic loci to the NE [65–67] were partly designed to test the effect of nuclear positioning on gene expression in a more controlled fashion. Two of these studies had a selectable marker inserted in the array and both found that transcription of this particular marker was reduced when the locus was at the periphery [65, 67]. Correspondingly they found that release of the locus from the periphery with IPTG restored the lost activity to the marker genes [65, 67]. One of the studies also tested endogenous genes close to the area where the lacO array was inserted, finding that repression was not general with only some genes being repressed when the locus was at the periphery [65]. In the third study no repressive effects or deficiencies in the induction dynamics of a reporter gene inserted by the array were observed between its internal and peripheral positioning [66]. However, in this study the reporter was strongly and actively induced from a promoter that could potentially overcome any repressive effects of the periphery. Therefore, the question of whether and how tethering a locus to the periphery directly results in its repression remained unresolved by these studies, but it is clear that changes in gene regulation can occur concomitantly with changes in gene positioning.

The general features of gene activity at loci in close proximity to the NE became apparent when the nuclear lamina associated chromatin domains were mapped on a genome-wide scale in *Drosophila* and mammalian cells by a technique known as DamID [93, 94]. The DamID method used in these studies employed a fusion of lamin B with a GATC sequence-specific bacterial DNA adenine methylase (Dam). As adenine methylation in GATC context does not exist in higher eukaryotes, any DNA carrying this mark in cells expressing the Dam-laminB fusion signifies proximity to the nuclear lamina. This approach defined the *l*amina-*a*ssociated *d*omains

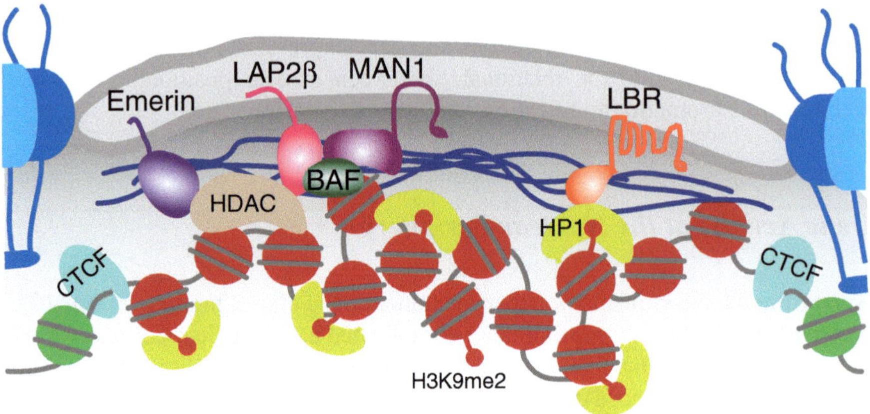

Fig. 2 Anchoring and silencing of chromatin at the nuclear envelope. Silenced chromatin is both recruited to and maintained at the NE through interactions of lamins and NETs with chromatin and chromatin modifying enzymes. Lamina-associated domains (LADs) are bounded by CTCF on chromatin. Silent marks such as histone H3 lysine 9 di and tri methylation (H3K9Me2 and Me3) recruit the additional silencing factor heterochromatin protein 1 (HP1), which in turn can bind to the NET lamin B receptor (LBR, so named because it in turn binds the lamin polymer). One part of the barrier-to-autointegration factor (BAF) can bind histones while another part binds to the NETs emerin, LAP2β and MAN1. LAP2β and emerin have also been found to bind histone deacetylases, which further promote and maintain silencing marks at the periphery

(LADs) as stretches of DNA 0.1–1 mega bases (Mb) in length. In human fibroblasts, there are more than 1,300 of such genomic regions in close contact with the nuclear lamina. LADs are present on all chromosomes, associated with repressive histone modifications (H3K9me2 and H3K27me3) and are relatively gene-poor (Fig. 2). Genes embedded in LADs have low levels of expression with very few active genes that escape silencing. Most LADs have sharp boundaries that are marked by insulator elements bound by CTCF [94] (Fig. 2). This organization of chromatin into LADs and inter-LAD domains is also conserved in *Drosophila* [93] with a different set of insulator elements marking the LAD boundaries [87].

Given the vast changes in gene expression that occur during differentiation and the observations that the radial positioning of chromosomes varies in a tissue-specific manner, one would expect LADs and CTCF binding sites also to change substantially during differentiation. Surprisingly, the LADs and H3K9me2-rich domains, which largely overlap with LADs, remain largely invariable during differentiation of embryonic stem cells into neuronal progenitors and further into mature neurons [96–98]. Instead of a global rearrangement of genome-nuclear lamina interactions, small local changes affecting individual genes were observed in these studies. Thus, LAD-embedded genes, which were upregulated during neuronal differentiation, were seen to dissociate from the lamina and lose silencing histone modifications [96, 97]. This may seem surprising, however, such global maps represent population average and cell-to-cell variations are lost in such studies. Thus, LADs can be viewed as a probabilistic map of genome-nuclear lamina interactions. Techniques that permit global genomic studies on a single-cell level are starting to emerge and promise to be instrumental in determining cell-to-cell variation in cultured cells and, excitingly, cells derived from specific tissues. As mentioned earlier, tracking individual cells through mitosis detected substantial rearrangement and dissociation of LADs from the nuclear lamina in daughter cells [27]. This, perhaps, allows rearrangement of LADs and activation of LAD-embedded genes. It also suggests that rearrangement of chromatin–lamina interactions may be proportional to the number of cell divisions undertaken by cells before they acquire the differentiated state. In the case of in vitro differentiation models using ES cells, the differentiated state is achieved within very few mitotic divisions, which may not allow sufficient time for significant LAD rearrangements to take place.

Is the dissociation of genes from the nuclear periphery sufficient to induce gene expression? In the lacO tethering experiments to the NE, it was found that the release of the locus from the periphery either by IPTG (disrupting lacO-lacI interactions) or treatment with histone deacetylase (HDAC) inhibitors could restore the lost activity to marker genes and few, but not all, endogenous genes in close proximity to the lacO array [65–67]. However, recruitment to peripheral chromatin of the strong transcriptional activator VP16, which interacts with histone acetyltransferases (HATs) and disrupts nuclear lamina–chromatin contacts, induces relatively few changes in gene expression on a global scale [27]. Taken together, these experiments suggest that movement of loci away from the nuclear lamina is perhaps permissive, but not sufficient for gene activation.

Recruitment and Silencing of Chromatin at the NE

The question of whether or not genes are silenced before they are anchored to the NE is not yet comprehensibly answered. Notably, many interactions between chromatin, the nuclear lamina and the NETs may require preexisting histone modifications. Thus, chromatin anchoring to the INM embedded lamin B receptor LBR via HP1α and HP1γ [10] requires H3K9me2 to allow HP1 binding (Fig. 2). Although there is some evidence that BAF (barrier-to-autointegration factor), which binds the NETs LAP2β, emerin, and MAN1 through a shared sequence motif called the LEM domain [81, 99, 100] interacts with histone H3 methylase G9a (EHMT2) [101], in most mammalian cell types G9a is distributed throughout the nucleoplasm and not exclusively anchored to the NE. However, additional evidence suggests that NETs emerin and LAP2ß interact with histone deacetylases, HDAC3 being one of them [102, 103] (Fig. 2). Therefore, it is possible that localized HDAC activity at the NE provides G9a with a suitable substrate for subsequent methylation of H3 tails generating H3K9me2 required for anchoring to LBR-interacting HP1 proteins.

The interactions of BAF with NETs raise another interesting issue regarding anchoring to the NE of silenced chromatin. The NET LAP2β has several soluble splice variants that also bind BAF [104–106]. One of these, LAP2ζ, principally resides in the cytoplasm and its upregulation compared to LAP2β causes BAF to be captured in the cytoplasm before nuclear import, thus reducing intranuclear pools and their corresponding functions in cross-linking chromatin and recruiting it to the

NE [106]. As many NETs have multiple splice variants, this type of competitive inhibition will likely prove to be used commonly as a regulatory mechanism for chromatin recruitment to the NE.

In addition to chromatin-mediated interactions, several DNA-binding factors have been found to interact with NETs and the nuclear lamina. One of these is the transcription factor Oct1, which interacts with lamin B and localizes to the nuclear periphery in a lamin B-dependent manner [107]. As Oct1 motifs are enriched within LADs, this suggests that Oct1 may contribute to tethering to the NE of genomic loci in a sequence-specific manner.

Activation of Chromatin at the Nuclear Periphery

Among the proteins identified in a proteomic study of rat liver NEs were several that can modify histones for not only repression, but also activation. One of these, NET43/hALP, is a histone acetyltransferase [16]. Interestingly, the membrane prediction for NET43/hALP only occurs in some organisms and was absent in the human homologue. Nonetheless, the human version localizes to mitotic chromosomes in mammalian cells by binding to another NET, SUN1 [108]. Depletion of SUN1 in human tissue culture cells resulted in delayed chromosome decondensation and a reduction in histone H2B and H4 acetylation in a manner dependent on hALP [108]. Interestingly NET43/hALP appears to be upregulated during lymphocyte activation when the large amount of dense peripheral chromatin of resting lymphocytes becomes decondensed [17].

Functions of the NPC in Spatial Genome Organization and Gene Regulation

Each NPC is a large protein assembly of >40 MDa in yeast and >60 MDa in mammalian cells made up of at least 30 core proteins [13, 15]. The assembly has an eightfold symmetry so that each protein is represented in a minimum of eight copies and in some cases in as many as 64 copies in a single NPC. Among these proteins is Tpr, a 270 kDa coiled-coil protein that extends roughly 100 nm into the nucleoplasm in a structure generally referred to as the nuclear basket [109], and Nup50, which also associates with the nuclear basket extending roughly 50 nm from the central plane of the NPC into the nucleoplasm [110]. These extensions allow interactions with chromatin of a different character to those at the level of the membrane. Electron micrographs and recent high-resolution imaging of mammalian nuclei [111] show that while heterochromatin contacts the areas of the NE which are lined with nuclear lamina, the NPCs are surrounded by less dense chromatin. Studies in yeast and higher eukaryotes demonstrate that NPCs contribute in several different ways to the spatial genome organization and regulation of gene expression.

The NPC in Spatial Genome Organization

Although in mammalian cells telomeres are tethered to the NE through interactions with SUN proteins, in yeast they are tethered by the NPC. In fact, the first specific interactions between telomeres and the NPC were determined in yeast with the yeast Tpr homologue, Mlp, as the anchoring site [34, 35]. The difference in telomere tethering between yeast and mammals together with the lack of yeast homologues for many NETs, particularly the more tissue-specific ones [18], suggests that many regulatory functions in yeast carried out by the NPC have been subsequently taken over by other NE proteins during evolution.

The recruitment of telomeres to the periphery in budding yeast is essential for silencing of subtelomeric genes. In addition to Mlp (Tpr), peripheral localization of telomeres also involves the soluble non-NPC protein Ku [112]. Mutation in Mlp and Ku proteins results in derepression of subtelomeric silenced genes [34, 35, 113, 114]. As a potential epigenetic mechanism, the heritability of this silencing was also addressed. Derepression of reporter genes integrated close to telomeres upon deletion of Mlp and Ku implied that silencing requires NE/NPC association [34, 114]. However, deletion of the NPC proteins also results in a redistribution of Sir3p fused to GFP [114]. The latter result argues that the observed derepression of genes could be a secondary consequence of NPC disruption, rather than due to relocation of the telomeres away from the nuclear periphery. Another study, using an elegant experimental setup to break the connection between the NPC and a silenced reporter after the silencing was established, found that the release from the periphery did not derepress the silent reporter [115]. This suggests that once established at the nuclear periphery the silenced chromatin state can be stably maintained without NPC association.

Active Chromatin at the NPC

In contrast to chromatin at the nuclear lamina, the NPCs in many species including yeast, flies and mammals associate with active chromatin [116–119]. Adaptor proteins that mediate these connections have been identified in a variety of systems. In yeast, the nuclear basket protein Mlp1 interacts with the chromatin-bound co-activator complex SAGA (Spt-Ada-Gcn5-Acetyltransferase), which is known to promote active transcription [120]. SAGA is a large protein assembly, containing two chromatin-modifying enzymes, the Gcn5 histone acetyltransferase and Ubp8 histone deubiquitinase [121, 122]. SAGA, via its Sus1 component, was also shown to interact with the NPC-bound TERX-2 complex, which plays important roles in transcript elongation and mRNA transport [123–125] (Fig. 3). Budding yeast TREX-2 consists of four proteins Sac3-Thp1, Cdc31, and Sus1) [126]. Thus, tethering of active acetylated chromatin to the NPC via Sus1, a shared component of SAGA and TREX-2 complexes, may help to coordinate transcription-coupled mRNA export [127]. This arrangement seems to be conserved in other species. In *Drosophila*, the Sus1 ortholog E(y)2 via the TREX-2 complex protein Xmas-2, the

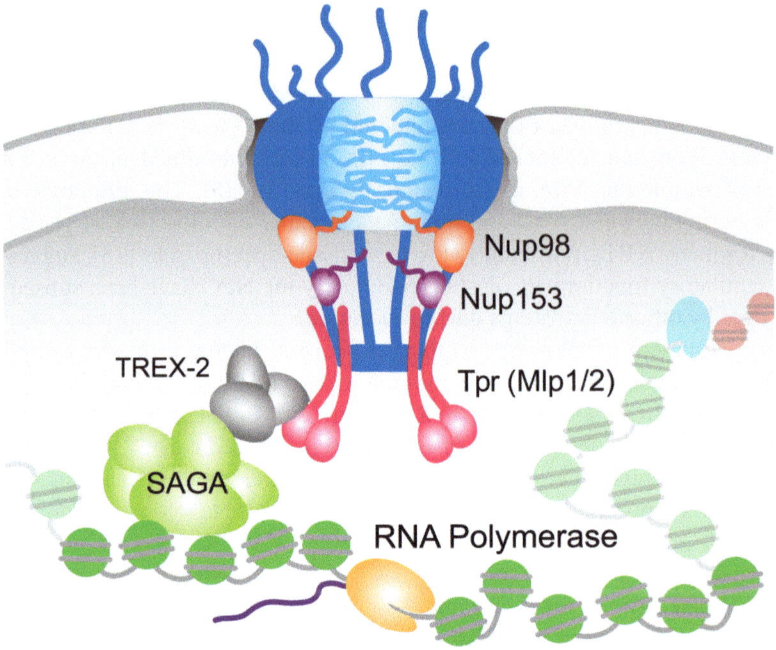

Fig. 3 Active chromatin at the nuclear pore. NPC proteins Nup98, Nup153, and Tpr form the nuclear basket, which tethers active chromatin through binding a complex of TREX-2 and SAGA. The SAGA complex promotes active transcription while the TREX complex is important for RNA export. Thus, recruitment of active genes to the periphery by this mechanism also facilitates rapid translocation of the mRNAs out of the nucleus so they can be translated on ribosomes in the cytoplasm. It is noteworthy that the average mammalian nucleus is estimated to have 2,000–3,000 NPCs and as they are 125 MDa complexes of greater than 100 nm diameter this translates to up to 40 % of the nuclear surface

equivalent of yeast Sac3, promotes transcription, mRNA export and positioning close to the NE of the *hsp70* gene cluster [128, 129]. Whether tethering of active chromatin follows a similar arrangement in mammalian cells is yet to be determined. Notably, the SAGA complex is highly conserved [122] and a Sac3-related protein GANP involved in mRNA export is present in human cells [130]. Therefore, it is likely that the positioning of active SAGA-bound chromatin at the NPC may operate in a similar manner across most eukaryotes.

Studies in yeast and *Drosophila* have also shown that several nucleoporins located at the nuclear side of the NPC are required for tethering of active chromatin to the nuclear periphery. Nup1, Nup2, Nup60, and Mlp1 (Tpr in metazoan species) have been implicated in mediating connections between active genes and the NPC [120, 131, 132]. One of the most interesting examples signifying the functional importance of such tethering is dosage compensation in *Drosophila*. Male flies, having one X chromosome, achieve balanced expression of X-linked genes relative to female flies with two X chromosomes by upregulating gene expression on the single male X. This upregulation requires *Drosophila* dosage compensation complex, which includes noncoding RNAs, histone acetyltransferase MOF and additional proteins, as well as the nuclear basket proteins Nup153 and Tpr [117, 133]. Taken together, these examples implicate nucleoporins located at the nuclear side of the NPC in spatial genome organization and regulation of gene activity.

It has been suggested that, topologically, positioning of active genes at the NPC may occur via formation of chromatin loops mediated by promoter–terminator interactions as well as the boundary elements in yeast and their equivalent, the insulator elements, in other metazoan species [92, 134, 135] (Fig. 1). Such elements are often found at the boundaries between active and repressed chromatin and serve to antagonize heterochromatin spreading. Consistent with such a function, studies in yeast and human cells have shown that chromatin in close proximity to the NPC carries modifications indicative of both, active and repressed, chromatin states [92, 136]. A screen for proteins involved in boundary activity in budding yeast identified Mlp, Nup2 (the yeast homologue of Nup153 in humans), the NPC-associated proteins Nup60p, and the Ran-GTP exchange factor Prp20p [134, 137]. The typically mobile Prp20 is bound to the core structure of the NPC through Nup2 on one side and on the other it binds H2A.Z (also called Htz1), a variant of histone H2A that is loaded by the SWR-C chromatin-remodeling complex [138]. H2A.Z marks relatively immobile nucleosomes in the yeast genome and can be found at most gene promoters and some intergenic regions of the budding yeast genome [139]. Interestingly, the SAGA complex may also contribute to boundary/insulator element function. It has been reported that the Ada2 component of SAGA is recruited to yeast telomeres and required to suppress the spreading of telomeric heterochromatin into subtelomeric regions [140]. Moreover, Sus1, a shared component of SAGA and TREX-2 complexes, functions together with Su(Hw) at insulator elements in *Drosophila* [129]. The overall similarity between the function of boundary elements in yeast and the insulators in *Drosophila* and mammalian cells indicates that the compartmentalization of chromatin into active and inactive domains at the NE is widely used in evolution.

Transcription Factors at the Nuclear Envelope

The nuclear surface provides a large scaffold on which the genome can be spatially organized and regulated; however, in relative terms it represents only ~1/30th of the total nuclear volume. If a spherical nucleus has a radius of ~5 μm, then the volume of the nucleus would be 523 μm^3 and the surface area would be 314 μm^2. However, the thickness of the NE from ONM to INM is only ~50 nm, and the penetration of NETs and lamins from the inner surface into the nucleoplasm is likely much less than this, while the NPC nuclear baskets have been measured to project roughly 100 nm into the nucleoplasm. Thus, if one considered an average penetration of 50 nm for the "volume" of the NE, the volume of the nuclear surface would be less than 16 μm^3 or roughly 30-fold smaller than the volume of the nucleoplasm. Thus, if, in addition to the genes tethered by the NE (discussed above), the NE also tethered transcriptional regulators for those genes, it would have the equivalent effect of increasing the local concentration of the transcriptional regulator by 30-fold.

The Function of NE in Direct and Indirect Repression of Transcription

Both LAP2β and emerin bind the transcriptional repressor germ cell-less (gcl) that is known to affect E2F/DP transcription factor heterodimers [141, 142]. Moreover, overexpression of LAP2β in tissue culture cells inhibited E2F-dependent transcription from a reporter construct [142]. Emerin also binds Btf, another transcriptional repressor with a different target specificity [143]. Of the lacO-lacI studies that used NE affinity tethering to recruit a lacO array to the periphery one fused the lacI to LAP2β and the other fused lacI to emerin [65, 67]. The overlap and discrepancies in results between these studies might be in part because of the partly shared and partly different specific transcriptional repressors binding to these two NETs. In the emerin study the amino-terminus was deleted to minimize this potential criticism [67]; however, the deleted region only partly overlaps with the binding site on emerin for germ cell-less [141], so this may not have been sufficient.

Whereas it has been demonstrated that multiple different NPC proteins can recruit transcription factors and their target genes to the same location to effectively increase the relative local concentration of the transcription factor, NETs and lamins that have been found to bind to transcription factors appear to function in the opposite fashion. They sequester the transcription factor at the periphery away from the gene target in the nuclear interior. Thus, NE binding of transcription factors appears to function to prevent gene activation.

The first demonstration of transcription factor binding to NE proteins was binding of lamin A to the retinoblastoma protein (Rb) [144]. However, the effect of lamins on Rb may be due mostly to the nucleoplasmic pool of lamins. This is supported by findings that a soluble splice variant of the NET LAP2β, LAP2α, forms a complex with lamin A and Rb in the nuclear interior [145]. Interestingly, this can have both positive and negative effects on gene regulation depending on cell type. Although this complex on the one hand sequesters Rb away from gene targets, on the other hand the complex stabilizes Rb, which, without this binding, turns over rapidly. As a consequence of this, in cells stimulated by phosphorylation of Rb to initiate progression into S-phase, the higher levels of stabilized available Rb enable much stronger activation. Thus, cells with more or less LAP2α and lamin would have different propensities to engage cell cycle progression upon the same activation stimulus. Consistent with this, LAP2α knockout mice exhibit hyperproliferation of both erythroid and epidermal lineage cells [146]. Lamins are ideal for sequestering transcriptional regulators as their abundance (~3,000,000 copies per mammalian nucleus; [147]) could easily saturate any transcription factor. Lamin A has also been shown to bind cFos [148].

In addition to lamins, NETs also repress gene function by sequestering transcription factors. Emerin interacts with the transcription factor Lmo7 [149]. Intriguingly, this interaction has been found to function in a tightly regulated feedback loop to regulate the emerin gene itself. Lmo7 activates the emerin gene (*EMD*) so that emerin protein binding of Lmo7 sequesters Lmo7 at the periphery away from the emerin gene; thus, the more emerin is produced, the more it can sequester Lmo7 to repress its own expression. Similarly, the NET MAN1 interacts with Smads and sequesters them away from target genes located in the nuclear interior [150, 151]. Very few of the many tissue-specific NETs recently identified have been characterized in detail, but it seems likely that some of these may bind to transcription factors for sequestration. The tissue specificity of these NETs would add a much greater complexity to NE regulation of gene expression.

Transcriptional Activation at the NE

In addition to binding Smad transcription factors, MAN1 directly binds DNA through a winged helix fold in its carboxyl-terminal domain [85]. Thus, though this has not been specifically demonstrated yet, it could potentially also tether both Smads and Smad-regulated genes to the NE to additionally activate transcription as shown for the NPC proteins.

While older studies showing the dissipation of dense chromatin at the nuclear periphery concomitant with activation of lymphocytes suggested a role for the NE in gene activation [26], several more recent studies show with certainty that transcriptional activation occurs from the NE. The *PLP* gene, involved in myelin production, becomes activated when already at the periphery [57]. It is interesting that this gene is active in glial cells, as it was noted earlier that brain cells tend to have minimal peripheral heterochromatin. It is thus reasonable to postulate that the ability to activate a gene at the NE may depend on cell type and the amount of peripheral heterochromatin. Other differentiation/cell state-associated genes observed to be active at the NE are the breast cancer *ERBB2* gene, the osteogenesis *COL1A1* gene and the interferon gamma *IFN-γ* locus [58–60].

Stabilization of the Genome by the Nuclear Envelope

Though only theoretical at this stage, the physical tethering of chromatin to the NE could play a major role in stabilizing the genome to protect against translocations and other factors that can lead to cancer. Proteins in the nuclear interior, even those associated with nucleoplasmic structures, tend to be relatively dynamic in FRAP studies [152]. In contrast, the movements of proteins at the NE are much less dynamic and indicate local constraints [153, 154]. One potential consequence of NE tethering could be to physically stabilize the genome, minimizing movement. Indeed, transgenes located near the nuclear periphery in mammalian cells have been shown to be less mobile than those residing in more internal positions [155]. In theory, tethering of chromatin to the NE could help maintain chromosome territories and prevent entanglement of chromosomes and potentially associated chromosome translocations that could lead to tumors. This could be particularly important during replication and even the reason why late-replicating DNA tends to be at the periphery [93, 156], i.e., because a peripheral tether combined with silent chromatin helps to stabilize chromosome territories so that chromosomes do not get entangled during replication.

The NE is a good tethering point for chromatin because it is a relatively stable structure due to the intermediate filament lamin polymer lining the inner surface of the NE [4]. It is striking that while the cytoplasm has actin filaments, microtubules and intermediate filaments, the NE has just the intermediate filament lamins. Unlike other cytoskeletal systems, intermediate filaments are highly elastic. Under compression or tension forces that would break actin filaments and microtubules the intermediate filaments are unaffected [157]. It is not surprising thus that spider's webs are made of intermediate filaments, tough yet elastic.

These properties are important because live cell microscopy indicates that nuclei move and exhibit frequent morphological aberrations while chromatin also moves dynamically. Thus, the NE needs to have a structural support that can bend, but not break connections. In this light it is not surprising that both lamins and INM NETs bind chromatin and each other, thus providing multiple contacts in an overlapping network embedded with a wide range of proteins into the membrane to create a very strong tether. Other INM NETs make connections across the lumen of the NE to outer membrane NETs [158] and these in turn connect the NE to the cytoskeleton providing an anchor for the nucleus in the cell [159]. Together these properties enable the NE to keep chromatin tethered while still being able to stretch in response to forces placed on the polymer by genome movements or the cytoskeleton. If the peripheral lamina nucleoskeleton were rigid like microtubules, it would likely break in response to such forces and genes and chromosomes would lose their tethering. Similarly, if tethered merely by transmembrane proteins, strong forces from chromosome movements might rip the tethering NETs out of the lipid bilayer. Thus, the use of both lamins and NETs is a sensible strategy to support the many dynamic movements of chromatin within the interphase nucleus.

Although much of the above is merely a hypothesis, it is clear that lamins contribute to the mechanical stability of the nucleus. Lamin depletion or mutant expression resulted in nuclear lobulation and increased deformation under mechanical stress [69, 160–163]. In theory the importance of this is manifold. Increased lobulation would also increase the ratio of NE to nucleoplasmic volume, thus enabling greater silencing and gene regulation from the NE. Due to the DNA of chromosomes being single rope-like molecules, connections to the NE could influence the ability of internal sections to engage in transcription factories.

Peripheral Chromatin Organization and Nuclear Envelope Disease

Though lamins, NETs, and NPC proteins have all been linked to cancer in a variety of ways, most of these are indirect, e.g., having functions in processes that are critical to cancer progression as opposed to mutations in a particular NE protein causing cancer. In contrast, mutations in NE proteins have been found to be causative of a wide variety of inherited disorders ranging from muscular dystrophies to the premature aging progeroid syndromes [164–166]. In fact lamin A is now the most mutated gene in the human genome. Lamin diseases include muscular dystrophies [167–169], lipodystrophy [170–172], neuropathy [173, 174], cardiomyopathy [175], dermopathy [176], and the aging disease progeria [177, 178]. Several NETs and associated proteins also cause diseases or syndromes affecting muscle [179–182], bone [183, 184], brain [185–189], skin [176] and immune cells [190]. Because most NETs tested bind to lamins, lamin mutations could also affect NET distribution and function. Thus, the pathology of lamin-based diseases could be as much due to secondary effects on NET function as to the loss of lamin function.

Chromatin Organization in Heritable Diseases Linked to the NE

The overall distribution of heterochromatin as defined by electron dense material in electron microscopy is altered in several NE-linked diseases. Normal fibroblasts typically have a reasonable amount of this dense peripheral chromatin generally distributed throughout the periphery. In fibroblasts from patients with Emery–Dreifuss muscular dystrophy (EDMD), the dense chromatin often seems to have broken away from the NE and resides about 500 nm in from the NE. This altered pattern is observed for two different NE-linked muscular dystrophies, EDMD and Limb-girdle muscular dystrophy, and for both NET-linked and lamin-linked disease [191–193]. It is also observed in cardiomyopathy linked to the NE [194]. In contrast, fibroblasts from patients with NE-linked progeroid diseases tended to lose all peripheral dense chromatin [195], and fibroblasts from NE-linked lipodystrophy patients exhibited an intermediate phenotype with partial loss of peripheral dense chromatin in some areas at the periphery and partial clumping in other areas [193]. The fact that the patterns are not just disrupted, but disrupted in reproducible ways for each disorder suggests that these spatial genome organizational patterns are functionally relevant.

These NE protein mutations had specific and reproducible effects on spatial genome organization also in controlled experimental systems. For example, a mutation in lamin A that causes Hutchinson–Gilford progeria syndrome (HGPS) yielded an abnormal distribution of telomeres and clustering of centromeres [196] while other mutations that cause variously a neuropathy, lipodystrophy and muscular dystrophy reposition chromosomes 13 and 18 away from the nuclear periphery [197]. However, the relevance of these changes to specific disease pathology is uncertain as different mutations that cause the same disease can yield different effects on chromosome positioning. For example, both E161K and D596N lamin A mutations cause cardiomyopathy, but only E161K causes chromosome 13 to lose its normal peripheral localization [198].

Epigenetic Changes in Heritable NE Diseases and Cancer

Some of the reorganization of chromosomes, observed by FISH, and altered distribution of peripheral dense chromatin, observed by electron microscopy, could be explained by changes to epigenetic heterochromatin marks. This appears to be the case for HGPS where primary fibroblasts from patients have a significant loss of silenced chromatin marks associated with facultative heterochromatin, such as H3K9me2 and H3K27me3, while marks of constitutive heterochromatin such as H4K20me3 were increased [199]. Similar changes could be induced in normal cells upon overexpression of lamin A carrying a progeria-causing mutation [199]. Fibroblasts from a female patient also lost silencing marks on the inactive X chromosome and this chromosome lost its tight association with the periphery. Similar changes in silent chromatin marks have also been reported in normal aging cells [200]. Most cases of HGPS are due to a splice site mutation that causes loss of an exon that contains a cleavage site close to the C-terminus of lamin A. similar to the small G proteins, Lamin A acquires a farnesyl group at its C-terminal CaaX box. This modification is transient as the last 18 amino acids are cleaved at this cleavage site in mature lamin A. Blocking lamin A farnesylation had a positive effect on HP1α accumulation [201] and some of the heterochromatin defects observed in progeroid cells in culture could be reversed by treatment of the cells with farnesyltransferase inhibitors [202].

A number of studies have detected altered expression of NE components in cancer cells. For example, reduced expression of Emerin in ovarian cancer [203] and Lamin A/C overexpression in colorectal, ovarian, and prostate cancer correlates with poor prognosis and advanced tumor stage [204, 205]. Although such examples are numerous [206], a direct causal relationship between either mutations or altered expression of NE proteins has been established in very few cases. Nevertheless, indications that NE–chromatin interactions play an important role in genome organization and gene expression in cancer have started to emerge. It is well known that tumors exhibit altered patterns of DNA methylation in comparison to normal tissues [207]. These alterations include localized gain of DNA methylation at normally methylation-free CpG-rich gene promoters (CpG islands), leading to stable silencing of tumor suppressor genes, and widespread loss of DNA methylation from extensive chromosomal regions potentially promoting genomic instability. Interestingly, recent reports have shown that loss of DNA methylation in colorectal and other types of cancer occurs in regions of the genome that coincide with LADs and the silencing H3K9me2 [98, 208]. Moreover, gain of DNA methylation at promoters also occurs at genes located in these domains [208]. Taken together, these findings suggest that genome organization mediated by chromatin–NE interactions is not only profoundly different in cancer cells, but that these rearrangements may have direct functional implications in cancer development.

Another striking connection between NE components and cancer has emerged when translocations involving the nucleoporin Nup98 were found in patients with acute myeloid leukemia (AML) [209]. The most commonly detected translocations generate a fusion between Nup98 and homeobox transcription factor HoxA9 genes [207, 210]. The resulting Nup98-HoxA9 chimeric protein is nucleoplasmic and contains the GLFG repeats of Nup98, normally involved in shuttling of nuclear transport receptors through the NPC, fused to the DNA binding domain of HoxA9. The mechanism by which the fusion protein induces AML is as yet unclear. It has been suggested that GLFG repeats bind transcriptional co-activator CBP/p300 histone acetyltransferase [211], thus maintaining expression of HoxA9 target genes, which are normally downregulated during differentiation of hematopoietic stem cells following downregulation of HoxA9 itself. Fusions of Nup98, all containing the Nup98 GLGF repeats, with a variety of other partners have also been identified in AML. These fusion partners include a variety of co-activator proteins, histone methyltransferases such as NSD1, DNA topoisomerases, and RNA helicases [209]. Similar to Nup98-HoxA9, the Nup98-NSD1 fusion protein was found to bind regulatory elements of polycomb target genes, normally silenced during myeloid differentiation, and induce high levels of H3K36me3 (a mark associated with active transcription elongation) and histone acetylation, thus maintaining the expression of several homeotic genes, which contribute to malignancy [212]. Mutations abolishing H3K36me3 activity of NSD1 as well as deletion of GLGF repeats of Nup98 within the Nup98-NSD1 fusion protein context abolish activation of homeotic genes and inhibit leukemogenesis [212]. By a broadly similar mechanism, fusion proteins between Nup98 GLGF repeat region and PHD (plant homeobox domain) zinc fingers, which bind to histone H3 tails methylated at K4, function as potent oncogenes and induce AML in mouse models [213]. These observations suggest that most chimeric Nup98 proteins function to promote AML by generating aberrant active chromatin, which counteracts polycomb-mediated gene repression and allows self-renewal and maintenance of the undifferentiated state of myeloid progenitors [214]. Thus, epigenetic changes caused by a mutant NE protein are an essential component of AML and, perhaps, other malignancies.

Conclusions

Research in the past few decades has led us to conclude that the NE is a complex compartment of the cell, with functions that extend beyond the separation of the genetic material from the cytoplasm of the cell and the control of nuclear trafficking. The NE contributes to tissue-specific genome organization and regulation of gene expression through many different mechanisms, the details of which remain to be fully elucidated. Lamins provide a structural framework for the nuclear periphery and interact with silenced chromatin. NETs bind both the lamin polymer and the membrane in addition to both general and specific chromatin proteins. Moreover, NETs and NPC proteins recruit transcriptional regulators, epigenetically modified chromatin, and separately recruit enzymes that add epigenetic modifications to chromatin. As only a fraction of the many novel NETs identified by proteomics have been characterized in detail, there are a great many possibilities for diverse mechanisms in the regulation of spatial genome architecture and chromatin modifications from the NE. The finding that many NETs are tissue specific further adds to this complexity.

There are many questions that remain to be elucidated. To what degree does the NE silence genes by bringing them into an environment that is rich with silencing factors that directly modify the chromatin versus bringing already silenced chromatin to the periphery by affinity interactions? Does NE tethering also contribute a steric effect to this regulation? To what degree are the NE influences on chromatin organization heritable? To what degree does NE tethering stabilize the genome to protect it from inappropriate recombination events? And how much do these various functions contribute to the initiation of tumorigenesis? The NE provides many layers for regulating genome function, and many important discoveries are certainly yet to be made in this exciting and rapidly moving field.

References

1. Callan HG, Tomlin SG (1950) Experimental studies on amphibian oocyte nuclei. I. Investigation of the structure of the nuclear membrane by means of the electron microscope. Proc R Soc Lond B Biol Sci 137(888):367–378
2. Tapley EC, Starr DA (2013) Connecting the nucleus to the cytoskeleton by SUN-KASH bridges across the nuclear envelope. Curr Opin Cell Biol 25(1):57–62, doi:S0955-0674(12)00177-9 [pii] 10.1016/j.ceb.2012.10.014
3. Suntharalingam M, Wente SR (2003) Peering through the pore: nuclear pore complex structure, assembly, and function. Dev Cell 4(6):775–789
4. Dechat T, Pfleghaar K, Sengupta K, Shimi T, Shumaker DK, Solimando L, Goldman RD (2008) Nuclear lamins: major factors in the structural organization and function of the nucleus and chromatin. Genes Dev 22(7):832–853
5. Stuurman N, Heins S, Aebi U (1998) Nuclear lamins: their structure, assembly, and interactions. J Struct Biol 122(1–2):42–66
6. Schirmer EC, Foisner R (2007) Proteins that associate with lamins: many faces, many functions. Exp Cell Res 313(10):2167–2179
7. Mattout-Drubezki A, Gruenbaum Y (2003) Dynamic interactions of nuclear lamina proteins with chromatin and transcriptional machinery. Cell Mol Life Sci 60(10):2053–2063
8. Brown CR, Kennedy CJ, Delmar VA, Forbes DJ, Silver PA (2008) Global histone acetylation induces functional genomic reorganization at mammalian nuclear pore complexes. Genes Dev 22:627–639
9. Makatsori D, Kourmouli N, Polioudaki H, Shultz LD, McLean K, Theodoropoulos PA, Singh PB, Georgatos SD (2004) The inner nuclear membrane protein lamin B receptor forms distinct microdomains and links epigenetically marked chromatin to the nuclear envelope. J Biol Chem 279(24):25567–25573
10. Ye Q, Worman HJ (1996) Interaction between an integral protein of the nuclear envelope inner membrane and human chromodomain proteins homologous to Drosophila HP1. J Biol Chem 271(25):14653–14656
11. Solovei I, Wang AS, Thanisch K, Schmidt CS, Krebs S, Zwerger M, Cohen TV, Devys D, Foisner R, Peichl L, Herrmann H, Blum H, Engelkamp D, Stewart CL, Leonhardt H, Joffe B (2013) LBR and Lamin A/C sequentially tether peripheral heterochromatin and inversely regulate differentiation. Cell 152(3):584–598, doi:S0092-8674(13)00012-3 [pii]10.1016/j.cell.2013.01.009
12. Zullo JM, Demarco IA, Pique-Regi R, Gaffney DJ, Epstein CB, Spooner CJ, Luperchio TR, Bernstein BE, Pritchard JK, Reddy KL, Singh H (2012) DNA sequence-dependent compartmentalization and silencing of chromatin at the nuclear lamina. Cell 149(7):1474–1487, doi:S0092-8674(12)00591-0 [pii]10.1016/j.cell.2012.04.035

13. Cronshaw J, Krutchinsky A, Zhang W, Chait B, Matunis M (2002) Proteomic analysis of the mammalian nuclear pore complex. J Cell Biol 158(5):915–927
14. Dreger M, Bengtsson L, Schoneberg T, Otto H, Hucho F (2001) Nuclear envelope proteomics: novel integral membrane proteins of the inner nuclear membrane. Proc Natl Acad Sci U S A 98(21):11943–11948
15. Rout MP, Aitchison JD, Suprapto A, Hjertaas K, Zhao Y, Chait BT (2000) The yeast nuclear pore complex: composition, architecture, and transport mechanism. J Cell Biol 148(4):635–651
16. Schirmer EC, Florens L, Guan T, Yates JR 3rd, Gerace L (2003) Nuclear membrane proteins with potential disease links found by subtractive proteomics. Science 301(5638):1380–1382
17. Korfali N, Wilkie GS, Swanson SK, Srsen V, Batrakou DG, Fairley EA, Malik P, Zuleger N, Goncharevich A, de Las Heras J, Kelly DA, Kerr AR, Florens L, Schirmer EC (2010) The leukocyte nuclear envelope proteome varies with cell activation and contains novel transmembrane proteins that affect genome architecture. Mol Cell Proteomics 9(12):2571–2585, doi:M110.002915 [pii]10.1074/mcp.M110.002915
18. Korfali N, Wilkie GS, Swanson SK, Srsen V, de Las Heras J, Batrakou DG, Malik P, Zuleger N, Kerr AR, Florens L, Schirmer EC (2012) The nuclear envelope proteome differs notably between tissues. Nucleus 3(6):552–564, doi:22257 [pii]10.4161/nucl.22257
19. Wilkie GS, Korfali N, Swanson SK, Malik P, Srsen V, Batrakou DG, de Las Heras J, Zuleger N, Kerr AR, Florens L, Schirmer EC (2011) Several novel nuclear envelope transmembrane proteins identified in skeletal muscle have cytoskeletal associations. Mol Cell Proteomics 10(1):M110.003129, doi:M110.003129 [pii]10.1074/mcp.M110.003129
20. Zuleger N, Boyle S, Kelly DA, de Las Heras JI, Lazou V, Korfali N, Batrakou DG, Randles KN, Morris GE, Harrison DJ, Bickmore WA, Schirmer EC (2013) Specific nuclear envelope transmembrane proteins can promote the location of chromosomes to and from the nuclear periphery. Genome Biol 14(2):R14, doi:gb-2013-14-2-r14 [pii]10.1186/gb-2013-14-2-r14
21. Fawcett DW (1981) The cell. Saunders, Philadelphia
22. Paddy MR, Belmont AS, Saumweber H, Agard DA, Sedat JW (1990) Interphase nuclear envelope lamins form a discontinuous network that interacts with only a fraction of the chromatin in the nuclear periphery. Cell 62(1):89–106
23. Belmont AS, Zhai Y, Thilenius A (1993) Lamin B distribution and association with peripheral chromatin revealed by optical sectioning and electron microscopy tomography. J Cell Biol 123(6 Pt 2):1671–1685
24. Bouvier D, Hubert J, Seve AP, Bouteille M (1985) Characterization of lamina-bound chromatin in the nuclear shell isolated from HeLa cells. Exp Cell Res 156(2):500–512
25. Mirsky AE, Allfrey V (1960) Biochemical activities of the cell nucleus. Dis Nerv Syst 21(2) Suppl:23–28
26. Hirschhorn R, Decsy MI, Troll W (1971) The effect of PHA stimulation of human peripheral blood lymphocytes upon cellular content of euchromatin and heterochromatin. Cell Immunol 2(6):696–701
27. Kind J, Pagie L, Ortabozkoyun H, Boyle S, de Vries SS, Janssen H, Amendola M, Nolen LD, Bickmore WA, van Steensel B (2013) Single-cell dynamics of genome-nuclear lamina interactions. Cell 153(1):178–192, doi:S0092-8674(13)00217-1 [pii]10.1016/j.cell.2013.02.028
28. Minc E, Allory Y, Worman HJ, Courvalin JC, Buendia B (1999) Localization and phosphorylation of HP1 proteins during the cell cycle in mammalian cells. Chromosoma 108(4):220–234, doi:91080220.412 [pii]
29. Rabl C (1885) Über Zelltheilung. Morphol Jahrb 10:214–330
30. Cowan CR, Carlton PM, Cande WZ (2001) The polar arrangement of telomeres in interphase and meiosis. Rabl organization and the bouquet. Plant Physiol 125(2):532–538
31. Bass HW, Marshall WF, Sedat JW, Agard DA, Cande WZ (1997) Telomeres cluster de novo before the initiation of synapsis: a three-dimensional spatial analysis of telomere positions before and during meiotic prophase. J Cell Biol 137(1):5–18
32. Scherthan H, Weich S, Schwegler H, Heyting C, Harle M, Cremer T (1996) Centromere and telomere movements during early meiotic prophase of mouse and man are associated with the onset of chromosome pairing. J Cell Biol 134(5):1109–1125

33. Aquiles Sanchez J, Karni RJ, Wangh LJ (1997) Fluorescent in situ hybridization (FISH) analysis of the relationship between chromosome location and nuclear morphology in human neutrophils. Chromosoma 106(3):168–177
34. Galy V, Olivo-Marin JC, Scherthan H, Doye V, Rascalou N, Nehrbass U (2000) Nuclear pore complexes in the organization of silent telomeric chromatin. Nature 403(6765):108–112
35. Scherthan H, Jerratsch M, Li B, Smith S, Hulten M, Lock T, de Lange T (2000) Mammalian meiotic telomeres: protein composition and redistribution in relation to nuclear pores. Mol Biol Cell 11(12):4189–4203
36. Schmitt J, Benavente R, Hodzic D, Hoog C, Stewart CL, Alsheimer M (2007) Transmembrane protein Sun2 is involved in tethering mammalian meiotic telomeres to the nuclear envelope. Proc Natl Acad Sci U S A 104(18):7426–7431
37. Bupp JM, Martin AE, Stensrud ES, Jaspersen SL (2007) Telomere anchoring at the nuclear periphery requires the budding yeast Sad1-UNC-84 domain protein Mps3. J Cell Biol 179(5):845–854, doi:jcb.200706040 [pii]10.1083/jcb.200706040
38. Chikashige Y, Tsutsumi C, Yamane M, Okamasa K, Haraguchi T, Hiraoka Y (2006) Meiotic proteins bqt1 and bqt2 tether telomeres to form the bouquet arrangement of chromosomes. Cell 125(1):59–69, doi:S0092-8674(06)00318-7 [pii]10.1016/j.cell.2006.01.048
39. Conrad MN, Lee CY, Wilkerson JL, Dresser ME (2007) MPS3 mediates meiotic bouquet formation in Saccharomyces cerevisiae. Proc Natl Acad Sci U S A 104(21):8863–8868
40. Hou H, Zhou Z, Wang Y, Wang J, Kallgren SP, Kurchuk T, Miller EA, Chang F, Jia S (2012) Csi1 links centromeres to the nuclear envelope for centromere clustering. J Cell Biol 199(5):735–744, doi:jcb.201208001 [pii]10.1083/jcb.201208001
41. Baricheva EA, Berrios M, Bogachev SS, Borisevich IV, Lapik ER, Sharakhov IV, Stuurman N, Fisher PA (1996) DNA from Drosophila melanogaster beta-heterochromatin binds specifically to nuclear lamins in vitro and the nuclear envelope in situ. Gene 171(2):171–176
42. Shoeman RL, Traub P (1990) The in vitro DNA-binding properties of purified nuclear lamin proteins and vimentin. J Biol Chem 265(16):9055–9061
43. Boveri T (1909) Die blastomerenkerne von Ascaris megalocephala und die Theorie der Chromosomen-indiviüalitat. Arch Zellforsch 3:181–268
44. Croft JA, Bridger JM, Boyle S, Perry P, Teague P, Bickmore WA (1999) Differences in the localization and morphology of chromosomes in the human nucleus. J Cell Biol 145(6):1119–1131
45. Bickmore WA, Teague P (2002) Influences of chromosome size, gene density and nuclear position on the frequency of constitutional translocations in the human population. Chromosome Res 10(8):707–715
46. Cremer M, Kupper K, Wagler B, Wizelman L, von Hase J, Weiland Y, Kreja L, Diebold J, Speicher MR, Cremer T (2003) Inheritance of gene density-related higher order chromatin arrangements in normal and tumor cell nuclei. J Cell Biol 162(5):809–820, doi:10.1083/jcb.200304096jcb.200304096 [pii]
47. Meaburn KJ, Newbold RF, Bridger JM (2008) Positioning of human chromosomes in murine cell hybrids according to synteny. Chromosoma 117(6):579–591
48. Kim SH, McQueen PG, Lichtman MK, Shevach EM, Parada LA, Misteli T (2004) Spatial genome organization during T-cell differentiation. Cytogenet Genome Res 105(2–4):292–301
49. Parada LA, McQueen PG, Misteli T (2004) Tissue-specific spatial organization of genomes. Genome Biol 5(7):R44
50. Parada LA, McQueen PG, Munson PJ, Misteli T (2002) Conservation of relative chromosome positioning in normal and cancer cells. Curr Biol 12(19):1692–1697
51. Solovei I, Kreysing M, Lanctot C, Kosem S, Peichl L, Cremer T, Guck J, Joffe B (2009) Nuclear architecture of rod photoreceptor cells adapts to vision in mammalian evolution. Cell 137(2):356–368, doi:S0092-8674(09)00137-8 [pii]10.1016/j.cell.2009.01.052
52. Marshall WF, Dernburg AF, Harmon B, Agard DA, Sedat JW (1996) Specific interactions of chromatin with the nuclear envelope: positional determination within the nucleus in Drosophila melanogaster. Mol Biol Cell 7(5):825–842

53. Kosak ST, Skok JA, Medina KL, Riblet R, Le Beau MM, Fisher AG, Singh H (2002) Subnuclear compartmentalization of immunoglobulin loci during lymphocyte development. Science 296(5565):158–162
54. Williams RR, Azuara V, Perry P, Sauer S, Dvorkina M, Jorgensen H, Roix J, McQueen P, Misteli T, Merkenschlager M, Fisher AG (2006) Neural induction promotes large-scale chromatin reorganisation of the Mash1 locus. J Cell Sci 119(Pt 1):132–140
55. Szczerbal I, Foster HA, Bridger JM (2009) The spatial repositioning of adipogenesis genes is correlated with their expression status in a porcine mesenchymal stem cell adipogenesis model system. Chromosoma 118(5):647–663. doi:10.1007/s00412-009-0225-5
56. Zink D, Amaral MD, Englmann A, Lang S, Clarke LA, Rudolph C, Alt F, Luther K, Braz C, Sadoni N, Rosenecker J, Schindelhauer D (2004) Transcription-dependent spatial arrangements of CFTR and adjacent genes in human cell nuclei. J Cell Biol 166(6):815–825
57. Nielsen JA, Hudson LD, Armstrong RC (2002) Nuclear organization in differentiating oligodendrocytes. J Cell Sci 115(Pt 21):4071–4079
58. Hewitt SL, High FA, Reiner SL, Fisher AG, Merkenschlager M (2004) Nuclear repositioning marks the selective exclusion of lineage-inappropriate transcription factor loci during T helper cell differentiation. Eur J Immunol 34(12):3604–3613
59. Park PC, De Boni U (1998) A specific conformation of the territory of chromosome 17 locates ERBB-2 sequences to a DNase-hypersensitive domain at the nuclear periphery. Chromosoma 107(2):87–95
60. Johnson C, Primorac D, McKinstry M, McNeil J, Rowe D, Lawrence JB (2000) Tracking COL1A1 RNA in osteogenesis imperfecta. splice-defective transcripts initiate transport from the gene but are retained within the SC35 domain. J Cell Biol 150(3):417–432
61. Morey C, Da Silva NR, Kmita M, Duboule D, Bickmore WA (2008) Ectopic nuclear reorganisation driven by a Hoxb1 transgene transposed into Hoxd. J Cell Sci 121(Pt 5):571–577, doi:jcs.023234 [pii]10.1242/jcs.023234
62. Albiez H, Cremer M, Tiberi C, Vecchio L, Schermelleh L, Dittrich S, Kupper K, Joffe B, Thormeyer T, von Hase J, Yang S, Rohr K, Leonhardt H, Solovei I, Cremer C, Fakan S, Cremer T (2006) Chromatin domains and the interchromatin compartment form structurally defined and functionally interacting nuclear networks. Chromosome Res 14(7):707–733
63. Morey C, Da Silva NR, Perry P, Bickmore WA (2007) Nuclear reorganisation and chromatin decondensation are conserved, but distinct, mechanisms linked to Hox gene activation. Development 134(5):909–919. doi:10.1242/dev.02779
64. Misteli T (2007) Beyond the sequence: cellular organization of genome function. Cell 128(4):787–800. doi:10.1016/j.cell.2007.01.028
65. Finlan LE, Sproul D, Thomson I, Boyle S, Kerr E, Perry P, Ylstra B, Chubb JR, Bickmore WA (2008) Recruitment to the nuclear periphery can alter expression of genes in human cells. PLoS Genet 4(3):e1000039
66. Kumaran RI, Spector DL (2008) A genetic locus targeted to the nuclear periphery in living cells maintains its transcriptional competence. J Cell Biol 180(1):51–65
67. Reddy KL, Zullo JM, Bertolino E, Singh H (2008) Transcriptional repression mediated by repositioning of genes to the nuclear lamina. Nature 452(7184):243–247
68. Clowney EJ, LeGros MA, Mosley CP, Clowney FG, Markenskoff-Papadimitriou EC, Myllys M, Barnea G, Larabell CA, Lomvardas S (2012) Nuclear aggregation of olfactory receptor genes governs their monogenic expression. Cell 151(4):724–737, doi:S0092-8674(12)01286-X [pii]10.1016/j.cell.2012.09.043
69. Schirmer EC, Gerace L (2004) The stability of the nuclear lamina polymer changes with the composition of lamin subtypes according to their individual binding strengths. J Biol Chem 279(41):42811–42817
70. Hoger TH, Krohne G, Kleinschmidt JA (1991) Interaction of Xenopus lamins A and LII with chromatin in vitro mediated by a sequence element in the carboxyterminal domain. Exp Cell Res 197(2):280–289
71. Taniura H, Glass C, Gerace L (1995) A chromatin binding site in the tail domain of nuclear lamins that interacts with core histones. J Cell Biol 131(1):33–44

72. Goldberg M, Harel A, Brandeis M, Rechsteiner T, Richmond TJ, Weiss AM, Gruenbaum Y (1999) The tail domain of lamin Dm0 binds histones H2A and H2B. Proc Natl Acad Sci U S A 96(6):2852–2857
73. Luderus ME, den Blaauwen JL, de Smit OJ, Compton DA, van Driel R (1994) Binding of matrix attachment regions to lamin polymers involves single-stranded regions and the minor groove. Mol Cell Biol 14(9):6297–6305
74. Luderus ME, de Graaf A, Mattia E, den Blaauwen JL, Grande MA, de Jong L, van Driel R (1992) Binding of matrix attachment regions to lamin B1. Cell 70(6):949–959
75. Rzepecki R, Bogachev SS, Kokoza E, Stuurman N, Fisher PA (1998) In vivo association of lamins with nucleic acids in Drosophila melanogaster. J Cell Sci 111(Pt 1):121–129
76. Haraguchi T, Koujin T, Hayakawa T, Kaneda T, Tsutsumi C, Imamoto N, Akazawa C, Sukegawa J, Yoneda Y, Hiraoka Y (2000) Live fluorescence imaging reveals early recruitment of emerin, LBR, RanBP2, and Nup153 to reforming functional nuclear envelopes. J Cell Sci 113(Pt 5):779–794
77. Polioudaki H, Kourmouli N, Drosou V, Bakou A, Theodoropoulos PA, Singh PB, Giannakouros T, Georgatos SD (2001) Histones H3/H4 form a tight complex with the inner nuclear membrane protein LBR and heterochromatin protein 1. EMBO Rep 2(10):920–925
78. Lachner M, O'Carroll D, Rea S, Mechtler K, Jenuwein T (2001) Methylation of histone H3 lysine 9 creates a binding site for HP1 proteins. Nature 410(6824):116–120. doi:10.1038/35065132
79. Bannister AJ, Zegerman P, Partridge JF, Miska EA, Thomas JO, Allshire RC, Kouzarides T (2001) Selective recognition of methylated lysine 9 on histone H3 by the HP1 chromo domain. Nature 410(6824):120–124. doi:10.1038/35065138
80. Foisner R, Gerace L (1993) Integral membrane proteins of the nuclear envelope interact with lamins and chromosomes, and binding is modulated by mitotic phosphorylation. Cell 73(7):1267–1279
81. Furukawa K (1999) LAP2 binding protein 1 (L2BP1/BAF) is a candidate mediator of LAP2-chromatin interaction. J Cell Sci 112(Pt 15):2485–2492
82. Cai M, Huang Y, Ghirlando R, Wilson KL, Craigie R, Clore GM (2001) Solution structure of the constant region of nuclear envelope protein LAP2 reveals two LEM-domain structures: one binds BAF and the other binds DNA. EMBO J 20(16):4399–4407
83. Zheng R, Ghirlando R, Lee MS, Mizuuchi K, Krause M, Craigie R (2000) Barrier-to-autointegration factor (BAF) bridges DNA in a discrete, higher-order nucleoprotein complex. Proc Natl Acad Sci U S A 97(16):8997–9002
84. Montes de Oca R, Lee KK, Wilson KL (2005) Binding of barrier to autointegration factor (BAF) to histone H3 and selected linker histones including H1.1. J Biol Chem 280(51):42252–42262
85. Caputo S, Couprie J, Duband-Goulet I, Konde E, Lin F, Braud S, Gondry M, Gilquin B, Worman HJ, Zinn-Justin S (2006) The carboxyl-terminal nucleoplasmic region of MAN1 exhibits a DNA binding winged helix domain. J Biol Chem 281(26):18208–18215
86. Gerasimova TI, Byrd K, Corces VG (2000) A chromatin insulator determines the nuclear localization of DNA. Mol Cell 6(5):1025–1035
87. Sun FL, Elgin SC (1999) Putting boundaries on silence. Cell 99(5):459–462
88. Ramos E, Ghosh D, Baxter E, Corces VG (2006) Genomic organization of gypsy chromatin insulators in Drosophila melanogaster. Genetics 172(4):2337–2349. doi:10.1534/genetics.105.054742
89. Yang J, Corces VG (2012) Insulators, long-range interactions, and genome function. Curr Opin Genet Dev 22(2):86–92. doi:10.1016/j.gde.2011.12.007
90. Bell AC, West AG, Felsenfeld G (1999) The protein CTCF is required for the enhancer blocking activity of vertebrate insulators. Cell 98(3):387–396
91. Loukinov DI, Pugacheva E, Vatolin S, Pack SD, Moon H, Chernukhin I, Mannan P, Larsson E, Kanduri C, Vostrov AA, Cui H, Niemitz EL, Rasko JE, Docquier FM, Kistler M, Breen JJ, Zhuang Z, Quitschke WW, Renkawitz R, Klenova EM, Feinberg AP, Ohlsson R, Morse HC 3rd, Lobanenkov VV (2002) BORIS, a novel male germ-line-specific protein associated with epigenetic reprogramming events, shares the same 11-zinc-finger domain with CTCF, the

insulator protein involved in reading imprinting marks in the soma. Proc Natl Acad Sci U S A 99(10):6806–6811. doi:10.1073/pnas.092123699

92. Handoko L, Xu H, Li G, Ngan CY, Chew E, Schnapp M, Lee CW, Ye C, Ping JL, Mulawadi F, Wong E, Sheng J, Zhang Y, Poh T, Chan CS, Kunarso G, Shahab A, Bourque G, Cacheux-Rataboul V, Sung WK, Ruan Y, Wei CL (2011) CTCF-mediated functional chromatin interactome in pluripotent cells. Nat Genet 43(7):630–638. doi:10.1038/ng.857
93. Pickersgill H, Kalverda B, de Wit E, Talhout W, Fornerod M, van Steensel B (2006) Characterization of the Drosophila melanogaster genome at the nuclear lamina. Nat Genet 38(9):1005–1014
94. Guelen L, Pagie L, Brasset E, Meuleman W, Faza MB, Talhout W, Eussen BH, de Klein A, Wessels L, de Laat W, van Steensel B (2008) Domain organization of human chromosomes revealed by mapping of nuclear lamina interactions. Nature 453(7197):948–951, doi:nature06947 [pii]10.1038/nature06947
95. Kind J, van Steensel B (2010) Genome-nuclear lamina interactions and gene regulation. Curr Opin Cell Biol 22(3):320–325, doi:S0955-0674(10)00054-2 [pii]10.1016/j.ceb.2010.04.002
96. Peric-Hupkes D, Meuleman W, Pagie L, Bruggeman SW, Solovei I, Brugman W, Graf S, Flicek P, Kerkhoven RM, van Lohuizen M, Reinders M, Wessels L, van Steensel B (2010) Molecular maps of the reorganization of genome-nuclear lamina interactions during differentiation. Mol Cell 38(4):603–613, doi:S1097-2765(10)00321-7 [pii]10.1016/j.molcel.2010.03.016
97. Lienert F, Mohn F, Tiwari VK, Baubec T, Roloff TC, Gaidatzis D, Stadler MB, Schubeler D (2011) Genomic prevalence of heterochromatic H3K9me2 and transcription do not discriminate pluripotent from terminally differentiated cells. PLoS Genet 7(6):e1002090. doi:10.1371/journal.pgen.1002090
98. Wen B, Wu H, Shinkai Y, Irizarry RA, Feinberg AP (2009) Large histone H3 lysine 9 dimethylated chromatin blocks distinguish differentiated from embryonic stem cells. Nat Genet 41(2):246–250. doi:10.1038/ng.297
99. Lee KK, Haraguchi T, Lee RS, Koujin T, Hiraoka Y, Wilson KL (2001) Distinct functional domains in emerin bind lamin A and DNA-bridging protein BAF. J Cell Sci 114(Pt 24):4567–4573
100. Mansharamani M, Wilson KL (2005) Direct binding of nuclear membrane protein MAN1 to emerin in vitro and two modes of binding to barrier-to-autointegration factor. J Biol Chem 280(14):13863–13870
101. Montes de Oca R, Shoemaker CJ, Gucek M, Cole RN, Wilson KL (2009) Barrier-to-autointegration factor proteome reveals chromatin-regulatory partners. PLoS One 4(9):e7050. doi:10.1371/journal.pone.0007050
102. Holaska JM, Wilson KL (2007) An emerin "proteome": purification of distinct emerin-containing complexes from HeLa cells suggests molecular basis for diverse roles including gene regulation, mRNA splicing, signaling, mechanosensing, and nuclear architecture. Biochemistry 46(30):8897–8908. doi:10.1021/bi602636m
103. Somech R, Shaklai S, Geller O, Amariglio N, Simon AJ, Rechavi G, Gal-Yam EN (2005) The nuclear-envelope protein and transcriptional repressor LAP2beta interacts with HDAC3 at the nuclear periphery, and induces histone H4 deacetylation. J Cell Sci 118(Pt 17):4017–4025. doi:10.1242/jcs.02521
104. Dechat T, Gajewski A, Korbei B, Gerlich D, Daigle N, Haraguchi T, Furukawa K, Ellenberg J, Foisner R (2004) LAP2alpha and BAF transiently localize to telomeres and specific regions on chromatin during nuclear assembly. J Cell Sci 117(Pt 25):6117–6128
105. Suzuki Y, Yang H, Craigie R (2004) LAP2alpha and BAF collaborate to organize the Moloney murine leukemia virus preintegration complex. EMBO J 23(23):4670–4678
106. Shaklai S, Somech R, Gal-Yam EN, Deshet-Unger N, Moshitch-Moshkovitz S, Hirschberg K, Amariglio N, Simon AJ, Rechavi G (2008) LAP2zeta binds BAF and suppresses LAP2beta-mediated transcriptional repression. Eur J Cell Biol 87(5):267–278
107. Malhas AN, Lee CF, Vaux DJ (2009) Lamin B1 controls oxidative stress responses via Oct-1. J Cell Biol 184(1):45–55. doi:10.1083/jcb.200804155

108. Chi YH, Haller K, Peloponese JM Jr, Jeang KT (2007) Histone acetyltransferase hALP and nuclear membrane protein hsSUN1 function in de-condensation of mitotic chromosomes. J Biol Chem 282(37):27447–27458
109. Jarnik M, Aebi U (1991) Toward a more complete 3-D structure of the nuclear pore complex. J Struct Biol 107(3):291–308
110. Guan T, Kehlenbach RH, Schirmer EC, Kehlenbach A, Fan F, Clurman BE, Arnheim N, Gerace L (2000) Nup50, a nucleoplasmically oriented nucleoporin with a role in nuclear protein export. Mol Cell Biol 20(15):5619–5630
111. Schermelleh L, Carlton PM, Haase S, Shao L, Winoto L, Kner P, Burke B, Cardoso MC, Agard DA, Gustafsson MG, Leonhardt H, Sedat JW (2008) Subdiffraction multicolor imaging of the nuclear periphery with 3D structured illumination microscopy. Science 320(5881):1332–1336. doi:10.1126/science.1156947
112. Laroche T, Martin SG, Gotta M, Gorham HC, Pryde FE, Louis EJ, Gasser SM (1998) Mutation of yeast Ku genes disrupts the subnuclear organization of telomeres. Curr Biol 8(11):653–656
113. Maillet L, Gaden F, Brevet V, Fourel G, Martin SG, Dubrana K, Gasser SM, Gilson E (2001) Ku-deficient yeast strains exhibit alternative states of silencing competence. EMBO Rep 2(3):203–210
114. Feuerbach F, Galy V, Trelles-Sticken E, Fromont-Racine M, Jacquier A, Gilson E, Olivo-Marin JC, Scherthan H, Nehrbass U (2002) Nuclear architecture and spatial positioning help establish transcriptional states of telomeres in yeast. Nat Cell Biol 4(3):214–221
115. Gartenberg MR, Neumann FR, Laroche T, Blaszczyk M, Gasser SM (2004) Sir-mediated repression can occur independently of chromosomal and subnuclear contexts. Cell 119(7):955–967
116. Akhtar A, Gasser SM (2007) The nuclear envelope and transcriptional control. Nat Rev Genet 8(7):507–517
117. Vaquerizas JM, Suyama R, Kind J, Miura K, Luscombe NM, Akhtar A (2010) Nuclear pore proteins nup153 and megator define transcriptionally active regions in the Drosophila genome. PLoS Genet 6(2):e1000846. doi:10.1371/journal.pgen.1000846
118. Capelson M, Liang Y, Schulte R, Mair W, Wagner U, Hetzer MW (2010) Chromatin-bound nuclear pore components regulate gene expression in higher eukaryotes. Cell 140(3): 372–383, doi:S0092-8674(09)01681-X [pii]10.1016/j.cell.2009.12.054
119. Kalverda B, Pickersgill H, Shloma VV, Fornerod M (2010) Nucleoporins directly stimulate expression of developmental and cell-cycle genes inside the nucleoplasm. Cell 140(3):360–371, doi:S0092-8674(10)00012-7 [pii]10.1016/j.cell.2010.01.011
120. Luthra R, Kerr SC, Harreman MT, Apponi LH, Fasken MB, Ramineni S, Chaurasia S, Valentini SR, Corbett AH (2007) Actively transcribed GAL genes can be physically linked to the nuclear pore by the SAGA chromatin modifying complex. J Biol Chem 282(5):3042–3049. doi:10.1074/jbc.M608741200
121. Baker SP, Grant PA (2007) The SAGA continues: expanding the cellular role of a transcriptional co-activator complex. Oncogene 26(37):5329–5340. doi:10.1038/sj.onc.1210603
122. Nagy Z, Tora L (2007) Distinct GCN5/PCAF-containing complexes function as co-activators and are involved in transcription factor and global histone acetylation. Oncogene 26(37): 5341–5357. doi:10.1038/sj.onc.1210604
123. Rodriguez-Navarro S, Fischer T, Luo MJ, Antunez O, Brettschneider S, Lechner J, Perez-Ortin JE, Reed R, Hurt E (2004) Sus1, a functional component of the SAGA histone acetylase complex and the nuclear pore-associated mRNA export machinery. Cell 116(1):75–86
124. Cabal GG, Genovesio A, Rodriguez-Navarro S, Zimmer C, Gadal O, Lesne A, Buc H, Feuerbach-Fournier F, Olivo-Marin JC, Hurt EC, Nehrbass U (2006) SAGA interacting factors confine sub-diffusion of transcribed genes to the nuclear envelope. Nature 441(7094):770–773. doi:10.1038/nature04752
125. Pascual-Garcia P, Govind CK, Queralt E, Cuenca-Bono B, Llopis A, Chavez S, Hinnebusch AG, Rodriguez-Navarro S (2008) Sus1 is recruited to coding regions and functions during transcription elongation in association with SAGA and TREX2. Genes Dev 22(20): 2811–2822. doi:10.1101/gad.483308

126. Klockner C, Schneider M, Lutz S, Jani D, Kressler D, Stewart M, Hurt E, Kohler A (2009) Mutational uncoupling of the role of Sus1 in nuclear pore complex targeting of an mRNA export complex and histone H2B deubiquitination. J Biol Chem 284(18):12049–12056. doi:10.1074/jbc.M900502200
127. Rodriguez-Navarro S (2009) Insights into SAGA function during gene expression. EMBO Rep 10(8):843–850. doi:10.1038/embor.2009.168
128. Kurshakova MM, Krasnov AN, Kopytova DV, Shidlovskii YV, Nikolenko JV, Nabirochkina EN, Spehner D, Schultz P, Tora L, Georgieva SG (2007) SAGA and a novel Drosophila export complex anchor efficient transcription and mRNA export to NPC. EMBO J 26(24):4956–4965. doi:10.1038/sj.emboj.7601901
129. Kurshakova M, Maksimenko O, Golovnin A, Pulina M, Georgieva S, Georgiev P, Krasnov A (2007) Evolutionarily conserved E(y)2/Sus1 protein is essential for the barrier activity of Su(Hw)-dependent insulators in Drosophila. Mol Cell 27(2):332–338. doi:10.1016/j.molcel.2007.05.035
130. Wickramasinghe VO, McMurtrie PI, Mills AD, Takei Y, Penrhyn-Lowe S, Amagase Y, Main S, Marr J, Stewart M, Laskey RA (2010) mRNA export from mammalian cell nuclei is dependent on GANP. Curr Biol 20(1):25–31. doi:10.1016/j.cub.2009.10.078
131. Jani D, Lutz S, Marshall NJ, Fischer T, Kohler A, Ellisdon AM, Hurt E, Stewart M (2009) Sus1, Cdc31, and the Sac3 CID region form a conserved interaction platform that promotes nuclear pore association and mRNA export. Mol Cell 33(6):727–737. doi:10.1016/j.molcel.2009.01.033
132. Casolari JM, Brown CR, Komili S, West J, Hieronymus H, Silver PA (2004) Genome-wide localization of the nuclear transport machinery couples transcriptional status and nuclear organization. Cell 117(4):427–439
133. Mendjan S, Taipale M, Kind J, Holz H, Gebhardt P, Schelder M, Vermeulen M, Buscaino A, Duncan K, Mueller J, Wilm M, Stunnenberg HG, Saumweber H, Akhtar A (2006) Nuclear pore components are involved in the transcriptional regulation of dosage compensation in Drosophila. Mol Cell 21(6):811–823. doi:10.1016/j.molcel.2006.02.007
134. Ishii K, Arib G, Lin C, Van Houwe G, Laemmli UK (2002) Chromatin boundaries in budding yeast: the nuclear pore connection. Cell 109(5):551–562
135. Gerasimova TI, Corces VG (2001) Chromatin insulators and boundaries: effects on transcription and nuclear organization. Annu Rev Genet 35:193–208. doi:10.1146/annurev.genet.35.102401.090349
136. Oki M, Valenzuela L, Chiba T, Ito T, Kamakaka RT (2004) Barrier proteins remodel and modify chromatin to restrict silenced domains. Mol Cell Biol 24(5):1956–1967
137. Dilworth DJ, Tackett AJ, Rogers RS, Yi EC, Christmas RH, Smith JJ, Siegel AF, Chait BT, Wozniak RW, Aitchison JD (2005) The mobile nucleoporin Nup2p and chromatin-bound Prp20p function in endogenous NPC-mediated transcriptional control. J Cell Biol 171(6):955–965
138. Luk E, Ranjan A, Fitzgerald PC, Mizuguchi G, Huang Y, Wei D, Wu C (2010) Stepwise histone replacement by SWR1 requires dual activation with histone H2A.Z and canonical nucleosome. Cell 143(5):725–736. doi:10.1016/j.cell.2010.10.019
139. Li B, Pattenden SG, Lee D, Gutierrez J, Chen J, Seidel C, Gerton J, Workman JL (2005) Preferential occupancy of histone variant H2AZ at inactive promoters influences local histone modifications and chromatin remodeling. Proc Natl Acad Sci U S A 102(51):18385–18390. doi:10.1073/pnas.0507975102
140. Jacobson S, Pillus L (2009) The SAGA subunit Ada2 functions in transcriptional silencing. Mol Cell Biol 29(22):6033–6045. doi:10.1128/MCB.00542-09
141. Holaska JM, Lee K, Kowalski AK, Wilson KL (2003) Transcriptional repressor germ cell-less (GCL) and barrier to autointegration factor (BAF) compete for binding to emerin in vitro. J Biol Chem 278(9):6969–6975
142. Nili E, Cojocaru GS, Kalma Y, Ginsberg D, Copeland NG, Gilbert DJ, Jenkins NA, Berger R, Shaklai S, Amariglio N, Brok-Simoni F, Simon AJ, Rechavi G (2001) Nuclear membrane protein LAP2beta mediates transcriptional repression alone and together with its binding partner GCL (germ-cell-less). J Cell Sci 114(18):3297–3307

143. Haraguchi T, Holaska JM, Yamane M, Koujin T, Hashiguchi N, Mori C, Wilson KL, Hiraoka Y (2004) Emerin binding to Btf, a death-promoting transcriptional repressor, is disrupted by a missense mutation that causes Emery-Dreifuss muscular dystrophy. Eur J Biochem 271(5):1035–1045
144. Ozaki T, Saijo M, Murakami K, Enomoto H, Taya Y, Sakiyama S (1994) Complex formation between lamin A and the retinoblastoma gene product: identification of the domain on lamin A required for its interaction. Oncogene 9(9):2649–2653
145. Markiewicz E, Dechat T, Foisner R, Quinlan R, Hutchison C (2002) Lamin A/C binding protein LAP2alpha is required for nuclear anchorage of retinoblastoma protein. Mol Biol Cell 13(12):4401–4413
146. Naetar N, Korbei B, Kozlov S, Kerenyi MA, Dorner D, Kral R, Gotic I, Fuchs P, Cohen TV, Bittner R, Stewart CL, Foisner R (2008) Loss of nucleoplasmic LAP2alpha-lamin A complexes causes erythroid and epidermal progenitor hyperproliferation. Nat Cell Biol 10(11):1341–1348, doi:ncb1793 [pii]10.1038/ncb1793
147. Gerace L, Burke B (1988) Functional organization of the nuclear envelope. Annu Rev Cell Biol 4:335–374
148. Ivorra C, Kubicek M, Gonzalez JM, Sanz-Gonzalez SM, Alvarez-Barrientos A, O'Connor JE, Burke B, Andres V (2006) A mechanism of AP-1 suppression through interaction of c-Fos with lamin A/C. Genes Dev 20(3):307–320, doi:20/3/307 [pii]10.1101/gad.349506
149. Holaska JM, Rais-Bahrami S, Wilson KL (2006) Lmo7 is an emerin-binding protein that regulates the transcription of emerin and many other muscle-relevant genes. Hum Mol Genet 15(23):3459–3472
150. Osada S, Ohmori SY, Taira M (2003) XMAN1, an inner nuclear membrane protein, antagonizes BMP signaling by interacting with Smad1 in Xenopus embryos. Development 130(9):1783–1794
151. Pan D, Estevez-Salmeron LD, Stroschein SL, Zhu X, He J, Zhou S, Luo K (2005) The integral inner nuclear membrane protein MAN1 physically interacts with the R-Smad proteins to repress signaling by the transforming growth factor-{beta} superfamily of cytokines. J Biol Chem 280(16):15992–16001
152. Misteli T (2001) Protein dynamics: implications for nuclear architecture and gene expression. Science 291(5505):843–847
153. Marshall WF, Straight A, Marko JF, Swedlow J, Dernburg A, Belmont A, Murray AW, Agard DA, Sedat JW (1997) Interphase chromosomes undergo constrained diffusional motion in living cells. Curr Biol 7(12):930–939, doi:S0960-9822(06)00412-X [pii]
154. Strickfaden H, Zunhammer A, van Koningsbruggen S, Kohler D, Cremer T (2010) 4D chromatin dynamics in cycling cells: Theodor Boveri's hypotheses revisited. Nucleus 1(3):284–297. doi:10.4161/nucl.1.3.11969
155. Chubb JR, Boyle S, Perry P, Bickmore WA (2002) Chromatin motion is constrained by association with nuclear compartments in human cells. Curr Biol 12(6):439–445
156. Zink D, Bornfleth H, Visser A, Cremer C, Cremer T (1999) Organization of early and late replicating DNA in human chromosome territories. Exp Cell Res 247(1):176–188
157. Janmey PA, Euteneuer U, Traub P, Schliwa M (1991) Viscoelastic properties of vimentin compared with other filamentous biopolymer networks. J Cell Biol 113(1):155–160
158. Crisp M, Liu Q, Roux K, Rattner JB, Shanahan C, Burke B, Stahl PD, Hodzic D (2006) Coupling of the nucleus and cytoplasm: role of the LINC complex. J Cell Biol 172(1):41–53
159. Starr DA, Fischer JA (2005) KASH 'n Karry: the KASH domain family of cargo-specific cytoskeletal adaptor proteins. Bioessays 27(11):1136–1146
160. Lammerding J, Schulze P, Takahashi T, Kozlov S, Sullivan T, Kamm R, Stewart C, Lee R (2004) Lamin A/C deficiency causes defective nuclear mechanics and mechanotransduction. J Clin Invest 113(3):370–378
161. Liu J, Ben-Shahar T, Riemer D, Treinin M, Spann P, Weber K, Fire A, Gruenbaum Y (2000) Essential roles for Caenorhabditis elegans lamin gene in nuclear organization, cell cycle progression, and spatial organization of nuclear pore complexes. Mol Biol Cell 11(11):3937–3947

162. Schirmer EC, Guan T, Gerace L (2001) Involvement of the lamin rod domain in heterotypic lamin interactions important for nuclear organization. J Cell Biol 153(3):479–489
163. Lenz-Bohme B, Wismar J, Fuchs S, Reifegerste R, Buchner E, Betz H, Schmitt B (1997) Insertional mutation of the Drosophila nuclear lamin Dm0 gene results in defective nuclear envelopes, clustering of nuclear pore complexes, and accumulation of annulate lamellae. J Cell Biol 137(5):1001–1016
164. Capell BC, Collins FS (2006) Human laminopathies: nuclei gone genetically awry. Nat Rev Genet 7(12):940–952
165. Foisner R, Aebi U, Bonne G, Gruenbaum Y, Novelli G (2007) 141st ENMC International Workshop inaugural meeting of the EURO-Laminopathies project "Nuclear Envelope-linked Rare Human Diseases: From Molecular Pathophysiology towards Clinical Applications", 10-12 March 2006, Naarden, The Netherlands. Neuromuscul Disord 17(8):655–660
166. Worman HJ, Bonne G (2007) "Laminopathies": a wide spectrum of human diseases. Exp Cell Res 313(10):2121–2133
167. Bonne G, Di Barletta MR, Varnous S, Becane HM, Hammouda EH, Merlini L, Muntoni F, Greenberg CR, Gary F, Urtizberea JA, Duboc D, Fardeau M, Toniolo D, Schwartz K (1999) Mutations in the gene encoding lamin A/C cause autosomal dominant Emery-Dreifuss muscular dystrophy. Nat Genet 21(3):285–288
168. Raffaele Di Barletta M, Ricci E, Galluzzi G, Tonali P, Mora M, Morandi L, Romorini A, Voit T, Orstavik KH, Merlini L, Trevisan C, Biancalana V, Housmanowa-Petrusewicz I, Bione S, Ricotti R, Schwartz K, Bonne G, Toniolo D (2000) Different mutations in the LMNA gene cause autosomal dominant and autosomal recessive Emery-Dreifuss muscular dystrophy. Am J Hum Genet 66(4):1407–1412
169. Muchir A, Bonne G, van der Kooi AJ, van Meegen M, Baas F, Bolhuis PA, de Visser M, Schwartz K (2000) Identification of mutations in the gene encoding lamins A/C in autosomal dominant limb girdle muscular dystrophy with atrioventricular conduction disturbances (LGMD1B). Hum Mol Genet 9(9):1453–1459
170. Cao H, Hegele RA (2000) Nuclear lamin A/C R482Q mutation in Canadian kindreds with Dunnigan-type familial partial lipodystrophy. Hum Mol Genet 9(1):109–112
171. Hegele RA, Cao H, Liu DM, Costain GA, Charlton-Menys V, Rodger NW, Durrington PN (2006) Sequencing of the reannotated LMNB2 gene reveals novel mutations in patients with acquired partial lipodystrophy. Am J Hum Genet 79(2):383–389
172. Shackleton S, Lloyd DJ, Jackson SN, Evans R, Niermeijer MF, Singh BM, Schmidt H, Brabant G, Kumar S, Durrington PN, Gregory S, O'Rahilly S, Trembath RC (2000) LMNA, encoding lamin A/C, is mutated in partial lipodystrophy. Nat Genet 24(2):153–156
173. De Sandre-Giovannoli A, Chaouch M, Kozlov S, Vallat J, Tazir M, Kassouri N, Szepetowski P, Hammadouche T, Vandenberghe A, Stewart C, Grid D, Levy N (2002) Homozygous defects in LMNA, encoding lamin A/C nuclear-envelope proteins, cause autosomal recessive axonal neuropathy in human (Charcot-Marie-Tooth disorder type 2) and mouse. Am J Hum Genet 70(3):726–736
174. Padiath QS, Saigoh K, Schiffmann R, Asahara H, Yamada T, Koeppen A, Hogan K, Ptacek LJ, Fu YH (2006) Lamin B1 duplications cause autosomal dominant leukodystrophy. Nat Genet 38(10):1114–1123
175. Fatkin D, MacRae C, Sasaki T, Wolff MR, Porcu M, Frenneaux M, Atherton J, Vidaillet HJ Jr, Spudich S, De Girolami U, Seidman JG, Seidman C, Muntoni F, Muehle G, Johnson W, McDonough B (1999) Missense mutations in the rod domain of the lamin A/C gene as causes of dilated cardiomyopathy and conduction-system disease [see comments]. N Engl J Med 341(23):1715–1724
176. Navarro CL, De Sandre-Giovannoli A, Bernard R, Boccaccio I, Boyer A, Genevieve D, Hadj-Rabia S, Gaudy-Marqueste C, Smitt HS, Vabres P, Faivre L, Verloes A, Van Essen T, Flori E, Hennekam R, Beemer FA, Laurent N, Le Merrer M, Cau P, Levy N (2004) Lamin A and ZMPSTE24 (FACE-1) defects cause nuclear disorganization and identify restrictive dermopathy as a lethal neonatal laminopathy. Hum Mol Genet 13(20):2493–2503

177. Eriksson M, Brown WT, Gordon LB, Glynn MW, Singer J, Scott L, Erdos MR, Robbins CM, Moses TY, Berglund P, Dutra A, Pak E, Durkin S, Csoka AB, Boehnke M, Glover TW, Collins FS (2003) Recurrent de novo point mutations in lamin A cause Hutchinson-Gilford progeria syndrome. Nature 423(6937):293–298
178. De Sandre-Giovannoli A, Bernard R, Cau P, Navarro C, Amiel J, Boccaccio I, Lyonnet S, Stewart C, Munnich A, Le Merrer M, Levy N (2003) Lamin A truncation in Hutchinson-Gilford progeria. Science 300(5628):2055
179. Bione S, Maestrini E, Rivella S, Mancini M, Regis S, Romeo G, Toniolo D (1994) Identification of a novel X-linked gene responsible for Emery-Dreifuss muscular dystrophy. Nat Genet 8(4):323–327
180. Taylor MR, Slavov D, Gajewski A, Vlcek S, Ku L, Fain PR, Carniel E, Di Lenarda A, Sinagra G, Boucek MM, Cavanaugh J, Graw SL, Ruegg P, Feiger J, Zhu X, Ferguson DA, Bristow MR, Gotzmann J, Foisner R, Mestroni L (2005) Thymopoietin (lamina-associated polypeptide 2) gene mutation associated with dilated cardiomyopathy. Hum Mutat 26(6):566–574
181. Zhang Q, Bethmann C, Worth NF, Davies JD, Wasner C, Feuer A, Ragnauth CD, Yi Q, Mellad JA, Warren DT, Wheeler MA, Ellis JA, Skepper JN, Vorgerd M, Schlotter-Weigel B, Weissberg PL, Roberts RG, Wehnert M, Shanahan CM (2007) Nesprin-1 and -2 are involved in the pathogenesis of Emery-Dreifuss Muscular Dystrophy and are critical for nuclear envelope integrity. Hum Mol Genet 16(23):2816–2833
182. Cronshaw JM, Matunis MJ (2003) The nuclear pore complex protein ALADIN is mislocalized in triple A syndrome. Proc Natl Acad Sci U S A 100(10):5823–5827
183. Hellemans J, Preobrazhenska O, Willaert A, Debeer P, Verdonk PC, Costa T, Janssens K, Menten B, Van Roy N, Vermeulen SJ, Savarirayan R, Van Hul W, Vanhoenacker F, Huylebroeck D, De Paepe A, Naeyaert JM, Vandesompele J, Speleman F, Verschueren K, Coucke PJ, Mortier GR (2004) Loss-of-function mutations in LEMD3 result in osteopoikilosis, Buschke-Ollendorff syndrome and melorheostosis. Nat Genet 36(11):1213–1218
184. Waterham H, Koster J, Mooyer P, Noort GG, Kelley R, Wilcox W, Wanders R, Hennekam R, Oosterwijk J (2003) Autosomal recessive HEM/Greenberg skeletal dysplasia is caused by 3beta-hydroxysterol delta14-reductase deficiency due to mutations in the lamin B receptor gene. Am J Hum Genet 72(4):1013–1017
185. Gros-Louis F, Dupre N, Dion P, Fox MA, Laurent S, Verreault S, Sanes JR, Bouchard JP, Rouleau GA (2007) Mutations in SYNE1 lead to a newly discovered form of autosomal recessive cerebellar ataxia. Nat Genet 39(1):80–85
186. Basel-Vanagaite L, Muncher L, Straussberg R, Pasmanik-Chor M, Yahav M, Rainshtein L, Walsh CA, Magal N, Taub E, Drasinover V, Shalev H, Attia R, Rechavi G, Simon AJ, Shohat M (2006) Mutated nup62 causes autosomal recessive infantile bilateral striatal necrosis. Ann Neurol 60(2):214–222
187. Gonzalez-Alegre P, Paulson HL (2004) Aberrant cellular behavior of mutant torsinA implicates nuclear envelope dysfunction in DYT1 dystonia. J Neurosci 24(11):2593–2601
188. Naismith TV, Heuser JE, Breakefield XO, Hanson PI (2004) TorsinA in the nuclear envelope. Proc Natl Acad Sci U S A 101(20):7612–7617
189. Goodchild RE, Dauer WT (2004) Mislocalization to the nuclear envelope: an effect of the dystonia-causing torsinA mutation. Proc Natl Acad Sci U S A 101(3):847–852
190. Hoffmann K, Dreger C, Olins A, Olins D, Shultz L, Lucke B, Karl H, Kaps R, Muller D, Vaya A, Aznar J, Ware R, Sotelo Cruz N, Lindner T, Herrmann H, Reis A, Sperling K (2002) Mutations in the gene encoding the lamin B receptor produce an altered nuclear morphology in granulocytes (Pelger-Huet anomaly). Nat Genet 31(4):410–414
191. Fidzianska A, Toniolo D, Hausmanowa-Petrusewicz I (1998) Ultrastructural abnormality of sarcolemmal nuclei in Emery-Dreifuss muscular dystrophy (EDMD). J Neurol Sci 159(1):88–93
192. Sewry CA, Brown SC, Mercuri E, Bonne G, Feng L, Camici G, Morris GE, Muntoni F (2001) Skeletal muscle pathology in autosomal dominant Emery-Dreifuss muscular dystrophy with lamin A/C mutations. Neuropathol Appl Neurobiol 27(4):281–290

193. Maraldi NM, Lattanzi G, Capanni C, Columbaro M, Mattioli E, Sabatelli P, Squarzoni S, Manzoli FA (2006) Laminopathies: a chromatin affair. Adv Enzyme Regul 46:33–49, doi:S0065-2571(06)00002-1 [pii]10.1016/j.advenzreg.2006.01.001
194. Verga L, Concardi M, Pilotto A, Bellini O, Pasotti M, Repetto A, Tavazzi L, Arbustini E (2003) Loss of lamin A/C expression revealed by immuno-electron microscopy in dilated cardiomyopathy with atrioventricular block caused by LMNA gene defects. Virchows Arch 443(5):664–671
195. Goldman RD, Shumaker DK, Erdos MR, Eriksson M, Goldman AE, Gordon LB, Gruenbaum Y, Khuon S, Mendez M, Varga R, Collins FS (2004) Accumulation of mutant lamin A causes progressive changes in nuclear architecture in Hutchinson-Gilford progeria syndrome. Proc Natl Acad Sci U S A 101(24):8963–8968
196. Taimen P, Pfleghaar K, Shimi T, Moller D, Ben-Harush K, Erdos MR, Adam SA, Herrmann H, Medalia O, Collins FS, Goldman AE, Goldman RD (2009) A progeria mutation reveals functions for lamin A in nuclear assembly, architecture, and chromosome organization. Proc Natl Acad Sci U S A 106(49):20788–20793, doi:0911895106 [pii]10.1073/pnas.0911895106
197. Meaburn KJ, Cabuy E, Bonne G, Levy N, Morris GE, Novelli G, Kill IR, Bridger JM (2007) Primary laminopathy fibroblasts display altered genome organization and apoptosis. Aging Cell 6(2):139–153
198. Mewborn SK, Puckelwartz MJ, Abuisneineh F, Fahrenbach JP, Zhang Y, MacLeod H, Dellefave L, Pytel P, Selig S, Labno CM, Reddy K, Singh H, McNally E (2010) Altered chromosomal positioning, compaction, and gene expression with a lamin A/C gene mutation. PLoS One 5(12):e14342. doi:10.1371/journal.pone.0014342
199. Shumaker DK, Dechat T, Kohlmaier A, Adam SA, Bozovsky MR, Erdos MR, Eriksson M, Goldman AE, Khuon S, Collins FS, Jenuwein T, Goldman RD (2006) Mutant nuclear lamin A leads to progressive alterations of epigenetic control in premature aging. Proc Natl Acad Sci U S A 103(23):8703–8708
200. Scaffidi P, Misteli T (2006) Lamin A-dependent nuclear defects in human aging. Science 312(5776):1059–1063
201. Lattanzi G, Columbaro M, Mattioli E, Cenni V, Camozzi D, Wehnert M, Santi S, Riccio M, Del Coco R, Maraldi NM, Squarzoni S, Foisner R, Capanni C (2007) Pre-Lamin A processing is linked to heterochromatin organization. J Cell Biochem 102(5):1149–1159
202. Columbaro M, Capanni C, Mattioli E, Novelli G, Parnaik VK, Squarzoni S, Maraldi NM, Lattanzi G (2005) Rescue of heterochromatin organization in Hutchinson-Gilford progeria by drug treatment. Cell Mol Life Sci 62(22):2669–2678
203. Capo-chichi CD, Cai KQ, Smedberg J, Ganjei-Azar P, Godwin AK, Xu XX (2011) Loss of A-type lamin expression compromises nuclear envelope integrity in breast cancer. Chin J Cancer 30(6):415–425
204. Willis ND, Cox TR, Rahman-Casans SF, Smits K, Przyborski SA, van den Brandt P, van Engeland M, Weijenberg M, Wilson RG, de Bruine A, Hutchison CJ (2008) Lamin A/C is a risk biomarker in colorectal cancer. PLoS One 3(8):e2988. doi:10.1371/journal.pone.0002988
205. Skvortsov S, Schafer G, Stasyk T, Fuchsberger C, Bonn GK, Bartsch G, Klocker H, Huber LA (2011) Proteomics profiling of microdissected low- and high-grade prostate tumors identifies Lamin A as a discriminatory biomarker. J Proteome Res 10(1):259–268. doi:10.1021/pr100921j
206. Chow KH, Factor RE, Ullman KS (2012) The nuclear envelope environment and its cancer connections. Nat Rev Cancer 12(3):196–209. doi:10.1038/nrc3219
207. Baylin SB, Jones PA (2011) A decade of exploring the cancer epigenome—biological and translational implications. Nat Rev Cancer 11(10):726–734. doi:10.1038/nrc3130
208. Berman BP, Weisenberger DJ, Aman JF, Hinoue T, Ramjan Z, Liu Y, Noushmehr H, Lange CP, van Dijk CM, Tollenaar RA, Van Den Berg D, Laird PW (2012) Regions of focal DNA hypermethylation and long-range hypomethylation in colorectal cancer coincide with nuclear lamina-associated domains. Nat Genet 44(1):40–46. doi:10.1038/ng.969

209. Romana SP, Radford-Weiss I, Ben Abdelali R, Schluth C, Petit A, Dastugue N, Talmant P, Bilhou-Nabera C, Mugneret F, Lafage-Pochitaloff M, Mozziconacci MJ, Andrieu J, Lai JL, Terre C, Rack K, Cornillet-Lefebvre P, Luquet I, Nadal N, Nguyen-Khac F, Perot C, Van den Akker J, Fert-Ferrer S, Cabrol C, Charrin C, Tigaud I, Poirel H, Vekemans M, Bernard OA, Berger R (2006) NUP98 rearrangements in hematopoietic malignancies: a study of the Groupe Francophone de Cytogenetique Hematologique. Leukemia 20(4):696–706. doi:10.1038/sj.leu.2404130
210. Nakamura T, Largaespada DA, Lee MP, Johnson LA, Ohyashiki K, Toyama K, Chen SJ, Willman CL, Chen IM, Feinberg AP, Jenkins NA, Copeland NG, Shaughnessy JD Jr (1996) Fusion of the nucleoporin gene NUP98 to HOXA9 by the chromosome translocation t(7;11)(p15;p15) in human myeloid leukaemia. Nat Genet 12(2):154–158. doi:10.1038/ng0296-154
211. Kasper LH, Brindle PK, Schnabel CA, Pritchard CE, Cleary ML, van Deursen JM (1999) CREB binding protein interacts with nucleoporin-specific FG repeats that activate transcription and mediate NUP98-HOXA9 oncogenicity. Mol Cell Biol 19(1):764–776
212. Wang GG, Cai L, Pasillas MP, Kamps MP (2007) NUP98-NSD1 links H3K36 methylation to Hox-A gene activation and leukaemogenesis. Nat Cell Biol 9(7):804–812. doi:10.1038/ncb1608
213. Wang GG, Song J, Wang Z, Dormann HL, Casadio F, Li H, Luo JL, Patel DJ, Allis CD (2009) Haematopoietic malignancies caused by dysregulation of a chromatin-binding PHD finger. Nature 459(7248):847–851. doi:10.1038/nature08036
214. Wang GG, Allis CD (2009) "Misinterpretation" of a histone mark is linked to aberrant stem cells and cancer development. Cell Cycle 8(13):1982–1983

Studying Lamins in Invertebrate Models

Roman Lyakhovetsky and Yosef Gruenbaum

Abstract Lamins are nuclear intermediate filament proteins that are conserved in all multicellular animals. Proteins that resemble lamins are also found in unicellular organisms and in plants. Lamins form a proteinaceous meshwork that outlines the nucleoplasmic side of the inner nuclear membrane, while a small fraction of lamin molecules is also present in the nucleoplasm. They provide structural support for the nucleus and help regulate many other nuclear activities. Much of our knowledge on the function of nuclear lamins and their associated proteins comes from studies in invertebrate organisms and specifically in the nematode *Caenorhabditis elegans* and the fruit fly *Drosophila melanogaster*. The simpler lamin system and the powerful genetic tools offered by these model organisms greatly promote such studies. Here we provide an overview of recent advances in the biology of invertebrate nuclear lamins, with special emphasis on their assembly, cellular functions and as models for studying the molecular basis underlying the pathology of human heritable diseases caused by mutations in lamins A/C.

Keywords *Caenorhabditis elegans* • *Drosophila melanogaster* • Intermediate filaments • Invertebrates • Lamin • Laminopathies • Muscular dystrophy • Nuclear envelope • Nuclear lamina • Hutchinson–Gilford progeria syndrome

Abbreviations

EDMD	Emery–Dreifuss muscular dystrophy
IF	Intermediate filament
INM	Inner nuclear membrane

R. Lyakhovetsky • Y. Gruenbaum (✉)
Department of Genetics, The Institute of Life Sciences, The Hebrew University of Jerusalem, Givat Ram, Jerusalem 91904, Israel
e-mail: gru@vms.huji.ac.il

E.C. Schirmer and J.I. de las Heras (eds.), *Cancer Biology and the Nuclear Envelope*, Advances in Experimental Medicine and Biology 773, DOI 10.1007/978-1-4899-8032-8_11,

LINC	Linker of nucleoskeleton and cytoskeleton
NLS	Nuclear localization signal
NPC	Nuclear pore complex
ONM	Outer nuclear membrane
PEV	Position effect variegation

The Nuclear Lamina: An Overview

The nuclear interior (nucleoplasm) is separated from the cytoplasm by a complex boundary called the nuclear envelope, which is comprised of outer and inner lipid bilayer membranes (ONM and INM, respectively), nuclear pore complexes (NPCs) and nuclear lamina. The ONM and INM are separated by a 20–40 nm-wide lumen and are fused at NPCs. The ONM is contiguous with the endoplasmic reticulum, while a filamentous meshwork termed the nuclear lamina is associated with the INM at the nucleoplasmic leaflet. The nuclear lamina is comprised of lamin filaments and lamin-binding proteins, including the integral proteins of the inner nuclear membrane. Proper assembly and function of the nuclear lamina is of utmost importance because it regulates many of the nuclear activities. The importance of the nuclear lamina and in particular lamins is manifested by the involvement of nuclear lamina defects in the pathology of many heritable diseases (reviewed in [1–3]). The study of the nuclear lamina in invertebrates, mainly in the nematode *Caenorhabditis elegans* and the fruit fly *Drosophila melanogaster*, provides major advantages. These include a relatively simple composition of the invertebrate nuclear lamina, efficient genetic manipulations of the animals and inexpensive culturing in laboratory conditions. We hereby summarize recent advances in invertebrate lamin research and highlight the contribution of these studies towards basic understanding of structure and functions of the nuclear lamina and of the mechanisms leading to laminopathies, many of which could reflect changes that occur in tumorigenesis.

The Structure of Lamin Proteins

Lamins are the nuclear members of the intermediate filament (IF) family of proteins and are the basic building blocks of the lamin filaments (reviewed in [2, 4]). Lamins, like all IF proteins, are composed of three distinct domains: an amino terminal head, central rod, and carboxyl tail domain (Fig. 1a). The lamin family is comprised of B-type and A-type lamins that differ in their biochemical properties, expression patterns, and activity [5]. The tail domain of all lamins contains an invariant nuclear localization signal (NLS) and an Ig-fold motif. In addition, except for in lamin C (an A-type lamin isoform), all lamins have a CaaX box at their C-terminus (a cysteine followed by two aliphatic amino acids and a fourth residue). This sequence is subjected to sequential posttranslational modifications: initially, the cysteine residue is farnesylated, followed by cleavage of the last three amino acid residues (aaX) and methyl-esterification of the newly exposed farnesylated cysteine (reviewed in [2]). In vertebrates, the C-terminus of

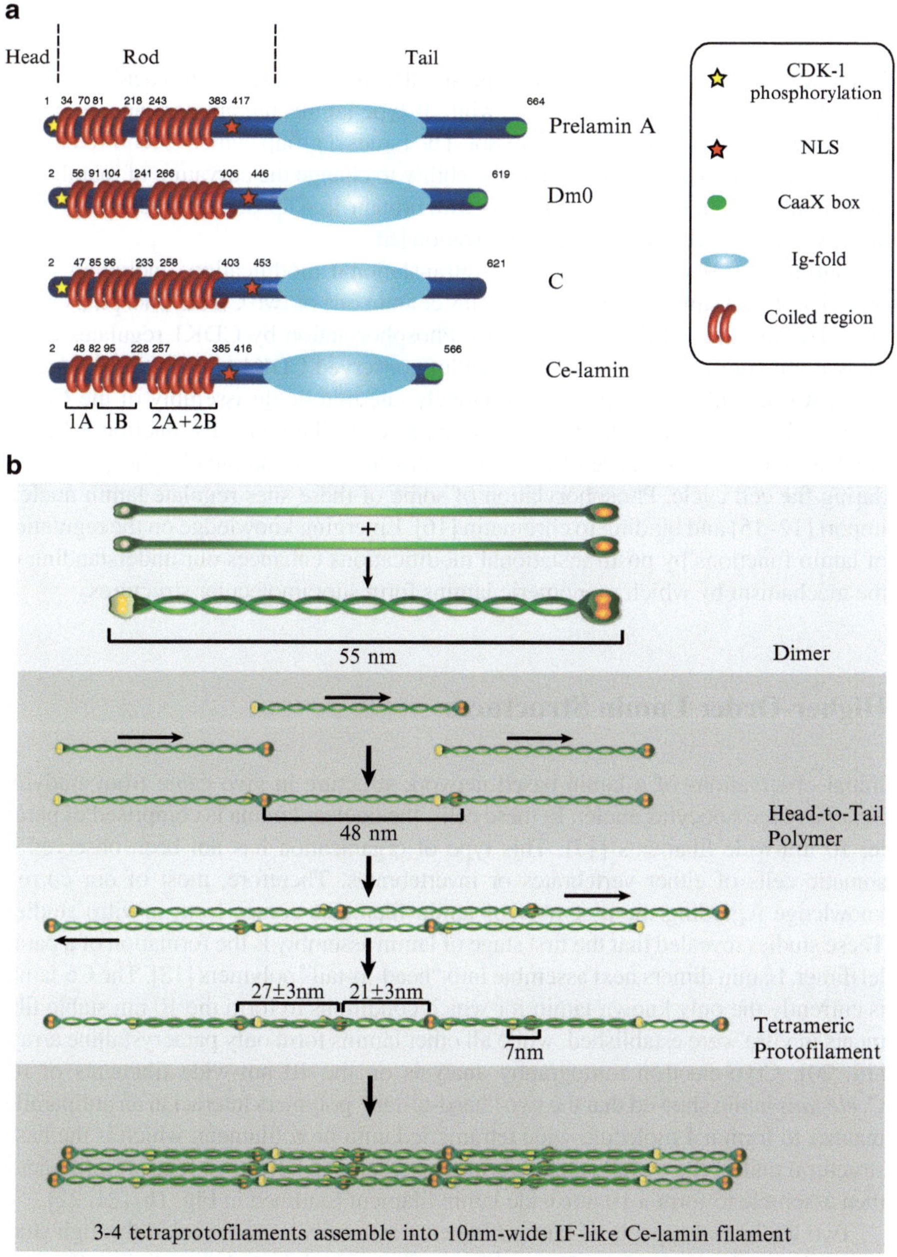

Fig. 1 Structure of invertebrate lamins and lamin filament assembly pathway. (**a**) Schematic representation of human prelamin A, *Drosophila* lamin Dm_0 (a B-type lamin), *Drosophila* lamin C, and Ce-lamin. All lamins display the tripartite structure of "head," "rod," and "tail" domains. The "head" domain contains a CDK-1 phosphorylation site (missing in *C. elegans* Ce-lamin). The "rod" domain is built from α-helical segments (labeled in the figure as 1A, 1B, and 2A+2B), separated by short linker peptides. The "tail" domain includes the conserved Ig-fold motif, the nuclear localization signal (NLS) and the CaaX box. The *D. melanogaster* lamin C lacks the CaaX box. (**b**). Assembly pathway of lamin filaments. Shown are the main stages of polymerization starting with the dimerization up to 10 nm intermediate filament. Detailed explanations of the process appear in the text

lamin A contains an additional cleavage site that is located 15 amino acids upstream from the farnesylated cysteine. Thus, while B-type lamins remain permanently farnesylated, the mature A-type lamins are not. The farnesyl group, which is required for the anchoring of lamins to the INM, has the ability to change the curvature of membranes as demonstrated in in vitro experiments with protein-free liposomes. Interestingly, this activity does not require lamin polymerization [6].

Lamins also undergo additional posttranslational modifications including phosphorylation and sumoylation. Most lamins contain one or two CDK1 phosphorylation sites flanking the rod domain (Fig. 1a). Phosphorylation by CDK1 regulates lamin disassembly during mitosis [7–10]. Lack of conserved CDK1 sites in the *C. elegans* lamin (Ce-lamin), may explain the relatively late mitotic disassembly of the *C. elegans* lamina [11]. The importance of lamin phosphorylation for its function is further emphasized by the presence of additional sites that are dynamically phosphorylated during the cell cycle. Phosphorylation of some of these sites regulate lamin nuclear import [12–15] and binding to chromatin [16]. Emerging knowledge on the regulation of lamin functions by posttranslational modifications enhances our understanding of the mechanism by which monomeric lamins form supramolecular structures.

Higher-Order Lamin Structures

Initial observations of a lamin-based network structure in vivo came from studying *Xenopus laevis* oocytes nuclei. In these cells, the nuclear lamina is comprised of parallel 10 nm-wide filaments [17]. This type of organization has not been observed in somatic cells of either vertebrates or invertebrates. Therefore, most of our current knowledge regarding the assembly of lamin filaments comes from in vitro studies. These studies revealed that the first stage of lamin assembly is the formation of a parallel dimer. Lamin dimers next assemble into "head-to-tail" polymers [18]. The Ce-lamin is currently the only known lamin for which conditions to form the 10 nm stable filaments in vitro were established, while all other lamins form only paracrystalline arrays [19, 20]. Cryo-electron tomography analysis on the 10 nm-wide filaments of the *C. elegans* lamin showed that the two "head-to-tail" polymers interact in an antiparallel manner to form a 4 molecule-wide tetrameric lamin protofilament, which is the basic structural unit of the cytoplasmic intermediate filaments. Three to four protofilaments then assemble to form a 10 nm-wide lamin filament (outlined in Fig. 1b) [21, 22].

Two additional aspects of lamin filaments assembly were revealed through studies of invertebrate models. Recombinant Ce-lamin injected into living *Xenopus oocytes* formed an intricate three-dimensional meshwork of 5–6 nm protofilaments. The difference between the 10 nm filaments observed in vitro and the 5–6 nm protofilaments observed in vivo suggests a role for lamin-binding proteins in lamin filament assembly [23]. A study in *Drosophila* showed that, when expressed in the background of lamin Dm_0 (a B-type lamin) knockout, lamin C does not associate with the preexisting maternal lamin Dm_0. This suggests that lamin Dm_0 and lamin C form mono-specific filament networks that occupy separate territories at the nuclear envelope [24]. Injection of Ce-lamin bearing the Q159K Hutchinson–Gilford progeria syndrome-linked mutation to living *Xenopus* oocytes induces the formation of bundled arrays composed of less isotropically oriented lamin protofilaments [23]. It is still in question whether the in vitro studies succeed to reconstitute the physiological lamin assembly. However, these studies propose that (1) distinct regions of the nuclear lamina are comprised of mono-specific (i.e., a single lamin type) filaments and (2) the assembly of the nuclear lamina is assisted by auxiliary factors. Successful modeling of lamina assembly in vivo will require further research, possibly assisted by state-of-the-art nanotechnology-based molecular imaging systems.

The Evolution of Lamin Gene Family

Except for the genomes of tunicates, that encode two B-type lamin genes, all other sequenced invertebrate genomes harbor a single B-type lamin gene (Fig. 2). *Drosophila* is the only known invertebrate organism to encode an additional A-type lamin gene (lamin C) [25]. The evolution pattern of the lamin gene family has been

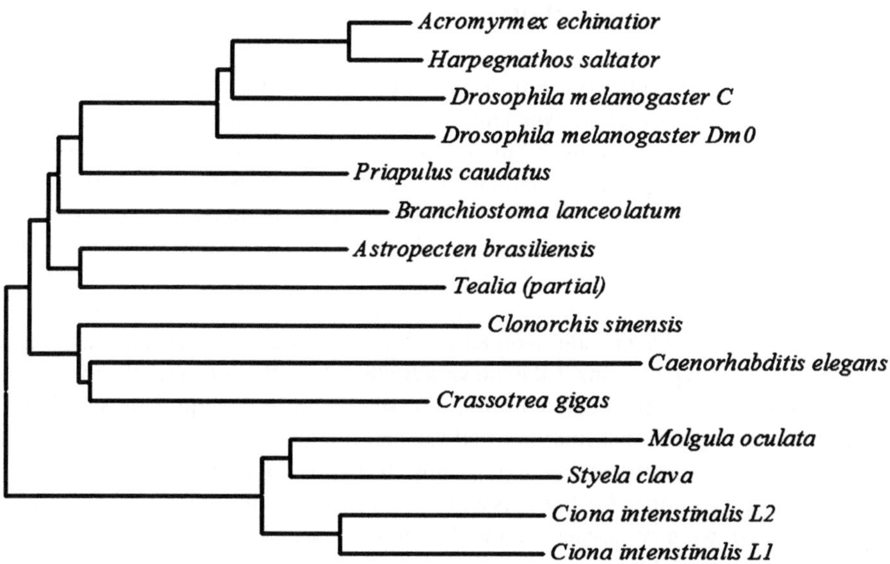

Fig. 2 Evolutionary relationships of invertebrate lamins. All the compared lamin sequences are of B-type except from *Drosophila melanogaster* lamin C, which is an A-type lamin. Evolutionary analyses were conducted in MEGA5 software [95]. The evolutionary history was inferred using the Minimum Evolution method [96]. The optimal tree with the sum of branch length = 6.40964395 is shown. The tree is drawn to scale, with branch lengths in the same units as those of the evolutionary distances used to infer the phylogenetic tree. The evolutionary distances were computed using the Poisson correction method [97] and are in the units of the number of amino acid substitutions per site. The ME tree was searched using the Close-Neighbor-Interchange (CNI) algorithm [98] at a search level of 1. The Neighbor-joining algorithm [99] was used to generate the initial tree. The analysis involved 15 amino acid sequences. All positions containing gaps and missing data were eliminated. There were a total of 281 positions in the final dataset

delineated through sequence analysis of lamin genes from various metazoans. The *Lmnb*1 gene, which encodes the mammalian lamin B1, is the most conserved vertebrate lamin gene; from *Nematostella vectensis* through *Xenopus laevis* to humans, the *Lmnb*1 homolog is always flanked by a gene that encodes for the membrane associated Ring-finger protein 2/3 [26]. In addition, the positions of eight of the nine introns of many invertebrate B-type lamin genes are conserved in human *LMNB1*. Interestingly, this conservation of introns also appears in genes encoding for vertebrate cytoplasmic IF proteins. The conserved intron-exon structure of lamin genes between invertebrates and vertebrate has led to the hypothesis that a B-type lamin gene was the ancestor to all lamin genes [27]. This hypothesis is further supported by the fact that in both invertebrate and vertebrate species, except for the sperm cells, B-type lamins are expressed in all cells. In contrast, in both *Drosophila* and vertebrate species, the A-type lamins are expressed only in a subset of tissues [2]. These results also led to the hypothesis that the vertebrate lamins diverged from an invertebrate B-type lamin through two rounds of duplication and that a B-type lamin is the ancestor gene for the whole IF multi-gene family [28, 29].

Lamins of Lower Metazoans

The most primitive metazoan with a characterized lamin is the cnidarian *Hydra vulgaris*, which has a single B-type lamin [30]. Similarly to other B-type lamins, the *Hydra* lamin contains an amino-terminal "head" domain, a coiled coil "rod" domain, a KRSR nuclear localization signal, and a CaaX box at the carboxyl-terminus. The positions of the three introns in the *Hydra* lamin gene are conserved in vertebrate B-type lamin genes. Lamin genes are found in other low metazoans including *Tealia* (*Cnidarian*), *Priapulus* (*Priapulid*), and the sea star *Asterias*. [30].

Caenorhabditis elegans lamin—The *C. elegans* genome contains a single lamin gene, termed *lmn-1* that encodes a 66 KDa protein [31]. Ce-lamin has characteristics of both A- and B-type lamins. Like the vertebrate A-type lamins, a small fraction of Ce-lamin is present in the nucleoplasm. It is essential for maintaining nuclear shape, required for the spatial distribution of NPCs, and it interacts with proteins that bind A-type lamins in vertebrates, including emerin, LEM-2 and BAF-1 [31–33]. Like vertebrate B-type lamins, Ce-lamin is expressed in all cells throughout development. It remains permanently farnesylated and interacts with proteins that in vertebrates bind B-type lamins. On the evolutionary scale, Ce-lamin displays an interesting divergence, since it lacks the conserved CDK1 mitotic phosphorylation site upstream to the "rod" domain, as well as 14 amino acids (two heptads) early in coil 2B in the rod domain [31, 34] (Fig. 1a). The head domain of Ce-lamin is 14-residue longer and the "tail" domain is 25-residues shorter compared to vertebrate B-type lamins [34].

Drosophila melanogaster lamins—*Drosophila melanogaster* stand out among the invertebrates with characterized lamins for expressing two types of lamin proteins encoded by two distinct genes. Lamin Dm_0 is essential and is expressed in all cells [35–37]. The additional *D. melanogaster* lamin, lamin C, resembles lamin Dm_0 more than a vertebrate A-type lamin. It also lacks the CaaX box, thus making it the only known invertebrate lamin that is unable to undergo farnesylation [38].

Are Lamins a Feature of Multicellular Organisms?

Unicellular organisms—Studied unicellular organisms such as yeast (*Saccharomyces cerevisiae*), *Tetrahymena thermophila* [39], *Dinoflagellates* [40], or *Physarum polycephalum* do not express functional lamin homologues (reviewed in [41]). Therefore, one of the most intriguing questions is whether lamins are a distinct feature of multicellular organisms. The study of unicellular organisms, that express lamin-like proteins, can undoubtedly clarify the evolutionary origins and physiological functions of the lamin proteins. For example, the nucleus of *Amoeba proteus* is surrounded by a honeycomb-structured layer resembling lamina, the first such structure ever to be defined [42]. This primordial lamina breaks down during mitosis as occurs in higher organisms. Nevertheless biochemical analysis and electron microscopy show that, unlike the metazoan nuclear lamina, this structure is not tightly anchored to the INM or to NPCs. Furthermore, when lamins from *Xenopus laevis* and *Drosophila melanogaster* are injected into *Amoeba* cells, they are effectively transported into the nucleus, but fail to localize to the nuclear periphery. The inability of the injected lamins to associate with the *Amoeba* nuclear envelope, in conjunction with the fact that the *Amoeba proteus* lamina-like structure is not associated with the INM, implies that the INM of the *Amoeba proteus* lacks essential proteins that tether lamins to the INM [43]. Strikingly, a different *Amoeba* species, *Dictyostelium discoideum*, expresses a protein termed NE81, which has structural and functional similarities to lamins [44]. Apart from the structural homology, NE81 shares several functional properties with lamin, such as association with the nuclear envelope of *Dictyostelium* during the entire cell cycle. In addition, similar to the metazoan lamins, it is immobile throughout interphase and becomes mobile in a short time window during mitosis when disintegration and reconstitution of the nuclear envelope occurs. Moreover, NE81 requires a carboxyl-terminal prenylation for proper integration into the nuclear envelope and is essential for mechanical stress resistance. Finally, when over-expressed in HeLa cells, the NE81-GFP fusion protein localizes to the nuclear envelope, further exhibiting lamin-like properties. Another lamin-like protein, termed NUP-1 is expressed at the nuclear periphery of *Trypanosomatid*s and exhibits nuclear envelope localization. In addition, it regulates the distribution of nuclear pore complexes, chromosome organization, maintenance of nuclear architecture, and regulation of gene expression [45].

Plants—Although plant cells do not contain *bona fide* lamin proteins, recent studies in plants revealed the involvement of the nuclear periphery in maintaining chromatin structure and intranuclear gene positioning. A family of coiled-coil plant proteins called *little nuclei* which were initially discovered in *Daucus carota* [46] and then in *Arabidopsis thaliana*, are proposed to be functional analogues of animal lamins. GFP-fused *little nuclei* proteins are localized at the nuclear periphery and mutations in these putative plant lamin-like genes cause a variety of defects including whole-plant dwarfing and alterations of nuclear structure that are reminiscent of nuclear defects observed in animal cells that express mutant lamins [47, 48].

Lamin-Binding Proteins

Identification of lamin-interacting proteins has become a powerful experimental tool to decipher the lamin biological functions. Among the known lamin-binding partners in *C. elegans* are the LEM-domain proteins Ce-emerin [49], LEM-2 (also called Ce-MAN1) [50], BAF-1 [51], UNC-84 [52], and core histones [16] (Table 1). Lamins also bind chromatin in mammalian cells, *Drosophila* and *C. elegans* [16, 53–55]. In addition, lamins associate with chromatin through interactions with lamin-associated proteins, such as the lamin B receptor (LBR), BAF-1, and the LEM-domain proteins (reviewed in [2, 56, 57]). An important question is whether lamin–chromatin association regulates gene expression.

A circumstantial answer to this question is provided by observations in nematodes and other organisms where many tissue-specific promoters are sequestered at the nuclear periphery when repressed, and shift to the nuclear interior when activated. The bacterial LacO/LacI system enables artificial tethering of a reporter gene to nuclear periphery. This allows one to follow the positioning of a tissue-specific promoter in living cells and to search for possible correlations between the nuclear positioning of a gene and its activity [58–60]. Employing a GFP-lacI/lacO recognition system in *C. elegans*, it was found that integrated large repetitive arrays containing developmental

Table 1 Lamin-binding proteins in invertebrate models

Protein type		Lamin-dependent localization[a]	Reference
C. elegans	LEM-domain proteins		
Ce-lamin	Ce-emerin[b]	Yes	[49]
	LEM-2/Ce-Man1[b]	Yes	[50]
	SUN-domain proteins		
	UNC-84[a]	Yes	[52]
	Matefin/SUN-1[b]	No	[73, 90]
	Other proteins		
	Titin[c,d]	Yes	[67]
	BAF-1	Yes	[51]
D. melanogaster	LEM-domain proteins		
Lamin Dm_0	Otefin[b,c]	Yes	[91, 92]
	Bocksbeutel-α	Yes	[49]
	dMAN1[b]	Yes	[49]
	Other proteins		
	dLBR[b]	No	[93]
	YA[c]	ND	[92]
	Histones H2A and H2B[b]	ND	[16]
	JIL-1 kinase[a,b,c]	ND	[94]
	Rab5[b]	ND	[69, 70]

Table 1 includes results of studies performed in *C. elegans* and *D. melanogaster*. *ND* not determined
[a]Analyzed by RNA interference and/or mutations
[b]Identified by co-immunoprecipitation, blot overlays or pull down experiments
[c]Identified by yeast two-hybrid
[d]Co-localization by immunohistochemistry

and tissue-specific promoters are positioned at the nuclear periphery in cells that do not activate the promoter. In contrast, in cells where a tissue-specific promoter is expressed, the array is shifted to the nuclear interior [61]. Lamin down regulation released the array from the nuclear periphery even in cells in which the array is not active. In contrast, worms expressing a lamin mutation that causes muscular dystrophy in humans retained the array at the nuclear periphery of muscle cells when under control of a muscle-specific promoter [62]. In *Drosophila*, lamin is required for specific regulation of testis-specific gene clusters. Ablation of Dm_0 through mutagenesis and RNAi-mediated silencing led to the detachment of testis-specific gene clusters from the nuclear envelope and to an increase in their expression in somatic cells. These findings are consistent with the observed coupling of transcriptional activation and detachment from the nuclear envelope of testis-specific genes in male germ lines [63].

In addition to mediating intranuclear activities, recent studies show that lamins play a key role in the determination of the spatial positioning of the nucleus within cells [64, 65]. This activity is mediated through lamin interactions with the SUN-KASH domain protein complexes (*li*nker of *n*ucleoskeleton and *c*ytoskeleton—LINC complexes). Rather than being static, the nucleus migrates in the cytoplasm of certain cell types, such as the photoreceptor cells in the *Drosophila* eye and this migration requires lamin Dm_0 activity [66]. In *C. elegans*, the correct positioning of UNC-84 (a SUN-domain protein) depends on lamin, suggesting that lamin–UNC-84 interaction plays a role in the spatial localization of nuclei [52]. Ce-lamin also interacts with nuclear Titin, a cytoskeletal protein, which has a role in chromosome maintenance in the nucleus [67].

Rab5 is a small GTPase and a known regulator of the assembly of early endosomes [68]. Surprisingly, in both *C. elegans* and *D. melanogaster* Rab5 also affects the disassembly of the nuclear envelope during mitosis, as well as chromosome alignment [69, 70]. The latter role requires Rab5 direct interaction with Mud, which is the *Drosophila* homologue of the vertebrate NuMa (*nuc*lear *m*itotic *a*pparatus) protein that is required for spindle formation and spindle maintenance. Rab5 and Mud act in the same pathway and downregulation of each protein leads to decreased tension at the kinetochore. Thus, Rab5 and lamin are both involved in progression through mitosis, which is relevant to progression of cancer.

Invertebrate Models in the Study of Lamin Functions and Lamin-Linked Disorders

Lamin mutations cause over 14 distinct hereditary disorders that are collectively referred to as “laminopathies”. These include both systemic and tissue-specific syndromes, most of which stem from mutations in A-type lamins. While invertebrates do not express a *bona fide* lamin A, the *Drosophila* lamin C is functionally defined as an A-type lamin and Ce-lamin displays features of both A- and B-type lamins [2, 31, 34, 71, 72]. To illustrate the use of invertebrate models in research of lamin-linked pathology we shall describe experiments in *C. elegans* and *D. melanogaster* testing the effects of lamin silencing and expression of lamins bearing disease-linked mutations.

C. elegans—Downregulation of lamin in *C. elegans* showed that embryos cease to develop when having up to several hundred cells [31]. However, the effect of lamin downregulation is observed already as early as the first zygotic division. In these embryos nuclear structure was altered and featured multiple pathological aberrations including nuclear membrane lobulation and invagination, clustering of NPCs, and several chromatin aberrations common in cancer cells such as inter-chromosome connections (bridges), abnormal chromatin condensation, and aberrant number of chromosomes due to abnormal segregation [31]. Some embryos manage to escape the lethal effects of *lmn-1* RNAi probably due to them being laid outside of the most effective time window of RNAi effect. These animals are sterile or semi-sterile displaying dramatic reduction in the number of germ cells and oocyte abnormalities including multiple nuclei and large vacuoles. Electron microscopy analysis shows that the densely stained heterochromatin was dramatically reorganized. These animals also display intranuclear vacuolation and sharing of the outer membrane by two adjacent nuclei [32]. Downregulation of lamin also affects the localization of other components of the nuclear envelope including emerin, MAN-1/LEM-2, and UNC-84, but not of matefin/SUN-1 [50, 52, 73, 74]. Interestingly, a shortened life span was detected in *C. elegans* where the lamin expression was either downregulated at adult stages or abolished by a deletion that contained the entire three N-terminal exons [75, 76]. Survival of the animals with the lamin deletion (tm1502) relied on the maternal supply of Ce-lamin protein and RNA [31, 75]. Interestingly, during the first 3 days of adulthood, the tm1502 animals display normal nuclear morphology and no apparent phenotypes. However, as early as the day four, a fraction of muscle nuclei start to show lobulation, and at days 8–10 muscle, gut, and hypodermal nuclei are highly convoluted, similar to what is normally observed in wild-type nuclei after 12–15 days. In addition, after day 4, the heterochromatin associated with the nuclear periphery is reduced and a small portion of these nuclei show additional layers of nuclear membrane and fragmentation/lobulation [76].

Expression in *C. elegans* of the disease-linked Ce-lamin Y59C mutation, which in humans (Y45C) causes Emery–Dreifuss muscular dystrophy (EDMD) affects the positioning of muscle-specific *myo-3* array [77]. Larvae expressing Ce-lamin Y59C show abnormal retention of the array at the nuclear envelope of muscle cells. The retention correlated with a tenfold reduction in the activation of the *myo-3* promoter activity in the array and alteration of expression of a number of muscle-specific genes in vivo. Consistent with a human lamin mutation that specifically affects muscle tissue, the effect in *C. elegans* was specific to muscle cells since an equivalent array under control of an intestine-specific *pha-4* promoter is expressed normally and shifts inward when activated in gut cells of Ce-lamin Y59C animals. Phenotypically, adult Ce-lamin Y59C expressing animals exhibit perturbations in body muscle ultrastructure and reduced muscle function (as in humans), suggesting that the function of lamins in muscle development is evolutionarily conserved [62]. Additional analysis of four disease-linked mutations, ΔK32 (EDMD), Q159K (progeria), T164P (EDMD), and L535P (EDMD), which affect lamin structure and nuclear localization and prevent proper assembly of Ce-lamin filaments and/or paracrystalline arrays, showed disease-like phenotypes in *C. elegans* strains as well as reduced fertility and motility coincident with muscle lesions. In addition, the Q159K- and T164P-expressing strains had a reduced life span [78, 79].

D. melanogaster—Downregulation of each of the two *D. melanogaster* lamin genes leads to developmental arrest. Insertion of a P element (a transposable element) into the first exon of lamin Dm_0 gene led to lethality at three distinct developmental stages: embryonic (20–30 %), pupal (50–60 %), and post-eclosion lethality (5–10 %) [80]. The variability in the stage of lethality is likely caused due to variations in the levels of maternal lamin RNA and protein [35, 81]. Absence of lamin in early embryos leads to lethality, with phenotypes that resemble loss of lamin in *C. elegans* [36]. The few animals that reach adulthood survive for up to 2 weeks after the eclosion. These flies are unable to fly and make only occasional small jumps [80]. Flies of both sexes are sterile. Strikingly, the most prominent difference in intensity and distribution of lamin Dm_0 signal between the wild-type and mutant animals is observed in the perinuclear regions of the central nervous system. Surprisingly, multiple cells in this region lack lamin completely implying that they manage to divide and differentiate into neurons either completely lacking or containing only minute, undetectable levels of lamin Dm_0 [82]. Furthermore, electron microscopy analysis of young adult fly heads shows multiple abnormalities at the subcellular level including incomplete nuclear envelopes, clustering of nuclear pore complexes and increased number of annulate lamellae. Another study presented a lamin Dm_0 knockout mutant that was modified by imprecise excision of the P element resulting in flies with deletions in the lamin Dm_0 gene. This modification prevented lamin protein expression. Multiple cells with no detectable lamin continued to proliferate with embryos reaching third instar developmental stage. A small fraction (3 %) of the animals made it through the eclosion but died thereafter exhibiting serious motility defects. Interestingly, *Drosophila* lamin C expression could not compensate for the loss of lamin Dm_0 [82]. In a recent study, analysis of a *Drosophila* lamin C knockout revealed a crucial role of this gene in the development of tendon cells. Flies knocked out for lamin C ceased to develop at pupal stage and displayed inability to do head eversion (a developmental stage of trans. Strikingly, this developmental arrest was rescued by expression of lamin C in tendon cells, thus suggesting the key role of lamin C expression in these cells for contraction of muscles that direct the turning of the head capsule outwards (head aversion) [83].

Ari3 is a mutant allele of lamin Dm_0 in which the carboxy part of the rod domain and the entire tail domain are missing [84]. Ari3 was originally discovered as an enhancer of position effect variegation (PEV). When Ari3 is expressed in the genetic background of In(1)w^{m4} (an inversion that brings the *white* locus next to centromeric heterochromatin, which leads to variegated eye color [85]), it significantly enhances the position effect and causes the eye to turn white [84]. The observed enhancement of PEV by the Ari3 allele suggests a role for lamin in the formation and/or maintenance of heterochromatin.

Drosophila also served as a model system to study EDMD through expression of lamin C mutant isoforms [86], and complete ablation of gene expression [83]. Animals expressing mutant lamin die in the pupal stages, due to defects in the muscle cells that appear as early as the third instar larval stage. These defects include:

Myonuclei clustering and mislocalization of lamin and Klaroid (a SUN-domain protein). These aberrations resemble the pathology in tissues of EDMD patients validating the authenticity of *D. melanogaster* as a model system to study this and possibly other human laminopathies [86]. Likewise, human disease-linked mutations were introduced into lamin Dm_0, including the EDMD-linked mutations R386K, R453W, W520S and L530P and the dilated cardiomyopathy-linked mutation N195K (numbers indicate the amino acid position in human lamin A). The mutant lamins either aggregate (N195K and R386K) or induce lethality (W520S) [87].

While no aberrant morphology of larval muscle or reduction in viability are observed in larval muscle cells overexpressing wild type lamin C, expression of N-terminally truncated lamin C leads to lethality with multiple defects. These defects include lamin C aggregation, deformed nuclear shape and aberrant chromatin condensation. The few animals that survive to adulthood display abnormal leg morphology [88]. Overexpression in *Drosophila* of either human lamin A or human progerin, (a truncated lamin A mutant that causes progeria), cause severe defects in nuclear morphology such as nuclear herniations, invaginations, fragmentation, and formation of micronuclei. These flies also showed a significantly shortened life span [89].

In conclusion, the investigations of lamins in *C. elegans* and *D. melanogaster* verify that it is possible to successfully reconstitute various aspects of lamin-linked pathology, both at the single cell and the whole organism level. Due to the many links between lamins and human cancers, these simple model systems should prove useful to investigating roles of lamins that contribute to tumor formation.

Conclusions and Future Perspectives

Lamins show a remarkable structural and functional conservation in metazoa evolution. Invertebrate species are used to determine the supramolecular structure of lamin filaments, as well as to decipher the roles of lamins in maintaining nuclear shape, regulating nuclear organization, spacing nuclear pore complexes, mitosis, heterochromatin organization, and spatial positioning of developmental genes, all of which can be linked to human cancers. Both *C. elegans* and *Drosophila* proved to be valid models to investigate the physiological roles of lamins. Since disease-linked mutations in these organisms at least partly recapitulate the muscle and aging phenotypes of human laminopathies, they are excellent models to study the pathology of these debilitating diseases and likely also cancer. There are still many open questions regarding the structure and functions of lamins. Some of the important questions yet to be answered include determination of the atomic structure of the lamin dimer, finding how lamin filaments are organized in somatic cells in vivo and assessing how their organization is affected by the lamin-binding partners. Another critical issue is the regulation, mechanism, and functional specificity of the lamin complexes. Answers to these questions will enhance our understanding of the molecular mechanism of laminopathies and lamin roles in cancer.

Acknowledgments We thank Yuval Reiss and other members of the Gruenbaum laboratory for critical comments on the manuscript. We also thank Nataly Levi for the assistance in preparation of the Fig. 1. We gratefully acknowledge funding from the Morasha Legacy 1798/10, Israel Ministry of Health (MOH 2965), the Muscular Dystrophy Association (MDA), the Israeli Science Foundation, the Binational Israel-USA Science Foundation (BSF 2007215), and the COST NANONET (BM1002).

References

1. Bank EM, Gruenbaum Y (2011) The nuclear lamina and heterochromatin: a complex relationship. Biochem Soc Trans 39(6):1705–1709. doi:10.1042/BST20110603
2. Prokocimer M, Davidovich M, Nissim-Rafinia M, Wiesel-Motiuk N, Bar DZ, Barkan R, Meshorer E, Gruenbaum Y (2009) Nuclear lamins: key regulators of nuclear structure and activities. J Cell Mol Med 13(6):1059–1085. doi:10.1111/j.1582-4934.2008.00676.x
3. Burke B, Stewart CL (2012) The nuclear lamins: flexibility in function. Nat Rev Mol Cell Biol 14(1):13–24. doi:10.1038/nrm3488
4. Dittmer TA, Misteli T (2011) The lamin protein family. Genome Biol 12(5):222. doi:10.1186/gb-2011-12-5-222
5. Stuurman N, Heins S, Aebi U (1998) Nuclear lamins: their structure, assembly, and interactions. J Struct Biol 122(1–2):42–66. doi:10.1006/jsbi.1998.3987
6. Polychronidou M, Hellwig A, Grosshans J (2010) Farnesylated nuclear proteins Kugelkern and lamin Dm0 affect nuclear morphology by directly interacting with the nuclear membrane. Mol Biol Cell 21(19):3409–3420. doi:10.1091/mbc.E10-03-0230
7. Dessev G, Iovcheva-Dessev C, Bischoff JR, Beach D, Goldman R (1991) A complex containing p34cdc2 and cyclin B phosphorylates the nuclear lamin and disassembles nuclei of clam oocytes in vitro. J Cell Biol 112(4):523–533
8. Rzepecki R, Fisher PA (2002) In vivo phosphorylation of Drosophila melanogaster nuclear lamins during both interphase and mitosis. Cell Mol Biol Lett 7(3):859–876
9. Heald R, McKeon F (1990) Mutations of phosphorylation sites in lamin A that prevent nuclear lamina disassembly in mitosis. Cell 61(4):579–589
10. Ward GE, Kirschner MW (1990) Identification of cell cycle-regulated phosphorylation sites on nuclear lamin C. Cell 61(4):561–577
11. Lee KK, Gruenbaum Y, Spann P, Liu J, Wilson KL (2000) C. elegans nuclear envelope proteins emerin, MAN1, lamin, and nucleoporins reveal unique timing of nuclear envelope breakdown during mitosis. Mol Biol Cell 11(9):3089–3099
12. Stuurman N, Maus N, Fisher PA (1995) Interphase phosphorylation of the Drosophila nuclear lamin: site-mapping using a monoclonal antibody. J Cell Sci 108(Pt 9):3137–3144
13. Schneider U, Mini T, Jeno P, Fisher PA, Stuurman N (1999) Phosphorylation of the major Drosophila lamin in vivo: site identification during both M-phase (meiosis) and interphase by electrospray ionization tandem mass spectrometry. Biochemistry 38(14):4620–4632. doi:10.1021/bi9827060
14. Zaremba-Czogalla M, Piekarowicz K, Wachowicz K, Koziol K, Dubinska-Magiera M, Rzepecki R (2012) The different function of single phosphorylation sites of Drosophila melanogaster lamin Dm and lamin C. PloS One 7(2):e32649. doi:10.1371/journal.pone.0032649
15. Zaremba-Czogalla M, Gagat P, Koziol K, Dubinska-Magiera M, Sikora J, Dadlez M, Rzepecki R (2011) Identification of new in vivo phosphosites on lamin Dm—the evidence of heterogeneity of phosphorylation sites in different Drosophila tissues. Nucleus 2(5):478–488. doi:doi:http://dx.doi.org/10.4161/nucl.2.5.1786410.4161/nucl.2.5.17
16. Mattout A, Goldberg M, Tzur Y, Margalit A, Gruenbaum Y (2007) Specific and conserved sequences in D. melanogaster and C. elegans lamins and histone H2A mediate the attachment of lamins to chromosomes. J Cell Sci 120(Pt 1):77–85. doi:10.1242/jcs.03325

17. Aebi U, Cohn J, Buhle L, Gerace L (1986) The nuclear lamina is a meshwork of intermediate-type filaments. Nature 323(6088):560–564. doi:10.1038/323560a0
18. Heitlinger E, Peter M, Haner M, Lustig A, Aebi U, Nigg EA (1991) Expression of chicken lamin B2 in Escherichia coli: characterization of its structure, assembly, and molecular interactions. J Cell Biol 113(3):485–495
19. Foeger N, Wiesel N, Lotsch D, Mucke N, Kreplak L, Aebi U, Gruenbaum Y, Herrmann H (2006) Solubility properties and specific assembly pathways of the B-type lamin from Caenorhabditis elegans. J Struct Biol 155(2):340–350. doi:10.1016/j.jsb.2006.03.026
20. Karabinos A, Schunemann J, Meyer M, Aebi U, Weber K (2003) The single nuclear lamin of Caenorhabditis elegans forms in vitro stable intermediate filaments and paracrystals with a reduced axial periodicity. J Mol Biol 325(2):241–247
21. Ben-Harush K, Wiesel N, Frenkiel-Krispin D, Moeller D, Soreq E, Aebi U, Herrmann H, Gruenbaum Y, Medalia O (2009) The supramolecular organization of the C. elegans nuclear lamin filament. J Mol Biol 386(5):1392–1402. doi:10.1016/j.jmb.2008.12.024
22. Wiesel N, Mattout A, Melcer S, Melamed-Book N, Herrmann H, Medalia O, Aebi U, Gruenbaum Y (2008) Laminopathic mutations interfere with the assembly, localization, and dynamics of nuclear lamins. Proc Natl Acad Sci U S A 105(1):180–185. doi:10.1073/pnas.0708974105
23. Grossman E, Dahan I, Stick R, Goldberg MW, Gruenbaum Y, Medalia O (2012) Filaments assembly of ectopically expressed Caenorhabditis elegans lamin within Xenopus oocytes. J Struct Biol 177(1):113–118. doi:10.1016/j.jsb.2011.11.002
24. Furukawa K, Ishida K, Tsunoyama TA, Toda S, Osoda S, Horigome T, Fisher PA, Sugiyama S (2009) A-type and B-type lamins initiate layer assembly at distinct areas of the nuclear envelope in living cells. Exp Cell Res 315(7):1181–1189. doi:10.1016/j.yexcr.2008.12.024
25. Bossie CA, Sanders MM (1993) A cDNA from Drosophila melanogaster encodes a lamin C-like intermediate filament protein. J Cell Sci 104(Pt 4):1263–1272
26. Zimek A, Weber K (2011) Flanking genes of an essential gene give information about the evolution of metazoa. Eur J Cell Biol 90(4):356–364. doi:10.1016/j.ejcb.2010.10.005
27. Peter A, Reimer S (2012) Evolution of the lamin protein family: what introns can tell. Nucleus 3(1):44–59
28. Ohno S (1970) Evolution by gene duplication. Springer, Berlin, NY
29. Putnam NH, Butts T, Ferrier DE, Furlong RF, Hellsten U, Kawashima T, Robinson-Rechavi M, Shoguchi E, Terry A, Yu JK, Benito-Gutierrez EL, Dubchak I, Garcia-Fernandez J, Gibson-Brown JJ, Grigoriev IV, Horton AC, de Jong PJ, Jurka J, Kapitonov VV, Kohara Y, Kuroki Y, Lindquist E, Lucas S, Osoegawa K, Pennacchio LA, Salamov AA, Satou Y, Sauka-Spengler T, Schmutz J, Shin IT, Toyoda A, Bronner-Fraser M, Fujiyama A, Holland LZ, Holland PW, Satoh N, Rokhsar DS (2008) The amphioxus genome and the evolution of the chordate karyotype. Nature 453(7198):1064–1071. doi:10.1038/nature06967
30. Erber A, Riemer D, Hofemeister H, Bovenschulte M, Stick R, Panopoulou G, Lehrach H, Weber K (1999) Characterization of the Hydra lamin and its gene: a molecular phylogeny of metazoan lamins. J Mol Evol 49(2):260–271
31. Liu J, Rolef Ben-Shahar T, Riemer D, Treinin M, Spann P, Weber K, Fire A, Gruenbaum Y (2000) Essential roles for Caenorhabditis elegans lamin gene in nuclear organization, cell cycle progression, and spatial organization of nuclear pore complexes. Mol Biol Cell 11(11):3937–3947
32. Cohen M, Tzur YB, Neufeld E, Feinstein N, Delannoy MR, Wilson KL, Gruenbaum Y (2002) Transmission electron microscope studies of the nuclear envelope in Caenorhabditis elegans embryos. J Struct Biol 140(1–3):232–240
33. Margalit A, Liu J, Fridkin A, Wilson KL, Gruenbaum Y (2005) A lamin-dependent pathway that regulates nuclear organization, cell cycle progression and germ cell development. Novartis Found Symp 264:231–240, discussion 240-235
34. Riemer D, Dodemont H, Weber K (1993) A nuclear lamin of the nematode Caenorhabditis elegans with unusual structural features; cDNA cloning and gene organization. Eur J Cell Biol 62(2):214–223

35. Gruenbaum Y, Landesman Y, Drees B, Bare JW, Saumweber H, Paddy MR, Sedat JW, Smith DE, Benton BM, Fisher PA (1988) Drosophila nuclear lamin precursor Dm0 is translated from either of two developmentally regulated mRNA species apparently encoded by a single gene. J Cell Biol 106(3):585–596
36. Guillemin K, Williams T, Krasnow MA (2001) A nuclear lamin is required for cytoplasmic organization and egg polarity in Drosophila. Nat Cell Biol 3(9):848–851. doi:10.1038/ncb0901-848
37. Harel A, Zlotkin E, Nainudel-Epszteyn S, Feinstein N, Fisher PA, Gruenbaum Y (1989) Persistence of major nuclear envelope antigens in an envelope-like structure during mitosis in Drosophila melanogaster embryos. J Cell Sci 94(Pt 3):463–470
38. Riemer D, Weber K (1994) The organization of the gene for Drosophila lamin C: limited homology with vertebrate lamin genes and lack of homology versus the Drosophila lamin Dm0 gene. Eur J Cell Biol 63(2):299–306
39. Chen B, Cai ST, Zhai ZH (1994) Investigation of nuclear lamina in Tetrahymena thermophila. Shi yan sheng wu xue bao 27(2):153–163
40. Minguez A, Franca S, Diaz M, de la Espina S (1994) Dinoflagellates have a eukaryotic nuclear matrix with lamin-like proteins and topoisomerase II. J Cell Sci 107(Pt 10):2861–2873
41. Melcer S, Gruenbaum Y, Krohne G (2007) Invertebrate lamins. Exp Cell Res 313(10): 2157–2166. doi:10.1016/j.yexcr.2007.03.004
42. Pappas GD (1956) The fine structure of the nuclear envelope of Amoeba proteus. J Biophys Biochem Cytol 2(4 Suppl):431–434
43. Schmidt M, Grossmann U, Krohne G (1995) The nuclear membrane-associated honeycomb structure of the unicellular organism Amoeba proteus: on the search for homologies with the nuclear lamina of metazoa. Eur J Cell Biol 67(3):199–208
44. Kruger A, Batsios P, Baumann O, Luckert E, Schwarz H, Stick R, Meyer I, Graf R (2012) Characterization of NE81, the first lamin-like nucleoskeleton protein in a unicellular organism. Mol Biol Cell 23(2):360–370. doi:10.1091/mbc.E11-07-0595
45. DuBois KN, Alsford S, Holden JM, Buisson J, Swiderski M, Bart JM, Ratushny AV, Wan Y, Bastin P, Barry JD, Navarro M, Horn D, Aitchison JD, Rout MP, Field MC (2012) NUP-1 Is a large coiled-coil nucleoskeletal protein in trypanosomes with lamin-like functions. PLoS Biol 10(3):e1001287. doi:10.1371/journal.pbio.1001287
46. Masuda K, Xu ZJ, Takahashi S, Ito A, Ono M, Nomura K, Inoue M (1997) Peripheral framework of carrot cell nucleus contains a novel protein predicted to exhibit a long alpha-helical domain. Exp Cell Res 232(1):173–181. doi:10.1006/excr.1997.3531
47. Dittmer TA, Stacey NJ, Sugimoto-Shirasu K, Richards EJ (2007) LITTLE NUCLEI genes affecting nuclear morphology in Arabidopsis thaliana. Plant Cell 19(9):2793–2803. doi:10.1105/tpc.107.053231
48. Dittmer TA, Richards EJ (2008) Role of LINC proteins in plant nuclear morphology. Plant Signal Behav 3(7):485–487
49. Gruenbaum Y, Lee KK, Liu J, Cohen M, Wilson KL (2002) The expression, lamin-dependent localization and RNAi depletion phenotype for emerin in C. elegans. J Cell Sci 115(Pt 5): 923–929
50. Liu J, Lee KK, Segura-Totten M, Neufeld E, Wilson KL, Gruenbaum Y (2003) MAN1 and emerin have overlapping function(s) essential for chromosome segregation and cell division in Caenorhabditis elegans. Proc Natl Acad Sci U S A 100(8):4598–4603. doi:10.1073/pnas.0730821100
51. Margalit A, Neufeld E, Feinstein N, Wilson KL, Podbilewicz B, Gruenbaum Y (2007) Barrier to autointegration factor blocks premature cell fusion and maintains adult muscle integrity in C. elegans. J Cell Biol 178(4):661–673. doi:10.1083/jcb.200704049
52. Lee KK, Starr D, Cohen M, Liu J, Han M, Wilson KL, Gruenbaum Y (2002) Lamin-dependent localization of UNC-84, a protein required for nuclear migration in Caenorhabditis elegans. Mol Biol Cell 13(3):892–901. doi:10.1091/mbc.01-06-0294
53. Goldberg M, Harel A, Brandeis M, Rechsteiner T, Richmond TJ, Weiss AM, Gruenbaum Y (1999) The tail domain of lamin Dm0 binds histones H2A and H2B. Proc Natl Acad Sci U S A 96(6):2852–2857

54. Glass CA, Glass JR, Taniura H, Hasel KW, Blevitt JM, Gerace L (1993) The alpha-helical rod domain of human lamins A and C contains a chromatin binding site. EMBO J 12(11):4413–4424
55. Burke B (1990) On the cell-free association of lamins A and C with metaphase chromosomes. Exp Cell Res 186(1):169–176
56. Maraldi NM, Lattanzi G, Cenni V, Bavelloni A, Marmiroli S, Manzoli FA (2010) Laminopathies and A-type lamin-associated signalling pathways. Adv enzyme Regul 50(1):248–261. doi:10.1016/j.advenzreg.2009.10.019
57. Wilson KL, Foisner R (2010) Lamin-binding proteins. Cold Spring Harbor Perspect Biol 2(4):a000554. doi:10.1101/cshperspect.a000554
58. Reddy KL, Zullo JM, Bertolino E, Singh H (2008) Transcriptional repression mediated by repositioning of genes to the nuclear lamina. Nature 452(7184):243–247. doi:10.1038/nature06727
59. Finlan LE, Sproul D, Thomson I, Boyle S, Kerr E, Perry P, Ylstra B, Chubb JR, Bickmore WA (2008) Recruitment to the nuclear periphery can alter expression of genes in human cells. PLoS Genet 4(3):e1000039. doi:10.1371/journal.pgen.1000039
60. Kumaran RI, Spector DL (2008) A genetic locus targeted to the nuclear periphery in living cells maintains its transcriptional competence. J Cell Biol 180(1):51–65. doi:10.1083/jcb.200706060
61. Meister P, Towbin BD, Pike BL, Ponti A, Gasser SM (2010) The spatial dynamics of tissue-specific promoters during C. elegans development. Genes Dev 24(8):766–782. doi:10.1101/gad.559610
62. Mattout A, Pike BL, Towbin BD, Bank EM, Gonzalez-Sandoval A, Stadler MB, Meister P, Gruenbaum Y, Gasser SM (2011) An EDMD mutation in C. elegans lamin blocks muscle-specific gene relocation and compromises muscle integrity. Curr Biol:CB 21(19):1603–1614. doi:10.1016/j.cub.2011.08.030
63. Shevelyov YY, Lavrov SA, Mikhaylova LM, Nurminsky ID, Kulathinal RJ, Egorova KS, Rozovsky YM, Nurminsky DI (2009) The B-type lamin is required for somatic repression of testis-specific gene clusters. Proc Natl Acad Sci U S A 106(9):3282–3287. doi:10.1073/pnas.0811933106
64. Starr DA (2007) Communication between the cytoskeleton and the nuclear envelope to position the nucleus. Mol bioSyst 3(9):583–589. doi:10.1039/b703878j
65. Burke B, Roux KJ (2009) Nuclei take a position: managing nuclear location. Dev Cell 17(5):587–597. doi:10.1016/j.devcel.2009.10.018
66. Patterson K, Molofsky AB, Robinson C, Acosta S, Cater C, Fischer JA (2004) The functions of Klarsicht and nuclear lamin in developmentally regulated nuclear migrations of photoreceptor cells in the Drosophila eye. Mol Biol Cell 15(2):600–610. doi:10.1091/mbc.E03-06-0374
67. Zastrow MS, Flaherty DB, Benian GM, Wilson KL (2006) Nuclear titin interacts with A- and B-type lamins in vitro and in vivo. J Cell Sci 119(Pt 2):239–249. doi:10.1242/jcs.02728
68. Somsel Rodman J, Wandinger-Ness A (2000) Rab GTPases coordinate endocytosis. J Cell Sci 113(Pt 2):183–192
69. Audhya A, Desai A, Oegema K (2007) A role for Rab5 in structuring the endoplasmic reticulum. J Cell Biol 178(1):43–56. doi:10.1083/jcb.200701139
70. Capalbo L, D'Avino PP, Archambault V, Glover DM (2011) Rab5 GTPase controls chromosome alignment through Lamin disassembly and relocation of the NuMA-like protein Mud to the poles during mitosis. Proc Natl Acad Sci U S A 108(42):17343–17348. doi:10.1073/pnas.1103720108
71. Liu B, Zhou Z (2008) Lamin A/C, laminopathies and premature ageing. Histol Histopathol 23(6):747–763
72. Burke B, Stewart CL (2002) Life at the edge: the nuclear envelope and human disease. Nat Rev Mol Cell Biol 3(8):575–585. doi:10.1038/nrm879
73. Fridkin A, Mills E, Margalit A, Neufeld E, Lee KK, Feinstein N, Cohen M, Wilson KL, Gruenbaum Y (2004) Matefin, a Caenorhabditis elegans germ line-specific SUN-domain nuclear membrane protein, is essential for early embryonic and germ cell development. Proc Natl Acad Sci U S A 101(18):6987–6992. doi:10.1073/pnas.0307880101

74. Tzur YB, Wilson KL, Gruenbaum Y (2006) SUN-domain proteins: 'Velcro' that links the nucleoskeleton to the cytoskeleton. Nat Rev Mol Cell Biol 7(10):782–788. doi:10.1038/nrm2003
75. Haithcock E, Dayani Y, Neufeld E, Zahand AJ, Feinstein N, Mattout A, Gruenbaum Y, Liu J (2005) Age-related changes of nuclear architecture in Caenorhabditis elegans. Proc Natl Acad Sci U S A 102(46):16690–16695. doi:10.1073/pnas.0506955102
76. Bar DZ, Neufeld E, Feinstein N, Gruenbaum Y (2009) Gliotoxin reverses age-dependent nuclear morphology phenotypes, ameliorates motility, but fails to affect lifespan of adult Caenorhabditis elegans. Cell Motil Cytoskel 66(10):791–797. doi:10.1002/cm.20347
77. Bonne G, Di Barletta MR, Varnous S, Becane HM, Hammouda EH, Merlini L, Muntoni F, Greenberg CR, Gary F, Urtizberea JA, Duboc D, Fardeau M, Toniolo D, Schwartz K (1999) Mutations in the gene encoding lamin A/C cause autosomal dominant Emery-Dreifuss muscular dystrophy. Nat Genet 21(3):285–288. doi:10.1038/6799
78. Bank EM, Ben-Harush K, Feinstein N, Medalia O, Gruenbaum Y (2012) Structural and physiological phenotypes of disease-linked lamin mutations in C. elegans. J Struct Biol 177(1): 106–112. doi:10.1016/j.jsb.2011.10.009
79. Bank EM, Ben-Harush K, Wiesel-Motiuk N, Barkan R, Feinstein N, Lotan O, Medalia O, Gruenbaum Y (2011) A laminopathic mutation disrupting lamin filament assembly causes disease-like phenotypes in Caenorhabditis elegans. Mol Biol Cell 22(15):2716–2728. doi:10.1091/mbc.E11-01-0064
80. Lenz-Bohme B, Wismar J, Fuchs S, Reifegerste R, Buchner E, Betz H, Schmitt B (1997) Insertional mutation of the Drosophila nuclear lamin Dm0 gene results in defective nuclear envelopes, clustering of nuclear pore complexes, and accumulation of annulate lamellae. J Cell Biol 137(5):1001–1016
81. Harel A, Goldberg M, Ulitzur N, Gruenbaum Y (1998) Structural organization and biological roles of the nuclear lamina. In: Boulikas T (ed) Textbook of gene therapy and molecular biology: from basic mechanism to clinical applications, vol 1. Gene Therapy Press, Palo Alto, CA
82. Osouda S, Nakamura Y, de Saint PB, McConnell M, Horigome T, Sugiyama S, Fisher PA, Furukawa K (2005) Null mutants of Drosophila B-type lamin Dm(0) show aberrant tissue differentiation rather than obvious nuclear shape distortion or specific defects during cell proliferation. Dev Biol 284(1):219–232. doi:10.1016/j.ydbio.2005.05.022
83. Uchino R, Nonaka YK, Horigome T, Sugiyama S, Furukawa K (2013) Loss of Drosophila A-type lamin C initially causes tendon abnormality including disintegration of cytoskeleton and nuclear lamina in muscular defects. Dev Biol 373(1):216–227. doi:10.1016/j.ydbio.2012.08.001
84. Bao X, Girton J, Johansen J, Johansen KM (2007) The lamin Dm0 allele Ari3 acts as an enhancer of position effect variegation of the wm4 allele in Drosophila. Genetica 129(3): 339–342. doi:10.1007/s10709-006-0012-7
85. Muller H (1930) Types of Visible Variations Induced by X-rays in Drosophila. J Genet 22(3):299–334
86. Dialynas G, Flannery KM, Zirbel LN, Nagy PL, Mathews KD, Moore SA, Wallrath LL (2012) LMNA variants cause cytoplasmic distribution of nuclear pore proteins in Drosophila and human muscle. Human Mol Genet 21(7):1544–1556. doi:10.1093/hmg/ddr592
87. Schulze SR, Curio-Penny B, Speese S, Dialynas G, Cryderman DE, McDonough CW, Nalbant D, Petersen M, Budnik V, Geyer PK, Wallrath LL (2009) A comparative study of Drosophila and human A-type lamins. PloS One 4(10):e7564. doi:10.1371/journal.pone.0007564
88. Dialynas G, Speese S, Budnik V, Geyer PK, Wallrath LL (2010) The role of Drosophila Lamin C in muscle function and gene expression. Development 137(18):3067–3077. doi:10.1242/dev.048231
89. Beard GS, Bridger JM, Kill IR, Tree DR (2008) Towards a Drosophila model of Hutchinson-Gilford progeria syndrome. Biochem Soc Trans 36(Pt 6):1389–1392. doi:10.1042/BST0361389
90. Tzur YB, Margalit A, Melamed-Book N, Gruenbaum Y (2006) Matefin/SUN-1 is a nuclear envelope receptor for CED-4 during Caenorhabditis elegans apoptosis. Proc Natl Acad Sci U S A 103(36):13397–13402. doi:10.1073/pnas.0604224103
91. Wagner N, Schmitt J, Krohne G (2004) Two novel LEM-domain proteins are splice products of the annotated Drosophila melanogaster gene CG9424 (Bocksbeutel). Eur J Cell Biol 82(12):605–616. doi:10.1078/0171-9335-00350

92. Goldberg M, Lu H, Stuurman N, Ashery-Padan R, Weiss AM, Yu J, Bhattacharyya D, Fisher PA, Gruenbaum Y, Wolfner MF (1998) Interactions among Drosophila nuclear envelope proteins lamin, otefin, and YA. Mol Cell Biol 18(7):4315–4323
93. Wagner N, Weber D, Seitz S, Krohne G (2004) The lamin B receptor of Drosophila melanogaster. J Cell Sci 117(Pt 10):2015–2028. doi:10.1242/jcs.01052
94. Bao X, Zhang W, Krencik R, Deng H, Wang Y, Girton J, Johansen J, Johansen KM (2005) The JIL-1 kinase interacts with lamin Dm0 and regulates nuclear lamina morphology of Drosophila nurse cells. J Cell Sci 118(Pt 21):5079–5087. doi:10.1242/jcs.02611
95. Tamura K, Peterson D, Peterson N, Stecher G, Nei M, Kumar S (2011) MEGA5: molecular evolutionary genetics analysis using maximum likelihood, evolutionary distance, and maximum parsimony methods. Mol Biol Evol 28(10):2731–2739. doi:10.1093/molbev/msr121
96. Rzhetsky A, Nei M (1992) A simple method for estimating and testing minimum evolution trees. Mol Biol Evol 9:945–967
97. Zuckercandl E, Pauling L (1965) Evolutionary divergence and convergence in proteins. In: Bryson V, Vogel HJ (eds) Evolving genes and proteins. Academic, New York, pp 97–166
98. Nei M, Kumar S (2000) Molecular evolution and phylogenetics. Oxford University Press, New York
99. Saitou N, Nei M (1987) The neighbor-joining method: a new method for reconstructing phylogenetic trees. Mol Biol Evol 4(4):406–425

The Non-random Repositioning of Whole Chromosomes and Individual Gene Loci in Interphase Nuclei and Its Relevance in Disease, Infection, Aging, and Cancer

Joanna M. Bridger, Halime D. Arican-Gotkas, Helen A. Foster, Lauren S. Godwin, Amanda Harvey, Ian R. Kill, Matty Knight, Ishita S. Mehta, and Mai Hassan Ahmed

Abstract The genomes of a wide range of different organisms are non-randomly organized within interphase nuclei. Chromosomes and genes can be moved rapidly, with direction, to new non-random locations within nuclei upon a stimulus such as a signal to initiate differentiation, quiescence or senescence, or also the application

---

J.M. Bridger (✉)
Laboratory of Nuclear and Genomic Health, Centre for Cell and Chromosome Biology, Biosciences, Brunel University, Kingston Lane, Uxbridge UB8 3PH, UK
e-mail: joanna.bridger@brunel.ac.uk

H.D. Arican-Gotkas • M.H. Ahmed
Laboratory of Nuclear and Genomic Health, Centre for Cell and Chromosome Biology, Biosciences, Brunel University, Kingston Lane, Uxbridge UB8 3PH, UK
e-mail: Halime.Arican@brunel.ac.uk; maieljeboury@hotmail.com

H.A. Foster • I.R. Kill
Biosciences, Brunel University, Kingston Lane, Uxbridge UB8 3PH, UK
e-mail: Helen.foster@brunel.ac.uk; Ian.kill@brunel.ac.uk

L.S. Godwin
Molecular Cell Biology Group, Biomedical Sciences Research Centre, Jenner Wing (box J2A), St. George's, University of London, Cranmer Terrace, London SW17 0RE, UK
e-mail: laurengodwin08@googlemail.com

A. Harvey
Brunel Institute for Cancer Genetics and Pharmacogenomics, Biosciences, Brunel University, Kingston Lane, Uxbridge UB8 3PH, UK
e-mail: Amanda.harvey@brunel.ac.uk

M. Knight
Department of Microbiology, Immunology & Tropical Medicine, School of Medicine & Health Sciences, The George Washington University, Ross Hall, 448, 2300 I (Eye) Street, NW, Washington, DC 20037, USA

I.S. Mehta
Tata Institute of Fundamental Research, Homi Bhabha Road, Mumbai 400005, India
e-mail: mehtaishi@gmail.com

E.C. Schirmer and J.I. de las Heras (eds.), *Cancer Biology and the Nuclear Envelope*, Advances in Experimental Medicine and Biology 773, DOI 10.1007/978-1-4899-8032-8_12,

of heat or an infection with a pathogen. It is now becoming increasingly obvious that chromosome and gene position can be altered in diseases such as cancer and other syndromes that are affected by changes to nuclear architecture such as the laminopathies. This repositioning seems to affect gene expression in these cells and may play a role in progression of the disease. We have some evidence in breast cancer cells and in the premature aging disease Hutchinson–Gilford Progeria that an aberrant nuclear envelope may lead to genome repositioning and correction of these nuclear envelope defects can restore proper gene positioning and expression in both disease situations.

Although spatial positioning of the genome probably does not entirely control expression of genes, it appears that spatio-epigenetics may enhance the control over gene expression globally and/or is deeply involved in regulating specific sets of genes. A deviation from normal spatial positioning of the genome for a particular cell type could lead to changes that affect the future health of the cell or even an individual.

Keywords Chromosome positioning • Gene positioning • Gene expression • Nuclear envelope • Nuclear lamins

Abbreviations

2D	Two-dimensional
3D	Three-dimensional
CML	Chronic myeloid leukemia
FISH	Fluorescence in situ hybridization
GFP	Green fluorescent protein
HGPS	Hutchinson–Gilford progeria syndrome

Introduction

The development of the technique of fluorescence in situ hybridization (FISH) and suitable probes to reveal whole chromosomes and individual genes for diagnostic purposes on mitotic chromosomes concomitantly allowed interphase nuclei to be analyzed by scientists interested in how the nucleus behaved functionally. The painting of whole chromosomes and individual gene loci led to the recognition that chromosomes and genes sit in individual locations within interphase nuclei. Indeed, chromosomes are found within their own nuclear territories and gene loci housed upon those chromosomes often sit at the edges of those chromosome territories [1], but can also be nearer the interior of the territories or at a distance away from the core individual chromosome territories, distended on chromatin loops (see Fig. 1) [2].

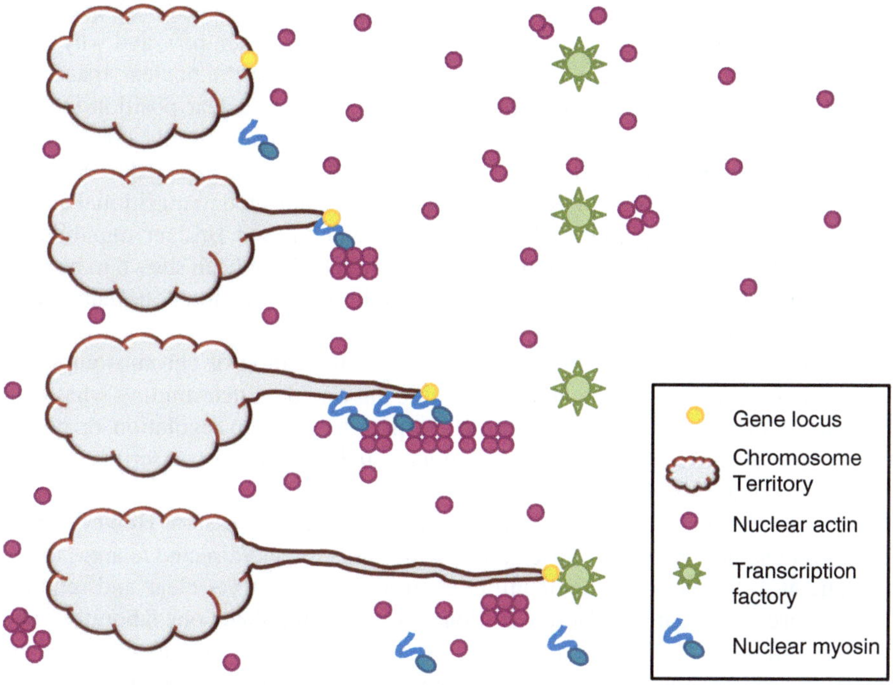

Fig. 1 The translocation of genes to transcription factories via nuclear motor activity. This cartoon shows how genes may be relocated to transcription factories at some distance from the main body of the chromosome territory that houses the gene. The nuclear myosin moves along actin filaments that polymerize where they are needed. There must be a signal from the chromatin to be moved, and this must unravel due to changes in chromatin modification, i.e., the histone code. If this process does not proceed correctly, then genome stability may be affected, genes may be over-expressed or under-expressed or become translocated with other chromosomes

The ability of FISH to reveal whole chromosomes and genes soon led to mapping endeavors whereby it was discovered that chromosomes reside in non-random radial locations within interphase nuclei with more gene-poor chromosomes such as 4, 13, 18, and X found at the nuclear periphery, whereas gene-rich chromosomes such as 17 and 19 were found towards the nuclear interior [3, 4]. It should be noted that this correlation with gene density was found in proliferating lymphoblasts and young proliferating primary fibroblasts [5, 6].

A non-random distribution of the genome that is maintained throughout interphase with all the dynamic processes that occur during this time, e.g., replication and transcription, must require energy and significant anchorage points that are dynamic in response to external stimuli. Indeed, when one looks further into cells that are no longer young and proliferating, diseased or subjected to an external stimulus, specific chromosomes and genes change nuclear location. The Bridger laboratory has put tremendous effort into finding situations where specific chromosomes and gene loci change location. This is so that we can ask questions about how and why the genome is spatially organized and then how and why individual genes and chromosomes become reorganized within the nuclear space. We have found that specific chromosomes and genes change nuclear position in aged senescent cells [7] in laminopathy patient cells [8, 9], cancer cells [10; Hassan Ahmed, Harvey, Karteris, and Bridger unpublished data], cells exposed to parasites [11; Arican, Bridger, Knight unpublished data], cells subjected to nutritional alterations [6, 12], and temperature change [Arican, Knight, and Bridger unpublished data]. All the changes in position that we have revealed have been shown to be non-random and even in some experiments reversible when the situation/treatment is removed/reversed.

The reason why the cell invests energy in the relocation of chromosomes and genes to new positions in the nucleus is being answered by determining what happens to them at their new location with respect to gene up-regulation or down-regulation. This is either done by techniques such as reverse transcriptase-PCR, quantitative real-time-PCR, microarray analysis, RNA FISH or by ChIP-seq and in many cases the repositioning correlates with changes in expression. However, how the chromosomes and genes move and why they are targeted/directed to areas of the nucleus at a distance from their initial environment is not yet clear and requires much more investigation. These questions are what stimulates our laboratory and we use a number of different situations, external stimuli and organisms to ask the questions where, how, and why are chromosomes and genes relocated. Here we describe several different experimental systems where such changes in spatial genome organization have been observed, ending with similar types of changes that we and others have observed in cancer cells that may be able to be taken advantage of for new therapies.

How We Map Genes and Chromosomes in Interphase Nuclei

Our laboratory has mapped many chromosomes and genes in many different cell types and organisms but we always use the same two ways of mapping for all situations for consistency and reproducibility.

We always employ both two-dimensional (2D) mapping that allows us to do lots of mapping relatively fast and three-dimensional mapping that takes longer but is important to confirm the 2D data. For the mapping to work, it is critical that the 2D sample is properly flattened, since we normalize and extrapolate out to three-dimensional (3D) with our findings and it is critical that the 3D sample has not undergone any structural changes since we take precise size and distance measurements from these samples.

For 2D samples imaging is performed, capturing 50 images of each chromosome/gene in each cell type. These images are run through a bespoke script that was devised by Dr Paul Perry in Prof Wendy Bickmore's group in the MRC Human Genetics Unit in Edinburgh [5]. This script outlines the entire nucleus based on a

DNA dye (such as DAPI) and erodes this mask, creating five shells of equal area. Within these five shells the intensity of the fluorescent signal from the DNA dye and the FISH probe is measured and recorded for each nucleus. In order to normalize for more DNA being in the interior shells when a spherical object is flattened, the probe signal is divided by the DNA signal. The data are plotted as a bar chart. This method does under record the signals that are at the periphery since they may appear interior if on the top or bottom of the nuclei but interiorly located signals always appear interior and since it is always used in a comparative way with other chromosomes, other cell types etc., it works exceptionally well as a method for mapping chromosomes and genes.

The 3D FISH method is based on one developed by Profs. Lichter and Cremer in Heidelberg to preserve the three-dimensionality of a nucleus while still allowing good penetration of the FISH probe [13]. We then use a confocal laser scanning microscope to collect optical sections and then the position of a chromosome or gene is measured in these images from the geometric center of its signal to the nearest nuclear edge, whether that be in the *x*, *y* or *z* axis. The results can be normalized to a measurement for the size of the nucleus but this does not often change the final outcome. The data are plotted as a frequency distribution. This method gives accurate measurements and we find that there are virtually no differences in general position when compared to the 2D method.

Live cell imaging for chromosome and gene movement is something we are presently working on. It is made complicated because we are often working in primary cells where transfection and selection of clones would make it impossible to collect proliferating cells at the end of the selection, i.e., they would have become senescent through the number of passages it would need to collect a colony of cells from a single cell. We also need to know what genes and chromosomes we are assessing—this is imperative since some chromosomes do not move at all and some move considerably. This can however be done using the GFP-lac repressor system [14, 15] stably transfected, but it is important when such sequences are added into a chromosome that they do not change its behavior, which could happen if the large number of repeats created a region of heterochromatin within a chromosome.

Alterations to Gene and Chromosome Position Using Growth Factor Addition and Removal

Addition of specific growth factors to cells in culture can induce cellular differentiation and removal of growth factors induces quiescence, a period of reversible growth arrest in cells. Both of these situations are controllable windows in which the cells have dramatic changes in their gene expression profiles. Thus, we have developed systems that can be controlled easily by the addition or removal of growth factors that have allowed us to analyze changes to genome organization in nuclei.

Porcine mesenchymal stem cells were isolated from fresh pig bone marrow and grown until there were copious numbers of cells and the culture was purely mesenchymal stem cells [16]. By adding human adipogenic growth factors to the medium the pig stem cells differentiated into adipocytes over a 2-week period, giving committed pre-adipocytes at 7 days. We were interested in what would happen from a spatial organizational perspective to genes involved in the adipogenic process during this in vitro differentiation. We studied seven genes involved in the adipogenesis pathway and six of them had moved to a more interior location after 14 days of treatment with growth factors, which was correlated with an up-regulation in gene expression in all these genes. The seventh gene was *GATA2,* a gene involved in pre-adipogenesis and this gene like the others was more peripheral at day 0 and then was found to be more interiorly located on day 7, but it had moved back to its original location towards the nuclear periphery by day 14. The movement of this gene to the interior also correlated with its up-regulation in expression at day 7 and its down-regulation by day 14 [17]. These were quite broad time points and we cannot determine from these experiments how fast genes respond after a stimulus. Other experiments in other experimental systems (see below) address this better.

During the induced adipogenesis of the mesenchymal stem cells the nuclear lamina was altered and the longer in adipogenesis induction medium the more cells became negative for A-type lamins, with the majority of cells being negative by day 11 (Foster and Bridger unpublished data). This is a major change in nuclear architecture and may be involved in allowing various regions of the genome to be freer to move into the interior of the nuclei, but as yet there is no direct evidence.

A follow-on study allowed us to ask the question—"where are these genes going?". In agreement with some other studies [18] we found that the gene loci were co-localized in significantly high numbers with SC35 splicing speckles [19]. Others have also found genes moving to transcription factories [20]; however, it is possible that these transcription factories were very close to splicing speckles and this is why both structures have been reported as a gene's destination. By analyzing three genes concomitantly that were each from different areas of the genome, it became clear that all six loci (in diploid cells) were found in the same splicing speckle in an individual nucleus much more often than could be considered a random occurrence. These data add to building body of evidence that genes from the same pathway may be transcribed together at common transcription factories or other nuclear structures [21]. This implies that some genes may have to travel large distances across the nucleus, avoiding the transcriptional structures in their locale and be directed to a specified nuclear location. We know from our studies inducing adipogenesis in porcine mesenchymal stem cells that whole chromosomes do not tend to move but we have seen genes loop out away from the core chromosomal territories on peninsulas that reach into the nucleoplasm.

In another series of experiments we reduced growth factors by placing cells into low serum. This makes proliferating primary cells, and some immortalized cells, enter a state called quiescence, a reversible growth arrest. As with differentiation this comes with a lot of gene expression changes and so makes an interesting inducible biological system in which to study changes to genome organization through the spatial positioning of chromosomes and genes. We have also been able to use this system to measure very precisely when the genome first responds to an external stimulus. We knew from proliferation marker staining that cells are not thought to be quiescent for at least 3 days after the removal of serum. We also knew that at 7 days after serum removal some whole chromosomes have a different nuclear location, such as chromosomes 13, 18 and 10 [6, 12]. A number of other chromosomes do not change their nuclear location, although some individual genes could, as in the adipogenesis system, still change their location through looping out. Most interestingly, when we investigated the specific timing of the chromosome shifts we found the response to the removal of serum to be much more rapid than 7 days or even 3 days. Indeed, after starting our time course at 7 days post-serum removal and working backwards, we found that whole chromosomes had become relocated within just 15 min after serum removal. This implies a directed repositioning requiring energy. In subsequent experiments where ATP and GTP were inhibited, the chromosomes would no longer relocate after serum removal. These data then begged the question what structures/entities that require energy could move chromosomes so rapidly? We followed a controversial line of thought that was based upon a small amount of evidence that in the nucleus there existed both actin and myosin isoforms that could work in concert to create a nuclear motor capable of moving chromatin around the nucleus [22–24]. Using immunofluorescence, some nuclear myosin 1β was observed in nuclei throughout the nucleoplasm, at the nuclear periphery and at nucleoli [12]. This distribution changed dramatically when serum was removed from the cells. The myosin 1β became located only in aggregates within the nucleoplasm. Using chemical inhibitors of both nuclear actin and myosin we also blocked the movement of the chromosomes upon serum withdrawal. Nuclear myosin 1β was also found to be a major player in chromosome relocation when we used short interference RNA protocols to remove it in >95 % of the cells. This study provides strong evidence to support that certain specific chromosomes are moved within the nucleus to new non-random locations by a nuclear motor (see Fig. 1) [23, 24].

When quiescence is induced in young proliferating primary fibroblasts chromosome 10 moves from an intermediate location to a peripheral location. If the hypothesis is absolute that the nuclear interior is for gene expression and the nuclear periphery is a region for gene silencing and down-regulation, then the movement of a whole chromosome should simultaneously alter the expression of many genes. We found that out of 10 genes on chromosome 10 only two were significantly down-regulated, whereas five were up-regulated when the chromosome moved to the periphery. Although this type of question requires global analysis, this small gene set already indicates that the nuclear periphery is not purely about gene silencing and down-regulation and the effects of repositioning depend on either individual characteristics or the local environment of specific genes.

Although many of the genes found associated with the nuclear lamina at the nuclear periphery are silenced or down-regulated, active genes can be moved towards the nuclear periphery on a chromosome and remain up-regulated [25]. A lot of interest is being focused on genes that tether to nuclear pore complex proteins as an area of the envelope that is associated with active genes [26]. This further shows that spatial positioning within nuclei is involved at some level with the regulation of gene expression, but it is much more complicated than a gene just being at the edge of the nucleus or the interior.

Alterations to Gene and Chromosome Position in Aging and Premature Aging

As chromosome and gene position was found to change specifically and reproducibly in primary human dermal fibroblasts as they become reversibly arrested, it was pertinent to also assess what happens to genome organization in human dermal fibroblasts that become silenced at the end of their replicative lives and become senescent. Cellular senescence does not mean cells die through apoptosis or necrosis. They sit within our tissues and probably send out signals that affect others cells around them in a negative capacity. We have mapped all chromosomes in senescent cells and have found for the most part their nuclear locations resemble those in quiescent human dermal fibroblasts [7]. For example, chromosome X remains at the nuclear periphery, and chromosomes 13 and 18 move to the nuclear interior, becoming associated (or more intensely associated) with the nucleolus. We also have some data that shows that chromosome 18 becomes embedded within and tightly attached to nucleoli in senescent human dermal fibroblasts (Bridger, unpublished data).

Most interestingly, there were two chromosomes that were found to be in different compartments in senescent cells when compared to quiescent human dermal fibroblasts. These were chromosome 15 and chromosome 10. Chromosome 10 was the most dramatic with it being at the nuclear periphery in quiescent cells and deeply in the nuclear interior in senescent cells—in fact, radially, at two opposing locations within the nuclei. This we postulate is so that different levels of control can be maintained over the chromosomes. When we looked at the same ten genes as we did for the quiescent human dermal fibroblasts, six of them were down-regulated, two did not change expression and two were up-regulated. Although this is again a small number of genes, it does show that there are measureable differences between the two arrested states. Quiescence and senescence are thus maybe not as similar as people believe and it is possible that the changes in gene expression are controlled by other means rather than just location. Loss of lamin B1 has been implicated in controlling genome behavior in senescence, i.e., allowing the creation of senescence associated heterochromatic foci and the relocation of genomic regions away from the nuclear periphery [27]. This also shows us that gene down-regulation does happen deep within the nucleus. This maybe a different type of silencing than is seen for genes at the periphery and this needs to be further investigated since any event that prevents proper silencing of genes at senescence could lead to re-expression and transformation of normal cells to cancer cells.

The chromosome and gene mappers are frequently asked how relevant is looking at chromosome positioning in tissue culture cells compared to real situations. When we map chromosomes in sections of tissue preserved for their three-dimensionality we find similar locations for chromosomes to the in vitro observations—this is has been seen both in the pig for a number of tissues from different cell sources in the pig [16] and in human skin (Mehta, Kill, Bridger, unpublished data). Indeed, in skin we have found chromosome 10 in three different locations depending on cell state—in proliferating nuclei at an intermediate location as it is in tissue culture cells and in non-proliferating cells at two opposing locations, at the nuclear periphery and in the nuclear interior, as was also seen in vitro. The cellular senescence field has been searching for a long time for a suitable biomarker that can properly differentiate between senescent and quiescent cells in vivo and perhaps the nuclear location of chromosome 10 could be this biomarker if exploited in the right way.

Chromosome Repositioning in Patient Cells with Mutations in the Lamin A Gene

We believe that the nuclear envelope and the proteins found there are responsible for organizing the genome in interphase nuclei: thus, we postulate that chromosomes and genes might reposition in cells from patients where proteins of the nuclear envelope are affected. We started with the nuclear lamina and the A-type lamins and used primary fibroblasts from a group of patients that have a laminopathy. These patients had different mutations along the *LMNA* gene, which encodes for nuclear lamins A and C. These laminopathies ranged from muscular dystrophies such as autosomal and X-linked Emery–Dreifuss muscular dystrophies to lipodystrophies such as Dunnigan's partial familial lipodystrophy to premature aging syndromes such as Mandibuloacral dysplasia and Hutchinson–Gilford Progeria Syndrome (HGPS). What we were expecting to do was to map the genome organizing regions of lamin A by mapping chromosome location in the different disease cells that were proliferating. Interestingly, we found that all the diseases had a completely altered chromosome location from the wild-type lamin A that was nonetheless similar between the cells from the different diseases [8]. Initially, with the chromosomes we were assessing this reorganization resembled the chromosome distribution of all non-proliferating cells, which would fit the premature aging aspect of some mutations in lamin A since chromosomes 13 and 18 were found in the nuclear interior. This was also seen by another group that also saw changes in gene expression associated with the chromosome repositioning [28]. It was not until we assessed chromosome 10 that we found, particularly in the proliferating HGPS cells, that it was not a senescent type pattern of chromosome positioning. Instead, it was a quiescent one, since territories of chromosome 10 were found at the nuclear periphery in the HGPS cells [9]. The effect that *LMNA* mutations have on these cells seems to uncouple the control over the chromosomes position in proliferating cells and allows them to take a resting state. This links lamin A to genome organization as has been shown recently by Solovei and colleagues [29] and others [30]. In HGPS cells chromosome position can be rescued by treating cells with a farnesyl transferase inhibitor that does not permit mutant lamin A with a uncleavable farnesyl group to be produced [9]. Thus, the toxic lamin A that retains a farnesyl group and associates with the nuclear membrane affects genome positioning with cells appearing as in a quiescent-like state when they display proliferating markers.

A genome-wide study of the sequences associated with the mutant toxic lamin A at the nuclear periphery in mouse confirms that A-type lamins are involved in chromatin and genome organization in nuclei, such that some genes have changed their location and are away from the nuclear periphery. On the other hand some genes have an enhanced association with the nuclear envelope in the mouse Progeria model [31]. In HGPS patient cells we have also shown that fewer telomeres are bound to the nuclear architecture which will inevitably affect genome stability and regulation (Godwin and Bridger unpublished).

It is interesting that nuclear myosin 1β, the myosin we believe is involved in moving some chromosomes around nuclei, is distributed quite differently in non-proliferating cells and in HGPS cells—as large aggregates in the nucleoplasm and absent at the nuclear periphery and nucleoli. Thus, we would predict that chromosomes and maybe even genes are not transposable around the nucleus in a resting state or a diseased state such as HGPS. This hypothesis is yet to be tested in normal cells, but we do know that chromosomes are not repositioned in restimulated quiescent human dermal fibroblasts until the cells have been through mitosis [6, 12], which takes more than 24 h from when the serum is readded. These timescales are similar to those seen for chromosome movement when specific nuclear envelope transmembrane proteins (NETs) as opposed to lamins are removed [32].

Alterations to Gene Position in Cells Exposed to a Pathogen

Pathogens are known to use their hosts for their own benefit and this may go as far as manipulating the hosts' genomes to alter host gene expression. We have been studying the effects on genes within the secondary host of the tropical disease schistosomiasis (bilharzia) which eventually leads to liver cancer in the human host. The host is the freshwater snail *Biomphalaria glabrata.* The snail is infected by miracidia found in water polluted with human feces. The miracidia burrow into the snail and develop into the next stage of their life-cycle. We have been looking at genes that are up-regulated in the snail after an infection with schistosoma. Two systems have been employed: an in vitro cell system whereby miracidia are only placed in the media with a snail cell line [11] and actual infected whole organisms. The schistosoma miracidia can be irradiated so that they are still alive and can burrow into the snails but are attenuated and do not progress further to an infection. In both these systems we have determined that exposure to the fully functioning parasite induces the genes that are up-regulated in these infections to be relocalized within the nuclear environment (Arican, Bridger, and Knight unpublished). By doing close time studies we have been able to show that the gene moves slightly prior to its being switched on and expressed. This helps answer an important question in the field as to whether genes and chromosomes move before they alter their transcriptional status as some have proposed that transcriptional activation can actually drive a gene away from the nuclear periphery.

The experiments with the attenuated parasite have been an important control since in the co-culture in vitro system the normally up-regulated genes were not relocated in interphase nuclei and remained stationary [11]. Further, in the whole organism experiments two genes did not move with attenuated parasite, remaining in their non-random location. However, one gene moved in the same direction under the same early time point in both attenuated and unirradiated miracidia samples. This shows that this gene is expressed due to the infiltration of the parasite into the host rather than any control or influence over the host genome. This species of snail also has a resistant laboratory bred strain in which the two genes that move in the susceptible strain are not expressed and remain stationary.

In order to work towards understanding where these genes are travelling to and what takes them to their new location upon a parasitic infection, we went back to the snail cells in culture and established a heat-shock stimulus, where cells were moved from 27 °C to 32 °C for 1 h. This allowed the gene *hsp70* to be expressed. From the literature we know that genes can move to PML bodies [33], SC35 speckles and transcription factories when they become active. Using 3D fixation and immuno-FISH in the snail we were able to see these structures with the gene loci of interest. Upon heat shock we found that there was a significantly increased number of gene loci associated with transcription factories as revealed by anti-RNA polymerase II antibodies. This association correlated with the increased gene expression of the *hsp70* gene.

Whilst staining the snail cells with antibodies that may have crossed the species we found that anti-nuclear myosin 1β that recognizes a nuclear myosin in human cells stained very strongly around the nuclear envelope and had foci throughout the nucleoplasm of the snail cells. We are already advocates of a nuclear motor system in cells moving chromosomes and possibly genes around functionally in the nucleoplasm. A drug that inhibits nuclear myosin polymerization was used and it removed all the internal foci of nuclear myosin 1β staining within the nucleoplasm. It also inhibited the genes moving to their new internal location and a produced much reduced expression of the *hsp70* gene. Thus, we believe, even in organisms such as molluscs they use the same system of moving around specific genes to regulate their expression and they do this via a nuclear motor system as has been shown in human cells.

We believe that the system we have found whereby a pathogen will influence genome reorganization in a host is a general mechanism benefitting the pathogen. This has been seen with a viral infection that elicited specific chromosome repositioning [34]. In a long term infection it may difficult for a host to regain control of its genome and cells may change their behavior through instability and become transformed, leading to either cellular premature senescence, death or immortalization.

Unlike other studies in the mammalian cells where we have shown genes moving more to the nuclear interior when they get up-regulated, in the snail cells we see genes that become activated move towards the nuclear periphery. This may have to do with the snail nuclear envelope being different to higher organisms and not such an area for down-regulation and silencing. Indeed, *B. glabrata* seems to have nuclear lamins, but they are more akin to *Drosophila* lamins than mammalian lamins (Town and Bridger, unpublished) and may instigate a different type of genome organization within this species/Genus.

Alterations to Genome Organization in Cancer

Genome organization is altered in cancer cells as has been shown nicely many years ago by the distribution of centromeres and telomeres being altered in bladder carcinoma cells [35]. A number of more recent studies now show abnormal chromosome positioning in cancer cells. Abnormal relocation of chromosome 18 from the nuclear periphery to the interior has been reported in several types of tumor cell lines, including those derived from melanoma, cervix carcinoma, colon carcinoma, Hodgkin's lymphoma, and metastasizing cells from a colon carcinoma [36]. Moreover, several reports support the idea of a functional correlation between non-random chromosome positioning and formation of specific chromosome translocations, for example human chromosomes 9 and 22 in chronic myeloid leukemia (CML) [37] as well as the correlation between tissue specific spatial organization and tissue specific translocations [38]. These findings are particularly compelling because chromosomes that tend to be adjacent to one another are much more likely to form particular fusion proteins from translocations that are prevalent in a particular tumor type. That this would be observed in large numbers of patients at the level of a specific fusion protein underscores that patterns of spatial genome organization are very highly conserved. Furthermore, the nuclear positions of chromosomes 10, 18 and 19 were assessed in normal thyroid tissue and compared to several types of thyroid cancers including adenomatous goiters, papillary carcinomas, and undifferentiated carcinomas. There was no difference in chromosome position in the normal and goiter tissue with chromosomes 10 and 18 positioned towards the nuclear periphery and chromosome 19 in a central location. However, in the papillary carcinoma tissue chromosome 19 was located more centrally. Furthermore, in undifferentiated carcinomas all the chromosomes assessed were mislocalized [37]. Marella et al. in 2009 [40], used normal human WI38 lung fibroblast and MCF10A epithelial breast cells and identified that similar levels of association were found in WI38 and MCF10A (both are non-tumorigenic) for chromosomes 1, 4, 11, 12, 14, and 16, whereas a nearly twofold increase in chromosomes 4 and 16 associations was found in a malignant breast cancer cell line (MCFCA1a) compared to the related normal epithelial cell line (MCF10A). This demonstrates that chromosome associations are cell-type specific and undergo alterations in cancer cells [40]. Furthermore, Wiech et al. 2005 analyzed chromosome 8 positions in wax embedded pancreatic cancer tissue samples. Their results obtained from non-neoplastic pancreatic cells indicated that the radial arrangement of the chromosome 8 territories did not significantly differ between normal individuals. However, in pancreatic tumors, the radial distance changes indicated the repositioning of chromosome 8 to the nuclear periphery. Positioning changes were also observed in breast cancer. In non-neoplastic ductal epithelium of the breast, there was a large distance between the position of the centromere 17 and *HER2* domains among individuals. In neoplastic epithelial breast cells, the distances between centromere and gene domains were smaller than in non-neoplastic cells. The centromere and the gene encoding *HER2* on chromosome 17 were shown to reposition to a more internal location [41, 42]. A later study by

Wiech et al. in 2009 looked at cervical carcinomas. They reported repositioning of chromosome 18 during cell differentiation of cervical squamous epithelium towards the nuclear center, whereas cervical squamous carcinomas showed a repositioning of chromosome 18 towards the nuclear periphery [43].

Therefore, changes in the radial position of specific gene loci in cancer cells could contribute to tumorigenesis, but further investigation is still needed. These observations strongly support the idea that the genomic regions influenced by states of gene activity and cell-type specific genome architecture are predisposed towards translocations that are characteristic to specific cell types and cancers.

All the aforementioned studies did not assess the status of the nuclear structure, especially those proteins involved in genome organization. Other studies only look at nuclear structure changes with respect to cancer and do not look at any genome behavioral changes [44]. Changing the nuclear architecture will have a direct effect on the genomes' stability and may then lead to cancer.

Alterations to Gene and Chromosome Position in Breast Cancer Cell Lines Can Possibly Be Manipulated to Reduce Cancer Phenotypes

A number of studies have shown that in cancer cells whole chromosomes and specific gene loci can change nuclear location away from the norm. Indeed, translocations prevalent in cancer cells can place genes in new locations in interphase nuclei that affect their behavior and expression profiles as has been shown with *HLXB9* in pediatric leukemia [45].

One of the best studies for looking at gene repositioning in cancer is that of Meaburn and Misteli for loci of genes that are involved in some of the important changes in breast cancer. These authors showed a number of genes were non-randomly located at new locations in a 3D culture model system and in tumor tissue sections. They showed altered positioning of cancer-associated genes such as *AKT1*, *BCL2*, *ERBB2*, and *VEGF* loci, although no correlation was found between this radial redistribution and gene activity levels [46].

Meaburn et al. in 2009 expanded on their study of the repositioning of genes that are involved in breast cancer. From 11 normal human breast and 14 invasive breast cancer tissue specimens, they identified eight genes (*HES5, ERBB2, MYC, FOSL2, HSP90AA1, AKT1, TGFB3,* and *CSF1R*) that had altered their position in breast cancer [47] Excitingly, the position of a specific gene, *HES5*, a transcription repressor that regulates cell differentiation, could distinguish between a cancerous tissue and a healthy one with almost 100 % accuracy. Alteration or repositioning of this gene has been associated with tumorigenesis and was observed in several types of breast cancer so that it could prove a useful diagnostic tool [47].

The studies by Meaburn and Misteli did not link any actual nuclear or chromosomal event or aberration in nuclear structure to this change in location. However, in a study using a panel of breast tumor epithelial cell lines we found that whole chromosomes were mis-positioned as well as individual genes. Most interestingly, when genes were mislocalized without repositioning of the whole chromosome they were found out on loops at some distance from the chromosomes. Interestingly, we also found that a number of the cells lacked or had reduced levels of lamin A and lamin B receptor that have been implicated in gene/chromosome/chromatin position by us and others. The cells also had large accumulations of B-type lamins in the center of their nuclei. The cell lines with the most pronounced loss or changes in nuclear envelope proteins had the most changes with respect to breast cancer gene relocation. In fact the three genes that were focused upon, *HER2*, *HSP90AA*, and *AKT1*, were found towards the nuclear periphery in these aberrant cells. When the cells were treated with a drug that restored lamin B receptor to the nuclear periphery and placed lamin B back at the nuclear periphery, one cell line had the genes *HER2*, *HSP90AA*, and *AKT1* become more internal with a corresponding up-regulation of all three genes. However, another cell line pulled the same genes more towards the nuclear periphery after treatment which correlated with a down-regulation of expression in *AKT1* and *HSP90AA* (Hassan Ahmed, Harvey, Karteris, and Bridger, unpublished data).

This is a very important finding because it links nuclear envelope aberrations with genome mislocalization in cancer. Though the reasons for the differences between cell lines need to be determined, our ability in this study to correct to a certain extent the nuclear envelope abnormalities and correspondingly restore proper gene location and expression may open the way for novel therapeutic treatments.

Summary

The non-random spatial positioning of the genome within nuclei appears to be highly relevant to controlling gene expression and silencing [49]. The gene-density distribution of chromosomes in proliferating cells requires energy and highly organized tethering to nuclear structures to be maintained. This organization is changed dramatically when cells become non-proliferating, perhaps some of the positioning is more relaxed requiring less energy to maintain. However, there are some noticeable differences between quiescent and senescent cells and we believe this is due differences in gene expression profiles but also the absolute need to silence irreversibly in senescent cells to prevent reactivation and these cells becoming cancerous. This silencing, we postulate, will be deep within nucleus.

It is not only chromosomes that move around nuclei after a stimulus but individual genes. These genes can move to new areas around the nucleus without the whole chromosome moving. We predict that these genes translocate across the nucleus in a directed manner, to structural entities such as transcription factories or splicing speckles for example using nuclear motor activity (see Fig. 1). We believe that the nuclear motors require nuclear envelope proteins such as emerin and the lamins to function correctly [48]. These genes will meet other genes at the transcription factories, and if this movement, co-occupation, transcription, and the return of the gene to its original location is not functioning correctly, then it is possible that chromosomal translocations are formed that are a hallmark of cancer [44].

Pathogen-led spatial reorganization of the genome is a newly discovered process and we need to discover how the pathogen controls specific selected gene expression but further we need to determine what irreversible alterations have been elicited in the cell that will affect its future.

Data from breast cancer cells and the premature aging disease HGPS demonstrates that the nuclear envelope is involved in chromosome and gene positioning, especially proteins such as A-type lamins, lamin B receptor and lamin B. Finally, and perhaps most importantly, these studies on gene repositioning and its consequences in cancer cells may pave the way for novel therapeutic interventions or combination treatments to inhibit tumor cell proliferation and restabilize the genome in cancer cells.

Acknowledgements The unpublished work referred to in this chapter complied by the Bridger group has been funded by NIH, EU-FP6, Brunel University Progeria Research Fund, The Gordon Memorial Foundation, Wellcome Trust, WestFocus, and The Malacological Society of London. The authors would like to thank colleagues at Brunel—Dr Christopher Eskiw, Dr Margaret Town, Dr Sabrina Tosi, Dr Emmanouil Karteris for collaboration, collegiality, and encouragement.

References

1. Cremer T, Kurz A, Zirbel R, Dietzel S, Rinke B, Schröck E, Speicher MR, Mathieu U, Jauch A, Emmerich P, Scherthan H, Ried T, Cremer C, Lichter P (1993) Role of chromosome territories in the functional compartmentalization of the cell nucleus. Cold Spring Harb Symp Quant Biol 58:777–792
2. Foster HA, Bridger JM (2005) The genome and the nucleus: a marriage made by evolution. Genome organisation and nuclear architecture. Chromosoma 114:212–229
3. Bridger JM, Bickmore WA (1998) Putting the genome on the map. Trends Genet 14:403–410
4. Boyle S, Gilchrist S, Bridger JM, Mahy NL, Ellis JA, Bickmore WA (2001) The spatial organization of human chromosomes within the nuclei of normal and emerin-mutant cells. Hum Mol Genet 10:211–219
5. Croft JA, Bridger JM, Boyle S, Perry P, Teague P, Bickmore WA (1999) Differences in the localization and morphology of chromosomes in the human nucleus. J Cell Biol 145:1119–1131
6. Bridger JM, Boyle S, Kill IR, Bickmore WA (2000) Re-modelling of nuclear architecture in quiescent and senescent human fibroblasts. Curr Biol 10:149–152
7. Mehta IS, Figgitt M, Clements CS, Kill IR, Bridger JM (2007) Alterations to nuclear architecture and genome behavior in senescent cells. Ann N Y Acad Sci 1100:250–263
8. Meaburn KJ, Cabuy E, Bonne G, Levy N, Morris GE, Novelli G, Kill IR, Bridger JM (2007) Primary laminopathy fibroblasts display altered genome organization and apoptosis. Aging Cell 6:139–153
9. Mehta IS, Eskiw CH, Arican HD, Kill IR, Bridger JM (2011) Farnesyltransferase inhibitor treatment restores chromosome territory positions and active chromosomes dynamics in Hutchinson-Gilford progeria syndrome cells. Genome Biol 12:R74
10. Wazir U, Ahmed MH, Bridger JM, Harvey A, Jiang WG, Sharma AK, Mokbel K (2013) The clinicopathological significance of lamin A/C, lamin B1 and lamin B receptor mRNA expression in human breast cancer. Cell Mol Biol Lett 18(4):595–611. doi:10.2478/s11658-013-0109-9
11. Knight M, Ittiprasert W, Odoemelam EC, Adema CM, Miller A, Raghavan N, Bridger JM (2011) Non-random organization of the Biomphalaria glabrata genome in interphase Bge cells and the spatial repositioning of activated genes in cells co-cultured with Schistosoma mansoni. Int J Parasitol 41:61–70

12. Mehta IS, Amira M, Harvey AJ, Bridger JM (2010) Rapid chromosome territory relocation by nuclear motor activity in response to serum removal in primary human fibroblasts. Genome Biol 11:R5
13. Bridger JM, Lichter P (1999) Analysis of mammalian interphase chromosomes by FISH and immunofluorescence. In: Chromosome Structural Analysis: a practical approach. Edited by Dr. Wendy Bickmore IRL Press.
14. Chuang CH, Carpenter AE, Fuchsova B, Johnson T, de Lanerolle P, Belmont AS (2006) Long-range directional movement of an interphase chromosome site. Curr Biol 16:825–831
15. Chubb JR, Boyle S, Perry P, Bickmore WA (2002) Chromatin motion is constrained by association with nuclear compartments in human cells. Curr Biol 12:439–445
16. Foster HA, Griffin DK, Bridger JM (2012) Interphase chromosome positioning in in vitro porcine cells and ex vivo porcine tissues. BMC Cell Biol 13:30
17. Szczerbal I, Foster HA, Bridger JM (2009) The Spatial Repositioning of Adipogenesis Genes Is Correlated with Their Expression Status in a Porcine Mesenchymal Stem Cell Adipogenesis Model System. Chromosoma 118:647–663
18. Brown JM, Green J, das Neves RP, Wallace HA, Smith AJ, Hughes J, Gray N, Taylor S, Wood WG, Higgs DR, Iborra FJ, Buckle VJ (2008) Association between active genes occurs at nuclear speckles and is modulated by chromatin environment. J Cell Biol 182: 1083–1097
19. Szczerbal I, Bridger JM (2010) Association of adipogenic genes with SC-35 domains during porcine adipogenesis. Chromosome Res 18:887–895
20. Osborne CS, Chakalova L, Mitchell JA, Horton A, Wood AL, Bolland DJ, Corcoran AE, Fraser P (2007) Myc dynamically and preferentially relocates to a transcription factory occupied by Igh. PLoS Biol 5:e192
21. Xu M, Cook PR (2008) The role of specialized transcription factories in chromosome pairing. Biochim Biophys Acta 1783(11):2155–2160
22. Hofmann WA, Johnson T, Klapczynski M, Fan JL, de Lanerolle P (2006) From transcription to transport: emerging roles for nuclear myosin I. Biochem Cell Biol 84:418–426
23. Bridger JM (2011) Chromobility: the rapid movement of chromosomes in interphase nuclei. Biochem Soc Trans 39:1747–1751
24. Bridger JM, Mehta IS (2011) Nuclear Molecular Motors for Active, Directed Chromatin Movement in Interphase Nuclei. Springer, Advances in Nuclear Architecture. Eds Niall Adams and Paul Freemont
25. Finlan LE, Sproul D, Thomson I, Boyle S, Kerr E, Perry P, Ylstra B, Chubb JR, Bickmore WA (2008) Recruitment to the nuclear periphery can alter expression of genes in human cells. PLoS Genet 4(3):e1000039
26. Arib G, Akhtar A (2011) Multiple facets of nuclear periphery in gene expression control. Curr Opin Cell Biol 23(3):346–353
27. Sadaie M, Salama R, Carroll T, Tomimatsu K, Chandra T, Young AR, Narita M, Pérez-Mancera PA, Bennett DC, Chong H, Kimura H, Narita M (2013) Redistribution of the Lamin B1 genomic binding profile affects rearrangement of heterochromatic domains and SAHF formation during senescence. Genes Dev 27:1800–1808
28. Mewborn SK, Puckelwartz MJ, Abuisneineh F, Fahrenbach JP, Zhang Y, MacLeod H, Dellefave L, Pytel P, Selig S, Labno CM, Reddy K, Singh H, McNally E (2010) Altered chromosomal positioning, compaction, and gene expression with a lamin A/C gene mutation. PLoS ONE 5:e14342
29. Solovei I, Wang AS, Thanisch K, Schmidt CS, Krebs S, Zwerger M, Cohen TV, Devys D, Foisner R, Peichl L, Herrmann H, Blum H, Engelkamp D, Stewart CL, Leonhardt H, Joffe B (2013) LBR and lamin A/C sequentially tether peripheral heterochromatin and inversely regulate differentiation. Cell 152(3):584–598
30. Taimen P, Pfleghaar K, Shimi T, Moller D, Ben-Harush K, Erdos MR, Adam SA, Herrmann H, Medalia O, Collins FS, Goldman AE, Goldman RD (2009) A progeria mutation reveals functions for lamin A in nuclear assembly, architecture, and chromosome organization. Proc Natl Acad Sci U S A 106:20788–20793

31. Kubben N, Adriaens M, Meuleman W, Voncken J, van Steensel B, Misteli T (2012) Mapping of lamin A- and progerin-interacting genome regions. Chromosoma 121:447–464
32. Zuleger N, Boyle S, Kelly DA, de Las Heras JI, Lazou V, Korfali N, Batrakou DG, Randles KN, Morris GE, Harrison DJ, Bickmore WA, Schirmer EC (2013) Specific nuclear envelope transmembrane proteins can promote the location of chromosomes to and from the nuclear periphery. Genome Biol 14(2):R14
33. Dundr M, Ospina JK, Sung MH, John S, Upender M, Ried T, Hager GL, Matera AG (2007) Actin-dependent Intranuclear Repositioning of an Active Gene Locus in Vivo. J Cell Biol 179:1095–1103
34. Li C, Shi Z, Zhang L, Huang Y, Liu A, Jin Y, Yu Y, Bai J, Chen D, Gendron C, Liu X, Fu S (2010) Dynamic changes of territories 17 and 18 during EBV-infection of human lymphocytes. Mol Biol Rep 37(5):2347–2354
35. Poddighe PJ, Ramaekers FC, Smeets AW, Vooijs GP, Hopman AH (1992) Structural chromosome 1 aberrations in transitional cell carcinoma of the bladder: interphase cytogenetics combining a centromeric, telomeric, and library DNA probe. Cancer Res 52(18):4929–4934
36. Cremer M, Kupper K, Wagler B, Wizelman L, von Hase J, Weiland Y, Kreja L, Diebold J, Speicher MR, Cremer T (2003) Inheritance of gene density-related higher order chromatin arrangements in normal and tumor cell nuclei. J Cell Biol 162:809–820
37. Lukasova E, Kozubek S, Kozubek M, Kjeronska J, Ryznar L, Horakova J, Krahulcova E, Horneck G (1997) Localisation and distance between ABL and BCR genes in interphase nuclei of bone marrow cells of control donors and patients with chronic myeloid leukaemia. Hum Genet 100:525–535
38. Parada LA, McQueen PG, Misteli T (2004) Tissue-specific spatial organization of genomes. Genome Biol 5:R44
39. Murata S, Nakazawa T, Ohno N, Terada N, Iwashina M, Mochizuki K, Kondo T, Nakamura N, Yamane T, Iwasa S, Ohno S, Katoh R (2007) Conservation and alteration of chromosome territory arrangements in thyroid carcinoma cell nuclei. Thyroid 17:489–496
40. Marella NV, Bhattacharya S, Mukherjee L, Xu J, Berezney R (2009) Cell type specific chromosome territory organization in the interphase nucleus of normal and cancer cells. J Cell Physiol 221:130–138
41. Wiech T, Timme S, Riede F, Stein S, Schuricke M, Cremer C, Werner M, Hausmann M, Walch A (2005) Human archival tissues provide a valuable source for the analysis of spatial genome organization. Histochem Cell Biol 123:229–238
42. Timme S, Schmitt E, Stein S, Schwarz-Finsterle J, Wagner J, Walch A, Werner M, Hausmann M, Wiech T (2011) Nuclear position and shape deformation of chromosome 8 territories in pancreatic ductal adenocarcinoma. Anal Cell Pathol (Amst) 34:21–33
43. Wiech T, Stein S, Lachenmaier V, Schmitt E, Schwarz-Finsterle J, Wiech E, Hildenbrand G, Werner M, Hausmann M (2009) Spatial allelic imbalance of BCL2 genes and chromosome 18 territories in nonneoplastic and neoplastic cervical squamous epithelium. Eur Biophys J 38:793–806
44. Bourne G, Moir C, Bikkul U, Hassan Ahmed M, Kill IR; Eskiw CH, Tosi S and Bridger JM (2013) Interphase chromosomes in disease. Human Interphase Chromosomes: the Biomedical Aspects Ed Yuri Yurov:Springer
45. Ballabio E, Cantarella CD, Federico C, Di Mare P, Hall G, Harbott J, Hughes J, Saccone S, Tosi S (2009) Ectopic Expression of the HLXB9 Gene Is Associated with an Altered Nuclear Position in T(7;12) Leukaemias. Leukemia 23:1179–1182
46. Meaburn KJ, Misteli T (2008) Locus-specific and activity-independent gene repositioning during early tumorigenesis. J Cell Biol 180:39–50
47. Meaburn KJ, Gudla PR, Khan S, Lockett SJ, Misteli T (2009) Disease-specific gene repositioning in breast cancer. J Cell Biol 187:801–812
48. Mehta IS, Amira M, Kill IR, Bridger JM (2008) Nuclear motors and nuclear structures containing A-type lamins and emerin: is there a functional link? Biochem Soc Trans 36:1384–1388
49. Elcock LS, Bridger JM (2010) Exploring the relationship between interphase gene positioning, transcriptional regulation and the nuclear matrix. Biochem Soc Trans 38:263–267

Part IV
Functions of the NPC in Cancer

Introduction

The nuclear pores are enormous complexes that perforate the nuclear envelope and direct the trafficking of macromolecules between the cytoplasm and nucleus [1]. There are typically between 2,000 and 3,000 nuclear pore complexes (NPCs) per mammalian nucleus [2], and each one is made up of over 30 core proteins in multiple copies [3, 4] that together with transport receptors and cargoes give an estimated total mass by cryo-electron microscopy of ~125 MDa [5, 6], making them among the largest protein structures in the cell. The transport receptors bind to sequences on cargo proteins called nuclear localization signals (NLSs) for transport into the nucleus and nuclear export signals (NESs) for transport out of the nucleus: several proteins with NESs bind to mRNAs to promote their export from the nucleus. The small GTPase Ran and its associated exchange proteins RCC1 and RanGAP are involved in cargo release after transport and recycling of transport receptors [1]. Although much of transport is unidirectional, many proteins have both NLSs and NESs that enable their shuttling between the nucleus and cytoplasm and the overall direction of transport at any given time is often directed by posttranslational modifications that either inactivate or make one of the localization signals inaccessible.

We have seen in Part II how several components of the NPCs have separate functions in mitosis involved in mitotic spindle and kinetochore function while others contribute to nuclear envelope reassembly at the end of mitosis. These functions are critical for the cell cycle and thus can play a central role in cancer development. However, there are even more ways that the NPCs can play roles in cancer development in interphase cells. Obviously direct transport roles can impact on the relative ratios of oncogenes in the nucleus versus in the cytoplasm to activate various transcriptional programs linked to cellular transformation, and likewise this same function could keep tumor suppressors out of the nucleus and thus ineffective. An example of this type of regulation was found in Yusuke Nakamura's laboratory for c-myc, which is translocated into the nucleus in a manner specifically dependent on the function of the NPC protein Nup205 together with the function of a nuclear envelope transmembrane protein NET31/Tmem209 [7]. Tmem209 is normally expressed highest in testis, but in this study they found that Tmem209 was strongly upregulated in lung cancer—both small-cell lung carcinoma and large-cell lung carcinoma. Searching for a function for Tmem209, they found its knockdown decreased the proliferation of lung cancer cells and expressing it exogenously promoted cell proliferation. Interestingly, a central effect of upregulating Tmem209 was to stabilize levels of the NPC protein Nup205 with a corresponding increase in the nuclear levels of c-myc, and this is thought to be the mechanism by which it plays a role in lung cancer. Moreover, individual NPC proteins have been found to play specific roles in tumorigenesis with some even being involved in specific chromosome translocations.

In this section, we first have Dan Simon and Mike Rout, who first identified the proteome of the NPC, of Rockefeller University providing a brief overview of the large number of NPC proteins that have been found to play roles in a wide range of tumors. They further raise the interesting possibility that the mechanisms viruses use for hijacking the NPCs parallel some of those used in tumorigenesis. Next, Chelsi Snow and Bryce Paschal, who discovered one of the first export factor families, of the University of Virginia give an in-depth view of the role of Tpr, a principal component of the nucleoplasmic face of the NPC called the nuclear basket, in both cancer and aging, specifically with regard to the premature aging progeroid syndromes directly caused by mutation of nuclear envelope proteins. This chapter moreover very nicely covers the relationship between aging and cancer, which is critical for understanding many aspects of tumorigenesis. This is followed by a detailed look at the Ran GTPase as a metastagene by Mohamed El-Tanani, who has led the investigation into Ran functions in cancer, together with Kyle Matchett and other colleagues from Queen's University Belfast, Trinity College Dublin, and University of Ulster. Ran is a small GTPase, similar in its characteristics to the many small GTPases linked to cancer such as the Ras, Rac, and Rho families. In this chapter the authors give evidence of how Ran can increase the metastatic potential of tumors as well as describing in detail its role in nuclear envelope reformation. Finally, in the last chapter of this section Beric Henderson, who has led much of the research into IQGAP and BRCA1, together with Manisha Sharma and other colleagues from the University of Sydney, discusses how the Wnt signaling pathway, often linked to tumorigenesis and metastasis, interacts with the NPC in its cancer roles.

1. Suntharalingam M, Wente SR (2003) Peering through the pore: nuclear pore complex structure, assembly, and function. Dev Cell 4(6):775–789
2. Maul GG, Deaven L (1977) Quantitative determination of nuclear pore complexes in cycling cells with differing DNA content. J Cell Biol 73(3):748–760
3. Cronshaw J, Krutchinsky A, Zhang W, Chait B, Matunis M (2002) Proteomic analysis of the mammalian nuclear pore complex. J Cell Biol 158(5):915–927
4. Rout MP, Aitchison JD, Suprapto A, Hjertaas K, Zhao Y, Chait BT (2000) The yeast nuclear pore complex: composition, architecture, and transport mechanism. J Cell Biol 148(4):635–651.

5. Hinshaw JE, Carragher BO, Milligan RA (1992) Architecture and design of the nuclear pore complex. Cell 69(7):1133–1141
6. Reichelt R, Holzenburg A, Buhle EL, Jr., Jarnik M, Engel A, Aebi U (1990) Correlation between structure and mass distribution of the nuclear pore complex and of distinct pore complex components. J Cell Biol 110(4):883–894
7. Fujitomo T, Daigo Y, Matsuda K, Ueda K, Nakamura Y (2012) Critical function for nuclear envelope protein TMEM209 in human pulmonary carcinogenesis. Cancer Res 72(16):4110–4118. doi:0008-5472.CAN-12-0159 [pii] 10.1158/0008-5472.CAN-12-0159

Cancer and the Nuclear Pore Complex

Dan N. Simon and Michael P. Rout

Abstract The nuclear pore complex (NPC) mediates trafficking between the cytoplasm and nucleoplasm. It also plays key roles in other nuclear processes such as chromatin silencing, transcriptional regulation, and DNA damage repair. Nucleoporins, the structural components of the NPC, have been linked to a multitude of cancers through chromosomal translocations generating fusion proteins, changes in protein expression levels, and single point mutations. Only a small number of nucleoporins have been linked to tumorigenesis thus far, and these proteins—Nup62, Nup88, Nup98, Nup214, Nup358/RanBP2, and Tpr—line the trafficking pathway and are particularly associated with mRNA export. Overexpression of several associated nuclear export factors, most also involved in various stages of mRNA export, has been linked to cancers as well. Some oncogenic nucleoporin mutants are mislocalized to either the cytoplasm or nucleoplasm while others are incorporated into the NPC, and in all these cases they are thought to misregulate signaling pathways and transcription through either altered or diminished nucleoporin functionality. Intriguingly, many viruses target the same cancer-linked nucleoporins, often causing their degradation or mislocalization, implying that these viruses exploit some of the same weaknesses as the oncogenic defects.

Keywords Nuclear envelope • Nuclear pore complex • Nucleoporin • Cancer • Leukemia • Tpr • Nup62 • Nup88 • Nup98 • Nup214 • Nup358/RanBP2 • Virus

D.N. Simon • M.P. Rout (✉)
The Laboratory of Cellular and Structural Biology,
The Rockefeller University, New York, NY 10065, USA
e-mail: dsimon01@rockefeller.edu; rout@rockefeller.edu

E.C. Schirmer and J.I. de las Heras (eds.), *Cancer Biology and the Nuclear Envelope*, Advances in Experimental Medicine and Biology 773, DOI 10.1007/978-1-4899-8032-8_13,

Abbreviations

NPC	Nuclear pore complex
NE	Nuclear envelope
Tpr	Translocated promoter region
NLS	Nuclear localization signal
HGFR	Hepatocyte growth factor receptor
NTrk1	Neurotrophic tyrosine receptor kinase 1
FGFR1	Fibroblast growth factor receptor 1
EMS	8p11 myeloproliferative syndrome
AML	Acute myeloid leukemia
AUL	Acute undifferentiated leukemia
MDS	Myelodysplastic syndrome
T-ALL	T-cell acute lymphoblastic leukemia
IMT	Inflammatory myofibroblastic tumor
Alk	Anaplastic lymphoma kinase
B-ALL	B-cell acute lymphoblastic leukemia
CML	Chronic myelogenous leukemia
CMML	Chronic myelomonocytic leukemia
JMML	Juvenile myelomonocytic leukemia
eIF4E	Eukaryotic initiation factor 4E
TMEV	Theiler's murine encephalomyelitis virus
VSV	Vesicular stomatitis virus
HIV-1	Human immunodeficiency virus

Introduction

The nuclear envelope (NE), a double membrane extension of the ER, separates the nucleoplasm from the cytoplasm. Embedded within pores in the NE (termed nuclear pores) are nuclear pore complexes (NPCs). The NPC is arguably the largest multiprotein complex in eukaryotic cells (60–120 MDa in human cells) with an evolutionarily conserved eightfold structural symmetry [1–4]. Each NPC is organized around a core composed of eight spokes joined by rings that surround the central transport channel (Fig. 1). A single NPC is comprised of multiple copies of ~30 different proteins, termed nucleoporins, which are made up from a limited set of structural domains that includes α-helices, β-propellers, (Phe-Gly) FG repeats, WD domains, and transmembrane domains [3–5]. Despite much progress, we still have neither a high-resolution structure for the NPC, nor a full picture of the many variants on this structure that are suspected between different organisms and tissues [4].

The primary, and perhaps best characterized, function of the NPC is to mediate the passive exchange of small molecules and the active transport of macromolecules between the nucleoplasm and cytoplasm. FG repeat-containing nucleoporins form an intrinsically disordered barrier in the central transport channel that, through a still

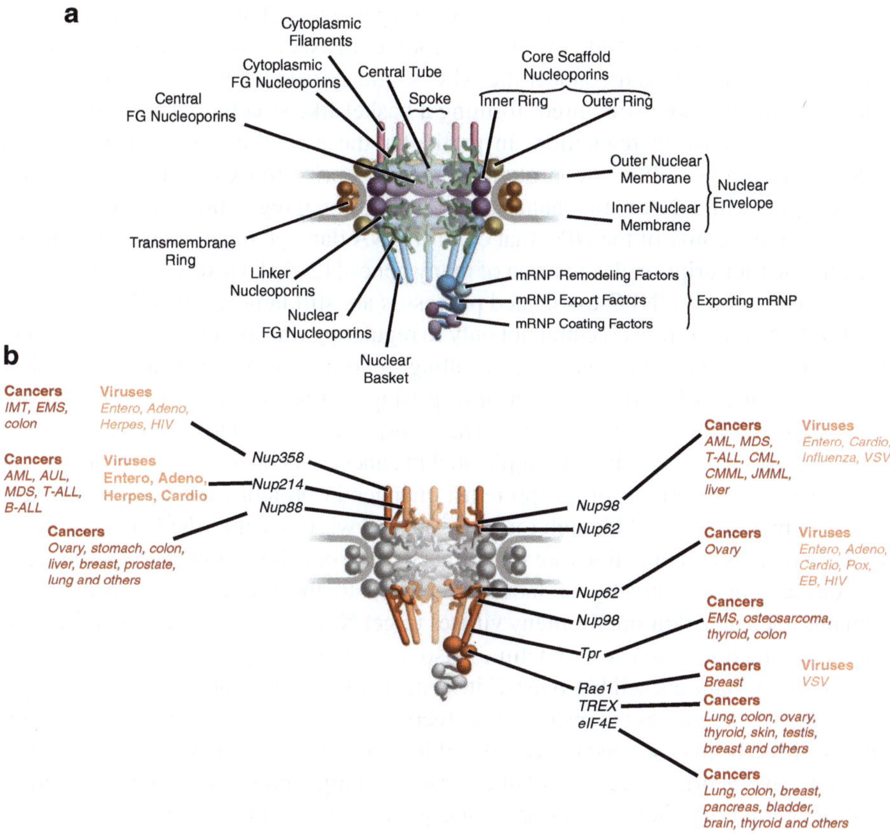

Fig. 1 Structure of the nuclear pore complex. (**a**) Each nuclear pore complex is a cylindrical structure comprised of eight spokes surrounding a central tube that connects the nucleoplasm and cytoplasm. The outer and inner nuclear membranes of the nuclear envelope join to form grommets, termed nuclear pores, in which the NPC sits. The NPC is anchored to the nuclear envelope by a transmembrane ring structure that connects to the core scaffold and comprises inner ring and outer ring elements. Linker nucleoporins help anchor the FG nucleoporins such that they line and fill the central tube. NPC-associated peripheral structures consist of cytoplasmic filaments and the nuclear basket which mediate assembly and export of mRNP complexes. (**b**) The nuclear pore complex highlighting nucleoporins and mRNP export factors linked to cancers and viral infections. Adapted from [3]

largely undetermined mechanism, impedes the passage of nonspecific macromolecules while mediating the passage of FG repeat-binding soluble transport factors (most falling into a family of chaperones termed karyopherins, or importins and exportins) carrying their specific cognate cargo macromolecules through the NPC [3, 6]. A gradient of the GTPase Ran, maintained by Ran cofactors, determines transport directionality by triggering the release of cargoes on the correct side of the NE [7]. Peripheral filaments, formed by a subset of nucleoporins, emanate from the core structure into both the nucleoplasm and cytoplasm—and though they too play a role in transport, they also connect the NPC to numerous other cellular processes. On the cytoplasmic side, the filaments are mostly disorganized and play a vital role in mediating traffic through the NPC [3, 6]; they couple mRNA export to translation initiation at ribosomes [8] and connect the NPC to the cytoskeleton [9–11]. The nuclear filaments appear more structured, forming a basket-like structure. This basket plays key roles in transport regulation, in particular the assembly and proofreading of mRNP (messenger ribonucleoprotein) complexes prior to export. It also connects RNA export with DNA maintenance and transcriptional regulation, organizing complexes in the vicinity of the NPC that control DNA damage repair, chromatin silencing and the transcriptional activation of many genes [12–15]. However, the molecular mechanisms of these NPC-associated processes are still largely unresolved.

The NPC is therefore essential not only to regulating transport between the nucleoplasm and cytoplasm but also to controlling genome organization and expression. These central cellular roles make it unsurprising—in retrospect—that the NPC has been linked with many diseases. These especially include cancers (Fig. 1). Nucleoporins have been directly implicated in cancers via three routes: chromosomal translocations generating fusion proteins; changes in protein expression levels; and single point mutations. Although found associated with cancers, whether or not many of these fusions and mutations are the primary cause or a downstream consequence of the disease, or are directly or indirectly linked to the processes of oncogenesis, remains unclear. Additionally, many viruses target NPC components, clearly facilitating viral infections and even sometimes also leading to oncogenesis. As viruses have been termed "nature's cell biologists," it seems likely that they are exploiting some of the same weaknesses as the oncogenic defects. The involvement of NPC components in cancer has been described in great detail in several recent reviews [4, 16–21], here, we will give a brief overview of these rapidly burgeoning areas of investigation, focussing on those nucleoporins particularly implicated in these diseases.

Tpr

Translocated Promoter Region (Tpr) is a ~270 kDa protein that forms the bulk of the NPC nuclear basket. With Nup153 and other partners, it forms NPC-linked filaments that extend into the nuclear interior [22]. Tpr and the various macromolecular complexes it recruits also maintain chromatin-free "channels" near the NPC, mediate export of proteins and mRNA, regulate telomere length and the mitotic spindle checkpoint, and help organize both the nuclear peripheral epigenetic silencing of some genes and the regulated transcriptional activation of others [13, 23–30]. The N-terminal coiled-coil domain of Tpr consists of multiple heptad repeat or leucine zipper motifs, while the acidic C-terminus is mostly unstructured (Fig. 2a) [31]. The coiled-coil domains are proposed to mediate dimerization and assembly of Tpr into the NPC, while the C-terminal domain contains a Nuclear Localization Sequence (NLS) [32].

Tpr was the first NPC component linked to cancer, as part of the MET oncogene in transformed osteogenic sarcoma cells, although at the time it was not yet known as a nucleoporin [33]. In this chromosomal translocation, the first 199 residues of Tpr are fused with the kinase domain of the Hepatocyte Growth Factor Receptor

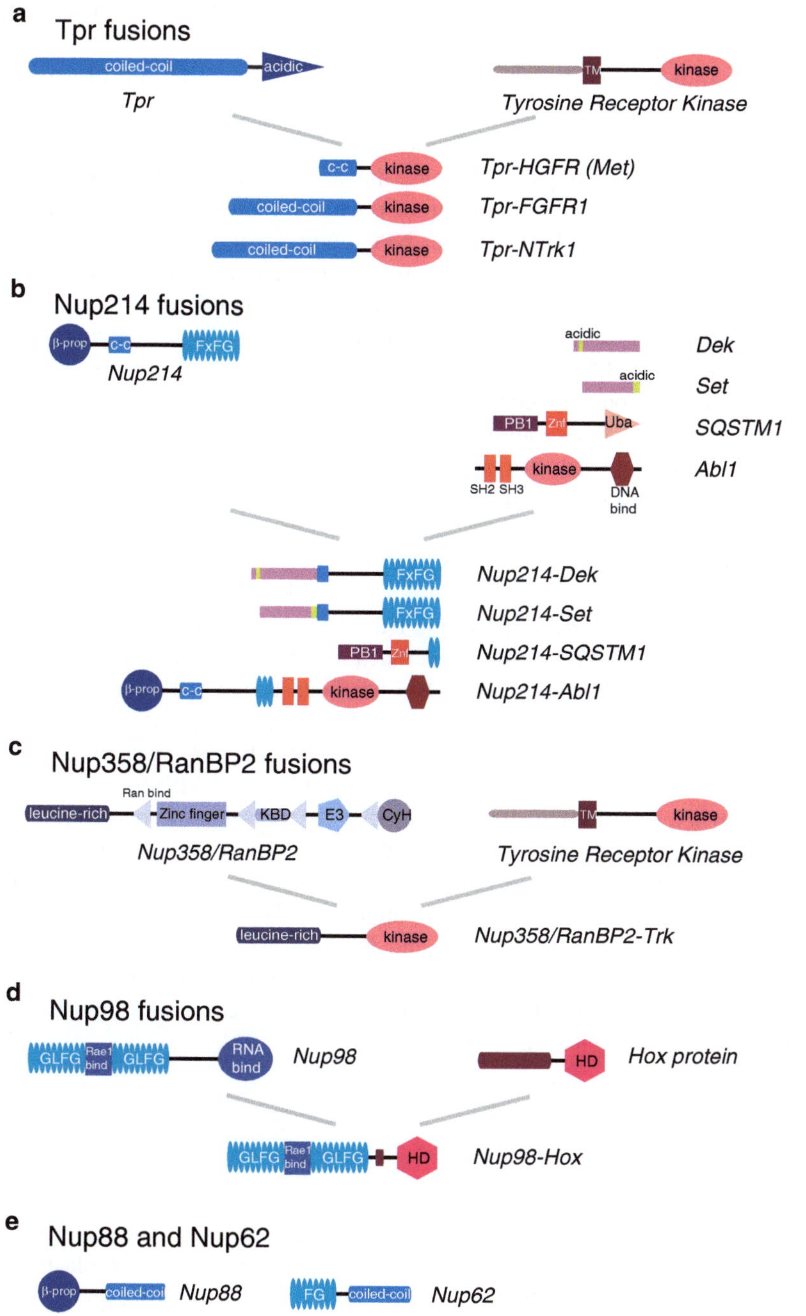

Fig. 2 Domain organization of cancer-linked nucleoporins and rearrangements following chromosomal translocations. The six cancer-linked nucleoporins and their translocation partners are shown with major structural domains, and the structure of the resulting fusion proteins found in cancer cells. (**a**) Tpr and fusions with HGFR, FGFR1, and NTrk1. (**b**) Nup214 and fusions with Dek, Set, SQSTM1, and Abl1. (**c**) Nup358/RanBP2 and fusions with tyrosine receptor kinases. (**d**) Nup98 and fusions with Hox proteins. (**e**) Nup88 and Nup62. *TM* transmembrane, *β-prop* β-propeller, *c-c* coiled-coil, *PB1* Phox and Bem 1, *Znf* zinc-finger, *Uba* ubiquitin associated, *SH2* Src homology 2, *SH3* Src homology 3, *KBD* kinesin binding domain, *E3* SUMO E3 domain, *CyH* cyclophilin homology, *HD* homeodomain. Adapted from [17]

(HGFR), generating a 65 kDa protein that localizes to the cytoplasm (Fig. 2a). The fusion protein dimerizes via the Tpr heptad domains and, since it lacks the HGFR ligand binding domain and the domain modulating its ubiquitin-mediated degradation, it acts as a constitutively active kinase that activates the Ras/MAPK and PI3K pathways [34–37]. Thus the mechanism of oncogenesis in this case seems at first glance to be incidental to Tpr being an NPC component. Despite being the first carcinogenic chromosomal translocation fusion of a tyrosine kinase to be discovered, Tpr translocations are actually rare in human tumors but Tpr-Met fusions are associated with gastric carcinomas where they are thought to represent an early step in carcinogenesis [38, 39].

Tpr has also been found fused with other tyrosine receptor kinases. Two different chromosomal translocations joining Tpr and Neurotrophic tyrosine receptor kinase 1 (NTrk1 or TrkA) were found in patients diagnosed with papillary thyroid carcinomas. In these translocations, either the first 199 or 1024 Tpr residues are fused with the kinase domain of NTrk1 generating oncogenic proteins with predicted masses of 66 kDa and 171 kDa, respectively (Fig. 2a) [40, 41]. As with Tpr-Met fusions, the Tpr-NTrk1 fusion creates a constitutively activated kinase that activates signaling via PLC-γ, SHC, FRS2, FRS3, IRS1, and IRS2 [17, 41, 42]. The Tpr-NTrk1 fusion containing the first 1024 Tpr residues includes the NPC-targeting site, but association of the fusion protein with NPCs has not been tested [42]. More recently a translocation generating a ~160 kDa fusion of the first 1032 residues of Tpr with the kinase domain of Fibroblast Growth Factor Receptor 1 (FGFR1) was identified in a patient diagnosed with 8p11 myeloproliferative syndrome (EMS) [43].

Notably, when six different human colorectal cancer tumors were examined, Tpr expression levels were found to be decreased fourfold to fivefold in all of them [44]. In yeast, disruption of the basket components (the MLP proteins) leads to a multitude of viable and disparate phenotypes, including alterations in gene expression, telomere tethering, epigenetic control, and RNA processing and export [28, 30]. Thus, as well as dimerization activation of the fusion protein described above, it is possible that the reduction of normal Tpr at the NPC may contribute significantly to the mechanism of oncogenesis.

Nup214

Nup214 (originally named CAN) is a developmentally essential nucleoporin localized to the cytoplasmic side of the NPC [45]; it is a component of cytoplasmic filaments, and forms a sub-complex with Nup88 that mediates mRNA export [46, 47]. The N-terminal portion of Nup214 contains β-propeller and coiled-coil domains, while the C-terminus is largely unstructured and contains degenerate FG and FxFG type repeats which bind to nuclear transport factors including the export receptor Crm1 (Fig. 2b) [48].

Nup214 was first identified as part of fusion proteins from chromosomal translocations in patients diagnosed with acute myeloid leukemia (AML), which have since also been found in acute undifferentiated leukemia (AUL), myelodysplastic syndrome (MDS), and T-cell acute lymphoblastic leukemia (T-ALL) [17, 20]. Many of these chromosomal translocations join Nup214 to either Dek (~165 kDa; [49]) or Set (~155 kDa; [50]). Dek and Set are both normally localized in the nucleoplasm where they are thought to play roles in chromatin organization and transcriptional regulation [20, 51]. Both fusions are structurally similar, joining almost the full Dek or Set proteins to the C-terminal two-thirds of Nup214, which includes a portion of the coiled-coil domain and the entire FG-repeat domain (Fig. 2b). Neither Nup214-Dek nor Nup214-Set fusion proteins have been found to associate with the NPC, but are instead located in distinct nucleoplasmic puncta [52].

The mechanisms by which Nup214-Set and Nup214-Dek fusions lead to cancers are unknown, but it has been proposed that the fusion proteins aberrantly alter transcription and chromatin organization [17, 20]. The *HOXA* genes are upregulated in five patients analyzed carrying the Nup214-Set fusion, suggesting that Nup214-Set may act by blocking differentiation of hematopoietic cells [53]. Overexpression of the Nup214-Dek fusion protein also increases mRNA translation in myeloid cells due to hyper-phosphorylation of the eukaryotic initiation factor 4E (eIF4E) [54]. The fusion proteins in the nucleoplasmic puncta may still bind to Nup88 via the partial coiled-coil domain of Nup214 [55] and can still bind to Crm1 via the FG repeat domain of Nup214. This in turn suggests that aberrant recruitment of nucleoporins and transport factors to these puncta may contribute to the mechanism of oncogenesis, compounded with reduction of normal Nup214 (and these aberrantly recruited proteins) at the NPCs [17, 55, 56].

A unique Nup214 chromosomal translocation was discovered in ~5 % of T-ALL patients where a small portion of chromosome 9 is circularized and episomally expressed generating a fusion of the Nup214 N-terminal domain with most of the Abl1 kinase [57]. The Nup214-Abl1 fusion is the second most common Abl1-containing cancer associated gene fusion, where only the circularized form of this fusion has been found to date [58]. Nine different Nup214-Abl1 fusions, ranging in size from ~239 to 333 kDa, have been observed in T-ALL patients. In all of these the N-terminus of Nup214 (including its β-propeller, coiled-coil, and varying amounts of the FG repeat regions) is fused to most of the Abl1 protein (including its SH2, SH3, and tyrosine kinase domains) (Fig. 2b) [58]. The Nup214-Abl1 protein is an active kinase [57]. Unlike Tpr-Met, the coiled-coil domains from Nup214 are not sufficient for dimerization and activation of the kinase in the Nup214-Abl1 protein. Instead, the coiled-coil domains bind to Nup88 and mediate association of the fusion protein with the NPC. Kinase activation then occurs though cross-phosphorylation of Nup214-Abl1 by neighboring fusion proteins [59]. Consequently, the activated Nup214-Abl1 fusion protein is tethered to the cytoplasmic filaments of the NPC from where it perhaps misregulates transport through the NPC by an unknown mechanism, or may aberrantly activate signaling cascades by phosphorylation of signaling proteins in the cytoplasm [17, 20]. The Nup214-Abl1 fusion protein is relatively weak as an oncogenic protein requiring 70–200 days to induce disease in mice compared to the 20 days necessary for the Bcr-Abl1 fusion to cause cancer in mouse T-cells [17, 59].

One T-ALL patient was found to carry a chromosomal translocation fusing a small portion of the Nup214 FG repeat domain (14 repeats out of 44) with the N-terminal half of Sequestosome 1 (SQSTM1/p62) (Fig. 2b) [60]. The localization of this fusion protein and mechanism of its activity are unknown. SQSTM1 is a ubiquitin binding adaptor protein that facilitates formation of signaling complexes in NFκB, ERK-1/2, p38 MAPK, and PKCζ pathways and thus functions in bone remodeling, adipocyte differentiation, and the inflammatory response [61–66]. SQSTM1 is also involved in targeting unfolded proteins for degradation [64, 67]. Interestingly, SQSTM1 knockout correlates with decreased growth of multiple myeloma cells in mice [62], but whether this occurs through its function in ubiquitination or another mechanism is unknown. Finally, in three B-cell acute lymphoblastic leukemia (B-ALL) patients a portion of chromosome 9 that included genes for SET, Nup214, and Nup188 was deleted, resulting in a ~2-fold decrease of mRNA levels of Nup214 and Nup188 [68]. Expression of several *HOXA* genes was upregulated in these patients, suggesting that normal function of SET, Nup214, and Nup188 is necessary for correct transcriptional regulation of *HOXA* genes, but whether via transport functions or loss of the SET oncogene or other functions is unknown.

Nup358/RanBP2

A multifunctional component of the cytoplasmic filaments, Nup358/RanBP2 is anchored at the NPC by the Nup88/Nup214 subcomplex [69]. It plays major roles in nuclear export [69, 70] and import [71] by providing a docking site for Ran and its cofactors, also mediating SUMOylation of the Ran cofactor RanGAP1 as well as of various cargo proteins [72–74]. Nup358/RanBP2 has multiple domains, including a leucine-rich domain, zinc-finger motif, four Ran binding domains, a cyclophilin homology domain, a SUMO E3 ligase domain, and several FG repeats (Fig. 2c) [75]. During mitosis Nup358/RanBP2 is found at kinetochores where it is involved in spindle formation and chromosome segregation [9, 76, 77]. Nup358/RanBP2 also links the NPC to the cytoskeleton by binding to the kinesin motors KIF5B and KIF5C [10, 11].

In one EMS patient the leucine-rich domain of Nup358/RanBP2 was fused with the tyrosine kinase domain of FGFR1 (Fig. 2c) [78]. The breakpoint in FGFR1 was the same as in the Tpr-FGFR1 fusion, which was also identified in an EMS patient [43]. However, four patients with inflammatory myofibroblastic tumors (IMTs) were found to carry chromosomal translocations that fuse the N-terminal leucine-rich domain of Nup358/RanBP2 with the tyrosine kinase domain of Anaplastic Lymphoma Kinase (Alk) (Fig. 2c) [79–81]. The resulting fusion protein has a predicted mass of ~160 kDa and localizes at the NE [79]. Whether it specifically interacts with the NPC and the potential mechanisms of oncogenesis are unknown. Single point mutations in Nup358/RanBP2 have been found in some human colorectal cancers [82] and, in mice, decreasing levels of Nup358/RanBP2 correlate with increased incidence of aneuploidy [83], suggesting that, as with the other nucleoporins discussed above, reduction of the protein at the NPC and disruption of normal transport function may contribute to oncogenesis. Overexpression of Nup358/RanBP2 is sufficient to inhibit eIF4E-induced oncogenic transformation [84]; and eIF4E has in turn been shown to function as an mRNA export factor (see below).

Nup98

Nup98 is a nucleoporin that localizes at both the cytoplasmic and nucleoplasmic sides of the NPC [85, 86] and also dynamically shuttles to chromatin throughout the nucleoplasm, and to poorly characterized FG repeat nucleoporin containing intranuclear "GLFG bodies" [87–89]. At the NPC, Nup98 interacts with transport factor Rae1 and mediates nuclear transport while in the nucleoplasm it acts as a transcriptional activator [15, 89, 90]. Nup98 consists of a large, mostly disordered region containing degenerate GLFG repeats with binding sites for Rae1, another transport factor TAP, and the transcriptional regulator CBP/p300 followed by a globular RNA-binding domain at the C-terminus (Fig. 2d) [91].

Nup98 was first linked to cancer when it was identified as part of fusions generated by chromosomal translocations in AML patients [92, 93]. The 59 kDa fusion protein contained all of the Nup98 GLFG repeats fused to the DNA-binding homeodomain of HoxA9 (Fig. 2d). This fusion protein principally localized at punctate foci in the nucleoplasm [94], although some foci appear to be near or at the NE [17]. Since the discovery of the Nup98-HoxA9 fusion, Nup98 has been found fused to at least 31 different proteins in a variety of myeloid malignancies, revealing itself to be something of an oncogenic hot spot [17–20, 95–97]. Cancers in which Nup98 fusion proteins have been found include AML, MDS, T-ALL, chronic myelogenous leukemia (CML), chronic myelomonocytic leukemia (CMML), and juvenile myelomonocytic leukemia (JMML). Structurally all of the fusion proteins are similar, consisting of the majority or all of the Nup98 GLFG repeats joined to the C-terminus of various partners. They also have similar localizations, mostly nucleoplasmic and in GLFG bodies, but excluded from nucleoli [18].

The fusion partners of Nup98 can be categorized as either homeodomain (HD) proteins or non-HD proteins [17–19]. HD proteins are transcription factors characterized by the DNA-binding HD domain that play crucial roles in embryonic development [98]. In Nup98 translocations involving HD proteins, the fusion protein always contains the DNA-binding HD domain (Fig. 2d). Several Nup98 fusions with non-HD proteins contain DNA- or chromatin-binding domains such as PHD finger, SET, PWWP, or AT-Hook. In other cases Nup98 is fused to a RNA helicase, DNA topoisomerases, or proteins with no known transcriptional roles [18, 19]. More than half of the fusion proteins contain coiled-coil domains that have been proposed to mediate oligomerization of the fusion proteins [99].

The mechanisms by which Nup98 fusion proteins lead to cancer are once again largely unknown, but recent studies in murine models of two non-HD fusions revealed some clues. The fusion of Nup98 with the histone methyltransferase NSD1 upregulates expression of *HOXA* genes by maintaining high levels of histone H3-K36 methylation and acetylation. Deletion of the Nup98 GLFG repeats or mutations inactivating NSD1 methyltransferase function both eliminated *HOXA* activation and leukemogenesis [100]. In the Nup98–JARID1A fusion, the PHD finger domain from JARID1A binds to trimethylated histone H3 at hematopoietic-specific promoters, leading to constitutive activation of these genes [101]. While informative, these studies do not explain how Nup98 fusions with so many different partners, some of which lack any DNA or chromatin binding ability, can cause similar disease phenotypes. This has led some to propose that leukemogenesis may arise due to aberrant function of Nup98 in either its transport or nucleoplasmic transcriptional roles. Since Nup98 is synthesized as a Nup98-Nup96 precursor molecule that is posttranslationally cleaved, any chromosomal translocation at the *NUP98-NUP96* locus would most likely decrease Nup96 protein levels leading to misregulation of nucleocytoplasmic transport [20]. Another possibility is that Nup98 normally functions as a transcriptional regulator during hematopoiesis. Chromosomal translocations involving Nup98 could lead to misregulation of its target genes and mislocalization of the wild-type Nup98 due to interactions between GLFG domains [19].

Recent findings indicate that Nup98 may have an important additional role: as a tumor suppressor. Nup98 binds to and prevents degradation of select p53 target mRNAs, supporting the tumor suppressing activity of p53; furthermore, the expression of Nup98 was found to be significantly decreased in one-quarter of surveyed hepatocellular carcinoma patients [102]. These results raise the intriguing idea that some nucleoporins, or the NPC itself, may function—directly or indirectly—as a suppressor of oncogenesis.

Nup88

Nup88 is a non-FG component of the cytoplasmic filaments. It consists of an N-terminal β-propeller domain and a C-terminal coiled-coil domain (Fig. 2e). Interestingly, in *Drosophila* larvae, Nup88 is expressed in a tissue-specific manner [103]. Within the NPC it forms a subcomplex with Nup214 that helps regulate export from the nucleus [46, 47, 104, 105]. Curiously, Nup88 also interacts with Nup98 and Nup358/RanBP2, nucleoporins that are also linked to cancers—and to mRNA export [69, 85].

Unlike the other oncogenic nucleoporins, Nup88 does not form fusion proteins but is instead often overexpressed in cancer cells. Nup88 was first found overexpressed in several human cancer cell lines and 75 % of ovarian tumors [106]. Subsequent studies revealed overexpression of Nup88 in a variety of tumors [107, 108]. In these tumors other nucleoporins such as Nup214, Nup153, or Nup107 were not overexpressed [108, 109], suggesting that overexpression is unique to Nup88 and not a result of a general increase in expression of NPC components or upregulation of certain subcomplexes [17]. However, no comprehensive analysis of all nucleoporins in these cancers has ever been performed, and alterations in expression of different nucleoporins have been observed in other cancers. A different study of breast cancers showed increased expression of Nup133 and decreased expression of Nup214 but no changes in Nup88 expression [110]. A study of ovarian cancers revealed downregulation of Nup62 in several tumors but did not examine any other nucleoporins [111]. Closer attention clearly needs to be paid in the future to changes in expression or localization of all nucleoporins in different tumor cell lines.

When it is overexpressed in cancer cells, the excess Nup88 is usually found in granular dots in the cytoplasm [17], though additional copies at the NPC cannot currently be excluded. The intensity of cytoplasmic Nup88 staining correlates with tumor grade: overexpression is typically seen in advanced tumors as opposed to benign tumors or mild hyperplasias [109, 112–114]. High overexpression correlates with tumor aggressiveness and poor differentiation [21, 109, 114–116], and is often seen at tumor edges suggesting a link to tumor invasivity [108, 113]. Due to these observations Nup88 is proposed to be a marker of tumor state and indicator of patient prognosis [17, 20].

The mechanistic link between Nup88 overexpression and cancer is unknown. Due to its known function in nuclear export it has been suggested that overexpression of Nup88 in cancer cells may cause mislocalization of its subcomplex partner Nup214 and export receptors causing misregulation of transport of signaling proteins and transcription factors [17, 20]. For example, loss of Nup88 causes accumulation of NF-κB in the cytoplasm due to increased nuclear export [103, 117]. Since NF-κB is seen mainly in the nucleus in many cancer types [118], it has been proposed that overexpression of Nup88 may lead to a decrease in NF-κB export from the nucleus leading to its accumulation and upregulation of target genes [17].

Nuclear Export Factors

All of the oncogenic nucleoporins described above play major roles in mediating nuclear export of mRNAs. It is therefore not surprising that several nuclear export factors (transport receptors) that associate with these proteins are also linked to cancers [119]. Rae1, an mRNA export factor that also plays a role in cell cycle regulation and binds to the FG repeat domain of Nup98 [120, 121], is upregulated in breast cancers [122]. Expression levels of two components of the transcriptional export (TREX) complex are altered in several cancer types. One component, THOC1, is upregulated in lung, colon, and ovarian cancers but is downregulated in thyroid, skin, and testis cancers [123–126]. Increased expression of THOC1 in breast cancers correlates with tumor size and metastatic state while siRNA-mediated downregulation of THOC1 in cancer cells leads to inhibition of mRNA export and decrease in cell proliferation [124]. Aly/REF, another component of the TREX complex, is upregulated in a wide variety of tumors [123, 127]. GANP, a component of the TREX-2 complex, is overexpressed in mantle cell, diffuse large B cell, and Hodgkin's lymphomas [128]. eIF4E mediates nuclear export of certain mRNAs via the Crm1 pathway [84, 129] and is overexpressed in ~30 % of analyzed tumors

[130]. Crm1 is also overexpressed in many tumors, such as cervical and pancreatic cancers as well as gliomas [131–134]. Along with Crm1, Karyopherin β1 and Karyopherin α2 are overexpressed in cervical cancers and also in transformed epithelial and fibroblast cells [134]. Importantly, downregulation of Crm1 and Karyopherin β1, but not of Karyopherin α2, in cancer cells causes cell cycle arrest and apoptosis, suggesting a link between tumorigenesis and increased expression levels of these export factors [119, 134]. Presumably as a result of Crm1 overexpression many cell cycle inhibitors and tumor suppressing proteins are mislocalized to the cytoplasm in a variety of cancers [133].

Viruses Exploit the NPC, Also Targeting the Oncogenic Nucleoporins

A remarkable number of viruses, including some major human pathogens, have specific mechanisms to alter the structure, composition, and function of the NPC. These viruses usually target the nucleoporins with links to cancer (Fig. 1) [135–137], suggesting they exploit some of the same weaknesses as the oncogenic defects. For example, the $2A^{pro}$ protease of enteroviruses poliovirus (PV) and human rhinovirus (HRV) sequentially cleaves Nup62 (2 h post-infection), Nup98 (4.5 h post-infection), and finally Nup153 (6 h post-infection) [138–140]. After all three are cleaved, nuclear import is inhibited and host nuclear proteins accumulate in the cytoplasm [140]. This could prevent tumor suppressor proteins, and apoptosis inducing proteins, from acting in the nucleus. Overexpression of the HRV $3C^{pro}$ protease leads to proteolysis of Nup153, Nup214, and Nup358/RanBP2 [141]. Enterovirus infected cells show significantly reduced staining with mAb414, which recognizes eight different nucleoporins, indicating defects in NPC organization and composition [138, 139].

Cardioviruses also target NPC components, but instead of degradation they are hyperphosphorylated in infected cells. The proteins affected are Nup62, Nup153, and Nup214 by *encephalomyocarditis* (EMCV) and Nup98 by Theiler's Murine Encephalomyelitis Virus (TMEV) [142–144]. Phosphorylation is mediated by the zinc finger domain of viral L protein, which does not possess kinase activity but is thought to recruit cellular kinases to the NPC [142, 145]. The BGLF4 kinase of Epstein-Barr virus associates with, and presumably phosphorylates, Nup62 and Nup153 [146].

Two types of human adenoviruses (Ad2 and Ad5) bind to Nup214 and the Nup358/RanBP2-associated kinesin-1 on the cytoplasmic side of the NPC [147]. This leads to disassembly of the viral coat, import of the viral genome into the nucleus, and mislocalization of Nup62, Nup214, and Nup358/RanBP2 into the cytoplasm. Nup153, found at the nuclear side of the NPC, was not affected. Why kinesin-1 is involved is unclear, but it is thought that binding of kinesin-1 to Nup358/RanBP2 activates its motor function, causing disassembly of the viral coat and the virus in turn acts on the NPC causing release of Nup62, Nup214, and Nup358/RanBP2 into the cytoplasm [135, 147]. Downregulation of either Nup358/RanBP2 or Nup214 using siRNA was sufficient to prevent viral uncoating and genome import into the nucleus. After release from the NPC the displaced nucleoporins, along with RanGAP1 and Crm1, colocalize with disassembled virus particles at the cell periphery [147]. Herpesviruses also attach at the NPC cytoplasmic filaments, binding specifically to Nup214, Nup358/RanBP2, and hCG1 (NupL2). As with adenoviruses, siRNA-mediated downregulation of either Nup214 or Nup358/RanBP2 was sufficient to prevent viral uncoating and genome import into the nucleus [148, 149].

The M protein of vesicular stomatitis virus (VSV) associates with the Nup98/Rae1 complex in order to inhibit host cell transcription and mitotic progression [150, 151]. Remarkably, VSV preferentially infects cancer cells rather than normal cells and therefore acts as an oncolytic agent [152, 153]. The influenza A virus affects nucleocytoplasmic trafficking by disrupting many mRNA export factors [137]. Nup98 expression is also significantly decreased in infected cells while Nup62 or Nup153 are unaffected [154], suggesting that influenza specifically targets and disrupts normal Nup98 function.

The human immunodeficiency virus (HIV-1) hijacks over 1,000 cellular proteins during its infection cycle [155, 156]. At the NPC it binds to the cyclophilin homology domain of Nup358/RanBP2 [157, 158]. siRNA-mediated downregulation of Nup358/RanBP2 but not of Nup98, Nup153, or Nup214 significantly reduced viral import [158]. However, Nup153 is necessary for the transport of HIV-1 through the NPC after it has docked at Nup358/RanBP2 [158–160]. Furthermore, in infected cells the heterogeneous nuclear ribonucleoprotein (hnRNP) A1 protein and its import receptor Karyopherin β2 are retained in the cytoplasm, due to downregulation and displacement of Nup62 into the cytoplasm [161]. Expression levels of many nucleoporins were also affected: Nup43, Nup45, Nup54, and Nup58 were downregulated, Nup35, Nup98, and Tpr were upregulated, and others were differentially regulated in different studies [162–164]. Surprisingly, Nup62 was also detected in budding virions, suggesting that it might play a role in assembly and infectivity of HIV-1 [135, 164]. Supporting this, siRNA-mediated downregulation of Nup62 caused decreased viral protein synthesis and viral production [164].

Emerging Themes

The tremendous amount of data now being accrued on the nature of the links between particular nucleoporins, tumorigenesis, and viral infection allows us to see a possible pattern of disease sensitivity at the NPC. Firstly, the nucleoporins linked to cancers and viral infections are all found in the same places in the NPC—flanking the entrance and exit of the central tube (Fig. 1). They are a discrete subset of proteins, representing less than a quarter of the total number of nucleoporins comprising the NPC, and are proximal to or components of the nuclear basket or the cytoplasmic filaments (Fig. 1). Secondly, these particular proteins—and most of the nuclear transport factors similarly implicated in disease—are especially important for the mechanism or regulation of mRNA export. Why exactly this should be so is still unclear. However, because many of the disease-causing defects or alterations lead to reduction of one or more of these nucleoporins at the NPC, it could be that an important additional function of these nucleoporins, and the NPC itself, is in tumor suppression. Such a function has already been demonstrated for Nup98, which supports p53 tumor suppressing activity [19, 102]. A possibility is that many of the oncogenic nucleoporin fusions represent a lethal double hit to the cell, as a consequence of both the fusion itself and the resulting depletion of normal nucleoporin from the NPC—the former altering the function of the fused protein, and the latter compromising the tumor suppressor activity of the NPC.

Of course, the complete pathway leading from a nucleoporin defect to full disease is surely multifactorial and will include oncogenic effects functionally unrelated to the fact that the protein involved is a nucleoporin, but nonetheless may combine with NPC-related alterations in function such as reduction of a key protein at the NPC or mis-targeting of nucleoporins or transport factors. What is certain is that, in recent years, the NPC has emerged as a major nexus of key human diseases, and that we now urgently need to determine the basic cell biology of the nucleocytoplasmic transport machinery if we are to understand how these diseases can be cured.

References

1. Alber F, Dokudovskaya S, Veenhoff LM, Zhang W, Kipper J, Devos D, Suprapto A, Karni-Schmidt O, Williams R, Chait BT, Sali A, Rout MP (2007) The molecular architecture of the nuclear pore complex. Nature 450(7170):695–701. doi:10.1038/nature06405
2. Bilokapic S, Schwartz TU (2012) 3D ultrastructure of the nuclear pore complex. Curr Opin Cell Biol 24(1):86–91. doi:10.1016/j.ceb.2011.12.011
3. Strambio-De-Castillia C, Niepel M, Rout MP (2010) The nuclear pore complex: bridging nuclear transport and gene regulation. Nat Rev Mol Cell Biol 11(7):490–501. doi:10.1038/nrm2928
4. Raices M, D'Angelo MA (2012) Nuclear pore complex composition: a new regulator of tissue-specific and developmental functions. Nat Rev Mol Cell Biol 13(11):687–699. doi:10.1038/nrm3461
5. Cronshaw JM, Krutchinsky AN, Zhang W, Chait BT, Matunis MJ (2002) Proteomic analysis of the mammalian nuclear pore complex. J Cell Biol 158(5):915–927. doi:10.1083/jcb.200206106
6. D'Angelo MA, Hetzer MW (2008) Structure, dynamics and function of nuclear pore complexes. Trends Cell Biol 18(10):456–466. doi:10.1016/j.tcb.2008.07.009
7. Quimby BB, Lamitina T, L'Hernault SW, Corbett AH (2000) The mechanism of ran import into the nucleus by nuclear transport factor 2. J Biol Chem 275(37):28575–28582. doi:10.1074/jbc.M005055200
8. Stewart M (2007) Ratcheting mRNA out of the nucleus. Mol Cell 25(3):327–330. doi:10.1016/j.molcel.2007.01.016
9. Joseph J, Dasso M (2008) The nucleoporin Nup358 associates with and regulates interphase microtubules. FEBS Lett 582(2):190–196. doi:10.1016/j.febslet.2007.11.087
10. Cai Y, Singh BB, Aslanukov A, Zhao H, Ferreira PA (2001) The docking of kinesins, KIF5B and KIF5C, to Ran-binding protein 2 (RanBP2) is mediated via a novel RanBP2 domain. J Biol Chem 276(45):41594–41602. doi:10.1074/jbc.M104514200

11. Singh BB, Patel HH, Roepman R, Schick D, Ferreira PA (1999) The zinc finger cluster domain of RanBP2 is a specific docking site for the nuclear export factor, exportin-1. J Biol Chem 274(52):37370–37378
12. Akhtar A, Gasser SM (2007) The nuclear envelope and transcriptional control. Nat Rev Genet 8(7):507–517. doi:10.1038/nrg2122
13. Nakano H, Funasaka T, Hashizume C, Wong RW (2010) Nucleoporin translocated promoter region (Tpr) associates with dynein complex, preventing chromosome lagging formation during mitosis. J Biol Chem 285(14):10841–10849. doi:10.1074/jbc.M110.105890
14. Towbin BD, Meister P, Gasser SM (2009) The nuclear envelope—a scaffold for silencing? Curr Opin Genet Dev 19(2):180–186. doi:10.1016/j.gde.2009.01.006
15. Kalverda B, Pickersgill H, Shloma VV, Fornerod M (2010) Nucleoporins directly stimulate expression of developmental and cell-cycle genes inside the nucleoplasm. Cell 140(3): 360–371. doi:10.1016/j.cell.2010.01.011
16. Capelson M, Hetzer MW (2009) The role of nuclear pores in gene regulation, development and disease. EMBO Rep 10(7):697–705. doi:10.1038/embor.2009.147
17. Xu S, Powers MA (2009) Nuclear pore proteins and cancer. Semin Cell Dev Biol 20(5): 620–630. doi:10.1016/j.semcdb.2009.03.003
18. Gough SM, Slape CI, Aplan PD (2011) NUP98 gene fusions and hematopoietic malignancies: common themes and new biologic insights. Blood 118(24):6247–6257. doi:10.1182/blood-2011-07-328880
19. Franks TM, Hetzer MW (2013) The role of Nup98 in transcription regulation in healthy and diseased cells. Trends Cell Biol 23(3):112–117. doi:10.1016/j.tcb.2012.10.013
20. Kohler A, Hurt E (2010) Gene regulation by nucleoporins and links to cancer. Mol Cell 38(1):6–15. doi:10.1016/j.molcel.2010.01.040
21. Chow KH, Factor RE, Ullman KS (2012) The nuclear envelope environment and its cancer connections. Nat Rev Cancer 12(3):196–209. doi:10.1038/nrc3219
22. Simon DN, Wilson KL (2011) The nucleoskeleton as a genome-associated dynamic 'network of networks'. Nat Rev Mol Cell Biol 12(11):695–708. doi:10.1038/nrm3207
23. Krull S, Dorries J, Boysén B, Reidenbach S, Magnius L, Norder H, Thyberg J, Cordes VC (2010) Protein Tpr is required for establishing nuclear pore-associated zones of heterochromatin exclusion. EMBO J 29(10):1659–1673. doi:10.1038/emboj.2010.54
24. Ben-Efraim I, Frosst PD, Gerace L (2009) Karyopherin binding interactions and nuclear import mechanism of nuclear pore complex protein Tpr. BMC Cell Biol 10:74. doi:10.1186/1471-2121-10-74
25. Frosst P, Guan T, Subauste C, Hahn K, Gerace L (2002) Tpr is localized within the nuclear basket of the pore complex and has a role in nuclear protein export. J Cell Biol 156(4):617–630. doi:10.1083/jcb.200106046
26. Dieppois G, Iglesias N, Stutz F (2006) Cotranscriptional recruitment to the mRNA export receptor Mex67p contributes to nuclear pore anchoring of activated genes. Mol Cell Biol 26(21):7858–7870. doi:10.1128/MCB.00870-06
27. Galy V, Gadal O, Fromont-Racine M, Romano A, Jacquier A, Nehrbass U (2004) Nuclear retention of unspliced mRNAs in yeast is mediated by perinuclear Mlp1. Cell 116(1): 63–73
28. Galy V, Olivo-Marin JC, Scherthan H, Doye V, Rascalou N, Nehrbass U (2000) Nuclear pore complexes in the organization of silent telomeric chromatin. Nature 403(6765):108–112. doi:10.1038/47528
29. Lee SH, Sterling H, Burlingame A, McCormick F (2008) Tpr directly binds to Mad1 and Mad2 and is important for the Mad1-Mad2-mediated mitotic spindle checkpoint. Genes Dev 22(21):2926–2931. doi:10.1101/gad.1677208
30. Luthra R, Kerr SC, Harreman MT, Apponi LH, Fasken MB, Ramineni S, Chaurasia S, Valentini SR, Corbett AH (2007) Actively transcribed GAL genes can be physically linked to the nuclear pore by the SAGA chromatin modifying complex. J Biol Chem 282(5): 3042–3049. doi:10.1074/jbc.M608741200

31. Byrd DA, Sweet DJ, Pante N, Konstantinov KN, Guan T, Saphire AC, Mitchell PJ, Cooper CS, Aebi U, Gerace L (1994) Tpr, a large coiled coil protein whose amino terminus is involved in activation of oncogenic kinases, is localized to the cytoplasmic surface of the nuclear pore complex. J Cell Biol 127(6 Pt 1):1515–1526
32. Bangs P, Burke B, Powers C, Craig R, Purohit A, Doxsey S (1998) Functional analysis of Tpr: identification of nuclear pore complex association and nuclear localization domains and a role in mRNA export. J Cell Biol 143(7):1801–1812
33. Cooper CS, Park M, Blair DG, Tainsky MA, Huebner K, Croce CM, Vande Woude GF (1984) Molecular cloning of a new transforming gene from a chemically transformed human cell line. Nature 311(5981):29–33
34. Hays JL, Watowich SJ (2003) Oligomerization-induced modulation of TPR-MET tyrosine kinase activity. J Biol Chem 278(30):27456–27463. doi:10.1074/jbc.M210648200
35. Park M, Dean M, Cooper CS, Schmidt M, O'Brien SJ, Blair DG, Vande Woude GF (1986) Mechanism of met oncogene activation. Cell 45(6):895–904
36. Peschard P, Park M (2007) From Tpr-Met to Met, tumorigenesis and tubes. Oncogene 26(9):1276–1285. doi:10.1038/sj.onc.1210201
37. Rodrigues GA, Park M (1993) Dimerization mediated through a leucine zipper activates the oncogenic potential of the met receptor tyrosine kinase. Mol Cell Biol 13(11):6711–6722
38. Soman NR, Correa P, Ruiz BA, Wogan GN (1991) The TPR-MET oncogenic rearrangement is present and expressed in human gastric carcinoma and precursor lesions. Proc Natl Acad Sci U S A 88(11):4892–4896
39. Yu J, Miehlke S, Ebert MP, Hoffmann J, Breidert M, Alpen B, Starzynska T, Stolte Prof M, Malfertheiner P, Bayerdorffer E (2000) Frequency of TPR-MET rearrangement in patients with gastric carcinoma and in first-degree relatives. Cancer 88(8):1801–1806
40. Greco A, Miranda C, Pagliardini S, Fusetti L, Bongarzone I, Pierotti MA (1997) Chromosome 1 rearrangements involving the genes TPR and NTRK1 produce structurally different thyroid-specific TRK oncogenes. Genes Chromosomes Cancer 19(2):112–123
41. Greco A, Miranda C, Pierotti MA (2010) Rearrangements of NTRK1 gene in papillary thyroid carcinoma. Mol Cell Endocrinol 321(1):44–49. doi:10.1016/j.mce.2009.10.009
42. Pierotti MA, Greco A (2006) Oncogenic rearrangements of the NTRK1/NGF receptor. Cancer Lett 232(1):90–98. doi:10.1016/j.canlet.2005.07.043
43. Li F, Zhai YP, Tang YM, Wang LP, Wan PJ (2012) Identification of a novel partner gene, TPR, fused to FGFR1 in 8p11 myeloproliferative syndrome. Genes Chromosomes Cancer 51(9):890–897. doi:10.1002/gcc.21973
44. Alfonso P, Canamero M, Fernandez-Carbonie F, Nunez A, Casal JI (2008) Proteome analysis of membrane fractions in colorectal carcinomas by using 2D-DIGE saturation labeling. J Proteome Res 7(10):4247–4255. doi:10.1021/pr800152u
45. van Deursen J, Boer J, Kasper L, Grosveld G (1996) G2 arrest and impaired nucleocytoplasmic transport in mouse embryos lacking the proto-oncogene CAN/Nup214. EMBO J 15(20):5574–5583
46. Bernad R, Engelsma D, Sanderson H, Pickersgill H, Fornerod M (2006) Nup214-Nup88 nucleoporin subcomplex is required for CRM1-mediated 60S preribosomal nuclear export. J Biol Chem 281(28):19378–19386. doi:10.1074/jbc.M512585200
47. Hutten S, Kehlenbach RH (2006) Nup214 is required for CRM1-dependent nuclear protein export in vivo. Mol Cell Biol 26(18):6772–6785. doi:10.1128/MCB.00342-06
48. Fornerod M, van Deursen J, van Baal S, Reynolds A, Davis D, Murti KG, Fransen J, Grosveld G (1997) The human homologue of yeast CRM1 is in a dynamic subcomplex with CAN/Nup214 and a novel nuclear pore component Nup88. EMBO J 16(4):807–816. doi:10.1093/emboj/16.4.807
49. von Lindern M, Fornerod M, van Baal S, Jaegle M, de Wit T, Buijs A, Grosveld G (1992) The translocation (6;9), associated with a specific subtype of acute myeloid leukemia, results in the fusion of two genes, dek and can, and the expression of a chimeric, leukemia-specific dek-can mRNA. Mol Cell Biol 12(4):1687–1697

50. von Lindern M, van Baal S, Wiegant J, Raap A, Hagemeijer A, Grosveld G (1992) Can, a putative oncogene associated with myeloid leukemogenesis, may be activated by fusion of its 3′ half to different genes: characterization of the set gene. Mol Cell Biol 12(8):3346–3355
51. Privette Vinnedge LM, Kappes F, Nassar N, Wells SI (2013) Stacking the DEK: from chromatin topology to cancer stem cells. Cell Cycle 12(1):51–66. doi:10.4161/cc.23121
52. Fornerod M, Boer J, van Baal S, Jaegle M, von Lindern M, Murti KG, Davis D, Bonten J, Buijs A, Grosveld G (1995) Relocation of the carboxyterminal part of CAN from the nuclear envelope to the nucleus as a result of leukemia-specific chromosome rearrangements. Oncogene 10(9):1739–1748
53. Van Vlierberghe P, van Grotel M, Tchinda J, Lee C, Beverloo HB, van der Spek PJ, Stubbs A, Cools J, Nagata K, Fornerod M, Buijs-Gladdines J, Horstmann M, van Wering ER, Soulier J, Pieters R, Meijerink JP (2008) The recurrent SET-NUP214 fusion as a new HOXA activation mechanism in pediatric T-cell acute lymphoblastic leukemia. Blood 111(9):4668–4680. doi:10.1182/blood-2007-09-111872
54. Ageberg M, Drott K, Olofsson T, Gullberg U, Lindmark A (2008) Identification of a novel and myeloid specific role of the leukemia-associated fusion protein DEK-NUP214 leading to increased protein synthesis. Genes Chromosomes Cancer 47(4):276–287. doi:10.1002/gcc.20531
55. Saito S, Miyaji-Yamaguchi M, Nagata K (2004) Aberrant intracellular localization of SET-CAN fusion protein, associated with a leukemia, disorganizes nuclear export. Int J Cancer 111(4):501–507. doi:10.1002/ijc.20296
56. Fornerod M, Boer J, van Baal S, Morreau H, Grosveld G (1996) Interaction of cellular proteins with the leukemia specific fusion proteins DEK-CAN and SET-CAN and their normal counterpart, the nucleoporin CAN. Oncogene 13(8):1801–1808
57. Graux C, Cools J, Melotte C, Quentmeier H, Ferrando A, Levine R, Vermeesch JR, Stul M, Dutta B, Boeckx N, Bosly A, Heimann P, Uyttebroeck A, Mentens N, Somers R, MacLeod RA, Drexler HG, Look AT, Gilliland DG, Michaux L, Vandenberghe P, Wlodarska I, Marynen P, Hagemeijer A (2004) Fusion of NUP214 to ABL1 on amplified episomes in T-cell acute lymphoblastic leukemia. Nat Genet 36(10):1084–1089. doi:10.1038/ng1425
58. De Braekeleer E, Douet-Guilbert N, Rowe D, Bown N, Morel F, Berthou C, Ferec C, De Braekeleer M (2011) ABL1 fusion genes in hematological malignancies: a review. Eur J Haematol 86(5):361–371. doi:10.1111/j.1600-0609.2011.01586.x
59. De Keersmaecker K, Rocnik JL, Bernad R, Lee BH, Leeman D, Gielen O, Verachtert H, Folens C, Munck S, Marynen P, Fornerod M, Gilliland DG, Cools J (2008) Kinase activation and transformation by NUP214-ABL1 is dependent on the context of the nuclear pore. Mol Cell 31(1):134–142. doi:10.1016/j.molcel.2008.05.005
60. Gorello P, La Starza R, Di Giacomo D, Messina M, Puzzolo MC, Crescenzi B, Santoro A, Chiaretti S, Mecucci C (2010) SQSTM1-NUP214: a new gene fusion in adult T-cell acute lymphoblastic leukemia. Haematologica 95(12):2161–2163. doi:10.3324/haematol.2010.029769
61. Duran A, Linares JF, Galvez AS, Wikenheiser K, Flores JM, Diaz-Meco MT, Moscat J (2008) The signaling adaptor p62 is an important NF-kappaB mediator in tumorigenesis. Cancer Cell 13(4):343–354. doi:10.1016/j.ccr.2008.02.001
62. Hiruma Y, Honjo T, Jelinek DF, Windle JJ, Shin J, Roodman GD, Kurihara N (2009) Increased signaling through p62 in the marrow microenvironment increases myeloma cell growth and osteoclast formation. Blood 113(20):4894–4902. doi:10.1182/blood-2008-08-173948
63. Martin P, Diaz-Meco MT, Moscat J (2006) The signaling adapter p62 is an important mediator of T helper 2 cell function and allergic airway inflammation. EMBO J 25(15):3524–3533. doi:10.1038/sj.emboj.7601250
64. Pursiheimo JP, Rantanen K, Heikkinen PT, Johansen T, Jaakkola PM (2009) Hypoxia-activated autophagy accelerates degradation of SQSTM1/p62. Oncogene 28(3):334–344. doi:10.1038/onc.2008.392
65. Rodriguez A, Duran A, Selloum M, Champy MF, Diez-Guerra FJ, Flores JM, Serrano M, Auwerx J, Diaz-Meco MT, Moscat J (2006) Mature-onset obesity and insulin resistance in mice deficient in the signaling adapter p62. Cell Metab 3(3):211–222. doi:10.1016/j.cmet.2006.01.011

66. Wooten MW, Geetha T, Seibenhener ML, Babu JR, Diaz-Meco MT, Moscat J (2005) The p62 scaffold regulates nerve growth factor-induced NF-kappaB activation by influencing TRAF6 polyubiquitination. J Biol Chem 280(42):35625–35629. doi:10.1074/jbc.C500237200
67. Seibenhener ML, Babu JR, Geetha T, Wong HC, Krishna NR, Wooten MW (2004) Sequestosome 1/p62 is a polyubiquitin chain binding protein involved in ubiquitin proteasome degradation. Mol Cell Biol 24(18):8055–8068. doi:10.1128/MCB.24.18.8055-8068.2004
68. Nowak NJ, Sait SN, Zeidan A, Deeb G, Gaile D, Liu S, Ford L, Wallace PK, Wang ES, Wetzler M (2010) Recurrent deletion of 9q34 in adult normal karyotype precursor B-cell acute lymphoblastic leukemia. Cancer Genet Cytogenet 199(1):15–20. doi:10.1016/j.cancergencyto.2010.01.014
69. Bernad R, van der Velde H, Fornerod M, Pickersgill H (2004) Nup358/RanBP2 attaches to the nuclear pore complex via association with Nup88 and Nup214/CAN and plays a supporting role in CRM1-mediated nuclear protein export. Mol Cell Biol 24(6):2373–2384
70. Forler D, Rabut G, Ciccarelli FD, Herold A, Kocher T, Niggeweg R, Bork P, Ellenberg J, Izaurralde E (2004) RanBP2/Nup358 provides a major binding site for NXF1-p15 dimers at the nuclear pore complex and functions in nuclear mRNA export. Mol Cell Biol 24(3):1155–1167
71. Hutten S, Walde S, Spillner C, Hauber J, Kehlenbach RH (2009) The nuclear pore component Nup358 promotes transportin-dependent nuclear import. J Cell Sci 122(Pt 8):1100–1110. doi:10.1242/jcs.040154
72. Matunis MJ, Wu J, Blobel G (1998) SUMO-1 modification and its role in targeting the Ran GTPase-activating protein, RanGAP1, to the nuclear pore complex. J Cell Biol 140(3):499–509
73. Pichler A, Gast A, Seeler JS, Dejean A, Melchior F (2002) The nucleoporin RanBP2 has SUMO1 E3 ligase activity. Cell 108(1):109–120
74. Reverter D, Lima CD (2005) Insights into E3 ligase activity revealed by a SUMO-RanGAP1-Ubc9-Nup358 complex. Nature 435(7042):687–692. doi:10.1038/nature03588
75. Wu J, Matunis MJ, Kraemer D, Blobel G, Coutavas E (1995) Nup358, a cytoplasmically exposed nucleoporin with peptide repeats, Ran-GTP binding sites, zinc fingers, a cyclophilin A homologous domain, and a leucine-rich region. J Biol Chem 270(23):14209–14213
76. Hamada M, Haeger A, Jeganathan KB, van Ree JH, Malureanu L, Walde S, Joseph J, Kehlenbach RH, van Deursen JM (2011) Ran-dependent docking of importin-beta to RanBP2/Nup358 filaments is essential for protein import and cell viability. J Cell Biol 194(4):597–612. doi:10.1083/jcb.201102018
77. Salina D, Enarson P, Rattner JB, Burke B (2003) Nup358 integrates nuclear envelope breakdown with kinetochore assembly. J Cell Biol 162(6):991–1001. doi:10.1083/jcb.200304080
78. Gervais C, Dano L, Perrusson N, Helias C, Jeandidier E, Galoisy AC, Ittel A, Herbrecht R, Bilger K, Mauvieux L (2013) A translocation t(2;8)(q12;p11) fuses FGFR1 to a novel partner gene, RANBP2/NUP358, in a myeloproliferative/myelodysplastic neoplasm. Leukemia 27(5):1186–1188. doi:10.1038/leu.2012.286
79. Ma Z, Hill DA, Collins MH, Morris SW, Sumegi J, Zhou M, Zuppan C, Bridge JA (2003) Fusion of ALK to the Ran-binding protein 2 (RANBP2) gene in inflammatory myofibroblastic tumor. Genes Chromosomes Cancer 37(1):98–105. doi:10.1002/gcc.10177
80. Chen ST, Lee JC (2008) An inflammatory myofibroblastic tumor in liver with ALK and RANBP2 gene rearrangement: combination of distinct morphologic, immunohistochemical, and genetic features. Hum Pathol 39(12):1854–1858. doi:10.1016/j.humpath.2008.04.016
81. Patel AS, Murphy KM, Hawkins AL, Cohen JS, Long PP, Perlman EJ, Griffin CA (2007) RANBP2 and CLTC are involved in ALK rearrangements in inflammatory myofibroblastic tumors. Cancer Genet Cytogenet 176(2):107–114. doi:10.1016/j.cancergencyto.2007.04.004
82. Gylfe AE, Kondelin J, Turunen M, Ristolainen H, Katainen R, Pitkanen E, Kaasinen E, Rantanen V, Tanskanen T, Varjosalo M, Lehtonen H, Palin K, Taipale M, Taipale J, Renkonen-Sinisalo L, Jarvinen H, Bohm J, Mecklin JP, Ristimaki A, Kilpivaara O, Tuupanen S, Karhu A, Vahteristo P, Aaltonen LA (2013) Identification of candidate oncogenes in human colorectal cancers with microsatellite instability. Gastroenterology. doi:10.1053/j.gastro.2013.05.015

83. Dawlaty MM, Malureanu L, Jeganathan KB, Kao E, Sustmann C, Tahk S, Shuai K, Grosschedl R, van Deursen JM (2008) Resolution of sister centromeres requires RanBP2-mediated SUMOylation of topoisomerase IIalpha. Cell 133(1):103–115. doi:10.1016/j.cell.2008.01.045
84. Culjkovic-Kraljacic B, Baguet A, Volpon L, Amri A, Borden KL (2012) The oncogene eIF4E reprograms the nuclear pore complex to promote mRNA export and oncogenic transformation. Cell Rep 2(2):207–215. doi:10.1016/j.celrep.2012.07.007
85. Griffis ER, Xu S, Powers MA (2003) Nup98 localizes to both nuclear and cytoplasmic sides of the nuclear pore and binds to two distinct nucleoporin subcomplexes. Mol Biol Cell 14(2):600–610. doi:10.1091/mbc.E02-09-0582
86. Radu A, Moore MS, Blobel G (1995) The peptide repeat domain of nucleoporin Nup98 functions as a docking site in transport across the nuclear pore complex. Cell 81(2):215–222
87. Griffis ER, Altan N, Lippincott-Schwartz J, Powers MA (2002) Nup98 is a mobile nucleoporin with transcription-dependent dynamics. Mol Biol Cell 13(4):1282–1297. doi:10.1091/mbc.01-11-0538
88. Griffis ER, Craige B, Dimaano C, Ullman KS, Powers MA (2004) Distinct functional domains within nucleoporins Nup153 and Nup98 mediate transcription-dependent mobility. Mol Biol Cell 15(4):1991–2002. doi:10.1091/mbc.E03-10-0743
89. Capelson M, Liang Y, Schulte R, Mair W, Wagner U, Hetzer MW (2010) Chromatin-bound nuclear pore components regulate gene expression in higher eukaryotes. Cell 140(3): 372–383. doi:10.1016/j.cell.2009.12.054
90. Liang Y, Franks TM, Marchetto MC, Gage FH, Hetzer MW (2013) Dynamic association of NUP98 with the human genome. PLoS Genet 9(2):e1003308. doi:10.1371/journal.pgen.1003308
91. Iwamoto M, Asakawa H, Hiraoka Y, Haraguchi T (2010) Nucleoporin Nup98: a gatekeeper in the eukaryotic kingdoms. Genes Cells 15(7):661–669. doi:10.1111/j.1365-2443.2010.01415.x
92. Borrow J, Shearman AM, Stanton VP Jr, Becher R, Collins T, Williams AJ, Dube I, Katz F, Kwong YL, Morris C, Ohyashiki K, Toyama K, Rowley J, Housman DE (1996) The t(7;11) (p15;p15) translocation in acute myeloid leukaemia fuses the genes for nucleoporin NUP98 and class I homeoprotein HOXA9. Nat Genet 12(2):159–167. doi:10.1038/ng0296-159
93. Nakamura T, Largaespada DA, Lee MP, Johnson LA, Ohyashiki K, Toyama K, Chen SJ, Willman CL, Chen IM, Feinberg AP, Jenkins NA, Copeland NG, Shaughnessy JD Jr (1996) Fusion of the nucleoporin gene NUP98 to HOXA9 by the chromosome translocation t(7;11) (p15;p15) in human myeloid leukaemia. Nat Genet 12(2):154–158. doi:10.1038/ng0296-154
94. Kasper LH, Brindle PK, Schnabel CA, Pritchard CE, Cleary ML, van Deursen JM (1999) CREB binding protein interacts with nucleoporin-specific FG repeats that activate transcription and mediate NUP98-HOXA9 oncogenicity. Mol Cell Biol 19(1):764–776
95. Gorello P, Nofrini V, Brandimarte L, Pierini V, Crescenzi B, Nozza F, Daniele G, Storlazzi CT, Di Giacomo D, Matteucci C, La Starza R, Mecucci C (2013) Inv(11)(p15q22)/NUP98-DDX10 fusion and isoforms in a new case of de novo acute myeloid leukemia. Cancer Genet 206(3):92–96. doi:10.1016/j.cancergen.2013.02.001
96. Lisboa S, Cerveira N, Bizarro S, Correia C, Vieira J, Torres L, Mariz JM, Teixeira MR (2013) POU1F1 is a novel fusion partner of NUP98 in acute myeloid leukemia with t(3;11)(p11;p15). Mol Cancer 12:5. doi:10.1186/1476-4598-12-5
97. Soler G, Kaltenbach S, Dobbelstein S, Broccardo C, Radford I, Mozziconacci MJ, Bernard OA, Penard-Lacronique V, Delabesse E, Romana SP (2013) Identification of GSX2 and AF10 as NUP98 partner genes in myeloid malignancies. Blood Cancer J 3:e124. doi:10.1038/bcj.2013.20
98. Shah N, Sukumar S (2010) The Hox genes and their roles in oncogenesis. Nat Rev Cancer 10(5):361–371. doi:10.1038/nrc2826
99. Hussey DJ, Dobrovic A (2002) Recurrent coiled-coil motifs in NUP98 fusion partners provide a clue to leukemogenesis. Blood 99(3):1097–1098
100. Wang GG, Cai L, Pasillas MP, Kamps MP (2007) NUP98-NSD1 links H3K36 methylation to Hox-A gene activation and leukaemogenesis. Nat Cell Biol 9(7):804–812. doi:10.1038/ncb1608

101. Wang GG, Song J, Wang Z, Dormann HL, Casadio F, Li H, Luo JL, Patel DJ, Allis CD (2009) Haematopoietic malignancies caused by dysregulation of a chromatin-binding PHD finger. Nature 459(7248):847–851. doi:10.1038/nature08036
102. Singer S, Zhao R, Barsotti AM, Ouwehand A, Fazollahi M, Coutavas E, Breuhahn K, Neumann O, Longerich T, Pusterla T, Powers MA, Giles KM, Leedman PJ, Hess J, Grunwald D, Bussemaker HJ, Singer RH, Schirmacher P, Prives C (2012) Nuclear pore component Nup98 is a potential tumor suppressor and regulates posttranscriptional expression of select p53 target genes. Mol Cell 48(5):799–810. doi:10.1016/j.molcel.2012.09.020
103. Uv AE, Roth P, Xylourgidis N, Wickberg A, Cantera R, Samakovlis C (2000) members only encodes a Drosophila nucleoporin required for rel protein import and immune response activation. Genes Dev 14(15):1945–1957
104. Sabri N, Roth P, Xylourgidis N, Sadeghifar F, Adler J, Samakovlis C (2007) Distinct functions of the Drosophila Nup153 and Nup214 FG domains in nuclear protein transport. J Cell Biol 178(4):557–565. doi:10.1083/jcb.200612135
105. Xylourgidis N, Roth P, Sabri N, Tsarouhas V, Samakovlis C (2006) The nucleoporin Nup214 sequesters CRM1 at the nuclear rim and modulates NFkappaB activation in Drosophila. J Cell Sci 119(Pt 21):4409–4419. doi:10.1242/jcs.03201
106. Martinez N, Alonso A, Moragues MD, Ponton J, Schneider J (1999) The nuclear pore complex protein Nup88 is overexpressed in tumor cells. Cancer Res 59(21):5408–5411
107. Gould VE, Martinez N, Orucevic A, Schneider J, Alonso A (2000) A novel, nuclear pore-associated, widely distributed molecule overexpressed in oncogenesis and development. Am J Pathol 157(5):1605–1613. doi:10.1016/S0002-9440(10)64798-0
108. Gould VE, Orucevic A, Zentgraf H, Gattuso P, Martinez N, Alonso A (2002) Nup88 (karyoporin) in human malignant neoplasms and dysplasias: correlations of immunostaining of tissue sections, cytologic smears, and immunoblot analysis. Hum Pathol 33(5):536–544
109. Agudo D, Gomez-Esquer F, Martinez-Arribas F, Nunez-Villar MJ, Pollan M, Schneider J (2004) Nup88 mRNA overexpression is associated with high aggressiveness of breast cancer. Int J Cancer 109(5):717–720. doi:10.1002/ijc.20034
110. Hernandez P, Sole X, Valls J, Moreno V, Capella G, Urruticoechea A, Pujana MA (2007) Integrative analysis of a cancer somatic mutome. Mol Cancer 6:13. doi:10.1186/1476-4598-6-13
111. Kinoshita Y, Kalir T, Rahaman J, Dottino P, Kohtz DS (2012) Alterations in nuclear pore architecture allow cancer cell entry into or exit from drug-resistant dormancy. Am J Pathol 180(1):375–389. doi:10.1016/j.ajpath.2011.09.024
112. Brustmann H, Hager M (2009) Nucleoporin 88 expression in normal and neoplastic squamous epithelia of the uterine cervix. Ann Diagn Pathol 13(5):303–307. doi:10.1016/j.anndiagpath.2009.05.005
113. Zhao ZR, Zhang LJ, Wang YY, Li F, Wang MW, Sun XF (2012) Increased serum level of Nup88 protein is associated with the development of colorectal cancer. Med Oncol 29(3):1789–1795. doi:10.1007/s12032-011-0047-1
114. Emterling A, Skoglund J, Arbman G, Schneider J, Evertsson S, Carstensen J, Zhang H, Sun XF (2003) Clinicopathological significance of Nup88 expression in patients with colorectal cancer. Oncology 64(4):361–369. doi:70294
115. Knoess M, Kurz AK, Goreva O, Bektas N, Breuhahn K, Odenthal M, Schirmacher P, Dienes HP, Bock CT, Zentgraf H, zur Hausen A (2006) Nucleoporin 88 expression in hepatitis B and C virus-related liver diseases. World J Gastroenterol 12(36):5870–5874
116. Zhang ZY, Zhao ZR, Jiang L, Li JC, Gao YM, Cui DS, Wang CJ, Schneider J, Wang MW, Sun XF (2007) Nup88 expression in normal mucosa, adenoma, primary adenocarcinoma and lymph node metastasis in the colorectum. Tumour Biol 28(2):93–99. doi:10.1159/000099154
117. Takahashi N, van Kilsdonk JW, Ostendorf B, Smeets R, Bruggeman SW, Alonso A, van de Loo F, Schneider M, van den Berg WB, Swart GW (2008) Tumor marker nucleoporin 88 kDa regulates nucleocytoplasmic transport of NF-kappaB. Biochem Biophys Res Commun 374(3):424–430. doi:10.1016/j.bbrc.2008.06.128
118. Kau TR, Way JC, Silver PA (2004) Nuclear transport and cancer: from mechanism to intervention. Nat Rev Cancer 4(2):106–117. doi:10.1038/nrc1274

119. Culjkovic-Kraljacic B, Borden KL (2013) Aiding and abetting cancer: mRNA export and the nuclear pore. Trends Cell Biol 23(7):328–335. doi:10.1016/j.tcb.2013.03.004
120. Pritchard CE, Fornerod M, Kasper LH, van Deursen JM (1999) RAE1 is a shuttling mRNA export factor that binds to a GLEBS-like NUP98 motif at the nuclear pore complex through multiple domains. J Cell Biol 145(2):237–254
121. Sitterlin D (2004) Characterization of the Drosophila Rae1 protein as a G1 phase regulator of the cell cycle. Gene 326:107–116
122. Chin K, DeVries S, Fridlyand J, Spellman PT, Roydasgupta R, Kuo WL, Lapuk A, Neve RM, Qian Z, Ryder T, Chen F, Feiler H, Tokuyasu T, Kingsley C, Dairkee S, Meng Z, Chew K, Pinkel D, Jain A, Ljung BM, Esserman L, Albertson DG, Waldman FM, Gray JW (2006) Genomic and transcriptional aberrations linked to breast cancer pathophysiologies. Cancer Cell 10(6):529–541. doi:10.1016/j.ccr.2006.10.009
123. Dominguez-Sanchez MS, Saez C, Japon MA, Aguilera A, Luna R (2011) Differential expression of THOC1 and ALY mRNP biogenesis/export factors in human cancers. BMC Cancer 11:77. doi:10.1186/1471-2407-11-77
124. Guo S, Hakimi MA, Baillat D, Chen X, Farber MJ, Klein-Szanto AJ, Cooch NS, Godwin AK, Shiekhattar R (2005) Linking transcriptional elongation and messenger RNA export to metastatic breast cancers. Cancer Res 65(8):3011–3016. doi:10.1158/0008-5472.CAN-04-3624
125. Guo S, Liu M, Godwin AK (2012) Transcriptional regulation of hTREX84 in human cancer cells. PLoS One 7(8):e43610. doi:10.1371/journal.pone.0043610
126. Yang J, Li Y, Khoury T, Alrawi S, Goodrich DW, Tan D (2008) Relationships of hHpr1/p84/Thoc1 expression to clinicopathologic characteristics and prognosis in non-small cell lung cancer. Ann Clin Lab Sci 38(2):105–112
127. Saito Y, Kasamatsu A, Yamamoto A, Shimizu T, Yokoe H, Sakamoto Y, Ogawara K, Shiiba M, Tanzawa H, Uzawa K (2013) ALY as a potential contributor to metastasis in human oral squamous cell carcinoma. J Cancer Res Clin Oncol 139(4):585–594. doi:10.1007/s00432-012-1361-5
128. Fujimura S, Xing Y, Takeya M, Yamashita Y, Ohshima K, Kuwahara K, Sakaguchi N (2005) Increased expression of germinal center-associated nuclear protein RNA-primase is associated with lymphomagenesis. Cancer Res 65(13):5925–5934. doi:10.1158/0008-5472.CAN-04-3259
129. Culjkovic B, Topisirovic I, Skrabanek L, Ruiz-Gutierrez M, Borden KL (2006) eIF4E is a central node of an RNA regulon that governs cellular proliferation. J Cell Biol 175(3):415–426. doi:10.1083/jcb.200607020
130. Borden KL, Culjkovic-Kraljacic B (2010) Ribavirin as an anti-cancer therapy: acute myeloid leukemia and beyond? Leuk Lymphoma 51(10):1805–1815. doi:10.3109/10428194.2010.496506
131. Huang WY, Yue L, Qiu WS, Wang LW, Zhou XH, Sun YJ (2009) Prognostic value of CRM1 in pancreas cancer. Clin Invest Med 32(6):E315
132. Shen A, Wang Y, Zhao Y, Zou L, Sun L, Cheng C (2009) Expression of CRM1 in human gliomas and its significance in p27 expression and clinical prognosis. Neurosurgery 65(1):153–159. doi:10.1227/01.NEU.0000348550.47441.4B, discussion 159–160
133. Turner JG, Dawson J, Sullivan DM (2012) Nuclear export of proteins and drug resistance in cancer. Biochem Pharmacol 83(8):1021–1032. doi:10.1016/j.bcp.2011.12.016
134. van der Watt PJ, Maske CP, Hendricks DT, Parker MI, Denny L, Govender D, Birrer MJ, Leaner VD (2009) The Karyopherin proteins, Crm1 and Karyopherin beta1, are overexpressed in cervical cancer and are critical for cancer cell survival and proliferation. Int J Cancer 124(8):1829–1840. doi:10.1002/ijc.24146
135. Cohen S, Etingov I, Pante N (2012) Effect of viral infection on the nuclear envelope and nuclear pore complex. Int Rev Cell Mol Biol 299:117–159. doi:10.1016/B978-0-12-394310-1.00003-5
136. Kobiler O, Drayman N, Butin-Israeli V, Oppenheim A (2012) Virus strategies for passing the nuclear envelope barrier. Nucleus 3(6):526–539. doi:10.4161/nucl.21979
137. Kuss SK, Mata MA, Zhang L, Fontoura BM (2013) Nuclear imprisonment: viral strategies to arrest host mRNA nuclear export. Viruses 5(7):1824–1849. doi:10.3390/v5071824

138. Gustin KE, Sarnow P (2001) Effects of poliovirus infection on nucleo-cytoplasmic trafficking and nuclear pore complex composition. EMBO J 20(1–2):240–249. doi:10.1093/emboj/20.1.240
139. Gustin KE, Sarnow P (2002) Inhibition of nuclear import and alteration of nuclear pore complex composition by rhinovirus. J Virol 76(17):8787–8796
140. Park N, Katikaneni P, Skern T, Gustin KE (2008) Differential targeting of nuclear pore complex proteins in poliovirus-infected cells. J Virol 82(4):1647–1655. doi:10.1128/JVI.01670-07
141. Ghildyal R, Jordan B, Li D, Dagher H, Bardin PG, Gern JE, Jans DA (2009) Rhinovirus 3C protease can localize in the nucleus and alter active and passive nucleocytoplasmic transport. J Virol 83(14):7349–7352. doi:10.1128/JVI.01748-08
142. Bardina MV, Lidsky PV, Sheval EV, Fominykh KV, van Kuppeveld FJ, Polyakov VY, Agol VI (2009) Mengovirus-induced rearrangement of the nuclear pore complex: hijacking cellular phosphorylation machinery. J Virol 83(7):3150–3161. doi:10.1128/JVI.01456-08
143. Porter FW, Palmenberg AC (2009) Leader-induced phosphorylation of nucleoporins correlates with nuclear trafficking inhibition by cardioviruses. J Virol 83(4):1941–1951. doi:10.1128/JVI.01752-08
144. Ricour C, Delhaye S, Hato SV, Olenyik TD, Michel B, van Kuppeveld FJ, Gustin KE, Michiels T (2009) Inhibition of mRNA export and dimerization of interferon regulatory factor 3 by Theiler's virus leader protein. J Gen Virol 90(Pt 1):177–186. doi:10.1099/vir.0.005678-0
145. Lidsky PV, Hato S, Bardina MV, Aminev AG, Palmenberg AC, Sheval EV, Polyakov VY, van Kuppeveld FJ, Agol VI (2006) Nucleocytoplasmic traffic disorder induced by cardioviruses. J Virol 80(6):2705–2717. doi:10.1128/JVI.80.6.2705-2717.2006
146. Chang CW, Lee CP, Huang YH, Yang PW, Wang JT, Chen MR (2012) Epstein-Barr virus protein kinase BGLF4 targets the nucleus through interaction with nucleoporins. J Virol 86(15):8072–8085. doi:10.1128/JVI.01058-12
147. Strunze S, Engelke MF, Wang IH, Puntener D, Boucke K, Schleich S, Way M, Schoenenberger P, Burckhardt CJ, Greber UF (2011) Kinesin-1-mediated capsid disassembly and disruption of the nuclear pore complex promote virus infection. Cell Host Microbe 10(3):210–223. doi:10.1016/j.chom.2011.08.010
148. Copeland AM, Newcomb WW, Brown JC (2009) Herpes simplex virus replication: roles of viral proteins and nucleoporins in capsid-nucleus attachment. J Virol 83(4):1660–1668. doi:10.1128/JVI.01139-08
149. Pasdeloup D, Blondel D, Isidro AL, Rixon FJ (2009) Herpesvirus capsid association with the nuclear pore complex and viral DNA release involve the nucleoporin CAN/Nup214 and the capsid protein pUL25. J Virol 83(13):6610–6623. doi:10.1128/JVI.02655-08
150. Chakraborty P, Seemann J, Mishra RK, Wei JH, Weil L, Nussenzveig DR, Heiber J, Barber GN, Dasso M, Fontoura BM (2009) Vesicular stomatitis virus inhibits mitotic progression and triggers cell death. EMBO Rep 10(10):1154–1160. doi:10.1038/embor.2009.179
151. Rajani KR, Pettit Kneller EL, McKenzie MO, Horita DA, Chou JW, Lyles DS (2012) Complexes of vesicular stomatitis virus matrix protein with host Rae1 and Nup98 involved in inhibition of host transcription. PLoS Pathog 8(9):e1002929. doi:10.1371/journal.ppat.1002929
152. Balachandran S, Barber GN (2004) Defective translational control facilitates vesicular stomatitis virus oncolysis. Cancer Cell 5(1):51–65
153. Barber GN (2005) VSV-tumor selective replication and protein translation. Oncogene 24(52):7710–7719. doi:10.1038/sj.onc.1209042
154. Satterly N, Tsai PL, van Deursen J, Nussenzveig DR, Wang Y, Faria PA, Levay A, Levy DE, Fontoura BM (2007) Influenza virus targets the mRNA export machinery and the nuclear pore complex. Proc Natl Acad Sci U S A 104(6):1853–1858. doi:10.1073/pnas.0610977104
155. Friedrich BM, Dziuba N, Li G, Endsley MA, Murray JL, Ferguson MR (2011) Host factors mediating HIV-1 replication. Virus Res 161(2):101–114. doi:10.1016/j.virusres.2011.08.001
156. Lever AM, Jeang KT (2011) Insights into cellular factors that regulate HIV-1 replication in human cells. Biochemistry 50(6):920–931. doi:10.1021/bi101805f

157. Schaller T, Ocwieja KE, Rasaiyaah J, Price AJ, Brady TL, Roth SL, Hue S, Fletcher AJ, Lee K, KewalRamani VN, Noursadeghi M, Jenner RG, James LC, Bushman FD, Towers GJ (2011) HIV-1 capsid-cyclophilin interactions determine nuclear import pathway, integration targeting and replication efficiency. PLoS Pathog 7(12):e1002439. doi:10.1371/journal.ppat.1002439
158. Di Nunzio F, Danckaert A, Fricke T, Perez P, Fernandez J, Perret E, Roux P, Shorte S, Charneau P, Diaz-Griffero F, Arhel NJ (2012) Human nucleoporins promote HIV-1 docking at the nuclear pore, nuclear import and integration. PLoS One 7(9):e46037. doi:10.1371/journal.pone.0046037
159. Di Nunzio F, Fricke T, Miccio A, Valle-Casuso JC, Perez P, Souque P, Rizzi E, Severgnini M, Mavilio F, Charneau P, Diaz-Griffero F (2013) Nup153 and Nup98 bind the HIV-1 core and contribute to the early steps of HIV-1 replication. Virology 440(1):8–18. doi:10.1016/j.virol.2013.02.008
160. Woodward CL, Prakobwanakit S, Mosessian S, Chow SA (2009) Integrase interacts with nucleoporin NUP153 to mediate the nuclear import of human immunodeficiency virus type 1. J Virol 83(13):6522–6533. doi:10.1128/JVI.02061-08
161. Monette A, Ajamian L, Lopez-Lastra M, Mouland AJ (2009) Human immunodeficiency virus type 1 (HIV-1) induces the cytoplasmic retention of heterogeneous nuclear ribonucleoprotein A1 by disrupting nuclear import: implications for HIV-1 gene expression. J Biol Chem 284(45):31350–31362. doi:10.1074/jbc.M109.048736
162. Chan EY, Qian WJ, Diamond DL, Liu T, Gritsenko MA, Monroe ME, Camp DG II, Smith RD, Katze MG (2007) Quantitative analysis of human immunodeficiency virus type 1-infected CD4+ cell proteome: dysregulated cell cycle progression and nuclear transport coincide with robust virus production. J Virol 81(14):7571–7583. doi:10.1128/JVI.00288-07
163. Chan EY, Sutton JN, Jacobs JM, Bondarenko A, Smith RD, Katze MG (2009) Dynamic host energetics and cytoskeletal proteomes in human immunodeficiency virus type 1-infected human primary CD4 cells: analysis by multiplexed label-free mass spectrometry. J Virol 83(18):9283–9295. doi:10.1128/JVI.00814-09
164. Monette A, Pante N, Mouland AJ (2011) HIV-1 remodels the nuclear pore complex. J Cell Biol 193(4):619–631. doi:10.1083/jcb.201008064

Roles of the Nucleoporin Tpr in Cancer and Aging

Chelsi J. Snow and Bryce M. Paschal

Abstract Tpr is a prominent architectural component of the nuclear pore complex that forms the basket-like structure on the nucleoplasmic side of the pore. Tpr, which stands for *t*ranslocated *p*romoter *r*egion, was originally described in the context of oncogenic fusions with the receptor tyrosine kinases Met, TRK, and Raf. Tpr has been since implicated in a variety of nuclear functions, including nuclear transport, chromatin organization, regulation of transcription, and mitosis. More recently, Tpr function has been linked to events including p53 signaling and premature aging in Hutchinson–Gilford Progeria Syndrome (HGPS). Here we provide an overview of the various processes that involve Tpr, and discuss how the levels and localization of a single protein can affect diverse pathways in the cell.

Keywords Tpr • Translocated promoter region • Nuclear pore • Met • TRK • Raf • Oncogenic fusion • Mad1 • Mad2

Abbreviations

APC	Anaphase promoting complex
EGF	Epidermal growth factor
HEZ	Heterochromatin exclusion zone
HGF	Hepatocyte growth factor
HOS	Human osteogenic sarcoma

C.J. Snow • B.M. Paschal (✉)
Center for Cell Signaling, University of Virginia, Charlottesville, VA, USA

Department of Biochemistry and Molecular Genetics, Health Sciences Center, University of Virginia, Room 7021 West Complex, 1400 Jefferson Park Ave, Box 800577, Charlottesville, VA 22908-0577, USA
e-mail: cjs8ed@gmail.com; bmp2h@cms.mail.virginia.edu

E.C. Schirmer and J.I. de las Heras (eds.), *Cancer Biology and the Nuclear Envelope*, Advances in Experimental Medicine and Biology 773, DOI 10.1007/978-1-4899-8032-8_14,

HGPS	Hutchinson–Gilford progeria syndrome
MSL	Male-specific lethal
MNNG	*N*-methyl-*N*′-nitronitrosoguanidine
NES	Nuclear export signal
NLS	Nuclear localization signal
NPC	Nuclear pore complex
NUP	Nucleoporin
SAC	Spindle assembly checkpoint
TPR	Translocated promoter region

The Nuclear Pore Complex

The nuclear envelope serves as a boundary that separates the nucleus from the cytoplasm. This simple arrangement allows compartment-based regulation of numerous cellular activities including transcription, cell cycle, DNA replication, and signal transduction. Movement of proteins and RNAs between these compartments is accomplished by highly specialized, macromolecular assemblies termed nuclear pore complexes (NPCs). The central channel of the NPC is thought to contain a meshwork of proteins that creates a selectivity barrier. This allows the free diffusion of ions and other small molecules, but restricts the translocation of proteins larger than approximately 40 kDa. Molecules larger than approximately 40 kDa require a transport signal (or binding to a partner that contains a transport signal) in order to gain entry into the nucleus via the NPC. Transport through the NPC is estimated to occur at a rate of approximately 1,000 translocation events per second [1, 2].

Detailed studies of the NPC over the past three decades have provided rich insights into its structure and probable mechanisms of nuclear transport. Two surprising outcomes of these studies are that the NPC is not a static channel, and that the NPC has roles beyond nuclear transport. There are approximately 30 different proteins that make up the NPC, termed nucleoporins (Nups), which are present in multiple copies per NPC and reflect the eightfold rotational symmetry of the NPC [3]. Many of the nucleoporins were discovered and characterized biochemically and genetically before a detailed understanding of NPC structure emerged. The NPC has a total mass estimated by cryo-electron microscopy of ~125 MDa in mammals and ~60 MDa in yeasts [4]. These values are larger than estimates based on proteomics [5], and the extent to which mass estimates are influenced by NPC-associated transport factors and their substrates is currently unknown. Several of the Nups are located exclusively on the cytoplasmic or nuclear sides of the NPC [6] an indication that certain nuclear transport events, or possibly other pathways involving the NPC, are regulated differently on the two sides of the NPC. This is clearly the case with the proteins that define the "terminal" structures of the NPC. Thus, on the cytoplasmic side of the NPC, there are 50 nm filaments composed of the nucleoporin RanBP2/Nup358 project into the cytoplasm [7, 8]. On the nucleoplasmic side of the NPC, 100 nm filaments built from the protein Tpr are organized into a nuclear basket [4].

Tpr at the NPC

The nuclear basket (Fig. 1a, b), composed of Tpr and possibly other proteins, is one of the most visually striking features of the NPC [9–11]. Tpr is a 267 kDa nucleoporin [11] which, in its purified form, appears as a mostly fibrillar protein [10]. Tpr contains 14 clusters of heptad repeats (identified in [10]) that are predicted to mediate coiled-coil interactions that give rise to a Tpr dimer, while the carboxy terminal region of Tpr is probably globular [12] (Fig. 1c). Electron microscopy indicates the basket contains eight filaments (Fig. 1b), which are believed to represent eight copies of the Tpr homodimer, each with a mass of approximately 534 kDa. The basket ends with a structure termed the terminal ring, the composition of which might simply be a part of Tpr, or it could involve another protein.

Tpr attachment to the NPC is mediated by the nucleoporin Nup153 [13]. Knockdown of Nup153 is sufficient to disrupt Tpr association with the NPC, resulting in an entirely nucleoplasmic distribution. Tpr amino acids 436–606, which includes heptad repeat 5, are sufficient for its NPC localization, and proline substitutions of amino acids 458 and 489 are sufficient to disrupt NPC binding (and Nup153 binding in vitro) but do not affect the ability of Tpr to homodimerize [10, 13]. Tpr also binds

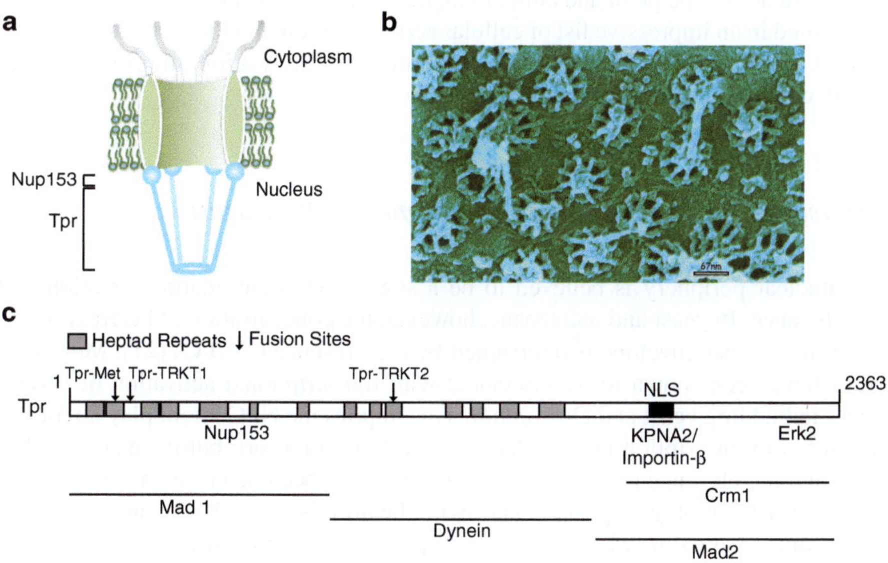

Fig. 1 Structure and properties of Tpr. (**a**) Cartoon of the NPC showing the location of Tpr and the NPC docking site (Nup153). (**b**) Scanning electron micrograph of the nuclear basket from the nuclear face of the NPC, from *Xenopus* oocytes. Spokes (Tpr coiled-coils) are pseudocolored *blue*. (**c**) Structural features and protein interactions associated with Tpr. Heptad repeat assignments were drawn based on a publication [11]. Electron micrograph image copyright Dr. Martin Goldberg, courtesy of the Biology Image Library (http://biologyimagelibrary.com/imageID=35958)

Nup98 [14]; however, the interaction with Nup98 does not regulate Tpr assembly into the NPC. The relative positions of Tpr, Nup153, and Nup98 and other Nups within the NPC have been delineated by elegant immunogold studies in mammals and yeast [11, 15]. It deserves mention that a small amount of soluble (triton-releasable) Tpr dimers can be detected within the nucleoplasm [10], leading to the suggestion that a pool of Tpr forms a "filamentous network" throughout the nucleus [16].

During mitosis, the NPC including the nuclear basket is disassembled in prophase and reassembled in telophase/early G1 in a stepwise and coordinated fashion. Experiments mapping temporal assembly of the NPC showed that Tpr is added at the end of NPC reassembly [17, 18]. Thus, a nearly complete and functional NPC exists prior to Tpr addition. Because the exclusion properties and transport abilities of the NPC are established prior to Tpr incorporation, its assembly on the nucleoplasmic side of the NPC requires signal dependent transport typical of other proteins that are imported into the nucleus. Tpr import relies on its bipartite 31 amino acid nuclear localization sequence (NLS; aa 1829–1860) located in its C-terminal domain, which binds the import receptors importin-α1 (also known as KPNA2) and importin-β [12, 19, 20].

Tpr has homologs in yeast, termed Mlp1 and Mlp2, in *D. melanogaster*, termed Megator, and *Arabidopsis* termed NUCLEAR PORE ANCHOR, or NUA. The yeast homologues designated Mlp1/2, stand for *m*yosin-*l*ike *p*rotein, due to the similar structure (globular head and fibrillar region). At least some functions of Tpr are conserved across the plant and animal kingdoms. Tpr and its homologues have been implicated in an impressive list of cellular activities, including transcription control, mRNA export/splicing control, protein export, deSUMOylation, mitosis, telomere length control, and senescence.

Transcription Regulation and Chromatin Organization

The nuclear periphery is believed to be a site of heterochromatin formation and maintenance. In yeast and metazoans; however, the concentration of heterochromatin at the nuclear envelope is interrupted by the presence of NPCs [21]. Moreover, Nups have been shown to be associated with transcriptional activation of certain genes in budding yeast and *Drosophila*. This implies the NPC might play an important role in transcription independent of its role in transport, but the exact mechanism and its relevance to higher eukaryotes have not been determined yet.

The yeast homolog Mlp1 associates with the promoters of GAL genes in S. cerevisiae upon induction with galactose [22]. In this setting, Mlp1 also undergoes inducible binding to the histone acetyl-transferase SAGA, a known regulator of gene activation. Human Tpr has been shown to associate with the HSP70 promoter upon heat shock induction. At 42 °C, Tpr interacts with HSF1, a transcription factor that undergoes recruitment to heat shock elements including those in the HSP70 gene (HSPA1). These examples suggest that Tpr is recruited to promoters of genes undergoing inducible activation, possibly to serve as a scaffold for assembly of transcription complexes [23].

Dosage compensation in *Drosophila*, in which the male X chromosome is hypertranscribed, is dependent upon association of the male-specific lethal (MSL) dosage compensation proteins with Megator (*Drosophila* Tpr) and Nup153 [23]. Both Megator and Nup153 co-precipitate with the dosage compensation complex. Knockdown of Megator and Nup153, but not other Nups, causes loss of MSL proteins from the X chromosome and reduced expression of X-linked genes [23]. These experiments provide strong evidence for Tpr and Nup153 involvement in gene expression in flies.

In human cells, Tpr also helps maintain heterochromatin-free zones in the vicinity of NPCs, termed heterochromatin exclusion zones (HEZs) [24]. In HeLa cells transfected with Poliovirus, which induces heterochromatin formation, cone-shaped HEZs can be observed proximal to NPCs by TEM (transmission electron microscopy). When Tpr is depleted from cells, HEZs are no longer observed, and dense heterochromatin appears in the NPC region. These experiments suggest that Tpr might play a role as a chromatin organizing factor [24].

Protein and RNA Export

Tpr participates in nuclear export of both protein and RNA. Reducing Tpr activity by antibody injection or by siRNA treatment has been shown to disrupt Crm1-dependent protein export [15, 25]. Tpr binds Crm1 directly in the presence of peptide containing a nuclear export signal (NES) [15, 19, 25]. Tpr function may be involved in export of p53 [25], a critical tumor suppressor protein. Thus far, there is no direct evidence that Tpr has a role in classical NLS-dependent import; however, Tpr is capable of binding the import receptor Importin-β [19].

Tpr is also required for efficient RNA export, and can regulate export of unspliced mRNA. Injection of antibodies to Tpr, or overexpression of Tpr causes poly(A)+RNA to accumulate in the nucleus [26, 27]. Knockdown of Tpr has also been shown to increase nuclear export of unspliced RNA, which normally should be retained for processing or degradation. Tpr knockdown enhances the export of mRNA containing a constitutive transport element (CTE) on the pathway mediated by Nxf1/Nxt1, but does not affect export of mRNA containing a Rev Response Element (RRE) mediated by Rev and Crm1 [28, 29]. Nuclear retention of unspliced RNA in this experimental setting requires NPC localization of Tpr. A wild-type siRNA resistant Tpr construct rescues mRNA retention, while a double mutant (L458P/M489P), which localizes to the nucleoplasm, does not [29]. Yeast Mlp1 and 2 are also implicated in retention of unspliced mRNA [30, 31]. Collectively, these observations suggest an important role for Tpr and the nuclear basket in mRNA quality control.

Oncogenic Fusions with Tpr

Tpr was originally discovered as part of the *tpr-met* oncogenic gene fusion product, and was termed *t*ranslocated *p*romoter *r*egion [32]. *Tpr* was later shown to form two separate fusion products with the oncogenic proteins TRK (termed TRK-T1 and TRK-T2) and another with Raf. Each of these oncogenes is a receptor tyrosine kinase. Although Tpr participates in several oncogenic fusions, the breakpoints used to generate the different fusions occur at distinct sites within the *Tpr* gene (Fig. 1c, note the sections of N-terminal Tpr in each human fusion).

The Met, TRK, and Raf Oncogenes

The Met receptor, also known as the hepatocyte growth factor (HGF) receptor, is normally localized to the plasma membrane. In response to HGF, also known as scatter factor, the Met receptor forms dimers and activates signal transduction cascades, including the PI3 Kinase and Ras pathways. Overexpressed or gain-of-function mutant forms of Met are oncogenic in a variety of cancers [33]. The *tpr-met* oncogene fusion transcript was first described in 1986. Treatment of a human osteogenic sarcoma (HOS) cell line with the carcinogen *N*-methyl-*N'*-nitronitrosoguanidine (MNNG) for 7 days resulted in formation of a hybrid mRNA encoding a 65 kDa gene product [32]. The untreated HOS cell line is non-tumorigenic (does not produce tumors when injected into mice), while the HOS-MNNG cell line is tumorigenic. Tpr-Met is the product of a genomic fusion between sections of chromosomes 1 and 7 (1q25:7q31). The 5′ end of the transcript is contributed by a section of the *Tpr* gene encoding the first two heptad repeats in Tpr, making the product capable of dimerization, while the 3′ end of the transcript encodes the kinase domain of Met (Tpr amino acids 1–142; Met amino acids 1010–1390). Significantly, transfection of this hybrid RNA is sufficient to transform cells and transgenic mice expressing the Tpr-Met fusion develop mammary tumors [34].

The fusion with Tpr has profound effects on Met localization and activity. The fusion protein does not localize to the NPC, which is not surprising given that NPC targeting of Tpr is mediated by a domain that is absent from the fusion product [13, 27]. The fusion product also lacks the transmembrane and juxtamembrane portions of Met that are required for plasma membrane localization. Thus, the Tpr-Met fusion localizes to the cytoplasm. Because the fusion retains two heptad repeats from Tpr, the Tpr-Met fusion product undergoes homodimerization and *trans-phosphorylation* that is independent of Met ligand binding [35]. Constitutive activation of Met induces proliferation, migration, and invasion in a process termed the invasive growth pathway [36]. This includes activation of the JNK/SAPK pathway, likely through PI-3-kinase signaling, which is required for transformation by Tpr-Met [37].

Recent evidence demonstrates that cytoplasmic Tpr-Met is able to escape receptor-mediated endocytosis [38]. This is due in part to the fact that the segment of Met required for binding the ubiquitylation machinery is missing from the Tpr-Met

Table 1 Properties of the Tpr-Met fusion

	Wild-type Met	Tpr-Met fusion
Promoter	Met	Tpr
Protein localization	Plasma membrane	Cytoplasm
Regulation by HGF	Yes	No
Dimerization	Regulated	Constitutive
Ubiquitinated	Yes	No
Regulation by endocytosis	Yes	No

fusion protein (see Table 1 for a summary of Met and Tpr-Met features) [38]. The signaling functions and tumorigenic effects of Tpr-Met are lost when amino acid substitutions are introduced into the dimerization domain of the Tpr fusion [35]. Thus, the oncogenic properties of Tpr-Met are driven by gain-of-functions that are dependent on dimerization.

The Tpr-Met rearrangement was first observed in human tumors in 1991, where it was detected in gastric carcinoma as well as precursor lesions [39]. It has been suggested that the *tpr-met* rearrangement occurs at an early stage in the disease process. In a small study of gastric carcinoma in nine patients with *tpr-met* fusions, five had *tpr-met* RNA in both tumors and noncancerous tissue, while two had the fusion in cancer tissue only, and two exhibited *tpr-met* only in noncancerous tissue [40]. This may indicate that the *tpr-met* translocation can occur as an early or late genetic event.

Two additional oncogenic fusion partners with Tpr have been described, TRK and Raf. TRK can form two different oncogenic fusions with Tpr (TRK-T1 and TRK-T2; 1q25:1q21-22) [41]. These involve fusion of *Tpr* with the gene encoding the NTRK1 kinase, which is a receptor for nerve growth factor. Both of these rearrangements are found in human papillary thyroid carcinomas. The rat homologue of Tpr can occur as a fusion with the Raf kinase, though it is unknown whether this fusion occurs in humans [42].

Other Nucleoporin Oncogenic Fusions

Intriguingly, Tpr is not the only Nup known to form oncogenic fusions. Nup98, perhaps the most widely studied Nup in oncogenic fusions, forms at least 28 genetic fusions associated with various leukemias [43]. Nup214 and Nup358 also occur in oncogenic chromosomal translocations [44, 45]. Why multiple Nups have emerged as oncogenic fusions is not clear, but the lessons learned from the Tpr-Met fusion suggests at least one contribution provided by the Nup is dimerization, which leads to kinase activation and oncogenesis. Self-association of Nups is a biochemical property that likely underpins NPC assembly; this might be the key gain-of-function provided by a fusion with Tpr (and other Nups) for kinases that rely on dimerization as part of their activation mechanism.

Tpr and the Spindle Assembly Checkpoint

During mitosis the NPC disassembles into Nup-containing subcomplexes that, unexpectedly, play roles in mitotic spindle assembly and kinetochore attachment [46]. It has been known that components of the mitotic *s*pindle *a*ssembly *c*heckpoint (SAC) localize to the nuclear periphery, and it was recently shown that Tpr acts as an anchor for SAC components in both mitotic and interphase cells.

The SAC functions during metaphase to protect cells from aneuploidy, one of the hallmarks of cancer. Mad1 and Mad2 (*m*itotic *a*rrest *d*eficient protein) are two of the central players in the SAC [47]. This checkpoint is activated if chromosomes fail to align at the metaphase plate. Consequently, Mad2 binds and inhibits the anaphase-promoting complex (APC). Tpr is necessary for the recruitment of both Mad1 and Mad2 to the kinetochore in mitosis and nuclear periphery in interphase [48]. Mad1 binds directly to an N-terminal region of Tpr in vitro (aa 1–774), while Mad2 independently binds the C-terminal region (aa 1700–2350). Tpr depletion by siRNA disrupts the association of Mad1 with Mad2 and results in aneuploidy and multinucleated cells. These data suggest that Tpr serves as a scaffold that helps regulate the SAC by controlling the localization and activity of Mad1 and Mad2. Inactivation of the SAC, which involves removal of Mad1 and Mad2 from the kinetochore, appears to involve an interaction between Tpr and the motor protein dynein [49].

Tpr Functions in Senescence and Aging

Recent studies have demonstrated several unanticipated roles for Tpr in senescence, autophagy, signal transduction, and aging.

Tpr and p53

Under normal conditions the tumor suppressor p53 is maintained at a low concentration in the cell. In response to cellular stress, p53 is modified, stabilized, and functions as a transcriptional activator for genes involved in cell cycle control, senescence and autophagy. Nuclear levels and activity of p53 are also regulated by nuclear export by the export receptor Crm1 [50]. Several studies suggest there is a pathway that links Tpr to p53 signaling and senescence (Fig. 2). Tpr knockdown increases the level and nuclear accumulation of p53. In turn, there is increased expression of the target gene p21, as well as upregulation of p16, both of which are tumor suppressor genes, along with reduced growth and proliferation [25, 51]. Tpr knockdown also arrests cells at the G0–G1 phase of the cell cycle, reducing the number of cells in S phase to less than half of controls after 2 days of knockdown. Eventually, after 6 days of siRNA treatment, Tpr depletion induces the expression of senescence-associated β-galactosidase, which is dependent on the presence of p53.

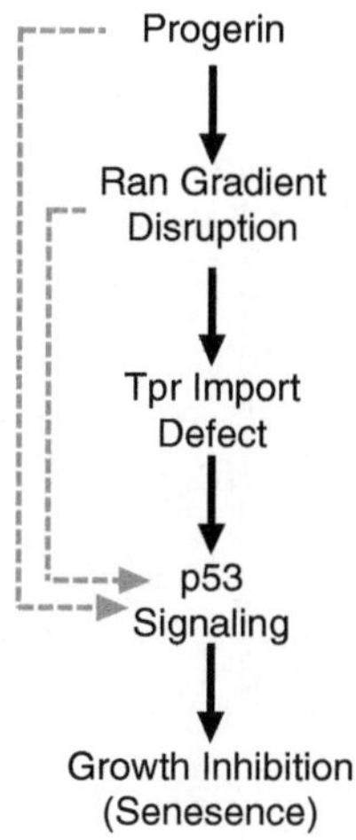

Fig. 2 Model linking Tpr to p53 signaling. The model is based on experiments performed in HeLa cells, human fibroblasts, and transgenic mice. Progerin induces a disruption of the Ran protein distribution in HGPS, which reduces Tpr import [20, 56]. Loss of Tpr from the nucleus increases p53 signaling [25, 51]. The induction of p53 can also occur as a result of the DNA damage caused by Progerin [58], and by simply reducing the nuclear concentration of Ran [57]. Growth inhibition and premature aging in mouse model of Progeria requires p53 function [58]

In addition, Tpr depletion induces expression of several autophagy markers. Interestingly, Nup153 knockdown (which causes Tpr to mislocalize to the nucleoplasm) also causes nuclear accumulation of p53, indicating p53 localization requires Tpr at the NPC [25]. These data suggest that Tpr plays a key role in regulating nuclear levels of p53 through a transport-based mechanism (Fig. 2). Tpr loss might also affect p53-related gene expression events via mRNA export pathways, though this is a point of speculation.

Tpr Is a Scaffold and Kinase Substrate for ERK2

Tpr has an interesting function in the context of signal transduction within the MAP kinase pathway. Addition of mitogens, such as epidermal growth factor (EGF), induces phosphorylation, activation, and nuclear translocation of ERK2 [52, 53]. In the nucleus, activated ERK2 regulates multiple genes involved in growth, invasion, and apoptosis. Indeed, the ERK signaling cascade is deregulated in one third of all human tumors [54]. Knockdown of Tpr results in ERK2 import defects, and ERK2 binds and phosphorylates Tpr and Tpr-associated proteins [52]. These data suggest that Tpr performs two distinct functions that help transduce the effects of EGF on the nucleus. The first is to facilitate ERK2 translocation into the nucleus, and the second is to promote ERK2 interactions with substrates on the nuclear side of the NPC.

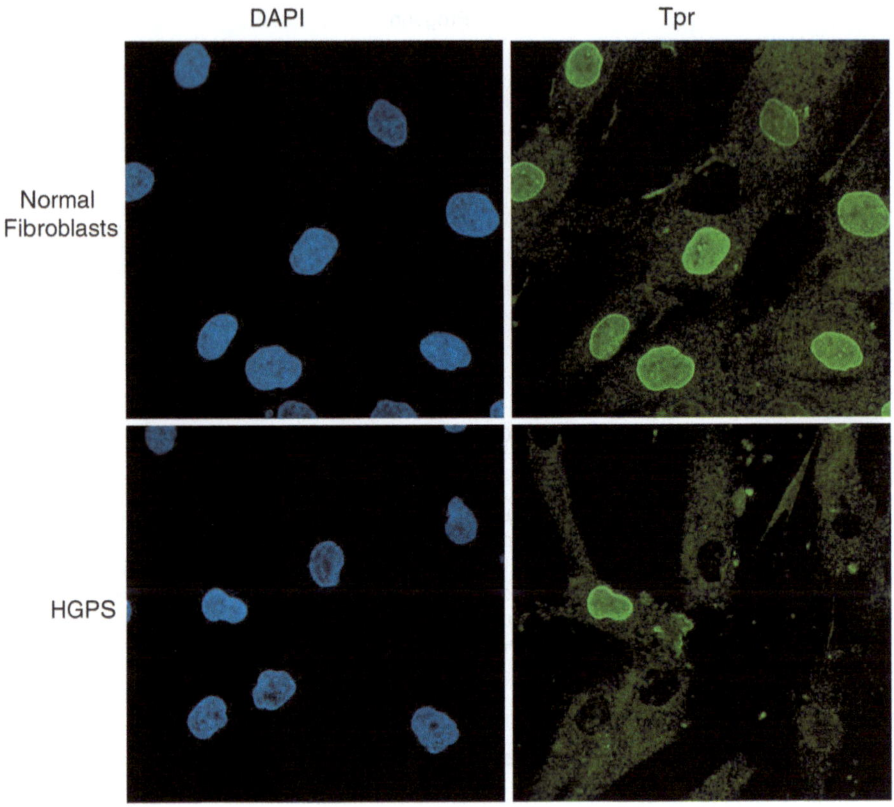

Fig. 3 Tpr is mislocalized in fibroblasts from patients with Hutchinson–Gilford Progeria Syndrome. In normal fibroblasts (*upper panels*), Tpr (*green*) is concentrated at the NPC and also found in the nucleoplasm. In HGPS patient fibroblasts (*lower panels*) Tpr is found primarily in the cytoplasm

Tpr and Premature Aging

The premature aging disease Hutchinson–Gilford Progeria Syndrome (HGPS) is caused by a mutation in the gene that encodes lamin A, a major component of the nuclear lamina. Mutant lamin A, termed progerin, acts in a dominant negative manner to cause changes in nuclear morphology and gene expression [55]. Fibroblasts from HGPS patients have defects in the Ran GTPase system that cause an import defect in Tpr [20, 56]. By immunofluorescence microscopy, Tpr in normal fibroblasts localizes to the NPC (Fig. 3, upper panels). In HGPS cells expressing Progerin, Tpr is detected in the cytoplasm (Fig. 3, lower panels). The Tpr import defects in HGPS cells are heterogeneous and reflect the extent to which the range of Ran gradient is disrupted in the cell population.

Progerin phenotypes can be induced in mice by deleting Zmpste24, which is the protease responsible for pre-lamin A processing. The accumulation of pre-lamin A in these mice leads to p53 activation and increased expression of p53 target genes whose functions contribute to aging phenotypes [57]. The premature aging phenotypes induced by the accumulation of pre-lamin A (which can be functionally equivalent to Progerin expression) in these mice is abrogated by deletion of the p53 gene [57]. These data, together with studies linking Tpr and Ran levels to p53 activation [25, 51, 58], and our data placing Progerin upstream of Ran gradient disruption and Tpr import, suggests there is a pathway that links the nuclear lamina to premature aging (Fig. 2). The pathway reflects a synthesis of observations made in cultured cells and mice, and as indicated, loss of Tpr is probably one of several triggers for induction of p53 activity. Given the variety of functions linked to Tpr and the nuclear basket, several of which were discussed in this review, we speculate that additional "outputs" affected by changes in Tpr and the Ran system might be important for p53 and development of growth inhibition and senescence associated with premature aging.

Conclusions

Tpr contributes to a variety of cellular pathways, several of which are clearly independent of the anticipated roles of Tpr in nuclear transport. Chromosomal translocations that generate fusions between Tpr and kinases are oncogenic, and Tpr can promote oncogenesis through mechanisms that involve cell cycle checkpoints, p53 and possibly MAP kinase signaling. The diverse biology and significant disease associations revealed from the studies on Tpr illustrate the multifunctional nature of Tpr and importance of developing a deep understanding of the NPC and its constituent proteins.

References

1. Ribbeck K, Gorlich D (2001) Kinetic analysis of translocation through nuclear pore complexes. EMBO J 20(6):1320–1330. doi:10.1093/emboj/20.6.1320
2. Yang W, Gelles J, Musser SM (2004) Imaging of single-molecule translocation through nuclear pore complexes. Proc Natl Acad Sci U S A 101(35):12887–12892. doi:10.1073/pnas.0403675101
3. Rabut G, Doye V, Ellenberg J (2004) Mapping the dynamic organization of the nuclear pore complex inside single living cells. Nat Cell Biol 6(11):1114–1121. doi:10.1038/ncb1184, ncb1184 [pii]
4. D'Angelo MA, Hetzer MW (2008) Structure, dynamics and function of nuclear pore complexes. Trends Cell Biol 18(10):456–466. doi:10.1016/j.tcb.2008.07.009, S0962-8924(08)00213-4 [pii]
5. Cronshaw JM, Krutchinsky AN, Zhang W, Chait BT, Matunis MJ (2002) Proteomic analysis of the mammalian nuclear pore complex. J Cell Biol 158(5):915–927. doi:10.1083/jcb.200206106

6. Hoelz A, Debler EW, Blobel G (2011) The structure of the nuclear pore complex. Annu Rev Biochem 80:613–643. doi:10.1146/annurev-biochem-060109-151030
7. Goldberg MW, Allen TD (1993) The nuclear pore complex: three-dimensional surface structure revealed by field emission, in-lens scanning electron microscopy, with underlying structure uncovered by proteolysis. J Cell Sci 106(Pt 1):261–274
8. Jarnik M, Aebi U (1991) Toward a more complete 3-D structure of the nuclear pore complex. J Struct Biol 107(3):291–308
9. Goldberg MW, Allen TD (1992) High resolution scanning electron microscopy of the nuclear envelope: demonstration of a new, regular, fibrous lattice attached to the baskets of the nucleoplasmic face of the nuclear pores. J Cell Biol 119(6):1429–1440
10. Hase ME, Kuznetsov NV, Cordes VC (2001) Amino acid substitutions of coiled-coil protein Tpr abrogate anchorage to the nuclear pore complex but not parallel, in-register homodimerization. Mol Biol Cell 12(8):2433–2452
11. Krull S, Thyberg J, Bjorkroth B, Rackwitz HR, Cordes VC (2004) Nucleoporins as components of the nuclear pore complex core structure and Tpr as the architectural element of the nuclear basket. Mol Biol Cell 15(9):4261–4277. doi:10.1091/mbc.E04-03-0165, E04-03-0165 [pii]
12. Cordes VC, Hase ME, Muller L (1998) Molecular segments of protein Tpr that confer nuclear targeting and association with the nuclear pore complex. Exp Cell Res 245(1):43–56. doi:10.1006/excr.1998.4246, S0014-4827(98)94246-X [pii]
13. Hase ME, Cordes VC (2003) Direct interaction with nup153 mediates binding of Tpr to the periphery of the nuclear pore complex. Mol Biol Cell 14(5):1923–1940. doi:10.1091/mbc.E02-09-0620, 14/5/1923 [pii]
14. Fontoura BM, Dales S, Blobel G, Zhong H (2001) The nucleoporin Nup98 associates with the intranuclear filamentous protein network of TPR. Proc Natl Acad Sci U S A 98(6):3208–3213. doi:10.1073/pnas.061014698, 98/6/3208 [pii]
15. Frosst P, Guan T, Subauste C, Hahn K, Gerace L (2002) Tpr is localized within the nuclear basket of the pore complex and has a role in nuclear protein export. J Cell Biol 156(4):617–630. doi:10.1083/jcb.200106046, jcb.200106046 [pii]
16. Paddy MR (1998) The Tpr protein: linking structure and function in the nuclear interior? Am J Hum Genet 63(2):305–310. doi:10.1086/301989, S0002-9297(07)61471-2 [pii]
17. Dultz E, Zanin E, Wurzenberger C, Braun M, Rabut G, Sironi L, Ellenberg J (2008) Systematic kinetic analysis of mitotic dis- and reassembly of the nuclear pore in living cells. J Cell Biol 180(5):857–865. doi:10.1083/jcb.200707026, jcb.200707026 [pii]
18. Bodoor K, Shaikh S, Salina D, Raharjo WH, Bastos R, Lohka M, Burke B (1999) Sequential recruitment of NPC proteins to the nuclear periphery at the end of mitosis. J Cell Sci 112(Pt 13):2253–2264
19. Ben-Efraim I, Frosst PD, Gerace L (2009) Karyopherin binding interactions and nuclear import mechanism of nuclear pore complex protein Tpr. BMC Cell Biol 10:74. doi:10.1186/1471-2121-10-74, 1471-2121-10-74 [pii]
20. Snow CJ, Dar A, Dutta A, Kehlenbach RH, Paschal BM (2013) Defective nuclear import of Tpr in Progeria reflects the Ran sensitivity of large cargo transport. J Cell Biol 201(4):541–557. doi:10.1083/jcb.201212117
21. Akhtar A, Gasser SM (2007) The nuclear envelope and transcriptional control. Nat Rev Genet 8(7):507–517. doi:10.1038/nrg2122, nrg2122 [pii]
22. Luthra R, Kerr SC, Harreman MT, Apponi LH, Fasken MB, Ramineni S, Chaurasia S, Valentini SR, Corbett AH (2007) Actively transcribed GAL genes can be physically linked to the nuclear pore by the SAGA chromatin modifying complex. J Biol Chem 282(5):3042–3049. doi:10.1074/jbc.M608741200, M608741200 [pii]
23. Mendjan S, Taipale M, Kind J, Holz H, Gebhardt P, Schelder M, Vermeulen M, Buscaino A, Duncan K, Mueller J, Wilm M, Stunnenberg HG, Saumweber H, Akhtar A (2006) Nuclear pore components are involved in the transcriptional regulation of dosage compensation in Drosophila. Mol Cell 21(6):811–823. doi:10.1016/j.molcel.2006.02.007, S1097-2765(06)00089-X [pii]
24. Krull S, Dorries J, Boysen B, Reidenbach S, Magnius L, Norder H, Thyberg J, Cordes VC (2010) Protein Tpr is required for establishing nuclear pore-associated zones of heterochromatin exclusion. EMBO J 29(10):1659–1673. doi:10.1038/emboj.2010.54, emboj201054 [pii]

25. David-Watine B (2011) Silencing nuclear pore protein Tpr elicits a senescent-like phenotype in cancer cells. PLoS One 6(7):e22423. doi:10.1371/journal.pone.0022423
26. Shibata S, Matsuoka Y, Yoneda Y (2002) Nucleocytoplasmic transport of proteins and poly(A)+RNA in reconstituted Tpr-less nuclei in living mammalian cells. Genes Cells 7(4):421–434, 525 [pii]
27. Bangs P, Burke B, Powers C, Craig R, Purohit A, Doxsey S (1998) Functional analysis of Tpr: identification of nuclear pore complex association and nuclear localization domains and a role in mRNA export. J Cell Biol 143(7):1801–1812
28. Coyle JH, Bor YC, Rekosh D, Hammarskjold ML (2011) The Tpr protein regulates export of mRNAs with retained introns that traffic through the Nxf1 pathway. RNA 17(7):1344–1356. doi:10.1261/rna.2616111, rna.2616111 [pii]
29. Rajanala K, Nandicoori VK (2012) Localization of nucleoporin Tpr to the nuclear pore complex is essential for Tpr mediated regulation of the export of unspliced RNA. PLoS One 7(1):e29921. doi:10.1371/journal.pone.0029921, PONE-D-11-17401 [pii]
30. Vinciguerra P, Iglesias N, Camblong J, Zenklusen D, Stutz F (2005) Perinuclear Mlp proteins downregulate gene expression in response to a defect in mRNA export. EMBO J 24(4): 813–823. doi:10.1038/sj.emboj.7600527, 7600527 [pii]
31. Galy V, Gadal O, Fromont-Racine M, Romano A, Jacquier A, Nehrbass U (2004) Nuclear retention of unspliced mRNAs in yeast is mediated by perinuclear Mlp1. Cell 116(1):63–73, S0092867403010262 [pii]
32. Park M, Dean M, Cooper CS, Schmidt M, O'Brien SJ, Blair DG, Vande Woude GF (1986) Mechanism of met oncogene activation. Cell 45(6):895–904, 0092-8674(86)90564-7 [pii]
33. Luraghi P, Schelter F, Kruger A, Boccaccio C (2012) The MET oncogene as a therapeutical target in cancer invasive growth. Front Pharmacol 3:164. doi:10.3389/fphar.2012.00164
34. Liang TJ, Reid AE, Xavier R, Cardiff RD, Wang TC (1996) Transgenic expression of tpr-met oncogene leads to development of mammary hyperplasia and tumors. J Clin Invest 97(12):2872–2877. doi:10.1172/JCI118744
35. Rodrigues GA, Park M (1993) Dimerization mediated through a leucine zipper activates the oncogenic potential of the met receptor tyrosine kinase. Mol Cell Biol 13(11):6711–6722
36. Peschard P, Park M (2007) From Tpr-Met to Met, tumorigenesis and tubes. Oncogene 26(9):1276–1285. doi:10.1038/sj.onc.1210201, 1210201 [pii]
37. Rodrigues GA, Park M, Schlessinger J (1997) Activation of the JNK pathway is essential for transformation by the Met oncogene. EMBO J 16(10):2634–2645. doi:10.1093/emboj/16.10.2634
38. Mak HH, Peschard P, Lin T, Naujokas MA, Zuo D, Park M (2007) Oncogenic activation of the Met receptor tyrosine kinase fusion protein, Tpr-Met, involves exclusion from the endocytic degradative pathway. Oncogene 26(51):7213–7221. doi:10.1038/sj.onc.1210522
39. Soman NR, Correa P, Ruiz BA, Wogan GN (1991) The TPR-MET oncogenic rearrangement is present and expressed in human gastric carcinoma and precursor lesions. Proc Natl Acad Sci U S A 88(11):4892–4896
40. Yu J, Miehlke S, Ebert MP, Hoffmann J, Breidert M, Alpen B, Starzynska T, Stolte Prof M, Malfertheiner P, Bayerdorffer E (2000) Frequency of TPR-MET rearrangement in patients with gastric carcinoma and in first-degree relatives. Cancer 88(8):1801–1806
41. Greco A, Miranda C, Pagliardini S, Fusetti L, Bongarzone I, Pierotti MA (1997) Chromosome 1 rearrangements involving the genes TPR and NTRK1 produce structurally different thyroid-specific TRK oncogenes. Genes Chromosomes Cancer 19(2):112–123. doi:10.1002/(SICI)1098-2264(199706)19:2<112::AID-GCC7>3.0.CO;2-1 [pii]
42. King HW, Tempest PR, Merrifield KR, Rance AJ (1988) tpr homologues activate met and raf. Oncogene 2(6):617–619
43. Gough SM, Slape CI, Aplan PD (2011) NUP98 gene fusions and hematopoietic malignancies: common themes and new biologic insights. Blood 118(24):6247–6257. doi:10.1182/blood-2011-07-328880
44. Kohler A, Hurt E (2010) Gene regulation by nucleoporins and links to cancer. Mol Cell 38(1):6–15. doi:10.1016/j.molcel.2010.01.040, S1097-2765(10)00172-3 [pii]
45. Xu S, Powers MA (2009) Nuclear pore proteins and cancer. Semin Cell Dev Biol 20(5): 620–630. doi:10.1016/j.semcdb.2009.03.003, S1084-9521(09)00041-X [pii]

46. Stukenberg PT, Macara IG (2003) The kinetochore NUPtials. Nat Cell Biol 5(11):945–947. doi:10.1038/ncb1103-945, ncb1103-945 [pii]
47. Musacchio A, Salmon ED (2007) The spindle-assembly checkpoint in space and time. Nat Rev Mol Cell Biol 8(5):379–393. doi:10.1038/nrm2163
48. Lee SH, Sterling H, Burlingame A, McCormick F (2008) Tpr directly binds to Mad1 and Mad2 and is important for the Mad1-Mad2-mediated mitotic spindle checkpoint. Genes Dev 22(21):2926–2931. doi:10.1101/gad.1677208, 22/21/2926 [pii]
49. Nakano H, Funasaka T, Hashizume C, Wong RW (2010) Nucleoporin translocated promoter region (Tpr) associates with dynein complex, preventing chromosome lagging formation during mitosis. J Biol Chem 285(14):10841–10849. doi:10.1074/jbc.M110.105890, M110.105890 [pii]
50. Freedman DA, Levine AJ (1998) Nuclear export is required for degradation of endogenous p53 by MDM2 and human papillomavirus E6. Mol Cell Biol 18(12):7288–7293
51. Funasaka T, Tsuka E, Wong RW (2012) Regulation of autophagy by nucleoporin Tpr. Sci Rep 2:878. doi:10.1038/srep00878
52. Vomastek T, Iwanicki MP, Burack WR, Tiwari D, Kumar D, Parsons JT, Weber MJ, Nandicoori VK (2008) Extracellular signal-regulated kinase 2 (ERK2) phosphorylation sites and docking domain on the nuclear pore complex protein Tpr cooperatively regulate ERK2-Tpr interaction. Mol Cell Biol 28(22):6954–6966. doi:10.1128/MCB.00925-08, MCB.00925-08 [pii]
53. Pouyssegur J, Volmat V, Lenormand P (2002) Fidelity and spatio-temporal control in MAP kinase (ERKs) signalling. Biochem Pharmacol 64(5–6):755–763
54. Dhillon AS, Hagan S, Rath O, Kolch W (2007) MAP kinase signalling pathways in cancer. Oncogene 26(22):3279–3290. doi:10.1038/sj.onc.1210421
55. Csoka AB, English SB, Simkevich CP, Ginzinger DG, Butte AJ, Schatten GP, Rothman FG, Sedivy JM (2004) Genome-scale expression profiling of Hutchinson-Gilford progeria syndrome reveals widespread transcriptional misregulation leading to mesodermal/mesenchymal defects andacceleratedatherosclerosis.AgingCell3(4):235–243.doi:10.1111/j.1474-9728.2004.00105.x, ACE105 [pii]
56. Kelley JB, Datta S, Snow CJ, Chatterjee M, Ni L, Spencer A, Yang CS, Cubenas-Potts C, Matunis MJ, Paschal BM (2011) The defective nuclear lamina in Hutchinson-Gilford progeria syndrome disrupts the nucleocytoplasmic Ran gradient and inhibits nuclear localization of Ubc9. Mol Cell Biol 31(16):3378–3395. doi:10.1128/MCB.05087-11, MCB.05087-11 [pii]
57. Varela I, Cadinanos J, Pendas AM, Gutierrez-Fernandez A, Folgueras AR, Sanchez LM, Zhou Z, Rodriguez FJ, Stewart CL, Vega JA, Tryggvason K, Freije JM, Lopez-Otin C (2005) Accelerated ageing in mice deficient in Zmpste24 protease is linked to p53 signalling activation. Nature 437(7058):564–568. doi:10.1038/nature04019
58. Nagai M, Yoneda Y (2013) Downregulation of the small GTPase ras-related nuclear protein accelerates cellular ageing. Biochim Biophys Acta 1830(3):2813–2819. doi:10.1016/j.bbagen.2012.11.001

Ran GTPase in Nuclear Envelope Formation and Cancer Metastasis

Kyle B. Matchett, Suzanne McFarlane, Sophie E. Hamilton, Yousef S. A. Eltuhamy, Matthew A. Davidson, James T. Murray, Ahmed M. Faheem, and Mohamed El-Tanani

Abstract Ran is a small ras-related GTPase that controls the nucleocytoplasmic exchange of macromolecules across the nuclear envelope. It binds to chromatin early during nuclear formation and has important roles during the eukaryotic cell cycle, where it regulates mitotic spindle assembly, nuclear envelope formation and cell cycle checkpoint control. Like other GTPases, Ran relies on the cycling between GTP-bound and GDP-bound conformations to interact with effector proteins and regulate these processes. In nucleocytoplasmic transport, Ran shuttles across the nuclear envelope through nuclear pores. It is concentrated in the nucleus by an active import mechanism where it generates a high concentration of RanGTP by nucleotide exchange. It controls the assembly and disassembly of a range of complexes that are formed between Ran-binding proteins and cellular cargo to maintain rapid nuclear transport. Ran also has been identified as an essential protein in nuclear envelope formation in eukaryotes. This mechanism is dependent on importin-β, which regulates the assembly of further complexes important in this process, such as Nup107–Nup160. A strong body of evidence is emerging implicating Ran as a key protein in the metastatic progression of cancer. Ran is overexpressed in a range of

K.B. Matchett • S. McFarlane • S.E. Hamilton • Y.S.A. Eltuhamy • M.A. Davidson
M. El-Tanani (✉)
Centre for Cancer Research and Cell Biology, Queen's University Belfast, Belfast BT9 7BL, UK
e-mail: k.matchett@qub.ac.uk; s.mcfarlane@qub.ac.uk; shamilton35@qub.ac.uk;
e-mail: ytuhamy@gmail.com; mdavidson17@qub.ac.uk; m.el-tanani@qub.ac.uk

J.T. Murray
Trinity Biomedical Sciences Institute, Trinity College Dublin, Dublin 2, Ireland
e-mail: james.murray@tcd.ie

A.M. Faheem
School of Pharmacy and Pharmaceutical Science, University of Ulster, Coleraine BT52 1SA, UK
e-mail: a.faheem@ulster.ac.uk

E.C. Schirmer and J.I. de las Heras (eds.), *Cancer Biology and the Nuclear Envelope*, Advances in Experimental Medicine and Biology 773, DOI 10.1007/978-1-4899-8032-8_15,

tumors, such as breast and renal, and these perturbed levels are associated with local invasion, metastasis and reduced patient survival. Furthermore, tumors with oncogenic KRAS or PIK3CA mutations are addicted to Ran expression, which yields exciting future therapeutic opportunities.

Keywords Ran GTPase • Nucleocytoplasmic transport • Mitotic spindle • Nuclear envelope • RCC1 • RanBP1 • CRM1 • TPX2 • Importin-β • Cell cycle checkpoint control • Osteopontin • Metastasis

Abbreviations

ARF	ADP ribosylation factor
CLL	Chronic lymphocytic leukemia
FEISEM	Field emission in-lens scanning electron microscopy
IMT	Inflammatory myofibroblastic tumors
NES	Nuclear export signal
NLS	Nuclear localization signal
NPC	Nuclear pore complex

Introduction

In every cell, a plethora of molecules are continually shuttling between the cytoplasm and nucleus in a complex system that ensures cellular homeostasis is maintained. This is necessitated by the compartmentalization within cells that separates transcription and translation. The separation by the nuclear envelope contributes to the ability of cells to regulate gene expression, but it also requires cells to develop mechanisms to transport cargoes between the nucleus and cytoplasm. Extensive research in eukaryotic model systems has identified RanGTP as a key protein in nucleocytoplasmic transport [1, 2]. Ran is a member of the Ras family of small G proteins, although highly unique due to a lack of the -CAAX membrane anchoring motif at its C-terminus (that is a characteristic feature of small GTPases) and its nuclear localization [3]. Ran contains a 20 kDa catalytic core domain and shares other similarities with Ras family members such as relying on a specific guanine nucleotide exchange factor (Regulator of Chromosome Condensation 1; RCC1 also known as RanGEF), and requiring a GTPase activating protein (Ran GTPase Activating Protein 1; RanGAP1) to contribute to its overall GTPase activity [4, 5]. Ran also possesses an unusually acidic tail at its C-terminus, stabilizing both GDP binding to Ran and its interaction with Ran Binding Protein 1 (RanBP1), which promotes GTP hydrolysis [6, 7]. The function of Ran has been studied in detail in interphase cells, where its regulator proteins RCC1, RanGAP1, and RanBP1 interact with RanGTP in the cytosol while RCC1 is located in the nucleus bound to chromatin [8–10]. This creates a RanGTP gradient across the nuclear envelope with a higher concentration of RanGTP in the nucleus than in the cytoplasm. The assembly and disassembly of cargo complexes for import and export is regulated by an asymmetrical distribution of various Ran factors, which direct nuclear-cytosolic exchange of these proteins [11]. RanGTP-binding receptors, such as chromosome region maintenance 1 (CRM1) and importin-α/β, control the transport of cargoes containing nuclear localization signals (NLS) or nuclear export signals (NES) [12]. When the nuclear envelope disassembles with the onset of mitosis, RCC1 continues to concentrate RanGTP near chromosomes, while RanGAP1/RanBP1 hydrolyzes RanGTP located further from chromosomal regions [13]. Spindle assembly factors are inhibited by importin-β in complexes located away from chromatin where RanGTP concentration is lower [14, 15]. It has been proposed that importin-β interacts with Ran on chromatin, highlighting the essential role importin-β plays in recruiting membrane vesicles to chromatin [16, 17]. However, an alternative model has also been reported, suggesting that importin-β inhibits NPC and nuclear envelope assembly [18]. When mitosis is completed and the mitotic spindle disintegrates, normal cellular structure must be resumed. The nuclear envelope and NPCs are reestablished and nucleocytoplasmic transport resumes. This process is dependent on a functional Ran system, as characterized by a number of specific in vitro and in vivo studies [19–23]. In addition to the intracellular functions of this small GTPase in nuclear transport and mitosis biology, Ran has been identified as a key protein in the induction of an invasive and metastatic phenotype in cancer [24]. Although no sporadic mutations in Ran have been reported in human cancer, extensive evidence has demonstrated the overexpression of Ran in various tumor types compared to normal tissue and increased Ran levels in vitro leads to malignant transformation [24–32]. Furthermore, Ran overexpression induces a metastatic phenotype through deregulation of its effector proteins with known oncogenic effects such as Aurora A [33], the microtubule associated protein HURP [34, 35], and BRCA1 [36]. Loss of Ran in normal cells confers minimal effects, whereas downregulation in cancer cells is associated with mitotic defects and increased apoptosis [37]. This provides the opportunity for the development of Ran inhibitors that selectively induce apoptosis in malignant cells as a potential future therapy.

Identification and Structure of Ran GTPase

Ran GTPase was identified while screening for genes homologous to *ras*-like coding sequences in a human teratocarcinoma cDNA library [3]. Of the ten clones that were initially discovered to hybridize to the conserved *ras* domain Asp-Thr-Ala-Gly-Gln-Glu (H-*ras* residues 57–62), four of these sequences (named TC4, TC10, TC21, and TC25) were identified as novel *ras*-like genes. Alignment of these clones with the H-*ras* amino acid sequence confirmed that four of the H-*ras* functional domains were present in the TC clones. Furthermore, a number of residues distinct from these domains, such as Phe-28 (interacts with the guanine ring), Thr-35 (important in Mg^{2+} ion regulation), Arg-68, Tyr-71, Pro-110, and Phe-156, were also conserved in the TC clones. This suggested that the proteins encoded by the novel sequences had biochemical characteristics similar to other *ras*-like proteins, such as guanine nucleotide binding and GTP hydrolysis. Despite the striking homology between the clones and H-*ras*, one of the four sequences was found to be unique from the others. The TC4 clone did not terminate with a CAAX coding sequence, a common feature of many *ras* superfamily proteins, whereas the other three clones did contain this motif. In addition to lacking a terminal CAAX sequence, the TC4 coding sequence did not end with a cysteine or YPT (*Y*east *P*rotein *T*ransport) group members, leading to the suggestion that TC4 was similar to ADP ribosylation factor (ARF), another low-molecular-weight GTP-binding protein [3]. However, the TC4 sequence, unlike ARF, encoded all four of the *ras* functional domains. In review of these findings, TC4 was classified as a *ras*-related gene. Subsequent studies on RCC1 identified TC4 as a protein that bound to an anti-RCC1 antibody and formed a nuclear complex with RCC1 [38]. Moreover, Bischoff and Ponstingl reported the 25 kDa *ras*-related protein was present in the nucleoplasm as a soluble monomer in 25-fold higher levels than the TC4-RCC1 complexed form [39]. Due to its homology to conserved *ras* sequences and nuclear localization, TC4 was named Ran (*Ras*-related *n*uclear protein) or Ran GTPase.

Ran contains a 20 kDa catalytic core domain similar in structure to GTP-binding domains of the Ras GTPases [4]. It shares other similarities with Ras family members as it relies on a specific guanine exchange factor (RCC1), and requires a GTPase activating protein (RanGAP) to contribute to its GTPase activity [5]. The first crystal structure of Ran, as part of the Ran•GDP•Mg^{2+} complex, demonstrated that despite the similarity with the Ras core, there were substantial differences in domains important in regulation of GDP and Mg^{2+} binding, suggesting there could be significant conformational alterations following GTP binding [6]. Furthermore, Ran lacks a -CAAX motif and is therefore not subject to lipid modification at its C-terminus [3]. It instead possesses an unusually acidic carboxyl terminus (^{211}DEDDDL^{216}), stabilizing both GDP binding to Ran and interaction with RanBP1, which facilitates GTP hydrolysis [7]. Analysis of the crystal structure also revealed that the amino-terminus (residues ^{1}MAAQGEP^{7}) and the extended tail are flexible [6].

The Roles of Ran in Nucleocytoplasmic Transport

Transport of molecules between the nucleus and cytoplasm is an essential activity of eukaryotic cells. It occurs through NPCs and is mediated by soluble receptors that shuttle between the nucleus and cytoplasm [40]. NPCs are approximately 170 nm in length and consist of more than 30 proteins called nucleoporins (or Nups) arranged in a symmetrical, octagonal structure that spans the nuclear envelope [41]. Nucleoporins form a semi-permeable barrier to macromolecules via an association of phenylalanine-glycine (FG) motifs, which can also act to mediate transport of cellular cargo [42]. Molecules less than 40 kDa are free to diffuse across the NPC, at a rate inversely related to their molecular size, but mRNA and other macromolecules rely on nucleocytoplasmic carrier proteins, called karyopherins, and other proteins to facilitate their transport [43]. The process is bidirectional, signal mediated, and energy dependent [44]. Karyopherins can be divided into two subgroups: importins (nuclear import) and exportins (nuclear export). These karyopherins recognize specific signals on the cargo to be transported, form complexes with them and attach onto the NPC through interactions with nucleoporins [41].

Ran was identified as a key factor for protein import into the nucleus and has since been demonstrated to have an essential role in the nucleocytoplasmic transport of macromolecules [1, 2, 45]. The nuclear envelope acts as a barrier to compartmentalize (and protect) the nuclear DNA from the cytoplasm and the enzymes that are found there, many of which could damage the genetic material. To maintain cellular homeostasis, cytoplasmic proteins are prevented from entering the nucleus by active transport, whereas nuclear proteins remain in the nucleoplasm by active import or by binding to nuclear structures [45]. Moreover, proteins that are important in nuclear/cytoplasmic communication transit rapidly in and out of the nucleus [45]. The directionality of this transport is largely supported by a concentration gradient of Ran GTP/GDP, which is characterized by a high nucleocytoplasmic concentration of RanGTP [11].

Ran, like other GTPases of the Ras family, switches between two conformations, GTP-bound and GDP-bound, which represent the active and inactive forms, respectively (Fig. 1). The conformation of Ran dictates how it interacts with effector proteins: for example the GTP/GDP bound state of Ran directs its recognition of the cargo to be transported, the directionality of transport and the timing of the transport [6, 46, 47]. In vitro disruption of the guanine nucleotide cycle can be achieved by introducing mutations in Ran that either block GTPase activity (e.g., RanQ69L) or inhibit nucleotide binding (e.g., RanT24N), resulting in the formation of an inhibitory Ran-RCC1 complex that prevents RanGTP production (Fig. 1a) [45, 48].

Both the GTP and GDP conformations of Ran are distributed asymmetrically in the cell. RanGTP is predominately located in a soluble form in the nucleus, whereas RanGDP is concentrated in the cytoplasm. Furthermore, more than 10^5 copies of Ran leave the average mammalian nucleus every second and the rapid nuclear efflux is consistent with the high nuclear concentration of Ran [8, 9]. Elevated Ran levels in the nucleus are maintained by a nuclear import mechanism involving nuclear transport factor-2 (NTF2), a cytosolic protein that interacts with RanGDP (Fig. 1b) [49, 50]. Moreover, the nucleocytoplasmic shuttling of NTF2 has recently been reported to be a regulated process in mammalian cells that may incorporate a tyrosine kinase-dependent signal transduction mechanism [51]. Loading of the Ran molecule with GTP during the GDP/GTP exchange reaction is necessary because intracellular concentrations of GTP are approximately tenfold higher than GDP [52]. Furthermore, RCC1 activates the RanGDP-RanGTP gradient by binding to the nucleosome (consisting of repeating units of chromatin), which recruits Ran to chromatin and activates Ran nucleotide exchange activity [53]. The binding of RCC1 to chromatin is important to this spatial signaling within the nucleus, and the process

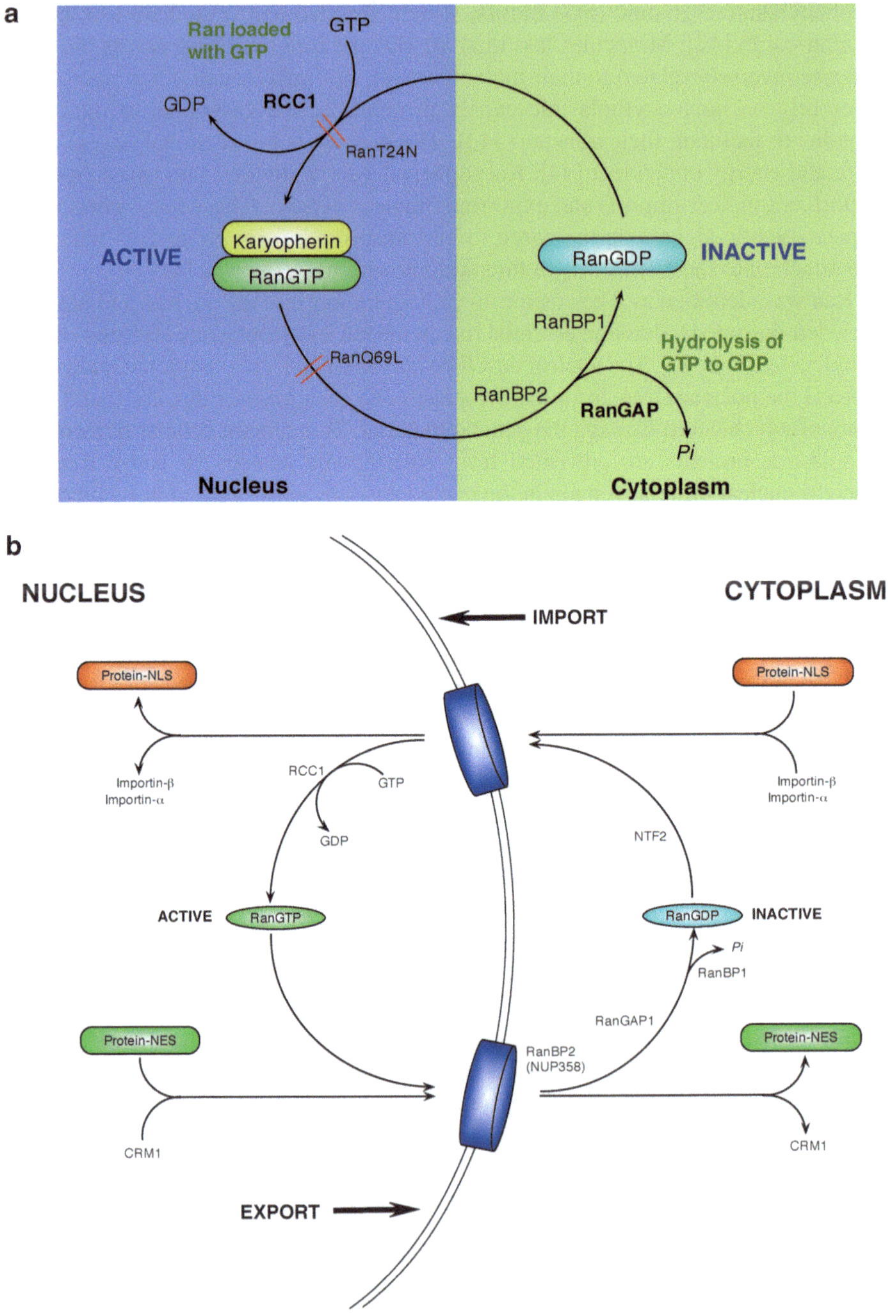

Fig. 1 Ran controls nucleocytoplasmic transport. (**a**) The guanine nucleotide cycle of Ran. The inactive RanGDP (lacking GTPase activity) is loaded with GTP by RCC1 in the cell nucleus to produce active RanGTP, which has GTPase activity. This reaction produces a conformational

depends on the N-terminal tail extension of RCC1, which is altered by an unusual amino methylation [10, 54]. Also, the RCC1 β-propeller motif associates with the histone H2A-H2B complex, which increases its nucleotide exchange activity [55].

In the cytoplasm, Ran mediated hydrolysis of GTP to GDP is stimulated by RanGAP1, and RanBP1, which interacts with Ran through a RanGTP-binding motif [56–59]. GTP hydrolysis by RanGAP can also be catalyzed by RanBP2 (also known as Nup358), a nucleoporin that localizes to cytoplasmic filaments and contains multiple binding domains for the nucleocytoplasmic transport machinery [60, 61]. Although RanBP2 is one of the few reported SUMO E3 ligases, it only exhibits its E3 ligase activity when complexed with RanGAP1•SUMO1/Ubc9, indicating that complexed RanBP2, rather than free RanBP2, is the E3 ligase [62].

Upon receiving GTP, Ran GTPase becomes activated and initiates the formation of nuclear import and export complexes. Karyopherins that mediate import, such as the importin-α/importin-β heterodimer, form complexes with proteins possessing a classical lysine-rich NLS in the cytoplasm [12]. The import complexes then translocate to the nucleus and are dissociated by RanGTP, which binds to importin-β and releases both the cargo protein and importin-α. The karyopherin is then recycled back to the cytoplasm to import the next cargo. Importin-α/β regulated import of NLS-carrying cargo is more effective than alternative Ran-dependent pathways that do not rely on importin-α [63]. Furthermore, digitonin permeabilized cell assays have shown that the association between importin-α and Nup153 (a NPC component) is required for nuclear import of proteins containing importin-β binding motifs but not other cargo substrates that bind directly to importin-β [63]. These assays work on the premise that digitonin permeabilization can permeabilize the plasma membrane specifically, leaving the nuclear membrane intact. This allows the differentiation between antigens on the nuclear and cytoplasmic sides of the nuclear envelope [64, 65]. Structural analysis of importin-β proteins have shown that they contain superhelices of HEAT repeats that can enclose an interacting protein or form a superhelix that can "spring" open to expose a potential binding site [66].

Fig. 1 (continued) alteration in Ran that facilitates interactions with effector molecules, such as karyopherins. In the cytoplasm, RanGTP is hydrolyzed back to RanGDP in a reaction dependent on RanGAP1, and promoted by RanBP1 and RanBP2. The in vitro mutants of the Ran GTP-GDP cycle disrupt GTPase activity by independent mechanisms—RanT24N blocks nucleotide binding and forms a stable inhibitory complex with RCC1, whereas RanQ69L lacks the ability to hydrolyze GTP and it remains in a fixed GTP-bound state. (**b**) Ran dynamics in macromolecular transport. Ran moves across the nuclear envelope in a bidirectional manner but is retained in the nucleus by an active import mechanism involving nuclear transport factor-2 (NTF2). Import complexes which incorporate proteins carrying a nuclear localization signal (NLS) are dissociated by RanGTP, and export is promoted by RanGTP binding with chromosome-region maintenance protein-1 (CRM1). Dissociation of the export complexes occurs in the cytoplasm when RanGTP associates with RanGAP and RanBP1 or RanBP2 (also known as Nup358), causing hydrolysis of GTP. Other importin family members are involved in the nucleocytoplasmic transport of certain cargoes

RanGTP can then form a stabilizing complex with these compacted structures. In contrast, the structure of CAS/CSE1, an export factor that recycles importin-α to the cytoplasm, opens up in response to RanGTP binding, permitting the binding of importin-α and the formation of a trimeric complex [45, 66].

In addition to nuclear import, Ran also controls the assembly and disassembly of complexes important for the export of macromolecules from the nucleus. To date, eight RanGTP-driven exporters have been identified in humans, differing widely in the cargo they transport [40]. Exportin 6, for example, specifically interacts with actin only and actively reverses its leakage into the nuclear space [40]. In contrast, RanGTP binds to the leptomycin-B-sensitive factor CRM1, an exportin with a variety of structurally diverse cargo [67]. This range of specificity of CRM1 is not only due to its highly adaptable cargo-binding site, but also a conformational change on CRM1 that occurs following RanGTP binding that increases its affinity for a short peptide sequence containing a canonical Leucine-rich NES [68]. When the RanGTP-CRM1-cargo export complex translocates across the nuclear pore, it is disassociated by Ran hydrolysis of GTP, which involves interaction with RanGAP and RanBP1 or RanBP2 [69]. CRM1 cargo proteins include several tumor suppressor proteins, such as FoxO3a, p53, and IκB [70]. CRM1 prevents the activation of these proteins, in the absence of DNA damage and other aberrations, by exporting them from the nucleus of normal cells.

Nucleocytoplasmic transport also regulates the compartmentalization of several of the effectors of the Ran GTP-GDP cycle. The amino terminal tail of RCC1 contains an NLS that associates with the importin-β-importin-α3 complex, and is translocated actively from the cytoplasm to the nucleus [71]. This pathway requires an established Ran gradient and is energy dependent [72]. Despite containing an NES, RanBP1, moves continuously between the nucleus and cytoplasm, but is largely a cytoplasmic protein [73]. In contrast, RanGAP1 is prevented from entering the nucleus because it carries an NES [74, 75]. Moreover, transport factors are also shuttled in both directions across the nuclear envelope by precise mechanisms, such as the CAS-mediated export of importin-α, highlighting the range of dynamic transport across the nuclear membrane at any one time [76]. Therefore, the spatial organization of Ran regulators may not be solely dictated by the localization of fixed structures, such as the NPC, but may also be due to the maintenance of protein gradients across the nuclear envelope [45].

The Roles of Ran in the Cell Cycle

In addition to the function of Ran in nucleocytoplasmic transport, it also has a distinct role in cell cycle regulation. The Ran system was first implicated in control of the cell cycle in 1990 when it was shown that a point mutation introduced in RCC1, which caused the amino acid change serine to phenylalanine, resulted in premature chromosome condensation [77, 78]. Further evidence that Ran is involved in cell cycle regulation has since been demonstrated in several organisms. In human cells, expression of a dominant active mutant of Ran disrupted cell cycle progression causing arrest primarily in G_2 but also in the G_1 phase of the cell cycle [79]. Likewise, disruption of the Ran system in yeast also resulted in similar cell cycle defects, whereas mutation of the RCC1 homologue Dcd1/Pim1 causes a defect at the end of mitosis which can be rescued by the Ran homologue spi1 [80]. In *Xenopus* egg extracts, generation of GTP bound Ran was required for activation of cyclin-dependent kinase-1 (CDK1)-cyclin B and entry into mitosis [81, 82]. However, the precise mechanism remains to be elucidated. In addition to its role in regulating cell cycle progression, Ran also controls formation of the mitotic spindle during mitosis.

Early experiments in *X. laevis* egg extracts suggested that RanGTP is required for the activation of CDK1-cyclin B and entry into mitosis. A dominant-inhibitory mutant form of Ran (T24N), which remains GDP-bound, promoted phosphorylation of CDK1 on tyrosine residues that inhibited its kinase activity thereby preventing entry into mitosis [81, 82]. Furthermore, when the extracts were supplemented with RCC1 protein together with Ran T24N, inhibition of CDK1-cyclin B was reduced, suggesting that Ran T24N sequesters RCC1 into an inactive form. In contrast, the GTP bound dominant activated mutant of Ran, Q69L, did not prevent CDK1-cyclin B activation, suggesting that RanGTP is required for entry into mitosis [82]. Experiments were also carried out on hamster tsBN2 cells, a temperature sensitive cell line with a point mutation in the RCC1 gene. When the cells were exposed to the nonpermissive temperature, RCC1 protein levels decreased significantly [82]. Moreover, cells prematurely entered S phase when RCC1 function was lost, probably as a consequence of impaired ability to actively exclude entry of mitotic inducers, such as cyclin B, from the nucleus. The presence of these mitotic inducers allows the activation of CDK1-cyclin B, and hence entry into mitosis. This discrepancy in the function of RCC1 during mitotic entry could be explained by an as yet undiscovered role of Ran in control of entry into mitosis. Alternatively, discrepancies in the studies could be explained by the experimental systems used in these studies: RCC1 is largely absent in the tsBN2 hamster kidney cell line while in the *X. laevis* egg extracts it is still present but in an inactive form. Ran also has an active role in the temporal control of progression through mitosis. Experiments in *X. laevis* egg extracts show that elevated levels of exogenous RCC1, and thus higher RanGTP levels, abrogate the spindle checkpoint [83]. Presumably, this is a consequence of delocalizing mitotic regulators, including Mad2, CENP-E, Bub1, and Bub3, from kinetochores. The displacement of these proteins prevents inhibition of APC/C, and so causes the cleaving of cohesion and transition from metaphase to anaphase. Reducing RanGDP levels (by depletion of RanGAP1 and RanBP1) abrogates the checkpoint arrest in a similar manner. In contrast, increasing RanGDP levels (by addition of RanGAP1 and RanBP1) in extracts with exogenous RCC1 restores the spindle checkpoint [83]. This suggests that the spindle checkpoint is directly linked to the RanGTP–GDP ratio. Therefore, changes in the production of RanGTP and release of cells from metaphase could be linked to the successful completion of spindle assembly.

The Roles of Ran in Mitotic Spindle Formation

The mitotic spindle is formed during mitosis from the reorganization of the microtubule network and functions to separate replicated chromosomes between the two daughter cells. There is considerable evidence that Ran plays a key role in mitotic spindle assembly during mitosis. In 1999, multiple groups demonstrated that increasing the amount of GTP bound Ran by addition of either RCC1 or dominant active Ran mutants promoted spindle formation in *Xenopus* egg extracts [84–87]. In contrast, inhibiting formation of GTP bound Ran caused abnormal spindle assembly and either RCC1 or dominant active Ran mutants promoted spindle formation [84–87]. Studies in somatic cells have shown similar mitotic defects when the Ran system is manipulated to those observed in *Xenopus* egg extracts. For example, expression of Q69L or T24N Ran mutants in human cells causes defects in mitotic spindle morphology and chromosome alignment [10]. Similarly, mis-localization of RCC1 through removal of the N-terminal region, which is required for chromosome interaction, causes comparable defects [10]. Furthermore, Ran has been implicated in various processes that are important in mitotic spindle assembly such as microtubule nucleation, dynamics, and motor activity [87–90].

RanGTP indirectly promotes microtubule nucleation in the vicinity of chromosomes through release of spindle assembly factors from inhibitory complexes with importins (Fig. 2) [14, 15, 91]. Spindle assembly factors are targeted to the nucleus during interphase through their NLS. Thus, the importin-α and importin-β dimer binds to the NLS of spindle assembly factors during mitosis and inhibits their activity. Subsequently, binding of RanGTP to importin-β displaces these factors and releases them from the inhibitory effect of importin binding [91]. Thus, RanGTP uses components of the nucleocytoplasmic transport system to localize active spindle assembly factors near to chromatin. Several spindle assembly factors, such as NuMA, XCTK2, and TPX2, are directly regulated by RanGTP [14, 92, 93]. TPX2 is one of the best-characterized downstream targets of RanGTP, and it is displaced from importin-α due to RanGTP activity (Fig. 3a) [91]. TPX2 mediates the binding of the motor protein XKLP2 to microtubules and activates Aurora A kinase which is associated with the centrosome (Fig. 3a) [33, 94, 95]. TPX2 has also been shown to form a multi protein complex with Aurora A kinase, Eg5 (a plus end-directed motor), and the microtubule associated proteins XMAP215 and HURP [34]. Formation and function of this complex is dependent on Aurora A kinase activity [34]. HURP is another key downstream target of RanGTP; it is an importin-β regulated protein that is involved in stabilizing and targeting K-fibers (microtubule bundles that attach the kinetochore to the spindle pole) to chromosomes [34, 35, 96]. More recently, the protein MCRS1 has also been identified as a spindle assembly factor that is released from interaction with importin-β by RanGTP binding [97]. MCRS1 is an essential protein for proper spindle assembly, as it stabilizes K-fibers by suppressing depolymerization at the minus end [97]. CDK11 is a further downstream target of RanGTP that is essential for microtubule stabilization around chromosomes [98].

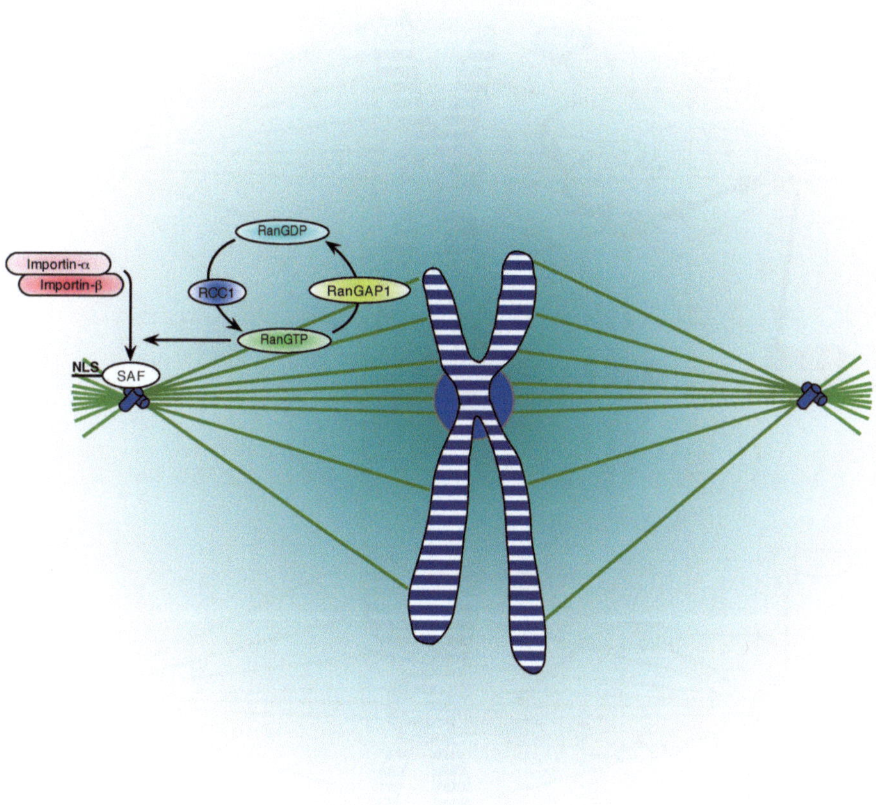

Fig. 2 Ran releases spindle assembly factors. Generation of RanGTP from the guanine nucleotide cycle forms a "cloud" of RanGTP around the mitotic spindle. RanGTP then acts to dissociate spindle assembly factors (SAF) from inhibitory complexes incorporating importin-α and importin-β. The importin dimer then binds to a nuclear localization sequence (NLS) of a SAF, which inhibits their activity

An indirect target of RanGTP is XRHAMM, a TPX2 and γ-tubulin associated factor that is required for microtubule nucleation and spindle pole formation [99]. It has also been suggested that the E3 ubiquitin ligase activity of the tumor suppressor complex BRCA1/BARD1 regulates Ran dependent spindle formation through attenuation of XRHAMM activity, thereby permitting accumulation of TPX2 on spindle poles (Fig. 3a) [100]. The nuclear export receptor CRM1 is another example of a RanGTP effector that is critical for mitotic spindle formation [101]. CRM1 and RanGTP are required for the recruitment of RanGAP1 and RanBP2 to kinetochores, for K-fiber assembly and proper chromosome segregation (Fig. 3b) [101]. The RanGTP-CRM complex may also have a role in regulating centrosome duplication during cell division (Fig. 3b) [102, 103].

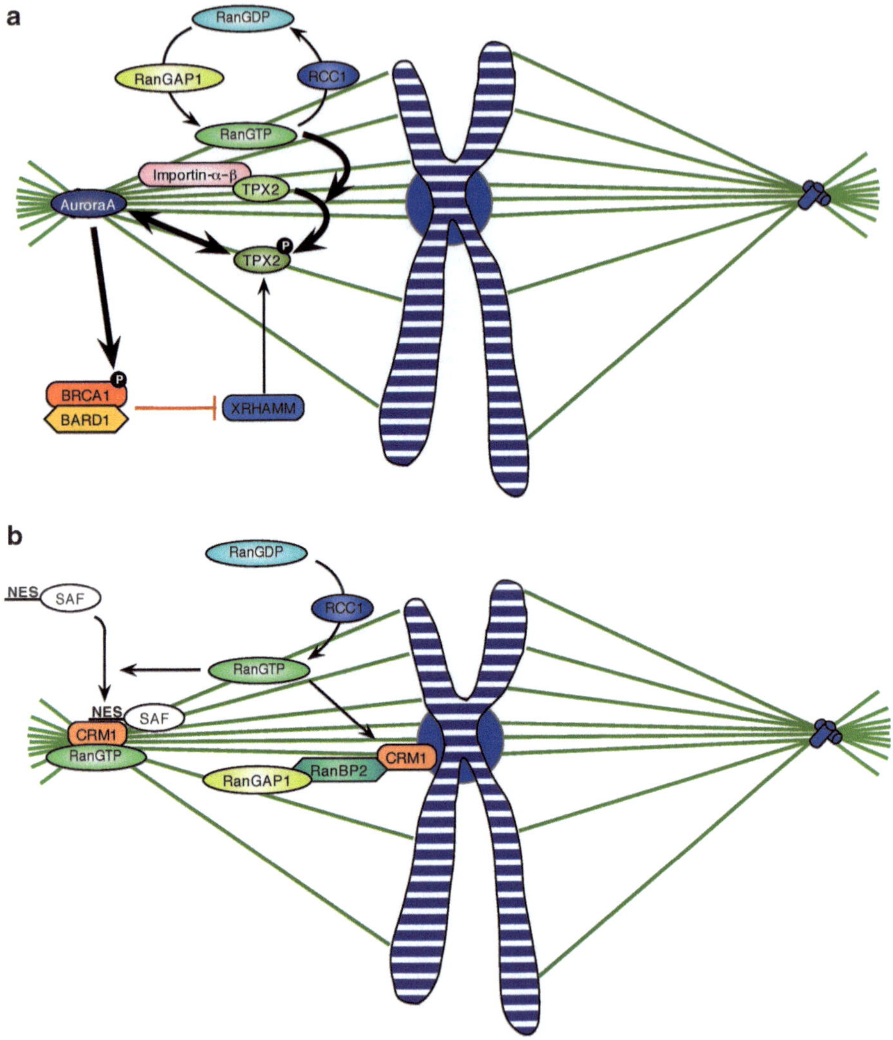

Fig. 3 Ran controls protein complexes during mitotic spindle formation. (**a**) RanGTP promotes the activation of Aurora A kinase (localized at spindle microtubules and centromeres) by catalyzing the release of TPX2 from a TPX2-importin complex. Free TPX2 activates Aurora A, which phosphorylates TPX2, producing a region of active TPX2 near chromosomes. XRHAMM can also activate TPX2, whereas the breast cancer type-1 susceptibility protein (BRCA1)-BARD1 (BRCA1-associated RING domain protein-1) ubiquitin ligase complex can inhibit XRHAMM. BRCA1 is also phosphorylated by Aurora A kinase. (**b**) Ran regulates mitotic spindle assembly through CRM1, which associates with RanBP1 and RanBP2 and interacts with kinetochores. The interaction between kinetochores and kinetochore fibers, which consist of microtubules that stretch from the centromeres to the kinetochore, is in part controlled by the RanBP2-RanGAP1 complex

The Roles of Ran in Nuclear Envelope Assembly

The completion of mitosis is characterized by disassembly of the mitotic spindle and formation of the nuclear envelope. Following the first report in 1995, the role of Ran in directing nuclear envelope assembly has been established in several eukaryotic organisms. In those seminal experiments, the introduction of a mutant form of RCC1 in *Schizosaccharomyces pombe* caused significant fragmentation of the nuclear envelope [104]. This also resulted in the aggregation of pore-containing membranes in the cytoplasm and "free" chromatin that was not compartmentalized. Subsequent studies using cell-free *Xenopus* extracts and demembraned sperm chromatin demonstrated that Ran was important in nuclear assembly and lamina formation around the nucleus [19–21]. However, it was suggested that these effects may be a result of disrupted nucleocytoplasmic transport, since the Ran mutants utilized in the experiments inhibited nuclear import of a protein containing a NLS [21]. Subsequently, Ran was shown to be directly required for the extensive vesicle fusion events that lead to membrane fusion and nuclear envelope formation in *Xenopus* extracts [22, 23]. The requirement of Ran in NPC assembly has since been further demonstrated [105–107] and a role for Ran in regulating the structure and function of the pore has been suggested [108]. Furthermore, Ran is essential for nuclear envelope assembly in *Drosophila melanogaster* [109] and *Caenorhabditis elegans* [110, 111], and for NPC formation in *Saccharomyces cerevisiae* [112], showing that these functions are conserved across eukaryotes.

In two independent studies, membrane vesicles, which fuse to form the lipid membrane of the nuclear envelope, were shown to rapidly accumulate around artificial beads coated with recombinant Ran protein [23, 113]. This lipid layer contained nucleoporins, formed NPCs and facilitated the active import of nuclear proteins through the formation of pseudo-nuclei [23]. Zhang and Clarke reported that RanGDP-coated beads require RCC1 in cell-free extracts to convert the RanGDP to RanGTP, which will promote vesicle binding and fusion. By contrast, RanGTP-coated beads can promote vesicle binding and fusion to form nuclear envelopes independently of RCC1. Inhibition of RCC1 and RanGAP using blocking antibodies decreased vesicle recruitment and fusion, indicating that RanGDP must be converted to RanGTP before vesicles can be recruited and subsequently GTP hydrolysis is required for vesicle fusion to form the nuclear envelope [113]. Elevated concentrations of RanGDP, but neither RanQ69L nor RanT24N, enhanced nuclear envelope formation around chromatin in *Xenopus* extracts, including the assembly of increased smoothed membranes and NPCs that can be observed by field emission in-lens scanning electron microscopy (FEISEM) [105]. In similar observations to RanBP1, RanT24N, a mutant with perturbed nucleotide binding, does not enhance nuclear envelope assembly, whereas RanQ69L, a mutant that disrupts GTPase activity, facilitates isolated vesicle recruitment to chromatin without membrane fusion or NPC formation, highlighting either decreased vesicle binding to chromatin or an inhibition in vesicle fusion [105].

Hetzer and colleagues demonstrated in other experiments that Ran is essential for nuclear membrane fusion events and integration of nucleoporins during nuclear envelope formation [22]. The incorporation of either Ran Q69L or T24N mutants into *Xenopus* extracts resulted in significant defects in nuclear membrane vesicle fusion, suggesting that both the generation and GDP-GTP cycling of RanGTP are essential for vesicle fusion events. Depletion of RCC1 also inhibited vesicle fusion around chromatin and this could be rescued by the re-addition of RCC1 or RanGTP, but not RanGDP [22]. Therefore, generation of RanGTP by RCC1 is essential for the events leading to nuclear envelope assembly. In contrast, extracts containing the GTP analogue GTPγS, which is non-hydrolyzable, are unable to rescue RCC1 reduction. This suggests Ran GTP hydrolysis is important for membrane assembly, but as we shall discuss later, this may not be entirely correct [22]. The inclusion of NPCs into existing nuclear membranes in *Xenopus* extracts is also RanGTP-dependent and can occur from both the nucleoplasmic and cytoplasmic pools [107].

The failure of GTPγS or RanQ69L to promote nuclear envelope formation around chromatin was initially thought to reflect the necessity for GTP hydrolysis in the fusion of vesicles [22]. However, the Q69L Ran mutant also prevents the chromatin recruitment of RCC1 thereby preventing generation of new RanGTP molecules on chromatin [105]. Typically, Ran in its GDP-bound conformation, associates with chromatin at an early stage in nuclear assembly, recruiting RCC1 which is essential for RanGTP generation at the chromatin surface [105]. This explains why RanGDP can support nuclear membrane formation—by catalyzing production of RanGTP on chromatin. It also explains why excess RanGTP (either through GTPγS or Q69L) inhibits viable nuclear envelope assembly—because the spatial organization is disrupted, rather than GTP/GDP cycling [105].

The mechanism of how Ran controls nuclear envelope formation is poorly understood, but the involvement of importin-β has been reported [16]. Zhang and colleagues demonstrated that the addition or removal of importin-β to Ran-coated beads inhibits the nuclear envelope formation initiated by Ran, whereas importin-β-coated beads were sufficient to initiate envelope assembly [16]. Importin-β is also essential for nuclear membrane formation in other eukaryotes such as *C. elegans* [23, 111] and *D. melanogaster* [109], and nuclear envelope assembly surrounding sperm chromatin in vitro [106]. Importin-α functions in concert with importin-β, and is also important in nuclear envelope assembly; formation of the envelope is dependent on membrane-bound importin-α, and this association is regulated by phosphorylation [114]. Lu and colleagues recently demonstrated that NLS-containing proteins bound to chromatin provide sites where nucleoporins and premature membrane vesicles can dock via importin-α and importin-β during nuclear membrane formation [115]. During this process, importin-β binds premature membrane nucleoporins and vesicles, while importin-α rapidly associates with NLS-containing proteins, such as histones and nucleoplasmin, which quickly accumulate around free chromatin in *Xenopus* extracts. Nucleoporins and membrane vesicles are subsequently recruited to the sperm chromatin surface by importin-β following importin-α–importin-β interactions on chromatin. Finally, Importin-β binds to RanGTP on demembraned chromatin and releases nucleoporins and membrane vesicles for nuclear envelope assembly. Experiments using reduced importin-β or perturbed levels of NLS-containing proteins can disrupt importin-α/

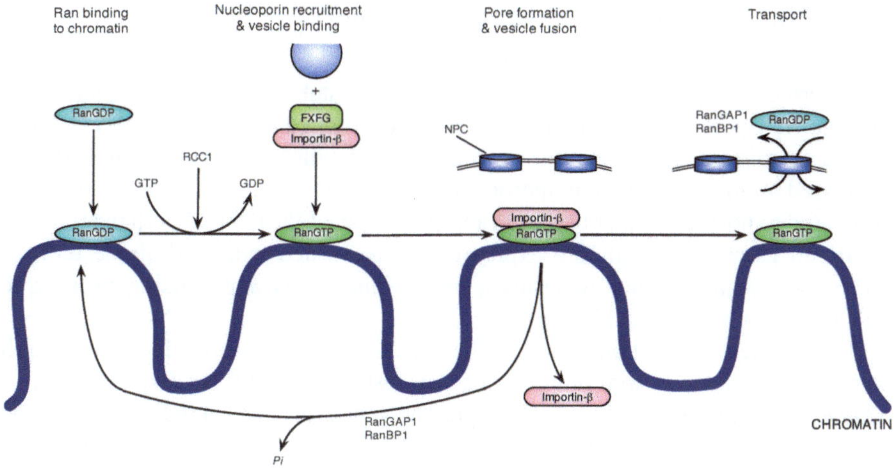

Fig. 4 Ran regulates nuclear-envelope assembly. RanGTP is recruited to the chromatin surface during telophase and its localized production recruits vesicles and nucleoporins through importin-β. Interactions between RanGTP and importin-β promote the assembly of nucleoporins and vesicles into nuclear pore complexes (NPCs), formation of the nuclear envelope and nuclear transport restart. The RanGTP-activating proteins RanGAP1 and RanBP1 or RanBP2 stimulate release of importin-β from Ran. FXFG is representative of a Phe-Gly (FG) repeat sequence

NLS-containing protein interactions, which result in defective nuclear membrane formation [115].

Importin-β regulation of membrane assembly probably requires RanGTP-mediated interactions with both FG and non-FG nucleoporins (Fig. 4). The Nup107–Nup160 complex, which is controlled by importin-β, may form a primary pre-pore on the chromatin surface where mature complexes are then formed [116]. Importin-β-driven inhibition of this pore complex is circumvented by RanGTP, indicating that RanGTP produced at chromosomes can release nucleoporins from inhibitory associations with importin-β [18, 106, 117].

The isolation of MEL-28/ELYS as a key factor in Nup107–Nup160 recruitment to chromatin has been a significant development in understanding how Ran directs nuclear envelope formation [118–121]. MEL-28/ELYS also associates with the MCM2-7 replication proteins on chromatin and ELY chromatin loading, and NPC formation, and nuclear membrane assembly are inhibited when MCM2-7 loading onto chromatin is inhibited [122]. Lamin B receptor (LBR), a lamin and chromatin binding nuclear membrane protein, is also involved in importin-β-mediated nuclear membrane formation [123]. LBR promotes membrane generation, membrane stack formation, nuclear envelope invagination and vesicle recruitment to chromatin. LBR has been identified in a mechanism that tethers heterochromatin to the nuclear envelope, in combination with lamin A/C [124–127]. These two tethers are utilized sequentially during development and have recently been found to regulate differentiation inversely. LBR-dependent heterochromatin tethers delay myogenic differentiation, whereas lamin A/C-dependent tethers promote myogenic differentiation [127].

Ran: The Effector of Metastatic Disease

As described earlier, Ran is best known for having established roles in nuclear cytoplasmic transport and mitosis. However, recent publications have clearly demonstrated an intriguing role for Ran in cancer development and progression. Ran is expressed at higher levels in numerous tumor derived cell lines and in tumor tissue compared to normal cells or tissues (Table 1) [25, 26, 128–133]. In both ovarian and renal cell carcinoma high Ran expression is associated with high-grade tumors, local invasion and tumor metastasis, establishing it as a promising prognostic indicator of poor survival [27, 28].

The spotlight was not on Ran from the outset. It was discovered through research carried out on an upstream regulator of Ran, osteopontin (OPN). Osteopontin is a 33 kDa glycophosphoprotein widely expressed in human cells, including osteoclasts, kidney cells, and endothelium, as well as breast and skin epithelial cells [134, 135]. It has been shown to enhance the transformative, migratory and invasive capacity of cancer cells in vitro [136, 137], and promote metastasis in vivo [138–140]. It does this by simultaneously inhibiting tumor suppressor genes and promoting oncogenic signaling, primarily via its downstream effector, Ran.

The Ran associated factor, RanBP2 has also been implicated in tumorigenesis in various studies. Dawlaty and colleagues demonstrated that transgenic mice with reduced RanBP2 expression developed severe aneuploidy and were predisposed to 7,12-dimethylbenz[a]anthracene (DMBA)-induced tumor formation [141]. Skin cancer incidence was significantly higher in RanBP2 hypomorphic mice compared

Table 1 Ran is over expressed in several tumor derived cell lines and tumor specimens compared to the equivalent normal cell lines or tissues, using various detection methods

Tumor type	Reference/detection method
Breast	[26] (WB)
	[128] (SSH, NH, qRT-PCR)
Lung	[25] (IHC)
Cervical	[129] (MA, qRT-PCR)
Prostate	[26] (WB)
	[130] (ISH)
Ovarian	[27] (IHC)
Colon	[25] (IHC)
	[26] (WB)
	[131] (MA)
	[132] (MA)
	[133] (IHC)
Renal	[28] (WB, IHC)
Nasopharyngeal	[29] (MA, qRT-PCR)
Mesothelioma	[30] (MA)
Mantle-Cell Lymphoma	[31] (qRT-PCR)
Myeloma	[32] (GEP)

WB western blotting, *IHC* immunohistochemistry, *SSH* suppressive subtractive hybridization, *qRT-PCR* quantitative RT-PCR, *ISH* in situ hybridization, *MA* microarray, *GEP* gene expression profile (custom), *NH* northern hybridization

to mice that carried at least one wild-type allele of RanBP2, with lung metastasis detectable in most of the hypomorphic mice. This suggests that RanBP2 has tumor suppressor activity so that when levels of this protein fall below a particular threshold, animals are more susceptible to carcinogen-induced tumorigenesis. In other studies, novel anaplastic lymphoma kinase (ALK) fusions with RanBP2 have been identified in inflammatory myofibroblastic tumors (IMT), rare cancers that are characterized by mesenchymal proliferation of myofibroblasts and numerous inflammatory cells. In these rearrangements, the amino terminal 867 residues of RanBP2 are fused to the cytoplasmic domain of ALK, generating a novel chimeric protein that has unique nuclear-membrane ALK staining compared to other ALK rearrangements in IMT [142]. Recently, a novel mutation in the ALK kinase domain (F1174L) has been identified in IMT patients that conferred resistance to the clinically approved ALK ATP-competitive inhibitor crizotinib in patients harboring the RanBP2-ALK translocation [143]. Overexpression of the *RanBP2* gene has also been associated with cisplatin resistance in ovarian cancer cell lines [144]. Furthermore, the RanGTP-associated export factor CAS/CSE1 has also been implicated in cancer. CAS/CSE1 has been identified in protein complexes that interact with p53 target promoters in vivo [145]. Downregulation of CAS/CSE1 was found to decrease apoptosis by reducing transcription from those p53 target promoters, including *PIG3*, a gene involved in early cellular DNA damage response [146]. Lately, CAS/CSE1 has also been reported as a pro-survival protein in ovarian cancer where it is regulated by the c-Met tyrosine kinase receptor, a receptor also implicated in Ran biology [147]. Another Ran-related protein that is perturbed in tumorigenesis is CRM1. Overexpression of CRM1 has been found in a number of solid and hematological tumors, and the upregulated expression correlates with reduced overall survival and therapy resistance [148–150]. Recently, nuclear export inhibitor studies have identified CRM1 as a therapeutic target in chronic lymphocytic leukemia (CLL) [70].

The decreased survival of cancer patients may depend on the overexpression of Ran, which is known to promote metastasis [24]. For instance, overexpression of Ran has been shown to induce a metastatic phenotype in an experimental model of breast cancer dissemination both in vitro and in vivo [24]. Stable transfection of benign mammary epithelial cells with Ran stimulated increased anchorage-independent growth, cell adhesion to fibronectin and invasion through Matrigel in vitro, and metastasis in a syngeneic rat model in vivo (Fig. 5) [24]. Conversely, this metastatic phenotype could be reversed both in vitro and in vivo by suppressing Ran expression through transfection of the cells with Ran-targeted siRNA [24]. Subsequent studies have corroborated these findings by further demonstrating that Ran expression increased the invasiveness of human lung carcinoma cells [151]. The GTPase activity of Ran is also required for efficient metastasis [24] and increased RanGTP levels have been associated with increased cell transformation and tumorigenicity [152].

As a consequence of the differential Ran expression between normal cells and tumor tissue, it has been suggested that Ran may represent a potential anticancer therapeutic target. Inhibition of Ran expression in several tumor cell lines causes abnormal mitotic spindle formation, mitochondrial dysfunction, and apoptosis [26, 153].

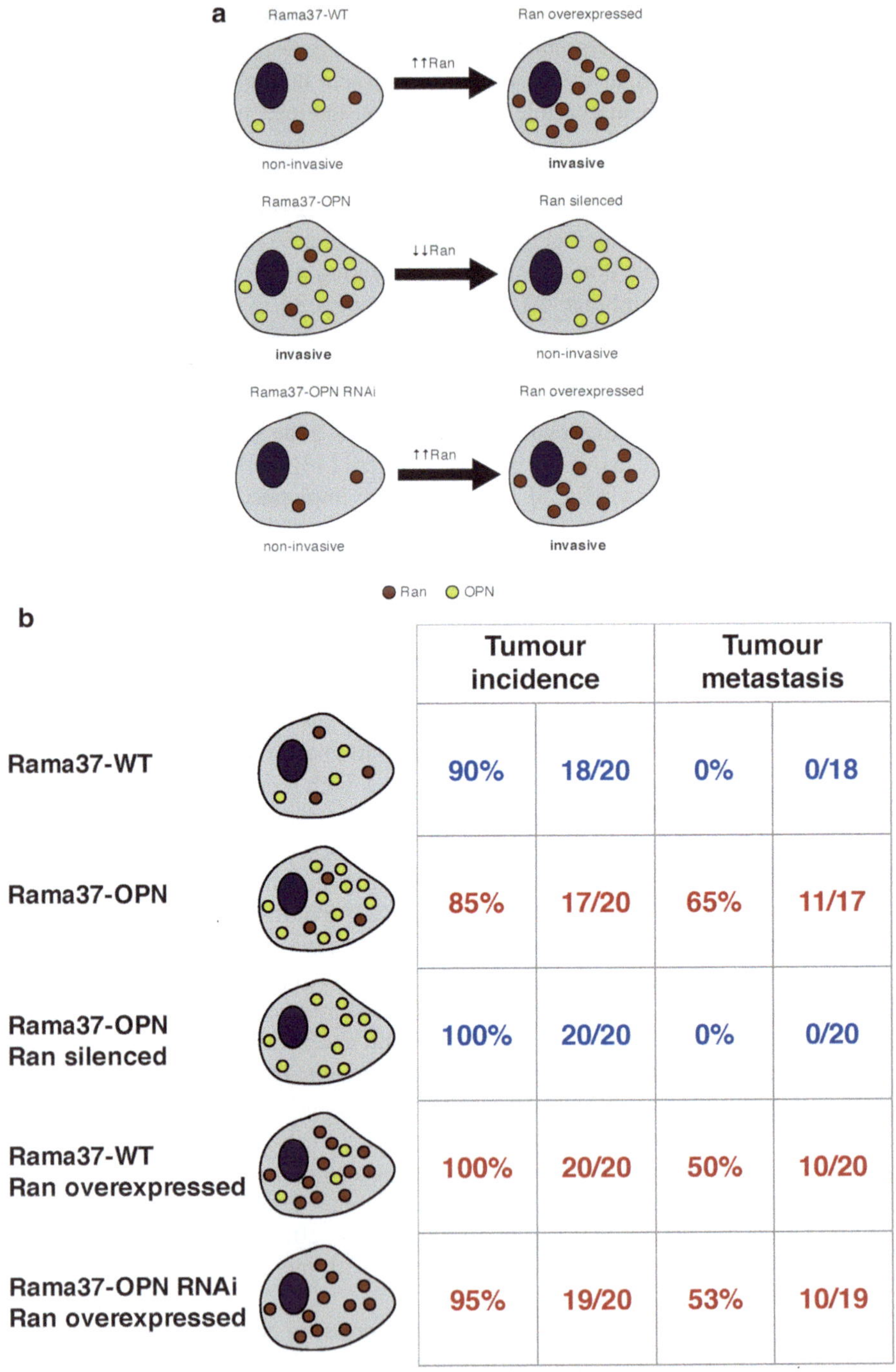

	Tumour incidence		Tumour metastasis	
Rama37-WT	90%	18/20	0%	0/18
Rama37-OPN	85%	17/20	65%	11/17
Rama37-OPN Ran silenced	100%	20/20	0%	0/20
Rama37-WT Ran overexpressed	100%	20/20	50%	10/20
Rama37-OPN RNAi Ran overexpressed	95%	19/20	53%	10/19

Fig. 5 Ran drives tumor invasion and metastasis. (**a**) In vitro analysis of Ran/OPN in cell invasion. Overexpression of Ran in benign noninvasive mammary Rama37 cells generated an invasive cell phenotype. Similarly, overexpression of Ran, together with OPN depletion, generated an invasive phenotype.

However, knock down of Ran has little effect on normal cells and does not result in decreased cell viability, indicating that tumor cells may become dependent on Ran signaling for survival [26]. Furthermore, inhibition of Ran expression may be a particularly effective anticancer therapeutic strategy in K-Ras mutant tumors since silencing Ran expression caused increased apoptotic cell death in K-Ras mutant cell lines in comparison to their isogenic counterparts [153].

Ran as a Novel Cancer Therapeutic Target

Nanomedicines have significant potential in enhancing delivery of cancer therapeutics. The proper understanding of the characteristics of tumor biology, coupled with proper tumor targeting using colloidal nanoparticles has been propelled to the forefront of cancer research. A nanoparticle constructed from a drug-loaded core with a peripheral targeting ligand is probably the design with the most potential to achieve these objectives. Such design has the potential to achieve tumor size reduction and elimination via its target-specific nature of action without damaging healthy tissue. However, the clinical worth of such nanoparticles is determined by their ability to disseminate in the body and reach target sites in therapeutically effective doses [154]. There is now a large overlap between tumor-targeting mechanisms and their utilization for diagnosis and therapy of cancer. This reflects the heterogeneity of tumor biology and the large potential for multiple targeting schemes using the same mechanism or ligand [155].

It is now well established that Ran is overexpressed in various cancers and this overexpression is correlated with increased aggressiveness of the cancer cells in vitro and in vivo [24, 151, 152]. Targeting the Ran pathway, through peptide blockades or siRNA molecules directed against human Ran and/or RCC1, represents a promising novel therapeutic modality for the treatment of complex diseases such as cancer [24]. However, delivering peptides or siRNA to tumors using clinically viable formulations remains the major technical hurdle. Over the past several years, significant effort has been devoted to explore novel delivery strategies, whereas relatively little attention has been paid to understand the impact of physiological constraints such as tumor vasculature on the efficiency of peptide/siRNA delivery [156]. Engineered virus particles can serve as multifunctional targeted, non-immunogenic,

Fig. 5 (continued) In contrast, overexpression of OPN and siRNA-mediated silencing of Ran rendered the cells noninvasive. This suggests the induction of an invasive phenotype is due to Ran expression and is induced independently of OPN. (**b**) In vivo analysis of Ran/OPN in tumor incidence and metastasis. In a syngeneic rat model of tumor formation and secondary metastasis, Rama37 generated tumors in 90 % of cases, but none of these progressed to metastasis. Overexpression of OPN (with concomitant increased Ran expression) in Rama37 cells induced both tumors and metastases. However, silencing of Ran abolished incidents of tumor metastasis, despite the formation of primary tumors in all cases. Overexpression of Ran was sufficient to generate tumor metastasis in low OPN expressed Rama37 or depleted by OPN-antisense. This suggests that Ran is the downstream effector of OPN-mediated invasion and metastasis

nanoscale devices with potential for a broad range of in vivo uses, especially for siRNA delivery. However, the limitations of viral vectors, particularly regarding safety concerns, have led to the development of colloidal nanoparticles based on biodegradable polymers and liposomes [156, 157]. A wide variety of synthetic nanoparticles have been shown to target tumor cells, enter cancer cells, and release therapeutic agents [158]. To reach cells, nanoparticles must readily diffuse through the capillary network, escape macrophage surveillance and interact with the cell membrane [159]. Their small size should allow penetration of cell membranes, binding and stabilization of proteins and lysosomal escape following endocytosis [160].

One interesting novel approach for nanoparticle delivery is a system using transferrin modified poly(ethylene) glycol (PEG) nanoparticles encapsulating the desired therapeutic agent [161]. This system combines both passive and active targeting mechanisms. Surface modification with hydrophilic polymers, such as PEG, has been used for steric stabilization and results in "stealth-like" nanoparticles that reduce serum protein interaction, thus prolonging drug residence time in the circulation and allowing nanoparticles, smaller than the fenestrations, to migrate and accumulate across leaky vasculature in tumor tissue by way of the enhanced permeation and retention effect (EPR). EPR is a form of passive targeting and can be also called selective targeting. Simultaneously, transferrin exhibits active targeting through transferrin-mediated endocytosis [155, 160]. In active targeting strategies, a targeting ligand is attached to the nanoparticles to allow site-specific delivery.

Conclusion

In summary, Ran is recognized as having a number of important roles in various cellular functions. Most notably, Ran is essential for proper nucleocytoplasmic transport, mitosis and nuclear envelope formation, with a further emerging role in tumor progression and metastasis. Ran mediates nucleocytoplasmic transport through the association of RanGTP with the importin-β family of karyopherins. This association with RanGTP regulates the interaction of karyopherins with the cargo proteins and NPC proteins. The localization of the RanGEF RCC1 to the nucleus, whilst RanGAP and Ran binding proteins are localized to the cytoplasm, causes the accumulation of RanGTP in the nucleus. This differential distribution of RanGTP guides the spatial orientation of the mitotic spindle and the nuclear envelope. Thus, defects in the Ran system may result in chromosomal instability, aneuploidy and tumorigenesis. Moreover, there is mounting evidence that Ran plays a crucial role in cancer development and progression. Ran is highly expressed in multiple tumor types and is associated with a more aggressive phenotype. It has been demonstrated that overexpression of Ran in benign mammary epithelium causes neoplastic transformation, dissemination and metastasis [24]. Current research suggests that Ran may not only be a useful prognostic biomarker but may also be a potential anticancer therapeutic target, which, when inhibited, selectively kills the more aggressive cancer cells. Ran's association with metastasis makes it an even more valuable therapeutic target since metastatic disease is often refractory to current treatments and remains the main cause of cancer related deaths. Whilst the discovery of Ran's involvement with cancer is a significant finding, many aspects of its role remain to be elucidated and it is not yet clear whether this knowledge will lead to an improvement in the diagnosis or treatment of cancer. However, the rate of progress in understanding the Ran system has been rapid to date, and it is an exciting area of interest for cancer biologists.

References

1. Moore MS, Blobel G (1993) The GTP-binding protein Ran/TC4 is required for protein import into the nucleus. Nature 365(6447):661–663. doi:10.1038/365661a0
2. Melchior F, Paschal B, Evans J, Gerace L (1993) Inhibition of nuclear protein import by nonhydrolyzable analogues of GTP and identification of the small GTPase Ran/TC4 as an essential transport factor. J Cell Biol 123(6 Pt 2):1649–1659
3. Drivas GT, Shih A, Coutavas E, Rush MG, D'Eustachio P (1990) Characterization of four novel ras-like genes expressed in a human teratocarcinoma cell line. Mol Cell Biol 10(4):1793–1798
4. Bourne HR, Sanders DA, McCormick F (1991) The GTPase superfamily: conserved structure and molecular mechanism. Nature 349(6305):117–127. doi:10.1038/349117a0
5. Joseph J (2006) Ran at a glance. J Cell Sci 119(Pt 17):3481–3484. doi:10.1242/jcs.03071, 119/17/3481 [pii]
6. Scheffzek K, Klebe C, Fritz-Wolf K, Kabsch W, Wittinghofer A (1995) Crystal structure of the nuclear Ras-related protein Ran in its GDP-bound form. Nature 374(6520):378–381. doi:10.1038/374378a0
7. Richards SA, Lounsbury KM, Macara IG (1995) The C terminus of the nuclear RAN/TC4 GTPase stabilizes the GDP-bound state and mediates interactions with RCC1, RAN-GAP, and HTF9A/RANBP1. J Biol Chem 270(24):14405–14411
8. Gorlich D, Seewald MJ, Ribbeck K (2003) Characterization of Ran-driven cargo transport and the RanGTPase system by kinetic measurements and computer simulation. EMBO J 22(5):1088–1100. doi:10.1093/emboj/cdg113
9. Smith AE, Slepchenko BM, Schaff JC, Loew LM, Macara IG (2002) Systems analysis of Ran transport. Science 295(5554):488–491. doi:10.1126/science.1064732
10. Moore W, Zhang C, Clarke PR (2002) Targeting of RCC1 to chromosomes is required for proper mitotic spindle assembly in human cells. Curr Biol 12(16):1442–1447
11. Kuersten S, Ohno M, Mattaj IW (2001) Nucleocytoplasmic transport: Ran, beta and beyond. Trends Cell Biol 11(12):497–503
12. Gorlich D, Kostka S, Kraft R, Dingwall C, Laskey RA, Hartmann E, Prehn S (1995) Two different subunits of importin cooperate to recognize nuclear localization signals and bind them to the nuclear envelope. Curr Biol 5(4):383–392
13. Li HY, Zheng Y (2004) Phosphorylation of RCC1 in mitosis is essential for producing a high RanGTP concentration on chromosomes and for spindle assembly in mammalian cells. Genes Dev 18(5):512–527. doi:10.1101/gad.1177304
14. Wiese C, Wilde A, Moore MS, Adam SA, Merdes A, Zheng Y (2001) Role of importin-beta in coupling Ran to downstream targets in microtubule assembly. Science 291(5504):653–656
15. Nachury MV, Maresca TJ, Salmon WC, Waterman-Storer CM, Heald R, Weis K (2001) Importin beta is a mitotic target of the small GTPase Ran in spindle assembly. Cell 104(1):95–106
16. Zhang C, Hutchins JR, Muhlhausser P, Kutay U, Clarke PR (2002) Role of importin-beta in the control of nuclear envelope assembly by Ran. Curr Biol 12(6):498–502

17. Clarke PR, Zhang C (2004) Spatial and temporal control of nuclear envelope assembly by Ran GTPase. Symp Soc Exp Biol 56:193–204
18. Harel A, Forbes DJ (2004) Importin beta: conducting a much larger cellular symphony. Mol Cell 16(3):319–330. doi:10.1016/j.molcel.2004.10.026
19. Dasso M, Seki T, Azuma Y, Ohba T, Nishimoto T (1994) A mutant form of the Ran/TC4 protein disrupts nuclear function in Xenopus laevis egg extracts by inhibiting the RCC1 protein, a regulator of chromosome condensation. EMBO J 13(23):5732–5744
20. Nicolas FJ, Zhang C, Hughes M, Goldberg MW, Watton SJ, Clarke PR (1997) Xenopus Ran-binding protein 1: molecular interactions and effects on nuclear assembly in Xenopus egg extracts. J Cell Sci 110(Pt 24):3019–3030
21. Hughes M, Zhang C, Avis JM, Hutchison CJ, Clarke PR (1998) The role of the ran GTPase in nuclear assembly and DNA replication: characterisation of the effects of Ran mutants. J Cell Sci 111(Pt 20):3017–3026
22. Hetzer M, Bilbao-Cortes D, Walther TC, Gruss OJ, Mattaj IW (2000) GTP hydrolysis by Ran is required for nuclear envelope assembly. Mol Cell 5(6):1013–1024
23. Zhang C, Clarke PR (2000) Chromatin-independent nuclear envelope assembly induced by Ran GTPase in Xenopus egg extracts. Science 288(5470):1429–1432
24. Kurisetty VV, Johnston PG, Johnston N, Erwin P, Crowe P, Fernig DG, Campbell FC, Anderson IP, Rudland PS, El-Tanani MK (2008) RAN GTPase is an effector of the invasive/metastatic phenotype induced by osteopontin. Oncogene 27(57):7139–7149. doi:10.1038/onc.2008.325, onc2008325 [pii]
25. Azuma K, Sasada T, Takedatsu H, Shomura H, Koga M, Maeda Y, Yao A, Hirai T, Takabayashi A, Shichijo S, Itoh K (2004) Ran, a small GTPase gene, encodes cytotoxic T lymphocyte (CTL) epitopes capable of inducing HLA-A33-restricted and tumor-reactive CTLs in cancer patients. Clin Cancer Res 10(19):6695–6702. doi:10.1158/1078-0432.ccr-04-0818
26. Xia F, Lee CW, Altieri DC (2008) Tumor cell dependence on Ran-GTP-directed mitosis. Cancer Res 68(6):1826–1833. doi:10.1158/0008-5472.can-07-5279
27. Ouellet V, Guyot MC, Le Page C, Filali-Mouhim A, Lussier C, Tonin PN, Provencher DM, Mes-Masson AM (2006) Tissue array analysis of expression microarray candidates identifies markers associated with tumor grade and outcome in serous epithelial ovarian cancer. Int J Cancer 119(3):599–607. doi:10.1002/ijc.21902
28. Abe H, Kamai T, Shirataki H, Oyama T, Arai K, Yoshida K (2008) High expression of Ran GTPase is associated with local invasion and metastasis of human clear cell renal cell carcinoma. Int J Cancer 122(10):2391–2397. doi:10.1002/ijc.23400
29. Li H, Ren CP, Tan XJ, Yang XY, Zhang HB, Zhou W, Yao KT (2006) Identification of genes related to nasopharyngeal carcinoma with the help of pathway-based networks. Acta Biochim Biophys Sin 38(12):900–910
30. Roe OD, Anderssen E, Sandeck H, Christensen T, Larsson E, Lundgren S (2010) Malignant pleural mesothelioma: genome-wide expression patterns reflecting general resistance mechanisms and a proposal of novel targets. Lung Cancer 67(1):57–68. doi:10.1016/j.lungcan.2009.03.016
31. Hartmann E, Fernandez V, Moreno V, Valls J, Hernandez L, Bosch F, Abrisqueta P, Klapper W, Dreyling M, Hoster E, Muller-Hermelink HK, Ott G, Rosenwald A, Campo E (2008) Five-gene model to predict survival in mantle-cell lymphoma using frozen or formalin-fixed, paraffin-embedded tissue. J Clin Oncol 26(30):4966–4972. doi:10.1200/jco.2007.12.0410
32. Harousseau JL, Shaughnessy J Jr, Richardson P (2004) Multiple myeloma. Hematology Am Soc Hematol Educ Program 237–256. http://www.ncbi.nlm.nih.gov/pubmed/15561686, doi:10.1182/asheducation-2004.1.237
33. Tsai MY, Wiese C, Cao K, Martin O, Donovan P, Ruderman J, Prigent C, Zheng Y (2003) A Ran signalling pathway mediated by the mitotic kinase Aurora A in spindle assembly. Nat Cell Biol 5(3):242–248. doi:10.1038/ncb936
34. Koffa MD, Casanova CM, Santarella R, Kocher T, Wilm M, Mattaj IW (2006) HURP is part of a Ran-dependent complex involved in spindle formation. Curr Biol 16(8):743–754. doi:10.1016/j.cub.2006.03.056

35. Sillje HH, Nagel S, Korner R, Nigg EA (2006) HURP is a Ran-importin beta-regulated protein that stabilizes kinetochore microtubules in the vicinity of chromosomes. Curr Biol 16(8):731–742. doi:10.1016/j.cub.2006.02.070
36. Joukov V, Groen AC, Prokhorova T, Gerson R, White E, Rodriguez A, Walter JC, Livingston DM (2006) The BRCA1/BARD1 heterodimer modulates ran-dependent mitotic spindle assembly. Cell 127(3):539–552. doi:10.1016/j.cell.2006.08.053
37. Yuen HF, Chan KK, Grills C, Murray JT, Platt-Higgins A, Eldin OS, O'Byrne K, Janne P, Fennell DA, Johnston PG, Rudland PS, El-Tanani M (2012) Ran is a potential therapeutic target for cancer cells with molecular changes associated with activation of the PI3K/Akt/mTORC1 and Ras/MEK/ERK pathways. Clin Cancer Res 18(2):380–391. doi:10.1158/1078-0432.ccr-11-2035
38. Bischoff FR, Ponstingl H (1991) Catalysis of guanine nucleotide exchange on Ran by the mitotic regulator RCC1. Nature 354(6348):80–82. doi:10.1038/354080a0
39. Bischoff FR, Ponstingl H (1991) Mitotic regulator protein RCC1 is complexed with a nuclear ras-related polypeptide. Proc Natl Acad Sci U S A 88(23):10830–10834
40. Guttler T, Gorlich D (2011) Ran-dependent nuclear export mediators: a structural perspective. EMBO J 30(17):3457–3474. doi:10.1038/emboj.2011.287
41. Lui K, Huang Y (2009) RanGTPase: a key regulator of nucleocytoplasmic trafficking. Mol Cell Pharmacol 1(3):148–156
42. Walde S, Kehlenbach RH (2010) The part and the whole: functions of nucleoporins in nucleocytoplasmic transport. Trends Cell Biol 20(8):461–469. doi:10.1016/j.tcb.2010.05.001
43. Macara IG (2001) Transport into and out of the nucleus. Microbiol Mol Biol Rev 65(4):570–594. doi:10.1128/MMBR.65.4.570-594.2001, table of contents
44. Kubitscheck U, Grunwald D, Hoekstra A, Rohleder D, Kues T, Siebrasse JP, Peters R (2005) Nuclear transport of single molecules: dwell times at the nuclear pore complex. J Cell Biol 168(2):233–243. doi:10.1083/jcb.200411005, jcb.200411005 [pii]
45. Clarke PR, Zhang C (2008) Spatial and temporal coordination of mitosis by Ran GTPase. Nat Rev Mol Cell Biol 9(6):464–477. doi:10.1038/nrm2410
46. Vetter IR, Nowak C, Nishimoto T, Kuhlmann J, Wittinghofer A (1999) Structure of a Ran-binding domain complexed with Ran bound to a GTP analogue: implications for nuclear transport. Nature 398(6722):39–46. doi:10.1038/17969
47. Vetter IR, Arndt A, Kutay U, Gorlich D, Wittinghofer A (1999) Structural view of the Ran-Importin beta interaction at 2.3 A resolution. Cell 97(5):635–646
48. Klebe C, Bischoff FR, Ponstingl H, Wittinghofer A (1995) Interaction of the nuclear GTP-binding protein Ran with its regulatory proteins RCC1 and RanGAP1. Biochemistry 34(2):639–647
49. Ribbeck K, Lipowsky G, Kent HM, Stewart M, Gorlich D (1998) NTF2 mediates nuclear import of Ran. EMBO J 17(22):6587–6598. doi:10.1093/emboj/17.22.6587
50. Smith A, Brownawell A, Macara IG (1998) Nuclear import of Ran is mediated by the transport factor NTF2. Curr Biol 8(25):1403–1406
51. Chafe SC, Pierce JB, Mangroo D (2012) Nuclear-cytoplasmic trafficking of NTF2, the nuclear import receptor for the RanGTPase, is subjected to regulation. PLoS One 7(8):e42501. doi:10.1371/journal.pone.0042501
52. Klebe C, Prinz H, Wittinghofer A, Goody RS (1995) The kinetic mechanism of Ran–nucleotide exchange catalyzed by RCC1. Biochemistry 34(39):12543–12552
53. Makde RD, England JR, Yennawar HP, Tan S (2010) Structure of RCC1 chromatin factor bound to the nucleosome core particle. Nature 467(7315):562–566. doi:10.1038/nature09321
54. Chen T, Muratore TL, Schaner-Tooley CE, Shabanowitz J, Hunt DF, Macara IG (2007) N-terminal alpha-methylation of RCC1 is necessary for stable chromatin association and normal mitosis. Nat Cell Biol 9(5):596–603. doi:10.1038/ncb1572
55. Nemergut ME, Mizzen CA, Stukenberg T, Allis CD, Macara IG (2001) Chromatin docking and exchange activity enhancement of RCC1 by histones H2A and H2B. Science 292(5521):1540–1543. doi:10.1126/science.292.5521.1540
56. Bischoff FR, Klebe C, Kretschmer J, Wittinghofer A, Ponstingl H (1994) RanGAP1 induces GTPase activity of nuclear Ras-related Ran. Proc Natl Acad Sci U S A 91(7):2587–2591

57. Seewald MJ, Korner C, Wittinghofer A, Vetter IR (2002) RanGAP mediates GTP hydrolysis without an arginine finger. Nature 415(6872):662–666. doi:10.1038/415662a
58. Bischoff FR, Krebber H, Smirnova E, Dong W, Ponstingl H (1995) Co-activation of RanGTPase and inhibition of GTP dissociation by Ran-GTP binding protein RanBP1. EMBO J 14(4):705–715
59. Coutavas E, Ren M, Oppenheim JD, D'Eustachio P, Rush MG (1993) Characterization of proteins that interact with the cell-cycle regulatory protein Ran/TC4. Nature 366(6455): 585–587. doi:10.1038/366585a0
60. Yokoyama N, Hayashi N, Seki T, Pante N, Ohba T, Nishii K, Kuma K, Hayashida T, Miyata T, Aebi U et al (1995) A giant nucleopore protein that binds Ran/TC4. Nature 376(6536): 184–188. doi:10.1038/376184a0
61. Lin DH, Zimmermann S, Stuwe T, Stuwe E, Hoelz A (2013) Structural and functional analysis of the C-terminal domain of Nup358/RanBP2. J Mol Biol 425(8):1318–1329. doi:10.1016/j.jmb.2013.01.021
62. Werner A, Flotho A, Melchior F (2012) The RanBP2/RanGAP1*SUMO1/Ubc9 complex is a multisubunit SUMO E3 ligase. Mol Cell 46(3):287–298. doi:10.1016/j.molcel.2012.02.017
63. Ogawa Y, Miyamoto Y, Oka M, Yoneda Y (2012) The interaction between importin-alpha and Nup153 promotes importin-alpha/beta-mediated nuclear import. Traffic 13(7):934–946. doi:10.1111/j.1600-0854.2012.01367.x
64. Liu J, Xiao N, DeFranco DB (1999) Use of digitonin-permeabilized cells in studies of steroid receptor subnuclear trafficking. Methods 19(3):403–409. doi:10.1006/meth.1999.0876
65. Niklas J, Melnyk A, Yuan Y, Heinzle E (2011) Selective permeabilization for the high-throughput measurement of compartmented enzyme activities in mammalian cells. Anal Biochem 416(2):218–227. doi:10.1016/j.ab.2011.05.039
66. Conti E, Muller CW, Stewart M (2006) Karyopherin flexibility in nucleocytoplasmic transport. Curr Opin Struct Biol 16(2):237–244. doi:10.1016/j.sbi.2006.03.010
67. Fukuda M, Asano S, Nakamura T, Adachi M, Yoshida M, Yanagida M, Nishida E (1997) CRM1 is responsible for intracellular transport mediated by the nuclear export signal. Nature 390(6657):308–311. doi:10.1038/36894
68. Petosa C, Schoehn G, Askjaer P, Bauer U, Moulin M, Steuerwald U, Soler-Lopez M, Baudin F, Mattaj IW, Muller CW (2004) Architecture of CRM1/Exportin1 suggests how cooperativity is achieved during formation of a nuclear export complex. Mol Cell 16(5):761–775. doi:10.1016/j.molcel.2004.11.018
69. Bischoff FR, Gorlich D (1997) RanBP1 is crucial for the release of RanGTP from importin beta-related nuclear transport factors. FEBS Lett 419(2–3):249–254
70. Lapalombella R, Sun Q, Williams K, Tangeman L, Jha S, Zhong Y, Goettl V, Mahoney E, Berglund C, Gupta S, Farmer A, Mani R, Johnson AJ, Lucas D, Mo X, Daelemans D, Sandanayaka V, Shechter S, McCauley D, Shacham S, Kauffman M, Chook YM, Byrd JC (2012) Selective inhibitors of nuclear export show that CRM1/XPO1 is a target in chronic lymphocytic leukemia. Blood 120(23):4621–4634. doi:10.1182/blood-2012-05-429506
71. Talcott B, Moore MS (2000) The nuclear import of RCC1 requires a specific nuclear localization sequence receptor, karyopherin alpha3/Qip. J Biol Chem 275(14):10099–10104
72. Nemergut ME, Macara IG (2000) Nuclear import of the ran exchange factor, RCC1, is mediated by at least two distinct mechanisms. J Cell Biol 149(4):835–850
73. Plafker K, Macara IG (2000) Facilitated nucleocytoplasmic shuttling of the Ran binding protein RanBP1. Mol Cell Biol 20(10):3510–3521
74. Matunis MJ, Coutavas E, Blobel G (1996) A novel ubiquitin-like modification modulates the partitioning of the Ran-GTPase-activating protein RanGAP1 between the cytosol and the nuclear pore complex. J Cell Biol 135(6 Pt 1):1457–1470
75. Hopper AK, Traglia HM, Dunst RW (1990) The yeast RNA1 gene product necessary for RNA processing is located in the cytosol and apparently excluded from the nucleus. J Cell Biol 111(2):309–321
76. Kutay U, Bischoff FR, Kostka S, Kraft R, Gorlich D (1997) Export of importin alpha from the nucleus is mediated by a specific nuclear transport factor. Cell 90(6):1061–1071

77. Nishimoto T, Eilen E, Basilico C (1978) Premature of chromosome condensation in a ts DNA-mutant of BHK cells. Cell 15(2):475–483
78. Uchida S, Sekiguchi T, Nishitani H, Miyauchi K, Ohtsubo M, Nishimoto T (1990) Premature chromosome condensation is induced by a point mutation in the hamster RCC1 gene. Mol Cell Biol 10(2):577–584
79. Ren M, Coutavas E, D'Eustachio P, Rush MG (1994) Effects of mutant Ran/TC4 proteins on cell cycle progression. Mol Cell Biol 14(6):4216–4224
80. Sazer S, Nurse P (1994) A fission yeast RCC1-related protein is required for the mitosis to interphase transition. EMBO J 13(3):606–615
81. Kornbluth S, Dasso M, Newport J (1994) Evidence for a dual role for TC4 protein in regulating nuclear structure and cell cycle progression. J Cell Biol 125(4):705–719
82. Clarke PR, Klebe C, Wittinghofer A, Karsenti E (1995) Regulation of Cdc2/cyclin B activation by Ran, a Ras-related GTPase. J Cell Sci 108(Pt 3):1217–1225
83. Arnaoutov A, Dasso M (2003) The Ran GTPase regulates kinetochore function. Dev Cell 5(1):99–111
84. Carazo-Salas RE, Guarguaglini G, Gruss OJ, Segref A, Karsenti E, Mattaj IW (1999) Generation of GTP-bound Ran by RCC1 is required for chromatin-induced mitotic spindle formation. Nature 400(6740):178–181. doi:10.1038/22133
85. Kalab P, Pu RT, Dasso M (1999) The ran GTPase regulates mitotic spindle assembly. Curr Biol 9(9):481–484
86. Ohba T, Nakamura M, Nishitani H, Nishimoto T (1999) Self-organization of microtubule asters induced in Xenopus egg extracts by GTP-bound Ran. Science 284(5418): 1356–1358
87. Wilde A, Zheng Y (1999) Stimulation of microtubule aster formation and spindle assembly by the small GTPase Ran. Science 284(5418):1359–1362
88. Carazo-Salas RE, Gruss OJ, Mattaj IW, Karsenti E (2001) Ran-GTP coordinates regulation of microtubule nucleation and dynamics during mitotic-spindle assembly. Nat Cell Biol 3(3):228–234. doi:10.1038/35060009
89. Wilde A, Lizarraga SB, Zhang L, Wiese C, Gliksman NR, Walczak CE, Zheng Y (2001) Ran stimulates spindle assembly by altering microtubule dynamics and the balance of motor activities. Nat Cell Biol 3(3):221–227. doi:10.1038/35060000
90. Zhang C, Hughes M, Clarke PR (1999) Ran-GTP stabilises microtubule asters and inhibits nuclear assembly in Xenopus egg extracts. J Cell Sci 112(Pt 14):2453–2461
91. Gruss OJ, Carazo-Salas RE, Schatz CA, Guarguaglini G, Kast J, Wilm M, Le Bot N, Vernos I, Karsenti E, Mattaj IW (2001) Ran induces spindle assembly by reversing the inhibitory effect of importin alpha on TPX2 activity. Cell 104(1):83–93
92. Ems-McClung SC, Zheng Y, Walczak CE (2004) Importin alpha/beta and Ran-GTP regulate XCTK2 microtubule binding through a bipartite nuclear localization signal. Mol Biol Cell 15(1):46–57. doi:10.1091/mbc.E03-07-0454
93. Schatz CA, Santarella R, Hoenger A, Karsenti E, Mattaj IW, Gruss OJ, Carazo-Salas RE (2003) Importin alpha-regulated nucleation of microtubules by TPX2. EMBO J 22(9): 2060–2070. doi:10.1093/emboj/cdg195
94. Wittmann T, Wilm M, Karsenti E, Vernos I (2000) TPX2, A novel xenopus MAP involved in spindle pole organization. J Cell Biol 149(7):1405–1418
95. Bayliss R, Sardon T, Vernos I, Conti E (2003) Structural basis of Aurora-A activation by TPX2 at the mitotic spindle. Mol Cell 12(4):851–862
96. Wong J, Fang G (2006) HURP controls spindle dynamics to promote proper interkinetochore tension and efficient kinetochore capture. J Cell Biol 173(6):879–891. doi:10.1083/jcb.200511132
97. Meunier S, Vernos I (2011) K-fibre minus ends are stabilized by a RanGTP-dependent mechanism essential for functional spindle assembly. Nat Cell Biol 13(12):1406–1414. doi:10.1038/ncb2372
98. Yokoyama H, Gruss OJ, Rybina S, Caudron M, Schelder M, Wilm M, Mattaj IW, Karsenti E (2008) Cdk11 is a RanGTP-dependent microtubule stabilization factor that regulates spindle assembly rate. J Cell Biol 180(5):867–875. doi:10.1083/jcb.200706189

99. Groen AC, Cameron LA, Coughlin M, Miyamoto DT, Mitchison TJ, Ohi R (2004) XRHAMM functions in ran-dependent microtubule nucleation and pole formation during anastral spindle assembly. Curr Biol 14(20):1801–1811. doi:10.1016/j.cub.2004.10.002
100. Clarke PR, Sanderson HS (2006) A mitotic role for BRCA1/BARD1 in tumor suppression? Cell 127(3):453–455. doi:10.1016/j.cell.2006.10.020
101. Arnaoutov A, Azuma Y, Ribbeck K, Joseph J, Boyarchuk Y, Karpova T, McNally J, Dasso M (2005) Crm1 is a mitotic effector of Ran-GTP in somatic cells. Nat Cell Biol 7(6):626–632. doi:10.1038/ncb1263
102. Wang W, Budhu A, Forgues M, Wang XW (2005) Temporal and spatial control of nucleophosmin by the Ran-Crm1 complex in centrosome duplication. Nat Cell Biol 7(8):823–830. doi:10.1038/ncb1282
103. Peloponese JM Jr, Haller K, Miyazato A, Jeang KT (2005) Abnormal centrosome amplification in cells through the targeting of Ran-binding protein-1 by the human T cell leukemia virus type-1 Tax oncoprotein. Proc Natl Acad Sci U S A 102(52):18974–18979. doi:10.1073/pnas.0506659103
104. Demeter J, Morphew M, Sazer S (1995) A mutation in the RCC1-related protein pim1 results in nuclear envelope fragmentation in fission yeast. Proc Natl Acad Sci U S A 92(5):1436–1440
105. Zhang C, Goldberg MW, Moore WJ, Allen TD, Clarke PR (2002) Concentration of Ran on chromatin induces decondensation, nuclear envelope formation and nuclear pore complex assembly. Eur J Cell Biol 81(11):623–633. doi:10.1078/0171-9335-00288
106. Walther TC, Askjaer P, Gentzel M, Habermann A, Griffiths G, Wilm M, Mattaj IW, Hetzer M (2003) RanGTP mediates nuclear pore complex assembly. Nature 424(6949):689–694. doi:10.1038/nature01898
107. D'Angelo MA, Anderson DJ, Richard E, Hetzer MW (2006) Nuclear pores form de novo from both sides of the nuclear envelope. Science 312(5772):440–443. doi:10.1126/science.1124196
108. Goldberg MW, Rutherford SA, Hughes M, Cotter LA, Bagley S, Kiseleva E, Allen TD, Clarke PR (2000) Ran alters nuclear pore complex conformation. J Mol Biol 300(3):519–529. doi:10.1006/jmbi.2000.3891
109. Timinszky G, Tirian L, Nagy FT, Toth G, Perczel A, Kiss-Laszlo Z, Boros I, Clarke PR, Szabad J (2002) The importin-beta P446L dominant-negative mutant protein loses RanGTP binding ability and blocks the formation of intact nuclear envelope. J Cell Sci 115(Pt 8):1675–1687
110. Askjaer P, Galy V, Hannak E, Mattaj IW (2002) Ran GTPase cycle and importins alpha and beta are essential for spindle formation and nuclear envelope assembly in living Caenorhabditis elegans embryos. Mol Biol Cell 13(12):4355–4370. doi:10.1091/mbc.E02-06-0346
111. Bamba C, Bobinnec Y, Fukuda M, Nishida E (2002) The GTPase Ran regulates chromosome positioning and nuclear envelope assembly in vivo. Curr Biol 12(6):503–507
112. Ryan KJ, McCaffery JM, Wente SR (2003) The Ran GTPase cycle is required for yeast nuclear pore complex assembly. J Cell Biol 160(7):1041–1053. doi:10.1083/jcb.200209116
113. Zhang C, Clarke PR (2001) Roles of Ran-GTP and Ran-GDP in precursor vesicle recruitment and fusion during nuclear envelope assembly in a human cell-free system. Curr Biol 11(3):208–212
114. Hachet V, Kocher T, Wilm M, Mattaj IW (2004) Importin alpha associates with membranes and participates in nuclear envelope assembly in vitro. EMBO J 23(7):1526–1535. doi:10.1038/sj.emboj.7600154
115. Lu Q, Lu Z, Liu Q, Guo L, Ren H, Fu J, Jiang Q, Clarke PR, Zhang C (2012) Chromatin-bound NLS proteins recruit membrane vesicles and nucleoporins for nuclear envelope assembly via importin-alpha/beta. Cell Res 22(11):1562–1575. doi:10.1038/cr.2012.113
116. Wozniak R, Clarke PR (2003) Nuclear pores: sowing the seeds of assembly on the chromatin landscape. Curr Biol 13(24):R970–R972
117. Harel A, Chan RC, Lachish-Zalait A, Zimmerman E, Elbaum M, Forbes DJ (2003) Importin beta negatively regulates nuclear membrane fusion and nuclear pore complex assembly. Mol Biol Cell 14(11):4387–4396. doi:10.1091/mbc.E03-05-0275

118. Fernandez AG, Piano F (2006) MEL-28 is downstream of the Ran cycle and is required for nuclear-envelope function and chromatin maintenance. Curr Biol 16(17):1757–1763. doi:10.1016/j.cub.2006.07.071
119. Galy V, Askjaer P, Franz C, Lopez-Iglesias C, Mattaj IW (2006) MEL-28, a novel nuclear-envelope and kinetochore protein essential for zygotic nuclear-envelope assembly in C. elegans. Curr Biol 16(17):1748–1756. doi:10.1016/j.cub.2006.06.067
120. Rasala BA, Orjalo AV, Shen Z, Briggs S, Forbes DJ (2006) ELYS is a dual nucleoporin/kinetochore protein required for nuclear pore assembly and proper cell division. Proc Natl Acad Sci U S A 103(47):17801–17806. doi:10.1073/pnas.0608484103
121. Franz C, Walczak R, Yavuz S, Santarella R, Gentzel M, Askjaer P, Galy V, Hetzer M, Mattaj IW, Antonin W (2007) MEL-28/ELYS is required for the recruitment of nucleoporins to chromatin and postmitotic nuclear pore complex assembly. EMBO Rep 8(2):165–172. doi:10.1038/sj.embor.7400889
122. Gillespie PJ, Khoudoli GA, Stewart G, Swedlow JR, Blow JJ (2007) ELYS/MEL-28 chromatin association coordinates nuclear pore complex assembly and replication licensing. Curr Biol 17(19):1657–1662. doi:10.1016/j.cub.2007.08.041
123. Ma Y, Cai S, Lv Q, Jiang Q, Zhang Q, Sodmergen ZZ, Zhang C (2007) Lamin B receptor plays a role in stimulating nuclear envelope production and targeting membrane vesicles to chromatin during nuclear envelope assembly through direct interaction with importin beta. J Cell Sci 120(Pt 3):520–530. doi:10.1242/jcs.03355
124. Ye Q, Worman HJ (1996) Interaction between an integral protein of the nuclear envelope inner membrane and human chromodomain proteins homologous to Drosophila HP1. J Biol Chem 271(25):14653–14656
125. Pyrpasopoulou A, Meier J, Maison C, Simos G, Georgatos SD (1996) The lamin B receptor (LBR) provides essential chromatin docking sites at the nuclear envelope. EMBO J 15(24):7108–7119
126. Makatsori D, Kourmouli N, Polioudaki H, Shultz LD, McLean K, Theodoropoulos PA, Singh PB, Georgatos SD (2004) The inner nuclear membrane protein lamin B receptor forms distinct microdomains and links epigenetically marked chromatin to the nuclear envelope. J Biol Chem 279(24):25567–25573. doi:10.1074/jbc.M313606200
127. Solovei I, Wang AS, Thanisch K, Schmidt CS, Krebs S, Zwerger M, Cohen TV, Devys D, Foisner R, Peichl L, Herrmann H, Blum H, Engelkamp D, Stewart CL, Leonhardt H, Joffe B (2013) LBR and lamin A/C sequentially tether peripheral heterochromatin and inversely regulate differentiation. Cell 152(3):584–598. doi:10.1016/j.cell.2013.01.009
128. Kurisetty VV, Johnston PG, Rudland PS, El-Tanani MK (2009) Identification of genes differentially expressed between benign and osteopontin transformed rat mammary epithelial cells. BMC Res Notes 2:15. doi:10.1186/1756-0500-2-15
129. Fitzpatrick MA, Funk MC, Gius D, Huettner PC, Zhang Z, Bidder M, Ma D, Powell MA, Rader JS (2006) Identification of chromosomal alterations important in the development of cervical intraepithelial neoplasia and invasive carcinoma using alignment of DNA microarray data. Gynecol Oncol 103(2):458–462. doi:10.1016/j.ygyno.2006.03.020
130. Li B, Zhang YL (2002) Identification of up-regulated genes in human uterine leiomyoma by suppression subtractive hybridization. Cell Res 12(3–4):215–221. doi:10.1038/sj.cr.7290127
131. Lin YM, Furukawa Y, Tsunoda T, Yue CT, Yang KC, Nakamura Y (2002) Molecular diagnosis of colorectal tumors by expression profiles of 50 genes expressed differentially in adenomas and carcinomas. Oncogene 21(26):4120–4128. doi:10.1038/sj.onc.1205518
132. Bertucci F, Salas S, Eysteries S, Nasser V, Finetti P, Ginestier C, Charafe-Jauffret E, Loriod B, Bachelart L, Montfort J, Victorero G, Viret F, Ollendorff V, Fert V, Giovaninni M, Delpero JR, Nguyen C, Viens P, Monges G, Birnbaum D, Houlgatte R (2004) Gene expression profiling of colon cancer by DNA microarrays and correlation with histoclinical parameters. Oncogene 23(7):1377–1391. doi:10.1038/sj.onc.1207262
133. Hung KE, Faca V, Song K, Sarracino DA, Richard LG, Krastins B, Forrester S, Porter A, Kunin A, Mahmood U, Haab BB, Hanash SM, Kucherlapati R (2009) Comprehensive pro-

teome analysis of an Apc mouse model uncovers proteins associated with intestinal tumorigenesis. Cancer Prev Res (Phila) 2(3):224–233. doi:10.1158/1940-6207.capr-08-0153
134. Johnston NI, Gunasekharan VK, Ravindranath A, O'Connell C, Johnston PG, El-Tanani MK (2008) Osteopontin as a target for cancer therapy. Front Biosci 13:4361–4372
135. Brown LF, Berse B, Van de Water L, Papadopoulos-Sergiou A, Perruzzi CA, Manseau EJ, Dvorak HF, Senger DR (1992) Expression and distribution of osteopontin in human tissues: widespread association with luminal epithelial surfaces. Mol Biol Cell 3(10):1169–1180
136. Standal T, Borset M, Sundan A (2004) Role of osteopontin in adhesion, migration, cell survival and bone remodeling. Exp Oncol 26(3):179–184
137. El-Tanani MK, Campbell FC, Kurisetty V, Jin D, McCann M, Rudland PS (2006) The regulation and role of osteopontin in malignant transformation and cancer. Cytokine Growth Factor Rev 17(6):463–474. doi:10.1016/j.cytogfr.2006.09.010
138. Oates AJ, Barraclough R, Rudland PS (1996) The identification of osteopontin as a metastasis-related gene product in a rodent mammary tumour model. Oncogene 13(1):97–104
139. Chen H, Ke Y, Oates AJ, Barraclough R, Rudland PS (1997) Isolation of and effector for metastasis-inducing DNAs from a human metastatic carcinoma cell line. Oncogene 14(13):1581–1588. doi:10.1038/sj.onc.1200993
140. El-Tanani MK (2008) Role of osteopontin in cellular signaling and metastatic phenotype. Front Biosci 13:4276–4284
141. Dawlaty MM, Malureanu L, Jeganathan KB, Kao E, Sustmann C, Tahk S, Shuai K, Grosschedl R, van Deursen JM (2008) Resolution of sister centromeres requires RanBP2-mediated SUMOylation of topoisomerase IIalpha. Cell 133(1):103–115. doi:10.1016/j.cell.2008.01.045
142. Ma Z, Hill DA, Collins MH, Morris SW, Sumegi J, Zhou M, Zuppan C, Bridge JA (2003) Fusion of ALK to the Ran-binding protein 2 (RANBP2) gene in inflammatory myofibroblastic tumor. Genes Chromosomes Cancer 37(1):98–105. doi:10.1002/gcc.10177
143. Sasaki T, Okuda K, Zheng W, Butrynski J, Capelletti M, Wang L, Gray NS, Wilner K, Christensen JG, Demetri G, Shapiro GI, Rodig SJ, Eck MJ, Janne PA (2010) The neuroblastoma-associated F1174L ALK mutation causes resistance to an ALK kinase inhibitor in ALK-translocated cancers. Cancer Res 70(24):10038–10043. doi:10.1158/0008-5472.can-10-2956
144. Solar P, Sytkowski AJ (2011) Differentially expressed genes associated with cisplatin resistance in human ovarian adenocarcinoma cell line A2780. Cancer Lett 309(1):11–18. doi:10.1016/j.canlet.2011.05.008
145. Tanaka T, Ohkubo S, Tatsuno I, Prives C (2007) hCAS/CSE1L associates with chromatin and regulates expression of select p53 target genes. Cell 130(4):638–650. doi:10.1016/j.cell.2007.08.001
146. Lee JH, Kang Y, Khare V, Jin ZY, Kang MY, Yoon Y, Hyun JW, Chung MH, Cho SI, Jun JY, Chang IY, You HJ (2010) The p53-inducible gene 3 (PIG3) contributes to early cellular response to DNA damage. Oncogene 29(10):1431–1450. doi:10.1038/onc.2009.438
147. Lorenzato A, Martino C, Dani N, Oligschlager Y, Ferrero AM, Biglia N, Calogero R, Olivero M, Di Renzo MF (2012) The cellular apoptosis susceptibility CAS/CSE1L gene protects ovarian cancer cells from death by suppressing RASSF1C. FASEB J 26(6):2446–2456. doi:10.1096/fj.11-195982
148. Yao Y, Dong Y, Lin F, Zhao H, Shen Z, Chen P, Sun YJ, Tang LN, Zheng SE (2009) The expression of CRM1 is associated with prognosis in human osteosarcoma. Oncol Rep 21(1):229–235
149. Shen A, Wang Y, Zhao Y, Zou L, Sun L, Cheng C (2009) Expression of CRM1 in human gliomas and its significance in p27 expression and clinical prognosis. Neurosurgery 65(1):153–159. doi:10.1227/01.neu.0000348550.47441.4b, discussion 159–160
150. Turner JG, Dawson J, Sullivan DM (2012) Nuclear export of proteins and drug resistance in cancer. Biochem Pharmacol 83(8):1021–1032. doi:10.1016/j.bcp.2011.12.016
151. Muniyappa MK, Dowling P, Henry M, Meleady P, Doolan P, Gammell P, Clynes M, Barron N (2009) MiRNA-29a regulates the expression of numerous proteins and reduces the inva-

siveness and proliferation of human carcinoma cell lines. Eur J Cancer 45(17):3104–3118. doi:10.1016/j.ejca.2009.09.014

152. Ly TK, Wang J, Pereira R, Rojas KS, Peng X, Feng Q, Cerione RA, Wilson KF (2010) Activation of the Ran GTPase is subject to growth factor regulation and can give rise to cellular transformation. J Biol Chem 285(8):5815–5826. doi:10.1074/jbc.M109.071886
153. Morgan-Lappe SE, Tucker LA, Huang X, Zhang Q, Sarthy AV, Zakula D, Vernetti L, Schurdak M, Wang J, Fesik SW (2007) Identification of Ras-related nuclear protein, targeting protein for xenopus kinesin-like protein 2, and stearoyl-CoA desaturase 1 as promising cancer targets from an RNAi-based screen. Cancer Res 67(9):4390–4398. doi:10.1158/0008-5472.can-06-4132
154. Byrne JD, Betancourt T, Brannon-Peppas L (2008) Active targeting schemes for nanoparticle systems in cancer therapeutics. Adv Drug Deliv Rev 60(15):1615–1626. doi:10.1016/j.addr.2008.08.005
155. McCarron PA, Faheem AM (2010) Nanomedicine-based cancer targeting: a new weapon in an old war. Nanomedicine (Lond) 5(1):3–5. doi:10.2217/nnm.09.89
156. Li L, Wang R, Wilcox D, Zhao X, Song J, Lin X, Kohlbrenner WM, Fesik SW, Shen Y (2012) Tumor vasculature is a key determinant for the efficiency of nanoparticle-mediated siRNA delivery. Gene Ther 19(7):775–780. doi:10.1038/gt.2011.146
157. Mansouri S, Lavigne P, Corsi K, Benderdour M, Beaumont E, Fernandes JC (2004) Chitosan-DNA nanoparticles as non-viral vectors in gene therapy: strategies to improve transfection efficacy. Eur J Pharm Biopharm 57(1):1–8
158. Peer D, Karp JM, Hong S, Farokhzad OC, Margalit R, Langer R (2007) Nanocarriers as an emerging platform for cancer therapy. Nat Nanotechnol 2(12):751–760. doi:10.1038/nnano.2007.387
159. Kuo PY, Saltzman WM (1996) Novel systems for controlled delivery of macromolecules. Crit Rev Eukaryot Gene Expr 6(1):59–73
160. De Jong WH, Borm PJ (2008) Drug delivery and nanoparticles:applications and hazards. Int J Nanomedicine 3(2):133–149
161. Hong M, Zhu S, Jiang Y, Tang G, Sun C, Fang C, Shi B, Pei Y (2010) Novel anti-tumor strategy: PEG-hydroxycamptothecin conjugate loaded transferrin-PEG-nanoparticles. J Control Release 141(1):22–29. doi:10.1016/j.jconrel.2009.08.024

Wnt Signaling Proteins Associate with the Nuclear Pore Complex: Implications for Cancer

Manisha Sharma, Michael Johnson, Mariana Brocardo, Cara Jamieson, and Beric R. Henderson

Abstract Several components of the Wnt signaling pathway have in recent years been linked to the nuclear pore complex. β-catenin, the primary transducer of Wnt signals from the plasma membrane to the nucleus, has been shown to transiently associate with different FG-repeat containing nucleoporins (Nups) and to translocate bidirectionally through pores of the nuclear envelope in a manner independent of classical transport receptors and the Ran GTPase. Two key regulators of β-catenin, IQGAP1 and APC, have also been reported to bind specific Nups or to locate at the nuclear pore complex. The interaction between these Wnt signaling proteins and different Nups may have functional implications beyond nuclear transport in cellular processes that include mitotic regulation, centrosome positioning and cell migration, nuclear envelope assembly/disassembly, and the DNA replication checkpoint. The broad implications of interactions between Wnt signaling proteins and Nups will be discussed in the context of cancer.

M. Sharma (✉) • M. Brocardo • C. Jamieson • B.R. Henderson (✉)
Westmead Institute for Cancer Research, Westmead Millennium Institute at Westmead Hospital, The University of Sydney, Darcy Road, PO Box 412, Westmead, NSW 2145, Australia
e-mail: manisha.sharma@sydney.edu.au; beric.henderson@sydney.edu.au

M. Johnson
Westmead Institute for Cancer Research, Westmead Millennium Institute at Westmead Hospital, The University of Sydney, Darcy Road, PO Box 412, Westmead, NSW 2145, Australia

Present Address: Kolling Institute of Medical Research, Royal North Shore Hospital, University of Sydney, St Leonards, NSW Australia

E.C. Schirmer and J.I. de las Heras (eds.), *Cancer Biology and the Nuclear Envelope*, Advances in Experimental Medicine and Biology 773, DOI 10.1007/978-1-4899-8032-8_16,

Keywords IQGAP1 • β-catenin • APC • Nuclear envelope • Wnt signaling • nucleus • Centrosome • Cancer

Abbreviations

APC	Adenomatous polyposis coli
IQGAP1	IQ motif containing GTPase activating protein
MTOC	Microtubule organizing center
NE	Nuclear envelope
NES	Nuclear export signal
NLS	Nuclear localization signal
NPC	Nuclear pore complex
NUP	Nucleoporin
PML	Promyelocytic leukemia

Introduction

Wnt signaling pathways play a critical role in regulating normal cell and tissue processes including embryonic development, stem cell self-renewal, tissue homeostasis, cell polarity and cell migration [1–4]. When constitutively activated by gene mutations, inappropriate Wnt signaling can drive cell transformation and the initiation of tumor progression, particularly of the colonic epithelium [1–3]. Therefore, aberrations in Wnt signaling can promote colorectal cancer [5]. In recent years, different protein components of the Wnt pathway, in particular the proteins β-catenin, adenomatous polyposis coli (APC), and IQ motif containing GTPase activating protein (IQGAP1), have been found to associate with the nuclear envelope (NE). These and other proteins of the Wnt pathway shuttle between nucleus and cytoplasm [6–8], and in some cases form complexes with nucleoporin (Nup) proteins that make up the nuclear pore complex (NPC). β-catenin is a unique example. In response to Wnt ligand activation of transmembrane Frizzled receptors at the cell surface, β-catenin is stabilized and rapidly translocates into the nucleus where it can transactivate proteins that promote cellular transformation and cancer [4]. We describe the unusual nuclear transport mechanism of β-catenin, which itself appears to act like a classical transport receptor in that it can traverse the NPC via multiple transient interactions with phenylalanine-glycine (FG)-containing Nups. Then the nuclear shuttling protein APC is discussed in the context of its binding to Nup153 and Nup358, and the potential impact on cytoskeletal regulation leading to changes in cell polarity and migration. Finally, the recent discovery of IQGAP1 in the nucleus and at the outer face of the NE is addressed, outlining how NE-associated IQGAP1 might tether and regulate the actin and microtubule cytoskeletons to control nuclear positioning or NE breakdown/assembly as cells enter mitosis.

β-Catenin Associates with the NPC: Implications for Nuclear Translocation and Transactivation

Wnt Signaling, β-Catenin Stabilization and Cancer

In the absence of Wnt signaling, β-catenin is immobilized by E-cadherin at the cell-cell junctions. It also binds to α-catenin at these junctions and indirectly modulates the actin cytoskeleton [2, 5, 9]. The accumulation of excess β-catenin is tightly controlled through its N-terminal phosphorylation by a destruction complex (comprising several factors including casein kinase 1, APC, glycogen synthase kinase 3-β(GSK3β) and axin) which marks it for proteosome-mediated degradation. Wnt signaling or gene mutations that disrupt specific members of the destruction complex result in increased levels of cytoplasmic β-catenin [5]. The stabilized β-catenin then enters the nucleus where it associates with members of the LEF-1/TCF family of transcription factors and drives transcription of diverse Wnt/β-catenin target genes [4, 10, 11]. The nuclear transactivation function of β-catenin is sufficient to induce cell immortalization and transformation [1]. Therefore, nuclear accumulation of β-catenin is linked to cellular transformation and cancers, especially those of the colon, breast, skin and liver [2, 3, 12]. More than 80 % of colorectal cancers have mutations in the *APC* gene and a majority of the remainder has mutations in other destruction complex proteins [13]. As a consequence, β-catenin stabilization is a common feature of colon cancers.

β-Catenin Moves Through the NPC via Direct Transient Association with FG-Rich Nups: The Nuclear Transport Process

The nuclear transport of proteins is a specific, rapid and highly active process mediated by chaperone proteins commonly referred to as nuclear import (importins) and export (exportins) receptors [14, 15]. The importins ferry protein cargo into the nucleus by interacting with Nuclear Localization Signals (NLSs) that are mostly positively charged recognition sequences. Conversely, proteins are exported from nucleus to the cytoplasm by exportins that bind to Nuclear Export Signals (NESs); the most common of these comprise closely spaced hydrophobic amino acids recognized by CRM1/exportin1. The majority of nuclear proteins are carried into the nucleus by the importin-β/importin-α complex. Importin-α binds directly to the protein NLS and importin-β mediates passage of the complex through the NPC by interacting with the FG rich repeats of certain Nups at the NPC. The import complex is dissociated in the nucleus when importin-β-bound Ran-GDP is exchanged for Ran-GTP, causing release of the cargo protein. Nuclear export of proteins by CRM1 requires Ran-GTP and release in the cytoplasm is achieved by its conversion to Ran-GDP; the reverse situation to import. Thus, the Ran-GTPase gradient across the NPC maintains directionality of movement. There are several mechanisms proposed for the actual movement of the importins/exportins and their protein-bound complexes through the NPC, which are discussed in detail elsewhere [15–21]. Of particular relevance here is that β-catenin appears to behave like a nuclear transport receptor, able to move back and forth rapidly across the NPC without need of other carriers and independently of the Ran-GTPase (see model in Fig. 1d).

β-Catenin Shuttles In and Out of the Nucleus Independently of Importins/Exportins

The link between nuclear accumulation of β-catenin and cancer has been known for many years, and yet we are only beginning to understand what are likely to be multiple mechanisms for β-catenin translocation into the nucleus. In normal cells not stimulated by Wnt signal, β-catenin is expressed in very small amounts and this is mostly found in complex with other proteins. Several of these partners, such as APC and axin, shuttle into and out of the nucleus through classical importin/exportin pathways and provide an indirect route for nuclear transport of β-catenin [6, 8]. In cancer cells or in Wnt-stimulated cells, however, β-catenin is highly expressed and its movement is not dictated by APC or other partners. Under these conditions it was discovered that β-catenin is unique in that it shuttles in and out of the nucleus independently of the importin/exportin pathway (β-catenin contains no typical NLS or NES sequence) [22–25]. Structurally, β-catenin comprises 12 Armadillo (Arm) repeat sequences that form a helical coil, and this central coil is flanked by unstructured N-and C-terminal tails [26, 27]. Arm repeats are ~43 amino acid motifs that form a super-helical structure with a major groove that mediates interaction with key protein partners [28]. The Arm repeats 9–12 are structurally similar to the HEAT repeats of importin-β [29]. These helicoidal repeats provide importins and exportins the flexibility with which to bind cargo proteins and the FG repeats of Nups simultaneously, allowing for translocation of the complex through the NPC [21, 30]. Recently, the β-catenin Arm repeats 10–12 have been shown to mediate rapid nuclear import/export of β-catenin and to bind directly to the FG repeats of NPC components Nup62, Nup98 and Nup153, and with the outer NPC cytoplasmic filament protein Nup358 [31]. The findings support a model wherein β-catenin translocation through the NPC is mediated by a series of transient and sequential interactions with multiple Nups. Of these, Nup358 has a crucial role given that it is located at the cytoplasmic filaments and is known to act as a docking platform for import complex formation at the entry point to the NPC [19]. Indeed, the loss of Nup358 was found to slow down the rate of nuclear entry of β-catenin in live cell assays (Fig. 1). By analogy to its role in regulating nuclear passage of importin-β [32], it is possible that Nup358 captures and concentrates β-catenin at the NPC to increase its import/export efficiency [19].

The interaction of Arm repeats 10–12 of β-catenin with components of the NPC has many implications. Firstly, most of the key partners of β-catenin (i.e., APC, LEF/TCFs, E-cadherin) bind within the first 8 Arm repeats, leaving Arm repeats 10–12

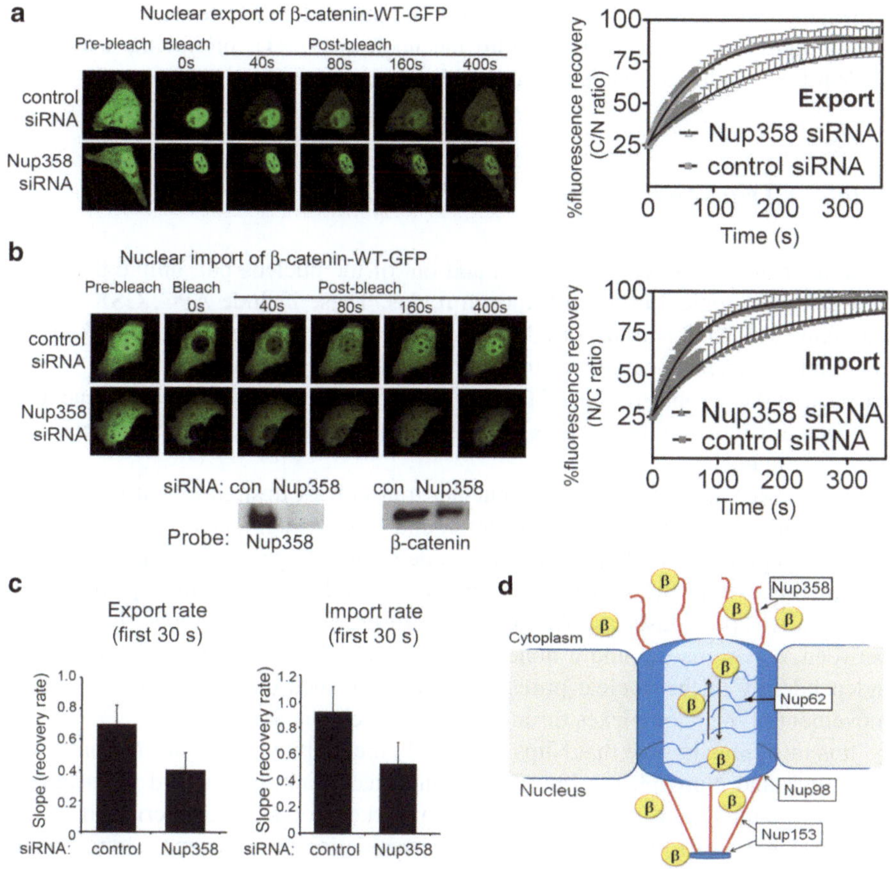

Fig. 1 Silencing Nup358 slows nuclear export and import of β-catenin. Analysis of GFP-tagged β-catenin transport in living cells by FRAP assay. (**a**) Mouse NIH 3T3 fibroblast cells were transfected with control siRNA or Nup358 siRNA and co-transfected with β-catenin-GFP for nuclear export FRAP analysis. Confocal cell images are shown before and after photobleaching the cytoplasm and fluorescence recovery was monitored over 400 s. Nuclear export rates (*right-hand* panel) were calculated for the first 30 s from initial recovery slopes using linear regression analysis (see [31] for details). (**b**) FRAP analysis of nuclear import was performed on β-catenin-GFP transfected cells by photobleaching the nucleoplasm. Nuclear import rates were calculated as in A revealing an inhibitory effect of Nup358 silencing (*right-hand* panel). (**c**) Western blot analysis to confirm knockdown of Nup358. (**d**) Schematic model for receptor-independent transport of β-catenin (β) through the NPC. β-catenin transiently binds to the FG-repeats of Nup358 (located at the cytoplasmic filaments), Nup62 (central channel of NPC), Nup98 and Nup153 (located at the nuclear basket of NPC). These data were originally published in [31] and reproduced with permission of the American Society for Biochemistry and Molecular Biology

free and accessible to transport the β-catenin-protein complexes [31]. Thus, β-catenin could function as a highly specialized transport receptor in cancer cells. Moreover, other proteins that contain Arm repeat domains such as the Wnt proteins APC, α-catenin and p120-catenin, might be capable of importin/exportin-independent movement through the NPC. In conclusion, it appears that β-catenin has evolved to transit rapidly and independently into the nucleus [7, 31], producing a dynamic nuclear pool readily available to transactivate gene expression.

Nuclear Transport of Other Wnt Proteins

A number of Wnt proteins shuttle in and out of the nucleus but, unlike β-catenin, they utilize canonical importin/exportin routes. These include APC, GSK-3β [33] and axin [34]. APC is transported by importin-α/β complexes into the nucleus and can bind to nuclear β-catenin to transport it out of the nucleus to stimulate its degradation in the cytoplasm [35–37]. Since all of the above factors are components of the β-catenin destruction complex, it is possible that β-catenin degradation may also occur in the nucleus.

In the more specialized case of nuclear transport of upstream Wnt receptors, a recent study of neuronal development found that a fragment of the *Drosophila* Wnt-1 receptor, named Dfrizzled2, was able to bind large mRNA particles and to presumably exit the nucleus through a process of budding [38]. This type of nuclear transport was previously ascribed only to viruses, and might provide the first link between Wnt signaling and a noncanonical form of nuclear transport that occurs independently of the nuclear pores. Alternatively, such data might also implicate movement of Wnt complexes through peripheral channels of the NPC.

It is intriguing to note that Nups, especially the mobile nucleoplasmic ones such as Nup98 and Nup153, are intimately connected to chromatin and contribute to gene regulation [39, 40], and are enriched in markers for active transcription such as RNA polymerase II [41] and the histone deacetylation and chromatin binding protein (CBP/p300) [42, 43]. These Nups were shown to regulate the expression of developmental and cell cycle related genes [39]. It is possible that β-catenin might regulate the transcription function of these Nups in the nucleus of cancer cells, or alternatively that these Nups could alter regulation of specific Wnt target genes.

The Role of Nup358 Sumo Ligase Activity in β-Catenin Transport and Function

Nup358, in addition to its role in protein docking and transport through the NPC, is a major constituent of the cytoplasmic filaments at the face of the NPC and acts as a Sumo E3-ligase involved in sumoylation of various target proteins. Interestingly, Nup358 has recently been found to sumoylate the Wnt protein TCF-4 [44]. TCF-4 is a member of the LEF-1/TCF family of transcription factors that are co-activated by β-catenin in colon cancers [45]. Sumoylated TCF-4 displayed an increased affinity for β-catenin and increased its transcriptional activity in colon cancer cell lines [44]. It is possible that sumoylated TCF-4 binds to β-catenin first in the cytoplasm and that the complex is then transported into the nucleus by β-catenin. At present there is no evidence that Nup358 can sumoylate β-catenin, although we note that the de-sumoylation activity of a different protein, Axam, was linked to the degradation of β-catenin [46]. Another Wnt-related sumo-substrate of Nup358 is promyelocytic leukemia protein (PML). Sumoylation of the PML tumor suppressor is essential for formation of PML nuclear bodies that regulate processes such as DNA replication and gene transcription. Wnt-stabilized β-catenin was found to inhibit Nup358 mediated sumoylation of PML and to prevent formation of PML nuclear bodies in cancer cells [47]. However, the mechanism of β-catenin action in this process has not yet been resolved. In future experiments it will be important to clarify whether Nup358 does sumo-modify β-catenin, and to explore whether β-catenin binding has a more global impact on Nup358 sumo-ligase activity, particular in relation to the nuclear transport process.

APC Associates with Specific Nups: Implications for Polarization of the Microtubule Cytoskeleton and Centrosome Orientation

Unlike other Wnt regulatory proteins, APC is a very large molecule and directly associates with the cellular cytoskeleton. Through its C-terminal BASIC domain, APC can bind directly to both microtubules [48] and to actin filaments [9, 49] in vitro and in vivo. Moreover, APC has recently been shown to associate (through its Arm domain) with vimentin, a component of intermediate filaments [50]. In interphase cells, APC binds and stabilizes the plus-end "cap" of microtubules, promoting assembly of microtubule bundles that drive formation of membrane protrusions during cell migration [48]. APC binds directly (via its extreme C-terminus) to end-binding protein-1 (EB1) which also stabilizes microtubule plus-ends. APC is most frequently detected by microscopy in clusters at the end of microtubules in membrane protrusions, and is thought to indirectly move along microtubules in a plus-end directed fashion with the aid of kinesin motor proteins and the adaptor KAP3A complex [51, 52]. At the membrane clusters, APC co-locates with other Wnt proteins including DLG, axin and β-catenin [53, 54]. APC is also present to some extent at actin-dependent membrane lamellipodia and ruffles of migrating cells where it interacts with IQGAP1, an effector molecule of Rac1-GTPase [55, 56]. At the plasma membrane, the APC–IQGAP1 interaction was proposed to mediate cross-linking between microtubules and actin to coordinate movement of the cell membrane during migration [55]. APC further regulates polarization of the centrosome during migration [57, 58]. Interestingly, two specific Nups, Nup153 and Nup358, were shown to interact with APC and regulate its role in cytoskeletal architecture.

APC, Nups and Microtubules

Microtubules are dynamic cytoskeletal tube-like structures that contribute to cell shape but also are linked to the NE where they contribute to the process of cell division through positioning of the centrosome and mitotic spindle and possibly through facilitation of NE breakdown and reassembly [59]. In interphase cells microtubules originate from the microtubule organizing center (MTOC) or centrosomes. They nucleate from γ-tubulin RING complexes at the centrosome (referred to as the minus-end of the microtubules) and extend as polymers outward to the cell cortex. Like APC, Nups can regulate microtubule stability and polymerization to control mitotic spindle assembly and disassembly [60] and if misregulated cause severe aneuploidy and missegregation of chromosomes [61, 62]. In interphase cells microtubules associate with, and were reported to be stabilized by, Nup358 [63] and Nup98 [64].

Microtubules and actin play an important role in cell migration and in cancer cell invasion, particularly in the direction of cell movement through generation of membrane protrusions (the relationship between Wnt proteins and actin will be discussed later). APC has been documented to associate with three distinct NPC-associated factors to date; the mobile Nups Nup358 [65], Nup153 [66] and the nuclear import receptor, importin-β [67]. The interaction with importin-β was proposed to stimulate microtubule stability and spindle formation, based mostly on in vitro assays with *Xenopus* egg extracts [67]. In contrast, the two Nups were each implicated in torsioning of the microtubule network for cell polarity and orientation of the centrosome (Fig. 2). In the case of Nup358, Murawala [65] reported that APC binds this Nup through its central β-catenin-binding domain, raising the possibility that β-catenin or axin might modulate the interaction between APC and Nup358. Curiously, and quite distinct from its role in the docking stages of nuclear transport, the cytoplasmic form of Nup358 was claimed to bind the kinesin-2 motor complex and promote movement of APC to the cell cortex and microtubule plus-ends, leading to centrosome polarization during cell migration [65]. This differs to the role of Nup153, which in fibroblasts and astrocytes was shown to bind APC and further postulated to anchor APC in the vicinity of the NPC at the NE [66]. The idea was that a subset of interphase microtubules are attached to the NE through an APC-Nup153 complex to assist in centrosome positioning during cell migration, although the evidence presented is mostly circumstantial. In future experiments it will be important to define more accurately the cellular localization of APC-Nup interactions, such as by proximity ligation assays, and also to determine whether the sumoylation function of Nup358 contributes to APC binding and/or cell polarity and migration. It is not yet known whether these APC-Nup interactions contribute to, or modulate, the rate of APC nuclear transport.

In the Wnt system, APC plays a key role in chromosome stability, and cancer associated mutations of APC lead to aneuploidy [61]. A similar outcome has been observed after loss of specific Nups such as Nup358, Nup98, and the Nup107-160 complex [60]. Thus, future research efforts can be envisaged to elucidate a more complex interplay or network between Wnt signaling and Nups in regulation of microtubules in cancer.

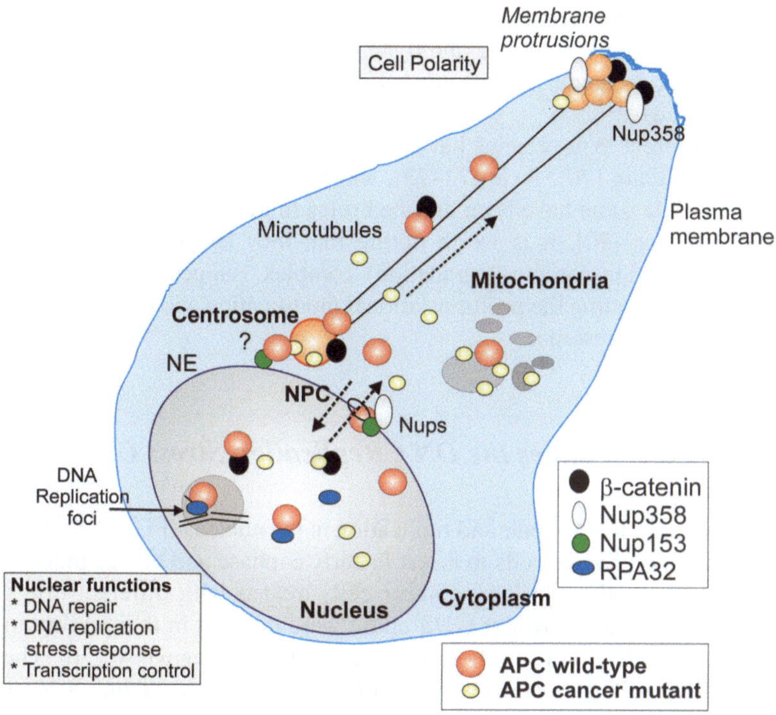

Fig. 2 APC localization, transport, and activities in interphase cells. APC localizes to different subcellular sites spanning from the nucleus to plasma membrane as shown. APC is thought to move to the plasma membrane along microtubules and driven by kinesin-motor proteins; at the membrane APC accumulates in clusters with other proteins including β-catenin and Nup358 to promote cell migration. Inside the cell, cytoplasmic APC (wild-type or cancer mutated forms) localize to the centrosome, a tiny structure that nucleates the microtubule network and primes assembly of the mitotic spindle in mitotic cells. The truncated cancer linked APC mutants accumulate at mitochondria in the cytoplasm with a predicted role in cancer cell survival. APC is known to shuttle in and out of the nucleus. At the nuclear pore complex (NPC) it associates with the nucleoporin Nup153 (and possibly Nup358). Inside the nucleus APC functions to regulate DNA repair, replication, and transcription. APC returns to the cytoplasm via the CRM1-dependent nuclear export pathway. Full-length APC is thought to act as a mobile scaffold that moves slowly between locations to regulate assembly of specific protein complexes [113]. Mutant truncated forms of APC are more dynamic but often display altered activity

APC and Other Wnt Components at the Centrosome

As described above, both β-catenin and APC form complexes with Nup358. In addition to the role of APC-Nups in orienting the centrosome to optimize microtubule dynamics for cell movement during interphase, Nup358 has recently been shown to bind to the microtubule motor-adaptor protein Bicaudal D2 [68] to regulate positioning of the nucleus and centrosome during late G2, prior to entry into mitosis. This mechanical process is important for integrity of NE breakdown and positioning of the mitotic spindle, and involves regulation of two counteracting forces along microtubules driven by kinesin-1 and dynein motors. At this stage it is not known whether APC or β-catenin contribute to this pre-mitotic orientation of the centrosome and nucleus.

Several different Wnt proteins have been detected at the centrosome, including APC [69], β-catenin [70], Axin [71–73], and the protein phosphatase 2A subunit B56α [74]. While some have been ascribed roles in microtubule nucleation [72] or centriole cohesion [70], it is worth noting that they are all components of the β-catenin destruction (or pre-degradation) complex, suggesting that they might share a role in regulating the posttranslational modification and possible turnover of β-catenin at the centrosome.

Nuclear APC Regulates the DNA Replication Stress Check Point

APC is involved in DNA repair and replication in the nucleus (Fig. 2). The silencing of APC expression causes cells to arrest in early S-phase and blocks their progression through the cell cycle [75]. Cancer cells treated with different drugs (e.g., hydroxyurea) to induce arrest at G1/S phase were delayed in their ability to reengage replication when APC levels were knocked down using RNAi [76]. The latter study found that APC binds to the replication protein A (RPA) subunit RPA32 in the nuclear chromatin fraction and that this interaction positioned APC as a regulator of the DNA replication stress response in cancer cells. The role of APC in safeguarding the replication process is underscored by the fact that cancer mutations reduce its binding to RPA protein [76]. Nup153 has independently been shown to regulate the DNA damage and repair response [77] and, like APC, to regulate the phosphorylation of CHK1 kinase after DNA damage, suggesting that APC-Nup153 complexes may cooperate in managing the cellular responses to DNA damage or to specific chemotherapeutic agents during colon cancer treatment.

IQGAP1 at the Outer Surface of the NE: Functional Implications

Regulation of Wnt Proteins at the Plasma Membrane by IQGAP1

IQGAP1 can bind to β-catenin and to APC [78]. It is a multi-domain actin-regulatory protein frequently detected at actin-dependent membrane structures such as lamellipodia and membrane ruffles involved in cell locomotion [79, 80]. Through its interaction with the plus-end binding proteins APC and CLIP-170, IQGAP1 tethers microtubules to the actin meshwork for cortical cell polarization cues [80].

Interruption of IQGAP1–APC linkages causes defective cell polarization [55]. IQGAP1 also regulates the internalization of β-catenin, APC and N-cadherin from membrane ruffles via macropinocytosis [56]. IQGAP1 expression is often high in cancers, and in some cases may promote cancer progression [78]. In this regard, IQGAP1 can destabilize membrane *adherens* junctions and indirectly coerce β-catenin from the plasma membrane to the nucleus where it activates the Wnt oncogenic transcription program.

Role of IQGAP1 in Cell Migration

There is evidence for a stimulatory role of IQGAP1 on cell migration [55, 81, 82]. At the leading edge of migrating cells, actin polymerization events induce membrane ruffles/lamellipodia that propel the cell across a substratum. IQGAP1 is unique in that it contributes to F-actin bundling/cross-linking capability [83] at membrane ruffles [83–85] and barbed-end capping of actin [86]. IQGAP1 targets two key actin nucleators, the N-WASp/Arp2/3 complex and Diaphanous1 (Dia1), to membrane ruffles [79, 87, 88], although at present it is not well understood how IQGAP1 coordinates the actin rearrangements that ensue. Prevailing evidence shows that Rac1 or Cdc42 regulate IQGAP1 activity [78]. These GTPases are also potent stimulators of N-WASp activity [89]. The ability of IQGAP1 to suppress the intrinsic GTPase activity of Rac1/Cdc42 could be critical for its ability to coordinate actin branching and nucleation at the membrane. In this context, for example, IQGAP1 may act as a scaffold linking active GTPase to N-WASP and actin filaments, and therefore have its strong effects on cell migration.

Implications for IQGAP1 at the NE

IQGAP1 has recently been found to localize to the NE in a range of cancer and non-tumorigenic cell lines [90]. Immunofluorescence staining and electron microscopy immune-gold labeling of IQGAP1 overlapped with cytoskeletal F-actin structures and, in part, microtubules at the NE (Fig. 3). This pattern is comparable to that observed at plasma membrane ruffles where IQGAP1 bridges the plus-ends of microtubules with the underlying actin meshwork. At the NE, actin and microtubules directly associate with outer nuclear membrane KASH and Nesprin proteins to assist nuclear positioning and cell polarization during cell migration. Several cytoskeletal proteins regulate the dynamics of NE-associated actin and/or nuclear positioning during cell migration [91–93]. Actin has been shown to polymerize directly from the cytoplasmic face of the NE [94]. Perinuclear actin polymerization would require the targeting of actin capping, bundling and nucleating proteins. Due to the scaffolding roles of IQGAP1, its ability to target and tether other cytoskeletal-associated proteins, and its F-actin regulatory properties, we speculate that IQGAP1 is a key mediator of cytoskeletal dynamics at the NE (Fig. 4).

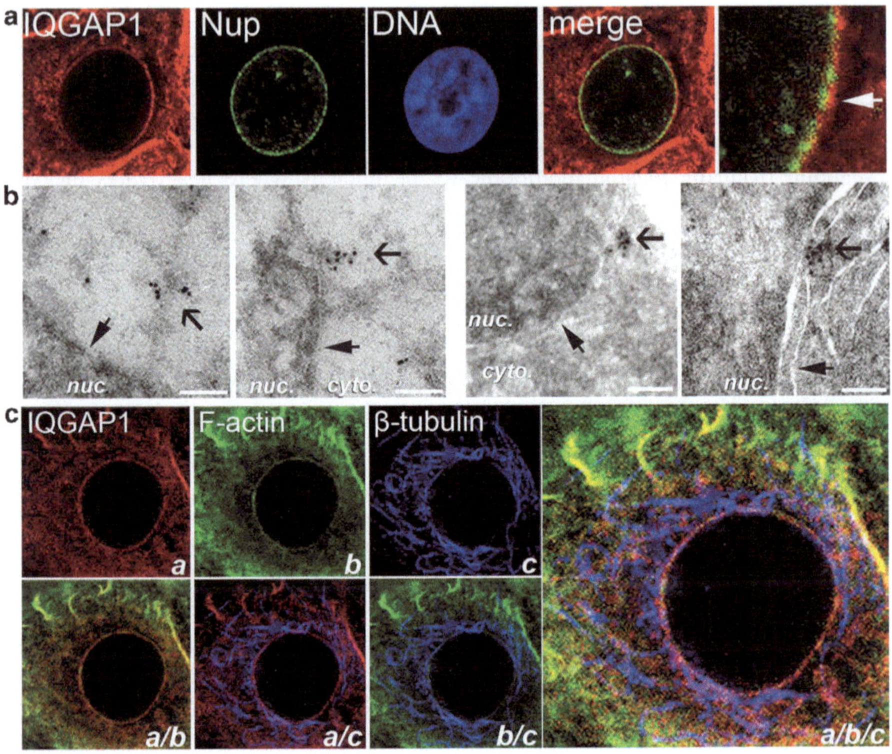

Fig. 3 Localization of IQGAP1 at the NE. (**a**) NE staining of IQGAP1. Deconvolution microscopy fluorescence images are shown of human MCF-7 epithelial cancer cells. IQGAP1 was detected at the cytoplasmic face of the outer nuclear membrane. Cells were stained with antibodies against IQGAP1 (*red*) and nups (mAb414; *green*). (**b**) Electron micrographs of ultrathin cryosections of MCF-7 cells immunolabeled with IQGAP1. *Thin closed arrow* indicates nuclear rim; *broad open arrow* indicates immunogold-labeling of IQGAP1. White scale bar, 200 nm. (**c**) Confocal fluorescence microscopy images of detergent-extracted MCF-7 cells labeled for detection of IQGAP1 (*a*; *red*), F-actin (*b*; *green*), and β-tubulin (*c*; *blue*). The *bottom* panel and enlarged image on the *right* show merged fluorescence micrographs, revealing overlap between IQGAP1 and the actin and microtubule networks at the outer edge of the nuclear envelope. These data were originally published in BioArchitecture [90] and reproduced with permission from Landes Bioscience

As discussed earlier, APC binds to Nup153 [66]. This was proposed to promote NE anchorage of proximal microtubules that radiate from the centrosome. It is conceivable that IQGAP1 assists in such events to aid positioning of the MTOC. This theory is supported by the observation that silencing of IQGAP1 in Vero fibroblasts inhibited MTOC reorientation as indicated by the formation of multiple leading edges [55]. We propose that this effect is caused not only by a deficiency in IQGAP1 cortical polarization cues, as suggested, but by defects in actin dynamics in the vicinity of the NE.

In migrating fibroblasts, repositioning of the nucleus is important during cell polarization [92], and can be facilitated by the retrograde movement of actin

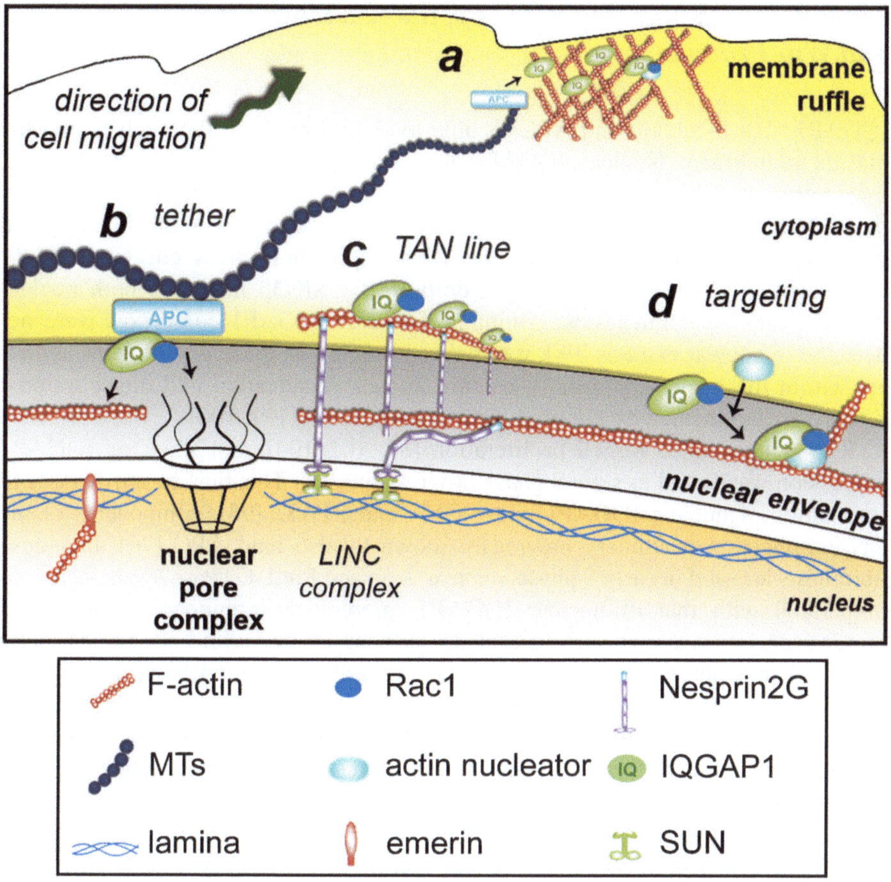

Fig. 4 Model proposing the roles of IQGAP1 at the cytoplasmic face of the nuclear envelope. At dynamic plasma membrane regions involved in cell migration (**a**), IQGAP1 tethers microtubule networks to the cortical actin network through a complex formed with APC. (**b**) At the cytoplasmic face of the nuclear envelope, IQGAP1 co-locates with microtubules and actin and may tether these cytoskeletal networks via APC or interaction with other NE proteins. (**c**) TAN lines in mesenchymal cells assist in nuclear repositioning during cell polarization and cell migration. (**d**) IQGAP1 recruits several actin-associated proteins to subcellular sites, and potentially targets proteins to the NE during cell migration or cell cycle events such as NE breakdown. This figure was originally published in BioArchitecture [90] and reproduced with permission from Landes Bioscience

filaments coupled to TAN lines [95, 96] (Fig. 4). IQGAP1 is involved in recruitment of different actin regulators such as APC [55], N-WASp [87, 88] and Dia1 [79] to membrane ruffles. Therefore, we speculate that IQGAP1 similarly recruits specific actin or microtubule-associated proteins to the outer NE to direct cytoskeletal rearrangements during nuclear repositioning. Alternatively, the NE-localized action of IQGAP1 in cytoskeletal dynamics and tethering might contribute to regulation of NE breakdown or reassembly, a process that commences late in G2 and in early mitosis and requires action of both actin and microtubule motors [97].

Links Between Nuclear IQGAP1, the Nucleoskeleton and Wnt Proteins

IQGAP1, like β-catenin and APC, is able to shuttle between the nucleus and the plasma membrane. Nuclear IQGAP1 may have roles comparable to those of β-catenin and APC in transcription and the DNA replication checkpoint, respectively. Nuclear IQGAP1 levels are regulated by two means, GSK-3β signaling and cell cycle-mediated cues, such as the DNA replication stress checkpoint [98]. Pharmacological or Wnt-mediated inhibition of GSK-3β increases both nuclear IQGAP1 and β-catenin levels. Although β-catenin-IQGAP1 complexes were not detected in nuclear extracts [98], it remains possible that these two proteins associate within the nucleus. Indeed, β-catenin nuclear localization and transcriptional activity are augmented by elevated IQGAP1 expression [99, 100].

IQGAP1 is linked to cell proliferation [82, 101–104], and silencing IQGAP1 expression was found to slow S phase progression [98]. This may be attributable to association of nuclear IQGAP1 with PCNA and RPA32 [98], components of the DNA replication machinery previously shown to also bind APC [76]. In mouse fibroblasts arrested in early S phase, nuclear actin and Rac1-GTPase levels increased in parallel with that of nuclear IQGAP1; moreover, immunoprecipitation and proximity-ligation assays suggest that IQGAP1 associates with both nuclear actin and Rac1 in S phase arrested cells (Johnson, Sharma and Henderson, unpublished results). Nuclear actin has been implicated in several nuclear processes, most notably transcription regulation and chromatin remodeling [105, 106]. The principal feature of actin is its ability to form filaments, which are readily detectable in the cytoplasm through fluorescently linked labels such as phalloidin. Despite any clear evidence that nuclear actin exists in a polymeric form, due to lack of nuclear phalloidin staining under physiological conditions, several studies implicate polymeric nuclear actin in transcription, chromatin remodeling and DNA repair [107–109]. We propose that nuclear-localized IQGAP1 stimulates polymerization of nuclear actin either through its actin-bundling/cross-linking activity or via actin nucleating machinery such as the ubiquitously nuclear N-WASp/Arp2/3 complex. Recent studies implicate nuclear Rac1 in DNA replication and repair processes [110–112]. Given that Rac1 is a key regulator of IQGAP1 activity at the plasma membrane, we postulate a similar role for Rac1 in mediating nuclear IQGAP1–actin regulation of gene transcription (e.g., linked to Wnt signaling responses) and/or DNA replication processes.

References

1. Fodde R, Smits R, Clevers H (2001) APC, signal transduction and genetic instability in colorectal cancer. Nat Rev Cancer 1(1):55–67. doi:10.1038/35094067
2. Polakis P (2007) The many ways of Wnt in cancer. Curr Opin Genet Dev 17(1):45–51. doi:10.1016/j.gde.2006.12.007, S0959-437X(06)00241-3 [pii]
3. Cadigan KM, Peifer M (2009) Wnt signaling from development to disease: insights from model systems. Cold Spring Harb Perspect Biol 1(2):a002881. doi:10.1101/cshperspect.a002881

4. Senda T, Iizuka-Kogo A, Onouchi T, Shimomura A (2007) Adenomatous polyposis coli (APC) plays multiple roles in the intestinal and colorectal epithelia. Med Mol Morphol 40(2):68–81. doi:10.1007/s00795-006-0352-5
5. MacDonald BT, Tamai K, He X (2009) Wnt/beta-catenin signaling: components, mechanisms, and diseases. Dev Cell 17(1):9–26. doi:10.1016/j.devcel.2009.06.016, S1534-5807(09)00257-3 [pii]
6. Henderson BR, Fagotto F (2002) The ins and outs of APC and beta-catenin nuclear transport. EMBO Rep 3(9):834–839. doi:10.1093/embo-reports/kvf181, 3/9/834 [pii]
7. Krieghoff E, Behrens J, Mayr B (2006) Nucleo-cytoplasmic distribution of beta-catenin is regulated by retention. J Cell Sci 119(Pt 7):1453–1463. doi:10.1242/jcs.02864, 119/7/1453 [pii]
8. Brocardo M, Henderson BR (2008) APC shuttling to the membrane, nucleus and beyond. Trends Cell Biol 18(12):587–596. doi:10.1016/j.tcb.2008.09.002, S0962-8924(08)00238-9 [pii]
9. Akiyama T, Kawasaki Y (2006) Wnt signalling and the actin cytoskeleton. Oncogene 25(57):7538–7544. doi:10.1038/sj.onc.1210063, 1210063 [pii]
10. Lustig B, Behrens J (2003) The Wnt signaling pathway and its role in tumor development. J Cancer Res Clin Oncol 129(4):199–221. doi:10.1007/s00432-003-0431-0
11. Jamieson C, Sharma M, Henderson BR (2011) Regulation of beta-catenin nuclear dynamics by GSK-3beta involves a LEF-1 positive feedback loop. Traffic 12(8):983–999. doi:10.1111/j.1600-0854.2011.01207.x
12. Brabletz T, Jung A, Reu S, Porzner M, Hlubek F, Kunz-Schughart LA, Knuechel R, Kirchner T (2001) Variable beta-catenin expression in colorectal cancers indicates tumor progression driven by the tumor environment. Proc Natl Acad Sci U S A 98(18):10356–10361. doi:10.1073/pnas.171610498, 98/18/10356 [pii]
13. Morin PJ, Sparks AB, Korinek V, Barker N, Clevers H, Vogelstein B, Kinzler KW (1997) Activation of beta-catenin-Tcf signaling in colon cancer by mutations in beta-catenin or APC. Science 275(5307):1787–1790
14. Strambio-De-Castillia C, Niepel M, Rout MP (2010) The nuclear pore complex: bridging nuclear transport and gene regulation. Nat Rev Mol Cell Biol 11(7):490–501. doi:10.1038/nrm2928, nrm2928 [pii]
15. Jamali T, Jamali Y, Mehrbod M, Mofrad MR (2011) Nuclear pore complex: biochemistry and biophysics of nucleocytoplasmic transport in health and disease. Int Rev Cell Mol Biol 287:233–286. doi:10.1016/B978-0-12-386043-9.00006-2, B978-0-12-386043-9.00006-2 [pii]
16. Terry LJ, Shows EB, Wente SR (2007) Crossing the nuclear envelope: hierarchical regulation of nucleocytoplasmic transport. Science 318(5855):1412–1416. doi:10.1126/science.1142204, 318/5855/1412 [pii]
17. Wente SR, Rout MP (2010) The nuclear pore complex and nuclear transport. Cold Spring Harb Perspect Biol 2(10):a000562. doi:10.1101/cshperspect.a000562, cshperspect.a000562 [pii]
18. D'Angelo MA, Hetzer MW (2008) Structure, dynamics and function of nuclear pore complexes. Trends Cell Biol 18(10):456–466. doi:10.1016/j.tcb.2008.07.009, S0962-8924(08)00213-4 [pii]
19. Walde S, Thakar K, Hutten S, Spillner C, Nath A, Rothbauer U, Wiemann S, Kehlenbach RH (2012) The nucleoporin Nup358/RanBP2 promotes nuclear import in a cargo- and transport receptor-specific manner. Traffic 13(2):218–233. doi:10.1111/j.1600-0854.2011.01302.x
20. Walde S, Kehlenbach RH (2010) The part and the whole: functions of nucleoporins in nucleocytoplasmic transport. Trends Cell Biol 20(8):461–469. doi:10.1016/j.tcb.2010.05.001, S0962-8924(10)00084-X [pii]
21. Stewart M (2007) Molecular mechanism of the nuclear protein import cycle. Nat Rev Mol Cell Biol 8(3):195–208. doi:10.1038/nrm2114, nrm2114 [pii]
22. Fagotto F, Gluck U, Gumbiner BM (1998) Nuclear localization signal-independent and importin/karyopherin-independent nuclear import of beta-catenin. Curr Biol 8(4):181–190. doi:10.1016/S0960-9822(98)70082-X, S0960-9822(98)70082-X [pii]
23. Yokoya F, Imamoto N, Tachibana T, Yoneda Y (1999) beta-catenin can be transported into the nucleus in a Ran-unassisted manner. Mol Biol Cell 10(4):1119–1131
24. Eleftheriou A, Yoshida M, Henderson BR (2001) Nuclear export of human beta-catenin can occur independent of CRM1 and the adenomatous polyposis coli tumor suppressor. J Biol Chem 276(28):25883–25888. doi:10.1074/jbc.M102656200, M102656200 [pii]

25. Wiechens N, Fagotto F (2001) CRM1- and Ran-independent nuclear export of beta-catenin. Curr Biol 11(1):18–27. doi:10.1016/S0960-9822(00)00045-2, S0960-9822(00)00045-2 [pii]
26. Huber AH, Nelson WJ, Weis WI (1997) Three-dimensional structure of the armadillo repeat region of beta-catenin. Cell 90(5):871–882. doi:10.1016/S0092-8674(00)80352-9, S0092-8674(00)80352-9 [pii]
27. Xing Y, Takemaru K, Liu J, Berndt JD, Zheng JJ, Moon RT, Xu W (2008) Crystal structure of a full-length beta-catenin. Structure 16(3):478–487. doi:10.1016/j.str.2007.12.021, S0969-2126(08)00050-6 [pii]
28. Tewari R, Bailes E, Bunting KA, Coates JC (2010) Armadillo-repeat protein functions: questions for little creatures. Trends Cell Biol 20(8):470–481. doi:10.1016/j.tcb.2010.05.003, S0962-8924(10)00096-6 [pii]
29. Lee SJ, Imamoto N, Sakai H, Nakagawa A, Kose S, Koike M, Yamamoto M, Kumasaka T, Yoneda Y, Tsukihara T (2000) The adoption of a twisted structure of importin-beta is essential for the protein-protein interaction required for nuclear transport. J Mol Biol 302(1): 251–264. doi:10.1006/jmbi.2000.4055, S0022-2836(00)94055-5 [pii]
30. Conti E, Muller CW, Stewart M (2006) Karyopherin flexibility in nucleocytoplasmic transport. Curr Opin Struct Biol 16(2):237–244. doi:10.1016/j.sbi.2006.03.010, S0959-440X(06)00048-0 [pii]
31. Sharma M, Jamieson C, Johnson M, Molloy MP, Henderson BR (2012) Specific armadillo repeat sequences facilitate beta-catenin nuclear transport in live cells via direct binding to nucleoporins Nup62, Nup153, and RanBP2/Nup358. J Biol Chem 287(2):819–831. doi:10.1074/jbc.M111.299099, M111.299099 [pii]
32. Hamada M, Haeger A, Jeganathan KB, van Ree JH, Malureanu L, Walde S, Joseph J, Kehlenbach RH, van Deursen JM (2011) Ran-dependent docking of importin-beta to RanBP2/Nup358 filaments is essential for protein import and cell viability. J Cell Biol 194(4):597–612. doi:10.1083/jcb.201102018, jcb.201102018 [pii]
33. Franca-Koh J, Yeo M, Fraser E, Young N, Dale TC (2002) The regulation of glycogen synthase kinase-3 nuclear export by Frat/GBP. J Biol Chem 277(46):43844–43848. doi:10.1074/jbc.M207265200, M207265200 [pii]
34. Wiechens N, Heinle K, Englmeier L, Schohl A, Fagotto F (2004) Nucleo-cytoplasmic shuttling of Axin, a negative regulator of the Wnt-beta-catenin Pathway. J Biol Chem 279(7):5263–5267. doi:10.1074/jbc.M307253200, M307253200 [pii]
35. Henderson BR (2000) Nuclear-cytoplasmic shuttling of APC regulates beta-catenin subcellular localization and turnover. Nat Cell Biol 2(9):653–660. doi:10.1038/35023605
36. Rosin-Arbesfeld R, Townsley F, Bienz M (2000) The APC tumour suppressor has a nuclear export function. Nature 406(6799):1009–1012. doi:10.1038/35023016
37. Neufeld KL, Zhang F, Cullen BR, White RL (2000) APC-mediated downregulation of beta-catenin activity involves nuclear sequestration and nuclear export. EMBO Rep 1(6):519–523. doi:10.1093/embo-reports/kvd117
38. Speese SD, Ashley J, Jokhi V, Nunnari J, Barria R, Li Y, Ataman B, Koon A, Chang YT, Li Q, Moore MJ, Budnik V (2012) Nuclear envelope budding enables large ribonucleoprotein particle export during synaptic Wnt signaling. Cell 149(4):832–846. doi:10.1016/j.cell.2012.03.032, S0092-8674(12)00420-5 [pii]
39. Kalverda B, Pickersgill H, Shloma VV, Fornerod M (2010) Nucleoporins directly stimulate expression of developmental and cell-cycle genes inside the nucleoplasm. Cell 140(3):360–371. doi:10.1016/j.cell.2010.01.011, S0092-8674(10)00012-7 [pii]
40. Capelson M, Liang Y, Schulte R, Mair W, Wagner U, Hetzer MW (2010) Chromatin-bound nuclear pore components regulate gene expression in higher eukaryotes. Cell 140(3):372–383. doi:10.1016/j.cell.2009.12.054, S0092-8674(09)01681-X [pii]
41. Vaquerizas JM, Suyama R, Kind J, Miura K, Luscombe NM, Akhtar A (2010) Nuclear pore proteins nup153 and megator define transcriptionally active regions in the Drosophila genome. PLoS Genet 6(2):e1000846. doi:10.1371/journal.pgen.1000846
42. Bai XT, Gu BW, Yin T, Niu C, Xi XD, Zhang J, Chen Z, Chen SJ (2006) Trans-repressive effect of NUP98-PMX1 on PMX1-regulated c-FOS gene through recruitment of histone deacetylase 1 by FG repeats. Cancer Res 66(9):4584–4590. doi:10.1158/0008-5472.CAN-05-3101, 66/9/4584 [pii]

43. Kasper LH, Brindle PK, Schnabel CA, Pritchard CE, Cleary ML, van Deursen JM (1999) CREB binding protein interacts with nucleoporin-specific FG repeats that activate transcription and mediate NUP98-HOXA9 oncogenicity. Mol Cell Biol 19(1):764–776
44. Shitashige M, Satow R, Honda K, Ono M, Hirohashi S, Yamada T (2008) Regulation of Wnt signaling by the nuclear pore complex. Gastroenterology 134(7):1961–1971. doi:10.1053/j.gastro.2008.03.010, 1971.e1961–1964, S0016-5085(08)00445-9 [pii]
45. Najdi R, Holcombe RF, Waterman ML (2011) Wnt signaling and colon carcinogenesis: beyond APC. J Carcinog 10:5. doi:10.4103/1477-3163.78111
46. Kadoya T, Yamamoto H, Suzuki T, Yukita A, Fukui A, Michiue T, Asahara T, Tanaka K, Asashima M, Kikuchi A (2002) Desumoylation activity of Axam, a novel Axin-binding protein, is involved in downregulation of beta-catenin. Mol Cell Biol 22(11):3803–3819
47. Satow R, Shitashige M, Jigami T, Fukami K, Honda K, Kitabayashi I, Yamada T (2012) beta-catenin inhibits promyelocytic leukemia protein tumor suppressor function in colorectal cancer cells. Gastroenterology 142(3):572–581. doi:10.1053/j.gastro.2011.11.041, S0016-5085(11) 01639-8 [pii]
48. Nathke I (2006) Cytoskeleton out of the cupboard: colon cancer and cytoskeletal changes induced by loss of APC. Nat Rev Cancer 6(12):967–974. doi:10.1038/nrc2010, nrc2010 [pii]
49. Moseley JB, Bartolini F, Okada K, Wen Y, Gundersen GG, Goode BL (2007) Regulated binding of adenomatous polyposis coli protein to actin. J Biol Chem 282(17):12661–12668. doi:10.1074/jbc.M610615200, M610615200 [pii]
50. Sakamoto Y, Boeda B, Etienne-Manneville S (2013) APC binds intermediate filaments and is required for their reorganization during cell migration. J Cell Biol 200(3):249–258. doi:10.1083/jcb.201206010, jcb.201206010 [pii]
51. Mimori-Kiyosue Y, Shiina N, Tsukita S (2000) Adenomatous polyposis coli (APC) protein moves along microtubules and concentrates at their growing ends in epithelial cells. J Cell Biol 148(3):505–518
52. Jimbo T, Kawasaki Y, Koyama R, Sato R, Takada S, Haraguchi K, Akiyama T (2002) Identification of a link between the tumour suppressor APC and the kinesin superfamily. Nat Cell Biol 4(4):323–327. doi:10.1038/ncb779, ncb779 [pii]
53. Bahmanyar S, Nelson WJ, Barth AI (2009) Role of APC and its binding partners in regulating microtubules in mitosis. Adv Exp Med Biol 656:65–74
54. Sharma M, Leung L, Brocardo M, Henderson J, Flegg C, Henderson BR (2006) Membrane localization of adenomatous polyposis coli protein at cellular protrusions: targeting sequences and regulation by beta-catenin. J Biol Chem 281(25):17140–17149. doi:10.1074/jbc.M513027200, M513027200 [pii]
55. Watanabe T, Wang S, Noritake J, Sato K, Fukata M, Takefuji M, Nakagawa M, Izumi N, Akiyama T, Kaibuchi K (2004) Interaction with IQGAP1 links APC to Rac1, Cdc42, and actin filaments during cell polarization and migration. Dev Cell 7(6):871–883. doi:10.1016/j.devcel.2004.10.017, S153458070400382X [pii]
56. Sharma M, Henderson BR (2007) IQ-domain GTPase-activating Protein 1 regulates beta-catenin at membrane ruffles and its role in macropinocytosis of N-cadherin and adenomatous polyposis coli. J Biol Chem 282(11):8545–8556, http://dx.doi.org/10.1074/jbc.M610272200
57. Etienne-Manneville S, Hall A (2003) Cdc42 regulates GSK-3beta and adenomatous polyposis coli to control cell polarity. Nature 421:753–756
58. Etienne-Manneville S, Manneville JB, Nicholls S, Ferenczi MA, Hall A (2005) Cdc42 and Par6-PKCzeta regulate the spatially localized association of Dlg1 and APC to control cell polarization. J Cell Biol 170(6):895–901. doi:10.1083/jcb.200412172, jcb.200412172 [pii]
59. de Forges H, Bouissou A, Perez F (2012) Interplay between microtubule dynamics and intracellular organization. Int J Biochem Cell Biol 44(2):266–274. doi:10.1016/j.biocel.2011.11.009, S1357-2725(11)00307-4 [pii]
60. Gomez-Cavazos JS, Hetzer MW (2012) Outfits for different occasions: tissue-specific roles of Nuclear Envelope proteins. Curr Opin Cell Biol 24(6):775–783. doi:10.1016/j.ceb.2012.08.008, S0955-0674(12)00133-0 [pii]
61. Caldwell CM, Kaplan KB (2009) The role of APC in mitosis and in chromosome instability. Adv Exp Med Biol 656:51–64

62. Dawlaty MM, Malureanu L, Jeganathan KB, Kao E, Sustmann C, Tahk S, Shuai K, Grosschedl R, van Deursen JM (2008) Resolution of sister centromeres requires RanBP2-mediated SUMOylation of topoisomerase IIalpha. Cell 133(1):103–115. doi:10.1016/j.cell.2008.01.045, S0092-8674(08)00210-9 [pii]
63. Joseph J, Dasso M (2008) The nucleoporin Nup358 associates with and regulates interphase microtubules. FEBS Lett 582(2):190–196. doi:10.1016/j.febslet.2007.11.087, S0014-5793(07)01241-0 [pii]
64. Cross MK, Powers MA (2011) Nup98 regulates bipolar spindle assembly through association with microtubules and opposition of MCAK. Mol Biol Cell 22(5):661–672. doi:10.1091/mbc.E10-06-0478, mbc.E10-06-0478 [pii]
65. Murawala P, Tripathi MM, Vyas P, Salunke A, Joseph J (2009) Nup358 interacts with APC and plays a role in cell polarization. J Cell Sci 122(Pt 17):3113–3122. doi:10.1242/jcs.037523, jcs.037523 [pii]
66. Collin L, Schlessinger K, Hall A (2008) APC nuclear membrane association and microtubule polarity. Biol Cell 100(4):243–252, http://dx.doi.org/10.1042/bc20070123
67. Dikovskaya D, Li Z, Newton IP, Davidson I, Hutchins JR, Kalab P, Clarke PR, Nathke IS (2010) Microtubule assembly by the Apc protein is regulated by importin-beta–RanGTP. J Cell Sci 123(Pt 5):736–746. doi:10.1242/jcs.060806, jcs.060806 [pii]
68. Splinter D, Tanenbaum ME, Lindqvist A, Jaarsma D, Flotho A, Yu KL, Grigoriev I, Engelsma D, Haasdijk ED, Keijzer N, Demmers J, Fornerod M, Melchior F, Hoogenraad CC, Medema RH, Akhmanova A (2010) Bicaudal D2, dynein, and kinesin-1 associate with nuclear pore complexes and regulate centrosome and nuclear positioning during mitotic entry. PLoS Biol 8(4):e1000350. doi:10.1371/journal.pbio.1000350
69. Louie RK, Bahmanyar S, Siemers KA, Votin V, Chang P, Stearns T, Nelson WJ, Barth AI (2004) Adenomatous polyposis coli and EB1 localize in close proximity of the mother centriole and EB1 is a functional component of centrosomes. J Cell Sci 117(Pt 7):1117–1128. doi:10.1242/jcs.00939, jcs.00939 [pii]
70. Huang P, Senga T, Hamaguchi M (2007) A novel role of phospho-beta-catenin in microtubule regrowth at centrosome. Oncogene 26(30):4357–4371. doi:10.1038/sj.onc.1210217, 1210217 [pii]
71. Kim SM, Choi EJ, Song KJ, Kim S, Seo E, Jho EH, Kee SH (2009) Axin localizes to mitotic spindles and centrosomes in mitotic cells. Exp Cell Res 315(6):943–954. doi:10.1016/j.yexcr.2009.01.013, S0014-4827(09)00030-5 [pii]
72. Fumoto K, Kadono M, Izumi N, Kikuchi A (2009) Axin localizes to the centrosome and is involved in microtubule nucleation. EMBO Rep 10(6):606–613. doi:10.1038/embor.2009.45, embor200945 [pii]
73. Ruan K, Ye F, Li C, Liou YC, Lin SC, Lin SY (2012) PLK1 interacts and phosphorylates Axin that is essential for proper centrosome formation. PLoS One 7(11):e49184. doi:10.1371/journal.pone.0049184, PONE-D-12-11766 [pii]
74. Flegg CP, Sharma M, Medina-Palazon C, Jamieson C, Galea M, Brocardo MG, Mills K, Henderson BR (2010) Nuclear export and centrosome targeting of the protein phosphatase 2A subunit B56alpha: role of B56alpha in nuclear export of the catalytic subunit. J Biol Chem 285(24):18144–18154. doi:10.1074/jbc.M109.093294, M109.093294 [pii]
75. Schneikert J, Behrens J (2006) Truncated APC is required for cell proliferation and DNA replication. Int J Cancer 119(1):74–79. doi:10.1002/ijc.21826
76. Brocardo MG, Borowiec JA, Henderson BR (2011) Adenomatous polyposis coli protein regulates the cellular response to DNA replication stress. Int J Biochem Cell Biol 43(9): 1354–1364. doi:10.1016/j.biocel.2011.05.013, S1357-2725(11)00159-2 [pii]
77. Lemaitre C, Fischer B, Kalousi A, Hoffbeck AS, Guirouilh-Barbat J, Shahar OD, Genet D, Goldberg M, Betrand P, Lopez B, Brino L, Soutoglou E (2012) The nucleoporin 153, a novel factor in double-strand break repair and DNA damage response. Oncogene 31(45): 4803–4809. doi:10.1038/onc.2011.638, onc2011638 [pii]
78. Johnson M, Sharma M, Henderson BR (2009) IQGAP1 regulation and roles in cancer. Cell Signal 21(10):1471–1478, http://dx.doi.org/10.1016/j.cellsig.2009.02.023

79. Brandt DT, Marion S, Griffiths G, Watanabe T, Kaibuchi K, Grosse R (2007) Dia1 and IQGAP1 interact in cell migration and phagocytic cup formation. J Cell Biol 178(2): 193–200, http://dx.doi.org/10.1083/jcb.200612071
80. Noritake J, Watanabe T, Sato K, Wang S, Kaibuchi K (2005) IQGAP1: a key regulator of adhesion and migration. J Cell Sci 118(10):2085–2092, http://dx.doi.org/10.1242/jcs.02379
81. Mataraza JM, Briggs MW, Li Z, Entwistle A, Ridley AJ, Sacks DB (2003) IQGAP1 promotes cell motility and invasion. J Biol Chem 278(42):41237–41245. doi:10.1074/jbc.M304838200
82. Yamaoka-Tojo M, Ushio-Fukai M, Hilenski L, Dikalov SI, Chen YE, Tojo T, Fukai T, Fujimoto M, Patrushev NA, Wang N, Kontos CD, Bloom GS, Alexander RW (2004) IQGAP1, a novel vascular endothelial growth factor receptor binding protein, is involved in reactive oxygen species–dependent endothelial migration and proliferation. Circ Res 95(3):276–283, http://dx.doi.org/10.1161/01.res.0000136522.58649.60
83. Bashour A-M, Fullerton AT, Hart MJ, Bloom GS (1997) IQGAP1, a Rac- and Cdc42-binding protein, directly binds and cross-links microfilaments. J Cell Biol 137(7):1555–1566. doi:10.1083/jcb.137.7.1555
84. Hart MJ, Callow MG, Souza B, Polakis P (1996) IQGAP1, a calmodulin-binding protein with a rasGAP-related domain, is a potential effector for cdc42Hs. EMBO J 15(12):2997–3005
85. Kuroda S, Fukata M, Kobayashi K, Nakafuku M, Nomura N, Iwamatsu A, Kaibuchi K (1996) Identification of IQGAP as a putative target for the small GTPases, Cdc42 and Rac1. J Biol Chem 271(38):23363–23367. doi:10.1074/jbc.271.38.23363
86. Pelikan-Conchaudron A, Le Clainche C, Didry D, Carlier MF (2011) The IQGAP1 protein is a calmodulin-regulated barbed end capper of actin filaments: possible implications in its function in cell migration. J Biol Chem 286(40):35119–35128. doi:10.1074/jbc.M111.258772, M111.258772 [pii]
87. Benseñor LB, Kan H-M, Wang N, Wallrabe H, Davidson LA, Cai Y, Schafer DA, Bloom GS (2007) IQGAP1 regulates cell motility by linking growth factor signaling to actin assembly. J Cell Sci 120(4):658–669, http://dx.doi.org/10.1242/jcs.03376
88. Le Clainche C, Schlaepfer D, Ferrari A, Klingauf M, Grohmanova K, Veligodskiy A, Didry D, Le D, Egile C, Carlier M-F, Kroschewski R (2007) IQGAP1 stimulates actin assembly through the N-WASP-Arp2/3 pathway. J Biol Chem 282(1):426–435, http://dx.doi.org/10.1074/jbc.M607711200
89. Mateer SC, Wang N, Bloom GS (2003) IQGAPs: integrators of the cytoskeleton, cell adhesion machinery, and signaling networks. Cell Motil Cytoskeleton 55(3):147–155, http://dx.doi.org/10.1002/cm.10118
90. Johnson MA, Henderson BR (2012) The scaffolding protein IQGAP1 co-localizes with actin at the cytoplasmic face of the nuclear envelope: implications for cytoskeletal regulation. Bioarchitecture 2(4):138–142, doi:10.4161/bioa.21182, 21182 [pii]
91. Cadot B, Gache V, Vasyutina E, Falcone S, Birchmeier C, Gomes ER (2012) Nuclear movement during myotube formation is microtubule and dynein dependent and is regulated by Cdc42, Par6 and Par3. EMBO Rep 13(8):741–749. doi:10.1038/embor.2012.89, embor201289 [pii]
92. Gomes ER, Jani S, Gundersen GG (2005) Nuclear movement regulated by Cdc42, MRCK, myosin, and actin flow establishes MTOC polarization in migrating cells. Cell 121(3): 451–463. doi:10.1016/j.cell.2005.02.022
93. Metzger T, Gache V, Xu M, Cadot B, Folker ES, Richardson BE, Gomes ER, Baylies MK (2012) MAP and kinesin-dependent nuclear positioning is required for skeletal muscle function. Nature 484(7392):120–124. doi:10.1038/nature10914, nature10914 [pii]
94. Munter S, Enninga J, Vazquez-Martinez R, Delbarre E, David-Watine B, Nehrbass U, Shorte S (2006) Actin polymerisation at the cytoplasmic face of eukaryotic nuclei. BMC Cell Biol 7(1):23, http://dx.doi.org/10.1186/1471-2121-7-23
95. Luxton GWG, Gomes ER, Folker ES, Vintinner E, Gundersen GG (2010) Linear arrays of nuclear envelope proteins harness retrograde actin flow for nuclear movement. Science 329(5994):956–959, http://dx.doi.org/10.1126/science.1189072

96. Luxton GWG, Gundersen GG (2011) Orientation and function of the nuclear–centrosomal axis during cell migration. Curr Opin Cell Biol 23(5):579–588, http://dx.doi.org/10.1016/j.ceb.2011.08.001
97. Guttinger S, Laurell E, Kutay U (2009) Orchestrating nuclear envelope disassembly and reassembly during mitosis. Nat Rev Mol Cell Biol 10(3):178–191, http://dx.doi.org/10.1038/nrm2641
98. Johnson M, Sharma M, Brocardo MG, Henderson BR (2011) IQGAP1 translocates to the nucleus in early S-phase and contributes to cell cycle progression after DNA replication arrest. Int J Biochem Cell Biol 43(1):65–73, http://dx.doi.org/10.1016/j.biocel.2010.09.014
99. Briggs MW, Li Z, Sacks DB (2002) IQGAP1-mediated stimulation of transcriptional co-activation by β-catenin is modulated by calmodulin. J Biol Chem 277(9):7453–7465. doi:10.1074/jbc.M104315200
100. Wang Y, Wang A, Wang F, Wang M, Zhu M, Ma Y, Wu R (2008) IQGAP1 activates Tcf signal independent of Rac1 and Cdc42 in injury and repair of bronchial epithelial cells. Exp Mol Pathol 85(2):122–128
101. Jadeski L, Mataraza JM, Jeong H-W, Li Z, Sacks DB (2008) IQGAP1 stimulates proliferation and enhances tumorigenesis of human breast epithelial cells. J Biol Chem 283(2):1008–1017. doi:10.1074/jbc.M708466200
102. Meyer RD, Sacks DB, Rahimi N (2008) IQGAP1-dependent signaling pathway regulates endothelial cell proliferation and angiogenesis. PLoS One 3(12):e3848
103. Wada A, Fukuda M, Mishima M, Nishida E (1998) Nuclear export of actin: a novel mechanism regulating the subcellular localization of a major cytoskeletal protein. EMBO J 17(6):1635–1641
104. Wang J-B, Sonn R, Tekletsadik YK, Samorodnitsky D, Osman MA (2009) IQGAP1 regulates cell proliferation through a novel CDC42-mTOR pathway. J Cell Sci 122(12):2024–2033. doi:10.1242/jcs.044644
105. de Lanerolle P, Serebryannyy L (2011) Nuclear actin and myosins: life without filaments. Nat Cell Biol 13(11):1282–1288. doi:10.1038/ncb2364, ncb2364 [pii]
106. Wilson KL, Berk JM (2010) The nuclear envelope at a glance. J Cell Sci 123(12):1973–1978, http://dx.doi.org/10.1242/jcs.019042
107. Andrin C, McDonald D, Attwood KM, Rodrigue A, Ghosh S, Mirzayans R, Masson JY, Dellaire G, Hendzel MJ (2012) A requirement for polymerized actin in DNA double-strand break repair. Nucleus 3(4):384–395. doi:10.4161/nucl.21055, 21055 [pii]
108. Castano E, Philimonenko VV, Kahle M, Fukalova J, Kalendova A, Yildirim S, Dzijak R, Dingova-Krasna H, Hozak P (2010) Actin complexes in the cell nucleus: new stones in an old field. Histochem Cell Biol 133(6):607–626. doi:10.1007/s00418-010-0701-2
109. Gieni RS, Hendzel MJ (2009) Actin dynamics and functions in the interphase nucleus: moving toward an understanding of nuclear polymeric actin. Biochem Cell Biol 87(1):283–306
110. Sandrock K, Bielek H, Schradi K, Schmidt G, Klugbauer N (2010) The nuclear import of the Small GTPase Rac1 is mediated by the direct interaction with karyopherin alpha2. Traffic 11(2):198–209
111. Huelsenbeck SC, Schorr A, Roos WP, Huelsenbeck J, Henninger C, Kaina B, Fritz G (2012) Rac1 signaling is required for DNA damage response stimulated by topoisomerase II poisons. J Biol Chem 287(46):38590–38599. doi:10.1074/jbc.M112.377903, M112.377903 [pii]
112. Yan Y, Greer PM, Cao PT, Kolb RH, Cowan KH (2012) Rac1 GTPase plays an important role in gamma-irradiation induced G2/M checkpoint activation. Breast Cancer Res 14(2):R60. doi:10.1186/bcr3164, bcr3164 [pii]
113. Lui C, Mills K, Brocardo MG, Sharma M, Henderson BR (2012) APC as a mobile scaffold: regulation and function at the nucleus, centrosomes, and mitochondria. IUBMB Life 64(3):209–214. doi:10.1002/iub.599

Part V
The Nuclear Envelope in DNA Damage and Stress Responses

Introduction

DNA damage is both a hallmark of and a major cause of cancer. When there is a lot of damage, repaired double strand breaks (DSBs) often result in chromosome translocations so that such translocations are used diagnostically as a reflection of the degree of damage that the progenitor cells suffered. Translocations during repair of breaks in gene sequences can yield gene fusions that, particularly when involving transcriptional regulators, result in changes to gene expression that can initiate or support tumorigenesis and metastasis (reviewed in [1]). The cell has checkpoints to prevent a nucleus with unrepaired damage from entering the cell cycle; however, the checkpoint can be overridden after about 15 h [2]. Unrepaired damage can result in loss of chromosome regions resulting in gene dosage and other pleiotropic effects. When the genes involved contribute to cell cycle regulation and cell migration or genes lost include tumor suppressors or genes directing error prone cells to an apoptotic pathway, the progeny cells may rapidly proliferate and/ or migrate to promote or enhance tumorigenesis and metastasis. At the same time, to proliferate tumor cells must maintain a balance between the limited DNA damage that supports such changes and too much damage that would direct the cells to apoptosis or necrosis. Therefore, the control and functioning of both DNA repair pathways and of DNA damage response checkpoints are critical for the host organism to avoid tumorigenesis and our ability to manipulate such processes in a tumor could promote apoptosis instead of proliferation. Indeed, driving irreparable DNA damage is the basis for most radiation and chemotherapy in cancer treatment [3].

The nuclear envelope has been lately linked to both processes of the DNA damage response and to stress pathways that can induce the DNA damage itself. First it was observed that nuclear envelope mutations in premature aging syndromes result in heightened accumulation of DNA damage [4, 5]. This initiated a number of studies more focused on DNA damage itself linked to the nuclear envelope and it was observed that if a DSB escapes repair and persists, it will be relocated to the nuclear envelope. Two types of nuclear envelope proteins have been clearly linked to this process: the nuclear envelope transmembrane (NET) SUN proteins and nuclear pore complex (NPC) proteins. In yeast the SUN domain-containing protein Mps3p was found to be responsible for recruiting DSBs to the nuclear periphery and notably this tethering delayed the recombination-repair process [6]. This might at first seem counterintuitive as one might have expected that tethering of breaks would facilitate mechanisms for their fusion; however, it has been shown by Susan Gasser's laboratory that DSBs are highly dynamic in the interphase nucleus [7, 8]. In mammals both the SUN1 and SUN2 proteins have been found to play redundant roles in DNA damage responses that, in addition to recruitment of breaks to the nuclear envelope, also seem to involve direction of signaling events for ATM activation and deposition of H2A.X at break sites [9]. NPC proteins have also been found to be involved in recruiting persistent DSBs to the nuclear envelope. In particular Nup84 has been reported to be required for the tethering of DSBs to the nuclear envelope [10]. One possibility for why persistent DSBs are relocated to the nuclear envelope would be that this allows for more effective recruitment of DNA repair factors or alternatively association of DSBs with the relatively stable nuclear envelope and lamin polymer might stabilize the genome sufficiently to overcome a checkpoint despite the break persisting. Regardless, there is clearly a mechanism that recognizes broken chromosome ends and relocates them to the nuclear envelope. Interestingly, there also appears to be a corresponding mechanism that can recruit the nuclear envelope to a DSB. Treatment with etoposide, which induces DSBs, results in lamin B1-containing invaginations of the membrane into the interior of the nucleus [11].

In this section Suzana Gonzalo, an expert in DNA repair pathways, of St Louis University comprehensively overviews aspects of DNA damage and genome instability associated with lamin A mutations and mutations in an enzyme that modifies lamins in premature aging syndromes. She also gives a detailed description of DNA damage and repair mechanisms and their links to telomere function and maintenance along with how these processes are affected by lamins. Next, Paola Vagnarelli, an expert in mitotic chromatin organization, of Brunel University discusses the protein phosphatase targeting subunit Repo-Man, which has both critical mitotic and interphase roles. In mitosis, Repo-Man counteracts Aurora B and facilitates nuclear envelope reassembly, relevant to cell cycle controls discussed in Part II. In interphase, Repo-Man targets in part to the nuclear envelope and negatively regulates DNA damage-induced signal transduction. As noted earlier, imbalanced DNA damage responses can increase metastasis and Repo-Man has now been found to be upregulated in more metastatic cancers for a number of tumor types. Finally, in this section Takeshi Shimi and Robert (Bob) Goldman, one of the historical discoverers of the lamins and leading intermediate filament researcher, of Northwestern University discuss how lamin B1 regulates the decision between cell proliferation and senescence through reactive oxygen species (ROS) signaling pathways. This is again a situation where finding the right balance, in this case for lamin B1 levels, could regulate choices for the cell relating to proliferation and the levels of the DNA damage causing ROS. Accordingly, the p53 tumor suppressor also plays a role in this mechanism.

1. Goodarzi AA, Jeggo PA (2013) The repair and signaling responses to DNA double-strand breaks. Adv Genet 82:1–45. doi:B978-0-12-407676-1.00001-9 [pii] 10.1016/B978-0-12-407676-1.00001-9
2. Toczyski DP, Galgoczy DJ, Hartwell LH (1997) CDC5 and CKII control adaptation to the yeast DNA damage checkpoint. Cell 90(6):1097–1106. doi:S0092-8674(00)80375-X [pii]
3. Cheung-Ong K, Giaever G, Nislow C (2013) DNA-damaging agents in cancer chemotherapy: serendipity and chemical biology. Chem Biol 20(5):648–659. doi:S1074-5521(13)00131-2 [pii] 10.1016/j.chembiol.2013.04.007
4. Liu B, Wang J, Chan KM, Tjia WM, Deng W, Guan X, Huang JD, Li KM, Chau PY, Chen DJ, Pei D, Pendas AM, Cadinanos J, Lopez-Otin C, Tse HF, Hutchison C, Chen J, Cao Y, Cheah KS, Tryggvason K, Zhou Z (2005) Genomic instability in laminopathy-based premature aging. Nat Med 11(7):780–785
5. Pegoraro G, Kubben N, Wickert U, Gohler H, Hoffmann K, Misteli T (2009) Ageing-related chromatin defects through loss of the NURD complex. Nat Cell Biol 11(10):1261–1267
6. Oza P, Jaspersen SL, Miele A, Dekker J, Peterson CL (2009) Mechanisms that regulate localization of a DNA double-strand break to the nuclear periphery. Genes Dev 23(8):912–927. doi:23/8/912 [pii] 10.1101/gad.1782209
7. Dion V, Kalck V, Horigome C, Towbin BD, Gasser SM (2012) Increased mobility of double-strand breaks requires Mec1, Rad9 and the homologous recombination machinery. Nat Cell Biol 14(5):502–509. doi:ncb2465 [pii] 10.1038/ncb2465
8. Neumann FR, Dion V, Gehlen LR, Tsai-Pflugfelder M, Schmid R, Taddei A, Gasser SM (2012) Targeted INO80 enhances subnuclear chromatin movement and ectopic homologous recombination. Genes Dev 26(4):369–383. doi:26/4/369 [pii] 10.1101/gad.176156.111
9. Lei K, Zhu X, Xu R, Shao C, Xu T, Zhuang Y, Han M (2012) Inner nuclear envelope proteins SUN1 and SUN2 play a prominent role in the DNA damage response. Curr Biol 22(17):1609–1615. doi:S0960-9822(12)00719-1 [pii] 10.1016/j.cub.2012.06.043
10. Nagai S, Dubrana K, Tsai-Pflugfelder M, Davidson MB, Roberts TM, Brown GW, Varela E, Hediger F, Gasser SM, Krogan NJ (2008) Functional targeting of DNA damage to a nuclear pore-associated SUMO-dependent ubiquitin ligase. Science 322(5901):597–602. doi:322/5901/597 [pii] 10.1126/science.1162790
11. Dellaire G, Kepkay R, Bazett-Jones DP (2009) High resolution imaging of changes in the structure and spatial organization of chromatin, gamma-H2A.X and the MRN complex within etoposide-induced DNA repair foci. Cell Cycle 8(22):3750–3769. doi:10065 [pii]

DNA Damage and Lamins

Susana Gonzalo

Abstract The spatial and temporal organization of the genome has emerged as an additional level of regulation of nuclear functions. Structural proteins associated with the nuclear envelope play important roles in the organization of the genome. The nuclear lamina, a polymeric meshwork formed by lamins (A- and B-type) and lamin-associated proteins, is viewed as a scaffold for tethering chromatin and protein complexes regulating a variety of nuclear functions. Alterations in lamins function impact DNA transactions such as transcription, replication, and repair, as well as epigenetic modifications that change chromatin structure. These data, and the association of defective lamins with a whole variety of degenerative disorders, premature aging syndromes, and cancer, provide evidence for these proteins operating as caretakers of the genome. In this chapter, we summarize current knowledge about the function of lamins in the maintenance of genome integrity, with special emphasis on the role of A-type lamins in the maintenance of telomere homeostasis and mechanisms of DNA damage repair. These findings have begun to shed some light onto molecular mechanisms by which alterations in A-type lamins induce genomic instability and contribute to the pathophysiology of aging and aging-related diseases, especially cancer.

Keywords Lamins • Laminopathies • Hutchinson–Gilford Progeria Syndrome (HGPS) • DNA repair • Cathepsins • Vitamin D • DNA damage response • Telomeres • Nonhomologous end-joining • Homologous recombination • Farnesyltransferase inhibitors • Reactive oxygen species • 53BP1 • Genomic instability

---

S. Gonzalo (✉)
Department of Biochemistry and Molecular Biology,
St Louis University School of Medicine, 1100 S Grand Ave, St. Louis, MO 63104, USA
e-mail: sgonzalo@slu.edu

E.C. Schirmer and J.I. de las Heras (eds.), *Cancer Biology and the Nuclear Envelope*, Advances in Experimental Medicine and Biology 773, DOI 10.1007/978-1-4899-8032-8_17,

Abbreviations

DDR	DNA damage response
DSB	Double-strand break
EDMD	Emery–Dreifuss muscular dystrophy
FTI	Farnesyltransferase inhibitor
HGPS	Hutchinson–Gilford progeria syndrome
HR	Homologous recombination
IR	Ionizing radiation
IRIF	Ionizing radiation-induced foci
LAD	Lamin-associated domain
MADA	Mandibuloacral dysplasia type A
MEF	Mouse embryonic fibroblast
NER	Nucleotide-excision repair
NHEJ	Nonhomologous end-joining
ROS	Reactive oxygen species
shRNA	Short hairpin RNA

Introduction

The human genome is organized into different levels of complexity. The DNA is packaged into nucleosomes to form the chromatin fiber, which in turn can acquire different structures and states of compaction. Chromatin compaction is fundamental for fitting 2 m of DNA in humans into the small volume of the nucleus. Another level of complexity is the 3D nuclear organization of the genome, which is responsible for maintaining the territories occupied by the different chromosomes, as well as the accumulation of repressive chromatin or heterochromatin at the nuclear periphery [1–3]. Alterations in nuclear architecture and chromatin structure are associated with disease, thus providing evidence of the importance of genome organization for genome function [4–6].

In recent years, the nuclear lamina has arisen as a key structure in the organization of genome function. The nuclear lamina is a filamentous structure under the inner nuclear membrane composed of A-type and B-type lamins and lamin-associated proteins, which provide a scaffold for tethering chromatin and protein complexes to specific subcompartments [7–11] (Fig. 1). This compartmentalization is brought about by the interaction of the nuclear lamina with large domains throughout the genome, known as lamina-associated domains (LADs), which vary in size between 0.1 and 10 Mb. LADs are usually gene-poor regions enriched in repressive chromatin marks and demarcated by boundaries including insulators, CpG islands, and gene promoters oriented away from the lamina [12]. The organization of the genome orchestrated by the nuclear lamina is thought to be essential for the proper regulation of nuclear functions, including gene transcription and silencing, DNA

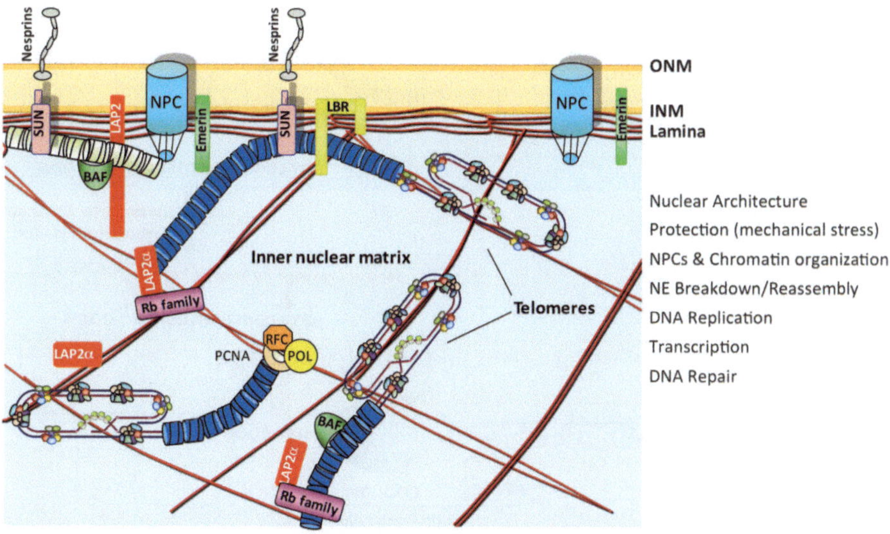

Fig. 1 A-type lamins function as a scaffold that ensures nuclear function. A-type lamins are a special class of intermediate filaments that form the nuclear lamina, a protein meshwork underlying the inner nuclear membrane, which can also extend throughout the nucleoplasm. They serve as a scaffold for tethering chromatin and proteins to specific nuclear subcompartments. A whole variety of functions have been ascribed to A-type lamins including processes such as DNA replication, transcription, and DNA repair. For the most part, the molecular mechanisms behind these functions remain poorly understood

replication and repair, positioning of nuclear pore complexes, chromatin remodeling, and nuclear envelope breakdown and reassembly during mitosis [6, 13, 14].

A-type and B-type lamins are type V intermediate filament proteins with a characteristic central α-helical coiled coil rod domain flanked by non-helical globular N-terminal "head" and C-terminal "tail" domains [15] (Fig. 2). While B-type lamins are constitutively expressed in all cell types and during all stages of development, A-type lamins are only expressed after the onset of cell differentiation. In the mouse, expression of A-type lamins is not observed until embryonic day 12. The B-type lamins class includes lamin B1, encoded by the *LMNB1* gene, and lamins B2 and B3 that result from alternative splicing of the *LMNB2* gene. In contrast, four A-type lamins (lamins A, C, AΔ10, and C2) proteins result from alternative splicing of a unique *LMNA* gene [16]. A-type and B-type lamins contain a -CAAX motif at their C-terminus, which undergoes farnesylation and carboxymethylation shortly after synthesis [17, 18]. This posttranslational modification is thought to facilitate anchoring of lamins to the inner nuclear membrane. In the case of lamin A, further processing of the protein takes place by the metalloprotease Zmpste24, which removes 15 residues at the C-terminus, including the farnesylated cysteine [19, 20], rendering mature lamin A (Fig. 2). This processing step is crucial for the proper function of lamin A, and alterations in this process result in severe nuclear abnormalities linked to disease, as discussed below. In addition, different studies have

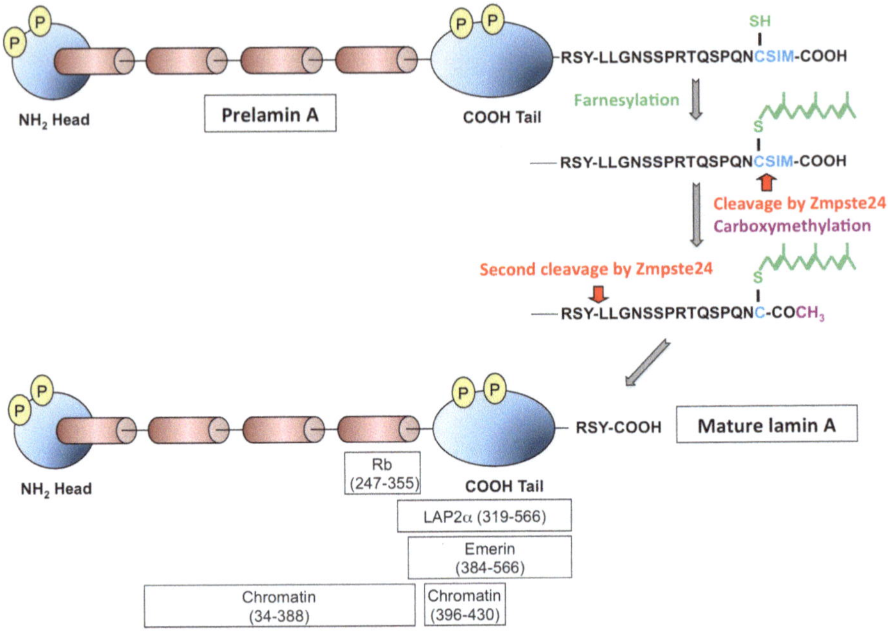

Fig. 2 Schematic representation of lamin A structure and posttranslational processing. Lamins consist of a central rod domain, flanked by a globular head and a globular tail domain. Lamin A is synthesized as a prelamin A precursor which undergoes processing of its C-terminus. The C-terminal -CAAX motif (*blue*) undergoes farnesylation (*green*), cleavage of the last three amino acids (*red*), followed by carboxymethylation (*purple*) of the new terminal cysteine. A second cleavage by the metalloprotease Zmpste24 (*red*) removes another 15 amino acids, including the farnesyl group, rendering mature lamin A. Sites of interaction with gene transcriptional regulators such as Rb family members, lamin-associated proteins such as LAP2α, inner nuclear membrane proteins such as emerin, or chromatin have been described that involve primarily the C-terminal part of the protein and the central rod domain

shown that the C-terminal domain of A-type lamins contains the binding sites for most lamin-binding proteins, as well as chromatin [13, 21, 22].

In recent years, interest in lamins has increased due to the association of hundreds of mutations in the LMNA gene with over a dozen degenerative disorders, broadly termed laminopathies, which include muscular dystrophies, neuropathies, lipodystrophies, and a variety of premature aging syndromes. In addition, alterations of A-type lamins function have been associated with physiological aging and cancer [4, 23, 24]. The connection between A-type lamins and aging came primarily from the discovery that the fatal premature aging disease Hutchinson Gilford Progeria Syndrome (HGPS) is caused by a mutation in the *LMNA* gene that results in the expression of a mutant dominant-negative prelamin A isoform known as "progerin" [25–27]. Progerin contains an internal deletion of 50 amino acids near the C-terminus, which removes the second cleavage site by Zmpste24. Thus, a farnesylated form of the protein accumulates, which is toxic for the cell.

Interestingly, the mouse knockout for Zmpste24 exhibits similar phenotypes as human patients with HGPS [20, 28], representing a good model to study progeria. The fact that progerin accumulates in fibroblasts from old individuals has implicated A-type lamins also in physiological aging [29, 30]. Furthermore, a link between A-type lamins and cancer has been established by studies showing that their expression is altered in many types of malignancies, which is often associated with increased aggressiveness [31–34]. Methylation-induced silencing of the *LMNA* gene is a major event in leukemia, lymphomas, and small cell lung cancer, while overexpression is associated with colon carcinoma. Thus, expression of mutant forms of A-type lamins as well as changes in their expression can lead to disease states. Although the specific molecular mechanisms affected by defects in lamins remain poorly understood, various lines of evidence have linked laminopathies with increased genomic instability. Here, we provide an overview of the advances made in recent years elucidating the roles of lamins, primarily A-type lamins, in mechanisms of DNA repair and maintenance of genome integrity.

Genomic Instability

Genomic instability is defined as the tendency of the genome to acquire mutations and epimutations as well as alterations in gene or chromosome dosage when processes involved in maintaining and replicating the genome become dysfunctional. Maintaining the stability and the correct sequence composition of the three billion bases that form our genome is critical for a faithful transmission of genomic information. Our genome is under constant attack by endogenous and exogenous agents [35]. As much as 10^5 lesions in DNA can occur per cell per day. DNA damage can result from side products of our normal metabolic activities such as free radicals and reactive oxygen (ROS) and reactive nitrogen species, as well as from environmental factors such as UV radiation, X-rays, and chemical compounds [36]. In addition, deficiencies in DNA replication or loss of telomere function can result in DNA lesions, primarily DNA double-strand breaks (Fig. 3). These dangerous lesions are repaired primarily by nonhomologous end-joining (NHEJ) or homologous recombination (see below). Improper or inefficient repair of DNA damage causes mutations, chromosomal translocations, or loss of genetic information that could ultimately lead to a large number of human syndromes, including premature aging, various cancer predispositions, and genetic abnormalities [36, 37]. A whole body of evidence has also implicated alterations in nuclear architecture and chromatin structure (epigenetic changes) in the acquisition of genomic instability. Furthermore, defects in the integrity of mitochondria profoundly impact on the stability of our genome, by generating ROS that damage DNA and oxidize proteins. Importantly, genomic instability increases with age and represents one of the highest risks for aging-related diseases, including cardiovascular diseases and cancer [38, 39]. Thus, understanding the molecular mechanisms behind the age-related acquisition of genomic instability is an active area of research, which will be fundamental to learn how to counteract its pathogenic effects.

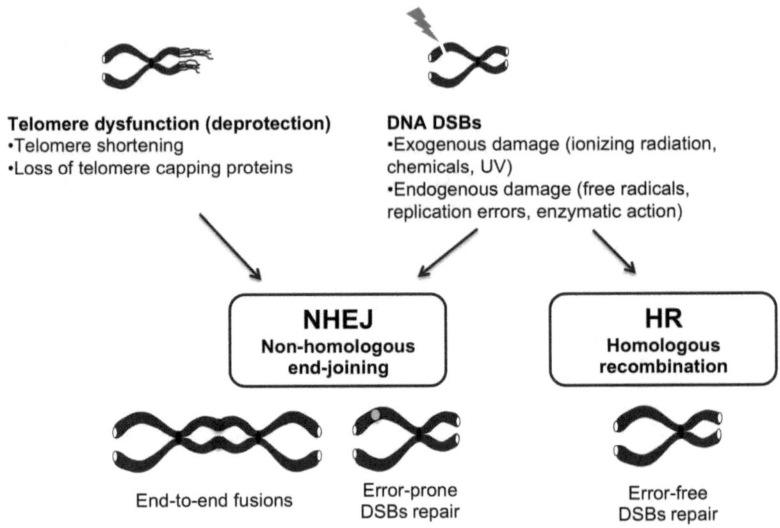

Fig. 3 Processes essential for maintenance of genome stability. Maintenance of telomere homeostasis and mechanisms of DNA DSB repair is critical for genome stability. DSBs resulting from telomere dysfunction or exogenous or endogenous insults are repaired primarily by two mechanisms: nonhomologous end-joining (NHEJ) and homologous recombination (HR)

The DNA Damage Response

The cell has evolved a complex pathway, termed DNA damage response (DDR), to sense, signal, and ultimately repair DNA lesions that arise from endogenous or exogenous insults [37, 38, 40–42]. A variety of DNA repair mechanisms, with specificity towards different categories of DNA damage have been identified. These include base-excision repair, nucleotide-excision repair (NER), transcription-coupled repair, homologous recombination (HR), and NHEJ. HR and NHEJ, the main mechanisms responsible for the repair of DNA double-strand breaks (DSBs), are considered to compete for repair substrates and be mutually exclusive [37, 43, 44]. These pathways have been extensively characterized due to the fact that DSBs are particularly dangerous to the cell [45]. HR repairs the damage with great fidelity and requires both resection of the 5′ ends around the break and the presence of a homologous template. In contrast, NHEJ is an error-prone mechanism of DNA repair, which involves end ligation of broken DNA. While the HR repair mechanism is active during S and G2 phases of the cell cycle, NHEJ seems to be active throughout the cell cycle [46].

A growing list of factors has been clearly implicated in the DDR. These include sensors of the lesions such as the MRN complex (made up of the factors MRE11,

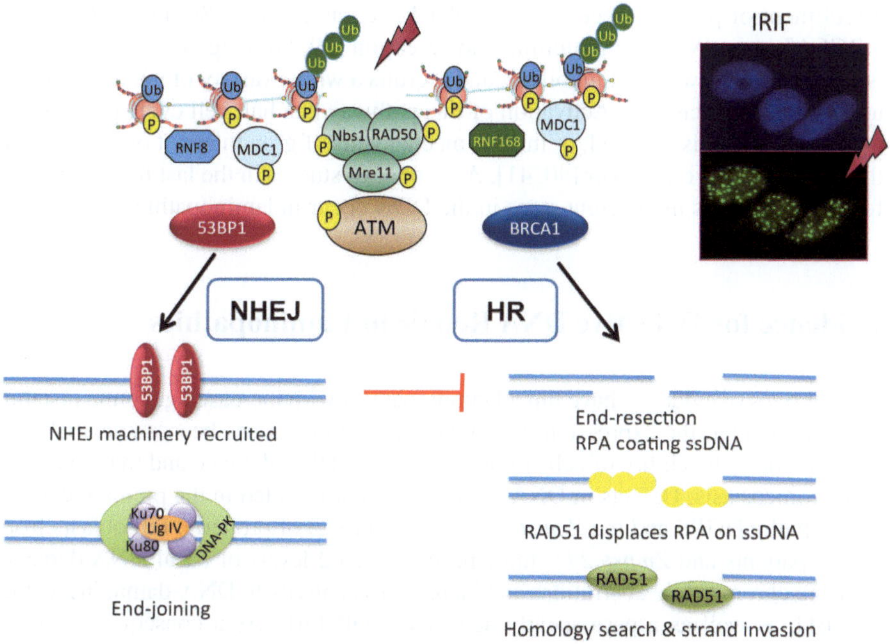

Fig. 4 Important factors in the DNA damage response (DDR). The presence of a DNA DSB is sensed by the MRN complex, which recruits and activates ATM, a master regulator of DDR. ATM phosphorylates H2AX, which targets MDC1 and other mediator factors to the break. These factors attract RNF8 and RNF168 which ubiquitylate histones around the break. These histone modifications facilitate recruitment of factors that participate in DNA DSB repair such as 53BP1 and BRCA1. 53BP1 promotes recruitment of the NHEJ machinery and leads to end-joining of the break. In contrast, BRCA1 promotes end-resection of the break, which initiates the complex process of recombination. The presence of DSBs can be easily visualized by performing immunofluorescence, as shown for γ-H2AX ionizing radiation-induced foci (IRIF)

NBS1, and RAD50), which recruits and activates the kinase ATM at sites of damage [47, 48]. ATM is considered a master regulator that phosphorylates a whole variety of factors involved in DDR and in the activation of cell cycle checkpoints [49, 50]. One of the earlier events in DDR is phosphorylation of the histone variant H2AX in the chromatin surrounding the break [51, 52] (Fig. 4). The presence of nuclear foci labeled with phospho-H2AX (γ-H2AX) indicates sites of DNA damage, and the kinetics of formation and resolution of these foci is widely used to identify defects in DNA repair. In turn, these events facilitate the recruitment of additional factors such as MDC1 [53], and ubiquitin and SUMO ligases that introduce further changes in the chromatin surrounding the break [54–58]. These changes facilitate the recruitment of downstream effectors of DNA DSB repair mechanisms. Two effectors have been recently in the spotlight due to their key role in regulating the balance between HR and NHEJ to repair DSBs: BRCA1 and 53BP1. BRCA1 promotes HR by facilitating end-resection and RAD51 recruitment, which is essential for homology search and strand invasion during recombination. In contrast, 53BP1 facilitates the recruitment of proteins necessary for NHEJ such as Ku70/Ku80, DNA-PKcs, and XRCC4/Ligase IV, while inhibiting end-resection [59, 60] (Fig. 4). Thus, the DDR is a complex pathway that mobilizes and recruits a whole variety of nuclear proteins to sites of DNA damage. Activation of this pathway can halt cell cycle progression until the damage is restored, or initiate mechanisms of growth arrest or cell death if the damage is beyond repair [40, 41]. A number of studies in the last few years have found that defects in different steps in the DDR occur in laminopathies.

Evidence for Defective DNA Repair in Laminopathies

Several lines of evidence have linked laminopathies with increased genomic instability. In particular, the expression of A-type lamin mutants has been associated with impairment in the ability of cells to properly repair DNA damage and maintain telomere homeostasis. Defects in DNA repair were first reported in the premature aging laminopathy HGPS and the *Zmpste24*$^{-/-}$ mouse model of progeria. Fibroblasts from HGPS patients and *Zmpste24*$^{-/-}$ mice have increased levels of basal DNA damage, showing chromosome aberrations and increased sensitivity to DNA-damaging agents [28, 61], as well as a permanently activated DDR [30]. As a consequence of the defects in DNA repair, *Zmpste24*$^{-/-}$ mice exhibit increased sensitivity to ionizing radiation (IR). When exposed to γ-irradiation, 80 % of knockout mice died within 12 days, in contrast to only 20 % of control mice [61]. Later studies indicated that the accumulation of unrepairable DNA damage in laminopathies such as HGPS and Restrictive Dermopathy is in part due to elevated levels of ROS and greater sensitivity to oxidative stress [62]. As such, the presence of ROS scavengers reduced the basal levels of DNA damage, thus suggesting that these types of compounds might help ameliorate the phenotype in some laminopathies. Similarly, the laminopathy mandibuloacral dysplasia type A (MADA), caused by the homozygous R527H mutation in the *LMNA* gene, is characterized by genomic instability. Fibroblasts from MADA patients exhibit defects in their ability to repair DNA damage, including reduced activation of p53 and its downstream targets, increased chromosome damage, and presence of residual γ-H2AX foci characteristic of unrepaired DNA lesions [63].

A whole variety of molecular mechanisms have been proposed to contribute to the defects in DNA repair and the increased genomic instability in progeria. The landmark study by Liu and colleagues [61] demonstrated that progeria cells exhibit delayed recruitment of 53BP1 and RAD51 to γ-H2AX-labeled DNA repair foci after IR. Later studies showed that HGPS and *Zmpste24*$^{-/-}$ fibroblasts have an aberrant accumulation of Xeroderma Pigmentosum group A (XPA), a protein involved in NER, at sites of DNA lesions [64]. Binding of XPA also activates ATM- and ATR-dependent signaling cascades that contribute to proliferation arrest, and thus could contribute to premature senescence in these cells. Interestingly, depletion of XPA by siRNA in progeria cells partially restored the recruitment of HR factors Rad50 and Rad51 to sites of DNA damage [61, 64]. However, the partial rescue indicated that additional mechanisms contribute to the DNA repair deficiencies in these cells. Some proposed mechanisms include the absence of the nuclear DNA-PK holoenzyme, a key factor in NHEJ repair that is linked to premature and physiological aging [65], as well as defects in chromatin-modifying activities. In particular, loss of components of the NuRD complex and decreased levels of the histone acetyltransferase Mof were observed in progeria cells and associated with impaired recruitment of DNA repair factors and increased basal DNA damage [66, 67]. The decrease in Mof levels in *Zmpste24*$^{-/-}$ cells correlates with lower global levels of H4K16 acetylation, a histone mark associated with chromatin compaction, and decreased recruitment of DNA repair factors such as 53BP1. The importance of this histone modification is demonstrated by the restoration of 53BP1 recruitment upon ectopic expression of Mof [67]. Overall, these reports provide a strong correlation between mutations or alterations in the processing of A-type lamins and a hindered ability to maintain the integrity of the genome. In addition, they have started to shed some light into putative molecular mechanisms behind the increased genomic instability in progeria. Elucidating the specific steps during these processes could bring about new possibilities for treatment of progeria, other laminopathies, and cancer.

Role of A-Type Lamins in Telomere Maintenance

Telomere dysfunction due to excessive attrition of telomeric DNA repeats or to defects in telomere-binding proteins known as the "shelterin" complex is sensed by the cell as DNA damage, which activates DDR and checkpoint pathways [68]. Dysfunctional telomeres, which can be easily visualized by the colocalization of γ-H2AX with a telomere DNA probe, are aberrantly repaired by the NHEJ pathway, which results in toxic chromosome end-to-end fusions [69] (Fig. 5). Telomere dysfunction leads to cell death or a permanent growth arrest, which can be bypassed by inactivation of tumor suppressor mechanisms, contributing to genomic instability.

There is evidence in the literature that A-type lamins associate with telomeres, putatively via telomere-binding proteins [70, 71]. This association has been shown to be important for the proper distribution of telomeres within the 3D nuclear space. A-type lamins-deficient cells exhibit accumulation of telomeres towards the nuclear periphery during interphase [72]. In addition, lamins have an impact on the mobility of telomeres in the nucleus [73]. Monitoring telomere dynamics in human fibroblasts revealed that HGPS cells have reduced mobility with respect to normal fibroblasts. Moreover, loss of A-type lamins increases telomere dynamics. These changes could contribute to the alterations in telomere metabolism observed in lamin-deficient cells [74]. HGPS fibroblasts are characterized by faster telomere attrition during proliferation [75, 76]. This was shown to be a direct consequence of the expression of progerin, as hematopoietic cells from HGPS patients, which do not express progerin, do not show this phenotype. In contrast, no differences in telomere length were observed in *Zmpste24*$^{-/-}$ fibroblasts, although whether or not *Zmpste24*$^{-/-}$ telomeres are dysfunctional due to uncapping remains to be tested. Furthermore, *Lmna*$^{-/-}$ mice exhibit a modest but consistent decreased telomere length when compared to wild-type littermates, as well as an increased frequency of chromosomes that lack telomere signals, presenting with signal-free ends [72].

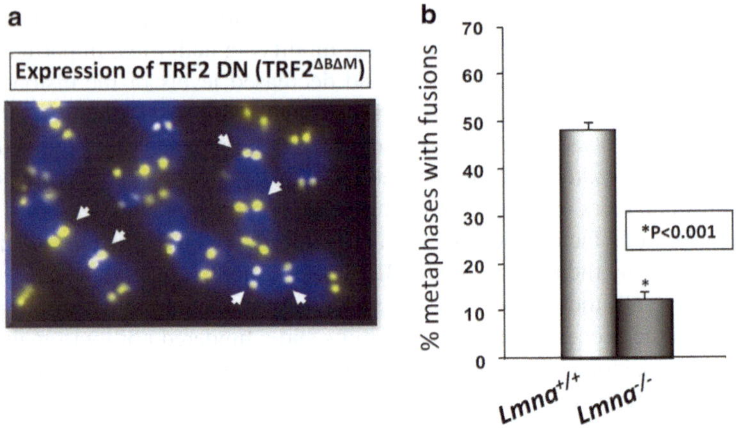

c

	Number of Metaphases	Total Fusions	% Metaphases with Fusions
Lmna$^{+/+}$ + TRF2$^{\Delta B\Delta M}$	170	170	48.4±2.3
Lmna$^{-/-}$ + TRF2$^{\Delta B\Delta M}$	164	50	12.4±1.7*

Fig. 5 A-type lamins are essential for NHEJ of dysfunctional telomeres. (**a**) Telomere dysfunction due to overexpression of a dominant-negative version of the telomere-binding protein TRF2 causes chromosome end-to-end fusions, a process that requires a functional NHEJ mechanism of repair. (**b**) Graph shows that loss of A-type lamins reduces significantly the frequency of chromosome end-to-end fusions by NHEJ. (**c**) Table shows the total number of fusions observed as well as the percentage of metaphases with fusions in *Lmna*$^{+/+}$ and *Lmna*$^{-/-}$ MEFs. These data demonstrated an unprecedented role for A-type lamins in NHEJ of dysfunctional telomeres

Further evidence for a role of A-type lamins in telomere biology was the finding that proliferative defects on human fibroblasts expressing lamin A mutants are rescued by telomerase [77].

The above data indicate that A-type lamins participate in the localization and mobility of telomeres within the nucleus, as well as in the maintenance of telomere homeostasis. Recent studies show that A-type lamins also participate indirectly in the aberrant processing of dysfunctional telomeres by NHEJ, which leads to chromosome end-to-end fusions. Expression of a dominant-negative version of the telomere-binding protein TRF2 (TRF2$^{\Delta B\Delta M}$) leads to uncapping of telomeres and massive appearance of chromosome end-to-end fusions in most cell types. However, in *Lmna*$^{-/-}$ mouse embryonic fibroblasts (MEFs), the extent of chromosome fusions in response to TRF2$^{\Delta B\Delta M}$ was greatly reduced [78] (Fig. 5). As discussed below, this defect was linked to the degradation of 53BP1, a protein that is essential for NHEJ of dysfunctional telomeres [79].

Novel Aspects of DNA Double-Strand Repair Revealed by Loss of A-Type Lamins

The generation of a mouse model deficient in A-type lamins (*Lmna*$^{-/-}$) revealed important information about the function of these structural nuclear proteins at the cellular and organismal level. However, a recent study showed that the *Lmna*$^{-/-}$ mouse model is not a complete knockout [80] and that low levels of a truncated form of the protein is produced lacking exons 8–11. These mice develop to term with no obvious abnormalities, but they exhibit severe postnatal growth retardation, dying at around 5–6 weeks of age. *Lmna*$^{-/-}$ mice develop defects in skeletal and cardiac muscles, a phenotype characteristic of muscular dystrophy. At the cellular level, perturbations of the nuclear envelope, loss of heterochromatin from the nuclear periphery, epigenetic alterations, and nuclear fragility are observed, concomitant with mislocalization of emerin, an inner nuclear membrane protein linked to Emery–Dreifuss muscular dystrophy (EDMD) [81]. The *Lmna*$^{-/-}$ mouse model has been utilized to improve our understanding of the role of A-type lamins in DNA repair, signaling through the DDR, and telomere maintenance.

Recent studies have shown that *Lmna*$^{-/-}$ MEFs exhibit signs of genomic instability illustrated by increased aneuploidy, higher incidence of chromosome and chromatid breaks, and basal levels of unrepaired DNA, as shown by the presence of γ-H2AX foci. As discussed above, *Lmna*$^{-/-}$ MEFs also present with decreased telomere length [72]. These data indicate that A-type lamins deficiency affects the ability of cells to properly repair DNA damage and maintain genome integrity. However, given the great variety of diseases associated with the different mutations/loss of A-type lamins, it is tempting to speculate that different types of alterations in these proteins would impact different aspects of the DDR and specific mechanisms of DNA repair. Establishing which steps in the DDR each type of alteration in lamins affects could provide new avenues for targeted treatment. In the case of cancer for example, knowing if tumor cells in which the *LMNA* promoter is silenced by DNA methylation are defective in DNA repair and more sensitive to DNA-damaging agents, would provide valuable information towards the use of specific cancer therapeutics.

Mechanisms Behind Genomic Instability in Lmna$^{-/-}$ *Mice*

One of the mechanisms behind genomic instability in A-type lamins-deficient cells is the loss of the DNA repair factor 53BP1. Studies monitoring the ability of A-type lamins-deficient cells to sense DNA damage did not find any defects in the activation of DDR [78]. ATM-dependent phosphorylation of H2AX (γ-H2AX) and phosphorylation of p53 at Ser15 were not affected in *Lmna*$^{-/-}$ MEFs, and the kinetics of formation and resolution of γ-H2AX ionizing radiation-induced foci (IRIF) was indistinguishable between *Lmna*$^{+/+}$ and *Lmna*$^{-/-}$ MEFs [78]. In contrast, this study

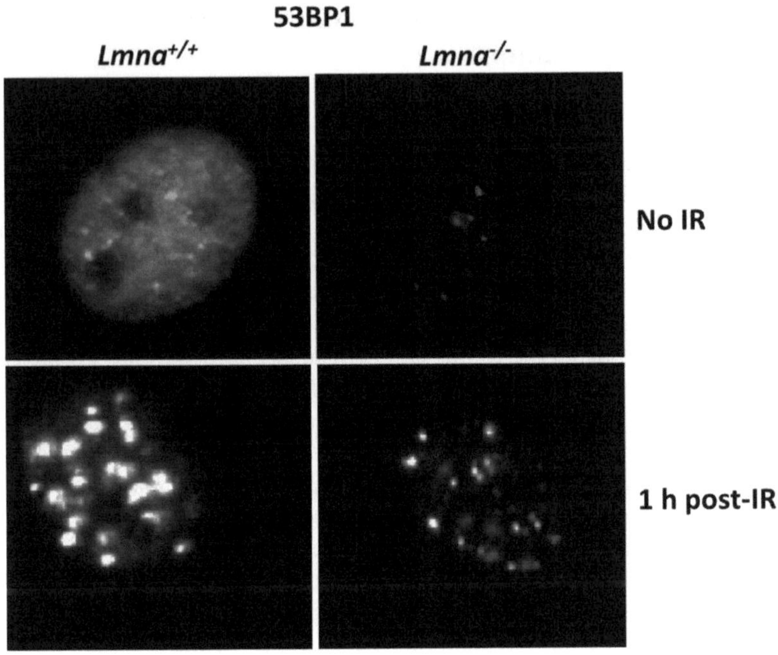

Fig. 6 Loss of A-type lamins impacts on 53BP1 stability and IRIF formation. Immunofluorescence showing a marked decrease in global levels of 53BP1 protein in *Lmna*$^{-/-}$ MEFs and reduced intensity of 53BP1 labeling of DNA repair foci at 1 h postirradiation. Thus, loss of A-type lamins decreases the accumulation of 53BP1 protein at DNA repair foci, which explains reduced NHEJ in lamin-deficient cells

demonstrated a marked decrease in the accumulation of 53BP1 at IRIF in *Lmna*$^{-/-}$ MEFs when compared to *Lmna*$^{+/+}$ at all postirradiation times tested. Importantly, this deficiency is due to a marked decrease in the global levels of 53BP1 protein, and not to failed recruitment, since a similar number of 53BP1 IRIF formed although at a much lower intensity [78] (Fig. 6). Further evidence for a role of A-type lamins in the stabilization of 53BP1 protein came from experiments showing that acute depletion of A-type lamins by lentiviral transduction with shRNAs leads to decreased levels of 53BP1 protein but not transcript levels. Interestingly, reconstitution of either lamin A, lamin C, or both, into *Lmna*$^{-/-}$ MEFs, rescued normal levels of 53BP1.

Loss of 53BP1 and Defects in NHEJ in Lmna$^{-/-}$ *Mice*

The DNA repair factor 53BP1 has been shown to be important for long-range NHEJ processes such as class-switch and *V*(*D*)*J* recombination, as well as processing of

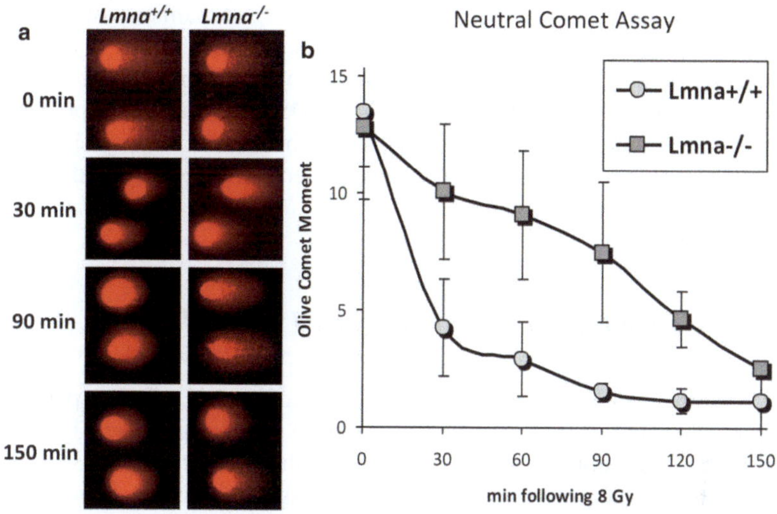

Fig. 7 Lamin-deficient cells exhibit defects in the repair of DNA DSBs induced by IR. (**a**) Images of neutral comet assays. *Lmna*$^{+/+}$ and *Lmna*$^{-/-}$ MEFs were irradiated and subjected to single cell electrophoresis at different times postirradiation. Note how the comet tail decreases over time due to repair of DSBs. (**b**) Olive moment measures unrepaired DNA damage. A bimodal form of DNA DSB repair is clearly observed in *Lmna*$^{+/+}$ fibroblasts, with the fast phase corresponding to NHEJ occurring within the first hour after DNA damage. *Lmna*$^{-/-}$ fibroblasts exhibited a greatly reduced rate of the fast component of DSB repair, indicating that these proteins play a role in the NHEJ repair mechanism

dysfunctional telomeres [79, 82–84]. In addition, 53BP1 participates in the repair of short-range DNA DSBs by inhibiting end-resection and facilitating the recruitment of the NHEJ DNA repair machinery [85, 86].

The decreased levels of 53BP1 protein in *Lmna*$^{-/-}$ MEFs suggest that these cells could be deficient in long-range and short-range NHEJ. The reduced ability of lamin A/C-deficient cells to promote chromosome end-to-end fusions upon telomere dysfunction is consistent with a defect in long-range NHEJ [78] (Fig. 5). In addition, studies were performed to assess putative defects in short-range NHEJ by monitoring the kinetics of repair of IR-induced DNA DSB by neutral comet assays. This assay examines the presence of unrepaired DNA by combining DNA gel electrophoresis with fluorescence microscopy to visualize migration of DNA fragments from individual agarose-embedded cells. The comet head contains high-molecular weight and intact DNA, and the comet tail contains the leading ends of migrating fragments. Olive moment, calculated as the product of the amount of DNA in the tail and the mean distance of migration in the tail, is a measure of unrepaired DNA [87] (Fig. 7). The comet assay showed a typical biphasic mode of repair for *Lmna*$^{+/+}$ MEFs, with a fast phase of repair corresponding to classical NHEJ, and a slower phase of repair that is associated with alternative mechanisms of repair [88]. In contrast, *Lmna*$^{-/-}$ MEFs exhibited profound defects in the fast phase of DNA DSB repair [78, 89]. Importantly, reconstitution of 53BP1 into *Lmna*$^{-/-}$ cells rescued the fast phase of repair of IR-induced DNA DSBs, as well as the processing of dysfunctional telomeres by NHEJ [78]. Altogether, these findings indicate that 53BP1 deficiency is a major contributor of the defects in long-range and short-range NHEJ observed in lamins-deficient cells.

How Do A-Type Lamins Regulate 53BP1 Levels?

A search for mechanisms by which A-type lamins affect DNA DSB repair revealed a role for cathepsin L (CTSL) in regulating 53BP1 protein stability. CTSL is a cysteine protease ubiquitously expressed in tissues that is synthesized as an inactive zymogen which undergoes autoproteolytic processing within the lysosomal/endosomal compartment to produce the mature active form [90]. In addition to its classical localization in the lysosomes, CTSL has been found in the nucleus [91] and secreted into the extracellular space [92]. Extracellular CTSL has been reported in numerous types of cancer and is often associated with increased invasiveness and metastasis [93–95].

The first association between CTSL and A-type lamins was established in *Zmpste24*$^{-/-}$ mice, which present with a marked increase in CTSL transcript levels [28]. Studies on *Lmna*$^{-/-}$ mice showed a marked increase in the levels of CTSL mRNA and protein, suggesting that loss of A-type lamins induces transcriptional upregulation of CTSL or increased stability of its transcripts [89]. Moreover, the increase in CTSL is directly responsible for the decrease in 53BP1 protein levels, since depletion of CTSL via shRNAs restored 53BP1 protein levels in *Lmna*$^{-/-}$ MEFs. Importantly, depletion of CTSL rescued the fast phase of repair of IR-induced DNA DSBs as well as NHEJ of dysfunctional telomeres. Interestingly, depletion of both 53BP1 and CTSL prevented restoration of NHEJ, demonstrating that the rescue is due to stabilization of 53BP1 (Fig. 8).

The regulatory effect of CTSL on 53BP1 and thus in DNA repair is not restricted to lamin A/C-deficient cells. A recent study shows that loss of the tumor suppressor DNA repair factor BRCA1 leads to upregulation of CTSL and degradation of 53BP1 protein [96]. Interestingly, activation of CTSL-mediated degradation of 53BP1 allows the bypass of the characteristic growth arrest induced by BRCA1 loss. Moreover, increased levels of nuclear CTSL were observed in human tumors with the poorest prognosis, including sporadic triple negative breast cancers and tumors from women carrying germline mutations in the BRCA1 gene. Importantly, upregulation of CTSL was correlated with decreased levels of 53BP1 in these tumors. These studies suggest that upregulation of CTSL activity is a novel mechanism contributing to genomic instability in some laminopathies and some types of cancer, which could be targeted with therapeutic purposes.

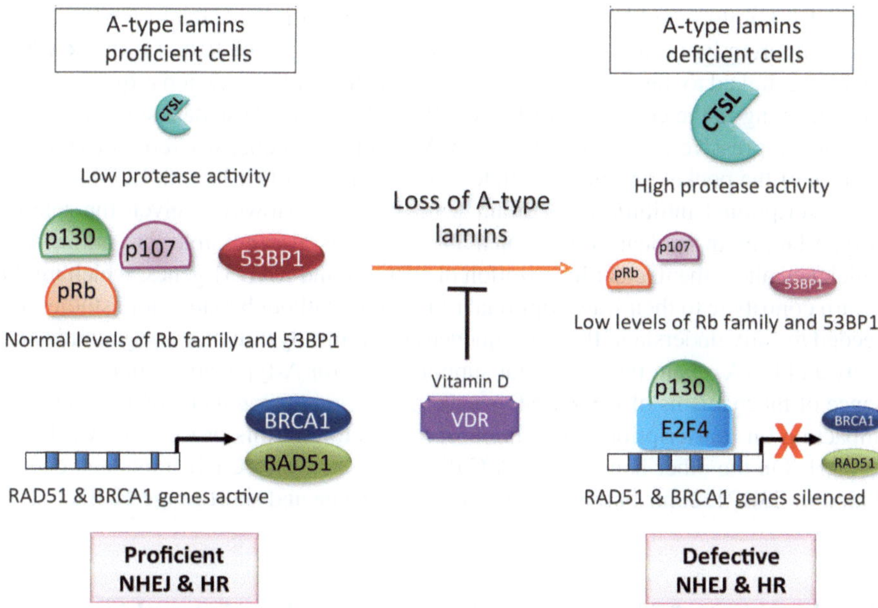

Fig. 8 Model of roles of A-type lamins in DNA repair. Loss of A-type lamins upregulates CTSL expression, resulting in elevated protein levels both in the nucleus and in the lysosomes. CTSL participates in the degradation of 53BP1, and the Rb family members pRb and p107, favoring the formation of p130/E2F4 repressor complexes, which in turn inhibit BRCA1 and RAD51 gene expression. The decrease in 53BP1 hinders NHEJ and the decrease in BRCA1 and RAD51 impairs HR repair. Other factors are likely to contribute to genomic instability in different laminopathies such as the accumulation of XPA at DSBs, the deficiency in DNA-PK, the increase in generation of ROS, epigenetic alterations, and loss of telomere homeostasis (not shown in the model). Importantly, vitamin D treatment inhibits CTSL-mediated degradation of 53BP1 and upregulates transcription of BRCA1 and RAD51, thus providing a potential therapeutic strategy

Defects in HR Upon Loss of A-Type Lamins

Based on the theory of competition between NHEJ and HR for the repair of DNA DSBs [86, 97], one could envision that loss of 53BP1-dependent NHEJ in lamin A/C-deficient cells is compensated by an increase in HR. However, studies in MCF7 cells carrying a reporter construct for monitoring HR demonstrated that depletion of A-type lamins leads to a 40 % decrease in HR, despite a marked downregulation of 53BP1 [78]. Experiments aiming to gain insight into mechanisms behind this defect revealed that depletion of lamins A/C results in transcriptional repression of the BRCA1 and RAD51 genes, which encode proteins essential for HR [78]. Previous reports had shown transcriptional repression of BRCA1 and RAD51 under stresses such as hypoxia, via formation of p130/E2F4 repressor complexes at their promoters [98, 99]. Interestingly, loss of A-type lamins is associated with increased degradation of the Rb family members pRb and p107, with only a minor effect on p130 [100, 101]. Importantly, in lamin A/C-deficient cells, repression of BRCA1 and RAD51 genes was linked to the status of Rb family members, with repression requiring p130 and occurring in the context of pRb and p107 deficiency. These data suggest activation of a repressive mechanism in lamin A/C-deficient cells, whereby altering the balance of the pocket proteins facilitates the association of p130 with E2F4, leading to transcriptional inhibition of certain genes (Fig. 8). However, given the role of A-type lamins in nuclear compartmentalization, it is also possible that lamin A/C deficiency alters the nuclear localization of BRCA1 and RAD51 genes, which might in turn contribute to their transcriptional repression. Although many more studies are needed to fully understand the consequences of lamins dysfunction for genomic stability and DNA repair, the data so far support a role for A-type lamins in the maintenance of mechanisms of DNA DSB repair. A-type lamins regulate HR indirectly by impacting on transcriptional regulation of key factors in this process, BRCA1 and RAD51. On the other hand, lamin A/C deficiency activates CTSL-mediated degradation of 53BP1, contributing to defects in long-range and short-range NHEJ.

Development of Therapeutic Strategies for Laminopathies

The progress made in recent years about the molecular mechanisms contributing to the pathogenesis of laminopathies has expedited the search for therapeutic strategies that could ameliorate these diseases. Preclinical studies on mouse models of laminopathies have identified farnesyltransferase inhibitors (FTIs) and prenylation inhibitors, such as statins and bisphosphonates, as beneficial treatments for certain lamin-related diseases [102, 103]. Cell-based assays have also suggested that rapamycin, scavengers of ROS, and vitamin D could have a protective role against increased genomic instability brought about by lamin dysfunction.

Farnesyltransferase Inhibitors, Statins and Bisphosphonates

Several lines of evidence indicate that the farnesylated forms of prelamin A and progerin cause the phenotypes associated with progeroid syndromes [104]. FTIs block prelamin A and progerin farnesylation and translocate these proteins from the nuclear envelope to the nucleoplasm [23, 105, 106]. FTI treatment of cells from HGPS knock-in and Zmpste24 knockout mouse models, as well as HGPS and Face1/Zmpste24-deficient human cells improved nuclear abnormalities, albeit not being efficient in reducing defects in DNA repair. In addition, treatment with FTIs ameliorates the disease phenotypes of *Zmpste24*$^{-/-}$ mice and improves their longevity, although there is considerable variation in the effect in different progeria mouse models [107, 108]. Based on these exciting results, the first clinical drug trial for children with progeria was started on May 2007. The FTI lonafarnib has proven some effectiveness for progeria, with patients showing improvement in rate of weight gain, hearing, bone structure, and flexibility of blood vessels [109]. For information see http://www.progeriaresearch.org/. However, the improvements were minor and the FTI treatment is far from curing the syndrome. Thus, the main challenge in the field of progeria is to find additional strategies to improve such dramatic phenotypes, which requires a detailed understanding of the molecular pathways affected.

In 2008, a study showed that some prelamin A undergoes geranylgeranylation when farnesylation is inhibited with FTIs [103]. This can explain the marginal efficiency of FTIs improving the overall progeria phenotype. Interestingly, a combination of two common drugs blocking several steps of the farnesyl pyrophosphate and geranylgeranyl pyrophosphate biosynthetic pathway, statins and aminobisphosphonates, have been shown to significantly improve aging-like symptoms and longevity in *Zmpste24*$^{-/-}$ mice [103]. Interestingly, these drugs improved nuclear morphology and also DNA damage, as assessed by decreased levels of γ-H2AX foci. These studies indicate that these drugs are likely to be more efficient than FTIs in ameliorating the phenotype of progeria patients. A follow-up trial is now in place in 45 children with HGPS being treated with lonafarnib in combination with a bisphosphonate (zoledronate) and a statin (pravastatin), with the hope that this regimen will be more efficient than FTI alone.

Novel Strategies: Vitamin D, Rapamycin, and ROS Scavengers

The identification of CTSL-mediated degradation of 53BP1 as a contributor to genomic instability in *Lmna*$^{-/-}$ fibroblasts prompted the exploration of strategies to inhibit this protease, with the goal of rescuing the DNA repair defects in these cells. As expected, treatment of A-type lamins-deficient cells with cathepsin inhibitors rescued the levels of 53BP1, as well as defects in DNA damage repair [89]. Studies in colon cancer models had previously shown that vitamin D exerts an inhibitory effect on cathepsins. The mechanism underlying this effect seems to be the transcriptional upregulation by vitamin D receptor (VDR) of cystatin D, an endogenous inhibitor of cathepsins, including CTSL [110]. Consistent with this notion, treatment of *Lmna*$^{-/-}$ MEFs with vitamin D inhibited CTSL-mediated degradation of 53BP1, rescued long-range and short-range NHEJ events, while causing a reduction in the percentage of cells presenting with nuclear morphological abnormalities and basal unrepaired DNA damage [89]. These cell-based studies suggest that vitamin D could represent a novel, safer, and cost-effective therapeutic strategy for lamin-related diseases that present with defects in DNA repair and genomic instability. Preclinical studies with vitamin D alone or in combination with FTIs or prenylation inhibitors in mouse models of laminopathies need to be performed to test the efficacy of such therapeutic regimens.

A recent report suggested that rapamycin, an immunosuppressant drug, reduces nuclear defects in fibroblasts from HGPS patients and prolongs cellular lifespan.

The mechanism behind this improvement is the increased clearance of toxic progerin by autophagy [111]. It will be interesting to determine whether rapamycin is capable of reducing disease phenotypes and extending life in mouse models of laminopathies.

New evidence has also linked lamin dysfunction with increased generation of ROS and reduced levels of antioxidant enzymes [112, 113]. In turn, persistent ROS can oxidize some cysteine residues in lamins, causing alterations in the structure of the nuclear lamina [113]. The accumulation of ROS could play a role in the increased levels of DNA damage and the genomic instability observed in lamin-deficient cells. This notion is supported by the fact that treatment of progeroid cells with the ROS scavenger *N*-acetyl cysteine reduces the amount of unrepairable DNA damage [62]. Thus, compounds that reduce the levels of ROS in the cell could represent yet another strategy to reduce genomic instability in lamin-related diseases [114].

Concluding Remarks

In recent years, interest in A-type lamins has increased due to the association of alterations of these structural nuclear proteins with a variety of human diseases, including severe premature aging syndromes and cancer. The molecular mechanisms contributing to the phenotypes of lamin-related diseases are beginning to be uncovered. New evidence indicates that A-type lamins play a key role maintaining telomere localization, structure, length and function, keeping in check the levels of ROS, as well as regulating the levels of factors participating in cell cycle regulation (Rb family members) and DNA repair (53BP1, BRCA1, RAD51, XPA). Understanding how lamins maintain genome integrity will surely provide new therapeutic strategies towards laminopathies and potentially cancer.

References

1. Cremer T et al (2006) Chromosome territories—a functional nuclear landscape. Curr Opin Cell Biol 18(3):307–316
2. Misteli T (2007) Beyond the sequence: cellular organization of genome function. Cell 128(4):787–800
3. Schneider R, Grosschedl R (2007) Dynamics and interplay of nuclear architecture, genome organization, and gene expression. Genes Dev 21(23):3027–3043
4. Broers JL et al (2006) Nuclear lamins: laminopathies and their role in premature ageing. Physiol Rev 86(3):967–1008
5. Dechat T et al (2008) Nuclear lamins: major factors in the structural organization and function of the nucleus and chromatin. Genes Dev 22(7):832–853
6. Taddei A et al (2004) The function of nuclear architecture: a genetic approach. Annu Rev Genet 38:305–345
7. Bridger JM et al (1993) Internal lamin structures within G1 nuclei of human dermal fibroblasts. J Cell Sci 104(Pt 2):297–306

8. Gerace L, Blum A, Blobel G (1978) Immunocytochemical localization of the major polypeptides of the nuclear pore complex-lamina fraction. Interphase and mitotic distribution. J Cell Biol 79(2 Pt 1):546–566
9. Goldman AE et al (1992) Pathway of incorporation of microinjected lamin A into the nuclear envelope. J Cell Biol 119(4):725–735
10. Hozak P et al (1995) Lamin proteins form an internal nucleoskeleton as well as a peripheral lamina in human cells. J Cell Sci 108(Pt 2):635–644
11. McKeon FD, Kirschner MW, Caput D (1986) Homologies in both primary and secondary structure between nuclear envelope and intermediate filament proteins. Nature 319(6053):463–468
12. Guelen L et al (2008) Domain organization of human chromosomes revealed by mapping of nuclear lamina interactions. Nature 453(7197):948–951
13. Goldman RD et al (2002) Nuclear lamins: building blocks of nuclear architecture. Genes Dev 16(5):533–547
14. Gruenbaum Y et al (2005) The nuclear lamina comes of age. Nat Rev Mol Cell Biol 6(1):21–31
15. Stuurman N, Heins S, Aebi U (1998) Nuclear lamins: their structure, assembly, and interactions. J Struct Biol 122(1–2):42–66
16. Lin F, Worman HJ (1993) Structural organization of the human gene encoding nuclear lamin A and nuclear lamin C. J Biol Chem 268(22):16321–16326
17. Beck LA, Hosick TJ, Sinensky M (1990) Isoprenylation is required for the processing of the lamin A precursor. J Cell Biol 110(5):1489–1499
18. Bergo MO et al (2002) Zmpste24 deficiency in mice causes spontaneous bone fractures, muscle weakness, and a prelamin A processing defect. Proc Natl Acad Sci U S A 99(20):13049–13054
19. Corrigan DP et al (2005) Prelamin A endoproteolytic processing in vitro by recombinant Zmpste24. Biochem J 387(Pt 1):129–138
20. Pendas AM et al (2002) Defective prelamin A processing and muscular and adipocyte alterations in Zmpste24 metalloproteinase-deficient mice. Nat Genet 31(1):94–99
21. Schirmer EC, Foisner R (2007) Proteins that associate with lamins: many faces, many functions. Exp Cell Res 313(10):2167–2179
22. Zastrow MS, Vlcek S, Wilson KL (2004) Proteins that bind A-type lamins: integrating isolated clues. J Cell Sci 117(Pt 7):979–987
23. Capell BC, Collins FS (2006) Human laminopathies: nuclei gone genetically awry. Nat Rev Genet 7(12):940–952
24. Prokocimer M et al (2009) Nuclear lamins: key regulators of nuclear structure and activities. J Cell Mol Med 13(6):1059–1085
25. De Sandre-Giovannoli A et al (2003) Lamin a truncation in Hutchinson-Gilford progeria. Science 300(5628):2055
26. Eriksson M et al (2003) Recurrent de novo point mutations in lamin A cause Hutchinson-Gilford progeria syndrome. Nature 423(6937):293–298
27. Hennekam RC (2006) Hutchinson-Gilford progeria syndrome: review of the phenotype. Am J Med Genet A 140(23):2603–2624
28. Varela I et al (2005) Accelerated ageing in mice deficient in Zmpste24 protease is linked to p53 signalling activation. Nature 437(7058):564–568
29. McClintock D et al (2007) The mutant form of lamin A that causes Hutchinson-Gilford progeria is a biomarker of cellular aging in human skin. PLoS One 2(12):e1269
30. Scaffidi P, Misteli T (2006) Lamin A-dependent nuclear defects in human aging. Science 312(5776):1059–1063
31. Agrelo R et al (2005) Inactivation of the lamin A/C gene by CpG island promoter hypermethylation in hematologic malignancies, and its association with poor survival in nodal diffuse large B-cell lymphoma. J Clin Oncol 23(17):3940–3947
32. Broers JL et al (1993) Nuclear A-type lamins are differentially expressed in human lung cancer subtypes. Am J Pathol 143(1):211–220

33. Prokocimer M, Margalit A, Gruenbaum Y (2006) The nuclear lamina and its proposed roles in tumorigenesis: projection on the hematologic malignancies and future targeted therapy. J Struct Biol 155(2):351–360
34. Willis ND et al (2008) Lamin A/C is a risk biomarker in colorectal cancer. PLoS One 3(8):e2988
35. Lindahl T (1993) Instability and decay of the primary structure of DNA. Nature 362(6422):709–715
36. Hoeijmakers JH (2001) Genome maintenance mechanisms for preventing cancer. Nature 411(6835):366–374
37. Jackson SP, Bartek J (2009) The DNA-damage response in human biology and disease. Nature 461(7267):1071–1078
38. Hoeijmakers JH (2009) DNA damage, aging, and cancer. N Engl J Med 361(15):1475–1485
39. Hanahan D, Weinberg RA (2011) Hallmarks of cancer: the next generation. Cell 144(5):646–674
40. Khanna KK, Jackson SP (2001) DNA double-strand breaks: signaling, repair and the cancer connection. Nat Genet 27(3):247–254
41. Zhou BB, Elledge SJ (2000) The DNA damage response: putting checkpoints in perspective. Nature 408(6811):433–439
42. Ciccia A, Elledge SJ (2010) The DNA damage response: making it safe to play with knives. Mol Cell 40(2):179–204
43. Hartlerode AJ, Scully R (2009) Mechanisms of double-strand break repair in somatic mammalian cells. Biochem J 423(2):157–168
44. Mao Z et al (2008) Comparison of nonhomologous end joining and homologous recombination in human cells. DNA Repair (Amst) 7(10):1765–1771
45. Wyman C, Kanaar R (2006) DNA double-strand break repair: all's well that ends well. Annu Rev Genet 40:363–383
46. Shibata A et al (2011) Factors determining DNA double-strand break repair pathway choice in G2 phase. EMBO J 30(6):1079–1092
47. Lee JH, Paull TT (2004) Direct activation of the ATM protein kinase by the Mre11/Rad50/Nbs1 complex. Science 304(5667):93–96
48. Lee JH, Paull TT (2005) ATM activation by DNA double-strand breaks through the Mre11-Rad50-Nbs1 complex. Science 308(5721):551–554
49. Bakkenist CJ, Kastan MB (2003) DNA damage activates ATM through intermolecular autophosphorylation and dimer dissociation. Nature 421(6922):499–506
50. Khanna KK et al (2001) ATM, a central controller of cellular responses to DNA damage. Cell Death Differ 8(11):1052–1065
51. Burma S et al (2001) ATM phosphorylates histone H2AX in response to DNA double-strand breaks. J Biol Chem 276(45):42462–42467
52. Fernandez-Capetillo O et al (2002) DNA damage-induced G2-M checkpoint activation by histone H2AX and 53BP1. Nat Cell Biol 4(12):993–997
53. Stewart GS et al (2003) MDC1 is a mediator of the mammalian DNA damage checkpoint. Nature 421(6926):961–966
54. Kolas NK et al (2007) Orchestration of the DNA-damage response by the RNF8 ubiquitin ligase. Science 318(5856):1637–1640
55. Mailand N et al (2007) RNF8 ubiquitylates histones at DNA double-strand breaks and promotes assembly of repair proteins. Cell 131(5):887–900
56. Doil C et al (2009) RNF168 binds and amplifies ubiquitin conjugates on damaged chromosomes to allow accumulation of repair proteins. Cell 136(3):435–446
57. Galanty Y et al (2009) Mammalian SUMO E3-ligases PIAS1 and PIAS4 promote responses to DNA double-strand breaks. Nature 462(7275):935–939
58. Morris JR et al (2009) The SUMO modification pathway is involved in the BRCA1 response to genotoxic stress. Nature 462(7275):886–890
59. Wang B et al (2002) 53BP1, a mediator of the DNA damage checkpoint. Science 298(5597):1435–1438

60. Scully R et al (1997) Association of BRCA1 with Rad51 in mitotic and meiotic cells. Cell 88(2):265–275
61. Liu B et al (2005) Genomic instability in laminopathy-based premature aging. Nat Med 11(7):780–785
62. Richards SA et al (2011) The accumulation of un-repairable DNA damage in laminopathy progeria fibroblasts is caused by ROS generation and is prevented by treatment with N-acetyl cysteine. Hum Mol Genet 20(20):3997–4004
63. di Masi A et al (2008) The R527H mutation in LMNA gene causes an increased sensitivity to ionizing radiation. Cell Cycle 7(13):2030–2037
64. Liu Y et al (2008) Involvement of xeroderma pigmentosum group A (XPA) in progeria arising from defective maturation of prelamin A. FASEB J 22(2):603–611
65. Liu GH et al (2011) Recapitulation of premature ageing with iPSCs from Hutchinson-Gilford progeria syndrome. Nature 472(7342):221–225
66. Pegoraro G et al (2009) Ageing-related chromatin defects through loss of the NURD complex. Nat Cell Biol 11(10):1261–1267
67. Krishnan V et al (2011) Histone H4 lysine 16 hypoacetylation is associated with defective DNA repair and premature senescence in Zmpste24-deficient mice. Proc Natl Acad Sci U S A 108(30):12325–12330
68. de Lange T (2002) Protection of mammalian telomeres. Oncogene 21(4):532–540
69. Smogorzewska A et al (2002) DNA ligase IV-dependent NHEJ of deprotected mammalian telomeres in G1 and G2. Curr Biol 12(19):1635–1644
70. Raz V et al (2008) The nuclear lamina promotes telomere aggregation and centromere peripheral localization during senescence of human mesenchymal stem cells. J Cell Sci 121(Pt 24):4018–4028
71. Shoeman RL et al (1988) The binding in vitro of the intermediate filament protein vimentin to synthetic oligonucleotides containing telomere sequences. J Biol Chem 263(35): 18744–18749
72. Gonzalez-Suarez I et al (2009) Novel roles for A-type lamins in telomere biology and the DNA damage response pathway. EMBO J 28(16):2414–2427
73. De Vos WH et al (2010) Increased plasticity of the nuclear envelope and hypermobility of telomeres due to the loss of A-type lamins. Biochim Biophys Acta 1800(4):448–458
74. Huang S et al (2008) Accelerated telomere shortening and replicative senescence in human fibroblasts overexpressing mutant and wild-type lamin A. Exp Cell Res 314(1):82–91
75. Allsopp RC et al (1992) Telomere length predicts replicative capacity of human fibroblasts. Proc Natl Acad Sci U S A 89(21):10114–10118
76. Decker ML et al (2009) Telomere length in Hutchinson-Gilford progeria syndrome. Mech Ageing Dev 130(6):377–383
77. Kudlow BA et al (2008) Suppression of proliferative defects associated with processing-defective lamin A mutants by hTERT or inactivation of p53. Mol Biol Cell 19(12): 5238–5248
78. Redwood AB et al (2011) A dual role for A-type lamins in DNA double-strand break repair. Cell Cycle 10(15):2549–2560
79. Dimitrova N et al (2008) 53BP1 promotes non-homologous end joining of telomeres by increasing chromatin mobility. Nature 456(7221):524–528
80. Jahn D et al (2012) A truncated lamin A in the Lmna-/- mouse line: implications for the understanding of laminopathies. Nucleus 3(5):463–474
81. Sullivan T et al (1999) Loss of A-type lamin expression compromises nuclear envelope integrity leading to muscular dystrophy. J Cell Biol 147(5):913–920
82. Difilippantonio S et al (2008) 53BP1 facilitates long-range DNA end-joining during V(D)J recombination. Nature 456(7221):529–533
83. Manis JP et al (2004) 53BP1 links DNA damage-response pathways to immunoglobulin heavy chain class-switch recombination. Nat Immunol 5(5):481–487
84. Ward IM et al (2004) 53BP1 is required for class switch recombination. J Cell Biol 165(4):459–464

85. Bothmer A et al (2010) 53BP1 regulates DNA resection and the choice between classical and alternative end joining during class switch recombination. J Exp Med 207(4):855–865
86. Bunting SF et al (2010) 53BP1 inhibits homologous recombination in Brca1-deficient cells by blocking resection of DNA breaks. Cell 141(2):243–254
87. Olive PL, Banath JP, Durand RE (1990) Heterogeneity in radiation-induced DNA damage and repair in tumor and normal cells measured using the "comet" assay. Radiat Res 122(1):86–94
88. Iliakis G et al (2004) Mechanisms of DNA double strand break repair and chromosome aberration formation. Cytogenet Genome Res 104(1–4):14–20
89. Gonzalez-Suarez I et al (2011) A new pathway that regulates 53BP1 stability implicates Cathepsin L and vitamin D in DNA repair. EMBO J 30(16):3383–3396
90. Katunuma N (1989) Mechanisms and regulation of lysosomal proteolysis. Revis Biol Celular 20:35–61
91. Goulet B et al (2004) A cathepsin L isoform that is devoid of a signal peptide localizes to the nucleus in S phase and processes the CDP/Cux transcription factor. Mol Cell 14(2):207–219
92. Reiser J, Adair B, Reinheckel T (2010) Specialized roles for cysteine cathepsins in health and disease. J Clin Invest 120(10):3421–3431
93. Gocheva V, Joyce JA (2007) Cysteine cathepsins and the cutting edge of cancer invasion. Cell Cycle 6(1):60–64
94. Jedeszko C, Sloane BF (2004) Cysteine cathepsins in human cancer. Biol Chem 385(11):1017–1027
95. Miyamoto K et al (2011) Cathepsin L is highly expressed in gastrointestinal stromal tumors. Int J Oncol 39(5):1109–1115
96. Grotsky DA et al (2013) BRCA1 loss activates cathepsin L-mediated degradation of 53BP1 in breast cancer cells. J Cell Biol 200(2):187–202
97. Bouwman P et al (2010) 53BP1 loss rescues BRCA1 deficiency and is associated with triple-negative and BRCA-mutated breast cancers. Nat Struct Mol Biol 17(6):688–695
98. Bindra RS, Glazer PM (2007) Repression of RAD51 gene expression by E2F4/p130 complexes in hypoxia. Oncogene 26(14):2048–2057
99. Hegan DC et al (2010) Inhibition of poly(ADP-ribose) polymerase down-regulates BRCA1 and RAD51 in a pathway mediated by E2F4 and p130. Proc Natl Acad Sci U S A 107(5):2201–2206
100. Johnson BR et al (2004) A-type lamins regulate retinoblastoma protein function by promoting subnuclear localization and preventing proteasomal degradation. Proc Natl Acad Sci U S A 101(26):9677–9682
101. Nitta RT, Smith CL, Kennedy BK (2007) Evidence that proteasome-dependent degradation of the retinoblastoma protein in cells lacking A-type lamins occurs independently of gankyrin and MDM2. PLoS One 2(9):e963
102. Fong LG et al (2006) Prelamin A and lamin A appear to be dispensable in the nuclear lamina. J Clin Invest 116(3):743–752
103. Varela I et al (2008) Combined treatment with statins and aminobisphosphonates extends longevity in a mouse model of human premature aging. Nat Med 14(7):767–772
104. Osorio FG et al (2009) Accelerated ageing: from mechanism to therapy through animal models. Transgenic Res 18(1):7–15
105. Glynn MW, Glover TW (2005) Incomplete processing of mutant lamin A in Hutchinson-Gilford progeria leads to nuclear abnormalities, which are reversed by farnesyltransferase inhibition. Hum Mol Genet 14(20):2959–2969
106. Toth JI et al (2005) Blocking protein farnesyltransferase improves nuclear shape in fibroblasts from humans with progeroid syndromes. Proc Natl Acad Sci U S A 102(36): 12873–12878
107. Yang SH et al (2005) Blocking protein farnesyltransferase improves nuclear blebbing in mouse fibroblasts with a targeted Hutchinson-Gilford progeria syndrome mutation. Proc Natl Acad Sci U S A 102(29):10291–10296

108. Fong LG et al (2006) A protein farnesyltransferase inhibitor ameliorates disease in a mouse model of progeria. Science 311(5767):1621–1623
109. Gordon LB et al (2012) Clinical trial of a farnesyltransferase inhibitor in children with Hutchinson-Gilford progeria syndrome. Proc Natl Acad Sci U S A 109(41):16666–16671
110. Alvarez-Diaz S et al (2009) Cystatin D is a candidate tumor suppressor gene induced by vitamin D in human colon cancer cells. J Clin Invest 119(8):2343–2358
111. Cao K et al (2011) Rapamycin reverses cellular phenotypes and enhances mutant protein clearance in Hutchinson-Gilford progeria syndrome cells. Sci Transl Med 3(89):89ra58
112. Sieprath T, Darwiche R, De Vos WH (2012) Lamins as mediators of oxidative stress. Biochem Biophys Res Commun 421(4):635–639
113. Pekovic V et al (2011) Conserved cysteine residues in the mammalian lamin A tail are essential for cellular responses to ROS generation. Aging Cell 10(6):1067–1079
114. Lattanzi G et al (2012) Nuclear damages and oxidative stress: new perspectives for laminopathies. Eur J Histochem 56(4):e45

Repo-Man at the Intersection of Chromatin Remodelling, DNA Repair, Nuclear Envelope Organization, and Cancer Progression

Paola Vagnarelli

Abstract Nuclear structure and chromatin changes are very useful biomarkers in cancer diagnosis. Despite this, their biological significance and relevance to cancer progression are still not well understood. The identification of new proteins that link the nuclear envelope to chromatin organization and the understanding of the molecular mechanisms underlying these connections have begun to provide some important clues. This review discusses the role of the nuclear protein Repo-Man (*CDCA2*) in the maintenance of genome stability. Repo-Man (*CDCA2*) is a targeting subunit for the protein phosphatase 1 involved in the dephosphorylation of histone H3 during mitotic exit. In this role, it is important for the chromatin organization in post-mitotic nuclei. Repo-Man (*CDCA2*) is also essential for proper nuclear envelope reformation and the regulation of DNA damage responses. The relevance of this complex for cancer biology is also corroborated by emerging evidence that provides a correlation between Repo-Man (*CDCA2*) expression levels and cancer progression; several studies now suggest that Repo-Man (*CDCA2*) represents a very strong prognostic marker for poor patient survival.

Keywords Mitosis • Chromosome instability • DNA repair • Cancer progression • NUP153

Abbreviations

CDCA2	Cell division cycle associated 2
CPC	Chromosomal passenger complex
Cisplatin	*Cis*-diamminedichloroplatinum(II)
DDR	DNA damage response

P. Vagnarelli (✉)
Division of Biosciences, Brunel University, London UB83PH, UK
e-mail: Paola.Vagnarelli@brunel.ac.uk

E.C. Schirmer and J.I. de las Heras (eds.), *Cancer Biology and the Nuclear Envelope*, Advances in Experimental Medicine and Biology 773, DOI 10.1007/978-1-4899-8032-8_18,

FRAP	Fluorescence recovery after photobleaching
MT	Microtubule
OSCC	Oral squamous cell carcinoma
PP1	Protein phosphatase 1
Repo-Man	Recruits PP1 onto mitotic chromatin at anaphase
RCA	Regulator of chromatin architecture
SS	Synovial sarcoma

Introduction

Nuclear structure changes are widely used by pathologists as they currently represent useful clinical biomarkers in cancer diagnosis for a number of cancer types. In fact, tumor cells are often distinguished by the presence of a lobulated nuclear envelope and abnormalities in chromatin organization [1]. Despite these criteria being of such general and important use in cancer diagnosis, we still do not have a clear picture of the causal-effect relationships of the observed changes and how or if any of the pathways that lead to such morphological changes could also be important as drug targets for cancer treatments.

Therefore, it is important to ultimately understand the molecular basis of alterations in nuclear structure that are associated with the clinical risk for disease recurrence and progression to metastasis. These changes may play an active role in cancer progression by contributing to many of the nuclear changes observed in the functionality of a cancer cell. Nuclear matrix changes appear to be a rich source for potential cancer biomarkers and may indeed reveal important cellular clues about the cancer process and its progression. Therefore the identification of the molecular pathways that link the nuclear envelope to chromatin organization and function has become of pivotal importance. Some components are well-known players in the process and currently under detailed investigation, while others have just emerged on the scene. In this review, we discuss the discovery and function of the novel chromatin binding and nuclear envelope interacting protein Repo-Man (*CDCA2*) and its connection with cancer progression.

Repo-Man Is a Cell-Cycle Associated Protein Phosphatase 1 Targeting Subunit

Cell division cycle associated 2 (CDCA2/Repo-Man) is localized on chromosome 8 (8p21.2) and was first identified by Walker [2] as a novel putative cell-cycle associated gene in a study that used co-expression analyses of several microarray databases to identify genes that were expressed in a similar way to known cell-cycle genes. However, the function for this protein remained unknown until Trinkle-Mulcahy and Lamond found CDCA2 in a proteomic analysis of protein

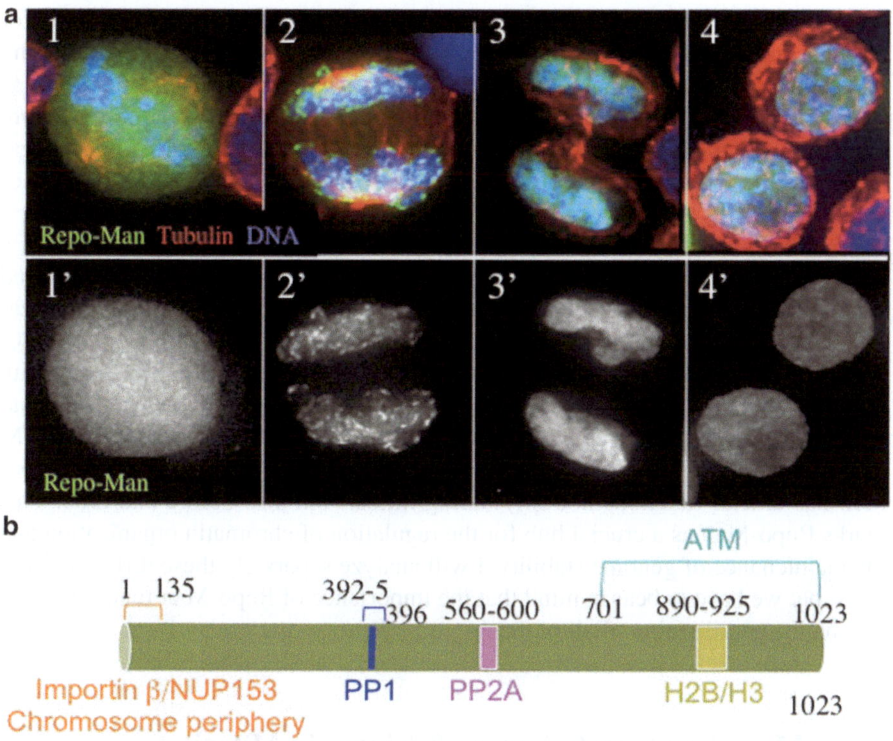

Fig. 1 Localization and interactions of Repo-Man (CDCA2). (**a**) Repo-Man is a nuclear protein in interphase (panel 4), it becomes dispersed in early mitosis (panel 1) and re-associates with the chromatin during mitotic exit (panels 2 and 3) where it localizes to the bulk of chromatin but has a separate population enriched at the periphery of the chromosomes. Repo-Man (*green*), α-Tubulin (*red*), DNA (*blue*). (**b**) Repo-Man schematic illustrating the binding domains for the interactors so far identified (see text for details)

phosphatase 1 (PP1) binding proteins. In this study, it was first shown that CDCA2 was a PP1 targeting subunit, responsible for the targeting of PP1 to chromatin in anaphase, and that it was essential for cell proliferation in vitro [3]. Because of these characteristics, it was renamed Repo-Man for "*Re*cruits *P*P1 *o*nto *M*itotic chromatin at *an*aphase."

Repo-Man is a nuclear protein in interphase that becomes dispersed in the cytoplasm at the onset of mitosis (Fig. 1a). In early mitosis (prometaphase/metaphase), it associates with chromatin in a highly dynamic manner as shown by fluorescence recovery after photobleaching analyses [4], and this dynamic behaviour is regulated by phosphorylation. At anaphase, Repo-Man becomes more tightly associated with chromatin, and it remains so during interphase where the chromosome-bound pool of Repo-Man has a very low turnover [4].

The biological importance of this protein was soon after revealed by a study of Vagnarelli and colleagues who showed that the Repo-Man/PP1 complex was a critical component of the chromatin reorganization machinery responsible for chromosome de-condensation at the transition from mitosis to G1 [5]. The targeting of Repo-Man/PP1 to anaphase chromatin is responsible for the inactivation of an as yet unknown factor, functionally termed Regulator of Chromosome Architecture (RCA), that acts in parallel with Condensin, Topoisomerase II, and Kif4 to organize/form the highly structured condensed form of chromatin in mitotic chromosomes [6–8].

In order to identify the molecular pathways that are controlled by Repo-Man, a few groups have used different approaches. The consensus from all their published studies is that Repo-Man is the phosphatase that regulates Aurora B kinase localization during mitotic progression and that dephosphorylates some chromosome-associated Aurora B substrates [9–12]. It was also found that Repo-Man interacts with nuclear membrane components, thus contributing to post-mitotic nuclear reformation [9]. Moreover, it was shown that the Repo-Man/PP1 complex modulates ATM activation, thereby setting the threshold for checkpoint activation [13]. Thus Repo-Man has crucial functions not only during mitosis but also before and after. This marks Repo-Man as a crucial hub for the regulation of chromatin organization and the maintenance of genome stability. I will analyze separately these different functions, but we have to bear in mind that the importance of Repo-Man function could rely on the coordination of all of them.

Repo-Man Counteracts Aurora B Kinase in Mitosis

Aurora B is a very important kinase for the regulation of mitotic progression. It is the catalytic subunit of the Chromosomal Passenger Complex (CPC) (for a recent review see ref. 14). In mitosis it modulates the strength of kinetochore attachments to microtubules (MTs) by differentially phosphorylating components of the kinetochore in response to tension and it is necessary for the establishment of bi-orientation and to allow correct chromosome segregation. The CPC is localized at the inner centromere of the mitotic chromosomes in metaphase, and this is due to the ability of the complex to recognize specifically the tail of Histone H3 when it is phosphorylated at Thr3. (This modification occurs only in mitosis at the inner centromeric region and it is accomplished by Haspin kinase [15–21].) This interaction is mediated by the CPC scaffolding protein Survivin. In fact, a sustained Haspin activity even after anaphase onset compromises the ability of the CPC to transfer to the spindle [18], while a premature removal of this phosphorylation by a hyperactive form of the Repo-Man/PP1 phosphatase causes a displacement of the CPC from the centromere and the increased dephosphorylation blocks the cells ability to correct inappropriate kinetochore-microtubules attachments [9, 10]. The complex interplay of this network of kinases and phosphatases at the kinetochore has become even clear based on some recent results from the Bollen laboratory. Repo-Man itself appears to be a substrate of Aurora B kinase: phosphorylation of Repo-Man at Ser893 decreases its affinity for chromatin. This site is then dephosphorylated by PP2A at anaphase onset which allows the targeting of the phosphatase back to the chromosomes where Repo-Man directs the dephosphorylation of H3Thr3ph by PP1 [12].

During mitotic exit the Repo-Man/PP1 complex also dephosphorylates Histone H3 at Ser10 and Ser28 (other Aurora B substrates) [9, 11]. This phosphatase has important functions also after cell division. Lack of Repo-Man/PP1 results in the inability to remove H3 mitotic phosphorylation which impairs binding of HP1. This in turn alters chromatin structural organization at the beginning of the new cell cycle [9]. All these results together clearly link Repo-Man to the maintenance of genome stability during cell division and they could explain why proper regulation of the expression levels of this PP1 targeting subunit is crucial for error-free mitoses.

Repo-Man Contributes to Nuclear Envelope Reassembly During Mitotic Exit

Beside its role as an important mitotic phosphatase, Repo-Man has a very peculiar localization during mitotic exit. As mentioned before, it targets to chromatin at anaphase onset where it localizes widely to the bulk of chromatin, but some becomes enriched at the periphery of the chromosomes (Figs. 1a and 2) in regions where both Importinβ and Nup153 are present. Mass spectrometry analyses of Repo-Man interactors have revealed that Repo-Man interacts directly with Importinβ and directly or indirectly with Nup153 [9] (Fig. 1b). These interactions are not present in early mitosis, but are established during mitotic exit and possibly maintained during much of interphase as well. Perturbation of these interactions by depleting Repo-Man causes major defects in nuclear envelope reorganization, thereby revealing yet another important function of the Repo-Man/PP1 complex in the establishment of a functional G1 nucleus after division.

The different chromatin sub-localizations are driven by two distinct chromatin-targeting domains in Repo-Man: the N-terminal domain is essential for the chromosome periphery targeting (aa 1–135), and the C-terminal domain (aa 560–9,250) is responsible for targeting to the bulk of chromatin (Fig. 1b). This enrichment of the protein in distinct compartments is established during mitotic exit but is maintained in interphase as well where a distinction in targeting is observed for the fragments between the nuclear periphery and the nucleoplasm (Fig. 2a, b). However, it is unclear whether the population of Repo-Man at the periphery of mitotic chromosomes directly corresponds to the same population/interactions responsible for targeting to the nuclear envelope/periphery in interphase. Another open question is whether these different enrichments in interphase have functional relevance. We presently do not know the chromatin binding sites for Repo-Man and the chromatin landscape that the binding of this phosphatase establishes; however, both aspects are pivotal toward the understanding of its complex functions in the cell cycle.

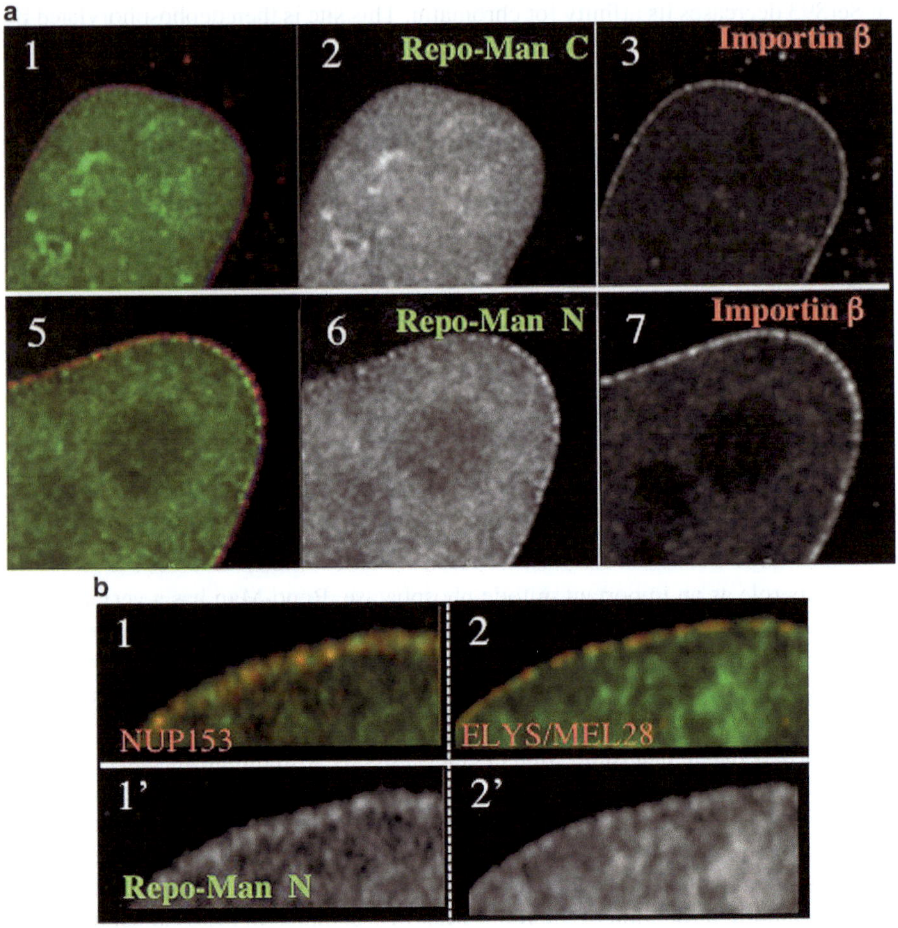

Fig. 2 Different chromatin-targeting domains localize Repo-Man in different nuclear compartments. (**a**) The C-terminal domain of Repo-Man targets the protein to the nuclear chromatin with exclusion of the nucleoli (panels 1–3), while the N-terminal domain (aa 1–135) targets the protein to the nuclear periphery (panels 5–7), as indicated by the proximity of Repo-Man (*green*) with Importinβ (*red*). (**b**) Co-staining of Repo-Man N-terminal domain (*green*) with the nucleoporins Nup153 and ELYS/MEL28

Repo-Man and the DNA Damage Response

The first link between Repo-Man/PP1 and DNA repair was provided by studies in *Xenopus* oocytes and egg extracts where it was shown to be required for setting the threshold for checkpoint activation after DNA damage. In this system, Repo-Man interacts with ATM (Fig. 1b) and co-localizes with ATM on chromatin [13]. However, the interaction between ATM and Repo-Man does not seem to be conserved in other systems as ATM was not found by proteomic analyses of the

Repo-Man interactome in DT40 and HeLa [9, 22], though it remains possible that other experimental conditions might reveal the interaction.

In the *Xenopus* system, Repo-Man appears to be responsible for targeting PP1γ to negatively regulate DNA damage-induced signal transduction. Further evidence indicates that Repo-Man overexpression reduces DNA damage-induced ATM activation, whereas a PP1 binding-deficient Repo-Man dominant-negative mutant enhances the response. By a mechanism that is not very clear at the moment (but possibly involves phosphorylation of Repo-Man itself) Repo-Man is released from the chromatin at DNA damage sites and dissociates from active ATM; this release presumably facilitates DNA damage response (DDR) activation.

These findings are very important since DDR is activated in early, pre-cancerous cells as a barrier to suppress cell proliferation and cancer progression [23], but it is reduced in late-stage cancer cells and the mechanism of this modulation in DNA damage responsiveness is unknown. In light of this, it is quite possible that overexpression of Repo-Man could result in desensitization of cells to DDR. Analyses of Repo-Man overexpression levels have revealed that indeed several but not all late-stage cancer cells have upregulated levels of Repo-Man. It has been proposed by J. Maller and collaborators that in the early stages of cancer progression DDR is activated in response to elevated genomic instability to prevent further cell proliferation [13]. If in some cells Repo-Man is upregulated, this will provide a selective growth advantage but also an increase of DNA damage, resulting in acquisition of additional mutations and further cancer progression. This implies that a reduction of Repo-Man levels should restore a normal DDR and this appears to be the case: Repo-Man depletion in advance stage cancer cells resensitizes them to the DDR and restrains their growth.

Additional support to these findings came from studies on oral squamous cell carcinoma (OSCC). Microarray analysis has shown that Repo-Man (*CDCA2*) is one of the genes upregulated in this cancer [24]. Depletion of Repo-Man in OSCC cell lines causes a decrease in cell proliferation due to cell-cycle arrest at the G1 phase. This is due to the upregulation of p21Cip1, p27Kip1, p15 INK4B, and p16INK4A and down-regulation of CDK4, CDK6, Cyclin D1, and Cyclin E [25]. In cells with wild-type p53, activated ATM phosphorylates p53 at Ser15 [26]. This stabilizes p53 so that it can induce transcription of p21Cip1 (CDK inhibitor) and prevent CDK4 and/or CDK6 and CDK2-mediated G1/S transition [27–30]. The activated ATM phosphorylates p53 at Ser46 as well, and this phosphorylation is essential for the induction of apoptosis after DNA damage [31].

The very important role of Repo-Man/PP1 in cancer progression and DNA damage signal transduction marks this complex a potentially very important target for cancer therapy. In fact, depletion of Repo-Man reactivates the DDR and blocks the abnormal proliferation of late-stage cancer cells [13]. This also suggests the possibility that lowering the levels of Repo-Man (in cancer where it is overexpressed) combined with the induction of DNA damage should drive the cells into the apoptotic pathway. This was proved to be correct, and it was shown that Repo-Man RNAi in OSCC (where Repo-Man is upregulated) combined with Cisplatin (cis-diamminedichloroplatinum(II); CDDP) treatments leads these cells into apoptosis while the control cells did not enter the apoptotic pathway [25]. This suggests that Repo-Man suppression might have a considerable potential in enhancing the therapeutic effects of irradiation and anticancer drugs that cause DNA damage, though more models must be examined in order to determine if this pathway could represent a general hit for future drug development.

Repo-Man and Cancer Progression

From the studies mentioned above, it is quite clear that the Repo-Man/PP1 complex could have a potential role in cancer progression. However, the question is what the relevance of these biological findings is in the landscape of human cancer types. Since the first identification of the gene it was clear that Repo-Man was associated with proliferative markers, but one of the first correlations between levels of expression of Repo-Man and malignant transformation came by studies of neuroblastomas [32]. Expression profiling of 103 neuroblastoma tumors revealed that Repo-Man (*CDCA2*) is among the top-scored genes that are upregulated in stage 4 neuroblastoma cancers. In collaboration with A. Sala (Brunel University, UK), we have investigated these findings further, demonstrating an increase in Repo-Man also at the protein level and showing a correlation between Repo-Man level and Myc-N expression (Fig. 3e). The Kaplain–Meyer curve for survival of neuroblastoma patients with low and high levels of Repo-Man also shows that upregulation of Repo-Man correlates with a bad prognosis (Vagnarelli and Sala, unpublished). Although we still do not know the molecular aspects that are behind this correlation, it becomes quite clear that Repo-Man profiling could add some more diagnostic and prognostic value for this particular tumor.

Another study has compared gene expression profiles from a series of melanoma cell lines representing discrete stages of malignant progression that recapitulate critical characteristics of the primary lesions from which they were derived [33]. The reported analyses have identified expression signatures associated with melanoma progression that include principally the upregulation of activators of cell-cycle progression and DNA replication/repair; Repo-Man (*CDCA2*) is part of this cohort of 18 signature genes. However, no further studies have been conducted to understand the biological significance of this correlation and its relevance with the progression of melanoma.

The theme that Repo-Man is overexpressed in late-stage cancers and correlates to a bad prognosis stands true also in other cancer types. A recent study was carried out to identify gene signatures that could be prognostic for the metastatic behavior of Synovial sarcoma (SS) [34]. SS occurs in both children and adults, although metastatic events are much more common in adults. Whereas the importance of the t(X;18) translocation in SS oncogenesis is well established, the genetic basis of SS metastasis is not clear. By comparing expression profiles of tumors with or without metastasis Repo-Man (*CDCA2*) was identified as one of the two top-ranked genes

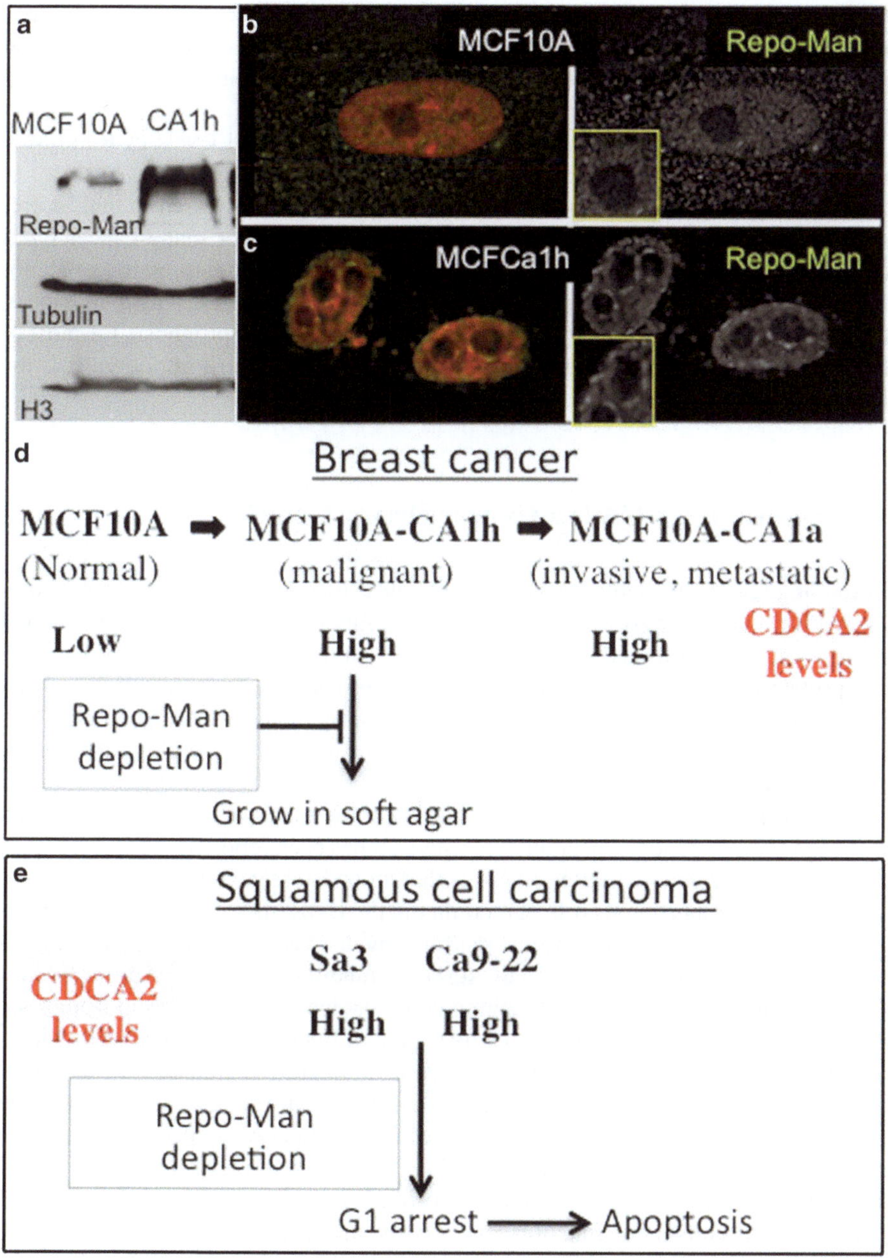

Fig. 3 Repo-Man functions in cancer progression. (**a**) Repo-Man is overexpressed in malignant breast cancer cell lines (CA1h) compared to normal breast cancer cells (MCF10A) in whole cell lysates. (**b**, **c**) In the malignant breast cancer cells, Repo-Man accumulates at the periphery of the nucleoli. (**d**) Repo-Man (*CDCA2*) levels are upregulated in breast cancer cells. Depletion of Repo-Man in malignant breast cancer cells restores the normal DNA damage response and blocks the ability of these cells to grow in soft agar [13]. (**e**) Depletion of Repo-Man in squamous cell carcinoma cell lines causes a G1 arrest and sensitizes the cells to DNA damaging agents [25]

(together with *KIF14*) for the metastatic prognosis of this cancer type. Kif14 belongs to the large family of kinesin proteins and appears to be a well-known oncogene that is overexpressed and associated with metastatic outcomes in lung [35], breast [36], ovary [37], and liver [38] carcinomas.

From the work presented above, it appears that increased levels of Repo-Man may represent a relatively common signature in human cancer biology. Analyses of Repo-Man (*CDCA2*) expression in different cancer types compared to their corresponding normal tissues clearly show this to be the case for many cancer types (Fig. 3a–e), though, interestingly, not all types show elevated expression of Repo-Man. These initial studies highlight Repo-Man as an important player in cancer, though much work is needed to test for a correlation with the prognosis for most of these cancer types, what is the molecular mechanism, and, most importantly, is the Repo-Man/PP1 complex a potential drug target for therapy?

Is Repo-Man (CDCA2) a Cancer Driver Gene?

Based on the data available so far it is quite clear that Repo-Man represents a biomarker for poor prognosis in cancer progression for at least some tumor types. What is less clear is why this is the case and which one of the functions so far revealed for this protein is the most relevant in cancer biology. The understanding of this aspect is extremely important if Repo-Man represents a drug target: we need to understand if the catalytic or the structural function of the protein is required for cancer progression.

Clearly all these cancer types and in particular their metastases are characterized by a high level of genome instability. However, it has been reported in several papers that Repo-Man can act at different levels that all promote genome instability. First, there are data supporting its role in modulating the DDR, therefore providing a platform for mutations and chromosome rearrangements. Second, it controls the regulation of chromosome segregation and Aurora B function. Impairing Aurora B leads to an increase of aneuploidy and lagging chromosomes. Lagging chromosomes are not lost in the cells but give rise to the formation of micronuclei. These phenomena, although very well known and widely used as a diagnostic signature for chromosome instability, have been recently shown to be the source of even greater genome instability [39, 40]. In this respect, increased Repo-Man levels could represent one of the means for generating the instability that characterizes late-stage cancers. Third, it has been clearly shown that Repo-Man is the phosphatase for histone H3 and that compromising its function causes problems in chromatin organization after cell division. Fourth, a change in the chromatin landscape for methylation and acetylation could well be another source of instability due to drastic changes in gene expression profiles. We do not know at the present time which type of chromatin Repo-Man binds to and how its expression levels alters the gene expression landscape of cells. It is clearly a question that needs to be addressed in the future.

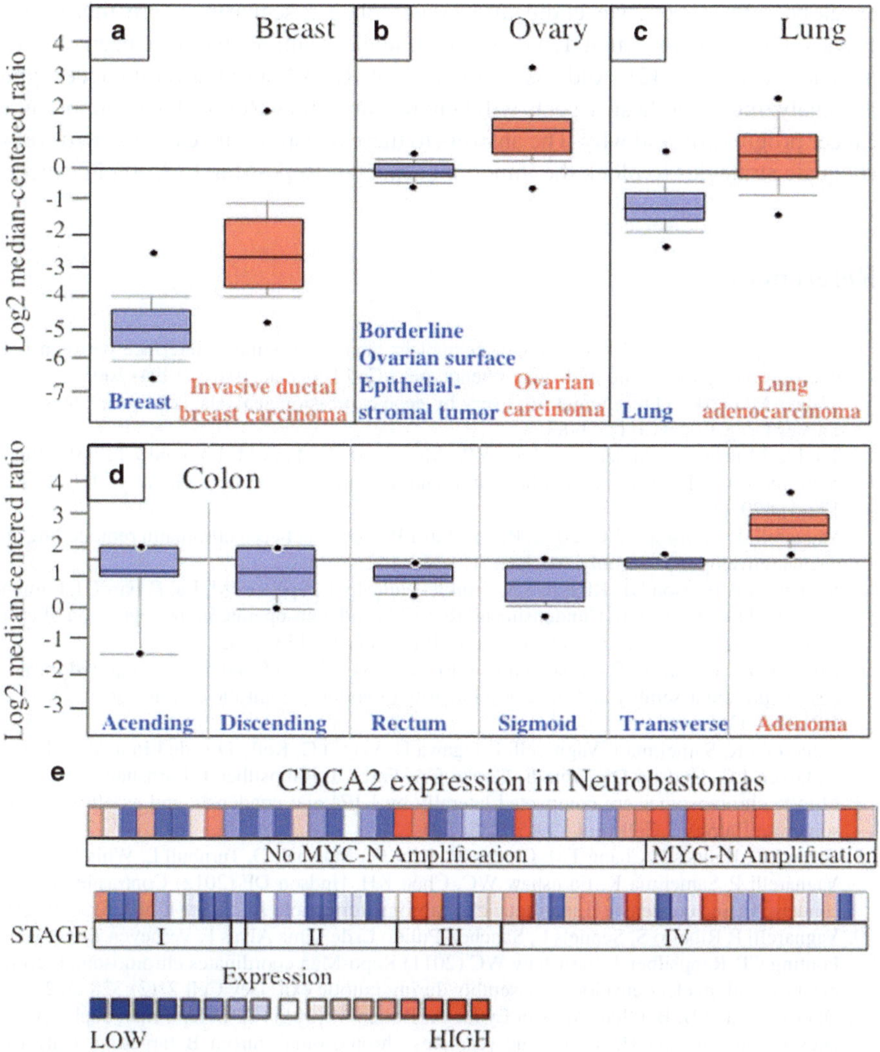

Fig. 4 Repo-Man overexpression in cancer. (**a**–**d**). Repo-Man is overexpressed in breast, ovary, lung, and colon cancers as shown by microarray data obtained in Oncomine. (**e**) Repo-Man is overexpressed in Neuroblastomas [32] and the expression levels correlate with Myc-N expression (Oncomine analyses)

Moreover, if we take a look at the localization of the protein in cancer cells where it is overexpressed, we can clearly see not just a general enhanced staining, but also a distinct pattern with accumulation to the nuclear periphery and at the periphery of the nucleoli (Fig. 4c). This is in contrast to normal interphase nuclei where it is evenly distributed in the nucleoplasm (Fig. 4b). Because it has been shown that Repo-Man binds Nup153 and Importinβ and is involved in aspects of nuclear envelope reformation, we could also contemplate a scenario where this protein plays a role in the abnormal dynamics of the nuclear lamina that have been reported in cancer cells [40–42]. Addressing the role of Repo-Man in several cancer types and analyzing all of these aspects will help to clarify how Repo-Man is important in cancer progression and why. The answers to these questions are essential to develop adequate strategies to block the important aspect of Repo-Man function in cancer.

References

1. True LD, Jordan CD (2008) The cancer nuclear microenvironment: interface between light microscopic cytology and molecular phenotype. J Cell Biochem 104(6):1994–2003
2. Walker MG (2001) Drug target discovery by gene expression analysis: cell cycle genes. Curr Cancer Drug Targets 1(1):73–83
3. Trinkle-Mulcahy L, Andersen J, Lam YW, Moorhead G, Mann M, Lamond AI (2006) Repo-Man recruits PP1 gamma to chromatin and is essential for cell viability. J Cell Biol 172(5):679–692
4. Vagnarelli P, Earnshaw WC (2012) Repo-Man-PP1: a link between chromatin remodelling and nuclear envelope reassembly. Nucleus 3(2):138–142
5. Vagnarelli P, Hudson DF, Ribeiro SA, Trinkle-Mulcahy L, Spence JM, Lai F, Farr CJ, Lamond AI, Earnshaw WC (2006) Condensin and Repo-Man-PP1 co-operate in the regulation of chromosome architecture during mitosis. Nat Cell Biol 8(10):1133–1142
6. Hudson DF, Vagnarelli P, Gassmann R, Earnshaw WC (2003) Condensin is required for non-histone protein assembly and structural integrity of vertebrate mitotic chromosomes. Dev Cell 5(2):323–336
7. Samejima K, Samejima I, Vagnarelli P, Ogawa H, Vargiu G, Kelly DA, de Lima Alves F, Kerr A, Green LC, Hudson DF, Ohta S, Cooke CA, Farr CJ, Rappsilber J, Earnshaw WC (2012) Mitotic chromosomes are compacted laterally by KIF4 and condensin and axially by topoisomerase IIalpha. J Cell Biol 199(5):755–770
8. Green LC, Kalitsis P, Chang TM, Cipetic M, Kim JH, Marshall O, Turnbull L, Whitchurch CB, Vagnarelli P, Samejima K, Earnshaw WC, Choo KH, Hudson DF (2012) Contrasting roles of condensin I and condensin II in mitotic chromosome formation. J Cell Sci 125(Pt 6):1591–1604
9. Vagnarelli P, Ribeiro S, Sennels L, Sanchez-Pulido L, de Lima Alves F, Verheyen T, Kelly DA, Ponting CP, Rappsilber J, Earnshaw WC (2011) Repo-Man coordinates chromosomal reorganization with nuclear envelope reassembly during mitotic exit. Dev Cell 21(2):328–342
10. Qian J, Lesage B, Beullens M, Van Eynde A, Bollen M (2011) PP1/Repo-man dephosphorylates mitotic histone H3 at T3 and regulates chromosomal aurora B targeting. Curr Biol 21(9):766–773
11. Wurzenberger C, Held M, Lampson MA, Poser I, Hyman AA, Gerlich DW (2012) Sds22 and Repo-Man stabilize chromosome segregation by counteracting Aurora B on anaphase kinetochores. J Cell Biol 198(2):173–183
12. Qian J, Beullens M, Lesage B, Bollen M (2013) Aurora B defines its own chromosomal targeting by opposing the recruitment of the phosphatase scaffold repo-man. Curr Biol 23(12):1136–1143
13. Peng A, Lewellyn AL, Schiemann WP, Maller JL (2010) Repo-man controls a protein phosphatase 1-dependent threshold for DNA damage checkpoint activation. Curr Biol 20(5):387–396. doi:10.1016/j.cub.2010.01.020
14. Carmena M, Wheelock M, Funabiki H, Earnshaw WC (2012) The chromosomal passenger complex (CPC): from easy rider to the godfather of mitosis. Nat Rev Mol Cell Biol 13(12):789–803

15. Dai J, Higgins JM (2005) Haspin: a mitotic histone kinase required for metaphase chromosome alignment. Cell Cycle 4(5):665–668
16. Dai J, Sullivan BA, Higgins JM (2006) Regulation of mitotic chromosome cohesion by Haspin and Aurora B. Dev Cell 11(5):741–750
17. Dai J, Sultan S, Taylor SS, Higgins JM (2005) The kinase haspin is required for mitotic histone H3 Thr 3 phosphorylation and normal metaphase chromosome alignment. Genes Dev 19(4):472–488
18. Kelly AE, Ghenoiu C, Xue JZ, Zierhut C, Kimura H, Funabiki H (2010) Survivin reads phosphorylated histone H3 threonine 3 to activate the mitotic kinase Aurora B. Science 330(6001):235–239
19. Varier RA, Outchkourov NS, de Graaf P, van Schaik FM, Ensing HJ, Wang F, Higgins JM, Kops GJ, Timmers HT (2010) A phospho/methyl switch at histone H3 regulates TFIID association with mitotic chromosomes. EMBO J 29(23):3967–3978
20. Wang F, Dai J, Daum JR, Niedzialkowska E, Banerjee B, Stukenberg PT, Gorbsky GJ, Higgins JM (2010) Histone H3 Thr-3 phosphorylation by Haspin positions Aurora B at centromeres in mitosis. Science 330(6001):231–235
21. Wang F, Ulyanova NP, van der Waal MS, Patnaik D, Lens SM, Higgins JM (2011) A positive feedback loop involving Haspin and Aurora B promotes CPC accumulation at centromeres in mitosis. Curr Biol 21(12):1061–1069
22. Prevost M, Chamousset D, Nasa I, Freele E, Morrice N, Moorhead G, Trinkle-Mulcahy L (2013) Quantitative fragmentome mapping reveals novel, domain-specific partners for the modular protein RepoMan (recruits PP1 onto mitotic chromatin at anaphase). Mol Cell Proteomics 12(5):1468–1486
23. Halazonetis TD, Gorgoulis VG, Bartek J (2008) An oncogene-induced DNA damage model for cancer development. Science 319(5868):1352–1355. doi:10.1126/science.1140735
24. Yamano Y, Uzawa K, Shinozuka K, Fushimi K, Ishigami T, Nomura H, Ogawara K, Shiiba M, Yokoe H, Tanzawa H (2008) Hyaluronan-mediated motility: a target in oral squamous cell carcinoma. Int J Oncol 32(5):1001–1009
25. Uchida F, Uzawa K, Kasamatsu A, Takatori H, Sakamoto Y, Ogawara K, Shiiba M, Bukawa H, Tanzawa H (2013) Overexpression of CDCA2 in human squamous cell carcinoma: correlation with prevention of G1 phase arrest and apoptosis. PLoS One 8(2):e56381
26. Bakkenist CJ, Kastan MB (2003) DNA damage activates ATM through intermolecular autophosphorylation and dimer dissociation. Nature 421(6922):499–506
27. Brew CT, Aronchik I, Hsu JC, Sheen JH, Dickson RB, Bjeldanes LF, Firestone GL (2006) Indole-3-carbinol activates the ATM signaling pathway independent of DNA damage to stabilize p53 and induce G1 arrest of human mammary epithelial cells. Int J Cancer 118(4): 857–868
28. He G, Siddik ZH, Huang Z, Wang R, Koomen J, Kobayashi R, Khokhar AR, Kuang J (2005) Induction of p21 by p53 following DNA damage inhibits both Cdk4 and Cdk2 activities. Oncogene 24(18):2929–2943
29. Harper JW, Elledge SJ, Keyomarsi K, Dynlacht B, Tsai LH, Zhang P, Dobrowolski S, Bai C, Connell-Crowley L, Swindell E et al (1995) Inhibition of cyclin-dependent kinases by p21. Mol Biol Cell 6(4):387–400
30. Tvrdik D, Djaborkhel R, Nagy A, Eckschlager T, Raska I, Muller J (2002) Cyclin D-cdk6 complex is targeted by p21(WAF) in growth-arrested lymphoma cells. J Struct Biol 140(1–3):49–56
31. Saito S, Goodarzi AA, Higashimoto Y, Noda Y, Lees-Miller SP, Appella E, Anderson CW (2002) ATM mediates phosphorylation at multiple p53 sites, including Ser(46), in response to ionizing radiation. J Biol Chem 277(15):12491–12494
32. Krasnoselsky AL, Whiteford CC, Wei JS, Bilke S, Westermann F, Chen QR, Khan J (2005) Altered expression of cell cycle genes distinguishes aggressive neuroblastoma. Oncogene 24(9):1533–1541
33. Ryu B, Kim DS, Deluca AM, Alani RM (2007) Comprehensive expression profiling of tumor cell lines identifies molecular signatures of melanoma progression. PLoS One 2(7):e594

34. Wozniak A, Schoffski P, Terrier P, Neuville A, Coindre JM, Italiano A, Orbach D, Debiec-Rychter M, Chibon F (2013) Chromosome instability accounts for reverse metastatic outcomes of pediatric and adult synovial sarcomas. J Clin Oncol 31(5):608–615
35. Corson TW, Zhu CQ, Lau SK, Shepherd FA, Tsao MS, Gallie BL (2007) KIF14 messenger RNA expression is independently prognostic for outcome in lung cancer. Clin Cancer Res 13(11):3229–3232
36. Corson TW, Gallie BL (2006) KIF14 mRNA expression is a predictor of grade and outcome in breast cancer. Int J Cancer 119(5):1088–1094
37. Theriault BL, Pajovic S, Bernardini MQ, Shaw PA, Gallie BL (2012) Kinesin family member 14: an independent prognostic marker and potential therapeutic target for ovarian cancer. Int J Cancer 130(8):1844–1854
38. Kim TM, Yim SH, Shin SH, Xu HD, Jung YC, Park CK, Choi JY, Park WS, Kwon MS, Fiegler H, Carter NP, Rhyu MG, Chung YJ (2008) Clinical implication of recurrent copy number alterations in hepatocellular carcinoma and putative oncogenes in recurrent gains on 1q. Int J Cancer 123(12):2808–2815
39. Crasta K, Ganem NJ, Dagher R, Lantermann AB, Ivanova EV, Pan Y, Nezi L, Protopopov A, Chowdhury D, Pellman D (2012) DNA breaks and chromosome pulverization from errors in mitosis. Nature 482(7383):53–58
40. Hatch EM, Fischer AH, Deerinck TJ, Hetzer MW (2013) Catastrophic nuclear envelope collapse in cancer cell micronuclei. Cell 154(1):47–60
41. Vargas JD, Hatch EM, Anderson DJ, Hetzer MW (2012) Transient nuclear envelope rupturing during interphase in human cancer cells. Nucleus 3(1):88–100
42. De Vos WH, Houben F, Kamps M, Malhas A, Verheyen F, Cox J, Manders EM, Verstraeten VL, van Steensel MA, Marcelis CL, van den Wijngaard A, Vaux DJ, Ramaekers FC, Broers JL (2011) Repetitive disruptions of the nuclear envelope invoke temporary loss of cellular compartmentalization in laminopathies. Hum Mol Genet 20(21):4175–4186

Nuclear Lamins and Oxidative Stress in Cell Proliferation and Longevity

Takeshi Shimi and Robert D. Goldman

Abstract In mammalian cells, the nuclear lamina is composed of a complex fibrillar network associated with the inner membrane of the nuclear envelope. The lamina provides mechanical support for the nucleus and functions as the major determinant of its size and shape. At its innermost aspect it associates with peripheral components of chromatin and thereby contributes to the organization of interphase chromosomes. The A- and B-type lamins are the major structural components of the lamina, and numerous mutations in the A-type lamin gene have been shown to cause many types of human diseases collectively known as the laminopathies. These mutations have also been shown to cause a disruption in the normal interactions between the A and B lamin networks. The impact of these mutations on nuclear functions is related to the roles of lamins in regulating various essential processes including DNA synthesis and damage repair, transcription and the regulation of genes involved in the response to oxidative stress. The major cause of oxidative stress is the production of reactive oxygen species (ROS), which is critically important for cell proliferation and longevity. Moderate increases in ROS act to initiate signaling pathways involved in cell proliferation and differentiation, whereas excessive increases in ROS cause oxidative stress, which in turn induces cell death and/or senescence. In this review, we cover current findings about the role of lamins in regulating cell proliferation and longevity through oxidative stress responses and ROS signaling pathways. We also speculate on the involvement of lamins in tumor cell proliferation through the control of ROS metabolism.

Keywords Nuclear lamins • Lamin A • Lamin B • Oxidative stress • Reactive oxygen species (ROS)

T. Shimi • R.D. Goldman (✉)
Department of Cell and Molecular Biology, Northwestern University Feinberg School of Medicine, Ward Building 11-145 303 E, Chicago Avenue, Chicago, IL 60611-3008, USA
e-mail: takeshi@northwestern.edu; r-goldman@northwestern.edu

E.C. Schirmer and J.I. de las Heras (eds.), *Cancer Biology and the Nuclear Envelope*, Advances in Experimental Medicine and Biology 773, DOI 10.1007/978-1-4899-8032-8_19,

Abbreviations

ADLD	Autosomal dominant leukodystrophy
DDR	DNA damage response
HDFs	Human diploid fibroblasts
HGPS	Hutchison-Gilford progeria syndrome
iPSCs	Inducible pluripotent stem cells
LA	Lamin A
LB1	Lamin B1
LB2	Lamin B2
LC	Lamin C
MSC	Mesenchymal stem cells
MEFs	Mouse embryonic fibroblasts
NE	Nuclear envelope
ROS	Reactive oxygen species
VSMCs	Vascular smooth muscle cells

Introduction

In mammalian cells, the nuclear lamina is a major determinant of nuclear architecture. The lamina is located at the inner membrane of the nuclear envelope (NE). The major structural components of the lamina in somatic cells are the A-type lamins (LA, LC) and the B-type lamins (LB1, LB2). LA and LC are derived from the single gene *LMNA* by alternative splicing, and LB1 and LB2 are encoded by two genes *LMNB1* and *LMNB2*, respectively [1]. In embryonic stem cells, the expression of LA and LC is low and they begin to increase at the onset of differentiation and continue to increase to relatively high levels in certain terminally differentiated cell types [2]. In contrast, LB1 and/or LB2 are expressed in all cells throughout development [2]. For example, T-cells and B-cells express only B-type lamins but not A-type lamins [1].

Lamins are type V intermediate filament proteins, which assemble into higher order filamentous structures within the peripheral lamina under the NE [3, 4]. All lamins contain a long central α-helical rod domain, flanked by globular N-terminal (head) and C-terminal (tail) domains. Many lamin subtypes are posttranslationally modified either transiently or permanently (see below). In particular, LA is transiently modified by C-terminal farnesylation and the failure to remove this farnesylation site results in nuclear defects, while B-type lamins tend to be permanently farnesylated. Electron microscopy has revealed that the lamina in *Xenopus* oocytes appears as a meshwork of ~10–15 nm filaments [5]. Lamin structures organized into meshworks have also been seen in nuclei of mouse cells by super resolution light microscopy [6]. Furthermore, it has been shown that A- and B-type lamin fibrils form separate but interacting meshworks within the lamina [7]. These lamin fibrils play important roles in assembling the lamina and contribute to the size, shape, and mechanical stability of the nucleus. Lamins are also involved in nuclear functions including chromatin organization, DNA replication, DNA repair, and transcription [7–10]. With respect to chromatin organization, the lamins provide anchorage sites for peripheral elements of heterochromatin, which are involved in the local regulation of gene expression [11–13]. Interestingly, silencing LB1 expression in HeLa cells dramatically alters the structure of the LA/C meshworks and induces LA/C-enriched NE blebs [7] that contain transcriptionally inactive gene-rich euchromatin in cancer cells [7].

The functional importance of lamins is further supported by the finding that structural changes in the lamina are among the most dramatic hallmarks of differentiation, cancer and aging and that numerous mutations in the *LMNA* gene are now known to be responsible for a wide range of genetic disorders called laminopathies. These combined studies suggest that lamins play important roles as key regulators of epigenetic events that may be critical in cellular stress responses. In particular, knowledge is accumulating to show an interdependence between oxidative stress and lamins. For example, oxidative stress modulates the expression and posttranslational modification of lamins. Conversely, mutations of lamin genes and depletion of lamins affect oxidative stress responses. Reactive oxygen species (ROS), major products of oxidative stress, are natural by-products of mitochondrial respiration which are normally eliminated in protective mechanisms such as antioxidant defenses [14–16]. Moderate increases in ROS act as a signaling mechanism to promote cell proliferation and differentiation [14–16]. However, excessive increases in ROS cause damage to DNA, proteins, and lipids, resulting in defects in proliferation and longevity that have been linked to cardiovascular and neurodegenerative diseases, as well as chronic inflammation [17]. Importantly, it is now becoming evident that lamins are involved in modulating ROS to regulate proliferation and longevity.

Here, we discuss current knowledge regarding the involvement of lamins in oxidative stress, cell proliferation, and longevity. Specifically, we focus our attention on the role of lamins in mediating cell proliferation and longevity through oxidative stress responses and ROS signaling pathways. We also consider the possible involvement of this nexus in tumor proliferation.

The Expression and Stability of Lamin Proteins Is Modulated by Oxidative Stress

Several studies have indicated that the expression and stability of lamin proteins is altered in response to oxidative stress, which in turn is tightly coupled to cell proliferation, cellular senescence, apoptosis, and autophagy.

Lamin expression is regulated by the tumor suppressors p53 and retinoblastoma protein (pRb) and by telomere functions; all master regulators of the cell cycle, apoptosis, replicative senescence, and autophagy. For example, LA/C expression is significantly upregulated upon the activation of p53 [18]. The LA mutant progerin, which causes the premature aging disease Hutchinson-Gilford Progeria Syndrome

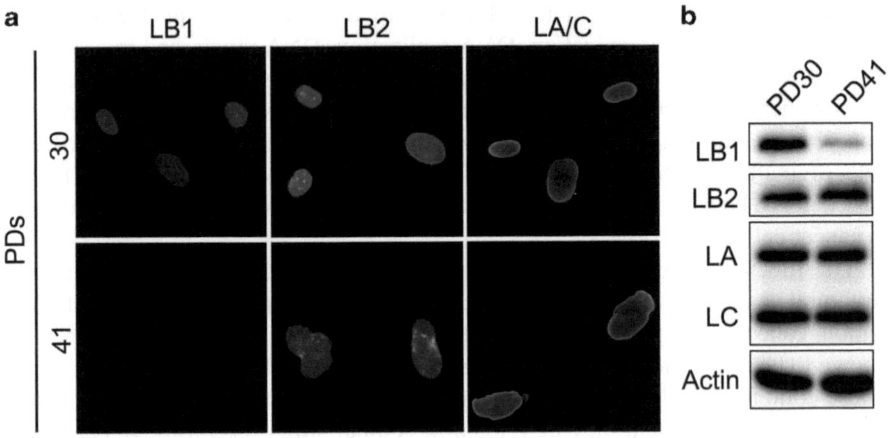

Fig. 1 The distribution and expression of lamin proteins were determined at population doublings (PD) 30 and 41 in WI-38 human embryonic lung fibroblasts. These cells are proliferating at PD 30 and senescent by PD 41. (**a**) LA/C, LB1, and LB2 are localized by immunofluorescence. (**b**) The expression of LA/C, LB1, and LB2 was determined by immunoblotting. The expression of LB1 but not LA/C or LB2 was significantly decreased during replicative senescence (permission to reproduce these data from Cold Spring Harbor Lab Press [22])

(HGPS) [19] is also expressed during normal aging [20]. Progerin expression is induced by telomere dysfunctions [21]. In contrast, the expression of LB1, but not LB2, is significantly down-regulated during senescence induced by replicative exhaustion, DNA damage, and oncogenic stress [22–24] (Fig. 1). A decrease in LB1 expression has also been observed in HGPS and in atypical progeroid syndromes caused by different mutations in the *LMNA* gene [25–27]. This decrease in LB1 expression is specifically coupled to senescence, since it does not occur in quiescence induced by serum depletion [22, 23]. The activation of pRb is required for the decrease in LB1 expression in senescence [22], and this is attributable to the fact that the *LMNB1* gene is a downstream target of the pRb–E2F pathway [28]. Based on these findings, it would be predicted that LA/C expression increases and LB1 expression decreases with no change in expression of progerin and LB2 when oxidative stress activates p53 and pRb [22] (Fig. 2). However, it has been reported that fibroblasts derived from patients with the recessive autosomal genetic disorder ataxia telangiectasia show an increase in LB1 expression in response to oxidative stress mediated by the activation of p38 mitogen-activated protein (MAP) kinase [29]. It is therefore, possible that the mutation causing ataxia telangiectasia alters the pathways required for regulating LB1 expression levels.

In addition to the transcriptional regulation of lamin expression described above, other studies have indicated that lamin levels are affected by directed degradation. Rapamycin induces autophagic protein degradation by inactivating the mammalian target of rapamycin (mTOR) pathway [30]. In this fashion, the non-farnesylated form of LA, premature LA (pre-LA), and progerin are degraded by rapamycin treatment [31, 32]. Therefore, it could be that oxygen tension which is upstream of

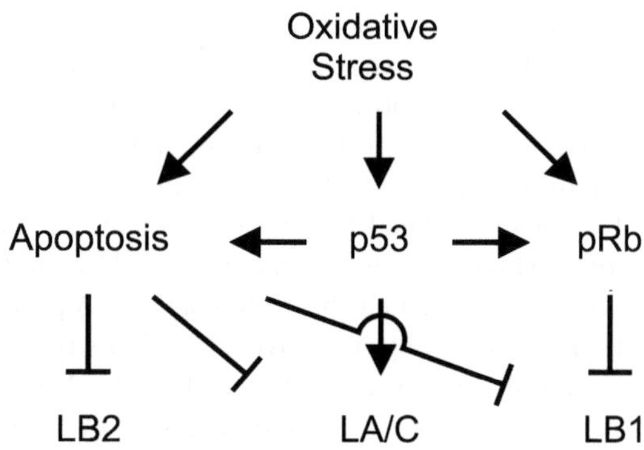

Fig. 2 A summary of the p53, pRb, and apoptotic pathways known to modulate the levels of lamin expression in response to oxidative stress. Oxidative stress induces an increase in LA/C expression, a decrease in LB1 expression, but no change in the expression of LB2 through the p53 and pRb pathways. Lamin proteins can be cleaved by oxidative-stress induced apoptosis

mTOR complex 1 (mTORC1) affects the protein levels of pre-LA and progerin through the mTOR pathway [33]. LB1 is degraded in transformed rat fibroblasts and a human cervical carcinoma cell line after these cells are exposed to a ROS inducer by the ubiquitin–proteasome pathway [34]. The lamina is also known to be broken down by caspases during apoptosis, and lamins are considered to be among the initial nuclear targets cleaved during the apoptotic process [35]. A- and B-type lamins are cleaved at their conserved VEID and VEVD sites by caspase-6 and 3, respectively [36–39]. Furthermore, both serine proteases, granzyme A and B, are known to cleave B-type lamins, whereas only granzyme A but not B appear to cleave A-type lamins [40]. Since oxidative stress can induce apoptosis [41], it is also likely that these cleavages of lamin proteins are induced by oxidative stress (Fig. 2). Together, these studies strongly support the idea that oxidative stress modulates the expression and stability of lamin proteins through transcription and proteolysis.

Posttranslational Modifications of Lamins in Response to Oxidative Stress

Lamins are known to be posttranslationally processed and the resulting modifications are likely to affect their functions, their interactions with each other, and their binding partners [42]. Several studies have indicated that lamins are posttranslationally modified by oxidation and enzymes in response to oxidative stress.

Lamins contain some amino acid residues that could be oxidized. During senescence, increased ROS results in the oxidation of cysteine residues in the LA tail domain, which in turn appears to inhibit the formation of LA inter- and intramolecular disulfide bonds [43]. Additionally, S-thiolation of A-type lamins is induced in isolated rat kidneys subjected to ischemia and reperfusion [44].

Phosphorylation of lamins has been the most extensively studied among many lamin posttranslational modifications. Though it is well known that hyperphosphorylation of lamins drives NE disassembly in mitosis [45–48], little is known regarding interphase phosphorylation of lamins. During interphase A-type lamins are known to be phosphorylated in response to oxidative stress in human neuroblastoma cells [49]. Since extracellular signal-regulated kinase 1/2 (ERK1/2) is activated by oxidative stress [50], it is possible that A-type lamins are phosphorylated by this kinase. In support of this, A-type lamins have been identified as among the most heavily phosphorylated proteins following activation of ERK1/2 [51, 52]. LB1 is also phosphorylated by p38 MAP kinase during senescence induced by oxidative stress, which leads to an increase in LB1 expression [29].

Furthermore, it has been shown that LA posttranslational processing by farnesylation is affected by oxidative stress. This is supported by the accumulation of pre-LA in old vascular smooth muscle cells (VSMCs) but not in young healthy blood vessels [53]. This accumulation of pre-LA correlates with the downregulation of the metallopeptidase Zmpste24/FACE-1, which is required for the processing of pre-LA into mature LA. Since both the mRNA and protein level of Zmpste24 are reduced in response to oxidative stress [53], this affects pre-LA levels. Posttranslational modifications and levels of B-type lamins are also altered by oxidative stress. For example, in transformed and cancer cells, LB1 protein is oxidized by ROS, which mediates the degradation of LB1 protein [34].

It still remains unclear how these posttranslational modifications of lamins induced by oxidative stress affect their structures and functions. However, one study shows that aggregates of LA/C and LB1 are observed in the nucleus during the early response to liver injury induced by oxidative stress [54]. This suggests, but certainly does not prove, that changes in posttranslational modifications of lamins caused by oxidative stress may inhibit lamin assembly into the lamina, which may lead to dysfunctions of the nucleus.

Lamin Functions in Cell Proliferation and Longevity

As mentioned above, the lamina and lamins are known to be involved in various nuclear functions including chromatin organization, DNA replication, DNA repair, and transcription (see "Introduction"). Some studies have indicated that the deregulation of these lamin functions by disease causing mutations and/or alterations in the expression levels of the different types of lamins inhibit cell proliferation and induce senescence or cell death.

The functions of A-type lamins in cell proliferation have been most extensively studied in the premature aging diseases, HGPS, and atypical progeroid syndromes. Dermal fibroblasts obtained from progeria patients are commonly found to proliferate slowly and become prematurely senescent [27, 55]. In several transgenic mouse models for HGPS, the mice also show marked reduction in body size and fibroblasts derived from the progeria mice exhibit slow proliferation and premature senescence [56, 57]. Mutations in the gene encoding a pre-LA processing enzyme, Zmpste24 and *Zmpste24* knockout mice (*Zmpste24*$^{-/-}$) cause a severe progeroid phenotype similar to HGPS [58, 59]. In support of this, the accumulation of pre-LA by silencing Zmpste24 expression or the overexpression of pre-LA accelerates VSMC

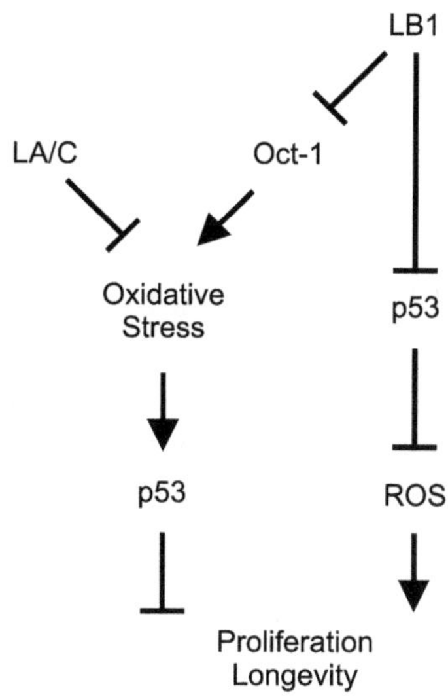

Fig. 3 A summary of p53- and Oct-1 dependent pathways linking lamin functions to oxidative stress responses, cell proliferation, and longevity. LA/C and LB1 suppress oxidative stress to promote proliferation and longevity through the p53 pathway. The suppression of oxidative stress by LB1 is mediated by Oct-1. LB1 also maintains ROS metabolism through the p53 pathway

senescence [53]. Similarly, silencing of LA/C expression slows cell proliferation and induces premature senescence in human diploid fibroblasts (HDFs) [60]. *Lmna* knockout mice (*Lmna*$^{-/-}$) also have proliferation defects within 4 weeks after birth and die by 8 weeks [61].

Alterations in LA/C either caused by silencing or defects in processing by Zmpste24 appear to increase the susceptibility of cells to DNA damage [62–65], most likely resulting in the activation of p53. In support of this, silencing LA/C expression or overexpression of LA and progerin slows proliferation and induces premature senescence in a p53-dependent but not pRb-dependent manner [60, 66] (Fig. 3). In *Zmpste24*$^{-/-}$ mice, p53 and the p53 target genes are upregulated and induce proliferation defects and premature death [67].

Recent studies have shown that B-type lamins are also involved in cell proliferation, differentiation, and longevity. Silencing the expression of LB1 slows cell proliferation in HDFs [22, 24] and induces premature senescence [22]. *Lmnb1* mutant mice (*Lmnb1*$^{\Delta/\Delta}$), *Lmnb1* knockout mice (*Lmnb1*$^{-/-}$), and *Lmnb1*/*Lmnb2* double knockout mice (*Lmnb1*$^{-/-}$*Lmnb2*$^{-/-}$) are born, but die immediately after birth due to developmental defects in specific differentiated cell types such as those comprising the lung and brain [68, 69]. *Lmnb2*$^{-/-}$ mice also have severe brain abnormalities, but these developmental defects are rather minor compared to those in *Lmnb1*$^{-/-}$ mice [69, 70]. Importantly, mouse embryonic fibroblasts (MEFs) derived from *Lmnb1*$^{\Delta/\Delta}$ mice and the mutant mice expressing a polymorphic variant of LB1 responsible for the curly tail phenotype show proliferation defects and prematurely senesce [68, 71]. On the other hand, mice with the conditional *Lmnb1*$^{-/-}$*Lmnb2*$^{-/-}$ restricted to their skin keratinocytes develop normally [72]. These findings support the idea that cell proliferation defects caused by the absence of B-type lamins are most likely to be tissue specific.

As in the case of the A-type lamins, there is evidence that p53 mediates the regulation of proliferation by LB1. Changes in proliferation rates caused by silencing LB1 expression and overexpression of LB1 can be restored by the inactivation of p53 [22, 24] (Fig. 3).

The Involvement of Lamins in Regulating Oxidative Stress

Oxidative stress activates several signaling pathways including the DNA damage response (DDR), the p38 MAP-kinase pathway, and the p53 pathway to induce slow proliferation, senescence, and cell death. Several studies have shown that lamins are involved in responses to oxidative stress.

There is significant evidence that in fibroblasts derived from HGPS patients, ROS levels are higher than those of normal fibroblasts, and these elevated levels of ROS are correlated with slow proliferation rates [73]. In addition, ROS-induced DNA double-strand breaks in HGPS fibroblasts are not repaired normally, and this appears to be related to their slow rate of proliferation [73]. This increase in ROS level in HGPS fibroblasts could be due to cysteine residues missing in progerin, which appear to be hyperoxidized during senescence and may contribute to the suppression of ROS-responsive genes [43]. Furthermore, mesenchymal stem cells (MSCs) overexpressing progerin or LA shows a decrease in expression of an antioxidant SOD2 and a SOD2-dependent increase in mitochondrial ROS, which leads to defects in chondrogenic differentiation potential [74]. Another study also shows that the viability of MSCs and VSMCs which have been differentiated from pluripotent stem cells (iPSCs) derived from HGPS fibroblasts are compromised by oxidative stress [75].

Other mutations in the *LMNA* gene have been reported to cause the accumulation of pre-LA and/or an increase in oxidative stress. Fibroblasts derived from patients with the *LMNA* mutations D47Y, L92F, L387V, R399H, L421P, and R482W, causing insulin resistance and/or lipodystrophy, accumulate pre-LA, and this is related to increased oxidative stress and the decreased expression of mitochondrial respiratory chain proteins that trigger premature cellular senescence [76]. Another study shows that cells from lipodystrophy patients with the LA mutations R439C, R482W, and H506D also accumulate pre-LA and show the expression of adipogenic proteins with brown fat-like features, an increased number of mitochondria and the overexpression of thermogenin (uncoupling protein 1, UCP1), which decreases the proton gradient generated in oxidative phosphorylation [77]. Higher levels of ROS are also induced by oxidative stress in fibroblasts from other lipodystrophy patients with a R439C *LMNA* mutation [78]. In addition, a homozygous *LMNA* mutation leads to expression of a mutated pre-LA with a deletion of 48 C-terminal amino acids, preventing its farnesylation and maturation. The resulting form of pre-LA is associated with increased oxidative stress and premature senescence [79]. *Zmpste24*$^{-/-}$ mice also accumulate pre-LA and show an increased mitochondrial response to oxidative stress [80]. Moreover, the mitochondrial proteins related to lipid metabolism, the tricarboxylic acid cycle, and oxidative phosphorylation are all upregulated in these mice. This supports the relationship between defective pre-LA processing and mitochondrial dysfunction, in addition to highlighting the relevance of pre-LA to oxidative damage in lipoatrophy and aging [80]. These results strongly support the idea that dysfunctions of LA/C cause oxidative stress (Fig. 3).

LB1 is also known to be involved in oxidative stress responses. The proliferation defects induced by silencing LB1 expression are accompanied by a p53-dependent reduction of mitochondrial ROS in HDFs, which can be rescued by growth under hypoxic conditions [22] (Fig. 3). RT-PCR analyses show that p53-target genes are altered under the experimental conditions used for silencing LB1 expression. For example, the antioxidant genes *SOD1*, *SOD2*, *SESN1*, *SESN2*, and *GPX1* are all upregulated, resulting in lowering the levels of ROS in LB1-silenced cells [22]. Other than p53, the POU-domain transcription factor Oct-1 also appears to mediate the LB1 regulation of the genes involved in oxidative stress responses [81]. Most importantly, Oct-1 has been shown to associate with the lamina through LB1 [11, 81]. This association is disrupted in the *Lmnb1*$^{\Delta/\Delta}$ fibroblasts, causing the elevation of ROS levels [81]. In autosomal dominant leukodystrophy (ADLD) fibroblasts, there is an increased expression of LB1 due to the duplication of the *LMNB1* gene, which is also coincident with an increase in the amount of Oct-1 associated with the lamina and a decrease in the nucleoplasmic fraction of Oct-1 by oxidative stress [82]. Since Oct-1 regulates the expression of genes involved in oxidative stress responses by binding to their putative octamer-binding DNA sequences [81], the sequestration of Oct-1 to the lamina by LB1 could be another mechanism by which LB1 modulates ROS levels (Fig. 3).

LB2 appears to regulate mitochondrial functions in neurons. For example, in *Lmnb2*$^{-/-}$ mice, a neuronal layering abnormality is caused by defective neuronal migration [70]. Inhibition of *LMNB2* mRNA translation in *Xenopus* retinal ganglion cell axons in vivo does not affect guidance but causes axonal degeneration [83]. This is attributable to the finding that a form of axonal LB2 associates with mitochondria and in LB2-deficient axons mitochondria are dysfunctional causing defects in axonal transport [83]. These results suggest that axonally synthesized LB2 plays a crucial role in axon maintenance by promoting mitochondrial functions to protect against oxidative stress.

Possible Pathways Related to Lamin Functions in Response to Oxygen Metabolism and Oxidative Stress Responses

Recent studies have indicated that lamins play important roles in regulating signaling pathways mediated by SIRT1, NF-κB, and mTORC1. These pathways are known to be involved in regulating oxygen metabolism and oxidative stress. Though in many cases, direct biochemical evidence has yet to be provided, these studies imply that lamins may regulate a cellular response to oxidative stress through these pathways.

In this regard, LA has been shown to interact with and activate an NAD-dependent deacetylase, Sirtuin 1 (SIRT1) [84]. SIRT1 deacetylates FOXO3, a member of the FOXO family of Forkhead transcription factors in response to oxidative stress, leading to cell cycle arrest, resistance to oxidative stress, and inhibition of cell death [85]. The presence of progerin or the accumulation of pre-LA in *Zmpste24*$^{-/-}$ mice induces a decrease in deacetylase activity of SIRT1, leading to rapid depletion of adult stem cells [84]. This scenario might also explain why a mitochondrial response to oxidative stress is increased in *Zmpste24*$^{-/-}$ mice and fibroblasts derived from HGPS patients [73, 80]. Therefore, it is possible that LA functions in oxidative stress responses are mediated by SIRT1.

It has also been shown that the accumulation of pre-LA activates an ATM- and NEMO-dependent signaling pathway, leading to the activation of NF-κB and secretion of proinflammatory cytokines in *Zmpste24*$^{-/-}$ and progeria mice (*Lmna*$^{G609G/}$G609G) [86]. The activation of NF-κB suppresses cell death by inducing the expression of genes encoding pro-oxidants [87]. In this fashion, the accumulation of pre-LA might increase the levels of ROS through the activation of NF-κB [80]. It is also possible that NF-κB mediates LA functions in oxidative stress responses.

Cellular oxygen sensing is upstream of the mTOR pathway [88, 89]. Interestingly, *Lmna*$^{-/-}$ mice show enhanced mTORC1 signaling in cardiac and skeletal muscle cells. Pharmacologic reversal of elevated mTORC1 signaling by rapamycin improves cardiac and skeletal muscle function and the longevity of the mice. In addition, this treatment also alleviates the defective autophagic-mediated degradation in *Lmna*$^{-/-}$ mice [90]. Based on these findings, it has been suggested that there is molecular cross talk between LA and oxygen sensing mechanisms.

The Prospective for Future Studies of Lamin Functions in Oxidative Stress

We have discussed current evidence for roles of lamins in regulating cell proliferation and longevity through the cellular response to oxidative stress and ROS signaling pathways. Though the lamin-related diseases that we have described are not directly involved in cancer, there are significant implications that lamins could be involved in tumor cell growth and in cancer progression. Many types of cancer cells are known to produce increased reactive ROS compared to normal cells [91]. Cancer cells also develop a unique way to control their proliferation as their increased glucose metabolism is coupled to fast proliferation [22] and their mitochondrial metabolism regulates ROS production which is essential for anchorage-independent growth [14]. Cancer cells are also more susceptible to oxidative stress compared to normal cells [91]. Therefore, progress in understanding how lamins control ROS metabolism in normal and cancer cells will provide new insights for cancer treatment. The modulation of ROS metabolism by changing lamin expression might provide a biochemical basis to design therapeutic strategies, including vectors for gene therapy and small molecule compounds for chemotherapy, to selectively slow cancer development, growth, and progression.

References

1. Adam SA, Goldman RD (2011) Insights into the differences between the A- and B-type nuclear lamins. Adv Enzyme Regul. doi:10.1016/j.advenzreg.2011.11.001
2. Eckersley-Maslin MA, Bergmann JH, Lazar Z, Spector DL (2013) Lamin A/C is expressed in pluripotent mouse embryonic stem cells. Nucleus 4(1):53–60. doi:10.4161/nucl.23384
3. Stuurman N, Heins S, Aebi U (1998) Nuclear lamins: their structure, assembly, and interactions. J Struct Biol 122(1–2):42–66
4. Herrmann H, Foisner R (2003) Intermediate filaments: novel assembly models and exciting new functions for nuclear lamins. Cell Mol Life Sci 60(8):1607–1612. doi:10.1007/s00018-003-3004-0
5. Aebi U, Cohn J, Buhle L, Gerace L (1986) The nuclear lamina is a meshwork of intermediate-type filaments. Nature 323(6088):560–564
6. Schermelleh L, Carlton PM, Haase S, Shao L, Winoto L, Kner P, Burke B, Cardoso MC, Agard DA, Gustafsson MG, Leonhardt H, Sedat JW (2008) Subdiffraction multicolor imaging of the nuclear periphery with 3D structured illumination microscopy. Science 320(5881):1332–1336
7. Shimi T, Pfleghaar K, Kojima S, Pack C, Solovei I, Goldman AE, Adam SA, Shumaker DK, Kinjo M, Cremer T, Goldman RD (2008) The A- and B-type nuclear lamin networks: microdomains involved in chromatin organization and transcription. Genes Dev 22(24):3409–3421, doi:22/24/3409 [pii]10.1101/gad.1735208
8. Moir RD, Spann TP, Herrmann H, Goldman RD (2000) Disruption of nuclear lamin organization blocks the elongation phase of DNA replication. J Cell Biol 149(6):1179–1192
9. Spann TP, Goldman AE, Wang C, Huang S, Goldman RD (2002) Alteration of nuclear lamin organization inhibits RNA polymerase II-dependent transcription. J Cell Biol 156(4):603–608
10. Shumaker DK, Solimando L, Sengupta K, Shimi T, Adam SA, Grunwald A, Strelkov SV, Aebi U, Cardoso MC, Goldman RD (2008) The highly conserved nuclear lamin Ig-fold binds to PCNA: its role in DNA replication. J Cell Biol 181(2):269–280
11. Guelen L, Pagie L, Brasset E, Meuleman W, Faza MB, Talhout W, Eussen BH, de Klein A, Wessels L, de Laat W, van Steensel B (2008) Domain organization of human chromosomes revealed by mapping of nuclear lamina interactions. Nature 453(7197):948–951, doi:nature06947 [pii]10.1038/nature06947
12. Reddy KL, Zullo JM, Bertolino E, Singh H (2008) Transcriptional repression mediated by repositioning of genes to the nuclear lamina. Nature 452(7184):243–7
13. Zullo JM, Demarco IA, Pique-Regi R, Gaffney DJ, Epstein CB, Spooner CJ, Luperchio TR, Bernstein BE, Pritchard JK, Reddy KL, Singh H (2012) DNA sequence-dependent compartmentalization and silencing of chromatin at the nuclear lamina. Cell 149(7):1474–1487. doi:10.1016/j.cell.2012.04.035
14. Weinberg F, Hamanaka R, Wheaton WW, Weinberg S, Joseph J, Lopez M, Kalyanaraman B, Mutlu GM, Budinger GR, Chandel NS (2010) Mitochondrial metabolism and ROS generation are essential for Kras-mediated tumorigenicity. Proc Natl Acad Sci U S A 107(19):8788–8793, doi:1003428107 [pii]10.1073/pnas.1003428107
15. Bell EL, Klimova TA, Eisenbart J, Schumacker PT, Chandel NS (2007) Mitochondrial reactive oxygen species trigger hypoxia-inducible factor-dependent extension of the replicative life span during hypoxia. Mol Cell Biol 27(16):5737–5745, doi:MCB.02265-06 [pii]10.1128/MCB.02265-06
16. Tormos KV, Anso E, Hamanaka RB, Eisenbart J, Joseph J, Kalyanaraman B, Chandel NS (2011) Mitochondrial complex III ROS regulate adipocyte differentiation. Cell Metab 14(4):537–544. doi:10.1016/j.cmet.2011.08.007
17. Uttara B, Singh AV, Zamboni P, Mahajan RT (2009) Oxidative stress and neurodegenerative diseases: a review of upstream and downstream antioxidant therapeutic options. Curr Neuropharmacol 7(1):65–74. doi:10.2174/157015909787602823
18. Rahman-Roblick R, Roblick UJ, Hellman U, Conrotto P, Liu T, Becker S, Hirschberg D, Jornvall H, Auer G, Wiman KG (2007) p53 targets identified by protein expression profiling. Proc Natl Acad Sci U S A 104(13):5401–5406. doi:10.1073/pnas.0700794104

19. Eriksson M, Brown WT, Gordon LB, Glynn MW, Singer J, Scott L, Erdos MR, Robbins CM, Moses TY, Berglund P, Dutra A, Pak E, Durkin S, Csoka AB, Boehnke M, Glover TW, Collins FS (2003) Recurrent de novo point mutations in lamin A cause Hutchinson-Gilford progeria syndrome. Nature 423(6937):293–298
20. Scaffidi P, Misteli T (2006) Lamin A-dependent nuclear defects in human aging. Science 312(5776):1059–1063
21. Cao K, Blair CD, Faddah DA, Kieckhaefer JE, Olive M, Erdos MR, Nabel EG, Collins FS (2011) Progerin and telomere dysfunction collaborate to trigger cellular senescence in normal human fibroblasts. J Clin Invest 121(7):2833–2844. doi:10.1172/JCI43578
22. Shimi T, Butin-Israeli V, Adam SA, Hamanaka RB, Goldman AE, Lucas CA, Shumaker DK, Kosak ST, Chandel NS, Goldman RD (2011) The role of nuclear lamin B1 in cell proliferation and senescence. Genes Dev 25(24):2579–2593. doi:10.1101/gad.179515.111
23. Freund A, Laberge RM, Demaria M, Campisi J (2012) Lamin B1 loss is a senescence-associated biomarker. Mol Biol Cell 23(11):2066–2075. doi:10.1091/mbc.E11-10-0884
24. Dreesen O, Chojnowski A, Ong PF, Zhao TY, Common JE, Lunny D, Lane EB, Lee SJ, Vardy LA, Stewart CL, Colman A (2013) Lamin B1 fluctuations have differential effects on cellular proliferation and senescence. J Cell Biol 200(5):605–617. doi:10.1083/jcb.201206121
25. Scaffidi P, Misteli T (2005) Reversal of the cellular phenotype in the premature aging disease Hutchinson-Gilford progeria syndrome. Nat Med 11(4):440–445
26. Kandert S, Luke Y, Kleinhenz T, Neumann S, Lu W, Jaeger VM, Munck M, Wehnert M, Muller CR, Zhou Z, Noegel AA, Dabauvalle MC, Karakesisoglou I (2007) Nesprin-2 giant safeguards nuclear envelope architecture in LMNA S143F progeria cells. Hum Mol Genet 16(23):2944–2959, doi:ddm255 [pii]10.1093/hmg/ddm255
27. Taimen P, Pfleghaar K, Shimi T, Moller D, Ben-Harush K, Erdos MR, Adam SA, Herrmann H, Medalia O, Collins FS, Goldman AE, Goldman RD (2009) A progeria mutation reveals functions for lamin A in nuclear assembly, architecture, and chromosome organization. Proc Natl Acad Sci U S A 106(49):20788–20793, doi:0911895106 [pii]10.1073/pnas.0911895106
28. Hallstrom TC, Mori S, Nevins JR (2008) An E2F1-dependent gene expression program that determines the balance between proliferation and cell death. Cancer Cell 13(1):11–22, doi:S1535-6108(07)00370-4 [pii]10.1016/j.ccr.2007.11.031
29. Barascu A, Le Chalony C, Pennarun G, Genet D, Imam N, Lopez B, Bertrand P (2012) Oxidative stress induces an ATM-independent senescence pathway through p38 MAPK-mediated lamin B1 accumulation. EMBO J 31(5):1080–1094. doi:10.1038/emboj.2011.492
30. Zoncu R, Efeyan A, Sabatini DM (2011) mTOR: from growth signal integration to cancer, diabetes and ageing. Nat Rev Mol Cell Biol 12(1):21–35. doi:10.1038/nrm3025
31. Cao K, Graziotto JJ, Blair CD, Mazzulli JR, Erdos MR, Krainc D, Collins FS (2011) Rapamycin reverses cellular phenotypes and enhances mutant protein clearance in Hutchinson-Gilford progeria syndrome cells. Sci Transl Med 3(89):89ra58, doi:10.1126/scitranslmed.3002346
32. Cenni V, Capanni C, Columbaro M, Ortolani M, D'Apice MR, Novelli G, Fini M, Marmiroli S, Scarano E, Maraldi NM, Squarzoni S, Prencipe S, Lattanzi G (2011) Autophagic degradation of farnesylated prelamin A as a therapeutic approach to lamin-linked progeria. Eur J Histochem 55(4):e36. doi:10.4081/ejh.2011.e36
33. Laplante M, Sabatini DM (2012) mTOR signaling in growth control and disease. Cell 149(2):274–293. doi:10.1016/j.cell.2012.03.017
34. Chiarini A, Whitfield JF, Pacchiana R, Armato U, Dal Pra I (2008) Photoexcited calphostin C selectively destroys nuclear lamin B1 in neoplastic human and rat cells—a novel mechanism of action of a photodynamic tumor therapy agent. Biochim Biophys Acta 1783(9):1642–1653. doi:10.1016/j.bbamcr.2008.03.014
35. Lazebnik YA, Takahashi A, Moir RD, Goldman RD, Poirier GG, Kaufmann SH, Earnshaw WC (1995) Studies of the lamin proteinase reveal multiple parallel biochemical pathways during apoptotic execution. Proc Natl Acad Sci U S A 92(20):9042–9046
36. Orth K, Chinnaiyan AM, Garg M, Froelich CJ, Dixit VM (1996) The CED-3/ICE-like protease Mch2 is activated during apoptosis and cleaves the death substrate lamin A. J Biol Chem 271(28):16443–16446

37. Takahashi A, Alnemri ES, Lazebnik YA, Fernandes-Alnemri T, Litwack G, Moir RD, Goldman RD, Poirier GG, Kaufmann SH, Earnshaw WC (1996) Cleavage of lamin A by Mch2 alpha but not CPP32: multiple interleukin 1 beta-converting enzyme-related proteases with distinct substrate recognition properties are active in apoptosis. Proc Natl Acad Sci U S A 93(16):8395–8400
38. Slee EA, Adrain C, Martin SJ (2001) Executioner caspase-3, -6, and -7 perform distinct, non-redundant roles during the demolition phase of apoptosis. J Biol Chem 276(10):7320–7326
39. Rao L, Perez D, White E (1996) Lamin proteolysis facilitates nuclear events during apoptosis. J Cell Biol 135(6 Pt 1):1441–1455
40. Zhang D, Beresford PJ, Greenberg AH, Lieberman J (2001) Granzymes A and B directly cleave lamins and disrupt the nuclear lamina during granule-mediated cytolysis. Proc Natl Acad Sci U S A 98(10):5746–5751
41. Chandra J, Samali A, Orrenius S (2000) Triggering and modulation of apoptosis by oxidative stress. Free Radic Biol Med 29(3–4):323–333
42. Simon DN, Wilson KL (2013) Partners and post-translational modifications of nuclear lamins. Chromosoma 122(1–2):13–31. doi:10.1007/s00412-013-0399-8
43. Pekovic V, Gibbs-Seymour I, Markiewicz E, Alzoghaibi F, Benham AM, Edwards R, Wenhert M, von Zglinicki T, Hutchison CJ (2011) Conserved cysteine residues in the mammalian lamin A tail are essential for cellular responses to ROS generation. Aging Cell 10(6):1067–1079. doi:10.1111/j.1474-9726.2011.00750.x
44. Eaton P, Jones ME, McGregor E, Dunn MJ, Leeds N, Byers HL, Leung KY, Ward MA, Pratt JR, Shattock MJ (2003) Reversible cysteine-targeted oxidation of proteins during renal oxidative stress. J Am Soc Nephrol 14(8 Suppl 3):S290–296
45. Dessev GN, Iovcheva-Dessev C, Goldman RD (1990) Lamin dimers. Presence in the nuclear lamina of surf clam oocytes and release during nuclear envelope breakdown. J Biol Chem 265(21):12636–12641
46. Heald R, McKeon F (1990) Mutations of phosphorylation sites in lamin A that prevent nuclear lamina disassembly in mitosis. Cell 61(4):579–589
47. Mall M, Walter T, Gorjanacz M, Davidson IF, Nga Ly-Hartig TB, Ellenberg J, Mattaj IW (2012) Mitotic lamin disassembly is triggered by lipid-mediated signaling. J Cell Biol 198(6):981–990. doi:10.1083/jcb.201205103
48. Peter M, Nakagawa J, Doree M, Labbe JC, Nigg EA (1990) In vitro disassembly of the nuclear lamina and M phase-specific phosphorylation of lamins by cdc2 kinase. Cell 61(4):591–602
49. Nakamura M, Yamada M, Ohsawa T, Morisawa H, Nishine T, Nishimura O, Toda T (2006) Phosphoproteomic profiling of human SH-SY5Y neuroblastoma cells during response to 6-hydroxydopamine-induced oxidative stress. Biochim Biophys Acta 1763(9):977–989
50. Finkel T, Holbrook NJ (2000) Oxidants, oxidative stress and the biology of ageing. Nature 408(6809):239–247. doi:10.1038/35041687
51. Lewis TS, Hunt JB, Aveline LD, Jonscher KR, Louie DF, Yeh JM, Nahreini TS, Resing KA, Ahn NG (2000) Identification of novel MAP kinase pathway signaling targets by functional proteomics and mass spectrometry. Mol Cell 6(6):1343–1354
52. Kosako H, Yamaguchi N, Aranami C, Ushiyama M, Kose S, Imamoto N, Taniguchi H, Nishida E, Hattori S (2009) Phosphoproteomics reveals new ERK MAP kinase targets and links ERK to nucleoporin-mediated nuclear transport. Nat Struct Mol Biol 16(10):1026–1035. doi:10.1038/nsmb.1656
53. Ragnauth CD, Warren DT, Liu Y, McNair R, Tajsic T, Figg N, Shroff R, Skepper J, Shanahan CM (2010) Prelamin A acts to accelerate smooth muscle cell senescence and is a novel biomarker of human vascular aging. Circulation 121(20):2200–2210. doi:10.1161/CIRCULATIONAHA.109.902056
54. Singla A, Griggs NW, Kwan R, Snider NT, Maitra D, Ernst SA, Herrmann H, Omary MB (2013) Lamin aggregation is an early sensor of porphyria-induced liver injury. J Cell Sci. doi:10.1242/jcs.123026
55. McClintock D, Gordon LB, Djabali K (2006) Hutchinson-Gilford progeria mutant lamin A primarily targets human vascular cells as detected by an anti-Lamin A G608G antibody. Proc Natl Acad Sci U S A 103(7):2154–2159

56. Mounkes LC, Kozlov S, Hernandez L, Sullivan T, Stewart CL (2003) A progeroid syndrome in mice is caused by defects in A-type lamins. Nature 423(6937):298–301. doi:10.1038/nature01631
57. Osorio FG, Navarro CL, Cadinanos J, Lopez-Mejia IC, Quiros PM, Bartoli C, Rivera J, Tazi J, Guzman G, Varela I, Depetris D, de Carlos F, Cobo J, Andres V, De Sandre-Giovannoli A, Freije JM, Levy N, Lopez-Otin C (2011) Splicing-directed therapy in a new mouse model of human accelerated aging. Sci Transl Med 3(106):106ra107, doi:10.1126/scitranslmed.3002847
58. Moulson CL, Go G, Gardner JM, van der Wal AC, Smitt JH, van Hagen JM, Miner JH (2005) Homozygous and compound heterozygous mutations in ZMPSTE24 cause the laminopathy restrictive dermopathy. J Invest Dermatol 125(5):913–919. doi:10.1111/j.0022-202X.2005.23846.x
59. Pendas AM, Zhou Z, Cadinanos J, Freije JM, Wang J, Hultenby K, Astudillo A, Wernerson A, Rodriguez F, Tryggvason K, Lopez-Otin C (2002) Defective prelamin A processing and muscular and adipocyte alterations in Zmpste24 metalloproteinase-deficient mice. Nat Genet 31(1):94–99
60. Moiseeva O, Bourdeau V, Vernier M, Dabauvalle MC, Ferbeyre G (2011) Retinoblastoma-independent regulation of cell proliferation and senescence by the p53-p21 axis in lamin A/C-depleted cells. Aging Cell 10(5):789–797. doi:10.1111/j.1474-9726.2011.00719.x
61. Sullivan T, Escalante-Alcalde D, Bhatt H, Anver M, Bhat N, Nagashima K, Stewart CL, Burke B (1999) Loss of A-type lamin expression compromises nuclear envelope integrity leading to muscular dystrophy. J Cell Biol 147(5):913–920
62. Gonzalez-Suarez I, Redwood AB, Perkins SM, Vermolen B, Lichtensztejin D, Grotsky DA, Morgado-Palacin L, Gapud EJ, Sleckman BP, Sullivan T, Sage J, Stewart CL, Mai S, Gonzalo S (2009) Novel roles for A-type lamins in telomere biology and the DNA damage response pathway. EMBO J 28(16):2414–2427. doi:10.1038/emboj.2009.196
63. Gonzalez-Suarez I, Redwood AB, Gonzalo S (2009) Loss of A-type lamins and genomic instability. Cell Cycle 8(23):3860–3865
64. Liu B, Wang J, Chan KM, Tjia WM, Deng W, Guan X, Huang JD, Li KM, Chau PY, Chen DJ, Pei D, Pendas AM, Cadinanos J, Lopez-Otin C, Tse HF, Hutchison C, Chen J, Cao Y, Cheah KS, Tryggvason K, Zhou Z (2005) Genomic instability in laminopathy-based premature aging. Nat Med 11(7):780–785. doi:10.1038/nm1266
65. Liu GH, Barkho BZ, Ruiz S, Diep D, Qu J, Yang SL, Panopoulos AD, Suzuki K, Kurian L, Walsh C, Thompson J, Boue S, Fung HL, Sancho-Martinez I, Zhang K, Yates J 3rd, Izpisua Belmonte JC (2011) Recapitulation of premature ageing with iPSCs from Hutchinson-Gilford progeria syndrome. Nature 472(7342):221–225. doi:10.1038/nature09879
66. Kudlow BA, Stanfel MN, Burtner CR, Johnston ED, Kennedy BK (2008) Suppression of proliferative defects associated with processing-defective lamin A mutants by hTERT or inactivation of p53. Mol Biol Cell 19(12):5238–5248. doi:10.1091/mbc.E08-05-0492
67. Varela I, Cadinanos J, Pendas AM, Gutierrez-Fernandez A, Folgueras AR, Sanchez LM, Zhou Z, Rodriguez FJ, Stewart CL, Vega JA, Tryggvason K, Freije JM, Lopez-Otin C (2005) Accelerated ageing in mice deficient in Zmpste24 protease is linked to p53 signalling activation. Nature 437(7058):564–568. doi:10.1038/nature04019
68. Vergnes L, Peterfy M, Bergo MO, Young SG, Reue K (2004) Lamin B1 is required for mouse development and nuclear integrity. Proc Natl Acad Sci U S A 101(28):10428–10433
69. Kim Y, Sharov AA, McDole K, Cheng M, Hao H, Fan CM, Gaiano N, Ko MS, Zheng Y (2011) Mouse B-type lamins are required for proper organogenesis but not by embryonic stem cells. Science 334(6063):1706–1710. doi:10.1126/science.1211222
70. Coffinier C, Chang SY, Nobumori C, Tu Y, Farber EA, Toth JI, Fong LG, Young SG (2010) Abnormal development of the cerebral cortex and cerebellum in the setting of lamin B2 deficiency. Proc Natl Acad Sci U S A 107(11):5076–5081. doi:10.1073/pnas.0908790107
71. De Castro SC, Malhas A, Leung KY, Gustavsson P, Vaux DJ, Copp AJ, Greene ND (2012) Lamin b1 polymorphism influences morphology of the nuclear envelope, cell cycle progression, and risk of neural tube defects in mice. PLoS Genet 8(11):e1003059. doi:10.1371/journal.pgen.1003059

72. Yang SH, Chang SY, Yin L, Tu Y, Hu Y, Yoshinaga Y, de Jong PJ, Fong LG, Young SG (2011) An absence of both lamin B1 and lamin B2 in keratinocytes has no effect on cell proliferation or the development of skin and hair. Hum Mol Genet 20(18):3537–3544, doi:ddr266 [pii]10.1093/hmg/ddr266
73. Richards SA, Muter J, Ritchie P, Lattanzi G, Hutchison CJ (2011) The accumulation of un-repairable DNA damage in laminopathy progeria fibroblasts is caused by ROS generation and is prevented by treatment with N-acetyl cysteine. Hum Mol Genet 20(20):3997–4004. doi:10.1093/hmg/ddr327
74. Mateos J, De la Fuente A, Lesende-Rodriguez I, Fernandez-Pernas P, Arufe MC, Blanco FJ (2013) Lamin A deregulation in human mesenchymal stem cells promotes an impairment in their chondrogenic potential and imbalance in their response to oxidative stress. Stem Cell Res 11(3):1137–1148. doi:10.1016/j.scr.2013.07.004
75. Zhang J, Lian Q, Zhu G, Zhou F, Sui L, Tan C, Mutalif RA, Navasankari R, Zhang Y, Tse HF, Stewart CL, Colman A (2011) A human iPSC model of Hutchinson Gilford Progeria reveals vascular smooth muscle and mesenchymal stem cell defects. Cell Stem Cell 8(1):31–45. doi:10.1016/j.stem.2010.12.002
76. Caron M, Auclair M, Donadille B, Bereziat V, Guerci B, Laville M, Narbonne H, Bodemer C, Lascols O, Capeau J, Vigouroux C (2007) Human lipodystrophies linked to mutations in A-type lamins and to HIV protease inhibitor therapy are both associated with prelamin A accumulation, oxidative stress and premature cellular senescence. Cell Death Differ 14(10):1759–1767. doi:10.1038/sj.cdd.4402197
77. Bereziat V, Cervera P, Le Dour C, Verpont MC, Dumont S, Vantyghem MC, Capeau J, Vigouroux C, Lipodystrophy Study G (2011) LMNA mutations induce a non-inflammatory fibrosis and a brown fat-like dystrophy of enlarged cervical adipose tissue. Am J Pathol 179(5):2443–2453. doi:10.1016/j.ajpath.2011.07.049
78. Verstraeten VL, Caputo S, van Steensel MA, Duband-Goulet I, Zinn-Justin S, Kamps M, Kuijpers HJ, Ostlund C, Worman HJ, Briede JJ, Le Dour C, Marcelis CL, van Geel M, Steijlen PM, van den Wijngaard A, Ramaekers FC, Broers JL (2009) The R439C mutation in LMNA causes lamin oligomerization and susceptibility to oxidative stress. J Cell Mol Med 13(5): 959–971. doi:10.1111/j.1582-4934.2009.00690.x
79. Le Dour C, Schneebeli S, Bakiri F, Darcel F, Jacquemont ML, Maubert MA, Auclair M, Jeziorowska D, Reznik Y, Bereziat V, Capeau J, Lascols O, Vigouroux C (2011) A homozygous mutation of prelamin-A preventing its farnesylation and maturation leads to a severe lipodystrophic phenotype: new insights into the pathogenicity of nonfarnesylated prelamin-A. J Clin Endocrinol Metab 96(5):E856–862. doi:10.1210/jc.2010-2234
80. Peinado JR, Quiros PM, Pulido MR, Marino G, Martinez-Chantar ML, Vazquez-Martinez R, Freije JM, Lopez-Otin C, Malagon MM (2011) Proteomic profiling of adipose tissue from Zmpste24$^{-/-}$ mice, a model of lipodystrophy and premature aging, reveals major changes in mitochondrial function and vimentin processing. Mol Cell Proteomics 10(11):M111.008094. doi:10.1074/mcp.M111.008094
81. Malhas AN, Lee CF, Vaux DJ (2009) Lamin B1 controls oxidative stress responses via Oct-1. J Cell Biol 184(1):45–55
82. Columbaro M, Mattioli E, Maraldi NM, Ortolani M, Gasparini L, D'Apice MR, Postorivo D, Nardone AM, Avnet S, Cortelli P, Liguori R, Lattanzi G (2013) Oct-1 recruitment to the nuclear envelope in adult-onset autosomal dominant leukodystrophy. Biochim Biophys Acta 1832(3):411–420. doi:10.1016/j.bbadis.2012.12.006
83. Yoon BC, Jung H, Dwivedy A, O'Hare CM, Zivraj KH, Holt CE (2012) Local translation of extranuclear lamin B promotes axon maintenance. Cell 148(4):752–764. doi:10.1016/j.cell.2011.11.064
84. Liu B, Ghosh S, Yang X, Zheng H, Liu X, Wang Z, Jin G, Zheng B, Kennedy BK, Suh Y, Kaeberlein M, Tryggvason K, Zhou Z (2012) Resveratrol rescues SIRT1-dependent adult stem cell decline and alleviates progeroid features in laminopathy-based progeria. Cell Metab 16(6):738–750. doi:10.1016/j.cmet.2012.11.007

85. Brunet A, Sweeney LB, Sturgill JF, Chua KF, Greer PL, Lin Y, Tran H, Ross SE, Mostoslavsky R, Cohen HY, Hu LS, Cheng HL, Jedrychowski MP, Gygi SP, Sinclair DA, Alt FW, Greenberg ME (2004) Stress-dependent regulation of FOXO transcription factors by the SIRT1 deacetylase. Science 303(5666):2011–2015. doi:10.1126/science.1094637
86. Osorio FG, Barcena C, Soria-Valles C, Ramsay AJ, de Carlos F, Cobo J, Fueyo A, Freije JM, Lopez-Otin C (2012) Nuclear lamina defects cause ATM-dependent NF-kappaB activation and link accelerated aging to a systemic inflammatory response. Genes Dev 26(20): 2311–2324. doi:10.1101/gad.197954.112
87. Morgan MJ, Liu ZG (2011) Crosstalk of reactive oxygen species and NF-kappaB signaling. Cell Res 21(1):103–115. doi:10.1038/cr.2010.178
88. Reiling JH, Hafen E (2004) The hypoxia-induced paralogs Scylla and Charybdis inhibit growth by down-regulating S6K activity upstream of TSC in Drosophila. Genes Dev 18(23):2879–2892. doi:10.1101/gad.322704
89. Brugarolas J, Lei K, Hurley RL, Manning BD, Reiling JH, Hafen E, Witters LA, Ellisen LW, Kaelin WG Jr (2004) Regulation of mTOR function in response to hypoxia by REDD1 and the TSC1/TSC2 tumor suppressor complex. Genes Dev 18(23):2893–2904. doi:10.1101/gad.1256804
90. Ramos FJ, Chen SC, Garelick MG, Dai DF, Liao CY, Schreiber KH, MacKay VL, An EH, Strong R, Ladiges WC, Rabinovitch PS, Kaeberlein M, Kennedy BK (2012) Rapamycin reverses elevated mTORC1 signaling in lamin A/C-deficient mice, rescues cardiac and skeletal muscle function, and extends survival. Sci Transl Med 4(144):144ra103. doi:10.1126/scitranslmed.3003802
91. Trachootham D, Alexandre J, Huang P (2009) Targeting cancer cells by ROS-mediated mechanisms: a radical therapeutic approach? Nat Rev Drug Discov 8(7):579–591. doi:10.1038/nrd2803

Part VI
The Nuclear Envelope Link to Cell Migration and Metastasis

Introduction

The nucleoskeleton is comprised of both the peripheral lamina and a less-characterized internal nuclear matrix. Protein components of both have been linked at various points to cancer and tumorigenesis. The nuclear matrix will only be touched on briefly in this section because its composition is less clearly delineated than the peripheral nucleoskeleton. However, work from Ronald Berezney and Jeffrey Nickerson has clearly demonstrated changes in the nuclear matrix in cancerous and proliferating cells and perhaps more importantly found that the AML transcription factors associated with acute myelogenous leukemia target to the nuclear matrix and require association with the matrix for function [1–4]. The lamina is made up of a polymer of the intermediate filament nuclear lamins, whose involvement in cancer was introduced in Part I, together with nuclear envelope transmembrane proteins (NETs) that connect it to the nuclear membrane. Some inner nuclear membrane NETs that bind the lamin polymer connect across the lumen of the double membrane nuclear envelope to NETs in the outer nuclear membrane. These outer nuclear membrane NETs in turn can connect to the cytoskeleton so that all the connections together connect the nucleoskeleton to the cytoskeleton. The main inner nuclear membrane NETs found thus far to be involved in these connections are the SUN proteins, and the outer nuclear membrane NETs involved are Nesprins/Syne proteins. The complex they form is generally referred to as the LINC complex for *LI*nker of *N*ucleoskeleton and *C*ytoskeleton [5]. The Nesprin/Syne proteins have been found to bind directly only to actin among the three major cytoskeletal systems [6, 7]; however, they likely connect at least indirectly to all three as they have been found also to bind microtubule motors [8, 9] and plectin [10] which can in turn connect to the intermediate filament, actin, and microtubule cytoskeletal systems.

Disruption of these connections has been found to affect the mechanical stability of cells, gene expression (likely through mechanotransduction), and cell polarity and migration. Moreover, specific links to cancer have been observed. In this section Celine Denais and Jan Lammerding of Cornell University, world experts on biophysical measurements of mechanical stability, discuss the overall role of the nuclear envelope in mechanical stability and mechanotransduction, touching on nuclear and cell mechanical stability, how this stability affects the nuclear shape changes observed in cancer, nuclear deformations leading to nuclear breaks in cancer, and mechanical connections involvement in cell migration. Then David Razafsky, Denis Wirtz, and Didier Hodzic of Washington University address the involvement of the LINC complex, discovered by Professor Hodzic, in cell polarity, cell migration, and nuclear migration. The role in nuclear migration is particularly intriguing as within different neuronal layers in the retina the spatial position of the nucleus changes in a highly reproducible manner. Next Sascha Neumann and Angelika Noegel from the University of Cologne, leading experts on the Nesprin proteins, discuss the roles of these proteins in nuclear size, centrosome positioning, and cell migration and the relationship of these functions to cancer. They also describe work that demonstrates a role for Nesprins in signal transduction mechanisms, likely more linked to mechanical strain than the types of mechanisms of inner nuclear membrane proteins described in Parts II and III. Finally, Daniel Osorio and Edgar Gomes of the University of Pierre et Marie Curie, who first described the role of TAN lines in nuclear migration, describe this more specific phenomenon and its connections with small GTPases involved in cell migration and metastasis.

1. Berezney R, Basler J, Hughes BB, Kaplan SC (1979) Isolation and characterization of the nuclear matrix from Zajdela ascites hepatoma cells. Cancer Res 39(8):3031–3039
2. Berezney R, Basler J, Kaplan SC, Hughes BB (1979) The nuclear matrix of slowly and rapidly proliferating liver cells. Eur J Cell Biol 20(2):139–142
3. Nickerson JA (1998) Nuclear dreams: the malignant alteration of nuclear architecture. J Cell Biochem 70(2):172–180. doi:10.1002/(SICI)1097-4644(19980801)70:2<172::AID-JCB3>3.0.CO;2-L [pii]
4. Zeng C, McNeil S, Pockwinse S, Nickerson J, Shopland L, Lawrence JB, Penman S, Hiebert S, Lian JB, van Wijnen AJ, Stein JL, Stein GS (1998) Intranuclear targeting of AML/CBFalpha regulatory factors to nuclear matrix-associated transcriptional domains. Proc Natl Acad Sci U S A 95(4):1585–1589
5. Crisp M, Liu Q, Roux K, Rattner JB, Shanahan C, Burke B, Stahl PD, Hodzic D (2006) Coupling of the nucleus and cytoplasm: role of the LINC complex. J Cell Biol 172(1):41–53
6. Starr DA, Han M (2002) Role of ANC-1 in tethering nuclei to the actin cytoskeleton. Science 298(5592):406–409
7. Zhang Q, Ragnauth C, Greener MJ, Shanahan CM, Roberts RG (2002) The nesprins are giant actin-binding proteins, orthologous to *Drosophila melanogaster* muscle protein MSP-300. Genomics 80(5):473–481
8. Roux KJ, Crisp ML, Liu Q, Kim D, Kozlov S, Stewart CL, Burke B (2009) Nesprin 4 is an outer nuclear membrane protein that can induce kinesin-mediated cell polarization. Proc Natl Acad Sci U S A 106(7):2194–2199

9. Schneider M, Lu W, Neumann S, Brachner A, Gotzmann J, Noegel AA, Karakesisoglou I (2011) Molecular mechanisms of centrosome and cytoskeleton anchorage at the nuclear envelope. Cell Mol Life Sci 68(9):1593–1610. doi:10.1007/s00018-010-0535-z
10. Wilhelmsen K, Litjens SH, Kuikman I, Tshimbalanga N, Janssen H, van den Bout I, Raymond K, Sonnenberg A (2005) Nesprin-3, a novel outer nuclear membrane protein, associates with the cytoskeletal linker protein plectin. J Cell Biol 171(5):799–810

Nuclear Mechanics in Cancer

Celine Denais and Jan Lammerding

Abstract Despite decades of research, cancer metastasis remains an incompletely understood process that is as complex as it is devastating. In recent years, there has been an increasing push to investigate the biomechanical aspects of tumorigenesis, complementing the research on genetic and biochemical changes. In contrast to the high genetic variability encountered in cancer cells, almost all metastatic cells are subject to the same physical constraints as they leave the primary tumor, invade surrounding tissues, transit through the circulatory system, and finally infiltrate new tissues. Advances in live cell imaging and other biophysical techniques, including measurements of subcellular mechanics, have yielded stunning new insights into the physics of cancer cells. While much of this research has been focused on the mechanics of the cytoskeleton and the cellular microenvironment, it is now emerging that the mechanical properties of the cell nucleus and its connection to the cytoskeleton may play a major role in cancer metastasis, as deformation of the large and stiff nucleus presents a substantial obstacle during the passage through the dense interstitial space and narrow capillaries. Here, we present an overview of the molecular components that govern the mechanical properties of the nucleus, and we

C. Denais
Department of Biomedical Engineering, Weill Institute for Cell and Molecular Biology, Cornell University, Ithaca, NY 14853, USA
e-mail: cmd326@cornell.edu

J. Lammerding (✉)
Department of Biomedical Engineering, Weill Institute for Cell and Molecular Biology, Cornell University, Ithaca, NY 14853, USA

Department of Biomedical Engineering, Weill Institute for Cell and Molecular Biology, Cornell University, Weill Hall, Room 235, Ithaca, NY 14853, USA
e-mail: jan.lammerding@cornell.edu

E.C. Schirmer and J.I. de las Heras (eds.), *Cancer Biology and the Nuclear Envelope*, Advances in Experimental Medicine and Biology 773, DOI 10.1007/978-1-4899-8032-8_20,

discuss how changes in nuclear structure and composition observed in many cancers can modulate nuclear mechanics and promote metastatic processes. Improved insights into this interplay between nuclear mechanics and metastatic progression may have powerful implications in cancer diagnostics and therapy and may reveal novel therapeutic targets for pharmacological inhibition of cancer cell invasion.

Keywords Cytoskeleton • LINC complex • Mechanotransduction • Mechanical stability • Nuclear lamina

Abbreviations

ER	Endoplasmic reticulum
KASH	Klarsicht, ANC-1, Syne Homology
LAPs	Lamina-associated polypeptides
LBR	Lamin B receptor
LINC	*L*inker of *N*ucleoskeleton and *C*ytoskeleton
NPC	Nuclear pore complex

Introduction

The cell nucleus was the first organelle discovered in the seventeenth century. In the oldest preserved depictions of the nucleus, Antonie van Leeuwenhoek described a central "clear area" in salmon blood cells that is now commonly acknowledged as the nucleus [1]. A more detailed description of the nucleus was subsequently provided by the botanist Robert Brown, who first articulated the concept of the nucleated cell as a structural unit in plants [1]. Today, the nucleus is recognized as the site of numerous essential functions in eukaryotes, including storage and organization of the genetic material, DNA synthesis, DNA transcription, transcriptional regulation, and RNA processing. In cancer biology, much of the research has traditionally been focused on this "DNA-centric view," starting with the identification of oncogenes and tumor-suppressor genes to the establishment of the multiple "hits" (i.e., mutations) concept now commonly accepted as a requirement for cancer initiation and progression [2]. Recently, however, it has become apparent that in addition to these genetic components, it is necessary to take the physical, i.e., biomechanical, factors of tumor cells and their microenvironment into consideration. Research conducted within the last 10 years has revealed that cancer cells have reduced stiffness [3–7], generate increased contractile forces [8], and are strongly influenced by their biomechanical environment [9, 10]. Furthermore, not only can cancer cells be mechanically distinguished from non-tumorigenic cells, but physical measurements also allow telling apart highly invasive cells from less invasive cells, for example, by their increased cell deformability [4] and increased traction forces [8], yielding the promise of future diagnostic and prognostic applications. Here, we focus on a particular aspect of cellular mechanics that has traditionally received less attention in cancer cell biology: the role of nuclear structure and mechanics in cancer progression.

Despite many advances in understanding the biology of cancer and its associated molecular changes, the most common and reliable diagnosis of cancer cells in tissue biopsies by pathologists still relies on the presence of morphological changes in nuclear structure, i.e., increased size, irregular shape and organization [11]. Nonetheless, the functional consequences of these characteristic changes have yet to be determined; thus, it remains unclear whether the observed morphological changes merely correlate with other, more difficult to observe cellular defects, or whether they can directly contribute to the disease progression.

In recent years, a growing number of studies have reported altered nuclear envelope composition in various cancers [12, 13]. The structure and composition of the nucleus, particularly the nuclear envelope, play an important role in cellular mechanics and function, ranging from determining nuclear deformability and fragility [14–17] to participating in mechanotransduction signaling, i.e., the sensing of biomechanical factors and the corresponding signaling response [15, 18]. One potential mechanism by which changes in nuclear envelope composition could contribute to cancer progression is that softer and more lobulated nuclei facilitate cancer cell invasion through dense tissues, where cells often have to pass through constrictions smaller than the nuclear diameter [19, 20]. Furthermore, the physical coupling between the nucleus and the cytoskeleton is critical for cytoskeletal organization and cell polarization [21–24], which could further affect cancer cell migration. In the following sections, we provide a brief review of normal nuclear structure and mechanics, highlight changes that occur during oncogenic transformation, and discuss recent findings suggesting an important role of nuclear mechanics and nucleo-cytoskeletal coupling in cancer progression.

Normal Nuclear Compartmentalization and Structure

The nucleus is a highly compartmentalized organelle that can be roughly subdivided into the nuclear envelope and the nuclear interior (Fig. 1), the latter representing most of the chromatin in diverse states of organization [25], the nucleolus, and diverse smaller subnuclear structures such as Cajal bodies and nuclear speckles [26–28]. In addition, the nuclear interior contains a still incompletely defined structural network (i.e., the nucleoskeleton or nuclear matrix), which may provide additional mechanical support and also act as scaffold for transcriptional complexes and other nuclear processes. The nuclear envelope forms the physical barrier between the nucleus and the cytoplasm. It consists of two phospholipid bilayers, the inner and the outer membranes, and the underlying nuclear lamina, a dense protein meshwork mostly comprising lamins. The inner and outer nuclear membranes are connected at the sites of nuclear pore complexes (NPCs) and encapsulate the perinuclear space or lumen.

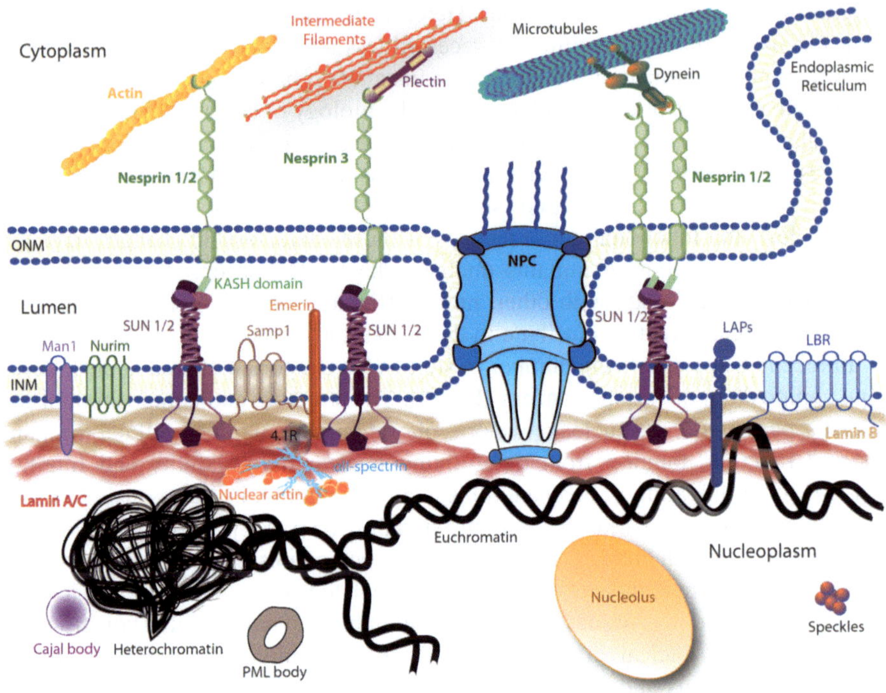

Fig. 1 Schematic overview of the nuclear structure and the LINC complex. The nuclear envelope is composed of the inner nuclear membrane (INM) and the outer nuclear membrane (ONM) punctuated by nuclear pore complexes (NPC). The ONM is continuous with the endoplasmic reticulum (ER). Several structures of the nuclear interior are depicted here, including the nucleolus, Cajal bodies, promyelocytic leukemia bodies (PML), and speckles. Chromatin is shown in its two states, very condensed (heterochromatin) and loosely organized (euchromatin). Only a subset of nuclear membrane proteins are portrayed in this picture: lamin B receptor (LBR), emerin, MAN1, and nurim. The schematic also illustrates some of the interactions between these proteins with the lamina meshwork (lamins B and A/C). The LINC complex is represented by nesprins, Sad1p/UNC-84 (SUN) proteins and Samp1. On the outer membrane, nesprin-1 and -2 can directly bind to actin filaments or indirectly interact with microtubules through motors proteins (dyneins or kinesin). Nesprin-3 is shown interacting with intermediate filaments via plectin

The Outer Nuclear Membrane

The outer nuclear membrane is continuous with the endoplasmic reticulum (ER); like the ER, its surface is scattered with ribosomes. The outer nuclear membrane exhibits a high degree of similarity to the ER membrane in terms of protein, enzyme, and lipid composition [29]. Nonetheless, recent studies have suggested that the outer nuclear membrane displays a certain degree of specialization [30] and participates in protein synthesis and processing [31]. The specialized protein composition of the outer nuclear membrane likely results from retention of specific proteins by direct interaction with inner nuclear membrane proteins across the lumen, thereby enriching them compared to the ER fraction [32, 33]. In mammals, one particularly important family of outer nuclear membrane proteins is the nesprins [34], which play a central role in connecting the nucleus to the cytoskeleton [35–39].

The Nuclear Lumen and Nuclear Pore Complexes

The nuclear lumen, also commonly termed the perinuclear space, is a 30–50 nm wide aqueous space separating the inner from the outer nuclear membrane that is continuous with the ER lumen [40]. It accommodates the luminal domains of integral nuclear membrane proteins [41]. The inner and outer nuclear membranes come together at sites of NPC insertion [42]. NPCs act as the main gateway for molecules between the cytoplasm and the nuclear interior (and also proteins of the inner nuclear membrane). Small molecules can diffuse freely through the NPC, while the exchange of macromolecules larger than ~40 kDa is mediated by a tightly controlled import and export mechanism requiring nuclear import and export signals and interaction with specific transport molecules [43–45].

The Inner Nuclear Membrane

The inner nuclear membrane contains at least 70–100 unique membrane-associated and integral membrane proteins that are retained at the inner nuclear membrane through interaction with nucleoplasmic proteins (e.g., lamins) and chromatin [13]. Most of these proteins have only been identified in recent proteomic studies [46–50], and the function of several of the nuclear envelope transmembrane proteins remains unclear. Some well-characterized inner nuclear membrane proteins include lamin B receptor (LBR), lamina-associated polypeptides (LAPs) [30], emerin, MAN1, nurim, nesprins, and Sad1p/UNC-84 (SUN) proteins [13]. Mislocalization or loss of these proteins due to mutations in nuclear envelope proteins causes a spectrum of diseases collectively known as laminopathies that include certain types of muscular dystrophies (e.g., Emery–Dreifuss muscular dystrophy and limb-girdle muscular dystrophy), dilated cardiomyopathy, and the premature aging disease Hutchinson–Gilford progeria syndrome [51].

The Nuclear Lamina

The lamina corresponds to a dense meshwork of proteins mainly composed of lamins underlying the inner nuclear membrane [52]. Lamins are type V intermediate filaments [53, 54] and display the characteristic tripartite molecular organization of all intermediate filaments, which consists of a central a-helical rod domain flanked by a short non-helical N-terminal "head" and a C-terminal "tail" domain that includes an Ig-like fold [55].

In vertebrates, lamins are classified into two major classes, A- and B-type lamins, depending on their sequence, expression pattern, and biochemical properties [56, 57]. A-type lamins, including lamins A, C, AΔ10, and C2, result from alternative splicing of the *LMNA* gene on chromosome 1. These proteins are expressed in a tissue-specific manner later in differentiation [58, 59], have neutral isoelectric points, and are dispersed upon phosphorylation of lamins during mitosis [60]. Lamin A and C can be distinguished by their unique C-terminal tail and processing: the C-terminus of prelamin A contains a CaaX motif, which is subject to a series of posttranslational modifications, including isoprenylation and proteolytic cleavage, to give rise to mature lamin A [61, 62]. In contrast, the shorter lamin C has a unique C-terminus that lacks the CaaX motif and does not require posttranslational processing. In addition to their localization at the nuclear lamina, A-type lamins are also present in the nuclear interior, where they form stable structures [63].

Unlike A-type lamins, B-type lamins are encoded by two separate genes: *LMNB1* for lamin B1 [64, 65] and *LMNB2* for lamin B2 and B3 [66, 67]. Only lamins B1 and B2 are found in somatic cells; expression of lamin B3 is restricted to germ cells. Unlike A-type lamins, at least one B-type lamin is expressed in all cells, including embryonic stem cells; B-type lamins are acidic and remain associated with membranes during mitosis [68]. The C-terminus of B-type lamins is also isoprenylated but, unlike prelamin A, does not undergo proteolytic cleavage. Consequently, B-type lamins remain permanently farnesylated, facilitating their attachment to the inner nuclear membrane.

The Nuclear Interior

In addition to DNA and histones, the nucleoplasm contains distinct structural and functional elements such as nucleoli [69], Cajal bodies [70], the Gemini of coiled bodies or gems [71], promyelocytic leukemia (PML) bodies [72], and splicing speckles [73]. The growing interest to decipher the detailed structure and composition of the nuclear interior has led to the recent discoveries that the nuclear interior contains actin [74, 75], myosin [76, 77], spectrin [78], and even titin [79]. It is now well established that actin oligomers or short polymers are present in the nucleus [80–82] and that all isoforms of actin contain nuclear export sequences [83], which may help prevent spontaneous assembly of actin filaments inside the nucleus. To date, many aspects of nuclear actin remain incompletely understood, including its precise structural organization [84]. Nonetheless, nuclear actin has been implicated in a number of functions highly relevant to tumorigenesis, including DNA organization, stabilization, and orientation during replication, determination of nuclear morphology, organization of gene regulatory complexes, and RNA synthesis [85]. The existence and function of the "nuclear matrix" or nucleoskeleton, typically defined as the insoluble structure remaining after nuclease, detergent, and high salt treatment of isolated nuclei [86], remains a matter of lively debate, but given the plethora of structural proteins present in the nucleus and their often low diffusional mobility, it is likely that some (possibly local) structural frameworks exist in the nuclear interior.

Nuclear Mechanics and Mechanotransduction

In recent years, it has emerged that physical factors, such as the biomechanical properties of the microenvironment and the mechanical forces acting between cells and their environment, play an important role in cellular function [87]. With regard to cancer cells, modulation of cytoskeletal tension by Rho inhibition alone can be sufficient to phenotypically revert epithelial morphogenesis of malignant cells [10]. Rho proteins belong to the family of small signaling G-proteins (GTPases) that can act as "molecular switches" in regulating actin cytoskeleton dynamics, while also playing important roles in cell polarity, migration vesicle trafficking, mitosis, proliferation and apoptosis [88]. Furthermore, recent studies found that aggressive cancer cells can be distinguished from less invasive and non-tumorigenic cancer cells based on their cytoskeletal stiffness [3] and their contractile force generation [8]. What is now becoming apparent is that in addition to cytoskeletal stiffness and force generation, nuclear deformability, as well as the physical coupling between the nucleus and the cytoskeleton, play a critical role in cell motility in three-dimensional (3D) environments [19, 20]. In this section, we discuss the molecular players governing normal nuclear mechanics, i.e., nuclear deformability and nucleo-cytoskeletal coupling, as well as their potential contribution to cellular mechanosensing. Their involvement in cancer progression is then described in the subsequent section.

Nuclear Deformability and Stability

Over the years, a variety of experimental techniques have been developed to probe the mechanical properties of the nucleus, particularly its deformability under applied forces. These approaches include micropipette aspiration [89–93], atomic force microscopy [91, 94–96], cell stretching [14, 97–99], tracking of particles within the nucleoplasm [100], and, most recently, optical stretching [101] and measuring transit times through microfluidic constriction channels [102, 103]. These experiments have revealed that the nucleus exhibits both elastic (the nuclear lamina) and viscoelastic (the nuclear interior) behavior and is typically ~2–10 times stiffer than the surrounding cytoplasm [93, 99, 104, 105]. The precise measurements for the apparent Young's modulus, a measure of material elasticity, range from ~0.1 to 10 kPa, depending on the experimental conditions and technique. This broad range of stiffness measures likely reflects a large degree of cell-to-cell variability, as well as different domains and mechanical behavior probed by the diverse experimental methods. For example, tracking of small particles within the nucleoplasm is sensitive to entanglement of the tracked particle within the nucleoskeleton/chromatin; in addition, the resulting measurements exclude contributions to nuclear stiffness from the nuclear envelope [90, 91]. In contrast, cell stretch experiments and other techniques that result in large nuclear deformations will yield "bulk" measurements that combine contributions from the nuclear interior and the nuclear envelope, but may also depend on the mechanical properties of the cytoskeleton and its connection to the nucleus [17].

Micropipette aspiration experiments [90–92] and computational modeling [105] indicate that the mechanical deformability of the nucleus is mainly governed by the nuclear lamina and the nuclear interior; the relative contribution of each component depends on diverse factors such as mechanical load (e.g., applied tension vs. compression), the specific cell type, differentiation state, and chromatin configuration. The contribution of the inner and outer nuclear membranes to the deformability of the nucleus is largely negligible [106], as lipid membranes exhibiting relatively low bending stiffness and a two-dimensional (2D) liquid-like behavior, i.e., they can flow in response to applied shear stress, with connections to a large membrane reservoir in the form of the ER [16, 106].

The importance of the nuclear lamina in providing structural support to the nucleus and controlling nuclear size is now well established [12, 17], with the nuclear lamina acting as a load-bearing, elastic shell surrounding a viscoelastic nuclear interior [90, 91, 107]. Experiments on cells from gene-modified mice lacking specific lamin isoforms [98] and *Xenopus* oocytes ectopically expressing human lamins [95] suggest that lamins A and C are the main contributors to nuclear stiffness, with loss of lamin A or C resulting in softer, more deformable nuclei, while increased expression of lamin A results in stiffer, less deformable nuclei. Given the structural similarities between A-type and B-type lamins, it may be somewhat surprising that these proteins have distinct roles in affecting nuclear deformability. However, recent findings suggest that A- and B-type lamins—and even lamins A and C—may form distinct but overlapping networks [108, 109], and that A-type lamins may form a thicker protein network at the nuclear envelope [110]; however, as imaging the nuclear lamina in intact somatic cells with sufficiently high resolution remains technically extremely challenging, the exact structure and organization of the lamina and the different lamin isoforms at the nuclear envelope remains unclear. Interaction of specific lamin isoforms with other nuclear (envelope) proteins may serve as additional explanation for the distinct roles of the diverse lamins in nuclear mechanics. For example, loss of the inner nuclear membrane protein emerin, which directly interacts with lamins A/C, results in more deformable nuclei, although to a lesser degree than functional loss of lamins A/C [92, 97]. In addition, functional loss of lamins due to mutations or (partial) deletion can also affects chromatin organization [111–114], which could affect nuclear deformability.

Further illustrating the importance of A-type lamins in nuclear mechanics, lamin A/C-deficient cells have more deformable nuclei that are more susceptible to rupture under mechanical stress [14, 115]. Of note, mutations in A-type lamins, as well as emerin, cause a spectrum of human diseases (laminopathies) that include

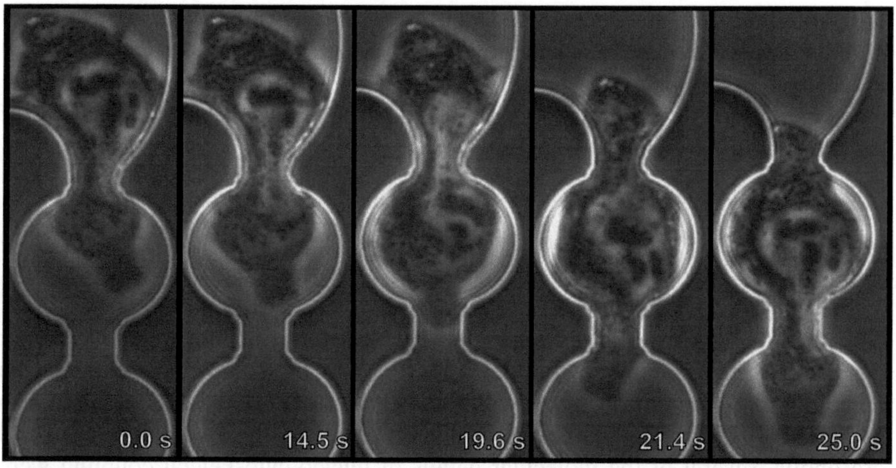

Fig. 2 Invasive cancer cell MDA-MB-231 squeezing into an 8 μm width constriction. Image sequences of a cancer cell being perfused through an 8 μm-wide constriction at a pressure difference (ΔP) of 10 psi. The viscoelastic deformation as the nucleus flows through the constriction is clearly visible

Emery–Dreifuss muscular dystrophy, limb-girdle muscular dystrophy, dilated cardiomyopathy, Dunnigan-type familial partial lipodystrophy, and Hutchinson–Gilford Progeria syndrome [51]. In many cases, cells from affected patients show characteristic features such as misshapen nuclei, increased nuclear fragility, and herniations [16]; furthermore, *LMNA* mutations resulting in disease affecting cardiac and skeletal muscle often cause defects in nuclear mechanics [116], providing a potential disease mechanism for the muscular laminopathies.

Importantly, lamins also interact with other inner nuclear membrane proteins (e.g., emerin, LAPs, and LBR), nuclear pore components, DNA, chromatin, and transcription factors (e.g., retinoblastoma protein [Rb], SREBPs, GCL, and MOK2), and structural proteins such as nuclear actin and titin [117]. These interactions could further modulate nuclear stiffness by forming nucleoskeletal structures or affecting chromatin organization and transcriptional regulation. For example, nuclear abnormalities have been observed in cells depleted of large repeat-domain proteins such as titin and αII-spectrin [118, 119]. On the other hand, the role of nuclear actin in providing structural support to the nucleus remains unclear [84]. Through their interaction with SUN proteins, nesprins, and Samp1, lamins also play an important role in connecting the nucleus to the surrounding cytoskeleton [120], as discussed in more detail below.

Besides the nuclear lamina, chromatin is an important contributor to nuclear stiffness. Unlike the mostly elastic nuclear lamina, chromatin exhibits more viscoelastic material behavior, i.e., it flows when subjected to forces (Fig. 2) and undergoes plastic deformations [106, 107]. Chromatin decondensation during initial lineage commitment of embryonic stem cells is associated with a significant softening of the nucleus [101]. Subsequently, the viscoelastic deformability of the cell nucleus in human embryonic stem cells changes during further cellular differentiation [107], becoming 6 times stiffer and also less fluid-like during terminal differentiation. It remains unclear, however, to what extent this behavior is caused by changes in chromatin organization, e.g., switching from loose euchromatin to more compacted heterochromatin, or results from the increased expression of A-type lamins in differentiated cells.

Nucleo-cytoskeletal Coupling

Over the last 10 years, it has become well established that the nucleus is physically coupled to the surrounding cytoskeleton [120]. Many of the molecular components are highly preserved throughout evolution, being present in unicellular organisms such as yeast all the way to mice and humans [121]. Building on work in yeast and drosophila, several of the molecular details of nucleo-cytoskeletal coupling were first unraveled in *Caenorhabditis elegans*, where UNC84 and ANC-1, in conjunction with Ce-lamin, participate in the actin-dependent anchorage and positioning of the nucleus [32, 122–125]. Subsequent studies have confirmed that closely related proteins are also responsible for nucleo-cytoskeletal coupling in mammalian cells; this physical connection is now commonly referred to as the *Li*nker of *N*ucleoskeleton and *C*ytoskeleton (LINC) complex [126]. In the strictest definition, the LINC complex contains two essential parts: (1) a member of the trimeric inner nuclear membrane SUN- [127] domain protein family, which engages with nucleoplasmic proteins such as lamins [121, 128, 129]; (2), KASH- (Klarsicht, ANC-1, Syne Homology) domain containing nesprins located on the outer nuclear membrane that bind across the perinuclear space to the SUN domain of Sun1/Sun2 trimers [130]. The cytoplasmic ends of nesprins interact directly or indirectly with various components of the cytoskeleton, including actin, intermediate filaments (via plectin) [131], and microtubules (via microtubule-binding motors such as dynein and kinesin), thereby completing the physical connection across the nuclear envelope [121]. In many cases, lamins are considered an extended part of the LINC complex, as they bind to SUN proteins and inner membrane variants of nesprins and help tether these proteins to the nuclear interior [132]. Since the cytoskeleton also connects to focal adhesion and cell-cell junctions, cells contain a continuous mechanical network linking the nuclear interior and the extracellular matrix and neighboring cells, thereby allowing forces exerted from the cellular environment or the cytoskeleton to be transmitted directly to the nuclear interior [39, 120, 133, 134].

SUN Domain Proteins

The characteristic feature of SUN domain family proteins is a 115–175 amino acid domain that shares homology with the Sad1 protein from *Schizosaccharomyces pombe* [135] and the UNC84 protein from *C. elegans* [122]. Mammalian cells have five SUN domain proteins, with two of these proteins (SUN1 and SUN2) present on the nuclear envelope in somatic cells (SUN3-5 are testis specific) [136]. SUN1 and SUN2 proteins consist of a helical N-terminal domain that can bind to lamins [137] and nuclear pore complex proteins [138, 139], a single pass transmembrane domain to anchor the protein in the inner nuclear membrane [140], a luminal helical domain required for trimerization of SUN proteins [130], and the C-terminal SUN domain, which interacts with the KASH domain of nesprins [126].

Nesprins and Other KASH Domain Proteins

Mammals have four nesprins (genes *SYNE 1–4*), with nesprins 1–3 having multiple isoforms resulting from alternative splicing, initiation, and termination [34, 120, 121]. Expression of various nesprin isoforms can be highly tissue-specific [34]. In skeletal muscle, levels of nesprin-1 (first described as Syne-1 for synaptic nuclear envelope protein-1) are highest in synaptic nuclei, suggesting that it might participate in the migration and anchoring of these specialized muscle nuclei [141]. Common to all nesprins is a central region containing multiple spectrin domains, whose number can greatly vary between isoforms [142]; all nesprins (but not all isoforms) contain a ~60 amino acid-long C-terminal KASH domain, consisting of a transmembrane domain and a short, highly conserved luminal domain, which is essential for anchoring nesprins to the nuclear envelop [59, 142]. The N-terminal domain of nesprins typically contains specific motifs to interact with different cytoskeletal proteins. For instance, the nesprin-1 and -2 "giant" isoforms (1,000 and 800 kDa in size, respectively) contain an actin-binding domain (ABD) composed of two calponin homology domains [35, 37, 143]; additionally, nesprins-1 and -2 can interact with the microtubule-associated motors dynein/dynactin and kinesin [120]. Nesprin-3 can connect to intermediate filaments via plectin [36]. Nesprin-4 binds the microtubule-associated motor kinesin, and ectopic expression of nesprin-4 induces dramatic changes in centrosome positioning in cells [144]. While localization of larger nesprin isoforms is restricted to the outer nuclear membrane, shorter isoforms can also be present at the inner nuclear membrane, where they can interact with lamins and emerin [38, 145–147]. Nesprin isoforms lacking the KASH domain may also be found in other cellular structures. In addition to nesprins 1–4, mammals express at least one additional KASH-domain protein, aptly named KASH5, which is found exclusively in spermatocytes and oocytes, where it plays a critical role in meiosis [148].

Other Molecules Involved in Nucleo-cytoskeletal Coupling

With the growing interest in understanding the mechanics of the nucleus and its connection to the cytoskeleton, several recent studies have focused on identifying additional molecular players involved in nucleo-cytoskeletal coupling. Based on experimental findings in emerin-deficient cells, one study has proposed that emerin binds to microtubules and that a subset of emerin located on the outer nuclear membrane is involved in coupling the centrosome to the nuclear envelope [149], but it remains unclear whether the emerin-microtubule interaction is direct or mediated through other proteins such as nesprins.

A more recent candidate to be involved in nucleo-cytoskeletal coupling is the inner nuclear membrane protein Samp1 [150], which associates with lamin A/C, emerin, Sun1, and Sun2 [150–152]. During mitosis, Samp1 is associated with the mitotic spindle [150]; during interphase, however, Samp1 is an important component of transmembrane actin-associated nuclear (TAN) lines [152], which promote rearward nuclear movement in polarizing fibroblasts by connecting the nucleus to retrograde actin flow via nesprin-2giant and SUN2 [153]. The involvement of lamins A/C in nucleo-cytoskeletal coupling is further illustrated by the finding that lamin mutants associated with muscular dystrophies can disrupt this retrograde nuclear movement [132] and that lamin A/C is required for retaining Samp1 at the nuclear envelope [152]. Another potential mediator of nucleo-cytoskeletal coupling is the luminal protein torsinA, part of the AAA + ATPase superfamily. TorsinA interacts with the KASH domains of nesprins 1–3, and loss of torsinA results in mislocalization of nesprin-3 from the nuclear envelope and impaired cell polarization and migration [131]. Given the promiscuous interaction of SUN domain proteins and nesprins [154], it is likely that tissue-specific expression of their isoforms, as well as potential interaction with other nuclear envelope proteins such as Samp1, play an important role in the spatial and temporal control of nucleo-cytoskeletal coupling.

Nucleo-cytoskeletal Coupling Is Critical for Many Cell Functions

Studies investigating molecules involved in connecting chromatin and cytoskeletal structures have often focused on processes during mitosis and meiosis. For instance, analysis of chromosome condensation during yeast prophase has unraveled a direct interaction between Sad1 (a Sun homologue protein) and meiotic-specific bouquet (Bqt) proteins [155]. Sad1 has also been linked to Kms1 protein [156] and this interaction is known to couple telomeres to microtubules and cytoplasmic dynein [157, 158]. Similar results were obtained in *C. elegans*, where selective inactivation of Sun1 protein or Kdp-1 (KASH domain protein-1) protein delays cell cycle progression [159, 160]. In mammalian cells, lamins, SUN proteins, KASH5, and Samp1 have all been implicated in specific roles during mitosis and/or meiosis [148, 161], and loss of A-type lamins causes telomere shortening defects and overall genomic instability [162].

In recent years, research has increasingly focused on the role of LINC complex proteins in interphase cells and consequences of LINC complex disruption. In *C. elegans*, deletion of the nesprin and SUN1 orthologues ANC-1 and UNC-84 result in impaired nuclear positioning and anchoring in muscle cells [32, 122]. In mammalian cells, LINC complex disruption causes defects in nuclear positioning, cell polarization, and migration [133] by impairing force transmission between the nucleus and cytoskeleton [24, 153]. LINC complex proteins are particularly important during cell migration in 3D environments, for example, inside collagen matrices or tissues. In particular, lamins A/C, nesprin-2giant, and nesprin3 modulate perinuclear actin organization and actin protrusions; consequently, deletion of lamins A/C or LINC complex disruption results in significantly impaired migration of cells in 3D collagen matrices [163]. The implications of impaired nucleo-cytoskeletal coupling in cancer progression are discussed in more detail below.

Nuclear Mechanics Stiffness and Nucleo-cytoskeletal Coupling in Mechanotransduction

As described above, the cytoskeleton physically connects the nucleus to the cellular microenvironment. Consequently, pulling on integrins on the surface of intact endothelial cells results not only in reorientation of cytoskeletal filaments, but also in distortion of the nucleus and spatial redistribution of subnuclear structures [134]. Similar results, including force-induced dissociation of nuclear protein complexes, have recently been obtained in HeLa cells subjected to forces applied via magnetic tweezers [164] and in human umbilical vein endothelial and osteosarcoma cells exposed to fluid shear stress [165]. It has long been speculated that such mechanically induced changes in nuclear structure and chromatin configuration could directly activate specific mechanosensitive genes, for example, by changing accessibility to transcription factors [18, 166]. This idea is further supported by studies that have found interactions between applied forces, Rho signaling, cell shape, and histone acetylation [167–169]. Nonetheless, direct evidence for such nuclear mechanosensing remains scarce, and the majority of data are rather correlative, making it difficult to discern whether mechanical forces acting on the nucleus are sufficient to directly induce changes in gene regulation, or whether the observed activation of mechanosensitive genes is the downstream result of signaling cascades originating in the cytoskeleton or the plasma membrane [15]. A recent study [24] addressing this question found that LINC complex disruption had no discernible effect on the mechanically induced expression of the mechanosensitive genes *Iex-1* and *Egr-1*, whose activation is impaired in lamin A/C-deficient cells [14, 170], even though LINC complex disruption resulted in substantially reduced nuclear deformation when the fibroblasts were subjected to substrate strain [24].

At the same time, changes in nuclear envelope composition undoubtedly affect cellular structure and function. For example, LINC complex disruption alters the mechanically induced proliferation of C2C12 myoblasts [171]; LINC complex depletion also causes impaired propagation of intracellular forces and disturbed organization of the perinuclear actin and intermediate filament networks, leading to defects in nuclear positioning and cell orientation [22, 24, 171]. In the case of impaired expression of mechanosensitive genes in lamin A/C- and emerin-deficient cells, it remains unclear whether this effect is due to direct mechanical defects or a consequence of altered interaction of lamins with specific transcriptional factors. An additional mechanism by which lamins and emerin can affect mechanotransduction signaling has recently been identified, revealing that the actin polymerization-promoting activity of emerin at the nuclear envelope can influence nuclear and cytoskeletal actin dynamics, thereby modulating localization and activity of the mechanosensitive transcription factor MKL1 (also known as MRTF-A or MAL), whose localization is dependent on interaction with monomeric G-actin [172].

Relevance of Nuclear Mechanics and Mechanotransduction in Cancer Progression

With growing advances in the understanding of the physics of cell motility, the mechanical properties of cancer cells have become an increasing area of interest [3]. As the nucleus is typically the largest and stiffest organelle, often occupying a large fraction of the cell's volume, the properties of the nucleus can dominate the overall cellular mechanical response when cells are subjected to large deformations [17]. Several lines of evidence suggest that the ability of the nucleus to deform can impose a rate-limiting step in non-proteolytic cell migration in 3D environments, when cells attempt to squeeze through narrow constrictions imposed by extracellular matrix fibers and other cells (Fig. 3) [19, 20]. In this section, we summarize changes in nuclear structure and morphology observed in various cancers and describe the role of nuclear deformability in cell motility. In addition, we discuss the intricate feedback between the mechanics of the cellular microenvironment and intracellular organization and function.

Altered Nuclear Structure and Morphology in Cancer Cells

With few exceptions, the nuclei of normal cells have an ellipsoid shape with smooth outlines; in contrast, many cancer cells are easily identifiable by increased nuclear size, irregular nuclear contours, and disturbed chromatin distribution, making nuclear morphology one of the oldest and most commonly used cancer markers [11]. The irregular nuclear outline in cancer cells is mainly the result of grooving, convolutions and invaginations of the nuclear envelope [173]. While the characteristic changes in nuclear morphology in cancer cells are well documented, their cause and consequence remain unclear. Interestingly, the irregular nuclear morphology of cancer cells often bears striking resemblance to the abnormal nuclear shapes observed in cells lacking or expressing mutant nuclear envelope proteins such as lamins A/C, lamin B1/B2, or LBR [174, 175], suggesting a possible involvement of dysregulated nuclear envelope proteins [173, 176].

This idea is supported by a growing number of publications that report altered expression of lamins in a variety of human tumors, often associated with particularly malignant phenotypes (Table 1). Interestingly, while some cancers frequently show downregulation of lamin A/C [177–179], other cancers have upregulated levels of lamins A/C [177, 180, 181], and for some cancers, such as colon cancer, both increased [182] and decreased [183] levels of lamin A/C have been reported.

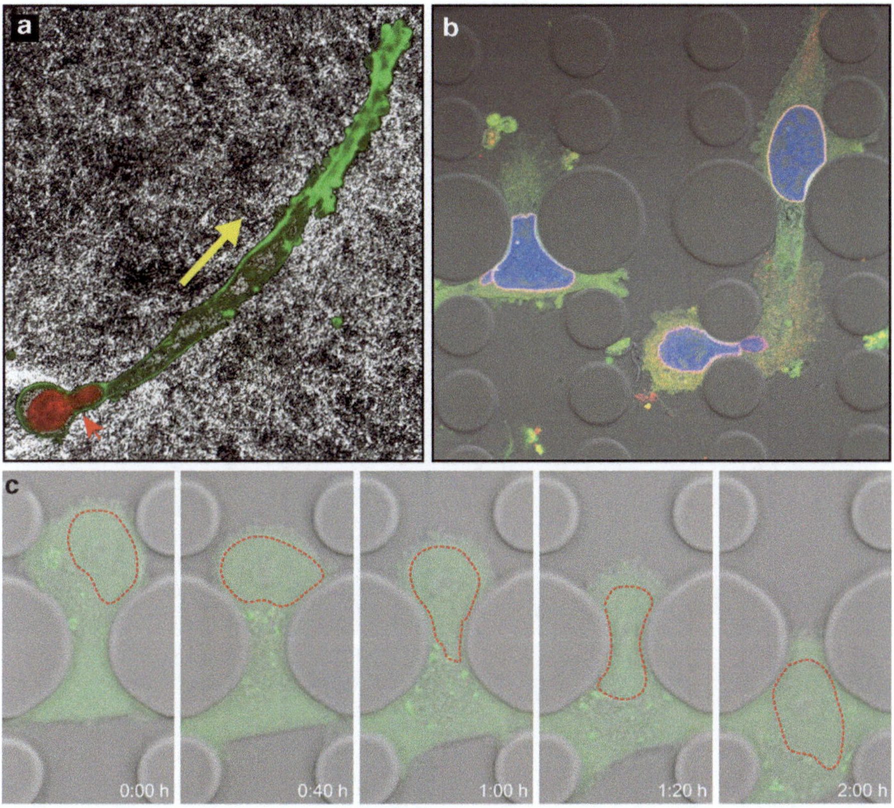

Fig. 3 Migration of cancer cell in a constrained environment. (**a**) Fibrosarcoma cell (HT1080 cell line) migrating through a dense collagen fiber matrix. The rat tail collagen matrix was imaged by reflection microscopy; the nucleus is visible in *red* (DAPI), F-actin in *green* (phalloidin). The cell body has already advanced in the direction of migration (*yellow arrow*), while the nucleus is still in the process of squeezing through constrictions in the collagen matrix (*red arrow head*). Image courtesy of Katarina Wolf, University of Nijmegen. (**b**) Fibrosarcoma cells (HT1080) migrating through 2 μm × 5 μm and 5 μm × 5 μm constrictions in a microfluidic channel. The cytoplasm is visible in *green*, the nucleus in *blue*, and the nuclear lamina (lamin B2) in *red*. (**c**) Time-lapse series of MDA-MB-231 breast cancer cell expressing a green fluorescent protein migrating through a 5 μm-wide constriction in a microfluidic channel. The nucleus is outlined in *red* (*dashed line*)

Furthermore, even within single tumors and individual cancer cell lines [184], highly heterogeneous expression levels of lamin A/C can be found [185]. Similarly, both high and low levels of lamins A/C have been considered poor prognostic markers for cancer patients, depending on the specific study and cancer subtype. For example, reduced lamin A/C expression is a sign of poor prognosis for patients with gastric carcinoma [186], and patients with stage II and III colon cancer have a significantly increased risk of cancer recurrence when their tumors are marked by loss of lamin A/C expression [183]. At the same time, another study found that patients with increased expression of lamins A/C in colorectal cancer tumors were almost

Table 1 Altered expression (and mutations) of nuclear envelope proteins in cancers

Protein	Cancer/tumor type	Reported change	Prognostic value	References
Emerin	Ovarian cancer	Loss of emerin		[222]
Lamins A/C	Lung cancer	Absence or very reduced expression in small cell lung carcinoma		[177]
	Colonic and gastric adenocarcinomas	Reduced levels and mislocalization (aberrant cytoplasmic immunolabelling)		[178]
	Other cancers: esophagus cancer, cervical and uterine cancer, breast cancer			
	Basal cell skin carcinoma	Low levels or absence of lamin A	Increased proliferation	[179]
	Basal cell skin carcinoma	Low levels of lamin C	Low proliferation	[179]
	Skin cancer	High levels of lamin A and C in the basal cell layer of the epidermis overlying basal cell carcinomas, squamous cell carcinomas, and actinic keratosis (AK)	Proliferative capacity	[180]
	Leukemia and lymphomas	Loss of gene expression by epigenetic silencing in nodal diffuse large B-cell lymphomas and acute lymphoblastic leukemias.	Poor outcome/overall survival	[223]
	Colorectal cancer	Increased expression (mainly lamin A)	Promote tumor invasiveness Poor prognosis (risk indicator of tumor related mortality)	[187]
	Ovarian serous cancer	High levels in all stages of ovarian serous carcinomas; increased immunoreactivity in the higher stage of tumor	Correlates with advanced stage	[224]
	Primary gastric carcinoma	Low levels	Poor histological differentiation; poor prognosis	[186]
	Prostate cancer	Low expression in lower grade; increased levels in higher grade	Correlates with advanced stage	[225]
	Colon cancer	Low expression in stage II and III patients	Correlates with increased relapse	[183]
	Ovarian cancer	Heterogeneous lamin A/C protein expression pattern or absence of lamin A/C and aneuploidy		[185]
	Breast cancer	Mutated		[190]

Lamins A/C	Breast cancer	Loss of lamins A/C	Higher lamin A/C expression was associated with better clinical outcomes and with better overall and disease-free survival	[243, 244]
Lamin B	Colon cancer	Reduced expression		[178]
	Colorectal carcinoma	Increased levels		[226]
	Ovarian cancer	Increased levels of lamin B1 and B2 in malignant cell compared to benign		[227]
	Hepatocellular carcinoma	Increased levels of Lamin B1 in cirrhotic tissue		[228]
	Liver cancer	Increased levels of lamin B1 in every stage (cirrhosis, early stage, late stage); presence of soluble lamin B1 in the circulation	Potential biomarker Correlate with the tumor development	[229]
	Prostate cancer	Increased levels of lamin B	Correlate with the tumor development	[230]
	Pancreatic cancer	Increased levels of lamin B1	Correlate with decreased levels of tumor differentiation, high metastatic potential and, poor overall survival	[231]
Lamin B1	Breast cancer	Reduced mRNA levels of lamin B1	Expression of LMNB1 declined with worsening clinical outcome	[244]
LAP2	Malignant lymphocytes	Increased levels	LAP2β correlates with highly proliferative malignant cells	[232]
Nesprins	Ovarian cancer	Nesprin 1 polymorphism Downregulation of a transcript (shorter isoform)	Associated with invasive ovarian cancer risk	[191]
	Colorectal cancer	Nesprin 1 is a candidate cancer gene (mutated in cancer)		[189]
	Breast cancer	Nesprin 1 (mutations)		[190]
	Breast cancer	Nesprin 2 (mutations)		[189, 190]

(continued)

Table 1 (continued)

Protein	Cancer/tumor type	Reported change	Prognostic value	References
NUP 88	Ovarian cancer	Increased levels		[233]
	Different type of cancers: sarcomas, lymphomas, mesotheliomas, and breast cancer	Increased levels	Correlates with high-grade malignancies	[234–237]
	Colorectal cancer and hepatocellular carcinoma	Increased levels	Correlates with poor differentiation	[238, 239]
NUP 98	Leukemia	Increased levels; may act as a component of a chromosomal translocation		[240]
NUP 214	Uterine, stomach and rectal tumors, leukemias, breast cancer	Increased levels; may act as a multifunctional oncogene and as a component of a chromosomal translocation		[189, 240–242]

twice as likely to die of the disease than patients with tumors negative for lamin A/C [187], possibly by lamin A/C promoting cell motility [188]. These apparently inconsistent findings point at the multiple roles lamin can play in cancer progression, which will be discussed in more detail below.

In addition to lamins, other nuclear (envelope) proteins have recently been implicated in a variety of cancers. A genome-wide scan in several patients with either breast, colorectal or ovarian cancer revealed genetic alterations in nesprin-1 [189], and another genome-wide study identified mutations in nesprin-1, -2 and lamin A/C in a panel of 100 breast cancer patients [190]. Furthermore, downregulation and mutations in nesprin have been associated with an increased risk of invasive ovarian cancer [191]. Lastly, several "nuclear matrix" or nucleoskeletal-associated proteins such as NuMA or nucleoporin proteins (NUP 88, NUP 98) have been correlated with aggressive tumor phenotypes [192] and used as prognostic markers of disease [193].

Implications of Altered Nuclear Envelope Composition in Cancer

What is the impact of altered nuclear envelope composition on nuclear mechanics? As lamin expression and chromatin organization determine nuclear deformability, it is expected that changes in nuclear architecture will alter the rigidity of the nucleus. In cancer, increased nuclear deformability may benefit metastatic cells that need to pass through narrow interstitial spaces or small capillaries, while defects in nucleo-cytoskeletal coupling may impair migration in 3D tissues [20]. In addition to these mechanical functions, the nuclear envelope and nuclear interior play important roles in the processing of genetic information [194–196]. Thus, changes in nuclear organization could have consequences on gene expression or DNA stability with important implications in cancer progression.

Nuclear Deformability and Cell Motility

The abnormal nuclear shapes observed in cancer cells and their resemblance to lamin-deficient or mutant cells, combined with the increasing reports of altered expression of nuclear envelope proteins in various cancers (Table 1), suggests that cancer cells may have altered nuclear mechanics. While direct measurements of nuclear deformability in cancer cells have not yet been reported, studies that measure whole-cell deformability consistently find that cancer cells, particularly highly invasive ones, have increased cellular deformability [3, 4, 7]. Why should (nuclear) deformability matter in cancer progression? During the metastatic process, cancer cells must undergo modifications and large elastic deformations to invade the tissue surrounding the primary tumor, intravasate blood vessels, survive the physical stresses during circulation in the blood stream, extravasate at new sites in the body, and eventually proliferate in a nutrient-deprived microenvironment [197].

Particularly during invasion and intravasation and extravasation, cells penetrate through interstitial spaces and openings ranging in size from 2 to 30 μm [198, 199]. Cytoskeletal shape is highly adaptive, owing to the rapid cytoskeletal remodeling and plasma membrane flexibility; consequently, cytoskeletal protrusions can invade spaces of less than 1 μm^2 in cross section [200, 201]. In contrast, the ability of the nucleus to pass through narrow constrictions is more limited due to its size and stiffness. Transient nuclear deformations, resulting in hourglass- and cigar-shaped nuclei, as well as nuclear protrusions indicative of attempts to pass through narrow constrictions, can be observed (at least transiently) during cancer cell migration in vivo [20]. Importantly, a recent report by Friedl, Wolf, and colleagues [19] found that deformation of the nucleus poses a rate-limiting step during proteolysis-independent cell migration. They found that in the absence of proteolysis, e.g., during matrix metalloprotease (MMP) inhibition or knockdown, migration of cancer cells through 3D collagen matrices and polycarbonate filters is limited by the available pore size: cell migration speed and migration efficiency gradually drops with decreasing cross-sectional areas of the constrictions until cell body movement is completely stalled [19]. A similar size-dependent effect was observed by Tong and colleagues [202] when studying cell migration in microchannels with varying width. Indeed, decreasing channel width below 20 μm (at a fixed channel height of 10 μm) resulted in increasing reduction in migration speed. At the extreme, cells in 3 μm-wide channels had a 70 % reduction in migration speed compared to 50 and 20 μm-wide channels. Interestingly, the minimum size requirement for (non-proteolytic) migration through 3D environments was found to be independent of the shape of the constriction and only depends on the available cross-sectional area [19].

While these studies illustrate the importance of nuclear deformability in cell migration in confined environments, the role of the nuclear lamina and nuclear stiffness in this process remains to be explored [20]. At least in neutrophil-like cells, which normally have extremely low levels of lamins A/C and which can migrate through constrictions only a few micrometers in diameter, overexpression of lamin A results in less deformable nuclei that have reduced efficiency at crossing narrow constrictions and that take significantly longer to transit narrow microfluidic channels mimicking capillaries [103]. Similarly, fibroblasts expressing a mutant form of lamin A (progerin) that is responsible for Hutchinson–Gilford progeria syndrome have difficulties migrating through an array of microfabricated pillars spaced 6 μm apart [203], likely due to the increased nuclear stiffness caused by progerin [204, 205], as migration on non-constricted surfaces was comparable to cells from healthy controls [203]. Although these findings suggest an important role of lamins A/C in moderating the ability of cells to pass through narrow constrictions, Wolf and colleagues [19] found that the maximal deformation the nucleus could achieve during passage through narrow constrictions, indicated as the ratio of the nuclear cross section in the constriction to the undeformed nuclear cross section, was consistently around 1:10, regardless of the cell type studied. These findings suggest that the size limit for nuclear passage through small constriction may be governed by the maximal compressibility of the nucleus. The theoretically maximal compression depends on the solid fraction of the nucleus, as the chromatin (and other nucleoplasmic proteins) can be no further compressed once all void spaces have been eliminated. This idea is consistent with the observed reduction in nuclear volume by up to 60 % during migration of skin fibroblasts through microfabricated constrictions [203] and with micropipette aspiration experiments that revealed that the nuclear volume can be compacted to about 20–40 % of its original size before reaching a state that resists further compression [92, 106].

But what about cancers in which increased, rather than decreased, levels of lamin A/C have been reported, which is expected to result in reduced nuclear deformability [98]? Cancer cells are highly plastic and heterogeneous in their gene expression, so it is likely that different subpopulations of cells with distinct roles in cancer progression exist. Increased lamin levels could help protect cells from mechanical stress caused by the high hydrostatic pressure inside solid tumors. At the same time, lamins are also involved in multiple signaling pathways [51, 117], which could modulate functions relevant to cancer progression. For example, increased levels of lamin A/C in prostate cancer cause changes in the PI3K/AKT/PTEN pathway [206], and upregulation of lamin A/C in colorectal cancer induces changes in cytoskeletal organization that promote cell motility [188]. As such, it is likely that different cells and tumors have found different approaches to find the best compromise between increasing nuclear deformability and activation of signaling pathways to increase cell motility and invasiveness.

Nuclear Rupture of Cancer Cells

As described earlier, the nuclear envelope forms a well-defined compartment that acts as a protective shield for the genetic material. In normal cells, nuclear envelope breakdown and reassembly is limited to mitosis and precisely regulated [207]. Recently, Vargas et al. [208] have reported that in many cancer cells, the nuclear envelope transiently ruptures and then reseals during interphase, resulting in temporary exchange between the nucleus and cytoplasm and the occasional entrapment of cytoplasmic organelles inside the nucleus. Nuclear envelope rupture was associated with the formation of micronuclei, portions of chromatin exiting the nuclear interior, and mislocalization of nucleoplasmic/cytoplasmic proteins. Importantly, the frequency of nuclear rupture events was increased in cells with small defects in the nuclear lamina [208]. These results are consistent with previous reports of increased nuclear fragility in lamin A/C-deficient mouse embryonic fibroblasts [14] and spontaneous (transient) nuclear rupture in these cells [209]. In our laboratory, we have frequently observed that cancer cells undergo transient nuclear rupture while migrating through narrow (~2 μm × 5 μm) microfluidic constrictions, with lamin-deficient cells displaying significantly increased rates of nuclear rupture (unpublished observations). Breakdown of the nuclear compartment during repetitive nuclear rupture could potentially result in increased genomic instability and chromatin rearrangements, which could further contribute to cancer progression, but this idea has not yet been experimentally tested.

Changes in Chromatin Organization in Cancer Cells

Epigenetic changes in chromatin configuration can directly impact nuclear stiffness. Therefore, the chromatin modifications frequently observed in cancer cells, including disturbed heterochromatin organization [11], could be associated with altered nuclear deformability and thereby affect 3D cell migration, in addition to their role in transcriptional activity. Importantly, there is a strong interplay between nuclear envelope proteins and chromatin organization. Lamin A regulates dynamics of heterochromatin proteins in early embryonic stem cells [25]; lamins A/C-deficiency and mutations in the *LMNA* gene result in loss of heterochromatin [111, 210]. Furthermore, lamins and lamin B receptor (LBR) play an important role in tethering specific chromatin regions to the nuclear periphery [211, 212], which typically serves as a transcriptionally repressive environment [195]. LBR also interacts with heterochromatin protein 1 [213] and histones H3/H4 [213]. Lamin-associated polypeptide-2β (LAP2β) can modulate gene expression by regulating higher order chromatin structure or binding the transcriptional repressors germ cell less (GCL) [214] and histone deacetylase 3 [215], resulting in deacetylation of histone H4 [215]. Emerin can directly associate with chromatin modifiers and transcriptional repressors such as the death promoting factor Btf [216], the splicing associated factor YT521-B [217], and the transcriptional repressor GCL [218]. Given these findings, it is tempting to speculate that the altered expression of nuclear envelope proteins found in various cancers (Table 1) can directly affect chromatin organization and gene expression. Of course, the observed changes in expression of nuclear envelope proteins could also be the consequence, rather than the cause of altered chromatin organization. In this case, the changes in nuclear envelope composition could still result in further modifications of nuclear structure and organization while also directly altering nuclear mechanics.

Conclusion and Future Perspectives

The field of cancer cell biology has dramatically changed since 1943, when George Papanicolaou published his book *Diagnosis of Uterine Cancer by the Vaginal Smear*, which laid the basis for the now abundant "pap smear" to detect early signs of cervical cancer. Since then, researchers and clinicians have learned not only to identify and assess cancer cells based on characteristic morphological changes, but also to peek inside the inner life of cancer cells, including their genetic changes, biochemical composition, and metabolic state. In recent years, these approaches have been complemented by a new research direction, focused on the biophysical changes in cancer cells and their microenvironment. This research has already led to striking discoveries, including the role of the extracellular matrix stiffness, composition and topology in cancer progression [219] and the characteristic difference in cell deformability of cancer cells, which may lead to new diagnostic and prognostic applications [3]. Motivated by research in other diseases (laminopathies), it is now emerging that the mechanical properties of the cell nucleus, particularly its deformability and connection to the cytoskeleton, may play a similarly important role in cancer metastasis. The idea that deformation of the large and stiff nucleus presents a rate-limiting factor during the passage of metastatic cancer cells through tight interstitial spaces or narrow capillaries has recently found increasing experimental support [19, 103, 165]. Given the increasing reports of altered expression and mutations in nuclear envelope proteins responsible for determining nuclear stiffness, it is intriguing to speculate that (a subset of) cancer cells may have acquired specific adaptations in their nuclear structure and mechanics to promote metastatic spreading. Nonetheless, experimental verification of this idea is still lacking. Additional experiments, using sophisticated combinations of live cell imaging and measurements of subcellular mechanics, including primary tumor (and metastatic) cells from cancer patients and complemented by in vivo studies in mouse models, will be required to firmly establish this hypothesis. These experiments will also have to address why some cancers frequently have increased lamin levels while others have decreased or unchanged levels, and whether such changes in nuclear envelope composition can serve as reliable prognostic markers. Given the diverse functions of lamins, it is likely that (varying) combinations of altered cellular mechanics, cell signaling, and stem cell differentiation contribute to the increasingly emerging role of lamins in cancer progression. Done correctly, such experiments have the potential to not only address these key questions but to also produce novel insights into the dynamic nature of cancer cells, which may switch between different morphological and mechanical modes depending on their current role in cancer progression. Novel technology developments to probe single cell mechanics at substantial higher throughput than traditional methods [5, 102, 220, 221] will enable detection of rare cell subpopulations, which could play a crucial role in cancer progression. Identifying key (mechanical) parameters that govern cancer cell metastasis may reveal novel therapeutic targets for pharmacological inhibition.

These clinical translation-driven experiments should be complemented by research to address some of the more fundamental questions in cancer cell biology, including the molecular mechanisms by which cells manage to squeeze the nucleus through constrictions only one tenth the diameter of the nucleus in size, and whether induced nuclear deformations can directly contribute to cellular mechanosensing. We are only at the beginning of a long road ahead, the destination a complete understanding of the physics of cancer progression und the underlying biology, but it will be exciting to see what is awaiting us around the next corner.

Acknowledgements We apologize to all authors whose work could not be cited due to space constraints. This work was supported by National Institutes of Health awards [R01 NS059348 and R01 HL082792], a National Science Foundation CAREER award to J. L. [CBET-1254846], the Department of Defense Breast Cancer Idea Award [BC102152], and an award from the Progeria Research Foundation [PRF2011-0035].

References

1. Fawcett DW (1966) An Atlas of fine structure. The cell, 2nd edn. W. B. Saunders Company, PA
2. Hanahan D, Weinberg RA (2011) Hallmarks of cancer: the next generation. Cell 144(5): 646–674. doi:10.1016/j.cell.2011.02.013
3. Suresh S (2007) Biomechanics and biophysics of cancer cells. Acta Biomater 3(4):413–438. doi:10.1016/j.actbio.2007.04.002
4. Guck J, Schinkinger S, Lincoln B, Wottawah F, Ebert S, Romeyke M, Lenz D, Erickson HM, Ananthakrishnan R, Mitchell D, Käs J, Ulvick S, Bilby C (2005) Optical deformability as an inherent cell marker for testing malignant transformation and metastatic competence. Biophys J 88(5):3689–3698. doi:10.1529/biophysj.104.045476
5. Remmerbach TW, Wottawah F, Dietrich J, Lincoln B, Wittekind C, Guck J (2009) Oral cancer diagnosis by mechanical phenotyping. Cancer Res 69(5):1728–1732. doi:10.1158/0008-5472.can-08-4073
6. Byun S, Son S, Amodei D, Cermak N, Shaw J, Kang JH, Hecht VC, Winslow MM, Jacks T, Mallick P, Manalis SR (2013) Characterizing deformability and surface friction of cancer cells. Proc Natl Acad Sci U S A 110(19):7580–7585. doi:10.1073/pnas.1218806110
7. Cross SE, Jin YS, Rao J, Gimzewski JK (2007) Nanomechanical analysis of cells from cancer patients. Nat Nanotechnol 2(12):780–783. doi:10.1038/nnano.2007.388
8. Kraning-Rush CM, Califano JP, Reinhart-King CA (2012) Cellular traction stresses increase with increasing metastatic potential. PLoS One 7(2):e32572. doi:10.1371/journal.pone.0032572
9. Baker EL, Lu J, Yu D, Bonnecaze RT, Zaman MH (2010) Cancer cell stiffness: integrated roles of three-dimensional matrix stiffness and transforming potential. Biophys J 99(7): 2048–2057. doi:10.1016/j.bpj.2010.07.051
10. Paszek MJ, Zahir N, Johnson KR, Lakins JN, Rozenberg GI, Gefen A, Reinhart-King CA, Margulies SS, Dembo M, Boettiger D, Hammer DA, Weaver VM (2005) Tensional homeostasis and the malignant phenotype. Cancer Cell 8(3):241–254. doi:10.1016/j.ccr.2005.08.010
11. Zink D, Fischer AH, Nickerson JA (2004) Nuclear structure in cancer cells. Nat Rev Cancer 4(9):677–687. doi:10.1038/nrc1430
12. Ho CY, Lammerding J (2012) Lamins at a glance. J Cell Sci 125(Pt 9):2087–2093. doi:10.1242/jcs.087288
13. de Las Heras JI, Batrakou DG, Schirmer EC (2013) Cancer biology and the nuclear envelope: a convoluted relationship. Semin Cancer Biol 23(2):125–137. doi:10.1016/j.semcancer.2012.01.008
14. Lammerding J, Schulze PC, Takahashi T, Kozlov S, Sullivan T, Kamm RD, Stewart CL, Lee RT (2004) Lamin A/C deficiency causes defective nuclear mechanics and mechanotransduction. J Clin Invest 113(3):370–378. doi:10.1172/jci19670
15. Dahl KN, Ribeiro AJ, Lammerding J (2008) Nuclear shape, mechanics, and mechanotransduction. Circ Res 102(11):1307–1318. doi:10.1161/circresaha.108.173989
16. Zwerger M, Ho CY, Lammerding J (2011) Nuclear mechanics in disease. Annu Rev Biomed Eng 13:397–428. doi:10.1146/annurev-bioeng-071910-124736
17. Lammerding J (2011) Mechanics of the nucleus. Compr Physiol 1(2):783–807. doi:10.1002/cphy.c100038
18. Wang N, Tytell JD, Ingber DE (2009) Mechanotransduction at a distance: mechanically coupling the extracellular matrix with the nucleus. Nat Rev Mol Cell Biol 10(1):75–82. doi:10.1038/nrm2594
19. Wolf K, Te Lindert M, Krause M, Alexander S, Te Riet J, Willis AL, Hoffman RM, Figdor CG, Weiss SJ, Friedl P (2013) Physical limits of cell migration: control by ECM space and nuclear deformation and tuning by proteolysis and traction force. J Cell Biol 201(7): 1069–1084. doi:10.1083/jcb.201210152
20. Friedl P, Wolf K, Lammerding J (2011) Nuclear mechanics during cell migration. Curr Opin Cell Biol 23(1):55–64. doi:10.1016/j.ceb.2010.10.015

21. Hale CM, Shrestha AL, Khatau SB, Stewart-Hutchinson PJ, Hernandez L, Stewart CL, Hodzic D, Wirtz D (2008) Dysfunctional connections between the nucleus and the actin and microtubule networks in laminopathic models. Biophys J 95(11):5462–5475. doi:10.1529/biophysj.108.139428
22. Chancellor TJ, Lee J, Thodeti CK, Lele T (2010) Actomyosin tension exerted on the nucleus through nesprin-1 connections influences endothelial cell adhesion, migration, and cyclic strain-induced reorientation. Biophys J 99(1):115–123. doi:10.1016/j.bpj.2010.04.011
23. Schneider M, Lu W, Neumann S, Brachner A, Gotzmann J, Noegel AA, Karakesisoglou I (2011) Molecular mechanisms of centrosome and cytoskeleton anchorage at the nuclear envelope. Cell Mol Life Sci 68(9):1593–1610. doi:10.1007/s00018-010-0535-z
24. Lombardi ML, Jaalouk DE, Shanahan CM, Burke B, Roux KJ, Lammerding J (2011) The interaction between nesprins and sun proteins at the nuclear envelope is critical for force transmission between the nucleus and cytoskeleton. J Biol Chem 286(30):26743–26753. doi:10.1074/jbc.M111.233700
25. Melcer S, Hezroni H, Rand E, Nissim-Rafinia M, Skoultchi A, Stewart CL, Bustin M, Meshorer E (2012) Histone modifications and lamin A regulate chromatin protein dynamics in early embryonic stem cell differentiation. Nat Commun 3:910. doi:10.1038/ncomms1915
26. Handwerger KE, Gall JG (2006) Subnuclear organelles: new insights into form and function. Trends Cell Biol 16(1):19–26. doi:10.1016/j.tcb.2005.11.005
27. Zhao R, Bodnar MS, Spector DL (2009) Nuclear neighborhoods and gene expression. Curr Opin Genet Dev 19(2):172–179. doi:10.1016/j.gde.2009.02.007
28. Dundr M, Misteli T (2010) Biogenesis of nuclear bodies. Cold Spring Harb Perspect Biol 2(12):a000711. doi:10.1101/cshperspect.a000711
29. Franke WW, Scheer U, Krohne G, Jarasch ED (1981) The nuclear envelope and the architecture of the nuclear periphery. J Cell Biol 91(3 Pt 2):39s–50s
30. Schirmer EC, Foisner R (2007) Proteins that associate with lamins: many faces, many functions. Exp Cell Res 313(10):2167–2179. doi:10.1016/j.yexcr.2007.03.012
31. Puddington L, Lively MO, Lyles DS (1985) Role of the nuclear envelope in synthesis, processing, and transport of membrane glycoproteins. J Biol Chem 260(9):5641–5647
32. Starr DA, Han M (2002) Role of ANC-1 in tethering nuclei to the actin cytoskeleton. Science 298(5592):406–409. doi:10.1126/science.1075119
33. Fridkin A, Mills E, Margalit A, Neufeld E, Lee KK, Feinstein N, Cohen M, Wilson KL, Gruenbaum Y (2004) Matefin, a Caenorhabditis elegans germ line-specific SUN-domain nuclear membrane protein, is essential for early embryonic and germ cell development. Proc Natl Acad Sci U S A 101(18):6987–6992. doi:10.1073/pnas.0307880101
34. Rajgor D, Mellad JA, Autore F, Zhang Q, Shanahan CM (2012) Multiple novel nesprin-1 and nesprin-2 variants act as versatile tissue-specific intracellular scaffolds. PLoS One 7(7):e40098. doi:10.1371/journal.pone.0040098
35. Padmakumar VC, Abraham S, Braune S, Noegel AA, Tunggal B, Karakesisoglou I, Korenbaum E (2004) Enaptin, a giant actin-binding protein, is an element of the nuclear membrane and the actin cytoskeleton. Exp Cell Res 295(2):330–339. doi:10.1016/j.yexcr.2004.01.014
36. Wilhelmsen K, Litjens SH, Kuikman I, Tshimbalanga N, Janssen H, van den Bout I, Raymond K, Sonnenberg A (2005) Nesprin-3, a novel outer nuclear membrane protein, associates with the cytoskeletal linker protein plectin. J Cell Biol 171(5):799–810. doi:10.1083/jcb.200506083
37. Zhang Q, Skepper JN, Yang F, Davies JD, Hegyi L, Roberts RG, Weissberg PL, Ellis JA, Shanahan CM (2001) Nesprins: a novel family of spectrin-repeat-containing proteins that localize to the nuclear membrane in multiple tissues. J Cell Sci 114(Pt 24):4485–4498
38. Zhang Q, Ragnauth CD, Skepper JN, Worth NF, Warren DT, Roberts RG, Weissberg PL, Ellis JA, Shanahan CM (2005) Nesprin-2 is a multi-isomeric protein that binds lamin and emerin at the nuclear envelope and forms a subcellular network in skeletal muscle. J Cell Sci 118(Pt 4):673–687. doi:10.1242/jcs.01642
39. Chen D, Zhao M, Mundy G (2004) Bone morphogenetic proteins. Growth Factors 22(4): 233–241, doi:HHV6108EX6P056CA [pii]10.1080/08977190412331279890

40. Wischnitzer S (1958) An electron microscope study of the nuclear envelope of amphibian oocytes. J Ultrastruct Res 1(3):201–222
41. Gerace L, Foisner R (1994) Integral membrane proteins and dynamic organization of the nuclear envelope. Trends Cell Biol 4(4):127–131
42. Grossman E, Medalia O, Zwerger M (2012) Functional architecture of the nuclear pore complex. Annu Rev Biophys 41:557–584. doi:10.1146/annurev-biophys-050511-102328
43. Nakielny S, Dreyfuss G (1999) Transport of proteins and RNAs in and out of the nucleus. Cell 99(7):677–690
44. D'Angelo MA, Anderson DJ, Richard E, Hetzer MW (2006) Nuclear pores form de novo from both sides of the nuclear envelope. Science 312(5772):440–443. doi:10.1126/science.1124196
45. D'Angelo MA, Hetzer MW (2008) Structure, dynamics and function of nuclear pore complexes. Trends Cell Biol 18(10):456–466. doi:10.1016/j.tcb.2008.07.009
46. Dreger M, Bengtsson L, Schöneberg T, Otto H, Hucho F (2001) Nuclear envelope proteomics: novel integral membrane proteins of the inner nuclear membrane. Proc Natl Acad Sci U S A 98(21):11943–11948. doi:10.1073/pnas.211201898
47. Korfali N, Wilkie GS, Swanson SK, Srsen V, Batrakou DG, Fairley EA, Malik P, Zuleger N, Goncharevich A, de Las HJ, Kelly DA, Kerr AR, Florens L, Schirmer EC (2010) The leukocyte nuclear envelope proteome varies with cell activation and contains novel transmembrane proteins that affect genome architecture. Mol Cell Proteomics 9(12):2571–2585. doi:10.1074/mcp.M110.002915
48. Schirmer EC, Florens L, Guan T, Yates JR, Gerace L (2003) Nuclear membrane proteins with potential disease links found by subtractive proteomics. Science 301(5638):1380–1382. doi:10.1126/science.1088176
49. Wilkie GS, Korfali N, Swanson SK, Malik P, Srsen V, Batrakou DG, de las Heras J, Zuleger N, Kerr AR, Florens L, Schirmer EC (2011) Several novel nuclear envelope transmembrane proteins identified in skeletal muscle have cytoskeletal associations. Mol Cell Proteomics 10(1):M110.003129. doi:10.1074/mcp.M110.003129
50. Korfali N, Wilkie GS, Swanson SK, Srsen V, de Las HJ, Batrakou DG, Malik P, Zuleger N, Kerr AR, Florens L, Schirmer EC (2012) The nuclear envelope proteome differs notably between tissues. Nucleus 3(6):552–564. doi:10.4161/nucl.22257
51. Schreiber KH, Kennedy BK (2013) When lamins go bad: nuclear structure and disease. Cell 152(6):1365–1375. doi:10.1016/j.cell.2013.02.015
52. Paddy MR, Belmont AS, Saumweber H, Agard DA, Sedat JW (1990) Interphase nuclear envelope lamins form a discontinuous network that interacts with only a fraction of the chromatin in the nuclear periphery. Cell 62(1):89–106
53. Strelkov S, Herrmann H, Aebi U (2003) Molecular architecture of intermediate filaments. Bioessays 25(3):243–251. doi:10.1002/bies.10246
54. Aebi U, Cohn J, Buhle L, Gerace L (1986) The nuclear lamina is a meshwork of intermediate-type filaments. Nature 323(6088):560–564. doi:10.1038/323560a0
55. Herrmann H, Aebi U (2004) Intermediate filaments: molecular structure, assembly mechanism, and integration into functionally distinct intracellular Scaffolds. Annu Rev Biochem 73:749–789. doi:10.1146/annurev.biochem.73.011303.073823
56. Peter A, Reimer S (2012) Evolution of the lamin protein family: what introns can tell. Nucleus 3(1):44–59
57. Batsios P, Peter T, Baumann O, Stick R, Meyer I, Gräf R (2012) A lamin in lower eukaryotes? Nucleus 3(3):237–243. doi:10.4161/nucl.20149
58. Stuurman N, Heins S, Aebi U (1998) Nuclear lamins: their structure, assembly, and interactions. J Struct Biol 122(1–2):42–66. doi:10.1006/jsbi.1998.3987
59. Gruenbaum Y, Margalit A, Goldman RD, Shumaker DK, Wilson KL (2005) The nuclear lamina comes of age. Nat Rev Mol Cell Biol 6(1):21–31. doi:10.1038/nrm1550
60. Goldman RD, Gruenbaum Y, Moir RD, Shumaker DK, Spann TP (2002) Nuclear lamins: building blocks of nuclear architecture. Genes Dev 16(5):533–547. doi:10.1101/gad.960502

61. Kitten GT, Nigg EA (1991) The CaaX motif is required for isoprenylation, carboxyl methylation, and nuclear membrane association of lamin B2. J Cell Biol 113(1):13–23
62. Zastrow MS, Vlcek S, Wilson KL (2004) Proteins that bind A-type lamins: integrating isolated clues. J Cell Sci 117(Pt 7):979–987. doi:10.1242/jcs.01102
63. Dechat T, Gesson K, Foisner R (2010) Lamina-independent lamins in the nuclear interior serve important functions. Cold Spring Harb Symp Quant Biol 75:533–543. doi:10.1101/sqb.2010.75.018
64. Höger TH, Krohne G, Franke WW (1988) Amino acid sequence and molecular characterization of murine lamin B as deduced from cDNA clones. Eur J Cell Biol 47(2):283–290
65. Höger TH, Zatloukal K, Waizenegger I, Krohne G (1990) Characterization of a second highly conserved B-type lamin present in cells previously thought to contain only a single B-type lamin. Chromosoma 99(6):379–390
66. Lin F, Worman HJ (1993) Structural organization of the human gene encoding nuclear lamin A and nuclear lamin C. J Biol Chem 268(22):16321–16326
67. Lin F, Worman HJ (1995) Structural organization of the human gene (LMNB1) encoding nuclear lamin B1. Genomics 27(2):230–236. doi:10.1006/geno.1995.1036
68. Moir RD, Spann TP, Lopez-Soler RI, Yoon M, Goldman AE, Khuon S, Goldman RD (2000) Review: the dynamics of the nuclear lamins during the cell cycle—relationship between structure and function. J Struct Biol 129(2–3):324–334. doi:10.1006/jsbi.2000.4251
69. Raska I, Koberna K, Malínský J, Fidlerová H, Masata M (2004) The nucleolus and transcription of ribosomal genes. Biol Cell 96(8):579–594. doi:10.1016/j.biolcel.2004.04.015
70. Matera AG (2003) Cajal bodies. Curr Biol 13(13):R503
71. Matera AG, Frey MR (1998) Coiled bodies and gems: Janus or gemini? Am J Hum Genet 63(2):317–321. doi:10.1086/301992
72. Dellaire G, Bazett-Jones DP (2004) PML nuclear bodies: dynamic sensors of DNA damage and cellular stress. Bioessays 26(9):963–977. doi:10.1002/bies.20089
73. Nyman U, Hallman H, Hadlaczky G, Pettersson I, Sharp G, Ringertz NR (1986) Intranuclear localization of snRNP antigens. J Cell Biol 102(1):137–144
74. Pederson T, Aebi U (2002) Actin in the nucleus: what form and what for? J Struct Biol 140(1–3):3–9
75. Hofmann WA (2009) Cell and molecular biology of nuclear actin. Int Rev Cell Mol Biol 273:219–263. doi:10.1016/s1937-6448(08)01806-6
76. Nowak G, Pestic-Dragovich L, Hozák P, Philimonenko A, Simerly C, Schatten G, de Lanerolle P (1997) Evidence for the presence of myosin I in the nucleus. J Biol Chem 272(27):17176–17181
77. de Lanerolle P, Serebryannyy L (2011) Nuclear actin and myosins: life without filaments. Nat Cell Biol 13(11):1282–1288. doi:10.1038/ncb2364
78. Young KG, Kothary R (2005) Spectrin repeat proteins in the nucleus. Bioessays 27(2): 144–152. doi:10.1002/bies.20177
79. Zastrow MS, Flaherty DB, Benian GM, Wilson KL (2006) Nuclear titin interacts with A- and B-type lamins in vitro and in vivo. J Cell Sci 119(Pt 2):239–249. doi:10.1242/jcs.02728
80. Rando OJ, Zhao K, Crabtree GR (2000) Searching for a function for nuclear actin. Trends Cell Biol 10(3):92–97
81. McDonald D, Carrero G, Andrin C, de Vries G, Hendzel MJ (2006) Nucleoplasmic beta-actin exists in a dynamic equilibrium between low-mobility polymeric species and rapidly diffusing populations. J Cell Biol 172(4):541–552. doi:10.1083/jcb.200507101
82. Gieni RS, Hendzel MJ (2009) Actin dynamics and functions in the interphase nucleus: moving toward an understanding of nuclear polymeric actin. Biochem Cell Biol 87(1):283–306. doi:10.1139/o08-133
83. Wada A, Fukuda M, Mishima M, Nishida E (1998) Nuclear export of actin: a novel mechanism regulating the subcellular localization of a major cytoskeletal protein. EMBO J 17(6):1635–1641. doi:10.1093/emboj/17.6.1635
84. Belin BJ, Cimini BA, Blackburn EH, Mullins RD (2013) Visualization of actin filaments and monomers in somatic cell nuclei. Mol Biol Cell 24(7):982–994. doi:10.1091/mbc.E12-09-0685

85. Visa N, Percipalle P (2010) Nuclear functions of actin. Cold Spring Harb Perspect Biol 2(4):a000620. doi:10.1101/cshperspect.a000620
86. Berezney R, Coffey DS (1977) Nuclear matrix. Isolation and characterization of a framework structure from rat liver nuclei. J Cell Biol 73(3):616–637
87. Jaalouk DE, Lammerding J (2009) Mechanotransduction gone awry. Nat Rev Mol Cell Biol 10(1):63–73. doi:10.1038/nrm2597
88. Heasman SJ, Ridley AJ (2008) Mammalian Rho GTPases: new insights into their functions from in vivo studies. Nat Rev Mol Cell Biol 9(9):690–701. doi:10.1038/nrm2476
89. Discher DE, Boal DH, Boey SK (1998) Simulations of the erythrocyte cytoskeleton at large deformation. II. Micropipette aspiration. Biophys J 75(3):1584–1597. doi:10.1016/s0006-3495(98)74076-7
90. Dahl KN, Kahn SM, Wilson KL, Discher DE (2004) The nuclear envelope lamina network has elasticity and a compressibility limit suggestive of a molecular shock absorber. J Cell Sci 117(Pt 20):4779–4786. doi:10.1242/jcs.01357
91. Dahl KN, Engler AJ, Pajerowski JD, Discher DE (2005) Power-law rheology of isolated nuclei with deformation mapping of nuclear substructures. Biophys J 89(4):2855–2864. doi:10.1529/biophysj.105.062554
92. Rowat AC, Lammerding J, Ipsen JH (2006) Mechanical properties of the cell nucleus and the effect of emerin deficiency. Biophys J 91(12):4649–4664. doi:10.1529/biophysj.106.086454
93. Guilak F, Tedrow JR, Burgkart R (2000) Viscoelastic properties of the cell nucleus. Biochem Biophys Res Commun 269(3):781–786. doi:10.1006/bbrc.2000.2360
94. Jiménez-García LF, Fragoso-Soriano R (2000) Atomic force microscopy of the cell nucleus. J Struct Biol 129(2–3):218–222. doi:10.1006/jsbi.2000.4233
95. Schäpe J, Prausse S, Radmacher M, Stick R (2009) Influence of lamin A on the mechanical properties of amphibian oocyte nuclei measured by atomic force microscopy. Biophys J 96(10):4319–4325. doi:10.1016/j.bpj.2009.02.048
96. Kaufmann A, Heinemann F, Radmacher M, Stick R (2011) Amphibian oocyte nuclei expressing lamin A with the progeria mutation E145K exhibit an increased elastic modulus. Nucleus 2(4):310–319. doi:10.4161/nucl.2.4.16119
97. Lammerding J, Hsiao J, Schulze PC, Kozlov S, Stewart CL, Lee RT (2005) Abnormal nuclear shape and impaired mechanotransduction in emerin-deficient cells. J Cell Biol 170(5):781–791. doi:10.1083/jcb.200502148
98. Lammerding J, Fong LG, Ji JY, Reue K, Stewart CL, Young SG, Lee RT (2006) Lamins A and C but not lamin B1 regulate nuclear mechanics. J Biol Chem 281(35):25768–25780. doi:10.1074/jbc.M513511200
99. Caille N, Thoumine O, Tardy Y, Meister JJ (2002) Contribution of the nucleus to the mechanical properties of endothelial cells. J Biomech 35(2):177–187
100. Tseng Y, Lee JS, Kole TP, Jiang I, Wirtz D (2004) Micro-organization and visco-elasticity of the interphase nucleus revealed by particle nanotracking. J Cell Sci 117(Pt 10):2159–2167. doi:10.1242/jcs.01073
101. Chalut KJ, Höpfler M, Lautenschläger F, Boyde L, Chan CJ, Ekpenyong A, Martinez-Arias A, Guck J (2012) Chromatin decondensation and nuclear softening accompany Nanog downregulation in embryonic stem cells. Biophys J 103(10):2060–2070. doi:10.1016/j.bpj.2012.10.015
102. Isermann P, Davidson PM, Sliz JD, Lammerding J (2012) Assays to measure nuclear mechanics in interphase cells. Curr Protoc Cell Biol Chapter 22:Unit22.16. doi:10.1002/0471143030.cb2216s56
103. Rowat AC, Jaalouk DE, Zwerger M, Ung WL, Eydelnant IA, Olins DE, Olins AL, Herrmann H, Weitz DA, Lammerding J (2013) Nuclear envelope composition determines the ability of neutrophil-type cells to passage through micron-scale constrictions. J Biol Chem 288(12):8610–8618. doi:10.1074/jbc.M112.441535
104. Kha HN, Chen BK, Clark GM, Jones R (2004) Stiffness properties for Nucleus standard straight and contour electrode arrays. Med Eng Phys 26(8):677–685. doi:10.1016/j.medengphy.2004.05.001

105. Vaziri A, Mofrad MR (2007) Mechanics and deformation of the nucleus in micropipette aspiration experiment. J Biomech 40(9):2053–2062. doi:10.1016/j.jbiomech.2006.09.023
106. Rowat AC, Lammerding J, Herrmann H, Aebi U (2008) Towards an integrated understanding of the structure and mechanics of the cell nucleus. Bioessays 30(3):226–236. doi:10.1002/bies.20720
107. Pajerowski JD, Dahl KN, Zhong FL, Sammak PJ, Discher DE (2007) Physical plasticity of the nucleus in stem cell differentiation. Proc Natl Acad Sci U S A 104(40):15619–15624. doi:10.1073/pnas.0702576104
108. Shimi T, Pfleghaar K, Kojima S, Pack CG, Solovei I, Goldman AE, Adam SA, Shumaker DK, Kinjo M, Cremer T, Goldman RD (2008) The A- and B-type nuclear lamin networks: microdomains involved in chromatin organization and transcription. Genes Dev 22(24):3409–3421. doi:10.1101/gad.1735208
109. Kolb T, Maass K, Hergt M, Aebi U, Herrmann H (2011) Lamin A and lamin C form homodimers and coexist in higher complex forms both in the nucleoplasmic fraction and in the lamina of cultured human cells. Nucleus 2(5):425–433. doi:10.4161/nucl.2.5.17765
110. Goldberg MW, Huttenlauch I, Hutchison CJ, Stick R (2008) Filaments made from A- and B-type lamins differ in structure and organization. J Cell Sci 121(Pt 2):215–225. doi:10.1242/jcs.022020
111. Sullivan T, Escalante-Alcalde D, Bhatt H, Anver M, Bhat N, Nagashima K, Stewart CL, Burke B (1999) Loss of A-type lamin expression compromises nuclear envelope integrity leading to muscular dystrophy. J Cell Biol 147(5):913–920
112. Puckelwartz MJ, Depreux FF, McNally EM (2011) Gene expression, chromosome position and lamin A/C mutations. Nucleus 2(3):162–167. doi:10.1083/jcb.201101046, 10.4161/nucl.2.3.16003
113. Mewborn SK, Puckelwartz MJ, Abuisneineh F, Fahrenbach JP, Zhang Y, MacLeod H, Dellefave L, Pytel P, Selig S, Labno CM, Reddy K, Singh H, McNally E (2010) Altered chromosomal positioning, compaction, and gene expression with a lamin A/C gene mutation. PLoS One 5(12):e14342. doi:10.1371/journal.pone.0014342
114. Bridger JM, Foeger N, Kill IR, Herrmann H (2007) The nuclear lamina. Both a structural framework and a platform for genome organization. FEBS J 274(6):1354–1361
115. Broers JL, Ramaekers FC (2004) Dynamics of nuclear lamina assembly and disassembly. Symp Soc Exp Biol 56:177–192
116. Zwerger M, Jaalouk DE, Lombardi ML, Isermann P, Mauermann M, Dialynas G, Herrmann H, Wallrath LL, Lammerding J (2013) Myopathic lamin mutations impair nuclear stability in cells and tissue and disrupt nucleo-cytoskeletal coupling. Hum Mol Genet 22(12):2335–2349. doi:10.1093/hmg/ddt079
117. Simon DN, Wilson KL (2013) Partners and post-translational modifications of nuclear lamins. Chromosoma 122(1–2):13–31. doi:10.1007/s00412-013-0399-8
118. Zhong Z, Wilson KL, Dahl KN (2010) Beyond lamins other structural components of the nucleoskeleton. Methods Cell Biol 98:97–119. doi:10.1016/s0091-679x(10)98005-9
119. Dahl KN, Kalinowski A (2011) Nucleoskeleton mechanics at a glance. J Cell Sci 124(Pt 5):675–678. doi:10.1242/jcs.069096
120. Gundersen GG, Worman HJ (2013) Nuclear positioning. Cell 152(6):1376–1389. doi:10.1016/j.cell.2013.02.031
121. Rothballer A, Kutay U (2013) The diverse functional LINCs of the nuclear envelope to the cytoskeleton and chromatin. Chromosoma. doi:10.1007/s00412-013-0417-x
122. Malone CJ, Fixsen WD, Horvitz HR, Han M (1999) UNC-84 localizes to the nuclear envelope and is required for nuclear migration and anchoring during C. elegans development. Development 126(14):3171–3181
123. Lee KK, Starr D, Cohen M, Liu J, Han M, Wilson KL, Gruenbaum Y (2002) Lamin-dependent localization of UNC-84, a protein required for nuclear migration in Caenorhabditis elegans. Mol Biol Cell 13(3):892–901. doi:10.1091/mbc.01-06-0294
124. Malone CJ, Misner L, Le Bot N, Tsai MC, Campbell JM, Ahringer J, White JG (2003) The C. elegans hook protein, ZYG-12, mediates the essential attachment between the centrosome and nucleus. Cell 115(7):825–836

125. Starr DA, Han M (2003) ANChors away: an actin based mechanism of nuclear positioning. J Cell Sci 116(Pt 2):211–216
126. Crisp M, Liu Q, Roux K, Rattner JB, Shanahan C, Burke B, Stahl PD, Hodzic D (2006) Coupling of the nucleus and cytoplasm: role of the LINC complex. J Cell Biol 172(1):41–53. doi:10.1083/jcb.200509124
127. Starr DA, Fridolfsson HN (2010) Interactions between nuclei and the cytoskeleton are mediated by SUN-KASH nuclear-envelope bridges. Annu Rev Cell Dev Biol 26:421–444. doi:10.1146/annurev-cellbio-100109-104037
128. Wagner N, Krohne G (2007) LEM-Domain proteins: new insights into lamin-interacting proteins. Int Rev Cytol 261:1–46. doi:10.1016/s0074-7696(07)61001-8
129. Méjat A, Misteli T (2010) LINC complexes in health and disease. Nucleus 1(1):40–52. doi:10.4161/nucl.1.1.10530
130. Sosa BA, Rothballer A, Kutay U, Schwartz TU (2012) LINC complexes form by binding of three KASH peptides to domain interfaces of trimeric SUN proteins. Cell 149(5):1035–1047. doi:10.1016/j.cell.2012.03.046
131. Nery FC, Zeng J, Niland BP, Hewett J, Farley J, Irimia D, Li Y, Wiche G, Sonnenberg A, Breakefield XO (2008) TorsinA binds the KASH domain of nesprins and participates in linkage between nuclear envelope and cytoskeleton. J Cell Sci 121(Pt 20):3476–3486. doi:10.1242/jcs.029454
132. Folker ES, Ostlund C, Luxton GW, Worman HJ, Gundersen GG (2011) Lamin A variants that cause striated muscle disease are defective in anchoring transmembrane actin-associated nuclear lines for nuclear movement. Proc Natl Acad Sci U S A 108(1):131–136. doi:10.1073/pnas.1000824108
133. Lombardi ML, Lammerding J (2011) Keeping the LINC: the importance of nucleocytoskeletal coupling in intracellular force transmission and cellular function. Biochem Soc Trans 39(6):1729–1734. doi:10.1042/bst20110686
134. Maniotis AJ, Chen CS, Ingber DE (1997) Demonstration of mechanical connections between integrins, cytoskeletal filaments, and nucleoplasm that stabilize nuclear structure. Proc Natl Acad Sci U S A 94(3):849–854
135. Hagan I, Yanagida M (1995) The product of the spindle formation gene sad1+ associates with the fission yeast spindle pole body and is essential for viability. J Cell Biol 129(4):1033–1047
136. Sosa BA, Kutay U, Schwartz TU (2013) Structural insights into LINC complexes. Curr Opin Struct Biol 23(2):285–291. doi:10.1016/j.sbi.2013.03.005
137. Haque F, Lloyd DJ, Smallwood DT, Dent CL, Shanahan CM, Fry AM, Trembath RC, Shackleton S (2006) SUN1 interacts with nuclear lamin A and cytoplasmic nesprins to provide a physical connection between the nuclear lamina and the cytoskeleton. Mol Cell Biol 26(10):3738–3751. doi:10.1128/mcb.26.10.3738-3751.2006
138. Liu Q, Pante N, Misteli T, Elsagga M, Crisp M, Hodzic D, Burke B, Roux KJ (2007) Functional association of Sun1 with nuclear pore complexes. J Cell Biol 178(5):785–798. doi:10.1083/jcb.200704108
139. Lussi YC, Hügi I, Laurell E, Kutay U, Fahrenkrog B (2011) The nucleoporin Nup88 is interacting with nuclear lamin A. Mol Biol Cell 22(7):1080–1090. doi:10.1091/mbc.E10-05-0463
140. Razafsky D, Hodzic D (2009) Bringing KASH under the SUN: the many faces of nucleocytoskeletal connections. J Cell Biol 186(4):461–472. doi:10.1083/jcb.200906068
141. Apel ED, Lewis RM, Grady RM, Sanes JR (2000) Syne-1, a dystrophin- and Klarsicht-related protein associated with synaptic nuclei at the neuromuscular junction. J Biol Chem 275(41):31986–31995. doi:10.1074/jbc.M004775200
142. Mellad JA, Warren DT, Shanahan CM (2011) Nesprins LINC the nucleus and cytoskeleton. Curr Opin Cell Biol 23(1):47–54. doi:10.1016/j.ceb.2010.11.006
143. Zhen YY, Libotte T, Munck M, Noegel AA, Korenbaum E (2002) NUANCE, a giant protein connecting the nucleus and actin cytoskeleton. J Cell Sci 115(Pt 15):3207–3222
144. Roux KJ, Crisp ML, Liu Q, Kim D, Kozlov S, Stewart CL, Burke B (2009) Nesprin 4 is an outer nuclear membrane protein that can induce kinesin-mediated cell polarization. Proc Natl Acad Sci U S A 106(7):2194–2199. doi:10.1073/pnas.0808602106

145. Libotte T, Zaim H, Abraham S, Padmakumar VC, Schneider M, Lu W, Munck M, Hutchison C, Wehnert M, Fahrenkrog B, Sauder U, Aebi U, Noegel AA, Karakesisoglou I (2005) Lamin A/C-dependent localization of Nesprin-2, a giant scaffolder at the nuclear envelope. Mol Biol Cell 16(7):3411–3424. doi:10.1091/mbc.E04-11-1009
146. Mislow JM, Kim MS, Davis DB, McNally EM (2002) Myne-1, a spectrin repeat transmembrane protein of the myocyte inner nuclear membrane, interacts with lamin A/C. J Cell Sci 115(Pt 1):61–70
147. Mislow JM, Holaska JM, Kim MS, Lee KK, Segura-Totten M, Wilson KL, McNally EM (2002) Nesprin-1alpha self-associates and binds directly to emerin and lamin A in vitro. FEBS Lett 525(1–3):135–140
148. Morimoto A, Shibuya H, Zhu X, Kim J, Ishiguro K, Han M, Watanabe Y (2012) A conserved KASH domain protein associates with telomeres, SUN1, and dynactin during mammalian meiosis. J Cell Biol 198(2):165–172. doi:10.1083/jcb.201204085
149. Salpingidou G, Smertenko A, Hausmanowa-Petrucewicz I, Hussey PJ, Hutchison CJ (2007) A novel role for the nuclear membrane protein emerin in association of the centrosome to the outer nuclear membrane. J Cell Biol 178(6):897–904. doi:10.1083/jcb.200702026
150. Buch C, Lindberg R, Figueroa R, Gudise S, Onischenko E, Hallberg E (2009) An integral protein of the inner nuclear membrane localizes to the mitotic spindle in mammalian cells. J Cell Sci 122(Pt 12):2100–2107. doi:10.1242/jcs.047373
151. Gudise S, Figueroa RA, Lindberg R, Larsson V, Hallberg E (2011) Samp1 is functionally associated with the LINC complex and A-type lamina networks. J Cell Sci 124(Pt 12):2077–2085. doi:10.1242/jcs.078923
152. Borrego-Pinto J, Jegou T, Osorio DS, Auradé F, Gorjánácz M, Koch B, Mattaj IW, Gomes ER (2012) Samp1 is a component of TAN lines and is required for nuclear movement. J Cell Sci 125(Pt 5):1099–1105. doi:10.1242/jcs.087049
153. Luxton GW, Gomes ER, Folker ES, Vintinner E, Gundersen GG (2010) Linear arrays of nuclear envelope proteins harness retrograde actin flow for nuclear movement. Science 329(5994):956–959. doi:10.1126/science.1189072
154. Stewart-Hutchinson PJ, Hale CM, Wirtz D, Hodzic D (2008) Structural requirements for the assembly of LINC complexes and their function in cellular mechanical stiffness. Exp Cell Res 314(8):1892–1905. doi:10.1016/j.yexcr.2008.02.022
155. Scherthan H (2001) A bouquet makes ends meet. Nat Rev Mol Cell Biol 2(8):621–627. doi:10.1038/35085086
156. Miki F, Kurabayashi A, Tange Y, Okazaki K, Shimanuki M, Niwa O (2004) Two-hybrid search for proteins that interact with Sad1 and Kms1, two membrane-bound components of the spindle pole body in fission yeast. Mol Genet Genomics 270(6):449–461. doi:10.1007/s00438-003-0938-8
157. Chikashige Y, Tsutsumi C, Yamane M, Okamasa K, Haraguchi T, Hiraoka Y (2006) Meiotic proteins bqt1 and bqt2 tether telomeres to form the bouquet arrangement of chromosomes. Cell 125(1):59–69. doi:10.1016/j.cell.2006.01.048
158. Shimanuki M, Miki F, Ding DQ, Chikashige Y, Hiraoka Y, Horio T, Niwa O (1997) A novel fission yeast gene, kms1+, is required for the formation of meiotic prophase-specific nuclear architecture. Mol Gen Genet 254(3):238–249
159. Penkner A, Tang L, Novatchkova M, Ladurner M, Fridkin A, Gruenbaum Y, Schweizer D, Loidl J, Jantsch V (2007) The nuclear envelope protein Matefin/SUN-1 is required for homologous pairing in C. elegans meiosis. Dev Cell 12(6):873–885. doi:10.1016/j.devcel.2007.05.004
160. McGee MD, Stagljar I, Starr DA (2009) KDP-1 is a nuclear envelope KASH protein required for cell-cycle progression. J Cell Sci 122(Pt 16):2895–2905. doi:10.1242/jcs.051607
161. Ottaviani A, Schluth-Bolard C, Rival-Gervier S, Boussouar A, Rondier D, Foerster AM, Morere J, Bauwens S, Gazzo S, Callet-Bauchu E, Gilson E, Magdinier F (2009) Identification of a perinuclear positioning element in human subtelomeres that requires A-type lamins and CTCF. EMBO J 28(16):2428–2436. doi:10.1038/emboj.2009.201
162. Gonzalez-Suarez I, Redwood AB, Gonzalo S (2009) Loss of A-type lamins and genomic instability. Cell Cycle 8(23):3860–3865

163. Khatau SB, Bloom RJ, Bajpai S, Razafsky D, Zang S, Giri A, Wu PH, Marchand J, Celedon A, Hale CM, Sun SX, Hodzic D, Wirtz D (2012) The distinct roles of the nucleus and nucleus-cytoskeleton connections in three-dimensional cell migration. Sci Rep 2:488. doi:10.1038/srep00488
164. Poh YC, Shevtsov SP, Chowdhury F, Wu DC, Na S, Dundr M, Wang N (2012) Dynamic force-induced direct dissociation of protein complexes in a nuclear body in living cells. Nat Commun 3:866. doi:10.1038/ncomms1873
165. Booth-Gauthier EA, Alcoser TA, Yang G, Dahl KN (2012) Force-induced changes in subnuclear movement and rheology. Biophys J 103(12):2423–2431. doi:10.1016/j.bpj.2012.10.039
166. Maniotis AJ, Bojanowski K, Ingber DE (1997) Mechanical continuity and reversible chromosome disassembly within intact genomes removed from living cells. J Cell Biochem 65(1):114–130
167. Ingber DE (2003) Mechanobiology and diseases of mechanotransduction. Ann Med 35(8):564–577
168. Ingber DE (2006) Cellular mechanotransduction: putting all the pieces together again. FASEB J 20(7):811–827. doi:10.1096/fj.05-5424rev
169. Moore KA, Polte T, Huang S, Shi B, Alsberg E, Sunday ME, Ingber DE (2005) Control of basement membrane remodeling and epithelial branching morphogenesis in embryonic lung by Rho and cytoskeletal tension. Dev Dyn 232(2):268–281. doi:10.1002/dvdy.20237
170. Lammerding J, Lee RT (2005) The nuclear membrane and mechanotransduction: impaired nuclear mechanics and mechanotransduction in lamin A/C deficient cells. Novartis Found Symp 264:264–273, discussion 273–268
171. Brosig M, Ferralli J, Gelman L, Chiquet M, Chiquet-Ehrismann R (2010) Interfering with the connection between the nucleus and the cytoskeleton affects nuclear rotation, mechanotransduction and myogenesis. Int J Biochem Cell Biol 42(10):1717–1728. doi:10.1016/j.biocel.2010.07.001
172. Ho CY, Jaalouk DE, Vartiainen MK, Lammerding J (2013) Lamin A/C and emerin regulate MKL1-SRF activity by modulating actin dynamics. Nature 497(7450):507–511. doi:10.1038/nature12105
173. Dey P (2009) Nuclear margin irregularity and cancer: a review. Anal Quant Cytol Histol 31(5):345–352
174. Vigouroux C, Auclair M, Dubosclard E, Pouchelet M, Capeau J, Courvalin JC, Buendia B (2001) Nuclear envelope disorganization in fibroblasts from lipodystrophic patients with heterozygous R482Q/W mutations in the lamin A/C gene. J Cell Sci 114(Pt 24):4459–4468
175. Muchir A, Medioni J, Laluc M, Massart C, Arimura T, van der Kooi AJ, Desguerre I, Mayer M, Ferrer X, Briault S, Hirano M, Worman HJ, Mallet A, Wehnert M, Schwartz K, Bonne G (2004) Nuclear envelope alterations in fibroblasts from patients with muscular dystrophy, cardiomyopathy, and partial lipodystrophy carrying lamin A/C gene mutations. Muscle Nerve 30(4):444–450. doi:10.1002/mus.20122
176. Friedl P, Alexander S (2011) Cancer invasion and the microenvironment: plasticity and reciprocity. Cell 147(5):992–1009. doi:10.1016/j.cell.2011.11.016
177. Broers JL, Raymond Y, Rot MK, Kuijpers H, Wagenaar SS, Ramaekers FC (1993) Nuclear A-type lamins are differentially expressed in human lung cancer subtypes. Am J Pathol 143(1):211–220
178. Moss SF, Krivosheyev V, de Souza A, Chin K, Gaetz HP, Chaudhary N, Worman HJ, Holt PR (1999) Decreased and aberrant nuclear lamin expression in gastrointestinal tract neoplasms. Gut 45(5):723–729
179. Venables RS, McLean S, Luny D, Moteleb E, Morley S, Quinlan RA, Lane EB, Hutchison CJ (2001) Expression of individual lamins in basal cell carcinomas of the skin. Br J Cancer 84(4):512–519. doi:10.1054/bjoc.2000.1632
180. Tilli CM, Ramaekers FC, Broers JL, Hutchison CJ, Neumann HA (2003) Lamin expression in normal human skin, actinic keratosis, squamous cell carcinoma and basal cell carcinoma. Br J Dermatol 148(1):102–109
181. Hudson ME, Pozdnyakova I, Haines K, Mor G, Snyder M (2007) Identification of differentially expressed proteins in ovarian cancer using high-density protein microarrays. Proc Natl Acad Sci U S A 104(44):17494–17499. doi:10.1073/pnas.0708572104

182. Willis ND, Wilson RG, Hutchison CJ (2008) Lamin A: a putative colonic epithelial stem cell biomarker which identifies colorectal tumours with a more aggressive phenotype. Biochem Soc Trans 36(Pt 6):1350–1353. doi:10.1042/bst0361350
183. Belt EJ, Fijneman RJ, van den Berg EG, Bril H, Delis-van Diemen PM, Tijssen M, van Essen HF, de Lange-de Klerk ES, Beliën JA, Stockmann HB, Meijer S, Meijer GA (2011) Loss of lamin A/C expression in stage II and III colon cancer is associated with disease recurrence. Eur J Cancer 47(12):1837–1845. doi:10.1016/j.ejca.2011.04.025
184. Wolf K, Wu YI, Liu Y, Geiger J, Tam E, Overall C, Stack MS, Friedl P (2007) Multi-step pericellular proteolysis controls the transition from individual to collective cancer cell invasion. Nat Cell Biol 9(8):893–904. doi:10.1038/ncb1616
185. Capo-chichi CD, Cai KQ, Simpkins F, Ganjei-Azar P, Godwin AK, Xu XX (2011) Nuclear envelope structural defects cause chromosomal numerical instability and aneuploidy in ovarian cancer. BMC Med 9:28. doi:10.1186/1741-7015-9-28
186. Wu Z, Wu L, Weng D, Xu D, Geng J, Zhao F (2009) Reduced expression of lamin A/C correlates with poor histological differentiation and prognosis in primary gastric carcinoma. J Exp Clin Cancer Res 28:8. doi:10.1186/1756-9966-28-8
187. Willis ND, Cox TR, Rahman-Casañs SF, Smits K, Przyborski SA, van den Brandt P, van Engeland M, Weijenberg M, Wilson RG, de Bruïne A, Hutchison CJ (2008) Lamin A/C is a risk biomarker in colorectal cancer. PLoS One 3(8):e2988. doi:10.1371/journal.pone.0002988
188. Foster CR, Robson JL, Simon WJ, Twigg J, Cruikshank D, Wilson RG, Hutchison CJ (2011) The role of Lamin A in cytoskeleton organization in colorectal cancer cells: a proteomic investigation. Nucleus 2(5):434–443. doi:10.4161/nucl.2.5.17775
189. Sjöblom T, Jones S, Wood LD, Parsons DW, Lin J, Barber TD, Mandelker D, Leary RJ, Ptak J, Silliman N, Szabo S, Buckhaults P, Farrell C, Meeh P, Markowitz SD, Willis J, Dawson D, Willson JK, Gazdar AF, Hartigan J, Wu L, Liu C, Parmigiani G, Park BH, Bachman KE, Papadopoulos N, Vogelstein B, Kinzler KW, Velculescu VE (2006) The consensus coding sequences of human breast and colorectal cancers. Science 314(5797):268–274. doi:10.1126/science.1133427
190. Stephens PJ, Tarpey PS, Davies H, Van Loo P, Greenman C, Wedge DC, Nik-Zainal S, Martin S, Varela I, Bignell GR, Yates LR, Papaemmanuil E, Beare D, Butler A, Cheverton A, Gamble J, Hinton J, Jia M, Jayakumar A, Jones D, Latimer C, Lau KW, McLaren S, McBride DJ, Menzies A, Mudie L, Raine K, Rad R, Chapman MS, Teague J, Easton D, Langerød A, Lee MT, Shen CY, Tee BT, Huimin BW, Broeks A, Vargas AC, Turashvili G, Martens J, Fatima A, Miron P, Chin SF, Thomas G, Boyault S, Mariani O, Lakhani SR, van de Vijver M, van't Veer L, Foekens J, Desmedt C, Sotiriou C, Tutt A, Caldas C, Reis-Filho JS, Aparicio SA, Salomon AV, Børresen-Dale AL, Richardson AL, Campbell PJ, Futreal PA, Stratton MR, OBCC (2012) The landscape of cancer genes and mutational processes in breast cancer. Nature 486(7403):400–404. doi:10.1038/nature11017
191. Doherty JA, Rossing MA, Cushing-Haugen KL, Chen C, Van Den Berg DJ, Wu AH, Pike MC, Ness RB, Moysich K, Chenevix-Trench G, Beesley J, Webb PM, Chang-Claude J, Wang-Gohrke S, Goodman MT, Lurie G, Thompson PJ, Carney ME, Hogdall E, Kjaer SK, Hogdall C, Goode EL, Cunningham JM, Fridley BL, Vierkant RA, Berchuck A, Moorman PG, Schildkraut JM, Palmieri RT, Cramer DW, Terry KL, Yang HP, Garcia-Closas M, Chanock S, Lissowska J, Song H, Pharoah PD, Shah M, Perkins B, McGuire V, Whittemore AS, Di Cioccio RA, Gentry-Maharaj A, Menon U, Gayther SA, Ramus SJ, Ziogas A, Brewster W, Anton-Culver H, Pearce C, Group AOCSM, Cancer ACS, OCAC (2010) ESR1/SYNE1 polymorphism and invasive epithelial ovarian cancer risk: an Ovarian Cancer Association Consortium study. Cancer Epidemiol Biomarkers Prev 19(1):245–250. doi:10.1158/1055-9965.epi-09-0729
192. Barboro P, Repaci E, D'Arrigo C, Balbi C (2012) The role of nuclear matrix proteins binding to matrix attachment regions (Mars) in prostate cancer cell differentiation. PLoS One 7(7):e40617. doi:10.1371/journal.pone.0040617
193. Malonia SK, Sinha S, Lakshminarasimhan P, Singh K, Jalota-Badhwar A, Rampalli S, Kaul-Ghanekar R, Chattopadhyay S (2011) Gene regulation by SMAR1: role in cellular homeostasis and cancer. Biochim Biophys Acta 1815(1):1–12. doi:10.1016/j.bbcan.2010.08.003

194. Stein GS, Lian JB, van Wijnen AJ, Stein JL, Javed A, Montecino M, Zaidi SK, Young D, Choi JY, Gutierrez S, Pockwinse S (2004) Nuclear microenvironments support assembly and organization of the transcriptional regulatory machinery for cell proliferation and differentiation. J Cell Biochem 91(2):287–302. doi:10.1002/jcb.10777
195. Bickmore WA, van Steensel B (2013) Genome architecture: domain organization of interphase chromosomes. Cell 152(6):1270–1284. doi:10.1016/j.cell.2013.02.001
196. Burke B, Stewart CL (2013) The nuclear lamins: flexibility in function. Nat Rev Mol Cell Biol 14(1):13–24. doi:10.1038/nrm3488
197. Valastyan S, Weinberg RA (2011) Tumor metastasis: molecular insights and evolving paradigms. Cell 147(2):275–292. doi:10.1016/j.cell.2011.09.024
198. Stoitzner P, Pfaller K, Stössel H, Romani N (2002) A close-up view of migrating Langerhans cells in the skin. J Invest Dermatol 118(1):117–125. doi:10.1046/j.0022-202x.2001.01631.x
199. Weigelin B, Bakker G-J, Friedl P (2012) Intravital third harmonic generation microscopy of collective melanoma cell invasion. Principles of interface guidance and microvesicle dynamics. IntraVital 1(1):32–43
200. Shankar J, Messenberg A, Chan J, Underhill TM, Foster LJ, Nabi IR (2010) Pseudopodial actin dynamics control epithelial-mesenchymal transition in metastatic cancer cells. Cancer Res 70(9):3780–3790. doi:10.1158/0008-5472.can-09-4439
201. Schoumacher M, Goldman RD, Louvard D, Vignjevic DM (2010) Actin, microtubules, and vimentin intermediate filaments cooperate for elongation of invadopodia. J Cell Biol 189(3):541–556. doi:10.1083/jcb.200909113
202. Tong Z, Balzer EM, Dallas MR, Hung WC, Stebe KJ, Konstantopoulos K (2012) Chemotaxis of cell populations through confined spaces at single-cell resolution. PLoS One 7(1):e29211. doi:10.1371/journal.pone.0029211
203. Booth-Gauthier EA, Du V, Ghibaudo M, Rape AD, Dahl KN, Ladoux B (2013) Hutchinson-Gilford progeria syndrome alters nuclear shape and reduces cell motility in three dimensional model substrates. Integr Biol (Camb) 5(3):569–577. doi:10.1039/c3ib20231c
204. Verstraeten VL, Ji JY, Cummings KS, Lee RT, Lammerding J (2008) Increased mechanosensitivity and nuclear stiffness in Hutchinson-Gilford progeria cells: effects of farnesyltransferase inhibitors. Aging Cell 7(3):383–393. doi:10.1111/j.1474-9726.2008.00382.x
205. Dahl KN, Scaffidi P, Islam MF, Yodh AG, Wilson KL, Misteli T (2006) Distinct structural and mechanical properties of the nuclear lamina in Hutchinson-Gilford progeria syndrome. Proc Natl Acad Sci U S A 103(27):10271–10276. doi:10.1073/pnas.0601058103
206. Kong L, Schäfer G, Bu H, Zhang Y, Klocker H (2012) Lamin A/C protein is overexpressed in tissue-invading prostate cancer and promotes prostate cancer cell growth, migration and invasion through the PI3K/AKT/PTEN pathway. Carcinogenesis 33(4):751–759. doi:10.1093/carcin/bgs022
207. Wandke C, Kutay U (2013) Enclosing chromatin: reassembly of the nucleus after open mitosis. Cell 152(6):1222–1225. doi:10.1016/j.cell.2013.02.046
208. Vargas JD, Hatch EM, Anderson DJ, Hetzer MW (2012) Transient nuclear envelope rupturing during interphase in human cancer cells. Nucleus 3(1):88–100. doi:10.4161/nucl.18954
209. De Vos WH, Houben F, Kamps M, Malhas A, Verheyen F, Cox J, Manders EM, Verstraeten VL, van Steensel MA, Marcelis CL, van den Wijngaard A, Vaux DJ, Ramaekers FC, Broers JL (2011) Repetitive disruptions of the nuclear envelope invoke temporary loss of cellular compartmentalization in laminopathies. Hum Mol Genet 20(21):4175–4186. doi:10.1093/hmg/ddr344
210. McCord RP, Nazario-Toole A, Zhang H, Chines PS, Zhan Y, Erdos MR, Collins FS, Dekker J, Cao K (2013) Correlated alterations in genome organization, histone methylation, and DNA-lamin A/C interactions in Hutchinson-Gilford progeria syndrome. Genome Res 23(2):260–269. doi:10.1101/gr.138032.112
211. Solovei I, Wang AS, Thanisch K, Schmidt CS, Krebs S, Zwerger M, Cohen TV, Devys D, Foisner R, Peichl L, Herrmann H, Blum H, Engelkamp D, Stewart CL, Leonhardt H, Joffe B (2013) LBR and lamin A/C sequentially tether peripheral heterochromatin and inversely regulate differentiation. Cell 152(3):584–598. doi:10.1016/j.cell.2013.01.009

212. Timp W, Feinberg AP (2013) Cancer as a dysregulated epigenome allowing cellular growth advantage at the expense of the host. Nat Rev Cancer 13(7):497–510. doi:10.1038/nrc3486
213. Polioudaki H, Kourmouli N, Drosou V, Bakou A, Theodoropoulos PA, Singh PB, Giannakouros T, Georgatos SD (2001) Histones H3/H4 form a tight complex with the inner nuclear membrane protein LBR and heterochromatin protein 1. EMBO Rep 2(10):920–925. doi:10.1093/embo-reports/kve199
214. Nili E, Cojocaru GS, Kalma Y, Ginsberg D, Copeland NG, Gilbert DJ, Jenkins NA, Berger R, Shaklai S, Amariglio N, Brok-Simoni F, Simon AJ, Rechavi G (2001) Nuclear membrane protein LAP2beta mediates transcriptional repression alone and together with its binding partner GCL (germ-cell-less). J Cell Sci 114(Pt 18):3297–3307
215. Somech R, Shaklai S, Geller O, Amariglio N, Simon AJ, Rechavi G, Gal-Yam EN (2005) The nuclear-envelope protein and transcriptional repressor LAP2beta interacts with HDAC3 at the nuclear periphery, and induces histone H4 deacetylation. J Cell Sci 118(Pt 17):4017–4025. doi:10.1242/jcs.02521
216. Haraguchi T, Holaska JM, Yamane M, Koujin T, Hashiguchi N, Mori C, Wilson KL, Hiraoka Y (2004) Emerin binding to Btf, a death-promoting transcriptional repressor, is disrupted by a missense mutation that causes Emery-Dreifuss muscular dystrophy. Eur J Biochem 271(5):1035–1045
217. Wilkinson FL, Holaska JM, Zhang Z, Sharma A, Manilal S, Holt I, Stamm S, Wilson KL, Morris GE (2003) Emerin interacts in vitro with the splicing-associated factor, YT521-B. Eur J Biochem 270(11):2459–2466
218. Holaska JM, Lee KK, Kowalski AK, Wilson KL (2003) Transcriptional repressor germ cell-less (GCL) and barrier to autointegration factor (BAF) compete for binding to emerin in vitro. J Biol Chem 278(9):6969–6975. doi:10.1074/jbc.M208811200
219. Lu P, Takai K, Weaver VM, Werb Z (2011) Extracellular matrix degradation and remodeling in development and disease. Cold Spring Harb Perspect Biol 3(12). doi:10.1101/cshperspect.a005058
220. Di Carlo D, Tse HT, Gossett DR (2012) Introduction: why analyze single cells? Methods Mol Biol 853:1–10. doi:10.1007/978-1-61779-567-1_1
221. Zhang W, Kai K, Choi DS, Iwamoto T, Nguyen YH, Wong H, Landis MD, Ueno NT, Chang J, Qin L (2012) Microfluidics separation reveals the stem-cell-like deformability of tumor-initiating cells. Proc Natl Acad Sci U S A 109(46):18707–18712. doi:10.1073/pnas.1209893109
222. Capo-chichi CD, Cai KQ, Testa JR, Godwin AK, Xu XX (2009) Loss of GATA6 leads to nuclear deformation and aneuploidy in ovarian cancer. Mol Cell Biol 29(17):4766–4777. doi:10.1128/mcb.00087-09
223. Agrelo R, Setien F, Espada J, Artiga MJ, Rodriguez M, Pérez-Rosado A, Sanchez-Aguilera A, Fraga MF, Piris MA, Esteller M (2005) Inactivation of the lamin A/C gene by CpG island promoter hypermethylation in hematologic malignancies, and its association with poor survival in nodal diffuse large B-cell lymphoma. J Clin Oncol 23(17):3940–3947. doi:10.1200/jco.2005.11.650
224. Wang Y, Wu R, Cho KR, Thomas DG, Gossner G, Liu JR, Giordano TJ, Shedden KA, Misek DE, Lubman DM (2009) Differential protein mapping of ovarian serous adenocarcinomas: identification of potential markers for distinct tumor stage. J Proteome Res 8(3):1452–1463. doi:10.1021/pr800820z
225. Skvortsov S, Schäfer G, Stasyk T, Fuchsberger C, Bonn GK, Bartsch G, Klocker H, Huber LA (2011) Proteomics profiling of microdissected low- and high-grade prostate tumors identifies Lamin A as a discriminatory biomarker. J Proteome Res 10(1):259–268. doi:10.1021/pr100921j
226. Alfonso P, Cañamero M, Fernández-Carbonié F, Núñez A, Casal JI (2008) Proteome analysis of membrane fractions in colorectal carcinomas by using 2D-DIGE saturation labeling. J Proteome Res 7(10):4247–4255. doi:10.1021/pr800152u
227. Bengtsson S, Krogh M, Szigyarto CA, Uhlen M, Schedvins K, Silfverswärd C, Linder S, Auer G, Alaiya A, James P (2007) Large-scale proteomics analysis of human ovarian cancer for biomarkers. J Proteome Res 6(4):1440–1450. doi:10.1021/pr060593y

228. Lim SO, Park SJ, Kim W, Park SG, Kim HJ, Kim YI, Sohn TS, Noh JH, Jung G (2002) Proteome analysis of hepatocellular carcinoma. Biochem Biophys Res Commun 291(4):1031–1037. doi:10.1006/bbrc.2002.6547
229. Sun S, Xu MZ, Poon RT, Day PJ, Luk JM (2010) Circulating Lamin B1 (LMNB1) biomarker detects early stages of liver cancer in patients. J Proteome Res 9(1):70–78. doi:10.1021/pr9002118
230. Coradeghini R, Barboro P, Rubagotti A, Boccardo F, Parodi S, Carmignani G, D'Arrigo C, Patrone E, Balbi C (2006) Differential expression of nuclear lamins in normal and cancerous prostate tissues. Oncol Rep 15(3):609–613
231. Li L, Du Y, Kong X, Li Z, Jia Z, Cui J, Gao J, Wang L, Xie K (2013) Lamin B1 is a novel therapeutic target of betulinic acid in pancreatic cancer. Clin Cancer Res. doi:10.1158/1078-0432.ccr-12-3630
232. Somech R, Gal-Yam EN, Shaklai S, Geller O, Amariglio N, Rechavi G, Simon AJ (2007) Enhanced expression of the nuclear envelope LAP2 transcriptional repressors in normal and malignant activated lymphocytes. Ann Hematol 86(6):393–401. doi:10.1007/s00277-007-0275-9
233. Martínez N, Alonso A, Moragues MD, Pontón J, Schneider J (1999) The nuclear pore complex protein Nup88 is overexpressed in tumor cells. Cancer Res 59(21):5408–5411
234. Agudo D, Gómez-Esquer F, Martínez-Arribas F, Núñez-Villar MJ, Pollán M, Schneider J (2004) Nup88 mRNA overexpression is associated with high aggressiveness of breast cancer. Int J Cancer 109(5):717–720. doi:10.1002/ijc.20034
235. Brustmann H, Hager M (2009) Nucleoporin 88 expression in normal and neoplastic squamous epithelia of the uterine cervix. Ann Diagn Pathol 13(5):303–307. doi:10.1016/j.anndiagpath.2009.05.005
236. Emterling A, Skoglund J, Arbman G, Schneider J, Evertsson S, Carstensen J, Zhang H, Sun XF (2003) Clinicopathological significance of Nup88 expression in patients with colorectal cancer. Oncology 64(4):361–369, doi:70294
237. Gould VE, Orucevic A, Zentgraf H, Gattuso P, Martinez N, Alonso A (2002) Nup88 (karyoporin) in human malignant neoplasms and dysplasias: correlations of immunostaining of tissue sections, cytologic smears, and immunoblot analysis. Hum Pathol 33(5):536–544
238. Knoess M, Kurz AK, Goreva O, Bektas N, Breuhahn K, Odenthal M, Schirmacher P, Dienes HP, Bock CT, Zentgraf H, zur Hausen A (2006) Nucleoporin 88 expression in hepatitis B and C virus-related liver diseases. World J Gastroenterol 12(36):5870–5874
239. Zhang ZY, Zhao ZR, Jiang L, Li JC, Gao YM, Cui DS, Wang CJ, Schneider J, Wang MW, Sun XF (2007) Nup88 expression in normal mucosa, adenoma, primary adenocarcinoma and lymph node metastasis in the colorectum. Tumour Biol 28(2):93–99. doi:10.1159/000099154
240. Xu S, Powers MA (2009) Nuclear pore proteins and cancer. Semin Cell Dev Biol 20(5):620–630. doi:10.1016/j.semcdb.2009.03.003
241. Cervoni N, Detich N, Seo SB, Chakravarti D, Szyf M (2002) The oncoprotein Set/TAF-1beta, an inhibitor of histone acetyltransferase, inhibits active demethylation of DNA, integrating DNA methylation and transcriptional silencing. J Biol Chem 277(28):25026–25031. doi:10.1074/jbc.M202256200
242. Hernández P, Solé X, Valls J, Moreno V, Capellá G, Urruticoechea A, Pujana MA (2007) Integrative analysis of a cancer somatic mutome. Mol Cancer 6:13. doi:10.1186/1476-4598-6-13
243. Capo-chichi CD, Cai KQ, Smedberg J, Ganjei-Azar P, Godwin AK, Xu XX. Loss of A-type lamin expression compromises nuclear envelope integrity in breast cancer. Chin J Cancer. 2011 Jun;30(6):415–425. http://www.ncbi.nlm.nih.gov/pubmed/21627864
244. Wazir U, Ahmed MH, Bridger JM, Harvey A, Jiang WG, Sharma AK, Mokbel K. The clinicopathological significance of lamin A/C, lamin B1 and lamin B receptor mRNA expression in human breast cancer. Cell Mol Biol Lett. 2013 Dec;18(4):595-611. doi: 10.2478/s11658-013-0109-9. Epub 2013 Nov 30. http://www.ncbi.nlm.nih.gov/pubmed/24293108

Nuclear Envelope in Nuclear Positioning and Cell Migration

David Razafsky, Denis Wirtz, and Didier Hodzic

Abstract Hauling and anchoring the nucleus within immobile or motile cells, tissues, and/or syncytia represents a major challenge. In the past 15 years, Linkers of the Nucleoskeleton to the Cytoskeleton (LINC complexes) have emerged as evolutionary-conserved molecular devices that span the nuclear envelope and provide interacting interfaces for cytoskeletal networks and molecular motors to the nuclear envelope. Here, we review the molecular composition of LINC complexes and focus on how their genetic alteration in vivo has provided a wealth of information related to the relevance of nuclear positioning during tissue development and homeostasis with a special emphasis on the central nervous system. As it may be relevant for metastasis in a range of cancers, the involvement of LINC complexes in migration of nonneuronal cells via its interaction with the perinuclear actin cap will also be developed.

Keywords LINC complexes • Sun protein • Nesprin • KASH domain • SUN domain • Nuclear lamina • Interkinetic nuclear migration • Neuronal migration • Nuclear anchorage • Actin cap • Skeletal muscle • Retina • Cell motility

---

D. Razafsky • D. Hodzic (✉)
Department of Ophthalmology and Visual Sciences, Washington University School of Medicine, 660 South Euclid Ave, St. Louis, MO 63110, USA
e-mail: Razafsky@vision.wustl.edu; hodzicd@vision.wustl.edu

D. Wirtz
Department of Chemical and Biomolecular Engineering, The Johns Hopkins University, 3400 North Charles St., Baltimore, MD 21218, USA
e-mail: wirtz@jhu.edu

E.C. Schirmer and J.I. de las Heras (eds.), *Cancer Biology and the Nuclear Envelope*, Advances in Experimental Medicine and Biology 773, DOI 10.1007/978-1-4899-8032-8_21,

Abbreviations

CNS	Central nervous system
DKO	Double knockout
ER	Endoplasmic reticulum
IKNM	Interkinetic nuclear migration
INM	Inner nuclear membrane
KO	Knockout
LINC	Linkers of the nucleoskeleton to the cytoskeleton
MTOC	Microtubule organizing center
NE	Nuclear envelope
NMJ	Neuromuscular junction
ONM	Outer nuclear membrane

Introduction

The nuclear envelope (NE) physically separates the genome from the cytoplasm (Fig. 1). It is composed of an inner and outer nuclear membrane (INM and ONM, respectively) that connect at nuclear pores and delineate the luminal compartment that is continuous with the lumen of the endoplasmic reticulum (ER). Whereas the ONM is an extension of the rough ER, the INM is devoid of ribosomes and displays a unique set of resident proteins that are immobilized within the INM by virtue of the

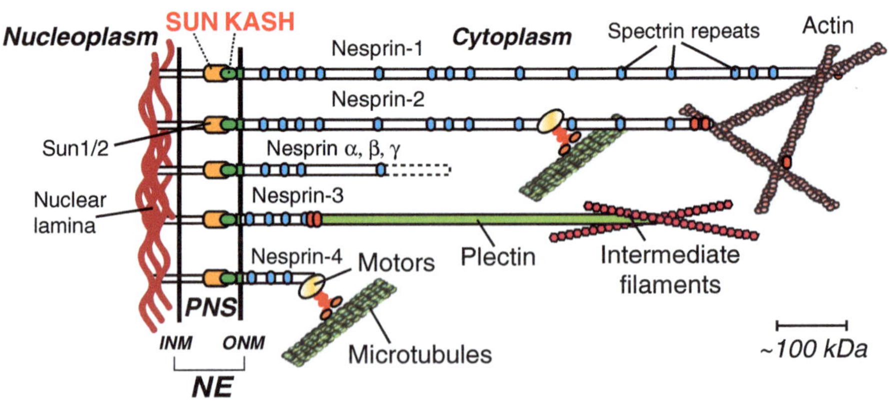

Fig. 1 Linkers of the nucleoskeleton to the cytoskeleton (LINC complexes) organization at the nuclear envelope in mammals (see text for more details). Nesprin α, β, γ refers to shorter spliced isoforms of Nesprin 1 and 2. *INM*, *ONM* inner and outer nuclear membrane, respectively, *NE* nuclear envelope, *SUN* Sad1 and Unc84 domain, *KASH* Klarsicht/ANC1/Syne homology domain

interaction of their nucleoplasmic domains with the nuclear lamina and/or chromatin [1]. The nuclear lamina is a meshwork of nuclear type-V intermediate filaments represented by A- and B-type lamins [2–4]. Whereas B-type lamins appear to be ubiquitously expressed both within progenitors and differentiated tissues, A-type lamins expression is restricted to subsets of differentiated cells and tissues [5–7].

LINC Complexes: Macromolecular Assemblies That Span the Nuclear Envelope and Mediate Nuclear Migration and Anchorage

The diversity of physiological functions exerted by resident proteins of the INM has completely redefined the functionality of the NE [8]. In particular, major progress has been made in the identification and physiological roles of macromolecular complexes that span the NE and mediate nuclear movements during development as well as nuclear anchorage in differentiated tissues. The term LInkers of the Nucleoskeleton to the Cytoskeleton (LINC) was coined to describe these molecular assemblies [9] (Fig. 1). They consist of interactions, within the lumen, between evolutionarily conserved motifs that characterize two families of integral transmembrane proteins of the NE: Sun (Sad1/Unc84) proteins and Nesprins (Nuclear Envelope SPectrin Repeat containing proteINS). In the following sections, we review the molecular nature of Sun proteins and Nesprins as well as their role as LINC complexes in nuclear dynamics and single cell migration.

Sun Proteins

Fifteen years ago, a seminal study by Malone et al. identified *unc84*, a gene whose mutation prevented both the migration and subsequent anchorage of nuclei populating the developing hypodermal syncytium of *C. elegans* [10]. UNC84 was identified as a transmembrane protein residing at the NE. One of its remarkable features was a C-terminal region of about 200 amino acids that displayed a strong homology with the C-terminal region of Sad1, a spindle pole body-associated protein in *S. pombe* [11]. This region, called the SUN domain (*S*ad1-*Un*c84 homology, PFAM family PF03856), was also identified in two human genes called *SUN1* and *SUN2*. Mammalian Sun1 and Sun2 proteins were later identified as type II integral membrane proteins of the INM with their conserved C-terminal SUN domain protruding within the luminal region between the INM and ONM of the NE [12–14]. Since then, orthologs have been identified in many phyla as well as in the plant kingdom [15] confirming the strong evolutionary conservation of SUN domains [16]. Recent crystallography studies have shown that SUN domains form homotrimeric structures displaying a cloverleaf-like arrangement [17, 18].

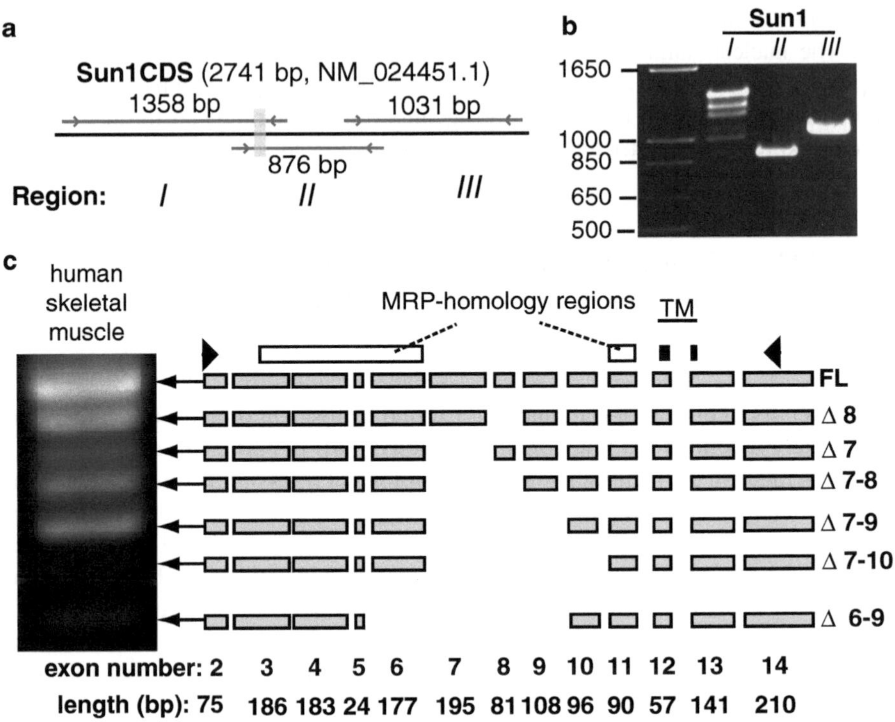

Fig. 2 The nucleoplasmic region of Sun1 is alternatively spliced and contains an MRP-homology region. (**a**) Primers location (*arrows*) and size of each amplicon predicted from indicated mRNA reference sequence of Sun1. *Gray shaded rectangle* corresponds to the transmembrane-encoding region. (**b**) RT-PCR amplification of Region I, II, and III of Sun1 transcripts carried out on total RNA of C2C12 mouse myoblasts. Note the amplification of multiple bands from Region I of Sun1 that encodes the nucleoplasmic domain. (**c**) PCR amplification of Sun1 transcripts from human skeletal muscle first-strand cDNAs with depiction of alternatively spliced isoforms. *White rectangles*: exonic region displaying MRP homology. *Black rectangles*: exonic region encoding the transmembrane domain of Sun1. *Arrowheads*: position of primers used in PCR amplification

The N-terminal nucleoplasmic region of Sun proteins is less well characterized. This region interacts with nuclear lamins [9, 13], but lamins are not required for human Sun proteins localization at the NE [19]. An emerging feature of Sun1 is the versatility of its primary sequence [20] due to the alternative splicing of exons 5–10 that encode the central nucleoplasmic region (Fig. 2). The nucleoplasmic region of Sun2, by contrast, does not display any comparable alternative splicing (data not shown). Interestingly, the extensively spliced region encodes a region reported to bind to nucleoplasmic binding partners of Sun1 [21], suggesting that specific Sun1 isoforms interact with distinct nucleoplasmic proteins. To date, the best-characterized nucleoplasmic variant of Sun1 was identified in testes and functionally defined by Gob et al. as being involved in mammalian sperm head formation [22]. Another intriguing feature of the nucleoplasmic region of Sun1 is the presence of a *M*itochondrial *R*NA binding *P*rotein domain (MRP, PFAM family PF09387) whose motif is highly conserved in mammalian Sun1 but not Sun2 (Fig. 2). In Trypanosoma, MRP1 and MRP2 belong to a complex machinery involved in mitochondrial RNA editing, a hallmark of kinetoplastids [23]. Whether the synthesis of MRP-like proteins by Sun1 genes actually occurs in mammals and their functional roles remains to be established.

Nesprins

Within the lumen, the SUN domain interacts directly with the evolutionary-conserved KASH (*K*larsicht/*A*nc-1, *S*yne *H*omology, PFAM family PF10541) domain, a stretch of ~60 amino acids made up of a transmembrane domain followed by a short stretch of ~30 luminal C-terminal amino acids (Fig. 1). To date, KASH domain-containing proteins have been functionally identified in *S. pombe*, *D. melanogaster*, *C. elegans*, and mammals [16]. In the latter, it is the typical molecular signature of a family of mammalian NE proteins called Nesprins that are encoded by five distinct genes (Nesprin-1 to -5) [24–28]. Whereas Nesprin-1, -2, and -3 are expressed in a wide variety of tissues; Nesprin-4 expression is more restricted and Nesprin-5 is a meiosis-specific KASH protein [28]. Nesprins harbor variable numbers of spectrin repeats along their cytoplasmic region that extend from ~50 kDa (Nesprin-4) to an astounding 1,000 kDa (giant isoform of Nesprin-1). Importantly, due to their gene size, a challenging plethora of Nesprin-1 and -2 isoforms (with and without KASH domains) are expressed to different degrees in different tissues and at different development times [29–31]. Next to these common structural features, giant isoforms of Nesprin-1 and -2 directly interact with actin through N-terminal actin-binding domains [32, 33] and Nesprin-3 with plectin [26] (Fig. 1). Nesprins also associate with molecular motors. In *C. elegans*, nuclear migration is mediated through direct interactions of the cytoplasmic region of UNC-83 with both kinesin-1 and dynein and their regulators [34, 35]. In mammals, Nesprin-2 coimmunoprecipitates with the dynein complex and Nesprin-4 with kinesin-1 [27, 36]. Together, these finding strongly support a model whereby LINC complexes connect the nucleus to the cytoskeleton and molecular motors.

LINC Complexes: Hubs for Force Transduction Across the NE

Interactions between SUN and KASH domain-containing proteins across the NE are direct and essential for the recruitment of KASH domain-containing proteins at the NE. Indeed, studies in different biological systems clearly demonstrate that the presence of SUN domains is strictly required for the ONM localization of KASH domain-containing proteins [9, 36, 37]. In addition, the KASH domain, by itself, is both sufficient to localize at the NE and strictly required to specify the NE localization of KASH proteins [38]. Accordingly, either the targeted expression of SUN domains within the perinuclear space or the overexpression of recombinant tagged KASH domains act in a dominant-negative manner by dislodging endogenous KASH domain-containing proteins from the ONM to the ER [39]. As we will see below, this property of recombinant KASH domain, such as EGFP-KASH, has multiple experimental applications to examine the role of LINC complexes in different species.

Several lines of evidence indicate that LINC complexes transduce forces across the NE. Physical coupling between the nucleus and the cytoplasm was directly demonstrated in harpooning experiments of the cytoplasm using microneedles. In this experimental setting, the NE clearly protrudes in the direction of an outward cytoplasmic pull, whereas it invaginates when the nucleoplasm itself is harpooned [40]. The NE is also distorted by manipulating microbeads attached to integrins, thereby suggesting that mechanical forces can be directly applied at the NE from the cytoplasmic membrane. As described above, the structural analysis and domain composition of Sun proteins and Nesprins strongly suggested a central role for LINC complexes in establishing such physical connections. Accordingly, disruption of SUN/KASH interactions drastically reduces nuclear deformation in microneedle manipulations [41] and further disrupts stretch-induced nuclear rotation [39]. Beside nuclear mechanics and dynamics, disruption of LINC complexes induces an overall loss of mechanical stiffness across the cytoskeleton [39], a phenotype that most likely reflects reported alterations of the perinuclear cytoskeleton [41] as well as decreased cellular migration and loss of polarization ensuing from LINC complex disruption [39].

Human SUN/KASH complexes have recently been crystallized [17]. Solved structures and biochemical approaches indicate that SUN domains physiologically assemble as trimers whose interacting interfaces provide large grooves for KASH domain binding. Because these interactions consist of an extensive network of noncovalent interactions between SUN and KASH triads and SUN and KASH domains interact covalently through disulfide bonds [17], SUN/KASH interactions appear well adapted as force-resistant coupling devices to move or still anchor nuclei within cells or syncytia [42]. Accumulating data indicate that forces transduced by LINC complexes are used in two important biological phenomena: nuclear positioning and chromosome movements. For more information on the role of SUN/KASH interaction in chromosome movements, readers are referred to a recent and thorough review by Kracklauer et al. [43]. Here, we further focus on the involvement of LINC complexes in nuclear positioning.

LINC Complexes and Nuclear Positioning in CNS Development

As described above, pioneering studies in *C. elegans* and *D. melanogaster* [44] clearly pointed out the role of SUN and KASH domain-containing proteins in nuclear migration and anchorage. More importantly, they paved the way to more recent studies aimed at understanding the physiological relevance of different types of nuclear movement observed during central nervous system (CNS) development [16].

The latter proceeds through the transformation of a pseudostratified layer of precursor cells into laminated layers of differentiated neurons whose interconnections establish the CNS circuitry. This transformation and its accompanying nuclear movements are well illustrated by the development of the mammalian retina. In the latter, the pseudostratified neuroepithelium, called the neuroblast layer, morphs into three distinct laminae of differentiated neurons. This process can be divided into different steps: (1) exit of retinal progenitor cells, which populate the neuroblast layer, from the cell cycle, (2) migration of post-mitotic newborn neurons towards their final laminar position, (3) anchorage of differentiated neurons to their specific laminar position. Below, we describe the various types of nuclear movements as well as what is known about the involvement of LINC complexes in these different developmental steps.

Interkinetic Nuclear Migration in Neuronal Progenitors

Interkinetic nuclear migration (IKNM, Fig. 3) consists of cell cycle-dependent oscillations of neuronal progenitors nuclei within pseudostratified neuroepithelia (recently reviewed in refs. 45, 46). Importantly, IKNM appears to be a universal property of pseudostratified epithelia. It has mostly been studied in developing neural tissues but, importantly, it is neither restricted to developing CNS tissues nor confined to vertebrates [46–48]. During IKNM, nuclei migrate towards the basal side of neuroepithelia during the G1 phase of the cell cycle and move back to the apical side during G2. As a result, mitoses take place exclusively at the apical side (Fig. 3a), while S-phase proceeds at the basal side of neuroepithelia. Anti-phospho-Histone H3, which specifically labels mitotic cells, and BrdU labeling, which labels S-phase cells, is commonly used to track IKNM in fixed samples.

Blocking nuclear oscillations during IKNM does not alter cell cycle progression of progenitors [49, 50], whereas cell cycle arrest blocks nuclear oscillations [51, 52]. These nuclear oscillations require an intact cytoskeleton (microtubule, actin, and centrosome) and molecular motors (actomyosin, kinesins, and the dynein complex). The identification of these molecular actors and their respective roles in apicobasal migration within neuroepithelial cells has been recently reviewed [45, 46]. By comparison to the wide amplitude of apicobasal migration of neuroblast nuclei, the centrosome is relatively stationary on the apical side of neuroepithelia [53] (Fig. 3a). However, recent time-lapse video microscopy experiments on chicken neural tube indicate that the centrosome migrates basally to "meet" apically migrating nuclei at late G2 [54].

To date, two reports indicate that LINC complexes directly mediate IKNM in mammals. Using time-lapse video microscopy, Zhang et al. showed that nuclear migration towards the apical side of the ventricular zone was significantly slower in brain slices of Nesprin-2 KO and Sun1/Sun2 double knockout (DKO) mouse embryos by comparison to wild-type brains [36]. Accordingly, mitotic cells labeled with phospho-Histone H3 antibodies mislocalized across the length of the

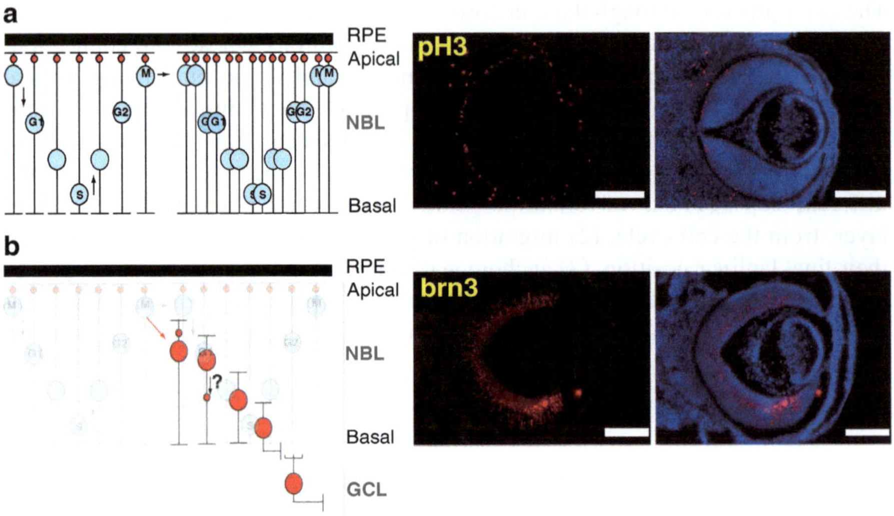

Fig. 3 Nuclear movements during embryonic retinal development. (**a**) Interkinetic nuclear migration consists of the basoapical migration of retinal progenitor cell nuclei (RPC, blue nuclei) in phase with the cell cycle. A symmetric mitotic division generating two progenitor cells is also illustrated. *Small red circles* depict the position of the centrosome. *Right panel*: Immunofluorescence of an E14.5 mouse ocular globe showing the apical localization of mitotic progenitors (labeled with anti-phosphoHistone3, pH3) on the apical side of the neuroretina. (**b**) In the case of an asymmetric division, the post-mitotic cells (in this case a retinal ganglion cell, red nuclei) migrate towards its final laminar position within the ganglion cell layer. *Right panel*: Immunofluorescence of an E14.5 mouse ocular globe showing newborn retinal ganglion cells (labeled with brn3) migrating towards the GCL. *M* M-phase nuclei, *S* S-phase nuclei, *RPE* retinal pigment epithelium, *NBL* neuroblast layer, *GCL* ganglion cell layer. Scale bar: 200 μm

ventricular zone rather than on its apical side. IKNM is also significantly altered in Sun1 and Nesprin-2 KO mouse retina [55]. In zebrafish retina, either the downregulation of Nesprin-2 or the expression of a recombinant dominant-negative EGFP-KASH protein during IKNM correlates with the accelerated genesis of earlier born neurons such as retinal ganglion cells at the expense of later-born photoreceptors [56]. These experiments did not directly examine nuclear positioning during IKNM. However, a similar phenotype of accelerated neurogenesis is also observed upon nonsense mutations of dynactin that directly affects nuclear positioning during IKNM. Because Nesprin-2 interacts with dynactin in mouse retina and brain lysates [36, 55] and these interactions are evolutionary conserved, these experimental results predict that LINC complex disruption should affect IKNM in Zebrafish retina. Furthermore they would suggest that intact LINC complexes are required to maintain an appropriate pool of neural progenitors while generating post-mitotic neurons in a timely fashion [50, 51, 56].

Together, these results strongly suggest that LINC complexes provide essential nucleocytoskeletal connections that directly mediate nuclear positioning during IKNM. Molecular motors-generated forces that move nuclei across the neuroepithelial cell length are most likely transduced to the NE through LINC complexes.

The identification of unc-83 binding domains to dyneins, kinesins, and their accessory proteins further suggest that the Nesprin interaction with molecular motors may be direct [35]. According to a model whereby basalmost migration of progenitor nuclei favors a subsequent asymmetric division into a newborn neuron and another retinal progenitor cells [49, 51, 56], faulty IKNM within retinal progenitor cells may affect the balance of asymmetric vs. symmetric division thereby affecting the relative abundance of retinal cell types in adult retinas. By contrast to the alteration of molecular motors that are involved in a diverse array of cellular functions, disruption of LINC complexes may provide a more specific mean to disrupt nuclear positioning during IKNM. Hence, the development of genetic tools to disrupt the SUN/KASH interactions in different species may provide additional insight into the possible physiological function(s) of IKNM.

Nuclear Translocation in Post-mitotic Neurons

Exit of progenitor cells from the cell cycle is followed by the migration of post-mitotic neurons from the apical side of the neuroepithelium towards their final laminar destination (Fig. 3b). In the developing neocortex, trekking of these newborn neurons towards the pial surface requires the active translocation of their nuclei [57, 58]. This actin- and microtubule-dependent process is initiated by the rapid extension and subsequent retraction of the leading neurite. The centrosome, which is invariably positioned ahead of the nucleus in these cells, moves away from the nucleus towards the leading process. The nucleus then translocates closer to the centrosome [59]. Hence, by contrast to IKNM where the centrosome remains relatively stationary, nuclear translocation within migrating post-mitotic neurons requires the coupling of the nucleus to the centrosome via a "fork-like" structure of microtubules wrapping the nucleus. The dynein/dynactin complex, LIS1, and other proteins that bind to microtubules and regulate dynein activity closely regulate nuclear translocation in these cells [57, 59–61]. Abnormal nuclear translocation underlies the erroneous lamination of cortical layers, a phenotype that characterizes a wide range of human pathologies of the CNS called lissencephalies [62]. Similar cortical lamination defects, caused by hypoglycosylation of basement membrane components, are also observed in a subgroup of congenital muscular dystrophies called dystroglycanopathies. These include muscle–eye–brain disease, Fukuyama congenital muscular dystrophy, and the Walker-Warburg syndrome [63, 64]. Severe retinal development abnormalities such as microphthalmia, optic nerve hypoplasia, and retinal dysgenetic stratification are also prominent in patients affected by lissencephalies [65–70]; and ocular defects are also observed in mouse models of human dystroglycanopathies [71, 72]. It is therefore likely that similar molecular mechanisms underlying nuclear migration in the neocortex are at play during retinogenesis. For these reasons, a better understanding of the role of nuclear movements during retinogenesis will most likely provide essential information about molecular mechanisms underlying both congenital retinopathies and congenital brain development disorders.

LINC complexes are required for nuclear translocation during the migration of newborn cortical neurons. Indeed, radial migration of newborn cortical neurons is severely hampered in the developing cerebral cortex of SUN1/2 DKO and Nesprin-2 KO embryos [36]. These defects are clearly associated to the failure of newborn cortical neurons to translocate their nuclei due to a loss of physical coupling between the nucleus and the centrosome. Indeed, whereas nuclei undergo robust translocation towards the pial surface in close association with the centrosome in wild-type brain slices, newborn cortical neuron nuclei remain at the ventricular side of SUN1/2 DKO or Nesprin-2 KO brain slices even though centrosomes keep moving towards the pial surface. These observations strongly suggest that transduction of apical forces on the migrating neuron nuclei are abolished upon LINC complex disruption. Whereas this mode of nuclear translocation has been clearly demonstrated for tangentially migrating cortical interneurons and radial cortical neurons [73, 74], other modes of neuronal nuclear translocation may exist within the CNS. Indeed, during radial migration of granule cells in cerebellar slices, the centrosome most often localizes between the anterior and posterior poles of the migrating nucleus rather than leading the nucleus [75]. Another example of "unconventional" nuclear translocation is provided by newborn retinal ganglion cells migrating towards their final laminar position at the beginning of zebrafish retinogenesis. In that case, the centrosome localizes at the trailing edge of migrating retinal ganglion cells [53]. The postnatal development of cone photoreceptors may provide another example of unconventional nuclear translocation within post-mitotic neurons. Whereas cone photoreceptors are specified early during embryonic retinal development [76], their nuclei remain on the apical surface of the neuroblast layer. However, during postnatal retinal development, mouse cone nuclei localize within the upper 2/3 of the developing outer plexiform layer before regaining their apical position [77]. Through the spatiotemporal control of EGFP-KASH2 expression during mouse cone photoreceptor maturation, we recently showed that LINC complexes mediate the apical migration leg of that oscillation, whereas basal migration appears unaffected. As a result, cone photoreceptor nuclei mislocalized at the basal edge of the outer nuclear layer as well as within the outer plexiform layer in adult retina [78]. Though less severe, a mislocalization phenotype of cone photoreceptor nuclei was also reported in adult retinas from Sun1 KO mice [55]. Even though the localization of the centrosome has not been formally examined during these oscillations, their apical location is required to elaborate the connecting cilium of presumptive photoreceptors. Hence, nuclear oscillations within maturing cone photoreceptors most likely take place without any significant movement of the centrosome thereby representing another case of unconventional nuclear translocation within a post-mitotic neuron.

Molecular mechanisms mediating photoreceptor nuclei positioning appear to be evolutionary conserved. Indeed, in *D. melanogaster* and zebrafish, genetic alterations of either the dynein/dynactin complex or SUN/KASH interactions clearly affect nuclear positioning during photoreceptor maturation [79–81].

As mentioned above, the nucleoplasmic region of Sun proteins interacts with nuclear lamins. Therefore, forces generated by molecular motors and transduced through LINC complexes should act on the lamina meshwork to pull the nucleus in the direction of cellular migration. This essential contribution of lamins is best exemplified by nuclear movements associated with the development of the compound eye of *D. melanogaster*. Indeed, genetic alteration of *lamin* Dm_0 prevents the apical migration of photoreceptor precursors nuclei [82]. Importantly, that same failure of apical migration is phenocopied by genetic alteration of *klaroid* (SUN protein), *Klarsicht* (KASH protein), or *glued* (Dynactin) mutants [80–83]. In mammals, A- and B-type lamins make up the nuclear lamina of most differentiated cells. However, mostly B-type lamins are expressed during CNS development. Accordingly, B-type lamins KO mice display severe neurodevelopmental defects and die at birth, whereas A-type lamin KO mice develop normally but die by 4–6 weeks and present muscular dystrophy phenotypes [84, 85].

Nuclear Anchorage

Mammalian Nesprin-1 was initially discovered as a direct binding partner of MuSK, a tyrosine kinase enriched at the post-synaptic membrane of the neuromuscular junction (NMJ), and further identified as an NE protein whose abundance was significantly higher in specialized nuclei that cluster beneath the NMJ (synaptic nuclei) by comparison to extrasynaptic nuclei [24]. Constitutive transgenic overexpression of the dominant-negative KASH domain of Nesprin-1 induced a phenotype where synaptic nuclei mislocalize away from the NMJ [86]. The development of Nesprin-1 KO mice confirmed that phenotype and further indicated that Nesprin-1 also mediates the spacing of extrasynaptic nuclei along skeletal muscle fibers. Because Nesprin-2 is dispensable for synaptic nuclei to localize at the NMJ [87], anchoring of synaptic nuclei at the NMJ is one of the few cellular functions of Nesprin-1 that does not overlap with Nesprin-2. Similar synaptic nuclei mispositioning phenotypes have been reported in different models of genetic alterations of the Nesprin-1 gene in mice [88, 89], and, in agreement with a central role of SUN/KASH interactions in synaptic nuclei anchoring, Sun1 and Sun2 function redundantly in myonuclear anchorage [90]. Synaptic nuclei mislocalization has also been reported in skeletal muscle biopsies from patients affected by autosomal recessive cerebellar ataxia type 1 (ARCA1) associated to mutations of Nesprin-1 [91]. In agreement with mouse models, NMJ structural organization is not affected by nuclear mislocalization. Whereas these results clearly indicate that genetic alteration of Nesprin-1 alters nuclear positioning in skeletal muscle, the phenotypical outcome in these different mouse models greatly varies from the lack of any overt phenotype [87] to Emery-Dreifuss muscular dystrophy-like [89] or growth retardation [88] phenotypes. The reader is referred to [88] for further discussion about the potential reasons underlying this phenotypical variability. Taken together, these results clearly indicate that LINC complexes, mostly made of Nesprin-1 and Sun1 in this case, mediate the anchorage of myonuclei within skeletal muscle fibers. Furthermore, results from different mouse models and ARCA1 patients suggest that genetic alterations of Nesprin-1 are clearly associated with mispositioning of synaptic nuclei away from the NMJ, but this mispositioning is not a good indicator of the clinical presentation.

By contrast, there is a clear clinical correlation between genetic alteration of Nesprin-4 and deafness. Indeed, a study by Horn et al. recently reported that patients affected by progressive high frequency hearing loss carry genetic alterations of the gene encoding Nesprin-4 [92], a phenotype that can be reproduced in Nesprin-4 KO mice. In agreement for a direct role of LINC complexes in hearing, Sun1KO mice also present with severe hearing loss. Within outer hair cells from both KO models, nuclei mislocalized on the apical side whereas wild-type outer hair cells maintain their nuclei at significantly more basal positions. This mispositioning is accompanied by a severe degeneration of outer hair cells that occurs in a basal to apical gradient across the cochlea. Disruption of LINC complexes upon expression of dominant-negative KASH proteins (under the control of a heat shock promoter) induced *after* the full differentiation of zebrafish photoreceptors also induce nuclear mislocalization, a phenotype that is also accompanied by severe photoreceptor degeneration. A challenging question regarding the consequences of LINC complex disruption in sensory neurons is the molecular mechanisms leading to their degeneration. Indeed, disruption of LINC complexes in cultured cells is accompanied by significant cytoskeletal and mechanical cellular defects. Hence, one can wonder whether either nuclear positioning per se or the alteration of cytoskeletal organization ensuing from LINC complex disruption underlies sensory neurons degeneration.

Role of the Nucleus and LINC Complexes in Single Cell Migration

Since the nucleus is topologically enclosed within the intracellular space of migrating eukaryotic cells, nuclear movements, and overall cell movements are highly correlated. But whether the interphase nucleus plays an active role in cell migration—and is not simply dragged along by the cell—is only beginning to be investigated. Recent discovery and characterization of LINC complexes suggests a much more tightly regulated functional connection between nuclear movements and positioning in the cell and cell migration than previously thought.

During planar cell migration in vitro, the nucleus is typically positioned near the myosin-rich contractile tail of the cell, while the microtubule-associated organizing center (MTOC) polarizes the cell and is positioned between the nucleus and the cell's leading edge, where active actin filament assembly and turnover occur [93]. Mesenchymal migrating cells dynamically switch back and forth between persistent moves and periods of migratory arrest, generating trajectories that are well approximated by a so-called persistent random walk. A persistent random walk is completely defined by just two parameters: cell speed and persistence time, the time it takes for the cell to deviate from a straight line, straightforwardly computed from fits of the mean squared displacements of cells. A long persistence time signifies a directionally persistent migration of the cell. During such migratory patterns, the cell dynamically changes its shape, from a mostly elongated morphology during persistent moves to a mostly rounded morphology during transient arrest and

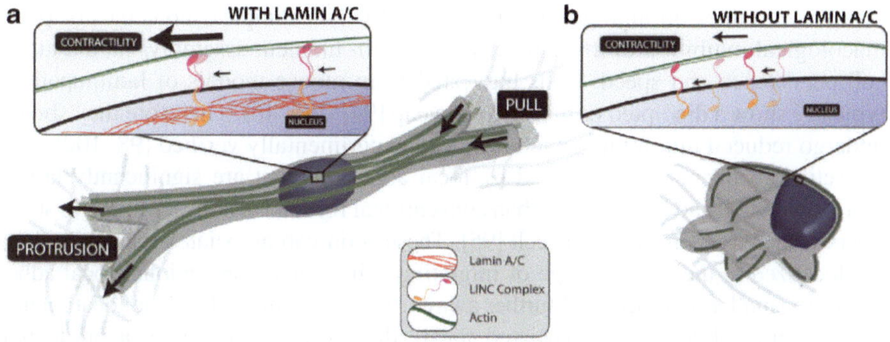

Fig. 4 The architecture and function of the LINC-anchored actin cap in 3D cell migration. (**a**) Cells with an intact actin cap, form thick pseudopodial protrusions that can pull on the surrounding matrix for net translocation. Actin-cap fibers (*green*) provide mechanical support to the protrusion thanks to LINC-mediated anchorage to the nuclear lamina. (**b**) Cells lacking an actin cap (caused by lamin A/C deficiency) cannot generate protrusions and, therefore, cannot translocate efficiently

repolarization of the MTOC for the cell's next persistent move in a direction uncorrelated to the previous persistent move.

Recent work has revealed the existence of a highly contractile acto-myosin filamentous structure, the perinuclear actin cap, which tightly wraps around the nucleus and regulates its shape [94]. The perinuclear actin cap is composed of thick and mostly parallel fibers that are rich in phosphorylated myosin II and F-actin crosslinker α-actinin and are dynamically anchored to the apical surface and lateral sides of the interphase nucleus and its lamina through LINC complexes [94, 95] (Fig. 4a). Hence, nucleus-anchored actin-cap fibers are topologically different from conventional stress fibers, which lie entirely at the bottom of adherent cells and are anchored to the plasma membrane. In interphase cells, the actin cap is not permanently affixed to the nucleus; rather it forms and dissolves dynamically due to both rapid turnover dynamics of F-actin fibers in the cap and the cap's sliding motion above the apical surface of the nucleus, which causes the actin-cap fibers to take on characteristics of basal stress fibers [95].

Cells forced to elongate on narrow adhesive (fibronectin- or collagen-coated) patterns flanked with nonadhesive PEG surfaces display an elongated nuclear morphology, while cells on round adhesive micropatterns display a round nucleus [94]. Accordingly, elongated cells show a prominent perinuclear actin cap, while rounded cells show no actin cap, suggesting that an important function of the actin cap is nuclear shaping and relating nuclear shape to cell shape. Cells on narrow patterns, which typically show an actin cap oriented along the long cell axis, are forced to undergo highly persistent migration, and cells on round patterns, which lack an actin cap, do not undergo net translocation. This global functional correlation between cell shape, nuclear shape, actin-cap status, and mode of migration predict that cells migrating on flat (unpatterned) substrates, which undergo persistent migration, will dynamically show an actin cap during directionally persistent moves and no or disrupted actin cap during transient migratory arrest and repolarization events [96, 97]. These results also predict that disruption of the actin cap and LINC complexes, either through genetic depletion of LINC complex molecules or pharmacological inhibition of myosin activity or actin filament assembly, should affect cell persistence and speed. Cells harvested from mouse models of laminopathies typically show a disrupted or missing actin cap [94] (Fig. 4b). Therefore they should undergo reduced migration, which is indeed experimentally verified [98–100].

Actin-cap fibers are terminated by focal adhesions that are significantly larger, more elongated, and longer-lived than conventional focal adhesions terminating stress fibers at the basal surface of the cell [95]. These actin-cap-associated focal adhesions are localized at the leading edge of migrating cells, while conventional focal adhesions are smaller and localized further away from the leading edge. The location and long lifetime of actin-cap-associated focal adhesions may maintain a productive lamellipodium in a given direction before retraction and cell repolarization. During random migration, upon spontaneous dissolution of the actin cap and of actin-cap-associated focal adhesions, the lamellipodium retracts, slowing down cell translocation. While the actin cap is present, the LINC interconnections between actin-cap fibers on the nuclear surface and nuclear lamina would prevent nuclear rotation and only allows for nuclear translocation, inducing persistent migration [95, 101]. When the actin cap (transiently) disappears, nuclear translocation stops, the brakes to nuclear rotation are released, and dynein-mediated nuclear rotation can occur. Accordingly, the MTOC can reorient as it is physically connected to the NE and the cell repolarizes to prepare for the next persistent move of the cell in a new direction.

Fibroblasts and post-epidermal-mesenchymal transition cancer cells migrate within a mostly three-dimensional (3D) collagen I-rich matrix. In 3D cell migration, the interphase nucleus is typically smaller and more elongated than the same cells on 2D flat collagen-coated surfaces and located in the middle of the cell [102–105]. The 3D equivalent of the actin cap for cells inside 3D collagen matrices is constituted of acto-myosin fibers that are now isotropically located all around the nucleus and prolong the nuclear region into long and thick protrusions [103]. These pseudopodial protrusions mediate traction forces on the collagen fibers surrounding the cell, inducing net cell translocation. Cell migration in 3D matrices is much more persistent than in 2D migration presumably because the 3D actin cap is longer-lived compared to its 2D counterpart. Moreover, activated fibroblasts and cancer cells will locally digest the matrix through the expression of membrane-bound matrix metalloproteinases (i.e., MT1-MMP), forming in their wake thin open channels of cross-sectional size smaller than the nuclear size [96, 97, 102, 106]. Disruption of LINC complexes significantly reduces the ability of cells to protrude and in turn reduces cell-induced matrix traction and therefore cell migration. Hence, the LINC-anchored contractile actin cap plays a central role in 3D cell migration by promoting the formation of protrusions and actively compressing the nucleus [96].

Lateral confinement of cells by microfabricated channels of cross-sectional size smaller than nuclear size typically induces actin cap formation, which may help migration of these highly confined cells even for microchannel sizes that are multiple-fold smaller than the natural size of the nucleus [105, 107]. How these in vitro systems mimic the in vivo context and whether actin cap formation is necessary for efficient cancer cell migration in the stromal matrix remains to be experimentally tested.

Concluding Remarks

For the past 15 years, our view of the NE has been radically transformed from a mere physical barrier between the nucleoplasm and the cytoplasm to a multifunctional compartment with "a life on its own." In particular, the identification of LINC complexes and their roles as force transmission hubs involved in the physical communication between cytoplasmic and nucleoplasmic networks across the NE is currently fueling new research avenues. Whereas the "core" of LINC complexes is now well defined, we are starting to have an appreciation of the variability of their cytoplasmic and nucleoplasmic interfaces. For these reasons, appropriate tools need to be developed to address the variable composition of Nesprins in different tissues at different development stages (Which nesprin? Which isoform thereof? When?). By the same token, the putative variability of LINC complexes nucleoplasmic interface(s) needs to be further addressed. Finally, whereas current knockout animal models have been central in our understanding of how the LINC complex affects CNS development, conditional models of LINC complex disruption need to be developed in order to examine their physiological roles in vivo in a cell-autonomous manner and at specific developmental stages. Recent advances in methods allowing for the expression of exogenous proteins in either organotypic mammalian tissue slices or primary cell cultures combined to time-lapse video microscopy also provide very attractive experimental options [108, 109]. Taken together, these approaches will most likely further fuel our expanding understanding of the role of the NE in normal and pathological biological processes. Given the central involvement of both LINC complexes and actin cap in cellular migration, examining new hypotheses about the relevance of these NE components in metastasis will be of particular interest.

Acknowledgements The authors are supported by the National Institute of Health (R01GM084204 to D.W. and D.H., R01CA174388 and U54CA143868 to D.W., R01EY022632 to D.H.) and by NIH training grant T32EY013360, core grant P30EY002687 and Unrestricted Grants from Research to Prevent Blindness, Inc. awarded to the Department of Ophthalmology and Visual Sciences at Washington University School of Medicine.

References

1. Schirmer EC, Florens L, Guan T, Yates JR III, Gerace L (2003) Nuclear membrane proteins with potential disease links found by subtractive proteomics. Science 301:1380–1382
2. Stuurman N, Heins S, Aebi U (1998) Nuclear lamins: their structure, assembly, and interactions. J Struct Biol 122:42–66
3. Hutchison CJ (2002) Lamins: building blocks or regulators of gene expression? Nat Rev Mol Cell Biol 3:848–858
4. Burke B, Stewart CL (2013) The nuclear lamins: flexibility in function. Nat Rev Mol Cell Biol 14:13–24
5. Stewart C, Burke B (1987) Teratocarcinoma stem cells and early mouse embryos contain only a single major lamin polypeptide closely resembling lamin B. Cell 51:383–392

6. Rober RA, Weber K, Osborn M (1989) Differential timing of nuclear lamin A/C expression in the various organs of the mouse embryo and the young animal: a developmental study. Development 105:365–378
7. Solovei I, Wang AS, Thanisch K, Schmidt CS, Krebs S et al (2013) LBR and lamin A/C sequentially tether peripheral heterochromatin and inversely regulate differentiation. Cell 152:584–598
8. Stewart CL, Roux KJ, Burke B (2007) Blurring the boundary: the nuclear envelope extends its reach. Science 318:1408–1412
9. Crisp M, Liu Q, Roux K, Rattner JB, Shanahan C et al (2006) Coupling of the nucleus and cytoplasm: role of the LINC complex. J Cell Biol 172:41–53
10. Malone CJ, Fixsen WD, Horvitz HR, Han M (1999) UNC-84 localizes to the nuclear envelope and is required for nuclear migration and anchoring during *C. elegans* development. Development 126:3171–3181
11. Hagan I, Yanagida M (1995) The product of the spindle formation gene sad1+ associates with the fission yeast spindle pole body and is essential for viability. J Cell Biol 129:1033–1047
12. Hodzic DM, Yeater DB, Bengtsson L, Otto H, Stahl PD (2004) Sun2 is a novel mammalian inner nuclear membrane protein. J Biol Chem 279:25805–25812
13. Haque F, Lloyd DJ, Smallwood DT, Dent CL, Shanahan CM et al (2006) SUN1 interacts with nuclear lamin A and cytoplasmic nesprins to provide a physical connection between the nuclear lamina and the cytoskeleton. Mol Cell Biol 26:3738–3751
14. Wang Q, Du X, Cai Z, Greene MI (2006) Characterization of the structures involved in localization of the SUN proteins to the nuclear envelope and the centrosome. DNA Cell Biol 25:554–562
15. Graumann K, Runions J, Evans DE (2010) Characterization of SUN-domain proteins at the higher plant nuclear envelope. Plant J 61:134–144
16. Razafsky D, Hodzic D (2009) Bringing KASH under the SUN: the many faces of nucleo-cytoskeletal connections. J Cell Biol 186:461–472
17. Sosa BA, Rothballer A, Kutay U, Schwartz TU (2012) LINC complexes form by binding of three KASH peptides to domain interfaces of trimeric SUN proteins. Cell 149:1035–1047
18. Zhou Z, Du X, Cai Z, Song X, Zhang H et al (2012) Structure of Sad1-UNC84 homology (SUN) domain defines features of molecular bridge in nuclear envelope. J Biol Chem 287:5317–5326
19. Hasan S, Guttinger S, Muhlhausser P, Anderegg F, Burgler S et al (2006) Nuclear envelope localization of human UNC84A does not require nuclear lamins. FEBS Lett 580:1263–1268
20. Gob E, Meyer-Natus E, Benavente R, Alsheimer M (2011) Expression of individual mammalian Sun1 isoforms depends on the cell type. Commun Integr Biol 4:440–442
21. Haque F, Mazzeo D, Patel JT, Smallwood DT, Ellis JA et al (2009) Mammalian SUN protein interaction networks at the inner nuclear membrane and their role in laminopathy disease processes. J Biol Chem 285:3487–3498
22. Gob E, Schmitt J, Benavente R, Alsheimer M (2010) Mammalian sperm head formation involves different polarization of two novel LINC complexes. PLoS One 5:e12072
23. Schumacher MA, Karamooz E, Zikova A, Trantirek L, Lukes J (2006) Crystal structures of *T. brucei* MRP1/MRP2 guide-RNA binding complex reveal RNA matchmaking mechanism. Cell 126:701–711
24. Apel ED, Lewis RM, Grady RM, Sanes JR (2000) Syne-1, a dystrophin- and Klarsicht-related protein associated with synaptic nuclei at the neuromuscular junction. J Biol Chem 275:31986–31995
25. Zhang Q, Ragnauth C, Greener MJ, Shanahan CM, Roberts RG (2002) The nesprins are giant actin-binding proteins, orthologous to *Drosophila melanogaster* muscle protein MSP-300. Genomics 80:473–481
26. Wilhelmsen K, Litjens SH, Kuikman I, Tshimbalanga N, Janssen H et al (2005) Nesprin-3, a novel outer nuclear membrane protein, associates with the cytoskeletal linker protein plectin. J Cell Biol 171:799–810

27. Roux KJ, Crisp ML, Liu Q, Kim D, Kozlov S et al (2009) Nesprin 4 is an outer nuclear membrane protein that can induce kinesin-mediated cell polarization. Proc Natl Acad Sci U S A 106:2194–2199
28. Morimoto A, Shibuya H, Zhu X, Kim J, Ishiguro K et al (2012) A conserved KASH domain protein associates with telomeres, SUN1, and dynactin during mammalian meiosis. J Cell Biol 198:165–172
29. Rajgor D, Mellad JA, Autore F, Zhang Q, Shanahan CM (2012) Multiple novel nesprin-1 and nesprin-2 variants act as versatile tissue-specific intracellular scaffolds. PLoS One 7:e40098
30. Cottrell JR, Borok E, Horvath TL, Nedivi E (2004) CPG2: a brain- and synapse-specific protein that regulates the endocytosis of glutamate receptors. Neuron 44:677–690
31. Randles KN, le Lam T, Sewry CA, Puckelwartz M, Furling D et al (2010) Nesprins, but not sun proteins, switch isoforms at the nuclear envelope during muscle development. Dev Dyn 239:998–1009
32. Zhen YY, Libotte T, Munck M, Noegel AA, Korenbaum E (2002) NUANCE, a giant protein connecting the nucleus and actin cytoskeleton. J Cell Sci 115:3207–3222
33. Padmakumar VC, Abraham S, Braune S, Noegel AA, Tunggal B et al (2004) Enaptin, a giant actin-binding protein, is an element of the nuclear membrane and the actin cytoskeleton. Exp Cell Res 295:330–339
34. Meyerzon M, Fridolfsson HN, Ly N, McNally FJ, Starr DA (2009) UNC-83 is a nuclear-specific cargo adaptor for kinesin-1 mediated nuclear migration. Development 136(16): 2725–2733
35. Fridolfsson HN, Ly N, Meyerzon M, Starr DA (2010) UNC-83 coordinates kinesin-1 and dynein activities at the nuclear envelope during nuclear migration. Dev Biol 338:237–250
36. Zhang X, Lei K, Yuan X, Wu X, Zhuang Y et al (2009) SUN1/2 and Syne/Nesprin-1/2 complexes connect centrosome to the nucleus during neurogenesis and neuronal migration in mice. Neuron 64:173–187
37. McGee MD, Rillo R, Anderson AS, Starr DA (2006) UNC-83 IS a KASH protein required for nuclear migration and is recruited to the outer nuclear membrane by a physical interaction with the SUN protein UNC-84. Mol Biol Cell 17:1790–1801
38. Padmakumar VC, Libotte T, Lu W, Zaim H, Abraham S et al (2005) The inner nuclear membrane protein Sun1 mediates the anchorage of Nesprin-2 to the nuclear envelope. J Cell Sci 118:3419–3430
39. Stewart-Hutchinson PJ, Hale CM, Wirtz D, Hodzic D (2008) Structural requirements for the assembly of LINC complexes and their function in cellular mechanical stiffness. Exp Cell Res 314:1892–1905
40. Maniotis AJ, Chen CS, Ingber DE (1997) Demonstration of mechanical connections between integrins, cytoskeletal filaments, and nucleoplasm that stabilize nuclear structure. Proc Natl Acad Sci U S A 94:849–854
41. Lombardi ML, Jaalouk DE, Shanahan CM, Burke B, Roux KJ et al (2011) The interaction between nesprins and sun proteins at the nuclear envelope is critical for force transmission between the nucleus and cytoskeleton. J Biol Chem 286:26743–26753
42. Rothballer A, Schwartz TU, Kutay U (2013) LINCing complex functions at the nuclear envelope: What the molecular architecture of the LINC complex can reveal about its function. Nucleus 4:29–36
43. Kracklauer MP, Link J, Alsheimer M (2013) LINCing the nuclear envelope to gametogenesis. Curr Top Dev Biol 102:127–157
44. Starr DA, Fischer JA (2005) KASH 'n Karry: the KASH domain family of cargo-specific cytoskeletal adaptor proteins. Bioessays 27:1136–1146
45. Taverna E, Huttner WB (2010) Neural progenitor nuclei IN motion. Neuron 67:906–914
46. Kosodo Y (2012) Interkinetic nuclear migration: beyond a hallmark of neurogenesis. Cell Mol Life Sci 69:2727–2738
47. Spear PC, Erickson CA (2012) Interkinetic nuclear migration: a mysterious process in search of a function. Dev Growth Differ 54:306–316

48. Lee HO, Norden C (2013) Mechanisms controlling arrangements and movements of nuclei in pseudostratified epithelia. Trends Cell Biol 23:141–150
49. Murciano A, Zamora J, Lopez-Sanchez J, Frade JM (2002) Interkinetic nuclear movement may provide spatial clues to the regulation of neurogenesis. Mol Cell Neurosci 21:285–300
50. Schenk J, Wilsch-Brauninger M, Calegari F, Huttner WB (2009) Myosin II is required for interkinetic nuclear migration of neural progenitors. Proc Natl Acad Sci U S A 106:16487–16492
51. Baye LM, Link BA (2007) Interkinetic nuclear migration and the selection of neurogenic cell divisions during vertebrate retinogenesis. J Neurosci 27:10143–10152
52. Ueno M, Katayama K, Yamauchi H, Nakayama H, Doi K (2006) Cell cycle progression is required for nuclear migration of neural progenitor cells. Brain Res 1088:57–67
53. Zolessi FR, Poggi L, Wilkinson CJ, Chien CB, Harris WA (2006) Polarization and orientation of retinal ganglion cells in vivo. Neural Dev 1:2
54. Spear PC, Erickson CA (2012) Apical movement during interkinetic nuclear migration is a two-step process. Dev Biol 370:33–41
55. Yu J, Lei K, Zhou M, Craft CM, Xu G et al (2011) KASH protein Syne-2/Nesprin-2 and SUN proteins SUN1/2 mediate nuclear migration during mammalian retinal development. Hum Mol Genet 20:1061–1073
56. Del Bene F, Wehman AM, Link BA, Baier H (2008) Regulation of neurogenesis by interkinetic nuclear migration through an apical-basal notch gradient. Cell 134:1055–1065
57. Ghashghaei HT, Lai C, Anton ES (2007) Neuronal migration in the adult brain: are we there yet? Nat Rev Neurosci 8:141–151
58. Ayala R, Shu T, Tsai LH (2007) Trekking across the brain: the journey of neuronal migration. Cell 128:29–43
59. Tsai LH, Gleeson JG (2005) Nucleokinesis in neuronal migration. Neuron 46:383–388
60. Hatten ME (2002) New directions in neuronal migration. Science 297:1660–1663
61. Solecki DJ, Trivedi N, Govek EE, Kerekes RA, Gleason SS et al (2009) Myosin II motors and F-actin dynamics drive the coordinated movement of the centrosome and soma during CNS glial-guided neuronal migration. Neuron 63:63–80
62. Kato M, Dobyns WB (2003) Lissencephaly and the molecular basis of neuronal migration. Hum Mol Genet 12(1):R89–R96
63. Labelle-Dumais C, Dilworth DJ, Harrington EP, de Leau M, Lyons D et al (2011) COL4A1 mutations cause ocular dysgenesis, neuronal localization defects, and myopathy in mice and Walker-Warburg syndrome in humans. PLoS Genet 7:e1002062
64. Godfrey C, Clement E, Mein R, Brockington M, Smith J et al (2007) Refining genotype phenotype correlations in muscular dystrophies with defective glycosylation of dystroglycan. Brain 130:2725–2735
65. Nabi NU, Mezer E, Blaser SI, Levin AA, Buncic JR (2003) Ocular findings in lissencephaly. J AAPOS 7:178–184
66. Dobyns WB, Pagon RA, Armstrong D, Curry CJ, Greenberg F et al (1989) Diagnostic criteria for Walker-Warburg syndrome. Am J Med Genet 32:195–210
67. Cormand B, Pihko H, Bayes M, Valanne L, Santavuori P et al (2001) Clinical and genetic distinction between Walker-Warburg syndrome and muscle–eye–brain disease. Neurology 56:1059–1069
68. Levine RA, Gray DL, Gould N, Pergament E, Stillerman ML (1983) Warburg syndrome. Ophthalmology 90:1600–1603
69. Dobyns WB, Truwit CL (1995) Lissencephaly and other malformations of cortical development: 1995 update. Neuropediatrics 26:132–147
70. Santavuori P, Somer H, Sainio K, Rapola J, Kruus S et al (1989) Muscle–eye–brain disease (MEB). Brain Dev 11:147–153
71. Gupta V, Kawahara G, Gundry SR, Chen AT, Lencer WI et al (2011) The zebrafish dag1 mutant: a novel genetic model for dystroglycanopathies. Hum Mol Genet 20:1712–1725
72. Chan YM, Keramaris-Vrantsis E, Lidov HG, Norton JH, Zinchenko N et al (2010) Fukutin-related protein is essential for mouse muscle, brain and eye development and mutation recapitulates the wide clinical spectrums of dystroglycanopathies. Hum Mol Genet 19:3995–4006

73. Bellion A, Baudoin JP, Alvarez C, Bornens M, Metin C (2005) Nucleokinesis in tangentially migrating neurons comprises two alternating phases: forward migration of the Golgi/centrosome associated with centrosome splitting and myosin contraction at the rear. J Neurosci 25:5691–5699
74. Schaar BT, McConnell SK (2005) Cytoskeletal coordination during neuronal migration. Proc Natl Acad Sci U S A 102:13652–13657
75. Umeshima H, Hirano T, Kengaku M (2007) Microtubule-based nuclear movement occurs independently of centrosome positioning in migrating neurons. Proc Natl Acad Sci U S A 104:16182–16187
76. Carter-Dawson LD, LaVail MM (1979) Rods and cones in the mouse retina. II. Autoradiographic analysis of cell generation using tritiated thymidine. J Comp Neurol 188:263–272
77. Rich KA, Zhan Y, Blanks JC (1997) Migration and synaptogenesis of cone photoreceptors in the developing mouse retina. J Comp Neurol 388:47–63
78. Razafsky D, Blecher N, Markov A, Stewart-Hutchinson PJ, Hodzic D (2012) LINC complexes mediate the positioning of cone photoreceptor nuclei in mouse retina. PLoS One 7:e47180
79. Tsujikawa M, Omori Y, Biyanwila J, Malicki J (2007) Mechanism of positioning the cell nucleus in vertebrate photoreceptors. Proc Natl Acad Sci U S A 104:14819–14824
80. Fan SS, Ready DF (1997) Glued participates in distinct microtubule-based activities in Drosophila eye development. Development 124:1497–1507
81. Kracklauer MP, Banks SM, Xie X, Wu Y, Fischer JA (2007) Drosophila klaroid encodes a SUN domain protein required for Klarsicht localization to the nuclear envelope and nuclear migration in the eye. Fly (Austin) 1:75–85
82. Patterson K, Molofsky AB, Robinson C, Acosta S, Cater C et al (2004) The functions of Klarsicht and nuclear lamin in developmentally regulated nuclear migrations of photoreceptor cells in the Drosophila eye. Mol Biol Cell 15:600–610
83. Mosley-Bishop KL, Li Q, Patterson L, Fischer JA (1999) Molecular analysis of the klarsicht gene and its role in nuclear migration within differentiating cells of the Drosophila eye. Curr Biol 9:1211–1220
84. Sullivan T, Escalante-Alcalde D, Bhatt H, Anver M, Bhat N et al (1999) Loss of A-type lamin expression compromises nuclear envelope integrity leading to muscular dystrophy. J Cell Biol 147:913–920
85. Coffinier C, Jung HJ, Nobumori C, Chang S, Tu Y et al (2011) Deficiencies in lamin B1 and lamin B2 cause neurodevelopmental defects and distinct nuclear shape abnormalities in neurons. Mol Biol Cell 22:4683–4693
86. Grady RM, Starr DA, Ackerman GL, Sanes JR, Han M (2005) Syne proteins anchor muscle nuclei at the neuromuscular junction. Proc Natl Acad Sci U S A 102:4359–4364
87. Zhang X, Xu R, Zhu B, Yang X, Ding X et al (2007) Syne-1 and Syne-2 play crucial roles in myonuclear anchorage and motor neuron innervation. Development 134:901–908
88. Zhang J, Felder A, Liu Y, Guo LT, Lange S et al (2009) Nesprin 1 is critical for nuclear positioning and anchorage. Hum Mol Genet 19:329–341
89. Puckelwartz MJ, Kessler E, Zhang Y, Hodzic D, Randles KN et al (2009) Disruption of nesprin-1 produces an Emery Dreifuss muscular dystrophy-like phenotype in mice. Hum Mol Genet 18:607–620
90. Lei K, Zhang X, Ding X, Guo X, Chen M et al (2009) SUN1 and SUN2 play critical but partially redundant roles in anchoring nuclei in skeletal muscle cells in mice. Proc Natl Acad Sci U S A 106:10207–10212
91. Gros-Louis F, Dupre N, Dion P, Fox MA, Laurent S et al (2007) Mutations in SYNE1 lead to a newly discovered form of autosomal recessive cerebellar ataxia. Nat Genet 39:80–85
92. Horn HF, Brownstein Z, Lenz DR, Shivatzki S, Dror AA, et al. (2013) The LINC complex is essential for hearing. J Clin Invest 123:740–750
93. Hale CM, Chen WC, Khatau SB, Daniels BR, Lee JS et al (2011) SMRT analysis of MTOC and nuclear positioning reveals the role of EB1 and LIC1 in single-cell polarization. J Cell Sci 124:4267–4285

94. Khatau SB, Hale CM, Stewart-Hutchinson PJ, Patel MS, Stewart CL et al (2009) A perinuclear actin cap regulates nuclear shape. Proc Natl Acad Sci U S A 106:19017–19022
95. Kim DH, Khatau SB, Feng Y, Walcott S, Sun SX et al (2012) Actin cap associated focal adhesions and their distinct role in cellular mechanosensing. Sci Rep 2:555
96. Khatau SB, Bloom RJ, Bajpai S, Razafsky D, Zang S et al (2012) The distinct roles of the nucleus and nucleus-cytoskeleton connections in three-dimensional cell migration. Sci Rep 2:488
97. Fraley SI, Feng Y, Giri A, Longmore GD, Wirtz D (2012) Dimensional and temporal controls of three-dimensional cell migration by zyxin and binding partners. Nat Commun 3:719
98. Hale CM, Shrestha AL, Khatau SB, Stewart-Hutchinson PJ, Hernandez L et al (2008) Dysfunctional connections between the nucleus and the actin and microtubule networks in laminopathic models. Biophys J 95:5462–5475
99. Lee JS, Hale CM, Panorchan P, Khatau SB, George JP et al (2007) Nuclear lamin A/C deficiency induces defects in cell mechanics, polarization, and migration. Biophys J 93:2542–2552
100. Khatau SB, Kim DH, Hale CM, Bloom RJ, Wirtz D (2010) The perinuclear actin cap in health and disease. Nucleus 1:337–342
101. Lee JS, Chang MI, Tseng Y, Wirtz D (2005) Cdc42 mediates nucleus movement and MTOC polarization in Swiss 3T3 fibroblasts under mechanical shear stress. Mol Biol Cell 16:871–880
102. Fraley SI, Feng Y, Krishnamurthy R, Kim DH, Celedon A et al (2010) A distinctive role for focal adhesion proteins in three-dimensional cell motility. Nat Cell Biol 12:598–604
103. Khatau SB, Kusuma S, Hanjaya-Putra D, Mali P, Cheng L et al (2012) The differential formation of the LINC-mediated perinuclear actin cap in pluripotent and somatic cells. PLoS One 7:e36689
104. Wirtz D, Konstantopoulos K, Searson PC (2011) The physics of cancer: the role of physical interactions and mechanical forces in metastasis. Nat Rev Cancer 11:512–522
105. Konstantopoulos K, Wu PH, Wirtz D (2013) Dimensional control of cancer cell migration. Biophys J 104:279–280
106. Fraley S, Feng Y, Wirtz D, Longmore G (2011) Reply: reducing background fluorescence reveals adhesions in 3D matrices. Nat Cell Biol 13:5-U254
107. Balzer EM, Tong Z, Paul CD, Hung WC, Stroka KM et al (2012) Physical confinement alters tumor cell adhesion and migration phenotypes. FASEB J 26:4045–4056
108. Karra D, Dahm R (2010) Transfection techniques for neuronal cells. J Neurosci 30:6171–6177
109. Taverna E, Haffner C, Pepperkok R, Huttner WB (2012) A new approach to manipulate the fate of single neural stem cells in tissue. Nat Neurosci 15:329–337

Nesprins in Cell Stability and Migration

Sascha Neumann and Angelika A. Noegel

Abstract Nesprins are a family of proteins that are primarily known for their localization along the nuclear envelope. Together with inner nuclear membrane SUN proteins, they form the core of the LINC (*Li*nker of *N*ucleoskeleton and *C*ytoskeleton) complex that traverses both nuclear membranes to connect the cytoplasm and the nuclear interior. Based on their structure and interactions, Nesprins integrate the nucleus into the cytoskeleton of a cell. Mutations in Nesprins have been identified in a group of human diseases that have been summarized as laminopathies. Cellular functions of the Nesprins and recent studies on different cancer types additionally draw interest on Nesprins in the field of cancer research. Here we summarize recent findings about the structural arrangements of Nesprins along the nuclear envelope, and highlight Nesprin functions in basic cellular processes like maintenance of nuclear shape and size, and of nuclear and cellular or cytoskeletal organization, centrosomal positioning, cell migration, and signal transduction. In summary, Nesprins are involved in critical cellular processes, which in case of malfunction contribute to the formation of cancer and might represent novel targets in cancer diagnosis or for therapeutic intervention.

S. Neumann (✉)
Medical Faculty, Center for Molecular Medicine Cologne (CMMC),
Cologne Excellence Cluster on Cellular Stress Responses in Aging-Associated
Diseases (CECAD), Institute for Biochemistry I, University of Cologne,
Joseph-Stelzmann-Str. 52, 50931 Cologne, Germany
e-mail: neumanns@uni-koeln.de

A.A. Noegel
Medical Faculty, Center for Molecular Medicine Cologne, University of Cologne,
Joseph-Stelzmann-Str. 52, 50931 Cologne, Germany

Cologne Excellence Cluster on Cellular Stress Responses in Aging-Associated
Diseases (CECAD), Joseph-Stelzmann-Str. 52, 50931 Cologne, Germany
e-mail: noegel@uni-koeln.de

E.C. Schirmer and J.I. de las Heras (eds.), *Cancer Biology and the Nuclear Envelope*,
Advances in Experimental Medicine and Biology 773, DOI 10.1007/978-1-4899-8032-8_22,

Keywords Nesprin • Nesprins • Nesprin-2 • Syne • Nuclear envelope • KASH • SUN • LINC • Centrosome • Cytoskeleton • Chromatin • Signaling • Nucleus • Signal transduction • Cancer

Abbreviations

ABD	Actin binding domain
ChIP	Chromatin Immuno*P*recipitation
INM	Inner nuclear membrane
IKNM	Interkinetic nuclear migration
KASH	*K*larsicht *A*NC-1, *S*yne *h*omology
LINC	*Li*nker of *N*ucleoskeleton and *C*ytoskeleton
MTOC	Microtubule organizing center
NE	Nuclear envelope
Nesprin	*N*uclear *e*nvelope *sp*ectrin *r*epeat prote*in*
ONM	Outer nuclear membrane
PML	Promyelocytic leukemia
SUN	*S*ad1p *Un*c84
SR	Spectrin repeat

Nesprins

The name Nesprin is an acronym of *N*uclear *e*nvelope *sp*ectrin *r*epeat prote*in* [1]. The name was chosen because of the predominant localization of these proteins along the nuclear envelope (NE) and the high content of spectrin repeats [2]. The NE is a double membrane system that encloses the genetic material in eukaryotes. It is composed of an inner nuclear membrane (INM), an outer nuclear membrane (ONM) and a lumen between both membranes. Four Nesprins have been described in mammals, namely, Nesprin-1, -2, -3, and -4. Each is encoded by a single gene (*SYNE-1*, *-2*, *-3*, *-4*) which gives rise to multiple isoforms based on different transcription initiation, termination, and alternative splicing [3]. For this reason size and domain composition differ among the isoforms. The largest, so-called giant isoforms of Nesprin-1 and -2 have molecular weights of 1,014 and 796 kDa, whereas Nesprin-3 and -4 are much smaller with molecular weights of less than 116 kDa. Nesprin-1 and -2 reach with about 10–20 isoforms, the highest degree of variations [3]. Nesprins have also been classified as KASH (*K*larsicht, *A*NC-1, *S*yne *h*omology) domain proteins based on a transmembrane domain at their C-termini. They are type II transmembrane proteins that can be integrated into the INM or the ONM. The KASH domain is a targeting motif, sufficient for anchoring the proteins to the NE (Fig. 1).

KASH domain protein orthologs have been characterized in eukaryotes like *Mus musculus* (Nesprin-1, -2, -3, -4 and KASH5), *Drosophila melanogaster*

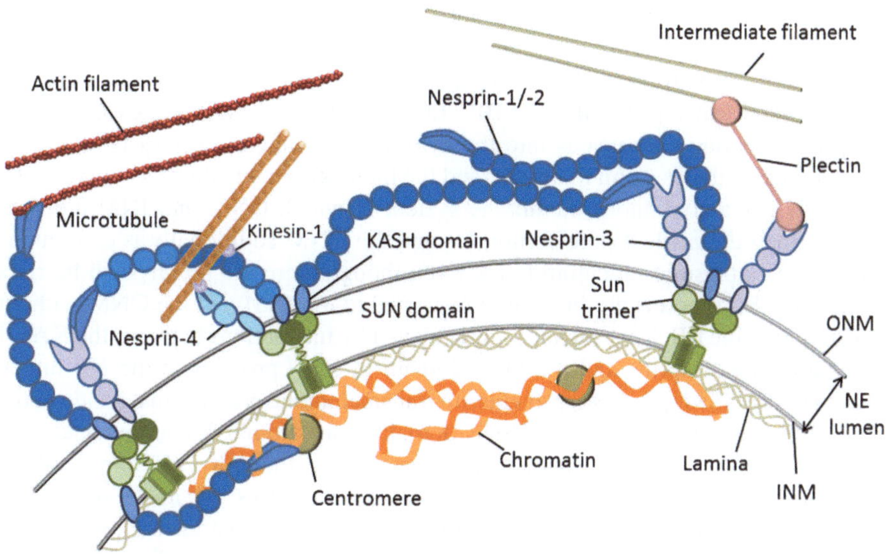

Fig. 1 Nuclear envelope protein assemblies. The scheme depicts an overview of Nesprin-1, -2, -3, -4 and some of their interaction partners along the nuclear envelope (NE). Nesprins reside at the inner (INM) and the outer nuclear membrane (ONM), where they are anchored through interactions with the SUN domain of the SUN proteins in the lumen of the NE. At the cytoplasmic site, Nesprins directly or indirectly interact with all cytoskeletal components of a cell. Nesprins form a lattice covering the NE rather than straightly projecting away from the NE. The giant isoforms of Nesprin-1 and -2 are brought into close proximity of the NE through interactions with the plectin-binding domain of Nesprin-3. Further stability is achieved through self-interactions among spectrin repeats of the Nesprins. Along the inner surface of the NE, Nesprins interact with the lamina or chromatin

(MSP-300, Klarsicht), *Caenorhabditis elegans* (ANC-1, ZYG-12, UNC-83, KDP-1), *Dictyostelium discoideum* (Interaptin), or the yeasts *Schizosaccharomyces pombe* (Kms1, Kms2) and *Saccharomyces cerevisiae* (Mps2, Csm4), and recently in the plant *Arabidopsis thaliana* (AtWIP1). The KASH domain consists of a 21-amino acid transmembrane region followed by a sequence of up to 30 amino acids that reaches into the NE lumen [4, 5]. At their N-termini Nesprins harbor binding sites for the different cytoskeletal filament systems. Nesprin-1 and -2 have at their N-termini a pair of calponin homology domains that mediate the binding to F-actin. Nesprin-3 possesses at its N-terminus a plectin-binding site. Plectin is a cytoskeletal linker protein that forms connections among all cytoskeletal systems of a cell, the microtubules, intermediate filaments, and the actin cytoskeleton [6]. Through the plectin-binding domain, Nesprin-3 can connect the NE to the intermediate filament system [7]. Nesprin-3 deficiency leads to reduced amounts of perinuclear intermediate filaments as shown in zebrafish [8] or human aortic endothelial cells [9]. Finally, Nesprin-2 and -4 interact with the microtubule network through interactions with the microtubule motor protein Kinesin-1 [10, 11]. Between their N- and C-terminus, Nesprins harbor an elongated domain composed of spectrin repeats that have been characterized in cytoskeletal organizers like spectrin, alpha-actinin, and dystrophin. The predicted number of more or less conserved spectrin repeats differs strongly among the Nesprins and their isoforms and reaches from 74 in Nesprin-1 giant and 56 in Nesprin-2 giant to a Nesprin-1 isoform or Nesprin-4 that contain only one spectrin repeat (SR) [3, 10]. Taken together, these findings argue that Nesprins integrate the nucleus into the cytoskeleton via connections between the NE and microtubules, actin filaments, and the intermediate filament system (Fig. 1).

The NE is a continuous membrane system in which ONM and INM are connected along the nuclear pore complexes and the ONM additionally is continuous with the endoplasmic reticulum (ER). Even though there are no physical barriers along the NE, Nesprins are in general restricted to the INM and the ONM without reaching into the ER. This anchoring is achieved by interactions between the KASH domain of the Nesprins and the SUN domain of SUN proteins in the NE lumen [12–14]. SUN (*S*ad1p *Un*c84) proteins are transmembrane proteins that reside in the INM. The eponymous C-terminal SUN domain spans around 200 amino acids and reaches into the perinuclear space where it interacts with the KASH domain of the Nesprins [15]. The N-terminus of the SUN proteins faces the nucleoplasm and binds to A- and B-type lamins and chromatin. Together these interactions form a nucleocytoplasmic bridge that is known as the LINC complex (*Li*nker of *N*ucleoskeleton and *C*ytoskeleton) [16] (Fig. 1).

In the following paragraphs we discuss the cellular functions of the Nesprins and their implications for tumorigenesis.

Nesprins Function as Guardians of Nuclear Shape

Tumor cells and nuclei are commonly characterized by architectural pleomorphisms in nuclear shape, size or chromatin organization [17]. Since changes of nuclear morphology have been used since a long time as a cytodiagnostic marker in cancer diagnosis, it is of particular importance to elucidate cellular entities that are involved in maintaining these structures. Nesprins as essential core components of the LINC complex represent one of the major components for these tasks. A hallmark of Nesprins is their ability to maintain proper nuclear architecture. Silencing of Nesprin-2 giant for example results in NE blebbing and severely misshapen nuclei [18]. A loss of Nesprins from the NE leads to a disruption of the LINC complex and yields a disorganization of the NE and an expansion of the otherwise evenly spaced lumen [19] (Fig. 2a). A similar effect is achieved by ectopic expression of a protein containing the SUN domain without the transmembrane region. This protein competes with the endogenous SUN proteins for the binding to the KASH domain of the Nesprins, resulting in a displacement of Nesprins from the NE to the ER and leading to nuclear morphology changes [20] (Fig. 2a).

Furthermore, Nesprins control nuclear shape and size by forming a protein lattice that covers the NE comparable to the meshwork of lamin proteins residing at the inner surface of the NE to stabilize nuclear morphology [21]. Based on their structure and size, Nesprins were originally depicted as stiff molecules that are connected to the NE by their C-terminus whereas their N-terminus reaches into the

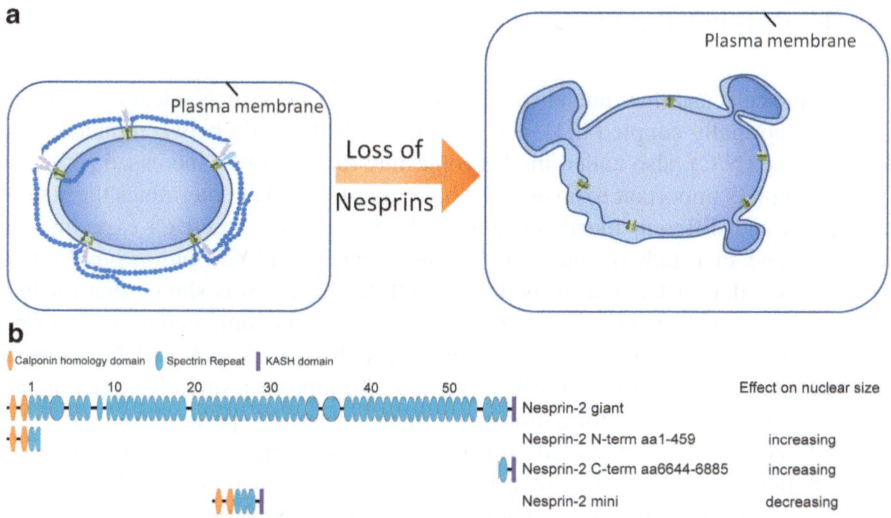

Fig. 2 Nesprins act as guardians of nuclear and cellular size and nuclear morphology. (**a**) The scheme summarizes the role of Nesprins in maintaining nuclear and cellular architecture. A loss of Nesprin leads to misshapen nuclei that are characterized by nuclear blebbing, expansions in the NE lumen, or a loss of the otherwise evenly shaped nuclear morphology. Beside morphological changes nuclear and cellular sizes are modulated as well and tend to be increased upon a loss of Nesprins. (**b**) Overview about Nesprin-2 giant, the biggest Nesprin-2 isoform and the polypeptides used to identify the role of Nesprin-2 in nuclear scaling [21]. Nesprin-2 N-term aa 1–459 form the calponin homology domains that act as actin binding domains. SR1 and most of SR2 are also part of this sequence. Nesprin-2 C-term aa 6,644–6,885 encompasses the KASH domain and the C-terminal SR56. Nesprin-2 mini is a combination of both that represents a shortened Nesprin-2 giant in which most of the central spectrin repeats are missing. The expression of these polypeptides affects nuclear size as summarized on the *right side*

cytoplasm over a distance of 300–500 nm. Compelling studies challenge this view and rather support the idea of cross connectivity among NE proteins to form a lattice encircling the nucleus (Figs. 1 and 2). Connections among Nesprins and further NE proteins have first been shown by identifying the ability of Nesprin-1 short isoforms to self-associate via their SRs [22]. Newer findings demonstrate an interaction between the actin binding domains (ABDs) of Nesprin-1 and -2 with Nesprin-3 [21, 23]. The binding site in Nesprin-3 was narrowed down to its N-terminal plectin-binding domain. Interestingly, these interactions do not interfere with the abilities of the Nesprins to interact with their cytoskeletal interaction partners. Based on these findings a novel model was predicted in which the giant Nesprin-1 and -2 molecules form a filamentous meshwork encasing the nucleus [21]. This is achieved through "anchorage" of the giant Nesprins in the NE at two sites, the C-terminus via the KASH domain and the N-terminus by the interaction with the ~100 kDa rather small Nesprin-3. These observations underline the importance of Nesprins for the NE interactome that acts in maintaining overt nuclear structure and shape and denotes them as a potential diagnostic marker to observe NE integrity and shape during cancer development.

Nuclear and Cellular Size

Neoplastic cells and their nuclei vary greatly in shape and size. Nuclear and cellular sizes are normally coupled [24], and in cancer cells the ratio between nuclear and cell volume (N/C), also called the karyoplasmic ratio, is often increased [25]. For this reason it is important to answer the basic question; what determines the size of a nucleus and different mechanisms have been proposed based on experimental data. For one, in a study on nuclei from *Xenopus laevis* and *Xenopus tropicalis*, two frog species that differ both in body and cellular size, it was shown that nuclear import affects nuclear size. The *X. laevis* nucleus is larger compared to the *X. tropicalis* nucleus and this is also reflected in their cell sizes as well as DNA content (tetraploid in *X. laevis*, diploid in *X. tropicalis*). Xenopus is a suitable model organism for studying nuclear formation since demembranated sperm chromatin is re-encapsulated by a NE after the addition of egg extracts [26]. It was shown by varying combinations of DNA and cytoplasmic extracts that nuclear scaling depended to greater extent on the transport factors importin-α and Ntf2 than on DNA content [27]. Similar results were obtained for yeast where nuclear size correlated with cell size rather than with ploidy [28].

A further mechanism may be provided by the action of Nesprins which also play an important role in nuclear and cellular scaling. Knockout mice lacking the giant isoform of Nesprin-2 (Nesprin-2 ΔABD) show a thickening of the dermis that was attributed to increased nuclei size [18]. A Nesprin-1α siRNA mediated knockdown also results in enlarged cell sizes [19]. Furthermore, morphometric analysis of cells expressing different Nesprin-2 domains shows alterations in nuclear areas as well as cellular sizes [21] (Fig. 2). The correlation between Nesprin-2 and nuclear and cellular size control was shown by the transient expression of a polypeptide containing the ABD, spectrin repeat 1 (SR1), and most of SR2 of Nesprin-2 (aa 1–459, Nesprin-2 N-term, Fig. 2b), a polypeptide encompassing aa 6,644–6,885 which form the KASH domain including the most C-terminal SR of Nesprin-2, SR56 (Nesprin-2 C-term, Fig. 2b), as defined by Simpson and Roberts [29], and Nesprin-2 aa 1–459, 6,644–6,885 (Nesprin-2 mini, Fig. 2b) in which the ABD and the KASH domain are combined. Nuclear areas of cells expressing Nesprin-2 N-term or C-term both show expansions of nuclear areas. This can be explained by interactions with endogenous Nesprin proteins. The anchoring sites along the NE are limited so that an overexpression of any interaction partner results in an occupation of interaction sites that might result in a dominant negative interruption of NE assemblies (Fig. 2) The overexpressed Nesprin-2 N-term occupies the binding sites for the Nesprin-2 ABD in Nesprin-3 and thus releases the endogenous protein from its close association with the NE. Vice versa Nesprin-2 C-term polypeptides compete for binding to SUN protein and disrupt the SUN–KASH bridge in the NE lumen. A well-defined example is the SUN–KASH interaction between the SUN domain of the SUN proteins and the KASH domain of the Nesprins in the lumen as it has been described earlier in this chapter.

Nesprin-2 mini acts in a different way. It replaces endogenous Nesprin-2 giant along the NE and acts as a belt to strengthen NE protein assemblies since it harbors both the ABD and the KASH domain but lacks nearly all of the central spectrin repeats. The conclusion is that the number of spectrin repeats is a determining factor in nuclear scaling (Fig. 2b). In addition to the changes in nuclear size, the cell size changed in the same manner. The overexpression of N- or C-term Nesprin-2 polypeptides led to enlarged cells, and the expression of Nesprin-2 mini to cellular compaction [21]. Whether Nesprins are also crucial factors for nuclear and cellular sizes or the ratio of both in different cancer types is not yet known.

Even though the role of Nesprins in determining cellular scaling and modeling has been widely studied, their role as diagnostic markers and targets for therapeutic intervention has only rarely been taken into account. Recent studies identified Nesprin-1 and -2 mutations that are involved in the formation of colorectal and breast cancers [30]. Further studies on gastrointestinal, lung, ovary, or prostate cancers demonstrate changes in the expression profile in which Nesprins-1 and -2 tend to be downregulated [31]. These findings are based on large-scale sequencing approaches of different tumor types, rather than specific studies on the role of a single gene. It remains to be determined in the future how Nesprins contribute to malignant transformations in these tumors.

Nesprins Control Nucleus and Centrosome Position and Cell Migration

In cancer, cells become metastatic when they acquire the ability to disseminate from the primary solid tumor to invade distant normal tissues. In the following part, we will discuss the role of Nesprins in the process of cell migration that includes aspects of cell polarization, centrosomal positioning, cell contact formation, and cytoskeletal architecture (Fig. 3).

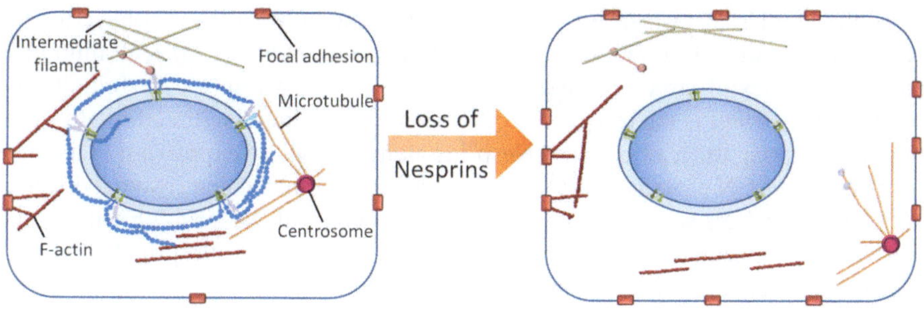

Fig. 3 Nesprins control cytoskeletal architecture and centrosomal positioning. Through their ability to simultaneously bind to the NE and all cytoskeletal components, Nesprins play an essential role in organizing cytoskeletal networks. An absence of Nesprins causes an interruption of F-actin, microtubule or intermediate filament arrangements and dislodges the centrosome from its close connection to the nucleus. Additionally cells in which the expression of Nesprins is silenced show faster focal adhesion formation. Detailed descriptions are given in the text

The nucleus is the largest organelle of a cell. During organogenesis, cells migrate and the nucleus undergoes spatial rearrangements. First hints for the role of Nesprins in nuclear positioning and anchoring came from studies in mammalian skeletal muscles. Muscle fibers are generated by the fusion of many hundreds of myoblasts to form a syncytium in which nuclei are evenly assembled. Exceptions are clusters of 3–6 synaptic nuclei that assemble beneath neuromuscular junctions, the site of signal transmission from neurons to muscle fibers. Overexpressing the Nesprin-1 KASH domain in transgenic mice largely dislodges Nesprins from the NE and causes a displacement of synaptic nuclei. Non-synaptic nuclei were not affected in these animals [32]. Mice in which either the KASH domain of Nesprin-1 or -2 was deleted showed more severe phenotypes probably due to the complete depletion of KASH domain containing isoforms. A deletion of the KASH domain of Nesprin-1 disrupted the spatial organization of synaptic and non-synaptic nuclei indicating that it is crucial for the positioning of synaptic and non-synaptic nuclei, whereas a deletion of the KASH domain of Nesprin-2 showed no effects on the uniform assemblies of muscle fiber nuclei pointing towards a less prominent role in myonuclei anchorage [33].

The role of LINC complex components in nuclear migration has also been studied in neuronal tissue morphogenesis [34]. Nuclear positioning and movement occurs in two distinct processes during neurogenesis, interkinetic nuclear migration (IKNM) and nucleokinesis. IKNM describes the movements of nuclei between apical and basal surfaces of epithelial cells, a common process in developing neuroepithelia. Nuclear movements occur in a cell cycle dependent manner with mitosis occurring in proximity to the apical site whereas for the S-phase nuclei have returned to the basal site. During IKNM, the centrosome does not move and remains at the basal site. Nucleokinesis is a part of the saltatory movement of neurons. In migrating neurons, first the leading edge extends, followed by a stable progression of the centrosome ahead of the nucleus and the nucleus is pulled towards the centrosome in saltatory steps referred to as nucleokinesis [35]. Both neurogenesis (IKNM) and neuronal migration (nucleokinesis) require functional LINC complexes [34].

With regard to metastasis we will focus on the role of Nesprins in connecting the centrosome to the NE to facilitate proper cellular migration (Fig. 3). Cell migration requires a series of structural and topological changes within a cell. Initially the most obvious event is the movement of the centrosome into the direction of migration. Interrupting the anchorage of Nesprins along the NE by expression of a dominant negative SUN protein increases the mean nucleosomal to centrosomal distance [11]. Similar effects were observed after treating cells with Nesprin-2 giant specific shRNAs and in primary cells derived from Nesprin-2 giant or Nesprin-1/-2 KASH domain or SUN-1/-2 knockout animals [11, 34] (Fig. 3).

Nucleosomal to centrosomal coupling is achieved through interactions between Nesprins and the cytoskeletal motor proteins kinesin and dynein. These motor proteins use microtubules as tracks for the transport of cargo. Microtubules have a minus and a plus end. The minus end of most microtubules is located near the centrosome also referred to as MTOC, *m*icro*t*ubule *o*rganizing *c*enter, and the plus end reaches into the cell. Kinesin mediates the plus end directed transport, whereas

Dynein transports cargo towards the minus end [36]. Populations of both kinesin and dynein localize to the NE [11, 37]. In terms of nuclear movement the nucleus can be considered a cargo for kinesin and dynein motor complexes in which minus end directed dynein carries the nucleus towards the centrosome and plus end directed kinesin moves the nucleus away from the centrosome. Defects in the LINC complex as well as loss of both dynein and kinesin result in a detachment of the centrosome away from the nucleus [10, 11, 38, 39].

The interplay between the centrosome and the NE is an important factor in understanding the potential role of Nesprins in migrating cancer cells. In migrating fibroblasts nuclear repositioning is an initial polarizing event and the Golgi complex and MTOC localize between the leading edge and the nucleus facing the direction of migration [40]. In scratch assay experiments where this response can be studied fibroblasts isolated from Nesprin-2 giant KO mice showed a severe polarization defect with both the Golgi complex and MTOC positioned away from the wound edge [18].

Dynamic processes in cell migration discussed so far are related to the microtubule network and the centrosome, but the cell harbors further cytoskeletal components that mediate cell migration and nuclear positioning. Recently TAN lines (*t*ransmembrane *a*ctin-associated *n*uclear lines) have been identified that form on the dorsal surface of a nucleus and which are anchored along the NE by lamin A/C [41, 42]. They have been observed in migrating NIH3T3 fibroblasts, in which nuclear movements are connected to retrograde actin flow. Nesprin-2 giant, SUN2 and lamin A/C connect the nucleus to the retrograde actin flow in linear arrays of TAN lines to ensure the movement of the nucleus to the rear of the centrosome which itself is maintained at the center of the cell. Taken together nuclear and cellular movements require an interplay between the centrosome, cytoskeletal components like microtubules or actin filaments, their associated proteins and factors that connect these components to the nucleus to ensure connectivity and force transmission. Based on their structures and interaction partners Nesprins adopt key roles in these processes. Interestingly it was also shown that a depletion of Nesprins results not only in disruption of the perinuclear actin cytoskeleton but also in topological changes in overt cellular microtubule organization [43] (Fig. 3). It remains to be determined how changes in Nesprin levels or distinct mutations in these proteins affect the organization of the cytoskeleton along the nucleus or the leading edge of migrating cancer cells.

A further important aspect in cell migration concerns the formation of cell–cell or cell–matrix connections. Invasiveness is characterized by the dissemination of cells from their original environment. Even though Nesprins are primarily known for their localization along the NE, they have also been shown to localize along the plasma membrane and cytoplasmic components [20, 44, 45] and their cellular functions reach to the border of a cell, namely, the plasma membrane that is the site of cell contact formation. The formation of cell adhesions can be studied by detaching and re-plating cells under cell culture conditions. A loss of Nesprin-2 results in faster focal adhesion formation after re-plating the cells [43], and a loss of Nesprin-1 has a similar effect [46] (Fig. 3). It remains to be tested how cancer related changes in Nesprin expression levels or mutations in these proteins affect the formation of cell–cell or cell–matrix contact.

Nesprins in Signal Transduction

Cancer results from malfunctions in basic cellular processes like increased cell survival or proliferation that are tightly controlled under normal conditions and in case of dysfunction trigger cells to form tumors and to disseminate into other organs. The NE is a hub in controlling signals that need to be transferred from the outside of a cell through the cytoplasm into the nucleus. Even though NE proteins reside at the border between the cytoplasm and the genetic material, only few aspects about their function in signaling have been explored so far. In the following paragraphs the growing role of Nesprins in various signal transduction processes will be discussed. Cellular signal transduction is based on two principles. First, signals can be triggered by mechanical strain, which causes cytoskeletal rearrangements that might be transferred to LINC complex components through the cytoskeleton to reach into the nucleus or they are converted into biochemical signals that need to be transferred across the NE by the nuclear pore complexes. Both mechanisms are collectively known as mechanotransduction. Second, signal transduction can include mobile messengers that are released from the plasma membrane upon activation and carry a signal across the NE to the chromatin to modulate gene expression (Fig. 4).

Research on the role of Nesprins in controlling signaling pathways is still at the very beginning and only few pieces of the mosaic have been revealed so far. However, it is already established that Nesprins are involved in both signal transduction routes. Mechanosensitive signaling pathways like the activation of NFκB signaling can be studied in cell culture by putting cultivated cells under cyclic strain

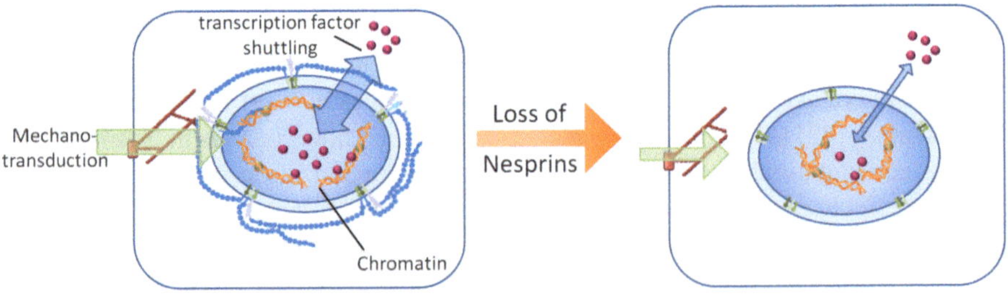

Fig. 4 Nesprins control signal transduction. Signals that reach the plasma membrane of a cell follow different routes into the nucleus where they finally affect the gene expression profile of a cell. Mechanically induced signals (*green arrow*) are transferred along cytoskeletal networks and reach the LINC complex along the NE, which transfers them into the nucleus. A disruption of the LINC complex releases the nucleoplasmic to cytoplasmic coupling and impairs the translocation of signals across the NE as indicated by the shorter *green arrow* in the *right figure*. Nesprins reside at the interface between cytoplasm and nucleoplasm. A loss of Nesprins results in impaired transport of transcription factors like c-Fos, SMAD, or β-Catenin into the nucleus (*blue arrow*). ChIP experiments demonstrate the interaction of Nesprin-2 with chromatin, and Nesprin-2 silencing leads to heterochromatin rearrangements

using a cell stretching device. NFκB signaling is based upon the translocation of NFκB from the cytoplasm into the nucleus after activation where it acts as a transcription factor. Disruption of the LINC complex by ectopic overexpression of dominant negative SUN or Nesprin proteins leads to elevated activation of NFκB signaling and increased levels of the transcription factors MyoD and myogenin [47]. Another study demonstrated that expression levels of selected mechanosensitive genes like *Egr-1* or *Iex-1* were similar between control cells and cells in which the LINC complex was disrupted after exposing them to mechanical strain [48]. These findings indicate that Nesprins are involved in regulating the expression of distinct rather than all mechanosensitive genes.

The translocation of transcription factors from the cytoplasm into the nucleus is a typical characteristic of signal transduction pathways. Nesprins reside at the NE that forms a selective barrier for the transport of proteins in or out of the nucleus as for instance in the absence of Nesprin-2 shuttling of the transcription factors SMAD2/3, β-Catenin, and c-Fos into the nucleus upon activation is reduced [20, 43]. Likewise KASH-less Nesprins affect signal transduction by acting as a scaffold for ERK1/2 to tether these kinases to promyelocytic leukemia protein (PML) nuclear bodies. Loss of Nesprin-2 leads to a sustained ERK1/2 activation and increased cell proliferation [49]. Finally Nesprins might affect the gene expression profile of a cell by associating with heterochromatic DNA as demonstrated by chromatin immunoprecipitation (ChIP) experiments. The altered heterochromatin organization in Nesprin-2 giant KO fibroblasts as revealed by an altered HP1β distribution might be a consequence [43]. In cancer the normal packaging and higher organization of heterochromatin is also often compromised and based on this property Nesprins have the potential to contribute to this phenotype [50].

Concluding Remarks

Irregular shaped cells and nuclei are cytological markers of neoplastic cells detectable by light microscopy. Mistakes in the tight control of cell cycle or transcriptional regulation or in cell contact formation release cells from their tissue environment to cause malignancy. The underlying cellular and molecular changes are closely related to Nesprin functions. Loss of Nesprins leads to enhanced adhesion to a substrate, perturbed cytoskeletal networks that affect cellular stiffness, integrity, shape and polarity of the cells and alterations of nuclear shape and size. Furthermore, a loss of Nesprins affects transcription factor shuttling across the NE and chromosomal organization. Mutations in Nesprins or changes in their expression profiles have been reported in different forms of cancer whereas the particular functional consequences remain to be explored [30, 31]. Taken together these data let us propose that Nesprins as pivotal NE components will take an important position in future questions in cancer research and might turn out to be diagnostic markers or targets for therapeutic interventions.

References

1. Zhang Q, Skepper JN, Yang F, Davies JD, Hegyi L, Roberts RG, Weissberg PL, Ellis JA, Shanahan CM (2001) Nesprins: a novel family of spectrin-repeat-containing proteins that localize to the nuclear membrane in multiple tissues. J Cell Sci 114(Pt 24):4485–4498
2. Apel ED, Lewis RM, Grady RM, Sanes JR (2000) Syne-1, a dystrophin- and Klarsicht-related protein associated with synaptic nuclei at the neuromuscular junction. J Biol Chem 275(41):31986–31995. doi:10.1074/jbc.M004775200
3. Rajgor D, Mellad JA, Autore F, Zhang Q, Shanahan CM (2012) Multiple novel nesprin-1 and nesprin-2 variants act as versatile tissue-specific intracellular scaffolds. PLoS One 7(7):e40098. doi:10.1371/journal.pone.0040098
4. Starr DA, Fischer JA (2005) KASH 'n Karry: the KASH domain family of cargo-specific cytoskeletal adaptor proteins. News Rev Mol Cell Dev Biol 27(11):1136–1146. doi:10.1002/bies.20312
5. Zhou X, Graumann K, Evans DE, Meier I (2012) Novel plant SUN-KASH bridges are involved in RanGAP anchoring and nuclear shape determination. J Cell Biol 196(2):203–211. doi:10.1083/jcb.201108098
6. Svitkina TM, Verkhovsky AB, Borisy GG (1996) Plectin sidearms mediate interaction of intermediate filaments with microtubules and other components of the cytoskeleton. J Cell Biol 135(4):991–1007
7. Wilhelmsen K, Litjens SH, Kuikman I, Tshimbalanga N, Janssen H, van den Bout I, Raymond K, Sonnenberg A (2005) Nesprin-3, a novel outer nuclear membrane protein, associates with the cytoskeletal linker protein plectin. J Cell Biol 171(5):799–810. doi:10.1083/jcb.200506083
8. Postel R, Ketema M, Kuikman I, de Pereda JM, Sonnenberg A (2011) Nesprin-3 augments peripheral nuclear localization of intermediate filaments in zebrafish. J Cell Sci 124(Pt 5):755–764. doi:10.1242/jcs.081174
9. Morgan JT, Pfeiffer ER, Thirkill TL, Kumar P, Peng G, Fridolfsson HN, Douglas GC, Starr DA, Barakat AI (2011) Nesprin-3 regulates endothelial cell morphology, perinuclear cytoskeletal architecture, and flow-induced polarization. Mol Biol Cell 22(22):4324–4334. doi:10.1091/mbc.E11-04-0287
10. Roux KJ, Crisp ML, Liu Q, Kim D, Kozlov S, Stewart CL, Burke B (2009) Nesprin 4 is an outer nuclear membrane protein that can induce kinesin-mediated cell polarization. Proc Natl Acad Sci U S A 106(7):2194–2199. doi:10.1073/pnas.0808602106
11. Schneider M, Lu W, Neumann S, Brachner A, Gotzmann J, Noegel AA, Karakesisoglou I (2011) Molecular mechanisms of centrosome and cytoskeleton anchorage at the nuclear envelope. Cell Mol Life Sci 68(9):1593–1610. doi:10.1007/s00018-010-0535-z
12. Wang W, Shi Z, Jiao S, Chen C, Wang H, Liu G, Wang Q, Zhao Y, Greene MI, Zhou Z (2012) Structural insights into SUN-KASH complexes across the nuclear envelope. Cell Res 22(10):1440–1452. doi:10.1038/cr.2012.126
13. Sosa BA, Rothballer A, Kutay U, Schwartz TU (2012) LINC complexes form by binding of three KASH peptides to domain interfaces of trimeric SUN proteins. Cell 149(5):1035–1047. doi:10.1016/j.cell.2012.03.046
14. Zhou Z, Du X, Cai Z, Song X, Zhang H, Mizuno T, Suzuki E, Yee MR, Berezov A, Murali R, Wu SL, Karger BL, Greene MI, Wang Q (2012) Structure of Sad1-UNC84 homology (SUN) domain defines features of molecular bridge in nuclear envelope. J Biol Chem 287(8):5317–5326. doi:10.1074/jbc.M111.304543
15. Padmakumar VC, Libotte T, Lu W, Zaim H, Abraham S, Noegel AA, Gotzmann J, Foisner R, Karakesisoglou I (2005) The inner nuclear membrane protein Sun1 mediates the anchorage of Nesprin-2 to the nuclear envelope. J Cell Sci 118(Pt 15):3419–3430. doi:10.1242/jcs.02471
16. Crisp M, Liu Q, Roux K, Rattner JB, Shanahan C, Burke B, Stahl PD, Hodzic D (2006) Coupling of the nucleus and cytoplasm: role of the LINC complex. J Cell Biol 172(1):41–53. doi:10.1083/jcb.200509124
17. Zink D, Fischer AH, Nickerson JA (2004) Nuclear structure in cancer cells. Nat Rev Cancer 4(9):677–687. doi:10.1038/nrc1430

18. Luke Y, Zaim H, Karakesisoglou I, Jaeger VM, Sellin L, Lu W, Schneider M, Neumann S, Beijer A, Munck M, Padmakumar VC, Gloy J, Walz G, Noegel AA (2008) Nesprin-2 Giant (NUANCE) maintains nuclear envelope architecture and composition in skin. J Cell Sci 121(Pt 11):1887–1898. doi:10.1242/jcs.019075
19. Zhang Q, Bethmann C, Worth NF, Davies JD, Wasner C, Feuer A, Ragnauth CD, Yi Q, Mellad JA, Warren DT, Wheeler MA, Ellis JA, Skepper JN, Vorgerd M, Schlotter-Weigel B, Weissberg PL, Roberts RG, Wehnert M, Shanahan CM (2007) Nesprin-1 and -2 are involved in the pathogenesis of Emery Dreifuss muscular dystrophy and are critical for nuclear envelope integrity. Hum Mol Genet 16(23):2816–2833. doi:10.1093/hmg/ddm238
20. Neumann S, Schneider M, Daugherty RL, Gottardi CJ, Eming SA, Beijer A, Noegel AA, Karakesisoglou I (2010) Nesprin-2 interacts with {alpha}-catenin and regulates Wnt signaling at the nuclear envelope. J Biol Chem 285(45):34932–34938. doi:10.1074/jbc.M110.119651
21. Lu W, Schneider M, Neumann S, Jaeger VM, Taranum S, Munck M, Cartwright S, Richardson C, Carthew J, Noh K, Goldberg M, Noegel AA, Karakesisoglou I (2012) Nesprin interchain associations control nuclear size. Cell Mol Life Sci 69(20):3493–3509. doi:10.1007/s00018-012-1034-1
22. Mislow JM, Holaska JM, Kim MS, Lee KK, Segura-Totten M, Wilson KL, McNally EM (2002) Nesprin-1alpha self-associates and binds directly to emerin and lamin A in vitro. FEBS Lett 525(1–3):135–140
23. Taranum S, Sur I, Muller R, Lu W, Rashmi RN, Munck M, Neumann S, Karakesisoglou I, Noegel AA (2012) Cytoskeletal interactions at the nuclear envelope mediated by nesprins. Int J Cell Biol 2012:736524. doi:10.1155/2012/736524
24. Neumann FR, Nurse P (2007) Nuclear size control in fission yeast. J Cell Biol 179(4): 593–600. doi:10.1083/jcb.200708054
25. Bhatia A, Dey P, Kakkar N, Srinivasan R, Nijhawan R (2006) Malignant atypical cell in urine cytology: a diagnostic dilemma. CytoJournal 3:28. doi:10.1186/1742-6413-3-28
26. Lohka MJ, Maller JL (1985) Induction of nuclear envelope breakdown, chromosome condensation, and spindle formation in cell-free extracts. J Cell Biol 101(2):518–523
27. Levy DL, Heald R (2010) Nuclear size is regulated by importin alpha and Ntf2 in Xenopus. Cell 143(2):288–298. doi:10.1016/j.cell.2010.09.012
28. Jorgensen P, Edgington NP, Schneider BL, Rupes I, Tyers M, Futcher B (2007) The size of the nucleus increases as yeast cells grow. Mol Biol Cell 18(9):3523–3532. doi:10.1091/mbc.E06-10-0973
29. Simpson JG, Roberts RG (2008) Patterns of evolutionary conservation in the nesprin genes highlight probable functionally important protein domains and isoforms. Biochem Soc Trans 36(Pt 6):1359–1367. doi:10.1042/BST0361359
30. Sjoblom T, Jones S, Wood LD, Parsons DW, Lin J, Barber TD, Mandelker D, Leary RJ, Ptak J, Silliman N, Szabo S, Buckhaults P, Farrell C, Meeh P, Markowitz SD, Willis J, Dawson D, Willson JK, Gazdar AF, Hartigan J, Wu L, Liu C, Parmigiani G, Park BH, Bachman KE, Papadopoulos N, Vogelstein B, Kinzler KW, Velculescu VE (2006) The consensus coding sequences of human breast and colorectal cancers. Science 314(5797):268–274. doi:10.1126/science.1133427
31. de Las Heras JI, Batrakou DG, Schirmer EC (2012) Cancer biology and the nuclear envelope: a convoluted relationship. Semin Cancer Biol. doi:10.1016/j.semcancer.2012.01.008
32. Grady RM, Starr DA, Ackerman GL, Sanes JR, Han M (2005) Syne proteins anchor muscle nuclei at the neuromuscular junction. Proc Natl Acad Sci U S A 102(12):4359–4364. doi:10.1073/pnas.0500711102
33. Zhang X, Xu R, Zhu B, Yang X, Ding X, Duan S, Xu T, Zhuang Y, Han M (2007) Syne-1 and Syne-2 play crucial roles in myonuclear anchorage and motor neuron innervation. Development 134(5):901–908. doi:10.1242/dev.02783
34. Zhang X, Lei K, Yuan X, Wu X, Zhuang Y, Xu T, Xu R, Han M (2009) SUN1/2 and Syne/Nesprin-1/2 complexes connect centrosome to the nucleus during neurogenesis and neuronal migration in mice. Neuron 64(2):173–187. doi:10.1016/j.neuron.2009.08.018
35. Tsai LH, Gleeson JG (2005) Nucleokinesis in neuronal migration. Neuron 46(3):383–388. doi:10.1016/j.neuron.2005.04.013

36. Tanenbaum ME, Akhmanova A, Medema RH (2011) Bi-directional transport of the nucleus by dynein and kinesin-1. Commun Integr Biol 4(1):21–25. doi:10.4161/cib.4.1.13780
37. Busson S, Dujardin D, Moreau A, Dompierre J, De Mey JR (1998) Dynein and dynactin are localized to astral microtubules and at cortical sites in mitotic epithelial cells. Curr Biol 8(9):541–544
38. Taranum S, Vaylann E, Meinke P, Abraham S, Yang L, Neumann S, Karakesisoglou I, Wehnert M, Noegel AA (2012) LINC complex alterations in DMD and EDMD/CMT fibroblasts. Eur J Cell Biol 91(8):614–628. doi:10.1016/j.ejcb.2012.03.003
39. Splinter D, Tanenbaum ME, Lindqvist A, Jaarsma D, Flotho A, Yu KL, Grigoriev I, Engelsma D, Haasdijk ED, Keijzer N, Demmers J, Fornerod M, Melchior F, Hoogenraad CC, Medema RH, Akhmanova A (2010) Bicaudal D2, dynein, and kinesin-1 associate with nuclear pore complexes and regulate centrosome and nuclear positioning during mitotic entry. PLoS Biol 8(4):e1000350. doi:10.1371/journal.pbio.1000350
40. Gomes ER, Jani S, Gundersen GG (2005) Nuclear movement regulated by Cdc42, MRCK, myosin, and actin flow establishes MTOC polarization in migrating cells. Cell 121(3): 451–463. doi:10.1016/j.cell.2005.02.022
41. Luxton GW, Gomes ER, Folker ES, Vintinner E, Gundersen GG (2010) Linear arrays of nuclear envelope proteins harness retrograde actin flow for nuclear movement. Science 329(5994):956–959. doi:10.1126/science.1189072
42. Folker ES, Ostlund C, Luxton GW, Worman HJ, Gundersen GG (2011) Lamin A variants that cause striated muscle disease are defective in anchoring transmembrane actin-associated nuclear lines for nuclear movement. Proc Natl Acad Sci U S A 108(1):131–136. doi:10.1073/pnas.1000824108
43. Rashmi RN, Eckes B, Glockner G, Groth M, Neumann S, Gloy J, Sellin L, Walz G, Schneider M, Karakesisoglou I, Eichinger L, Noegel AA (2012) The nuclear envelope protein Nesprin-2 has roles in cell proliferation and differentiation during wound healing. Nucleus 3(2):172–186. doi:10.4161/nucl.19090
44. Warren DT, Zhang Q, Weissberg PL, Shanahan CM (2005) Nesprins: intracellular scaffolds that maintain cell architecture and coordinate cell function? Expert Rev Mol Med 7(11):1–15. doi:10.1017/S1462399405009294
45. Padmakumar VC, Abraham S, Braune S, Noegel AA, Tunggal B, Karakesisoglou I, Korenbaum E (2004) Enaptin, a giant actin-binding protein, is an element of the nuclear membrane and the actin cytoskeleton. Exp Cell Res 295(2):330–339. doi:10.1016/j.yexcr.2004.01.014
46. Chancellor TJ, Lee J, Thodeti CK, Lele T (2010) Actomyosin tension exerted on the nucleus through nesprin-1 connections influences endothelial cell adhesion, migration, and cyclic strain-induced reorientation. Biophys J 99(1):115–123. doi:10.1016/j.bpj.2010.04.011
47. Brosig M, Ferralli J, Gelman L, Chiquet M, Chiquet-Ehrismann R (2010) Interfering with the connection between the nucleus and the cytoskeleton affects nuclear rotation, mechanotransduction and myogenesis. Int J Biochem Cell Biol 42(10):1717–1728. doi:10.1016/j.biocel.2010.07.001
48. Lombardi ML, Jaalouk DE, Shanahan CM, Burke B, Roux KJ, Lammerding J (2011) The interaction between nesprins and sun proteins at the nuclear envelope is critical for force transmission between the nucleus and cytoskeleton. J Biol Chem 286(30):26743–26753. doi:10.1074/jbc.M111.233700
49. Warren DT, Tajsic T, Mellad JA, Searles R, Zhang Q, Shanahan CM (2010) Novel nuclear nesprin-2 variants tether active extracellular signal-regulated MAPK1 and MAPK2 at promyelocytic leukemia protein nuclear bodies and act to regulate smooth muscle cell proliferation. J Biol Chem 285(2):1311–1320. doi:10.1074/jbc.M109.032557
50. Carone DM, Lawrence JB (2012) Heterochromatin instability in cancer: from the Barr body to satellites and the nuclear periphery. Semin Cancer Biol. doi:10.1016/j.semcancer.2012.06.008

Connecting the Nucleus to the Cytoskeleton for Nuclear Positioning and Cell Migration

Daniel S. Osorio and Edgar R. Gomes

Abstract The position of the nucleus in the cytoplasm is a highly regulated process and is required for multiple cellular and developmental processes. Defects on different nuclear positioning events are associated with several pathologies such as muscle and nervous system disorders. In this chapter we describe the current knowledge on the mechanism of nuclear positioning. We discuss how the nucleus connects to the cytoskeleton by nesprins and SUN proteins, how this connection is regulated by Samp1, and how this connection is required for proper nuclear positioning. Furthermore, we discuss how nesprins, SUN, and Samp1 form transmembrane actin-associated nuclear (TAN) lines, novel nuclear envelope structures involved in force transduction during nuclear movement. Finally, we describe the recent evidences suggesting a role for the connection between the nucleus and the cytoskeleton in cancer.

Keywords Nuclear positioning • Centrosome reorientation • Cell migration • LINC complex • Nucleo-cytoskeletal connections • Nuclear envelope • Actin cytoskeleton

D.S. Osorio (✉)
INSERM UMR S 787-Groupe Myologie, Faculté de Médecine - Pitié-Salpêtrière
Université Pierre et Marie Curie, 105 boulevard de l'Hôpital, 75634 Paris, Cedex 13, France

PhD Program in Biomedicine, Faculty of Medicine, University of Porto, Porto, Portugal
e-mail: daniel.s.osorio@gmail.com

E.R. Gomes (✉)
INSERM UMR S 787-Groupe Myologie, Faculté de Médecine - Pitié-Salpêtrière,
Université Pierre et Marie Curie, 105 boulevard de l'Hôpital, 75634 Paris, Cedex 13, France

Groupe Hospitalier Pitié-Salpêtrière, Institut de Myologie, Paris, France
e-mail: edgar.gomes@upmc.fr

E.C. Schirmer and J.I. de las Heras (eds.), *Cancer Biology and the Nuclear Envelope*, Advances in Experimental Medicine and Biology 773, DOI 10.1007/978-1-4899-8032-8_23,

Abbreviations

CHIP	Chromatin immunoprecipitation
EMT	Epithelial–mesenchymal transition
INM	Inner nuclear membrane
LINC	*LI*nker of *N*ucleoskeleton and *C*ytoskeleton
LPA	Lysophosphatidic acid
MTOC	Microtubule organizing center
MAS	MTOC attachment site
MRCK	Myotonic dystrophy kinase-related Cdc42 binding protein
N2G	Nesprin-2 giant
refilin	*RE*gulator of *FIL*amin prote*IN*
SPB	Spindle pole body
TAN lines	*T*ransmembrane *A*ctin-associated *N*uclear lines

Introduction

The traditional depiction of the nucleus standing passively in the center of the cell has led to an underappreciation of the importance of its localization and the mechanisms that govern nuclear positioning. Like any other organelle, the position of the nucleus is dynamically regulated in space and time, and indeed many cellular and developmental processes involve positioning of the nucleus inside the cell. Extreme examples are skeletal muscle cells that have their nuclei positioned at the periphery and epithelial tissues where nuclei are usually away from the apical membrane. Multiple nuclear movement events have been described during early (zygote formation and establishment of organism polarity axes) and late (nervous system and skeletal muscle formation) developmental stages, and these nuclear positioning events have been found to be required for proper development [1]. Moreover, mispositioning of the nucleus is associated with pathological conditions such as muscular disorders and neuronal pathologies [2, 3]. Nuclear position is probably also involved in the migration of immune and metastatic tumor cells since the migration of these cells requires the squeezing of the nucleus through tight gaps in the surrounding matrix [4–6]. Thus, cells and tissues actively regulate nuclear positioning and nuclear position is involved in diverse events during development, homeostasis, and regeneration.

Nuclear positioning mechanisms have been studied in different organisms, most of them non-vertebrates, and under different stages of cell differentiation or development. Examples are mitosis and karyogamy in *S. cerevisiae* and *S. pombe*, fertilization in *C. elegans* and Sea Urchins, hypodermal syncytium and P-cell formation in *C. elegans*, and finally oogenesis in *Drosophila* [2, 7–12]. Nuclear positioning is usually mediated by connections between the nucleus and the cytoskeleton, involving cytoskeleton-associated proteins, motors and nuclear envelope proteins. The spindle pole body (SPB) in fungi mediates the connection between the nucleus and the microtubule cytoskeleton. The SPB is directly connected to the nuclear envelope and nucleates microtubules. Cortically anchored dynein pulls the microtubules connected to the nucleus resulting in nuclear movement [10, 13]. In metazoans, the function of the SPB is usually mediated by the centrosome, the main microtubule organizing center (MTOC). The microtubules that emanate from the centrosome connect to the nuclear envelope probably by microtubule motors (dynein and kinesins). Then, as observed in SPB-dependent nuclear movements, cortically anchored dynein pulls on the microtubules. In some other situations, such as pronuclear migration during zygote formation or skeletal muscle myotube formation, dynein anchored to the nucleus also contributes to nuclear movement by pulling the nucleus directly on the microtubules, similarly to transport of vesicles along microtubules [14, 15].

Connections between the nucleus and the cytoskeleton can also be mediated by other mechanisms. One class of nuclear envelope proteins containing a transmembrane KASH domain was originally identified in *C. elegans* (ANC-1) for its involvement in the connection between the nucleus and the cytoskeleton. Some KASH-containing proteins are integral to the outer nuclear membrane and connect with the cytoskeleton either directly or by interactions with cytoskeleton-binding proteins. Moreover, some KASH domain-containing proteins might be considered themselves to be cytoskeletal proteins as most contain spectrin repeats. On the other hand, another class of nuclear envelope proteins containing a SUN domain has also been identified for their involvement in targeting KASH-containing proteins to the nuclear envelope [16]. These proteins are conserved in different species and are integral to the inner nuclear membrane (INM). In vertebrates the KASH domain-containing proteins are known as nesprins or Syne proteins and the SUN-domain-containing proteins are known as SUN proteins. In *S. pombe*, they are known as Kms1 (with the KASH domain) and Sad1p (with the SUN domain). The KASH-SUN proteins complex, named *LI*nker of *N*ucleoskeleton and *C*ytoskeleton (LINC), provides a direct connection between the cytoskeleton and the nuclear lamina [17].

Recently, a novel structure composed of LINC complex proteins has been identified in migrating cells [18, 19]. This structure, named TAN lines (for *T*ransmembrane *A*ctin-associated *N*uclear Lines), directly connects the nucleus to the actin cytoskeleton. In this chapter we describe the current knowledge on the mechanisms of nuclear positioning in migrating cells and on the processes involving the formation and maintenance of TAN lines during nuclear movement.

Nuclear Positioning in Migrating Cells

Cell migration is fundamental for different physiological and pathological conditions. Cell polarization prior to migration is required for proper cell migration. The future leading edge starts to exhibit protrusive activity, filopodia and lamellipodia formation, and polarized changes in the actin cytoskeleton and adhesion machinery are observed [20]. In multiple cell types, the centrosome and the Golgi apparatus become positioned between the nucleus and the future leading edge [21–23]. This process, named centrosome and Golgi reorientation, has been proposed to be important for the polarized delivery of intracellular factors towards the leading edge and has a role in the maintenance of cell polarity during cell migration [24, 25]. The fibroblast wound healing assay has been used extensively as a system to study cell polarization. In this assay, a confluent monolayer of cells is stimulated to migrate by a scratch-wound. The cells positioned on the wound edge acquire a synchronized polarized morphology, with formation of protrusions, filopodia, and lamellipodia in the side facing the wound edge (that becomes the leading edge of the cell), and the reorientation of the centrosome and Golgi apparatus towards the wound edge. However, if the monolayer of cells is starved prior to wounding, no polarization is observed upon wounding. In this situation, polarization can be triggered by the addition of serum or the serum-component lysophosphatidic acid (LPA) [26]. When stimulated with LPA, cells polarize without migrating allowing the uncoupling of these two processes [27].

The small GTPase Cdc42, the motor protein dynein/dynactin and the polarity protein Par6 and aPKCζ were the first proteins to be linked to centrosome reorientation [23, 28, 29]. Upon serum or LPA stimulation, the levels of active Cdc42-GTP increased and constitutively active Cdc42 is sufficient to induce centrosome reorientation in serum-starved cells. Inhibition of dynein or dynactin caused the inhibition of reorientation in the presence of LPA [29]. These results, together with the observation that dynein accumulates at the leading edge of polarized migrating cells, suggested a mechanism for centrosome reorientation where the centrosome moves towards the leading edge to become positioned between the leading edge and the nucleus [23, 28–30]. Real-time analysis of centrosome reorientation by video microscopy of GFP-tubulin expressing cells provided an unexpected observation: during centrosome reorientation, the centrosome stayed mostly static in the center of the cell, whereas the nucleus moved rearwards, away from the future leading edge, resulting in centrosome reorientation without centrosome movement [27] (Fig. 1).

This nuclear movement was regulated by Cdc42 and Myotonic Dystrophy Kinase-Related Cdc42 binding protein (MRCK), a Cdc42 effector, and driven by actin retrograde flow. On the other hand, microtubules, dynein/dynactin and Par6 were required for the maintenance of the centrosome in the cell center. The activation of Cdc42 by LPA or serum activates two pathways that lead to centrosome reorientation. One that regulates nuclear movement and another that regulates centrosome positioning at the cell center. In the nuclear movement pathway, Cdc42 activates MRCK that directly regulates myosin II activity and probably actin retrograde flow. On the other hand, in the centrosome positioning pathway, Cdc42 activates Par-6 and aPKCζ that acts together with dynein/dynactin, probably anchored to the cell cortex [30]. Anchored dynein keeps centrosomal microtubules under tension, thus maintaining the centrosome in the center of the cell. Par-3 is also implicated in this process through an interaction with dynein light intermediate chain 2 (LIC2) [31]. Inhibition of nuclear movement by disruption of different components of the nuclear movement pathway impairs cell migration. It is still unknown why positioning the nucleus away from the leading edge is important for cell migration. Probably, keeping the nucleus away from the leading edge of a migrating cell allows the front of the cell to penetrate through tissues. After initial penetration, the nucleus and the cell body are pulled, so

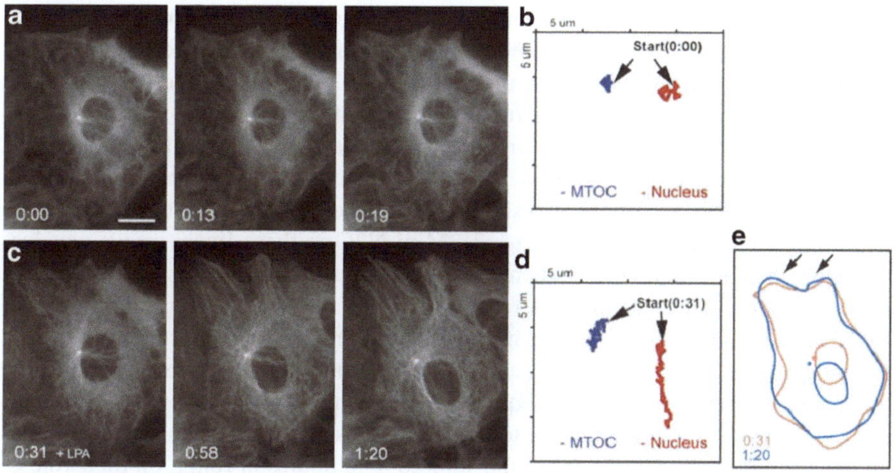

Fig. 1 Centrosome reorientation in fibroblasts occurs through rearwards nuclear movement. (**a** and **c**) Frames from a time-lapse recording of a starved wound-edge NIH3T3 cell expressing GFP-Tubulin before (**a**) and after (**c**) LPA treatment (time in h:min). (**b** and **d**) Traces of the MTOC (*blue*) and nucleus (*red*) centroid positions before (**b**) and after (**d**) addition of LPA. (**e**) Superimposition of the cell outline, nucleus, and MTOC from frame 0:31 (*orange*; beginning of nuclear movement) and frame 1:20 (*blue*; end of nuclear movement) shows rearward movement of the nucleus relative to the leading edge (*arrows*). Adapted from [27] by permission from Elsevier Ltd. Copyright © 2005

then the cell can migrate more efficiently within a tissue. If the nucleus would remain in the front of the cell, it would be more difficult for the cell to penetrate a tissue, since the nucleus is the stiffest organelle in the cell [32].

*T*ransmembrane *A*ctin-Associated *N*uclear (TAN) Lines and Nuclear Movement

Recent work has identified how actin retrograde flow drives nuclear movement. In migrating fibroblasts, the KASH-containing protein nesprin-2 giant (N2G) and SUN2 were found to be required for nuclear movement prior to cell migration [18]. Accordingly, the use of dominant negative KASH constructs or depletion of N2G (the only giant nesprin isoform expressed in these cells that is able to bind to actin) or SUN2 was sufficient to inhibit centrosome reorientation and nuclear movement upon LPA stimulation. The inhibition of nuclear movement by N2G RNAi could be rescued by expressing a mini-N2G construct containing the KASH domain, two spectrin repeats, and the actin-binding domain, and this rescue depended on functional actin binding and KASH domains. Analysis of the dorsal nuclear surface of LPA stimulated cells demonstrated that SUN2 and N2G formed linear arrays. Furthermore, these arrays co-localized with actin cables, as assessed by using both

GFP fusions and immunofluorescence against endogenous proteins. These dorsal actin cables oriented parallel to the leading edge of the cell and were independent entities from ventral stress fibers, mostly oriented along the major cell axis. The formation of linear arrays was disrupted when acto-myosin cables were perturbed by cytochalasin D or blebbistatin treatments, indicating that TAN lines depend on acto-myosin cable engagement by N2G. Using the mini-N2G construct and live-imaging analysis, the authors also showed that these linear arrays co-migrated with actin filaments and the nucleus (Fig. 2). Mini-N2G in these linear arrays was less dynamic than mini-N2G elsewhere in the nucleus, indicating its inclusion in a molecularly defined structure. Therefore these actin-dependent molecular arrays of SUN2 and N2G form TAN lines, a new nuclear envelope supra structure responsible for moving the nucleus [18]. These actin-dependent arrays that are formed at the nuclear envelope are functionally equivalent to plasma membrane focal adhesions in their role connecting a membrane to the actin cytoskeleton. In addition, TAN lines are also functionally equivalent to the spindle pole bodies in fungi, on their role of directly connecting the nucleus to the cytoskeleton to drive nuclear movement.

The LINC complex connects to the nuclear lamina via SUN proteins. As expected, lamin A/C, the main components of the nuclear lamina encoded by the *LMNA* gene, are also required for nuclear movement during cell migration. Although lamin A/C do not form linear arrays on the cell surface, they have a role in anchoring TAN lines at the nuclear envelope. In *Lmna*$^{-/-}$ mouse embryonic fibroblast (MEF) cells, the TAN lines still formed and moved together with actin but just slid over the non-moving nucleus, indicating an anchoring defect of the complex to the nucleus. Additionally, TAN lines were less persistent in these cells. Similar results were obtained for SUN2 siRNA knockdown. Interestingly, expression of lamin A/C with mutations that are involved in muscle dystrophy impaired nuclear movement in fibroblasts, whereas mutations involved in lipodystrophies and related phenotypes impaired centrosome reorientation but not nuclear movement. This result not only suggests a role for nuclear positioning in lamin-related muscle dystrophies but also provides some mechanistic insight on how different lamin mutations generate such diverse clinical phenotypes [33].

Regulation of Actin Dynamics at the Nuclear Surface

Besides the actin cables in TAN lines, the existence of an actin cap regulating nuclear shape has been observed in several cell types [34]. This set of filaments associates with the dorsal side of the nucleus and terminates in focal adhesions forming a mesh that holds the nucleus to the ventral surface [35]. Depolymerization of actin by latrunculin-B disrupted the nuclear cap at concentrations that did not affect ventral stress fibers. In this condition, nuclear shape regulation by cell shape was abrogated as assessed by plating on patterned substrates. Phosphorylated myosin also localizes to this cap and actomyosin contractility was shown to have a role in the formation and regulation of the cap, albeit smaller than its role in actin

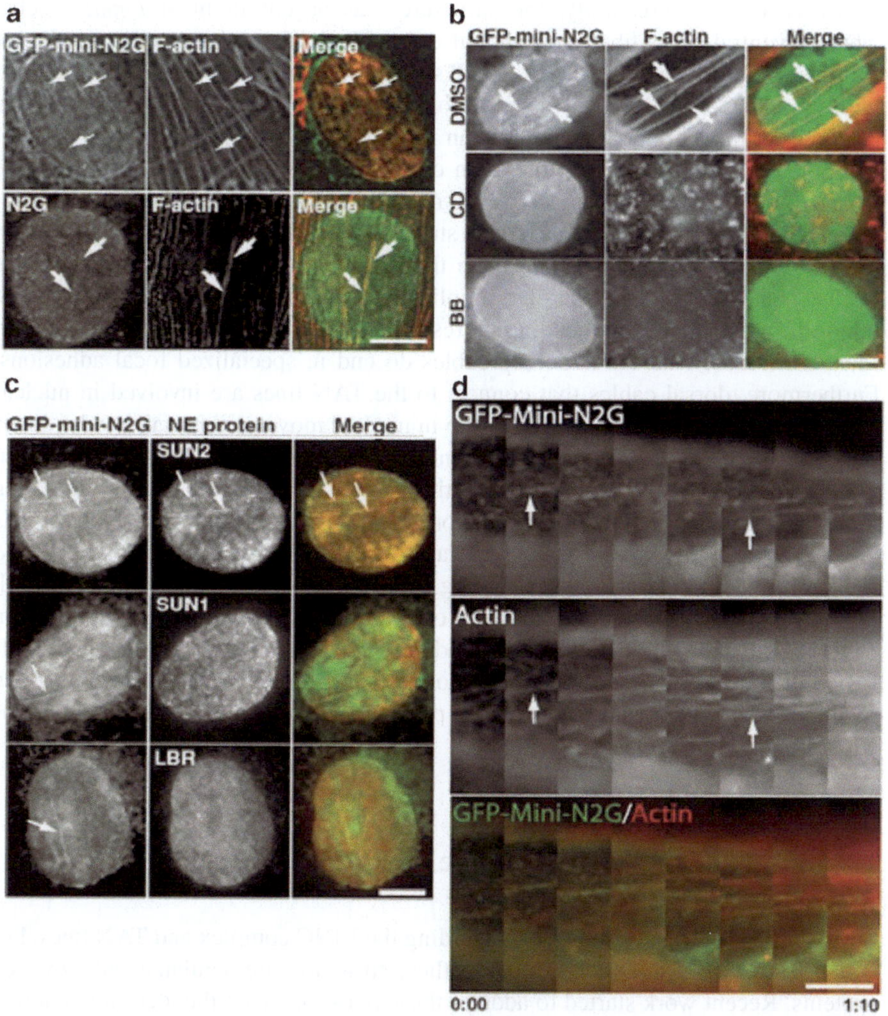

Fig. 2 SUN2 and N2G form actin associated linear arrays (TAN lines) in the nuclear surface. (**a**) Fluorescence images of a nucleus from a nesprin2G depleted cell expressing GFP-mini-N2G and stained with rhodamine-phalloidin (F-actin). (*Bottom*) Fluorescent images of a nucleus in a cell stained with nesprin2G antibody (N2G) and rhodamine-phalloidin (F-actin). *Arrows*, colocalization of N2G and dorsal actin cables. (**b**) Fluorescence images of nuclei in nesprin2G-depleted cells expressing GFP-mini-N2G. The cells were stained with GFP antibody (GFP-mini-N2G) and rhodamine-phalloidin (F-actin). The cells were treated with dimethyl sulfoxide (DMSO), 50 mM blebbistatin (BB), or 0.5 mM cytochalasin D (CD) for 1 h before and during LPA treatment. (**c**) Fluorescence images of nuclei in nesprin2G depleted cells. Staining from SUN1, SUN2, LBR, and GFP (GFP-mini-N2G) antibodies. *Arrows* show N2G colocalizing with SUN2 but not SUN1 or LBR. (**d**) Fluorescence kymograph of GFP-miniN2G and Lifeact-mCherry on the dorsal nuclear surface of a nucleus upon LPA stimulation. *Arrows* indicate examples of TAN lines forming on actin cables. Images are for every 10 min and oriented with the wound edge at the top. Scale bars in A-D, 5 μm. Time is h:min for (**d**). Adapted from [18] by permission from the American Association for the Advancement of Science, Copyright © 2010

polymerization. Interestingly, this structure was absent in most *Lmna*$^{-/-}$ cells, whereas ventral stress fibers were not affected. Nuclei from these cells were larger and their shape was uncoupled from cell shape in patterned substrates. Interference with the LINC complex expressing a dominant-negative construct containing the last 65 c-terminal amino acids of human nesprin2 that includes the KASH domain (KASH2) [36] also disrupted the actin cap. These results indicate a role for the LINC complex not only in attaching the nuclear envelope to actin but also in regulating the formation of perinuclear actin structures. The precise relationship between the dorsal actin cables that connect to the TAN lines and the actin cap remains unclear, but the molecular partners implicated in both seem to be the same. The major difference between these structures is that these dorsal cables do not end on focal adhesions, whereas actin cap cables do end in specialized focal adhesions. Furthermore, dorsal cables that connect to the TAN lines are involved in nuclear movement, whereas the role of actin cap in nuclear movement remains to be determined [18, 35]. Additionally, it is still unclear how actin dynamics are regulated at the nuclear surface. A recent study identified a unique family of F-actin regulating proteins designated refilins (for *RE*gulator of *FIL*amin prote*IN*) whose expression is induced upon epithelial–mesenchymal transition (EMT). Refilins dimerize and bind to filamins, actin binding and scaffolding proteins that mediate actin bundling [37]. Filamin A localizes to the nuclear actin cap together with RefilinB upon induction of EMT. Knockdown of RefilinB resulted in the disruption of the actin cap and diffusion of filamin A. Therefore, refilin-bound filamin proteins might have a role in regulating the structure and dynamics of perinuclear actin, in particular in situations where migration is induced.

Regulation of the LINC Complex and TAN Lines

Several unresolved questions exist regarding the LINC complex and TAN lines. For instance we still do not understand how these structures are regulated and by which proteins. Recent work started to address these questions with the identification of a new nuclear envelope protein enriched at the sites of SPB attachment to microtubules in *S. pombe*. This protein, named Ima1 in *S. pombe*, is conserved in all *metazoa* but not present in *S. cerevisiae* (Fig. 3). In mammals this protein is encoded by the *TMEM201* gene and was named NET5 as one of the nuclear envelope proteins identified by subtractive proteomics, and Samp1 (for Spindle associated membrane protein 1) due to its distribution in membrane compartments along the microtubules of the mitotic spindle [38, 39].

Fig. 3 (continued) The first predicted transmembrane domain has experimentally been shown to not be integral to the membrane but instead to associate with the INM [42, 45]. (**b**) Alignment of PFAM DUF2349/Ima-1 domain, the most conserved region of the protein, highlighting the existence of four conserved CXXC motifs that potentially form two zinc-fingers in the N-terminal region of this protein [42]. Figure created using Geneious version 6.1.6 from Biomatters. Available from http://www.geneious.com

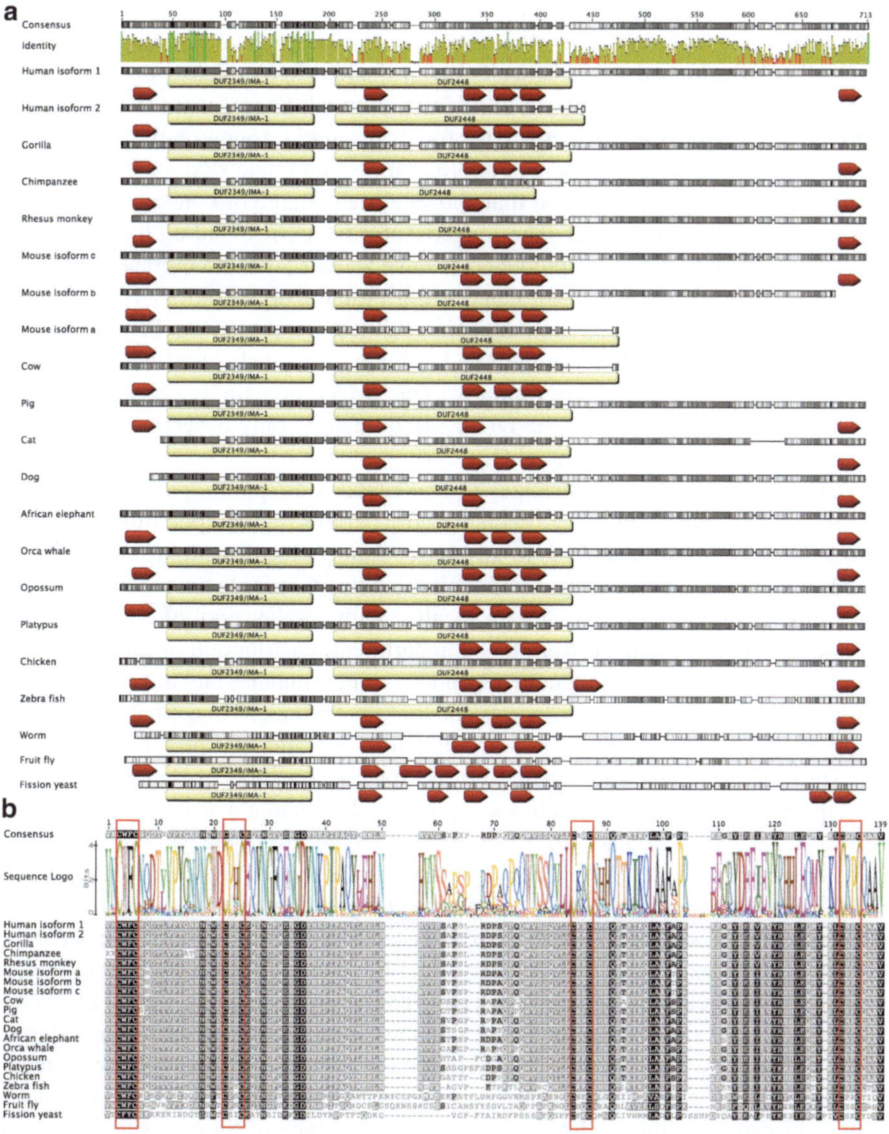

Fig. 3 Samp1 orthologs. (**a**) Multiple sequence alignment of NET-5/Samp1 proteins from 20 representative species (*Homo sapiens* isoform 1, NP_001124396.2; *Homo sapiens* isoform 2, NP_001010866.1; *Gorilla gorilla*, XP_004024661.1|; *Pan troglodytes*, XP_003307844.1; *Macaca mulatta*, XP_002802208.1; *Mus musculus* isoform c, A2A8U2; *Mus musculus* isoform b, A2A8U2-3; *Mus musculus* isoform a, A2A8U2-2; *Felis catus*, XP_003989620.1; *Canis lupus* familiaris, XP_546763.3; *Bos taurus*, NP_001032528.1; *Loxodonta africana*, XP_003413529.1; *Sus scrofa*, XP_003127587.1; *Orcinus orca*, XP_004272413.1; *Monodelphis domestica*, XP_001505803.2; *Gallus gallus*, XP_001376083.1; *Danio rerio*, XP_417601.3; *Drosophila melanogaster*, CG7744; *Caenorhabditis elegans*, T24F1.2; *Schizosaccharomyces pombe*, SPCC737.03c). Consensus and identity are plotted on *top* and the *shading* level represents conservation. Mouse isoform nomenclature is according to Borrego-Pinto et al. [41], mouse isoform a corresponding to isoform 2 in human. PFAM domains are indicated in *yellow* and predicted transmembrane domains (TMHMM) are in *red*.

There are three predicted Samp1 isoforms in mouse, which share the same N-terminus and different C-termini (Fig. 3). The shorter isoform (Samp1a) was originally characterized in human cells [39, 40] and mRNA expression of the three isoforms was confirmed in mouse fibroblasts although Samp1a was not detectable at the protein level [41]. As is the case for its *S. pombe* counterpart, Samp1 is an inner nuclear membrane protein with four transmembrane domains and a long N-terminal nucleoplasmic region that shows a high degree of conservation among metazoans and *S. pombe* [42]. The longer isoforms, corresponding to human isoform 2 or mouse Samp1c are predicted to have an extra transmembrane domain in the C-terminus (Figs. 3 and 4). Concerning domain organization, Ima1/NET-5/Samp1 proteins have a common domain architecture (based on PFAM domain server) that includes the N-terminal IMA1 domain (previously known as DUF2349) and a DUF2448 localized c-terminally. The IMA1 domain is conserved in all eukaryotes analyzed, whereas the DUF2448 domain is only observed in tetrapods and fish, but not in insects or fungi (Fig. 3a).

During mitosis, Samp1a localizes to and remains at spindle poles, an atypical behavior for INM proteins, usually displaced from this region upon mitotic spindle formation [39]. From metaphase to anaphase, Samp1a localizes to a specific membrane component that associates with spindle poles that is distinct from bulk ER, since other INM proteins like emerin localize with another pool of Samp1a in the mitotic ER but are excluded from this compartment. Distribution of Samp1a to the spindle pole membrane component depended on microtubules and not on actin. The conserved N-terminus of Samp1a was shown to be sufficient to mediate both interphase INM and mitotic spindle pole localizations in a chimeric protein. The role of this spindle membrane component and the specific function of Samp1a within it remain elusive [39]. Recent studies suggest that the spindle matrix provides a conserved strategy to segregate genetic material from the rest of the cell during mitosis [43]; however, the relationship between the spindle matrix and this spindle membrane component remains unclear. The existence of eight conserved cysteines distributed as four CxxC motifs in the Samp1 N-terminus led the authors to analyze the potential formation of two zinc fingers (Fig. 3b). Mutations of conserved cysteines impaired INM localization of constructs and nuclei expressing this mutant displayed shape anomalies reminiscent of lamin mutants. Furthermore, emerin, Sun1, lamin A/C, and partially SUN2 were displaced from the nuclear envelope in these mutants. These functional connections were further explored by RNAi depletion and emerin was displaced from the nuclear envelope in the absence of Samp1, suggesting the requirement of Samp1a for emerin localization. Furthermore, a zinc-finger-dependent interaction between emerin and Samp1a was reported, as emerin co-immunoprecipitated with YFP-Samp1 only in the presence of zinc and the interaction was disrupted in the presence of EDTA [42]. Samp1 also interacts with lamin A/C since it was co-immunoprecipitated with lamin A/C and was found to be in the vicinity of lamin A/C using proximity-dependent biotin identification (BioID) [41, 44]. Taken together these results suggest that the N-terminus of Samp1 constitutes a nucleoplasmic domain containing two potential zinc fingers that are important for Samp1 nuclear envelope localization. High-resolution microscopy analysis

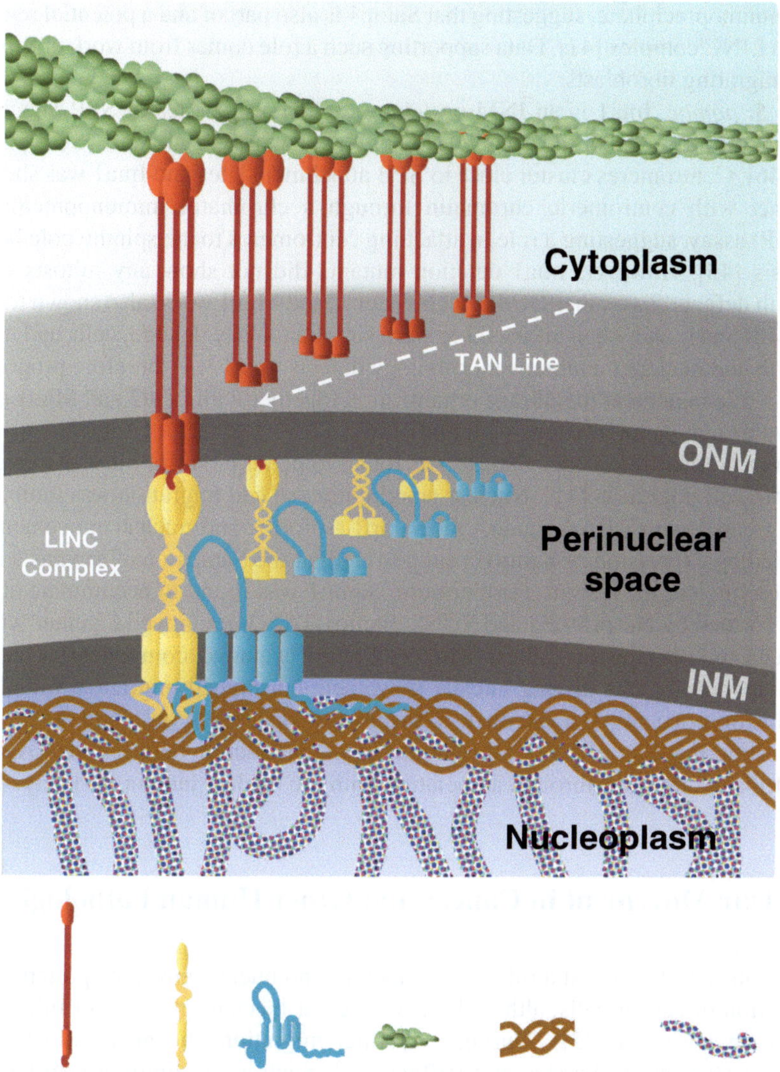

Fig. 4 Schematic representation of a TAN line at the nuclear envelope. TAN lines composed of Nesprin-2G in the outer nuclear membrane (ONM) and SUN2 and Samp1 in the inner nuclear membrane (INM) connect the actin dorsal cables to the nuclear lamina and chromatin. Adapted from [41] by permission from the Company of Biologists, Copyright © 2012

demonstrated that Samp1a protein is not uniformly distributed along the nuclear envelope but seems to concentrate in microdomains. Furthermore, it partially colocalized with SUN1, but showed only a minor colocalization with NPC, Lamins, emerin, and Sun2, suggesting a functional association with SUN1 in human cells [42]. In migrating fibroblasts, Samp1b and SUN2 were shown to co-immunoprecipitate, suggesting that Samp1 is also part of and a potential regulator of the LINC complex [41]. Data supporting such a role comes from work in *S. pombe* and migrating fibroblasts.

In *S. pombe*, Ima1 is an INM protein, enriched at the sites of SPB attachment during a short period of the cell cycle, where it colocalizes with Sad1 (SUN protein) [45, 46]. Centromeres cluster close to SPB attachment sites and Ima1 was shown to interact with centromeric chromatin through a chromatin immunoprecipitation (CHIP) assay, suggesting a role in attaching centromeres to the spindle pole body in mitosis [45]. However, Ima1 deletion mutants did not show any mitosis or cell growth defects. Two other INM proteins, Lem2 and Man1, were also shown to interact with Sad1, and when all the three were simultaneously deleted, cells had mitotic growth and nuclear membrane morphology defects. Ima1 was therefore proposed to have a role in nuclear membrane organization together with Lem2 and Man1 potentially through its interactions with Sad1 [46].

In migrating fibroblasts, Samp1 was found to be required for nuclear movement prior to cell migration [41]. Nuclear positioning and cell migration was inhibited in cells treated with siRNA against all Samp1 isoforms and nuclear movement was rescued by expression of Samp1b, suggesting that the Samp1c isoform has a minor role on nuclear movement. Furthermore, Samp1 was found to accumulate at TAN lines formed by Nesprin-2G and SUN2. Samp1 TAN lines moved together with the nucleus and dorsal actin cables, therefore Samp1 is a novel component of the TAN lines, responsible for driving nuclear movement during cell migration. Finally, the localization of Samp1 at the nuclear envelope is dependent on lamin A/C, overall suggesting that Samp1 has a role stabilizing the interaction between the TAN lines and LINC complexes through association with the nuclear lamina [41] (Fig. 4).

Nuclear Movement in Cancer and Other Human Pathologies

Different reports suggest a role for the nucleus and nuclear envelope proteins in the migration of cancer cells, although the molecular mechanisms are not fully understood (reviewed in [32]). During interstitial migration, the nucleus is the rate-limiting step due to its size and stiffness [5]. Such a rate-limiting step has been observed during the migration of glioma cells and is dependent on myosin II [6]. In these cells, myosin II activation is proposed to have a role squeezing the nucleus through holes within the interstitial regions. Interestingly, nesprin-1 is down regulated in multiple carcinomas [47] and both nesprin-1 and -2 accumulate mutations in breast and colorectal cancers [48]. In addition, nesprin-1 polymorphisms have been associated with ovarian cancer [49]. Finally, changes on gene expression and mutations in lamin A/C have been found in multiple cancers [32]. All these proteins are involved in nuclear positioning and nuclear stiffness, suggesting an important role for the correct positioning of the nucleus during cell migration in cancer.

Other examples of pathologies linked to nuclear positioning have been described. Defects in brain formation and development occur in lissencephaly patients, which is caused by a mutation in the *LIS1* gene, a dynein-associated protein [50]. Impairment of *LIS1* interferes with neuronal nuclear movements that occur during interkinetic movement and neuronal cell migration, originating the defects in brain development. More recently, mutations in *SYNE1* that encodes Nesprin-1, a KASH-containing protein involved in nuclear positioning, have been associated with autosomal recessive arthrogryposis [51], cerebellar ataxia [52], and Emery–Dreifuss muscular dystrophy [53]. Mutations in *SYNE4*, that encodes Nesprin-4 cause hereditary hearing loss [54]. Mutations in torsinA, an endoplasmic reticulum protein involved in the regulation of the Nesprin-SUN (LINC) complex [55, 56] is also associated with torsion dystonia. The extensive list of pathologies associated with *LMNA* mutations (laminopathies) might partially be due to the role of lamin A/C in nuclear movement [3, 33]. Finally, in skeletal muscle where nuclear positioning is important for muscle function, some muscular disorders cause an aberrant displacement of the nuclei [57, 58]. How mislocalized nuclei lead to muscle dysfunction is still unknown. With the exception of lissencephaly, the molecular mechanisms impaired in these diseases are largely unknown and the role of nuclear positioning in these pathologies also remains to be determined.

References

1. Burke B, Roux KJ (2009) Nuclei take a position: managing nuclear location. Dev Cell 17: 587–597. doi:10.1016/j.devcel.2009.10.018
2. Starr DA, Han M (2003) ANChors away: an actin based mechanism of nuclear positioning. J Cell Sci 116:211–216
3. Dauer WT, Worman HJ (2009) The nuclear envelope as a signaling node in development and disease. Dev Cell 17:626–638. doi:10.1016/j.devcel.2009.10.016
4. Guilak F, Tedrow JR, Burgkart R (2000) Viscoelastic properties of the cell nucleus. Biochem Biophys Res Commun 269:781–786. doi:10.1006/bbrc.2000.2360
5. Lammermann T, Bader BL, Monkley SJ, Worbs T, Wedlich-Soldner R, Hirsch K, Keller M, Forster R, Critchley DR, Fassler R, Sixt M (2008) Rapid leukocyte migration by integrin-independent flowing and squeezing. Nature 453:51–55. doi:10.1038/nature06887
6. Beadle C, Assanah MC, Monzo P, Vallee R, Rosenfeld SS, Canoll P (2008) The role of myosin II in glioma invasion of the brain. Mol Biol Cell 19:3357–3368. doi:10.1091/mbc.E08-03-0319
7. Morris NR (2003) Nuclear positioning: the means is at the ends. Curr Opin Cell Biol 15:54–59
8. Malone CJ, Fixsen WD, Horvitz HR, Han M (1999) UNC-84 localizes to the nuclear envelope and is required for nuclear migration and anchoring during C. elegans development. Development 126:3171–3181
9. Reinsch S, Gönczy P (1998) Mechanisms of nuclear positioning. J Cell Sci 111:2283–2295
10. Adames NR, Cooper JA (2000) Microtubule interactions with the cell cortex causing nuclear movements in Saccharomyces cerevisiae. J Cell Biol 149:863–874
11. Gibeaux R, Politi AZ, Nédélec F, Antony C, Knop M (2013) Spindle pole body-anchored Kar3 drives the nucleus along microtubules from another nucleus in preparation for nuclear fusion during yeast karyogamy. Genes Dev 27:335–349. doi:10.1101/gad.206318.112
12. Yu J, Starr DA, Wu X, Parkhurst SM, Zhuang Y, Xu T, Xu R, Han M (2006) The KASH domain protein MSP-300 plays an essential role in nuclear anchoring during Drosophila oogenesis. Dev Biol 289:336–345

13. Malone CJ, Misner L, Le Bot N, Tsai M-C, Campbell JM, Ahringer J, White JG (2003) The C. elegans hook protein, ZYG-12, mediates the essential attachment between the centrosome and nucleus. Cell 115:825–836. doi:10.1016/S0092-8674(03)00985-1
14. Reinsch S, Karsenti E (1997) Movement of nuclei along microtubules in Xenopus egg extracts. Curr Biol 7:211–214
15. Cadot B, Gache V, Vasyutina E, Falcone S, Birchmeier C, Gomes ER (2012) Nuclear movement during myotube formation is microtubule and dynein dependent and is regulated by Cdc42, Par6 and Par3. EMBO Rep 13:741–749. doi:10.1038/embor.2012.89
16. Troy A, Cadwallader AB, Fedorov Y, Tyner K, Tanaka KK, Olwin BB (2012) Coordination of satellite cell activation and self-renewal by Par-complex-dependent asymmetric activation of p38α/β MAPK. Cell Stem Cell 11:541–553. doi:10.1016/j.stem.2012.05.025
17. Crisp M, Liu Q, Roux K, Rattner JB, Shanahan C, Burke B, Stahl P, Hodzic D (2006) Coupling of the nucleus and cytoplasm: role of the LINC complex. J Cell Biol 172:41–53. doi:10.1083/jcb.200509124
18. Luxton GWG, Gomes ER, Folker ES, Vintinner E, Gundersen GG (2010) Linear arrays of nuclear envelope proteins harness retrograde actin flow for nuclear movement. Science 329:956–959. doi:10.1126/science.1189072
19. Luxton GWG, Gomes ER, Folker ES, Worman H, Gundersen GG (2011) TAN lines: a novel nuclear envelope structure involved in nuclear positioning. Nucleus 2:173–181. doi:10.4161/nucl.2.3.16243
20. Ridley AJ, Schwartz MA, Burridge K, Firtel RA, Ginsberg MH, Borisy G, Parsons JT, Horwitz AR (2003) Cell migration: integrating signals from front to back. Science 302:1704–1709. doi:10.1126/science.1092053
21. Kupfer A, Louvard D, Singer SJ (1982) Polarization of the Golgi apparatus and the microtubule-organizing center in cultured fibroblasts at the edge of an experimental wound. Proc Natl Acad Sci U S A 79:2603–2607
22. Gundersen GG, Bulinski JC (1988) Selective stabilization of microtubules oriented toward the direction of cell migration. Proc Natl Acad Sci U S A 85:5946–5950
23. Nobes CD, Hall A (1999) Rho GTPases control polarity, protrusion, and adhesion during cell movement. J Cell Biol 144:1235–1244. doi:10.1083/jcb.144.6.1235
24. Bergmann JE, Kupfer A, Singer SJ (1983) Membrane insertion at the leading edge of motile fibroblasts. Proc Natl Acad Sci U S A 80:1367–1371
25. Schmoranzer J, Kreitzer G, Simon SM (2003) Migrating fibroblasts perform polarized, microtubule-dependent exocytosis towards the leading edge. J Cell Sci 116:4513–4519
26. Cook TA, Nagasaki T, Gundersen GG (1998) Rho guanosine triphosphatase mediates the selective stabilization of microtubules induced by lysophosphatidic acid. J Cell Biol 141:175–185. doi:10.1083/jcb.141.1.175
27. Gomes ER, Jani S, Gundersen GG (2005) Nuclear movement regulated by Cdc42, MRCK, myosin, and actin flow establishes MTOC polarization in migrating cells. Cell 121:451–463
28. Etienne-Manneville S, Hall A (2001) Integrin-mediated activation of Cdc42 controls cell polarity in migrating astrocytes through PKCzeta. Cell 106:489–498
29. Palazzo AF, Joseph HL, Chen YJ, Dujardin DL, Alberts AS, Pfister KK, Vallee RB, Gundersen GG (2001) Cdc42, dynein, and dynactin regulate MTOC reorientation independent of Rho-regulated microtubule stabilization. Curr Biol 11:1536–1541
30. Dujardin DL, Barnhart LE, Stehman SA, Gomes ER, Gundersen GG, Vallee RB (2003) A role for cytoplasmic dynein and LIS1 in directed cell movement. J Cell Biol 163:1205–1211. doi:10.1083/jcb.200310097
31. Schmoranzer J, Fawcett JP, Segura M, Tan S, Vallee RB, Pawson T, Gundersen GG (2009) Par3 and dynein associate to regulate local microtubule dynamics and centrosome orientation during migration. Curr Biol 19:1065–1074. doi:10.1016/j.cub.2009.05.065
32. Friedl P, Wolf K, Lammerding J (2011) Nuclear mechanics during cell migration. Curr Opin Cell Biol 23:55–64. doi:10.1016/j.ceb.2010.10.015
33. Folker ES, Östlund C, Luxton GWG, Worman HJ, Gundersen GG (2011) Lamin A variants that cause striated muscle disease are defective in anchoring transmembrane actin-associated nuclear lines for nuclear movement. Proc Natl Acad Sci U S A 108:131–136. doi:10.1073/pnas.1000824108

34. Khatau SB, Hale CM, Stewart-Hutchinson PJ, Patel MS, Stewart CL, Searson PC, Hodzic D, Wirtz D (2009) A perinuclear actin cap regulates nuclear shape. Proc Natl Acad Sci U S A 106:19017–19022. doi:10.1073/pnas.0908686106
35. Khatau SB, Bloom RJ, Bajpai S, Razafsky D, Zang S, Giri A, Wu P-H, Marchand J, Celedon A, Hale CM, Sun SX, Hodzic D, Wirtz D (2012) The distinct roles of the nucleus and nucleus-cytoskeleton connections in three-dimensional cell migration. Sci Rep. doi:10.1038/srep00488
36. Stewart-Hutchinson PJ, Hale CM, Wirtz D, Hodzic D (2008) Structural requirements for the assembly of LINC complexes and their function in cellular mechanical stiffness. Exp Cell Res 314:1892–1905. doi:10.1016/j.yexcr.2008.02.022
37. Gay O, Gilquin B, Nakamura F, Jenkins ZA, McCartney R, Krakow D, Deshiere A, Assard N, Hartwig JH, Robertson SP, Baudier J (2011) RefilinB (FAM101B) targets FilaminA to organize perinuclear actin networks and regulates nuclear shape. Proc Natl Acad Sci U S A 108:11464–11469. doi:10.1073/pnas.1104211108
38. Schirmer EC, Florens L, Guan T, Yates JR, Gerace L (2003) Nuclear membrane proteins with potential disease links found by subtractive proteomics. Science 301:1380–1382
39. Buch C, Lindberg R, Figueroa R, Gudise S, Onischenko E, Hallberg E (2009) An integral protein of the inner nuclear membrane localizes to the mitotic spindle in mammalian cells. J Cell Sci 122:2100–2107. doi:10.1242/jcs.047373
40. Figueroa R, Gudise S, Larsson V, Hallberg E (2010) A transmembrane inner nuclear membrane protein in the mitotic spindle. Nucleus 1:249–253. doi:10.4161/nucl.1.3.11740
41. Borrego-Pinto J, Jegou T, Osorio D, Aurade F, Gorjánácz M, Koch B, Mattaj I, Gomes ER (2012) Samp1 is a new component of the TAN lines and is required for nuclear movement. J Cell Sci 125:1099–1105
42. Gudise S, Figueroa RA, Lindberg R, Larsson V, Hallberg E (2011) Samp1 is functionally associated with the LINC complex and A-type lamina networks. J Cell Sci 124:2077–2085
43. Zheng Y (2010) A membranous spindle matrix orchestrates cell division. Nat Rev Mol Cell Biol 11:529–535. doi:10.1038/nrm2919
44. Roux KJ, Kim DI, Raida M, Burke B (2012) A promiscuous biotin ligase fusion protein identifies proximal and interacting proteins in mammalian cells. J Cell Biol 196:801–810. doi:10.1083/jcb.201112098
45. King MC, Drivas TG, Blobel G (2008) A network of nuclear envelope membrane proteins linking centromeres to microtubules. Cell 134:427–438. doi:10.1016/j.cell.2008.06.022
46. Hiraoka Y, Maekawa H, Asakawa H, Chikashige Y, Kojidani T, Osakada H, Matsuda A, Haraguchi T (2011) Inner nuclear membrane protein Ima1 is dispensable for intranuclear positioning of centromeres. Genes Cells 16:1000–1011. doi:10.1111/j.1365-2443.2011.01544.x
47. Marmé A, Zimmermann H, Moldenhauer G, Schorpp-Kistner M, Müller C, Keberlein O, Giersch A, Kretschmer J, Seib B, Spiess E, Hunziker A, Merchán F, Möller P, Hahn U, Kurek R, Marmé F, Bastert G, Wallwiener D, Ponstingl H (2008) Loss of Drop1 expression already at early tumor stages in a wide range of human carcinomas. Int J Cancer 123:2048–2056. doi:10.1002/ijc.23763
48. Sjöblom T, Jones S, Wood LD, Parsons DW, Lin J, Barber TD, Mandelker D, Leary RJ, Ptak J, Silliman N, Szabo S, Buckhaults P, Farrell C, Meeh P, Markowitz SD, Willis J, Dawson D, Willson JKV, Gazdar AF, Hartigan J, Wu L, Liu C, Parmigiani G, Park BH, Bachman KE, Papadopoulos N, Vogelstein B, Kinzler KW, Velculescu VE (2006) The consensus coding sequences of human breast and colorectal cancers. Science 314:268–274. doi:10.1126/science.1133427
49. Doherty JA, Rossing MA, Cushing-Haugen KL, Chen C, Van Den Berg DJ, Wu AH, Pike MC, Ness RB, Moysich K, Chenevix-Trench G, Beesley J, Webb PM, Chang-Claude J, Wang-Gohrke S, Goodman MT, Lurie G, Thompson PJ, Carney ME, Hogdall E, Kjaer SK, Hogdall C, Goode EL, Cunningham JM, Fridley BL, Vierkant RA, Berchuck A, Moorman PG, Schildkraut JM, Palmieri RT, Cramer DW, Terry KL, Yang HP, Garcia-Closas M, Chanock S, Lissowska J, Song H, Pharoah PDP, Shah M, Perkins B, McGuire V, Whittemore AS, Di Cioccio RA, Gentry-Maharaj A, Menon U, Gayther SA, Ramus SJ, Ziogas A, Brewster W, Anton-Culver H, The Australian Ovarian Cancer Study Management Group, The Australian Cancer Study (Ovarian Cancer), Pearce CL, on behalf of the Ovarian Cancer Association Consortium (OCAC) (2010) ESR1/SYNE1 Polymorphism and invasive epithelial ovarian cancer risk: an ovarian cancer association consortium study. Cancer Epidemiol Biomarkers Prev 19:245–250. doi:10.1158/1055-9965.EPI-09-0729

50. Vallee RB, Tsai J-W (2006) The cellular roles of the lissencephaly gene LIS1, and what they tell us about brain development. Genes Dev 20:1384–1393. doi:10.1101/gad.1417206
51. Attali R, Warwar N, Israel A, Gurt I, McNally E, Puckelwartz M, Glick B, Nevo Y, Ben-Neriah Z, Melki J (2009) Mutation of SYNE-1, encoding an essential component of the nuclear lamina, is responsible for autosomal recessive arthrogryposis. Hum Mol Genet 18:3462–3469. doi:10.1093/hmg/ddp290
52. Gros-Louis F, Dupre N, Dion P, Fox MA, Laurent S, Verreault S, Sanes JR, Bouchard J-P, Rouleau GA (2007) Mutations in SYNE1 lead to a newly discovered form of autosomal recessive cerebellar ataxia. Nat Genet 39:80–85. doi:10.1038/ng1927
53. Zhang Q, Bethmann C, Worth N, Davies J, Wasner C, Feuer A, Ragnauth C, Yi Q, Mellad J, Warren D, Wheeler M, Ellis J, Skepper J, Vorgerd M, Schlotter-Weigel B, Weissberg P, Roberts R, Wehnert M, Shanahan C (2007) Nesprin-1 and -2 are involved in the pathogenesis of Emery-Dreifuss Muscular Dystrophy and are critical for nuclear envelope integrity. Hum Mol Genet. doi:10.1093/hmg/ddm238
54. Horn HF, Brownstein Z, Lenz DR, Shivatzki S, Dror AA, Dagan-Rosenfeld O, Friedman LM, Roux KJ, Kozlov S, Jeang K-T, Frydman M, Burke B, Stewart CL, Avraham KB (2013) The LINC complex is essential for hearing. J Clin Invest. doi:10.1172/JCI66911
55. Vander Heyden AB, Naismith TV, Snapp EL, Hodzic D, Hanson PI (2009) LULL1 Retargets TorsinA to the nuclear envelope revealing an activity that is impaired by the DYT1 dystonia mutation. Mol Biol Cell 20(11):2661–2672. doi:10.1091/mbc.E09-01-0094
56. Nery FC, Zeng J, Niland BP, Hewett J, Farley J, Irimia D, Li Y, Wiche G, Sonnenberg A, Breakefield XO (2008) TorsinA binds the KASH domain of nesprins and participates in linkage between nuclear envelope and cytoskeleton. J Cell Sci 121:3476–3486. doi:10.1242/jcs.029454
57. Metzger T, Gache V, Xu M, Cadot B, Folker ES, Richardson BE, Gomes ER, Baylies MK (2012) MAP and kinesin-dependent nuclear positioning is required for skeletal muscle function. Nature 484:120–124. doi:10.1038/nature10914
58. Jungbluth H, Wallgren-Pettersson C, Laporte J (2008) Centronuclear (myotubular) myopathy. Orphanet J Rare Dis 3:26. doi:10.1186/1750-1172-3-26

Part VII
Towards a Molecular Explanation of Prognostic Links to the Nuclear Envelope

Introduction

Although there is still a long way to go, considerable progress has been made in recent years in understanding the mechanisms of control of nuclear size and shape and the function of the apparent loss of lamin A in certain tumor types. As noted in Part I, nuclear shape and size aberrations have long been used in both cancer diagnosis and prognosis. In this section, Ashraf Malhas and David Vaux, one of the most creative minds to hit the nuclear envelope field who first clearly described the nucleoplasmic reticulum, of Oxford University note that, with higher resolution microscopy than typically used in cancer diagnostics, many of the nuclear shape changes observed likely reflect large-scale invaginations of the nuclear envelope that can penetrate most or even all of the way through the nucleus. They further discuss a variety of possible functional benefits these invaginations could give to a normal cell and how they could contribute to the cancer pathology.

Cells tend to maintain a roughly constant ratio of nuclear to cell volume, the karyoplasmic ratio [1, 2], and changes in this have been a standard criterion used in cancer diagnostics and prognostics for at least the last half century [3, 4]. However, the seeming paradox that malignancy was associated with increased nuclear size for some tumors (e.g., invasive meningiomas and bladder carcinoma [5, 6]) and reduced nuclear size for others (e.g., squamous cell carcinoma of the lung [7]) diffused the search for the molecular mechanism behind these changes. Next in this section Predrag Jevtic and Daniel Levy, an expert on intracellular scaling, of the University of Wyoming describe a wide variety of cancer types for which nuclear size plays an important role. They then discuss the roles of DNA ploidy, nuclear structural components, cytoplasmic factors, nucleocytoplasmic transport, cytoskeleton, and extracellular matrix in nuclear size and speculate on how these size changes could contribute to cancer development and progression. This is followed by another chapter on nuclear size focused specifically on the role of the NPC by Masatoshi Takagi and Naoko Imamoto, one of the leading experts in NPC assembly and the world's top expert in nucleoporin dynamics, of RIKEN. They further discuss the aspects of NPC roles in nuclear size in a wide range of organisms and then discuss specific changes of the NPC in cancer. This latter area is very exciting for cancer biology as three structural components of the NPC are involved in fusions from chromosome translocations that promote tumorigenesis.

Measurements of the karyoplasmic ratio in epidermal layers revealed that nuclear volume was largely maintained, but cytoplasmic volume increased fourfold as cells passed from basal to granular layers [8]. This suggested that the characteristic ratios observed in various tumors might reflect the stage of the progenitor cell that originated the tumor and could be used as prognostic markers. This idea has now been found to apply also to lamin A levels, often used in prognosis of different tumor types. In the last chapter Christopher Hutchison of Durham University, one of the world's leading lamin researchers whose work has potentially resolved the great paradox of lamin levels in cancer, reviews the roles of lamins in a variety of cancer types and presents the hypothesis that similar changes in lamin abundance in different levels of an epithelium as observed for nuclear size could explain the fact that increased lamin A levels are a poor prognostic for some tumor types while reduced levels are a poor prognostic for other tumor types. Professor Hutchison uses several modern studies to give a very insightful view to the question of whether lamins truly influence cancer progression.

1. Cavalier-Smith T (2005) Economy, speed and size matter: evolutionary forces driving nuclear genome miniaturization and expansion. Ann Bot 95(1):147–175. doi:95/1/147 [pii] 10.1093/aob/mci010
2. Wilson EB (1925) The karyoplasmic ratio. In: The cell in development and heredity, 3rd edn. The Macmillan Company, New York, pp 727–733
3. Ishizuka Y, Oota K, Masubachi K (1972) Practical cytodiagnosis. Lippincott, Philadelphia
4. Weibel ER (1972) The value of stereology in analysing structure and function of cells and organs. J Microsc 95(1):3–13
5. Fukuzawa S, Hashimura T, Sasaki M, Yamabe H, Yoshida O (1995) Nuclear morphometry for improved prediction of the prognosis of human bladder carcinoma. Cancer 76(10):1790–1796
6. Madsen C, Schroder HD (1996) Stereological estimation of nuclear mean volume in invasive meningiomas. APMIS 104(2):103–107
7. Ladekarl M, Boek-Hansen T, Henrik-Nielsen R, Mouritzen C, Henriques U, Sorensen FB (1995) Objective malignancy grading of squamous cell carcinoma of the lung. Stereologic estimates of mean nuclear size are of prognostic value, independent of clinical stage of disease. Cancer 76(5):797–802
8. Rowden G (1975) Ultrastructural studies of keratinized epithelia of the mouse. III. Determination of the volumes of nuclei and cytoplasm of cells in murine epidermis. J Invest Dermatol 64(1):1–3

Nuclear Envelope Invaginations and Cancer

Ashraf N. Malhas and David J. Vaux

Abstract The nuclear envelope (NE) surrounds the nucleus and separates it from the cytoplasm. The NE is not a passive structural component, but rather contributes to various cellular processes such as genome organization, transcription, signaling, and stress responses. Although the NE is mostly a smooth surface, it also forms invaginations that can reach deep into the nucleoplasm and may even traverse the nucleus completely. Cancer cells are generally characterized by irregularities and invaginations of the NE that are of diagnostic and prognostic significance. In the current chapter, we describe the link between nuclear invaginations and irregularities with cancer and explore possible mechanistic roles they might have in tumorigenesis.

Keywords Nuclear envelope • Nucleoplasmic reticulum • Nuclear invaginations • Gene expression

Abbreviations

CCT-a	CTP:phosphocholine-cytidylyltransferase-α
HCC	Hepatocellular carcinoma
HGPS	Hutchinson–Gilford Progeria Syndrome
INM	Inner nuclear membrane
MARs	Matrix attachment regions
NE	Nuclear envelope
NR	Nucleoplasmic reticulum
PaCa	Pancreatic carcinoma

A.N. Malhas • D.J. Vaux (✉)
Sir William Dunn School of Pathology, South Parks Road, Oxford, OX1 3RE, UK
e-mail: Ashraf.malhas@path.ox.ac.uk; david.vaux@path.ox.ac.uk

E.C. Schirmer and J.I. de las Heras (eds.), *Cancer Biology and the Nuclear Envelope*, Advances in Experimental Medicine and Biology 773, DOI 10.1007/978-1-4899-8032-8_24,

The Nuclear Envelope and Nucleoplasmic Reticulum

The nuclear envelope (NE) is composed of two phospholipid bilayers, the inner nuclear membrane (INM) and outer nuclear membrane, with a lumenal space between them and a nuclear lamina which underlies the INM [1]. Our current understanding of the NE is that it contributes to nuclear structural integrity, genome organization, selective bidirectional transport of ions and macromolecular cargo, regulation of gene expression, DNA repair, mechano-transduction and acts as a platform for signaling [2–4]. The structural and functional importance of the NE is highlighted by diseases that result from mutations in genes coding for NE components. These diseases are collectively referred to as laminopathies and can affect a range of differentiated tissue types, including striated muscle, adipose tissue, peripheral nerve or multiple tissues [5, 6].

Although nuclei are often represented in schematics as circular or oval in shape, they actually vary in different cell types and under different pathological conditions, and thus nuclear shape has been described as a "differentiation-related phenotype" [7]. There is currently plenty of evidence that confirms that the NE from many cell types and tissues can contain deep, branching invaginations [8–10]. These invaginations are collectively referred to as the nucleoplasmic reticulum (NR) because of its morphological resemblance to the endoplasmic reticulum. NR invaginations are classified into two main classes, type I and type II. Type I invaginations are those where the inner nuclear membrane alone invaginates into the nucleoplasm, while type II involve the invagination of both the inner nuclear membrane and outer nuclear membrane. Type II invaginations may contain microtubules, microfilaments, and mitochondria in their cytoplasmic core [11, 12] (see Fig. 1). The NR is observed in nuclei from various normal and abnormal tissues [9] as well as cells grown in 2D and 3D cultures including many tumor cell types, such as brain, breast, kidney, bladder, prostate, and ovary [13, 14]. The NR may arise during NE reassembly after mitosis or de novo in interphase cell nuclei without mitosis. It usually persists throughout interphase and can show heritable patterns in specific cell types [8, 15, 16]. Although the NR is now widely recognized as a nuclear structural component, its exact functional roles in normal and cancer cells are not fully understood. There is evidence for roles in calcium signaling gene expression, transport, and nuclear lipid metabolism [17]. The composition of the NR, its formation mechanisms and regulation have been reviewed recently [13] and will not be discussed in detail here. The changes in NE composition that are observed in many cancers have also been reviewed elsewhere [18]. In the current review, we discuss the NR connection to cancer and what functional roles it is thought to play in tumorigenesis.

Invaginations of the NE are a common feature of some laminopathies [19]. One of the most striking laminopathies is Hutchinson–Gilford progeria syndrome (HGPS). HGPS cells express a mutant form of lamin A called progerin that persists as a farnesylated prelamin at the nuclear periphery and as a result have an increased abundance of invaginated nuclei [20, 21] which share similar morphological

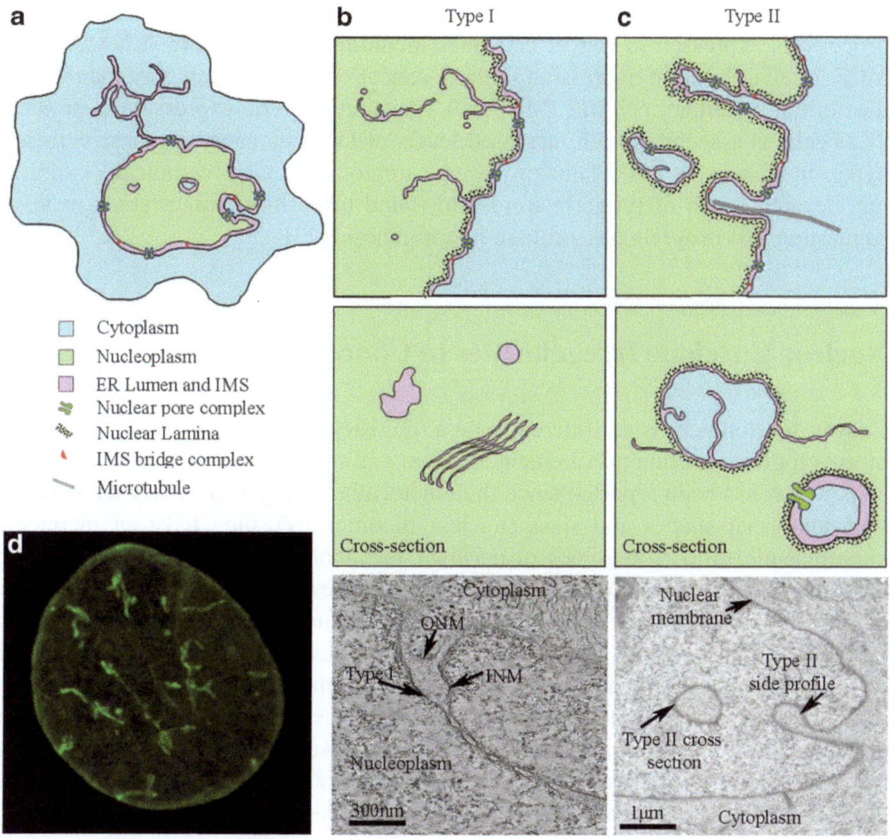

Fig. 1 Schematic representations of two types of membrane invaginations into the nucleoplasm. (**a**) Whole cell schematic defining the color coding of relevant compartments and structures. (**b**) Schematic of type I NR, including the cross-sectional appearance (*middle panel*) and an annotated TEM view (*bottom panel*). (**c**) Schematic of type II NR, including the cross-sectional appearance (*middle panel*) and an annotated TEM view (*bottom panel*). (**d**) 3D reconstruction of a confocal z-stack showing GFP-lamin B1 expressed in a HeLa cell treated with a farnesyltransferase inhibitor that causes NR proliferation. While type I NR brings the lumen to deep nuclear locations it does not provide the cytoplasmic conduit characteristic of the type II NR channels. In either case a complete distinction between transverse sections of reticular channels (*panels top right*) and long slots slicing into the nucleus can only be achieved by three dimensional analysis (for example confocal z-stacks, serial section EM, or electron tomography). Such 3D analysis confirms that type I and type II structures may coexist, and indeed interconnect within the same nucleus, and that isolated NR features with no apparent connection to the nuclear envelope may be frequent. Cross-sections of NR channels (schematic *middle panel* in (**c**)) show the complex topology of the type II channels, with a cytoplasmic interior that may be invaded by folds of outer nuclear membrane and often contains vesicles, cytoskeletal elements, and, in some cell types, mitochondria . Reproduced from [13] with permission of Elsevier

features to nuclei from normal aged cells [22]. It has also been shown that progerin accumulates in cells from normal older individuals suggesting a possible role for it in the normal ageing process [22]. How a pathologically invaginated nucleus leads to segmental accelerated ageing is yet to be determined, and it is unclear if the NR that develops serves any functional role. A recent study however showed that progerin is expressed in cancer cell lines including PC-3, DU145, LN-CAP, and MCF-7 cells. Progerin expression in these cancer cell lines was higher than that in the normal mammary cell line (MCF-10). Elevated progerin expression in prostate PC-3 cells is associated with increased levels of DNA damage and these cells are hypersensitive to the DNA damaging agent camptothecin. Cells ectopically expressing elevated levels of progerin were also found to be more tumorigenic in vivo, suggesting that progerin has a role in tumorigenesis [23].

Nuclear Envelope Irregularities in Cancer Diagnostics

Cancer development is considered to be a multistep process that is related to both morphological and functional changes. A general morphological feature of cancer cells which has been reported since the nineteenth century is that they have nuclei with abnormal shapes and sizes (nuclear pleomorphy), they have an increased karyoplasmic ratio (sometimes also called nucleoplasmic to cytoplasmic ratio), have increased DNA content, and have irregularities in their NE [14, 24]. These aspects of nuclear structure are so striking that they are used for grading of cancers by pathologists. In breast cancer, for example, nuclear pleomorphisms are graded and correlate with clinical aggressiveness and prognosis [25, 26]. These changes occur early and are even detectable in pre-cancerous stages making them a valuable diagnostic tool [27]. Bloom and Richardson [28] observed that there is variation in progress of cases of breast cancer even in patients of the same age and found that clinical staging is not an accurate indicator of the likelihood and speed of metastasis. They found however that when NE irregularities are taken into account in grading, a more accurate prognosis is achieved. NE irregularities are of prognostic significance in a number of other cancers including head and neck [29], kidney [30], bladder [31], prostate [32–34], and ovary [35].

Nuclear shape is also used in the diagnosis of papillary thyroid carcinoma. Staining of the NE component emerin identifies irregularities in the NE and is considered a useful tool for a “definitive and objective diagnosis” [36, 37]. Diagnosis based on emerin staining of thyroid fine needle aspiration biopsies was found to have a sensitivity of 77 % and accuracy of 84 % compared to only 36 % and 62 % achieved by using the conventional H and E staining method. Better consistency in diagnosis is also achieved using emerin staining with inter-observer concordance reported as moderate in the case of H and E staining but substantial in the case of emerin staining [37].

Bussolati and coworkers reported similar results in the diagnosis of breast cancer [25]. They found a marked increase in NE irregularities when comparing normal mammary cells and breast cancer cells using lamin B and emerin staining. Staining using NE markers enabled the authors to recognize tumors that are more prone to metastasis even among low-grade cancers. In routine practice H and E staining is used to assess nuclear pleomorphism of cancers. The above studies indicate that determination of nuclear pleomorphisms based on NE markers generates better sensitivity and accuracy than those achieved by the conventional H and E stain [38, 39]. It is important to note at this point that there are exceptions to the rule regarding nuclear morphology and cancer. Nuclear pleomorphisms and changes in chromatin distribution are not observed in the transition zone variant of prostate adenocarcinoma and nuclei of lobular breast carcinoma [14].

It is clear from the aforementioned studies that NR formation can be prevalent in cancer cells. It is still unclear whether the irregularities in the NE and the NR that is formed in such cells is a deleterious consequence of dysregulation or a beneficial response altering nuclear architecture and gene expression.

Formation and Possible Functions of Nuclear Envelope Irregularities

A well-characterized example of NE irregularities in cancer is that of papillary thyroid carcinoma which can be caused by oncogenic translocation events resulting in RET/PTC1 fusion, which generates a chimeric tyrosine kinase. This translocation is found in over half of papillary thyroid carcinomas [40]. The translocation can be induced by radiation exposure, and has been detected at elevated frequency in a group of Chernobyl-associated papillary thyroid carcinomas [41]. Both papillary thyroid carcinomas and follicular thyroid neoplasms are derived from thyroid epithelial cells. The latter however lack irregularities in the NE [16]. When normal thyroid epithelial cells are microinjected with the RET/PTC1 oncogene, they develop irregularities of the NE within hours and without the cells going through mitosis [42, 43]. This indicates that NR formation can occur in the intact nuclei of interphase cells and does not require post-mitotic nuclear reassembly. It is important to note here that NE irregularities are not observed when normal thyroid epithelial cells are microinjected with oncogenic RAS which leads to follicular thyroid neoplasms [44]. The mechanism of NR formation in papillary thyroid carcinoma and the roles that RET/PTC1 play in it are not fully understood. Posttranslational modifications of NE components have been excluded as a cause since no such modifications were observed in analyses of papillary thyroid carcinomas [45]. Another possibility is that the RET/PTC1 product leads to signaling events that generate new forces that act on the NE leading to a new characteristic NR. These forces might come from the chromatin side or the cytoskeletal side of the NE.

Cytoskeletal changes—There is a well-characterized link between the NE and cytoskeleton. Muscle cell nuclei have irregular shapes and this is probably because of the physical stress that the cells experience [46, 47]. Irregular NE and NR formation in cancer cells might also be a consequence of extranuclear forces by cytoskeletal components. Pancreatic carcinoma (PaCa) cells are characterized by irregularities of the NE reported as nuclear lobulations. PaCa cells have intermediate filament bundles composed of vimentin and keratins distributed around the nucleus. These bundles are thicker in nuclear invaginations where they form rings. These rings were found to cut into the nucleus in such a way that they aggregate at the base of the NR invagination, run through the invagination and back to the aggregate. Similar structures are observed in bladder carcinoma (T24), melanoma (G-361), and cervical carcinoma (HeLa) cells, although the intermediate filament bundles in these cells were thinner than those observed in PaCa cells. There are however other cancer cell lines such as adrenal adenocarcinoma (SW13) cells which have an NR but lack these intermediate filament bundles [7]. These studies collectively suggest that the cytoskeleton might play a role in NR formation in some cancer cells, but similarly also suggest that this cannot be a universal mechanism. Since these observations are of preexisting NR they do not address the potentially separable roles of the cytoskeleton in the formation and maintenance of a stable NR network.

Chromatin changes—Irregular thickening of chromatin at the nuclear periphery is observed in cancer nuclei due to formation of peripheral heterochromatin [48]. This change in chromatin structure, which maybe a result of NE irregularities, can influence gene expression. NE components interact directly with chromatin and chromatin-associated proteins such as emerin, LAP2b, BAF1, the retinoblastoma protein Rb, histone deacetylases, DNA methyltransferases, histone methyltransferases, and HP-1 [24, 49–52]. Many of these influence heterochromatin assembly at the NE and hence have an effect on gene expression. Changes to nuclear peripheral shape by NR formation not only affect heterochromatin distribution but also are often associated with changes of gene positioning and expression.

Interphase nuclei have an ordered structure where chromosomes occupy distinct nonoverlapping territories [53]. The NE can affect this 3D organization of the genome within the nucleus by acting as a docking platform for certain loci [54, 55]. Hence, any change in NE shape is expected to influence the distribution and organization of at least some chromosomes within the nucleus. Many cancers exhibit structural and functional changes that are consequences of interchromosomal or intrachromosomal rearrangements. Both of these mechanisms are influenced by the spatial proximity of the relevant chromosomal regions within the 3D space of the nucleus [56, 57]. Loci participating in interchromosomal exchanges are positioned at the edges of chromosome territories more frequently than loci that are involved in intrachromosomal inversions [58]. This suggests that the formation of additional platforms for chromosomal attachment sites when an irregular and extensive NR is formed in cancer cells may be a cause for altering the 3D arrangement of chromosome territories and hence play a role in chromosomal translocations.

Selected portions of DNA loops are attached to the NE through sequences known as matrix attachment regions (MARs) [59]. MAR-binding proteins have a role in regulating gene expression and some have been linked to certain cancers. For example, MAR-binding of p114 in tissues correlates inversely with histological grade of breast carcinomas. Another MAR-binding protein is the AT-rich sequence binding protein 1 (SATB1) which is associated with the nuclear matrix surrounding dense heterochromatin [60]. Loss of SATB1 results in relocalization of binding of genes that normally bind SATB1 and dysregulation of genes involved in signal transduction, apoptosis, and tumor suppression [60]. This shows the importance of nuclear architecture in normal cell function and the potential role its loss may play in cancer. Some drugs which target AT-rich regions including MARS, such as the cyclopropylpyrroloindole, bizelesin, which alkylates MARS, have been used in phase I [61] and phase II clinical trials [48].

Regulation of NR membranes—We have recently shown that NR formation requires the activity of CTP:phosphocholine-cytidylyltransferase-α (CCT-a), an enzyme that is known to be involved in regulating the rate of phosphatidylcholine synthesis, and thus new membrane formation [62]. The CCT-a inhibitor Perifosine is used in clinical trials as an anticancer drug and is thought to act through different pathways such as inhibiting Akt phosphorylation [63, 64]. It is possible that an additional pathway through which it might act could be via NR alterations caused by the inhibition of CCT-a in cancer cells.

A Potential Mechanism from NR to Tumorigenesis

Prostate tissue cells from patients with prostate cancer have chromosome translocations as well as alterations to chromatin structure and nuclear shape. These cells have nuclear blebs that are enriched in lamin A/C and activated Polymerase II and preferentially associate with chromosomal regions enriched in mutations that have been linked to increased prostate cancer risks, such as 8q and 17q [65–67]. This suggests that nuclear blebs are involved in prostate cancer development and progression and that the NR maybe involved in similar mechanisms. Nuclear blebs were found in five human prostate cancer cell lines, RWPE-1, LNCaP, DU145, PC-3, and PC-3m. All nuclear blebs in these cells were found to be enriched in lamin A/C, deficient in lamin B and had reduced Hoechst staining indicating chromatin alterations. The frequency of these domains was found to be related to cell motility and invasiveness. Epigenetic analysis indicates that chromosomes within these domains are euchromatic and transcriptionally active. It is worth noting that in cell lines nuclear blebs were only observed following mitosis suggesting that they require lamin reassembly in order to form [65]. This is in contrast with nuclear invagination, which can form in interphase nuclei without the need for post-mitotic reassembly.

We and others have shown that alterations to the composition of the NE can affect gene expression by altering the position of specific chromosomes within the nucleus [24, 68–71] and by affecting the nucleoplasmic levels of specific transcription factors. The latter include SMAD, SREBP1, and Oct-1 [72–74]. We confirmed that lamin B1 can sequester Oct-1 and that loss of this sequestration leads to the released Oct-1 binding to its target sequences. As a result, Oct-1 targets, a large subset of which is involved in cellular responses to oxidative stress, are dysregulated. Cells in which this occurs exhibit a phenotype of harboring elevated levels of reactive oxygen species and are sensitive to oxidative stress. These results provide a mechanism that has implications for normal ageing and tumorigenesis. Furthermore, we found that miRNA-31 is one of the targets of Oct-1 that are upregulated when the Oct-1 sequestration at the NE is affected. As a result miRNA-31 is dysregulated [75]. Functional analysis showed that these targets are significantly enriched in genes involved in cell cycle regulation, such as the tumor suppressor *Cdkn2a.* We found that miRNA-31 can regulate the products of *Cdkn2a*; p16Ink4a and p19Arf

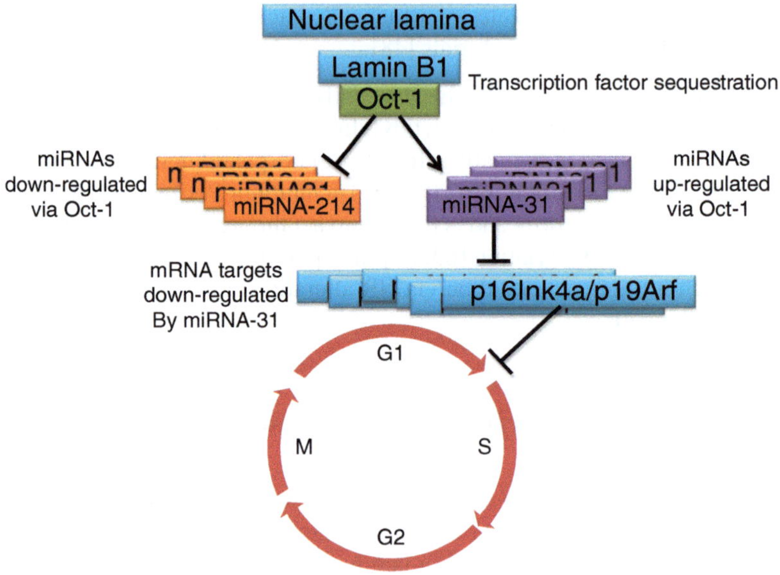

Fig. 2 Sequestration of transcription factors by the nuclear lamina regulates downstream processes, including cell cycle progression. The schematic shows the effect of a failure of nuclear lamina-dependent sequestration of Oct-1. Loss of lamin B1-Oct-1 interactions releases excess Oct-1 into the nucleoplasm, resulting in changes in mRNA transcription (not shown), and also altered expression of regulatory miRNAs. This results in the upregulation and downregulation of distinct sets of miRNAs. The most up-regulated miRNA is miRNA-31, which normally acts to reduce the level of a diverse set of target mRNAs, including gene products involved in cell cycle regulation. These targets include multiple splice forms generated from the cdkN2a gene, including the regulators p16Ink4a and p19arf. Since the normal function of these regulators is to block G1-S transition, loss of this regulation results in increased cell proliferation (see [75] for details)

(see schematic in Fig. 2). The link between the NE, Oct-1, and miRNA-31 levels is of interest since altered levels of miRNA-31 have been reported in a number of tumors including colorectal [76, 77], breast [78, 79], hepatocellular (HCC), and pancreatic ductal carcinomas [80]. Hence, an irregular NE in the nuclei of these cancer cells can lead to disruption of Oct-1 sequestration and dysregulation of Oct-1 targets, such as genes involved in oxidative stress responses and miRNA-31, with subsequent consequences for cell cycle regulation.

Conclusions

Although NE invaginations and irregularities are a common feature of many cancer cells, little is known about the exact mechanisms of their formation or the roles they play in tumorigenesis and metastasis. The NE is now seen as an active regulator of vital processes such as genome organization, gene expression, signaling, apoptosis and response to stress. The NE regulatory functions are often thought of in the context of the nuclear periphery, but the existence of an extensive NR offers a new platform that protrudes deep into the nucleoplasm and can therefore affect the aforementioned processes and may play a key role in tumorigenesis.

References

1. Prunuske AJ, Ullman KS (2006) The nuclear envelope: form and reformation. Curr Opin Cell Biol 18(1):108–116. doi:10.1016/j.ceb.2005.12.004, S0955-0674(05)00185-7 [pii]
2. Dauer WT, Worman HJ (2009) The nuclear envelope as a signaling node in development and disease. Dev Cell 17(5):626–638. doi:10.1016/j.devcel.2009.10.016, S1534-5807(09)00438-9 [pii]
3. Mekhail K, Moazed D (2010) The nuclear envelope in genome organization, expression and stability. Nat Rev Mol Cell Biol 11(5):317–328. doi:10.1038/nrm2894, nrm2894 [pii]
4. Redwood AB, Perkins SM, Vanderwaal RP, Feng Z, Biehl KJ, Gonzalez-Suarez I, Morgado-Palacin L, Shi W, Sage J, Roti-Roti JL, Stewart CL, Zhang J, Gonzalo S (2011) A dual role for A-type lamins in DNA double-strand break repair. Cell Cycle 10(15):2549–2560. doi:16531 [pii]
5. Worman HJ (2012) Nuclear lamins and laminopathies. J Pathol 226(2):316–325. doi:10.1002/path.2999
6. Broers JL, Ramaekers FC, Bonne G, Yaou RB, Hutchison CJ (2006) Nuclear lamins: laminopathies and their role in premature ageing. Physiol Rev 86(3):967–1008. doi:10.1152/physrev.00047.2005, 86/3/967 [pii]
7. Kamei H (1994) Relationship of nuclear invaginations to perinuclear rings composed of intermediate filaments in MIA PaCa-2 and some other cells. Cell Struct Funct 19(3):123–132
8. Fricker M, Hollinshead M, White N, Vaux D (1997) The convoluted nucleus. Trends Cell Biol 7(5):181. doi:S0962-8924(97)84084-6, [pii]10.1016/S0962-8924(97)84084-6
9. Fricker M, Hollinshead M, White N, Vaux D (1997) Interphase nuclei of many mammalian cell types contain deep, dynamic, tubular membrane-bound invaginations of the nuclear envelope. J Cell Biol 136(3):531–544
10. Schermelleh L, Carlton PM, Haase S, Shao L, Winoto L, Kner P, Burke B, Cardoso MC, Agard DA, Gustafsson MG, Leonhardt H, Sedat JW (2008) Subdiffraction multicolor imaging of the nuclear periphery with 3D structured illumination microscopy. Science 320(5881):1332–1336. doi:10.1126/science.1156947, 320/5881/1332 [pii]
11. Clubb BH, Locke M (1998) 3T3 cells have nuclear invaginations containing F-actin. Tissue Cell 30(6):684–691
12. Brandes D, Schofield BH, Anton E (1965) Nuclear mitochondria? Science 149(3690):1373–1374
13. Malhas A, Goulbourne C, Vaux DJ (2011) The nucleoplasmic reticulum: form and function. Trends Cell Biol 21(6):362–373. doi:10.1016/j.tcb.2011.03.008, S0962-8924(11)00050-X [pii]
14. True LD, Jordan CD (2008) The cancer nuclear microenvironment: interface between light microscopic cytology and molecular phenotype. J Cell Biochem 104(6):1994–2003. doi:10.1002/jcb.21478
15. Gupton SL, Collings DA, Allen NS (2006) Endoplasmic reticulum targeted GFP reveals ER organization in tobacco NT-1 cells during cell division. Plant Physiol Biochem 44(2–3):95–105. doi:10.1016/j.plaphy.2006.03.003, S0981-9428(06)00033-7 [pii]
16. Fischer AH, Taysavang P, Jhiang SM (2003) Nuclear envelope irregularity is induced by RET/PTC during interphase. Am J Pathol 163(3):1091–1100. doi:10.1016/S0002-9440(10)63468-2, S0002-9440(10)63468-2 [pii]
17. Bootman MD, Fearnley C, Smyrnias I, MacDonald F, Roderick HL (2009) An update on nuclear calcium signalling. J Cell Sci 122(Pt 14):2337–2350. doi:10.1242/jcs.028100, 122/14/2337 [pii]
18. Chow KH, Factor RE, Ullman KS (2012) The nuclear envelope environment and its cancer connections. Nat Rev Cancer 12(3):196–209. doi:10.1038/nrc3219, nrc3219 [pii]

19. Mounkes L, Kozlov S, Burke B, Stewart CL (2003) The laminopathies: nuclear structure meets disease. Curr Opin Genet Dev 13(3):223–230, S0959437X03000583 [pii]
20. McClintock D, Gordon LB, Djabali K (2006) Hutchinson-Gilford progeria mutant lamin A primarily targets human vascular cells as detected by an anti-Lamin A G608G antibody. Proc Natl Acad Sci U S A 103(7):2154–2159. doi:10.1073/pnas.0511133103, 0511133103 [pii]
21. De Sandre-Giovannoli A, Bernard R, Cau P, Navarro C, Amiel J, Boccaccio I, Lyonnet S, Stewart CL, Munnich A, Le Merrer M, Levy N (2003) Lamin a truncation in Hutchinson-Gilford progeria. Science 300(5628):2055. doi:10.1126/science.1084125, 1084125 [pii]
22. Scaffidi P, Misteli T (2006) Lamin A-dependent nuclear defects in human aging. Science 312(5776):1059–1063. doi:10.1126/science.1127168, 1127168 [pii]
23. Tang Y, Chen Y, Jiang H, Nie D (2010) Promotion of tumor development in prostate cancer by progerin. Cancer Cell Int 10:47. doi:10.1186/1475-2867-10-47, 1475-2867-10-47 [pii]
24. Zink D, Fischer AH, Nickerson JA (2004) Nuclear structure in cancer cells. Nat Rev Cancer 4(9):677–687. doi:10.1038/nrc1430, nrc1430 [pii]
25. Bussolati G, Marchio C, Gaetano L, Lupo R, Sapino A (2008) Pleomorphism of the nuclear envelope in breast cancer: a new approach to an old problem. J Cell Mol Med 12(1):209–218. doi:10.1111/j.1582-4934.2007.00176.x, JCMM176 [pii]
26. Elston CW, Ellis IO (1991) Pathological prognostic factors in breast cancer. I. The value of histological grade in breast cancer: experience from a large study with long-term follow-up. Histopathology 19(5):403–410
27. Boyd J, Pienta KJ, Getzenberg RH, Coffey DS, Barrett JC (1991) Preneoplastic alterations in nuclear morphology that accompany loss of tumor suppressor phenotype. J Natl Cancer Inst 83(12):862–866
28. Bloom HJ, Richardson WW (1957) Histological grading and prognosis in breast cancer; a study of 1409 cases of which 359 have been followed for 15 years. Br J Cancer 11(3): 359–377
29. Giardina C, Caniglia DM, D'Aprile M, Lettini T, Serio G, Cipriani T, Ricco R, Pesce Delfino V (1996) Nuclear morphometry in squamous cell carcinoma (SCC) of the tongue. Eur J Cancer B Oral Oncol 32B(2):91–96
30. Nativ O, Sabo E, Bejar J, Halachmi S, Moskovitz B, Miselevich I (1996) A comparison between histological grade and nuclear morphometry for predicting the clinical outcome of localized renal cell carcinoma. Br J Urol 78(1):33–38
31. Partin AW, Gearhart JP, Leonard MP, Leventhal BG, Yoo JK, Crooks D, Epstein JI, Beckwith JB (1993) The use of nuclear morphometry to predict prognosis in pediatric urologic malignancies: a review. Med Pediatr Oncol 21(3):222–229
32. Epstein JI, Berry SJ, Eggleston JC (1984) Nuclear roundness factor. A predictor of progression in untreated Stage A2 prostate cancer. Cancer 54(8):1666–1671
33. Mohler JL, Figlesthaler WM, Zhang XZ, Partin AW, Maygarden SJ (1994) Nuclear shape analysis for the assessment of local invasion and metastases in clinically localized prostate carcinoma. Cancer 74(11):2996–3001
34. Partin AW, Steinberg GD, Pitcock RV, Wu L, Piantadosi S, Coffey DS, Epstein JI (1992) Use of nuclear morphometry, gleason histologic scoring, clinical stage, and age to predict disease-free survival among patients with prostate cancer. Cancer 70(1):161–168
35. Liu CQ, Sasaki H, Fahey MT, Sakamoto A, Sato S, Tanaka T (1999) Prognostic value of nuclear morphometry in patients with TNM stage T1 ovarian clear cell adenocarcinoma. Br J Cancer 79(11–12):1736–1741. doi:10.1038/sj.bjc.6690276
36. Asioli S, Bussolati G (2009) Emerin immunohistochemistry reveals diagnostic features of nuclear membrane arrangement in thyroid lesions. Histopathology 54(5):571–579. doi:10.1111/j.1365-2559.2009.03259.x, HIS3259 [pii]
37. Asioli S, Maletta F, Pacchioni D, Lupo R, Bussolati G (2010) Cytological detection of papillary thyroid carcinomas by nuclear membrane decoration with emerin staining. Virchows Arch 457(1):43–51. doi:10.1007/s00428-010-0910-z
38. Batistatou A, Scopa CD (2009) Pathogenesis and diagnostic significance of nuclear grooves in thyroid and other sites. Int J Surg Pathol 17(2):107–110. doi:10.1177/1066896908316071, 1066896908316071 [pii]

39. Bussolati G (2008) Proper detection of the nuclear shape: ways and significance. Rom J Morphol Embryol 49(4):435–439, 490408435439 [pii]
40. Elisei R, Romei C, Vorontsova T, Cosci B, Veremeychik V, Kuchinskaya E, Basolo F, Demidchik EP, Miccoli P, Pinchera A, Pacini F (2001) RET/PTC rearrangements in thyroid nodules: studies in irradiated and not irradiated, malignant and benign thyroid lesions in children and adults. J Clin Endocrinol Metab 86(7):3211–3216
41. Pisarchik AV, Ermak G, Fomicheva V, Kartel NA, Figge J (1998) The ret/PTC1 rearrangement is a common feature of Chernobyl-associated papillary thyroid carcinomas from Belarus. Thyroid 8(2):133–139
42. Santoro M, Sabino N, Ishizaka Y, Ushijima T, Carlomagno F, Cerrato A, Grieco M, Battaglia C, Martelli ML, Paulin C et al (1993) Involvement of RET oncogene in human tumours: specificity of RET activation to thyroid tumours. Br J Cancer 68(3):460–464
43. Fischer AH, Bond JA, Taysavang P, Battles OE, Wynford-Thomas D (1998) Papillary thyroid carcinoma oncogene (RET/PTC) alters the nuclear envelope and chromatin structure. Am J Pathol 153(5):1443–1450. doi:10.1016/S0002-9440(10)65731-8, S0002-9440(10)65731-8 [pii]
44. Wynford-Thomas D (1997) Origin and progression of thyroid epithelial tumours: cellular and molecular mechanisms. Horm Res 47(4–6):145–157
45. Fischer AH, Taysavang P, Weber CJ, Wilson KL (2001) Nuclear envelope organization in papillary thyroid carcinoma. Histol Histopathol 16(1):1–14
46. Franke WW, Schinko W (1969) Nuclear shape in muscle cells. J Cell Biol 42(1):326–331
47. Lane BP (1965) Alterations in the cytologic detail of intestinal smooth muscle cells in various stages of contraction. J Cell Biol 27(1):199–213
48. Dey P (2010) Cancer nucleus: morphology and beyond. Diagn Cytopathol 38(5):382–390. doi:10.1002/dc.21234
49. Luo RX, Postigo AA, Dean DC (1998) Rb interacts with histone deacetylase to repress transcription. Cell 92(4):463–473, S0092-8674(00)80940-X [pii]
50. Robertson KD, Ait-Si-Ali S, Yokochi T, Wade PA, Jones PL, Wolffe AP (2000) DNMT1 forms a complex with Rb, E2F1 and HDAC1 and represses transcription from E2F-responsive promoters. Nat Genet 25(3):338–342. doi:10.1038/77124
51. Nielsen SJ, Schneider R, Bauer UM, Bannister AJ, Morrison A, O'Carroll D, Firestein R, Cleary M, Jenuwein T, Herrera RE, Kouzarides T (2001) Rb targets histone H3 methylation and HP1 to promoters. Nature 412(6846):561–565. doi:10.1038/35087620, 35087620 [pii]
52. Vandel L, Nicolas E, Vaute O, Ferreira R, Ait-Si-Ali S, Trouche D (2001) Transcriptional repression by the retinoblastoma protein through the recruitment of a histone methyltransferase. Mol Cell Biol 21(19):6484–6494
53. Marshall WF (2002) Order and disorder in the nucleus. Curr Biol 12(5):R185–R192, S0960982202007248 [pii]
54. Pickersgill H, Kalverda B, de Wit E, Talhout W, Fornerod M, van Steensel B (2006) Characterization of the Drosophila melanogaster genome at the nuclear lamina. Nat Genet 38(9):1005–1014. doi:10.1038/ng1852, ng1852 [pii]
55. Guelen L, Pagie L, Brasset E, Meuleman W, Faza MB, Talhout W, Eussen BH, de Klein A, Wessels L, de Laat W, van Steensel B (2008) Domain organization of human chromosomes revealed by mapping of nuclear lamina interactions. Nature 453(7197):948–951. doi:10.1038/nature06947, nature06947 [pii]
56. Parada L, Misteli T (2002) Chromosome positioning in the interphase nucleus. Trends Cell Biol 12(9):425–432, S0962892402023516 [pii]
57. Cremer T, Cremer M, Dietzel S, Muller S, Solovei I, Fakan S (2006) Chromosome territories–a functional nuclear landscape. Curr Opin Cell Biol 18(3):307–316. doi:10.1016/j.ceb.2006.04.007, S0955-0674(06)00056-1 [pii]
58. Gandhi MS, Stringer JR, Nikiforova MN, Medvedovic M, Nikiforov YE (2009) Gene position within chromosome territories correlates with their involvement in distinct rearrangement types in thyroid cancer cells. Genes Chromosomes Cancer 48(3):222–228. doi:10.1002/gcc.20639
59. Heng HH, Goetze S, Ye CJ, Liu G, Stevens JB, Bremer SW, Wykes SM, Bode J, Krawetz SA (2004) Chromatin loops are selectively anchored using scaffold/matrix-attachment regions. J Cell Sci 117(Pt 7):999–1008. doi:10.1242/jcs.00976, 117/7/999 [pii]

60. Cai S, Han HJ, Kohwi-Shigematsu T (2003) Tissue-specific nuclear architecture and gene expression regulated by SATB1. Nat Genet 34(1):42–51. doi:10.1038/ng1146 ng1146 [pii]
61. Schwartz GH, Patnaik A, Hammond LA, Rizzo J, Berg K, Von Hoff DD, Rowinsky EK (2003) A phase I study of bizelesin, a highly potent and selective DNA-interactive agent, in patients with advanced solid malignancies. Ann Oncol 14(5):775–782
62. Goulbourne CN, Malhas AN, Vaux DJ (2011). The induction of a nucleoplasmic reticulum by prelamin A accumulation requires CTP:phosphocholine cytidylyltransferase-alpha. J Cell Sci 124 (Pt 24):4253–4266. doi:jcs.091009 [pii]10.1242/jcs.091009
63. Leighl NB, Dent S, Clemons M, Vandenberg TA, Tozer R, Warr DG, Crump RM, Hedley D, Pond GR, Dancey JE, Moore MJ (2008) A Phase 2 study of perifosine in advanced or metastatic breast cancer. Breast Cancer Res Treat 108(1):87–92. doi:10.1007/s10549-007-9584-x
64. Su JS, Woods SM, Ronen SM (2012) Metabolic consequences of treatment with AKT inhibitor perifosine in breast cancer cells. NMR Biomed 25(2):379–388. doi:10.1002/nbm.1764
65. Helfand BT, Wang Y, Pfleghaar K, Shimi T, Taimen P, Shumaker DK (2012) Chromosomal regions associated with prostate cancer risk localize to lamin B-deficient microdomains and exhibit reduced gene transcription. J Pathol 226(5):735–745. doi:10.1002/path.3033
66. Helfand BT, Loeb S, Cashy J, Meeks JJ, Thaxton CS, Han M, Catalona WJ (2008) Tumor characteristics of carriers and noncarriers of the deCODE 8q24 prostate cancer susceptibility alleles. J Urol 179(6):2197–2201. doi:10.1016/j.juro.2008.01.110, discussion 2202. S0022-5347(08)00257-7 [pii]
67. Staebler A, Karberg B, Behm J, Kuhlmann P, Neubert U, Schmidt H, Korsching E, Burger H, Lelle R, Kiesel L, Bocker W, Shih Ie M, Buchweitz O (2006) Chromosomal losses of regions on 5q and lack of high-level amplifications at 8q24 are associated with favorable prognosis for ovarian serous carcinoma. Genes Chromosomes Cancer 45(10):905–917. doi:10.1002/gcc.20356
68. Hewitt SL, High FA, Reiner SL, Fisher AG, Merkenschlager M (2004) Nuclear repositioning marks the selective exclusion of lineage-inappropriate transcription factor loci during T helper cell differentiation. Eur J Immunol 34(12):3604–3613. doi:10.1002/eji.200425469
69. Chuang CH, Carpenter AE, Fuchsova B, Johnson T, de Lanerolle P, Belmont AS (2006) Long-range directional movement of an interphase chromosome site. Curr Biol 16(8):825–831. doi:10.1016/j.cub.2006.03.059, S0960-9822(06)01356-X [pii]
70. Malhas A, Lee CF, Sanders R, Saunders NJ, Vaux DJ (2007) Defects in lamin B1 expression or processing affect interphase chromosome position and gene expression. J Cell Biol 176(5):593–603. doi:10.1083/jcb.200607054, jcb.200607054 [pii]
71. Ruault M, Dubarry M, Taddei A (2008) Re-positioning genes to the nuclear envelope in mammalian cells: impact on transcription. Trends Genet 24(11):574–581. doi:10.1016/j.tig.2008.08.008, S0168-9525(08)00228-X [pii]
72. Heessen S, Fornerod M (2007) The inner nuclear envelope as a transcription factor resting place. EMBO Rep 8(10):914–919. doi:10.1038/sj.embor.7401075, 7401075 [pii]
73. Gonzalez JM, Navarro-Puche A, Casar B, Crespo P, Andres V (2008) Fast regulation of AP-1 activity through interaction of lamin A/C, ERK1/2, and c-Fos at the nuclear envelope. J Cell Biol 183(4):653–666. doi:10.1083/jcb.200805049, jcb.200805049 [pii]
74. Malhas AN, Lee CF, Vaux DJ (2009) Lamin B1 controls oxidative stress responses via Oct-1. J Cell Biol 184(1):45–55. doi:10.1083/jcb.200804155, jcb.200804155 [pii]
75. Malhas A, Saunders NJ, Vaux DJ (2010) The nuclear envelope can control gene expression and cell cycle progression via miRNA regulation. Cell Cycle 9(3):531–539, 10511 [pii]
76. Bandres E, Cubedo E, Agirre X, Malumbres R, Zarate R, Ramirez N, Abajo A, Navarro A, Moreno I, Monzo M, Garcia-Foncillas J (2006) Identification by Real-time PCR of 13 mature microRNAs differentially expressed in colorectal cancer and non-tumoral tissues. Mol Cancer 5:29. doi:10.1186/1476-4598-5-29, 1476-4598-5-29 [pii]
77. Slaby O, Svoboda M, Fabian P, Smerdova T, Knoflickova D, Bednarikova M, Nenutil R, Vyzula R (2007) Altered expression of miR-21, miR-31, miR-143 and miR-145 is related to clinicopathologic features of colorectal cancer. Oncology 72(5–6):397–402. doi:10.1159/000113489, 000113489 [pii]

78. Valastyan S, Reinhardt F, Benaich N, Calogrias D, Szasz AM, Wang ZC, Brock JE, Richardson AL, Weinberg RA (2009) A pleiotropically acting microRNA, miR-31, inhibits breast cancer metastasis. Cell 137(6):1032–1046. doi:10.1016/j.cell.2009.03.047, S0092-8674(09)00390-0 [pii]
79. Valastyan S, Chang A, Benaich N, Reinhardt F, Weinberg RA (2011) Activation of miR-31 function in already-established metastases elicits metastatic regression. Genes Dev 25(6): 646–659. doi:10.1101/gad.2004211, 25/6/646 [pii]
80. Szafranska AE, Davison TS, John J, Cannon T, Sipos B, Maghnouj A, Labourier E, Hahn SA (2007) MicroRNA expression alterations are linked to tumorigenesis and non-neoplastic processes in pancreatic ductal adenocarcinoma. Oncogene 26(30):4442–4452. doi:10.1038/sj.onc.1210228, 1210228 [pii]

Mechanisms of Nuclear Size Regulation in Model Systems and Cancer

Predrag Jevtić and Daniel L. Levy

Abstract Changes in nuclear size have long been used by cytopathologists as an important parameter to diagnose, stage, and prognose many cancers. Mechanisms underlying these changes and functional links between nuclear size and malignancy are largely unknown. Understanding mechanisms of nuclear size regulation and the physiological significance of proper nuclear size control will inform the interplay between altered nuclear size and oncogenesis. In this chapter we review what is known about molecular mechanisms of nuclear size control based on research in model experimental systems including yeast, *Xenopus*, *Tetrahymena*, *Drosophila*, plants, mice, and mammalian cell culture. We discuss how nuclear size is influenced by DNA ploidy, nuclear structural components, cytoplasmic factors and nucleocytoplasmic transport, the cytoskeleton, and the extracellular matrix. Based on these mechanistic insights, we speculate about how nuclear size might impact cell physiology and whether altered nuclear size could contribute to cancer development and progression. We end with some outstanding questions about mechanisms and functions of nuclear size regulation.

Keywords Nuclear size • Cancer nucleus • Nuclear scaling • Developmental scaling • *Xenopus* • DNA ploidy • Nucleocytoplasmic transport • Nuclear lamina • Linker of nucleoskeleton and cytoskeleton (LINC) complexes • Endoplasmic reticulum • Reticulons • Extracellular matrix

P. Jevtić • D.L. Levy (✉)
Department of Molecular Biology, University of Wyoming,
1000 E. University Avenue, Laramie, WY 82071, USA
e-mail: pjevtic@uwyo.edu; dlevy1@uwyo.edu

E.C. Schirmer and J.I. de las Heras (eds.), *Cancer Biology and the Nuclear Envelope*, Advances in Experimental Medicine and Biology 773, DOI 10.1007/978-1-4899-8032-8_25,

Abbreviations

ABD	Actin binding domain
CRC	Colorectal cancer
ECM	Extracellular matrix
ER	Endoplasmic reticulum
HCC	Hepatocellular carcinoma
INM	Inner nuclear membrane
kuk	Kugelkern
LINC	Linker of nucleoskeleton and cytoskeleton
MAC	Macronucleus
MBT	Midblastula transition
MIC	Micronucleus
NE	Nuclear envelope
NLS	Nuclear localization signal
NPC	Nuclear pore complex
Rtn	Reticulon
SCLC	Small-cell lung carcinoma

Introduction

Cancers are heterogeneous and exhibit diverse biological properties that usually arise from and are characterized by genetic instability. This heterogeneity includes variations in function (pleiotropism) and structure (pleomorphism) [1]. Among structural changes, the most common observed in tumor cells are alterations in nuclear size and shape, chromatin organization, nucleolar size and number, and perinucleolar space [2, 3]. Changes in nuclear size have long been used by cytopathologists as an important parameter to stage and prognose cancer of the prostate [4–8], breast [9–12], lung [13, 14], skin [15], ovary [16], pancreas [17] bladder [18, 19], liver [20], and thyroid [21] (Fig. 1a–c). Aggressive metastatic tumors generally exhibit enlarged nuclei and routine preoperative treatment of breast cancer with anti-estrogen therapy led to nuclear size reduction in tumors [22]. However, it is not clear if increased nuclear size can promote tumorigenesis nor how nuclear size and malignant phenotype might be functionally linked. Changes in nuclear size are

Fig. 1 (continued) visualized with a nuclear membrane marker Cut11-GFP (*left panel*) or by DNA staining with DAPI (*right panel*). In this example, a massive increase in DNA ploidy does not affect the size of the nucleus. Image adapted with permission from [48]. (**g**) Nuclei were isolated from different stage *X. laevis* embryos and visualized by immunofluorescence using an antibody against the NPC (mAb414). Stage 3=4-cell embryo, stage 6=32-cell embryo, stage 8=MBT embryo (~4,000 cells), stage 12=gastrula, stage 20=neurula. Dramatic reductions in nuclear size are observed as development proceeds without any changes in DNA content. Image adapted with permission from [51]

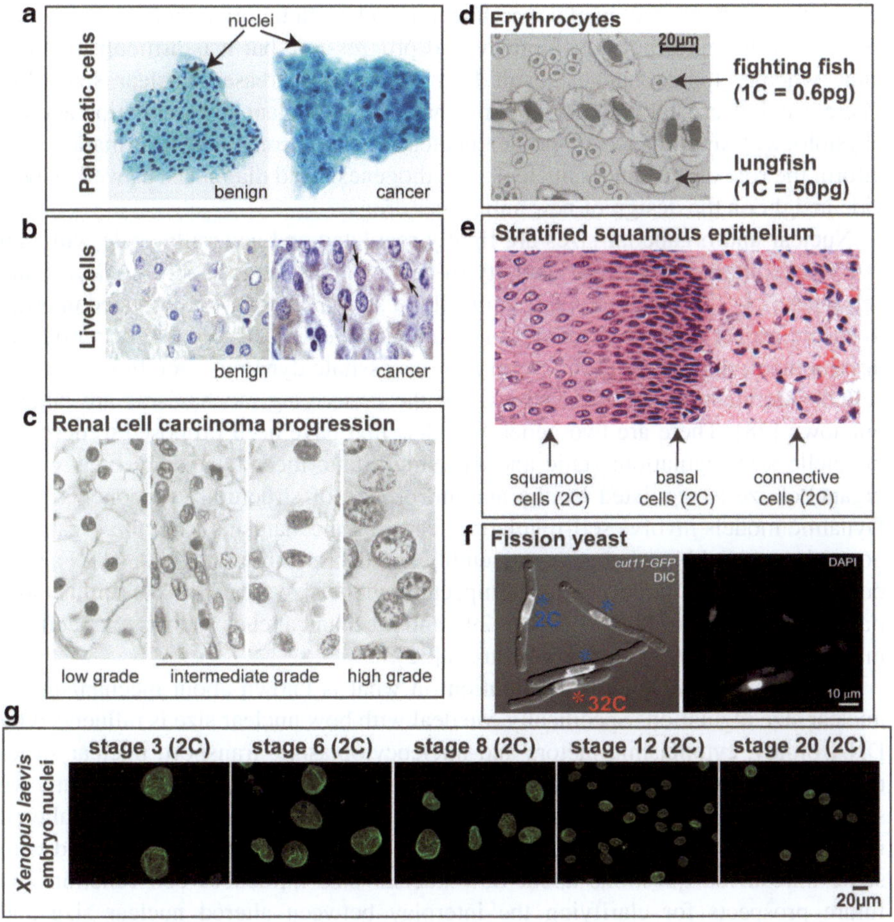

Fig. 1 Relationships between DNA ploidy and nuclear size. (**a**) Benign and cancerous pancreatic cells are shown at the same magnification, stained using the Papanicolaou technique. Nuclei (darker staining) are clearly enlarged in the cancer, though the ploidy status was not documented. Image adapted with permission from [17]. (**b**) Cells from normal human liver and hepatocellular cancer tissue are shown at the same magnification using hematoxylin and eosin (H and E) stain. Again, cancer cells exhibit enlarged nuclei. Image adapted with permission from [233]. (**c**) Cells from different stages of renal cell carcinoma are shown at the same magnification using H and E stain. Progressive nuclear enlargement is evident. Image adapted with permission from [234]. (**d**) Photomicrographs of Feulgen-stained erythrocytes from Siamese fighting fish (*Betta splendens*) and Australian lungfish (*Neoceratodus forsteri*). The lungfish genome is approximately 100 times larger than that of the fighting fish and has larger cells and nuclei, providing an example where DNA content and nuclear size correlate. Image adapted with permission from [235]. (**e**) Esophageal stratified squamous epithelium was stained with H and E. Different layers of the tissue exhibit dramatic differences in nuclear size despite having the same DNA content. Image adapted from http://commons.wikimedia.org/wiki/File%3APancreatic_acinar_metaplasia_-_high_mag.jpg (Copyright © 2011 Nephron, made available under the terms of the GNU Free Documentation License Version 1.2 at http://www.gnu.org/copyleft/fdl.html). (**f**) Fission yeast cells were either arrested with 2C DNA content (*blue asterisk*) or induced to over-replicate their DNA 16-fold (*red asterisk*). Nuclei were

associated with altered nuclear architecture, chromatin organization, and gene expression during tumor development and progression, but it is difficult to determine whether these changes result from or cause increased nuclear size [23]. Therefore, it is crucial to understand mechanisms of nuclear size regulation and the physiological significance of proper nuclear size control. Such information will inform the role of the cancer nucleus in pathogenesis and diagnosis, as well as provide insight on the design of new anticancer drugs.

Nuclear and organellar sizes are tightly regulated and generally scale with cell size to accommodate a great diversity of cell volumes ranging across species and cell types [24–27]. Size scaling of intracellular structures must be regulated in order to ensure proper physiological cell function. Changes in cell size that occur during development, division, and differentiation necessitate dynamic alterations of intracellular architecture and dimensions, but the underlying mechanisms are largely unknown [28]. There are two general models that have been proposed to describe organelle size regulation: static and dynamic. Static mechanisms are ones where organelle size is regulated by the amount or size of structural components [29]. Dynamic models involve self-regulating feedback mechanisms from the organelle, governing assembly or balancing assembly and disassembly rates [30, 31]. Factors outside the general models that can impact organelle size are cell cycle timing, spatial limitations imposed by cell size [29], and mechanical cues such as extracellular matrix (ECM) stiffness, strain, and stress [32].

In this review we focus our attention on what is known about mechanisms of nuclear size regulation. Specifically, we deal with how nuclear size is influenced by DNA ploidy, cytoplasmic factors and nucleocytoplasmic transport, nuclear structural components, and the cytoskeleton. Based on these mechanistic insights, we speculate about how nuclear size might impact cell physiology and whether altered nuclear size could contribute to cancer development and progression. We end with some unresolved questions about how nuclear size influences cell function and future prospects for clarifying the interplay between altered nuclear size and oncogenesis.

Ploidy and Nuclear Size

Cells from different species range over a 1,000-fold in size. Recognized over a century ago and supported by abundant data is the strong positive correlation between genome and cell size [26, 27, 33, 34]. Across the five orders of magnitude spanning eukaryotic genome contents, organisms with larger genomes tend to have larger cells. Interestingly, no relationship exists between genome size and the number of coding genes or organismal complexity. This observation, termed the "C-value paradox," and possible explanations for how it arose have been reviewed elsewhere [25, 33].

Nuclear size scales with cell size over a wide range of species and cell types [24, 25, 27], and cell shape also influences nuclear shape via a perinuclear actin cap [35]. Since genome and cell size tend to correlate, it is generally true that species with larger genomes possess larger nuclei (Fig. 1d). While this suggests that genome size [25, 36, 37] or ploidy [38, 39] might determine nuclear size, it is important to note that these are only correlations. Diverse cell types within the same multicellular organism exhibit cell and nuclear size differences despite possessing the same nuclear DNA [40] (Fig. 1e). Embryonic development and cell differentiation are also associated with dramatic cell and nuclear size changes that are independent of alterations in genome content [41–44]. Nuclear volume in cardiac myocytes increases during normal physiological growth without any changes in DNA content [45], and uncoupling of nuclear size and DNA content is observed during cell differentiation in some plants [46]. In addition to these empirical examples, studies experimentally manipulating DNA content in yeast and *Xenopus* also support the idea that ploidy is not the only determinant of nuclear size.

Yeast Nuclear Size and DNA Content

Yeast offers a useful genetic system for investigating nuclear size regulation. In growing *Saccharomyces cerevisiae* and *Schizosaccharomyces pombe* cells, it was observed that nuclear size increases with cell size throughout the cell cycle maintaining roughly the same karyoplasmic ratio, the ratio of nuclear to cytoplasmic volumes [47, 48]. Importantly there was no abrupt increase in nuclear volume during the S phase of the cell cycle as might be expected if bulk DNA directly accounted for nuclear size, and a similar result was observed in growing HeLa cells [49]. Though diploid yeast cells and nuclei were proportionately larger than haploids, the impact of ploidy on nuclear size is likely indirect, as evidenced by the fact that nuclear size was unaffected in a mutant fission yeast strain induced to over-replicate its DNA 16-fold [48] (Fig. 1f). Taken together, DNA content does not appear to directly determine yeast nuclear size but may set a minimum to the size of the nucleus.

Xenopus Extracts Provide a Model System to Study Nuclear Size Regulation

Xenopus egg extracts represent a powerful biochemical system to investigate mechanisms of nuclear size regulation. These extracts constitute undiluted cytoplasms that are amenable to biochemical manipulation and have been extensively used to study nuclear assembly and import [50]. Since intact cells are not present, essential processes can be studied in an in vitro setting where proteins of interest are immunodepleted or inhibited with specific antibodies or compounds, and recombinant proteins can be added. Nuclei are easily visualized by adding fluorescent membrane dyes or labeled nuclear proteins, and live time-lapse microscopy enables functional assays such as monitoring nuclear import and expansion kinetics. Moreover, sufficient material can be obtained for biochemical experiments and protein purification.

Nuclear size regulation can be effectively studied using two related *Xenopus* frog species: *Xenopus laevis* and *Xenopus tropicalis*. *X. laevis* frogs, cells, eggs, and nuclei are larger than those of *X. tropicalis*. *Xenopus* egg extracts contain all the cytoplasmic components necessary to assemble nuclei in vitro, but lack the egg chromosomes, so assembly is initiated by addition of an exogeneous chromatin source, usually demembranated *Xenopus* sperm. Cell cycle control during meiosis in *Xenopus* differs significantly from lower eukaryotes in that laid amphibian eggs are arrested in metaphase-II of the meiotic cell cycle and are therefore easily manipulated to generate cell-free extracts arrested in defined stages of the cell cycle. *X. tropicalis* egg extracts are also capable of reconstituting the fundamental cell cycle events that have made *X. laevis* egg extracts so useful in elucidating molecular mechanisms of development and cell biology in the absence of the cell. We use the term "interspecies scaling" to refer to comparisons between *X. laevis* and *X. tropicalis*.

Nuclear scaling was recapitulated using *X. laevis* and *X. tropicalis* egg extracts [51]. Addition of the same *X. laevis* chromatin to each extract formed nuclei that expanded more rapidly in *X. laevis* egg cytoplasm than in *X. tropicalis* cytoplasm, generating nuclei with over twice the nuclear envelope (NE) area. Mixing experiments using extracts from these two frog species produced a graded effect on nuclear size, demonstrating that titratable cytoplasmic factors determine the size of the nucleus in this system. Importantly, when nuclei were assembled with *X. tropicalis* sperm that have approximately half the DNA content of *X. laevis* sperm, nuclear size was only minimally smaller, showing that cytoplasm has a greater effect on nuclear size than bulk DNA content.

Xenopus Development Provides a Model System to Study Nuclear Size Regulation

Early embryo development represents another robust cellular scaling system because cell divisions are rapid with no overall growth in the size of the embryo itself. In *X. laevis* the initially ~1.2 mm diameter fertilized egg undergoes 12 rapid, synchronous cell divisions (each approximately 30 min), to produce about 4,000 50 μm cells at the midblastula transition (MBT), also referred to as stage 8.5 [52]. As the embryo proceeds through gastrulation (stages 10–12), further reductions in cell size occur, reaching 12 μm in the tadpole [53]. While *Xenopus* egg extracts represent the traditional model system, *Xenopus* embryo extracts can be manipulated and analyzed similarly to egg extracts [51]. Different stage embryo extracts can be mixed and fractionated, and endogenous embryonic nuclei can be examined in their native cytoplasm. Changes in scaling factor abundance, localization, and activity can be monitored by protein immunoblot and immunofluorescence. A transcriptomics study comparing *X. laevis* and *X. tropicalis* development facilitates analysis of experiments using different stage embryo extracts [54]. Furthermore, microinjected *Xenopus* embryos enable in vivo functional studies of organelle size regulation. Single-cell fertilized *X. laevis* embryos can be microinjected with mRNA to ectopically express proteins of interest, morpholino oligonucleotides to inhibit target mRNA expression, or inhibitory antibodies, as well as with fluorescent markers to visualize cells and organelles.

Nuclear size was observed to decrease throughout early embryonic development in both *X. laevis* and *X. tropicalis*, providing another system to investigate nuclear scaling in the absence of DNA ploidy changes [51, 55] (Fig. 1g). Experimentally halving the DNA content in *X. laevis* embryonic nuclei only reduced NE surface area by 10 %, demonstrating that, like egg cytoplasm, embryo cytoplasm determines nuclear size to a greater extent than ploidy. Thus, evidence from many systems supports the notion that DNA amount is not the primary determinant of nuclear size, yet it remains to be seen what role chromatin compaction plays in nuclear size regulation.

Cancer Cell Nuclear Size and Ploidy

As already discussed, ploidy and nuclear size generally correlate across species, and when comparing the same cell type, cells with higher ploidy are generally larger and contain larger nuclei [56, 57]. Many diploid organisms possess cell types that undergo endoreplication during their normal development, producing polyploid nuclei that tend to be larger than their diploid counterparts [58, 59], though ploidy and cell size are not necessarily coupled in plants [58, 60, 61]. Consistent with these observations, increased ploidy is frequently observed in cancer cells with enlarged nuclei, including cancer of the prostate [6], breast [9], ovary [16], thyroid [62, 63], and a variety of malignant small round-cell tumors [64].

However, it is clear that ploidy is not the sole determinant of nuclear size in a variety of different experimental systems. We have discussed relevant examples in yeast and *Xenopus*, as well as cited nuclear size variation in different cell types from the same species. Ploidy-independent increases in nuclear size are also observed in cancer cells, and in fact the correlation between nuclear volume and ploidy in cancer cells is frequently weak or nonexistent [65]. For example, many malignant tumors are diploid but still exhibit nuclear morphological changes [2, 66]. A lack of correlation between nuclear size and ploidy has been reported in cancers of the bladder [18, 19, 67], colon [68], breast [69], lung [70], skin [71], cervix [72], and prostate [73], as well as in tissue culture cells derived from colon cancers and T-cell leukemia [74]. Furthermore, oncogene activation in cultured cells is sufficient to cause nuclear enlargement without changes in DNA ploidy, for instance by RET tyrosine kinase activation in thyroid carcinoma [75] and p300 expression in prostate cancer cells [76].

Chromatin organization is frequently altered in malignant nuclei, and altered ploidy is one possible explanation. Although the underlying causes are largely unknown and could be varied, increased nuclear size, even in diploid tumor cells, might arise from changes in chromatin structure [2]. We propose that when it comes to nuclear size and ploidy in cancer, it may be informative to distinguish between cancers where aneuploidy impacts nuclear size versus those where nuclear size increases independently of ploidy. The latter category may be elucidated by the types of mechanistic studies on nuclear size regulation reviewed in this chapter. Furthermore, whether nuclear size changes in a given cancer are ploidy-mediated may influence strategies for diagnosis and treatment.

Regulation of Nuclear Size by Cytoplasmic Factors

It has long been observed that the karyoplasmic ratio is maintained at a constant value in normal cells [24, 77], and classic transplantation experiments support models where cytoplasmic volume and components regulate nuclear size. Heterokaryons formed by fusing hen erythrocytes with HeLa cells resulted in expansion of the erythrocyte nucleus, changes in its chromatin organization, and reactivation of DNA synthesis and transcription [78]. Similarly, somatic nuclei grew when injected into *X. laevis* eggs or oocytes [79, 80]. Experimental manipulation of sea snail, *Crepidula plana*, embryos demonstrated that cytoplasmic volume, and not cell size per se, determined nuclear size, as nuclei exposed to larger volumes of cytoplasm grew bigger than nuclei within less cytoplasm [81]. In some cell types it has been observed that nuclear size correlates with the ratio of RNA–DNA, expression of ribosomal genes, and general transcription rate [82, 83]. Models invoking limiting pools of cytoplasmic components might be sufficient to account for organelle size, number, and scaling [29]. More recent studies offer mechanistic insight into the cytoplasmic factors involved in nuclear scaling, further supporting the notion that cytoplasm is the predominant determinant of nuclear size, with DNA bulk setting a minimum nuclear size [47, 48, 51, 84].

Cytoplasmic Volume Regulates Yeast Nuclear Size

Yeast cell size can be readily altered by mutation or varied growth conditions. Taking this approach, it was demonstrated that budding yeast *S. cerevisiae* cells with increased size possessed large nuclei while nuclei were smaller in small cells, such that a constant karyoplasmic ratio was maintained [47]. Interestingly, the size of the nucleolus also scaled with cell and nuclear size. Varying growth rate and blocking nuclear export both failed to uncouple cell and nuclear size regulation. In another study, mutations upregulating phospholipid synthesis caused proliferation of NE and endoplasmic reticulum (ER) membrane, but a wild-type karyoplasmic ratio was maintained [85]. Mechanisms that account for these observed scaling relationships remain to be uncovered, nevertheless, these data suggest that nuclear size in budding yeast is determined by cytoplasmic volume.

Similar studies were undertaken in fission yeast. Across a 35-fold range of *S. pombe* cell sizes, a constant karyoplasmic volume ratio was maintained (Fig. 2a), and nucleolar size also scaled proportionately with nuclear size [48]. To show how cytoplasmic volume affects nuclear size, a cytokinesis mutant was used to generate multinucleate cells. Within the same cell, nuclei surrounded by a greater proportional amount of cytoplasm were larger than nuclei exposed to smaller cytoplasmic volumes (Fig. 2b). To uncouple cell and nuclear growth, nuclei in multinucleate cells were displaced by centrifugation and then cells were allowed to divide,

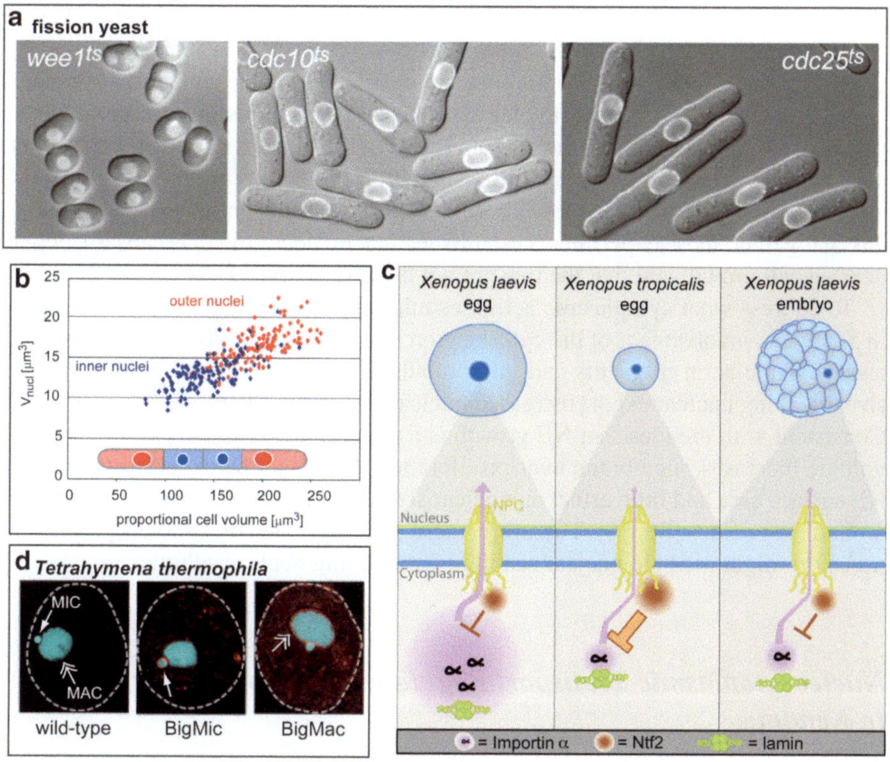

Fig. 2 Cytoplasmic volume and nuclear import contribute to nuclear size regulation in a variety of experimental systems. (**a**) Nuclear size was examined in different fission yeast cell size mutants using the nuclear envelope marker Cut11-GFP. The Wee1 kinase inhibits cyclin-dependent kinase and mitotic entry, so the temperature sensitive mutant *wee1*^{ts} accelerates mitotic entry resulting in small cells. The Cdc10 protein acts early in G_1 to regulate commitment to the mitotic cell cycle, so the temperature-sensitive mutant *cdc10*^{ts} delays progression past G_1 leading to larger cells. The Cdc25 phosphatase activates cyclin-dependent kinase by removing inhibitory phosphates, so the temperature-sensitive mutant *cdc25*^{ts} exhibits a mitotic delay resulting in increased cell size. Larger cells have correspondingly larger nuclei such that a constant karyoplasmic ratio is maintained. Image adapted with permission from [48]. (**b**) Multinucleate fission yeast cells were generated using cytokinesis mutant *cdc11*^{ts}. Nuclear volumes were plotted relative to proportional cell volumes, defined as cytoplasmic volumes separated by the midpoints of two adjacent nuclear centers or the cell end. Nuclei exposed to larger volumes of cytoplasm are proportionately larger. Image adapted with permission from [48]. (**c**) Interspecies and developmental nuclear scaling in *Xenopus* are summarized. Nuclei assembled in *X. laevis* egg extract are larger than nuclei assembled in *X. tropicalis* egg extract, correlating with differences in nuclear import between the two extracts. Importin α levels positively regulate nuclear size by controlling bulk import rates, while Ntf2 levels negatively regulate nuclear size by slowing large cargo translocation through the nuclear pore complex. These two factors are sufficient to account for interspecies nuclear scaling in *Xenopus*, and lamin B3 is one importin α cargo required for regulating nuclear size. During early *Xenopus* development, nuclear size scales smaller. Reductions in bulk import and importin α levels correlate with reductions in nuclear size up to the MBT, and ectopic expression of importin α by mRNA microinjection in the early embryo leads to increased nuclear size at the MBT. Image adapted with permission from [51]. (**d**) Wild type and nucleoporin (Nup98A)-chimera-expressing *T. thermophila* cells are shown at the same magnification. DNA was stained with DAPI (*blue*). A *single arrowhead* indicates the MIC (micronucleus) and a *double arrowhead* indicates the MAC (macronucleus). The cell in the *middle panel* is expressing micronucleus-localized mCherry-BigMic (*red*). The cell in the far *right panel* is expressing macronucleus-localized mCherry-BigMac (*red*). This example demonstrates how variable NPC composition and import within the same cell can dictate nuclear size. Image adapted with permission from [95]

generating cells with variable karyoplasmic ratios. In cells with proportionately larger amounts of cytoplasm, nuclei expanded rapidly to restore a normal karyoplasmic ratio, indicating that the rate of growth of the nucleus can exceed that of the cell. Nuclear growth was arrested in cells with high karyoplasmic ratios until cell size increased sufficiently, showing nuclei do not shrink under these conditions. Taken together, these experiments established that nuclear growth is causally dependent on cell growth and that the two are not directly coupled [48].

To address what cytoplasmic activities might be responsible for nuclear scaling in *S. pombe*, involvement of the cytoskeleton and nucleocytoplasmic transport were tested. While actin filaments and microtubules were not required to scale nuclear size, blocking nuclear export increased nuclear size and the karyoplasmic ratio [48]. Consistent with the idea that NE growth is a result rather than the cause of nuclear volume increases, membrane over-proliferation that might be expected to increase NE surface area had little effect on nuclear size, and instead NE sheets accumulated around the nucleus [86, 87]. Therefore, these data suggested that nucleocytoplasmic transport might mediate nuclear scaling by regulating nuclear volume.

Nucleocytoplasmic Transport Regulates Nuclear Size in Xenopus

Consistent with findings in yeast, cytoplasm and nucleocytoplasmic transport also scale nuclear size in metazoans. Studies of interspecies scaling in *Xenopus* again proved fruitful, demonstrating that differential nuclear import rates can determine nuclear size. Nuclear transport pathways have been extensively reviewed elsewhere [88, 89]. Nuclear import rates were found to differ between *X. laevis* and *X. tropicalis* egg extracts, with *X. laevis* nuclei exhibiting faster import than *X. tropicalis* nuclei. Notably, the levels of importin α and Ntf2 differed between the two extracts, and modulating the levels of these two proteins was nearly sufficient to account for the differences in import and nuclear size between the two species [51]. While importin α levels positively modulated bulk import rates, Ntf2 acted by reducing import of large cargo molecules (Fig. 2c). Lamin B was demonstrated to be one of the imported cargos mediating these nuclear size differences, consistent with lamin B3 depletion from egg extracts blocking nuclear growth [51, 90]. The nuclear lamins are intermediate filament proteins that form a meshwork on the NE nucleoplasmic face and are important for chromatin organization and for providing mechanical strength to the nucleus [91]. These experiments provided mechanistic insight into an example of interspecies nuclear scaling and highlighted a physiologically relevant role for nuclear import and components of nuclear structure like the lamina.

As with interspecies scaling, developmental nuclear scaling in *Xenopus* was also shown to involve changes in nuclear import capacity and the lamins. Reductions in bulk import and cytoplasmic importin α levels correlated with reductions in nuclear size at the MBT, and nuclear size in the embryo was sensitive to importin α levels as its ectopic expression in MBT embryos increased nuclear size [51] (Fig. 2c). At later developmental stages expression of both importin α and lamin B was necessary to increase nuclear size, demonstrating that lamins also become limiting. It is worth noting that this import pathway cannot fully account for nuclear size changes in the post-MBT embryo. One possibility is that nuclear pore complex (NPC) number limits nuclear growth during *Xenopus* developmental progression. However, comparisons between different organisms and cell types suggested that NPC number is regulated independently of nuclear volume and surface area [92], revealing an inverse relationship between nuclear volume and NPC density [93]. Furthermore, NPC number and nuclear size could be uncoupled in mammalian cells where inhibition of cyclin-dependent kinase blocked interphase NPC assembly but not nuclear growth [49]. Open questions remain about what roles nuclear export or membrane availability might play in *Xenopus* post-MBT nuclear size regulation, as well as other novel scaling mechanisms.

Differential Nuclear Import of Histones Regulates **Tetrahymena thermophila** *Nuclear Size*

Tetrahymena thermophila offers a unique system in which to investigate mechanisms of nuclear size control within the same cell. *Tetrahymena* is a ciliated protozoan that possesses two morphologically and functionally distinct nuclei. The micronucleus (MIC) is small, transcriptionally inert, and contains a diploid genome originating from the zygote. The macronucleus (MAC) on the other hand is much larger, being generated by programmed DNA rearrangements and amplifications. Linker histone H1 was shown to be one factor determining nuclear size in this system. Different H1 isoforms are specifically targeted to each nucleus, and deleting one nucleus-specific isoform enlarged that nucleus without affecting the size of the other [94]. It was demonstrated that NPC differences in the MAC and MIC nuclei determine correct H1 targeting. Nucleoporin domain swapping experiments showed that MAC nuclei contain NPCs composed of nucleoporin isoforms that mediate transport of a certain subset of importin α transport receptors responsible for importing MAC-specific H1. On the other hand, NPCs in MIC nuclei recognize a different set of importin α isoforms that import MIC-specific H1 [95, 96] (Fig. 2d). How H1 itself affects nuclear size, whether through altered chromatin structure or gene expression, is an open question. Interestingly, this example illustrates a different way in which nuclear import can contribute to nuclear scaling, through altered NPC composition, and is consistent with other studies in which manipulating the composition of the NPC caused concomitant changes in nuclear size [97, 98].

Might Altered Nucleocytoplasmic Transport in Cancer Cells Account for Increased Nuclear Size?

Nuclear import plays a central role in both cell differentiation and transformation. Switching expression of importin α isoforms and alterations in NPC composition are involved in the differentiation of embryonic stem cells into disparate cell lineages [99–101]. Lymphocyte activation is also characterized by changes in importin α expression and subcellular localization [102]. Altered nuclear transport is frequently observed in transformed cells. For instance, importin α expression is a reliable biomarker for aggressive cancers as its levels are elevated in metastatic breast [103, 104] and non-small-cell lung cancer [105]. In addition, changes in the nuclear transport machinery or NPC can mislocalize transcription factors or oncoproteins contributing to cellular transformation [106–109]. Nevertheless, while trends have been observed, the cause and effect relationships between altered nuclear size and transport in cancer remain to be elucidated.

Nuclear import is emerging as a common mechanism for regulating nuclear size. In some systems, control is mediated by the import factors themselves, determined by which isoforms are expressed and at what levels. In other systems, the composition of the NPC is important, modulating the identity and size of cargos that are imported. Given these mechanistic insights into nuclear size regulation, a particularly intriguing hypothesis is that altered nuclear import in some cancers might account for their increased nuclear size. Although this idea awaits direct experimental testing, it has been proposed that viable therapeutic approaches may be to target the nuclear transport machinery [107, 110] and nuclear size [111].

The Nuclear Lamina in Nuclear Size Regulation and Cancer

Structural components of the NE likely play an important role in defining nuclear size. While the outer NE is continuous with the ER, the inner NE is lined with the nuclear lamina and chromatin on the nucleoplasmic side. The nuclear lamina is a meshwork of lamin intermediate filaments important for chromatin organization, providing mechanical strength to the nucleus [112, 113], and regulating transcription and DNA metabolism [114]. Expression of different lamin isoforms is developmentally regulated [115], with the four major vertebrate lamin proteins being lamins B1 and B2 (coded by two separate genes, *LMNB1* and *LMNB2*) and lamins A and C (alternatively spliced products of the lamin A gene, *LMNA*). Lamins C2 and B3 are germ-cell-specific lamins alternatively spliced from *LMNA* and *LMNB2*, respectively [116].

Lamin and Lamin-Associated Proteins Regulate Nuclear Envelope Growth

Lamin proteins are good candidates as regulators of nuclear size. We have already described nuclear size control in *Xenopus* that is regulated by lamin B3 import, the major egg lamin required for NE growth [51, 90, 117]. Lamins also regulate nuclear size during early *Xenopus* development, where ectopic expression of importin α and lamin B3 increased nuclear size at the MBT. During frog, chicken, and mouse development, lamin expression patterns change [118–120] and different isoforms are expressed in different cell types [121, 122]. It will be interesting to test the idea that the type of lamin expressed might differentially control nuclear size. A tantalizing clue comes from experiments comparing the effects on NE growth of overexpressing lamins with and without a C-terminal CaaX box, a motif targeted for post-translational farnesylation and localization to the inner nuclear membrane (INM). Expression of lamins containing a CaaX motif in *Xenopus* A6 cells induced NE proliferation, but not if this site was mutated or when lamins naturally lacking this motif were expressed [123].

INM proteins like lamina-associated polypeptide 2 (LAP2), emerin, and MAN1 (LEM-2) can contribute to nuclear size regulation due to their interactions with lamins and influence on lamina organization. Knocking down INM and lamin proteins in *Caenorhabditis elegans* resulted in small nuclei and a concomitant defect in centrosome attachment [124]. The LAP2 family includes seven alternatively spliced mammalian proteins [125] and LAP2 isoforms contribute to chromatin and lamina organization [126]. Microinjection of the lamin-binding domain of LAP2β into HeLa cells strongly inhibited nuclear expansion, suggesting that LAP2 functions in growth of the nuclear lamina and NE [127]. Similar effects of LAP2β fragments on NE growth were observed in an in vitro *Xenopus* system [128], and nuclear size was also reduced upon immunodepletion of a LAP2-associated protein, TPX2 [129]. Lamin B receptor has also been implicated in NE membrane growth, through its sterol reductase activity and/or lamina interactions [130]. These data further support the idea that nuclear lamina dynamics regulate NE growth.

Lamin-Like Proteins Regulate Nuclear Envelope Growth

Though not all organisms possess canonical nuclear lamins, many have lamin-like proteins that contain lamin structural motifs (including coiled coil, nuclear localization signal [NLS], and C-terminal farnesylation sequences) and can influence nuclear size. In early *Drosophila melanogaster* development, significant nuclear growth occurs during cellularization with a 2.5-fold increase in nuclear length. A NE component similar to lamins, kugelkern (kuk), was shown to be responsible for this nuclear elongation and expansion [131]. The expression of kuk is upregulated at the time of cellularization, coinciding with nuclear expansion. The kuk protein exclusively localizes to the NE, and its overexpression in embryos resulted in ruffled and abnormally shaped nuclei likely due to extensive NE expansion. Ectopic kuk expressed in *Xenopus* tissue culture cells localized to the NE and caused significant nuclear growth, which was diminished when the coiled coil, NLS, or farnesylation sequences were deleted.

Although there are no obvious yeast lamin orthologs, there are yeast proteins that may serve a homologous function to the nuclear lamina in higher eukaryotes. Fission yeast LEM-domain protein (LEM, Lap2/emerin/Man1) Lem2 is an integral nuclear membrane protein important for NE integrity and chromatin organization, and Lem2 overexpression induces NE proliferation [132, 133]. Budding yeast Heh1 and Heh2 are also LEM-family proteins that localize to the INM [134]. Lamin-like proteins have been identified in *Dictyostelium* [135] and *Trypanosoma brucei* [136], and two homologous *Arabidopsis thaliana* proteins, encoded by *LITTLE NUCLEI* genes (*LINC1/2*), localize to the nuclear periphery. Reducing *LINC1/2* expression levels resulted in smaller nuclei, with the double mutant having an additive effect producing nuclei with half the volume of wild-type [137]. *A. thaliana* seeds possess characteristically small nuclei, dependent on the seed maturation regulator *ABI3*, and increases in nuclear size that occur during germination require *LINC1/2* [138]. Thus, these lamin-like proteins behave similarly to bona fide lamins and play similar roles in regulating nuclear size.

Do Lamins Regulate Nuclear Size in Cancer Cells?

In many tumors, aberrant lamin expression or localization correlate with a vast range of cellular processes involved in tumor progression, such as control of nuclear size and shape, chromatin organization, gene expression, apoptosis, and senescence [139]. The level of altered lamin expression depends on the subtype of cancer, its aggressiveness, proliferative capacity, and degree of differentiation [140]. For example, some lung, intestinal, and skin squamous cell carcinomas exhibit reduced B-type lamin expression [141–143], while lamin B1 expression is frequently elevated in hepatocellular carcinoma (HCC) and prostate cancer [139]. In HCC, lamin B1 expression levels correlated with tumor stage, and circulating levels of lamin B1 could be successfully used as a marker for detection of early stage disease [144].

A-type lamins are also differentially expressed in cancers. In general, lamin A/C expression is low or absent in poorly differentiated, highly proliferative tumors, for instance in basal and squamous cell carcinomas [140, 143], although elevated levels of lamin A/C were detected in ovarian cancers [145] and were associated with increased mortality in colorectal cancer [146]. Another example is small-cell lung carcinoma (SCLC) exhibiting reduced lamin A/C levels relative to non-SCLC, which might contribute to the nuclear structural differences used to differentiate between these two cancers [142, 147]. Expression of lamina-associated proteins can also vary in tumors. As one example, LAP2β levels were significantly increased in colorectal adenocarcinoma, ovarian carcinoma, and SCLC, which may account for the increased size of these cancerous nuclei [148].

Given established mechanistic links between nuclear size and lamins, it is possible that changes in lamin expression levels and isoforms might determine the size and shape of cancer nuclei. This, in turn, could influence nuclear structure, chromatin organization, gene positioning, and gene expression, contributing to malignant progression. Future research is necessary to directly test these hypotheses. Nonetheless, lamin levels and localization patterns in tumors are often useful as biomarkers for diagnosis and prognosis of different cancers [23]. For instance, immunofluorescence staining for lamin B and emerin was used to assess NE pleomorphism in breast cancer and found to be superior to classical histological grading procedures (Fig. 3a) [149].

Perinuclear Structures that Influence Nuclear Size

The Cytoskeleton and Nuclear Size

Linker of nucleoskeleton and cytoskeleton (LINC) complexes consist of families of proteins that connect the nuclear lamina with the cytoskeleton through the NE [150]. These connections are mediated by interactions between INM SUN-domain containing proteins, including Sun1/2, and outer nuclear membrane KASH-domain proteins that directly or indirectly bind actin, such as Nesprins [151]. LINC complexes perform diverse cellular functions, including roles in cell division, regulating nucleus–centrosome attachments, and nuclear migration and anchorage [152]. There is growing evidence that these proteins are also important for regulating nuclear size.

Nesprin-2 Giant regulates nuclear size and epidermal thickness in mice [153]. Knockout epidermis lacking the N-terminal actin binding domain (ABD) of Nesprin-2 Giant showed no difference in proliferation and differentiation compared to wild-type; however, fibroblasts and keratinocytes from knockout mice exhibited large and severely misshapen nuclei with mislocalized emerin, reduced function in a wound healing assay, and slower rates of cell migration. Significant thickening of the epidermis was also observed, possibly as a consequence of increased nuclear size. Therefore, it appears that Nesprin-2 Giant function relies on its ability to maintain NE morphology and regulate nuclear size through interactions with the lamina and actin cytoskeleton.

Various Nesprin isoforms are able to associate with one another, and these interactions contribute to nuclear size regulation. The ABDs of Nesprin-1/-2 interact with Nesprin-3 in human keratinocyte cells (HaCaT cells). Overexpressing either Nesprin-2 ABD or Nesprin-2 C-terminal KASH domain increased nuclear size. Conversely, overexpression of Nesprin-2 mini, a construct composed of both the ABD and KASH domain but lacking most of the centrally located spectrin repeats, decreased nuclear size in HaCaT cells [154] (Fig. 3b). This confirmed the idea that the number of spectrin repeats within the Nesprin-2 rod domain is important in regulating nuclear size. The observed Nesprin-2 mediated nuclear size changes were accompanied by analogous cellular size changes suggesting that nuclear and cell size are concomitantly regulated. Nuclear size reductions caused by Nesprin-2 mini overexpression were further enhanced by Nesprin-3 co-expression or actin depolymerization. The model proposed from these data was that Nesprin interchain interactions and their links with the cytoskeleton form a belt-like filamentous network that covers the outer surface of the nucleus to regulate size (Fig. 3c). A similar model for nuclear size regulation can be invoked for the nuclear lamina acting on the nucleoplasmic face of the NE.

LINC complex proteins have been implicated in oncogenesis. In a screen for genes that are mutated at high frequency in cancer, *SYNE2* (*NESPRIN-2*) was altered in 3 out of 11 breast cancers examined, and *SYNE1* (*NESPRIN-1*) was mutated in 5 out of 11 colorectal cancers (CRCs) [155]. CRC progression is characterized by transcriptional silencing of a set of genes by promoter CpG island hypermethylation. *SYNE1* is one such gene that exhibited promoter methylation in all 102

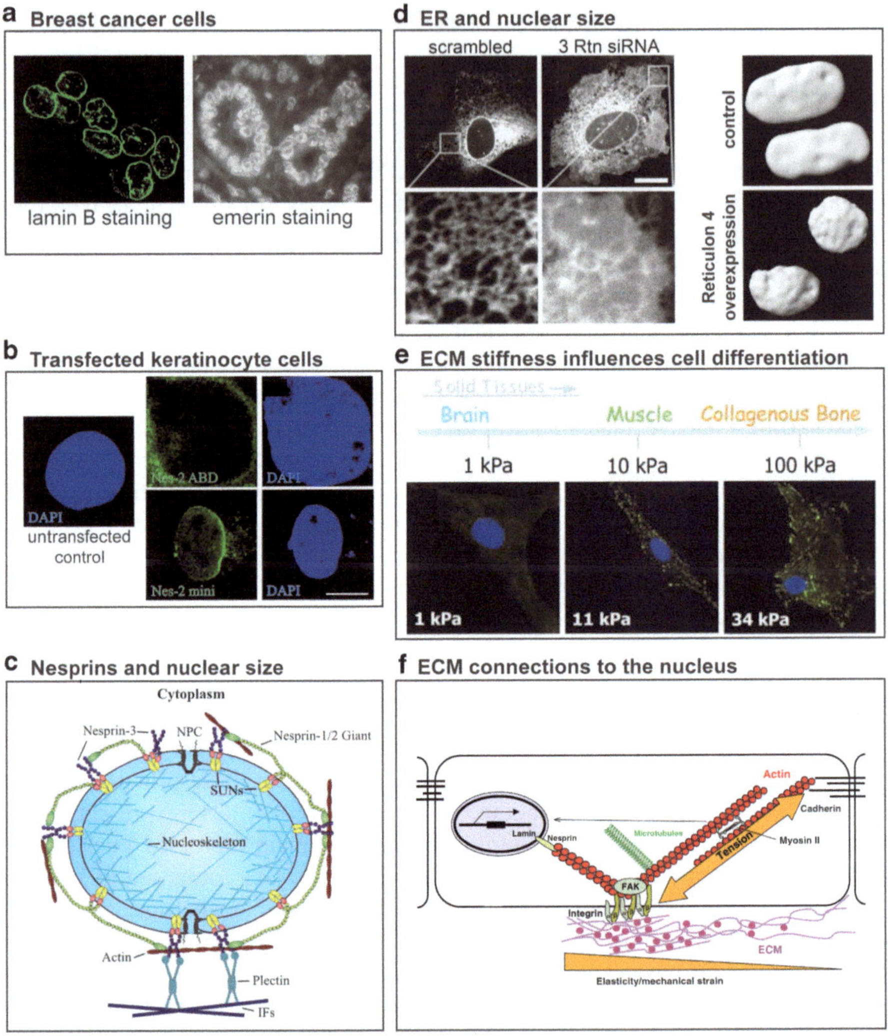

Fig. 3 Components of the nuclear envelope and perinuclear structures influence nuclear size and morphology. (**a**) The *left panel* shows nuclei visualized by fluorescent tagging of membrane-associated lamin B from BT474 breast cancer cells, revealing NE irregularities and intranuclear tubules. The *right panel* shows nuclei stained by immunofluorescence for emerin in a histologically low-grade ductal carcinoma, with NE invaginations and indentations that could not be observed by H and E stain. Note that these images are intended to depict NE pleomorphism in cancer cells, and as such the two images are not shown at the same magnification. Images adapted with permissions from [149]. (**b**) HaCaT cells transiently transfected with the actin-binding domain of Nesprin-2 fused to GFP (Nes-2 ABD; *top two right panels*) exhibit larger nuclei than untransfected controls (*left panel*). Transient transfection of Nesprin-2 lacking the spectrin repeats and fused to GFP (Nes-2 mini; *bottom panels*) reduces nuclear size. GFP (*green*) and DNA DAPI (*blue*) staining are shown. Scale bar, 10 μm. Images adapted with permission from [154]. (**c**) The model depicts how different Nesprin isoforms interact with each other and the cytoskeleton. Alignment of Nesprin-1/-2 Giant isoforms at the surface of the NE might form an

analyzed CRC samples, indicating epigenetic modification of *SYNE1* is important in CRC development [156]. Genotyping single nucleotide polymorphisms in a large population-based case-control study identified a correlation between *SYNE1/NESPRIN-1* gene polymorphisms and increased risk of invasive ovarian cancer [157]. Furthermore, Nesprin-1 expression was downregulated 20- to 180-fold in a majority of ovarian and mammary tumors, as well as in early stage carcinomas and metastases of the uterus, cervix, kidney, lung, thyroid, and pancreas. Thus, reduced Nesprin expression could serve as a biomarker for early tumor detection [158]. Taken together, these studies suggest possible links between Nesprin alterations, nuclear morphological changes, and cancer cell migration and invasiveness.

The Endoplasmic Reticulum and Nuclear Size

In addition to the cytoskeleton, another perinuclear structure that may play a role in regulating nuclear size is the ER. The ER is an interconnected network of lipid bilayer tubules and sheets that is continuous with the NE. Molecules responsible for shaping the ER include proteins in the reticulon (Rtn) and REEP families that form ER tubules, likely by inserting a hydrophobic wedge into lipid bilayers to induce membrane curvature [159, 160]. These proteins were sufficient in vitro to shape liposomes into tubules [161]. Rtns are also important for stabilizing high curvature membranes at the edges of ER sheets, with Climp63 acting as a spacer to set sheet width [162]. Atlastins fuse ER tubules to generate the three-way junctions characteristic of ER morphology [163, 164].

Fig. 3 (continued) extranuclear filamentous network that functions as a molecular "belt" to set nuclear size. IFs refer to intermediate filaments. Image adapted with permission from [154]. (**d**) U2OS cells were transfected with Sec61-GFP (to visualize the ER and NE) and scrambled RNA oligos ("scrambled") or siRNA oligos directed against Rtn1, Rtn3, and Rtn 4 ("3 Rtn siRNA"). Reticulon knockdown reduces ER tubule formation in favor of ER sheets, accelerates NE formation, and increases nuclear size. Scale bar, 20 μm. The panels to the right show nuclei in U2OS cells, visualized with GFP-NLS. Nuclear surfaces were reconstructed from confocal z stacks. Control cells (top) and cells overexpressing V5-Rtn4 (*bottom*) are shown at the same magnification. Images adapted with permission from [166]. (**e**) Embryonic mesenchymal stem cells were grown on neurogenic, soft gels (1 kPa), myogenic gels (11 kPa), and stiff, osteogenic gels (34 kPa). Focal adhesions were visualized with an anti-paxillin antibody (*green*), nuclei are labeled (*blue*), and images are shown at the same magnification. Adhesions grow from undetectable diffuse "contacts" on soft gels to punctate adhesions on stiffer gels, becoming long and thin on the stiffest gels. Nuclear size reductions correlate with increased matrix stiffness. Image adapted with permission from [186]. (**f**) The model depicts the protein connections that link the NE, cytoskeleton, cell periphery, neighboring cells, and ECM. Changes in mechanical forces are transmitted from the ECM and neighboring cells through integrin and cadherin surface receptors, respectively. The actin and microtubule cytoskeleton along with associated motor proteins transduce these forces to the nucleus through connections with NE proteins such as the Nesprins. In turn, the nuclear lamina can modulate nuclear architecture and gene expression. Image adapted with permission from *Development* [185]

The NE is a specialized domain of the ER and as such Rtn levels can impact nuclear size by affecting the relative proportion of ER tubules to NE sheets. Membrane responsible for NE formation and expansion is derived from the ER, so it is possible that ER could become limiting for nuclear growth [165]. In support of this idea, overexpression of Rtns in tissue culture cells increased ER tubulation and concomitantly reduced NE surface area, demonstrating how size control of two organelles can be interconnected [166] (Fig. 3d). The model proposed from these experiments is that a tug-of-war relationship exists between membranes of the ER and NE [166]. This relationship could be modulated by INM proteins that bridge the lamina and chromatin [167] as well as SUN-KASH proteins that link the NE and cytoskeleton [168].

Rtn4a was found to be overexpressed in malignant brain tumor (glioma) cells [169], while Rtn4 Interacting Protein1 (Rtn4IP1) was downregulated in thyroid cancers, correlating with larger primary tumor size and other malignant phenotypes [170]. Rtn4IP1 may influence Rtn4 function and its low expression in thyroid cancer might alter the ability of Rtn4 to shape ER tubules. *RTN1* was upregulated in malignant pancreatic carcinoma [169], diffusely infiltrating gliomas [171], and neuroendocrine tumors [172, 173], correlating with histological grade and prognosis. Increased expression of reticulons may affect nuclear size in these neural and endocrine tumors [174]. For example, high-level reticulon expression could have a dominant negative effect that reduces ER tubule formation in favor of ER sheet and NE expansion. Alternatively, altered reticulon expression might not be linked to increased nuclear size in these cases [21, 175, 176]. Examination of ER morphology in these cancer cells could prove quite informative.

Extracellular Matrix Stiffness, Nuclear Size, and Cancer Progression

Mechanical force from the ECM is transmitted intracellularly from focal adhesions to nuclear LINC complexes and the lamina, via the actin-myosin cytoskeleton [177–181]. Due to the stiffness and elasticity of the nuclear lamina, which functions as a "molecular shock absorber," the nucleus changes its size and shape in response to transduced force, potentially altering chromatin structure, organization, and gene expression [182]. Tissue stiffness regulates cell physiology and morphology during cell migration [183], cell contraction [184], differentiation, tissue and organ development [181, 185], and carcinogenesis [179]. For instance, naive embryonic mesenchymal stem cells transform into neurogenic cells if cultured in vitro using soft collagen-coated gels that mimic brain. Stiffer gels that mimic muscle promote myogenesis, and rigid gels similar to collagenous bone lead to osteogenesis [186] (Fig. 3e).

Integrins and cadherins are major components of focal adhesions that are sometimes altered in cancer. In one example, cell surface and total β1 integrin levels were shown to be higher in a malignant breast cancer cell line compared to the breast epithelial cell line from which it was derived [187]. Upon β1 integrin inhibition, tumor reversal was observed in 3D cultures characterized by the formation of acinar structures, establishment of normal cell-cell contacts, cellular polarization, cell differentiation and growth arrest, and reduced nuclear size. Furthermore, these cells exhibited reduced tumorigenicity when inoculated into nude mice. Thus, cancer cell interactions with the ECM may directly or indirectly influence nuclear size and cancer severity.

Qualitative and quantitative changes in ECM components, focal adhesions, and gene expression impact the bidirectional flow of signal transduction information between the ECM and the nucleus (Fig. 3f). These changes are interconnected and can affect tumor progression and invasiveness. Cancer cells generally display decreased cell-cell contacts and ECM attachments, corresponding to increased proliferation, motility, infiltrative potential, and metastasis [188]. These cells are therefore exposed to reduced ECM stiffness and mechanical force, potentially contributing to the increased nuclear size usually observed in cancer. Future work is necessary to directly test this hypothesis.

Future Directions

A common mechanistic theme in nuclear size regulation is the role of nuclear import and NE structural components, especially the nuclear lamins. Though not all organisms possess canonical lamins, proteins with structural homology to the lamins often function in nuclear size regulation. Screening for proteins with lamin structural motifs, including coiled-coils, NLS, and C-terminal farnesylation sequences, may uncover additional regulators of nuclear size. We have discussed examples where nuclear import factors, NPC and NE components, the ER, and the cytoskeleton contribute to nuclear size. Many questions remain regarding mechanisms and functions of nuclear size control and their relevance to oncogenesis.

Mechanisms of Nuclear Size Regulation

Evidence from many systems supports the notion that DNA amount is not the primary determinant of nuclear size, but it remains to be seen what role chromatin compaction plays in nuclear size regulation. While limiting lamin import can restrict nuclear size, it is unknown how precise concentrations of the import factors and lamins are determined [29]. Another open question is how a steady-state, equilibrium nuclear size is maintained. Perhaps there exist activities that balance NE expansion, such as the tug-of-war relationship between membranes of the ER and NE discussed earlier [166]. Besides nucleocytoplasmic transport and ER/NE balance, lipid homeostasis may also impinge on nuclear size regulation.

Changes in nuclear shape are frequently observed in cancer cells in addition to increased nuclear size. Invaginations of the NE, termed the nucleoplasmic reticulum, are features of normal cells that can become exaggerated in cancer nuclei [189]. Several studies suggest a role for lipid homeostasis in nuclear size and shape regulation. Mutations in chromatin-remodeling [190], integral ER [191], and INM [192, 193] proteins all resulted in NE proliferation that could be reversed by addition of chemicals that increase membrane fluidity, such as benzyl alcohol. A role for vesicular trafficking in nuclear size and shape regulation has also been demonstrated [85]. Nuclear shape regulation has been the subject of several excellent reviews [114, 194] that propose the following particularly intriguing hypothesis linking the regulation of nuclear size and shape. The karyoplasmic ratio is critically important for cell function as it is maintained in response to cell and nuclear size changes and is altered in disease states. Changes in nuclear size may manifest as nuclear shape changes in order for the cell to maintain a constant karyoplasmic ratio [85, 114]. Thus, changes in the shape of cancer cell nuclei may be a manifestation of altered nuclear size.

Nuclear size defects in cancer may result from changes in nuclear composition. For instance, tumor-specific nuclear matrix proteins are expressed in certain bone, prostate, and bladder cancers [195–199], and these proteins are sometimes used as cancer-specific proteomic markers [2, 200]. Mutations in NE components associated with various types of cancer have been thoroughly reviewed elsewhere [23]; however, it remains to be directly tested if and how these changes in NE proteins affect nuclear size. Another outstanding question is how nuclear size is sensed and regulated relative to the size of the cell. Some cancers are associated with changes in the karyoplasmic ratio [201–203], while in other cancers cell and nuclear size increase concomitantly, maintaining a normal karyoplasmic ratio [204–209]. Thus, elucidating mechanisms responsible for scaling nuclear size with cell size may have important implications for disease diagnosis and treatment.

Functions of Nuclear Size Regulation

Another significant question is the functional relevance of nuclear size regulation. We have already discussed several examples where functions for nuclear size control have been demonstrated or proposed (Fig. 4). For instance, during early *C. elegans* development, centrosome attachment to the male pronucleus is critical, and defects in this process have developmental consequences. It was shown that NE surface area dictates centrosome attachment, likely by regulating the access of centrosome microtubules to dynein localized at the NE [124]. During early development in many organisms there are dramatic changes in the karyoplasmic ratio that correlate with the timing of developmental transitions, like the MBT [210–212]. It is tempting to speculate that nuclear size might play a functional role in this process.

What are the functional consequences of altering nuclear size? One hypothesis is that nuclear size directly affects chromatin organization and gene expression [213]. Enlarged nuclei are often observed in cells adjacent to a tumor, and these cells appear otherwise normal by histology [214–218]. As one example, morphologically normal lobules of the mammary gland adjacent to breast cancer were found to possess genetic aberrations characteristic of hyperplastic, premalignant, and malignant

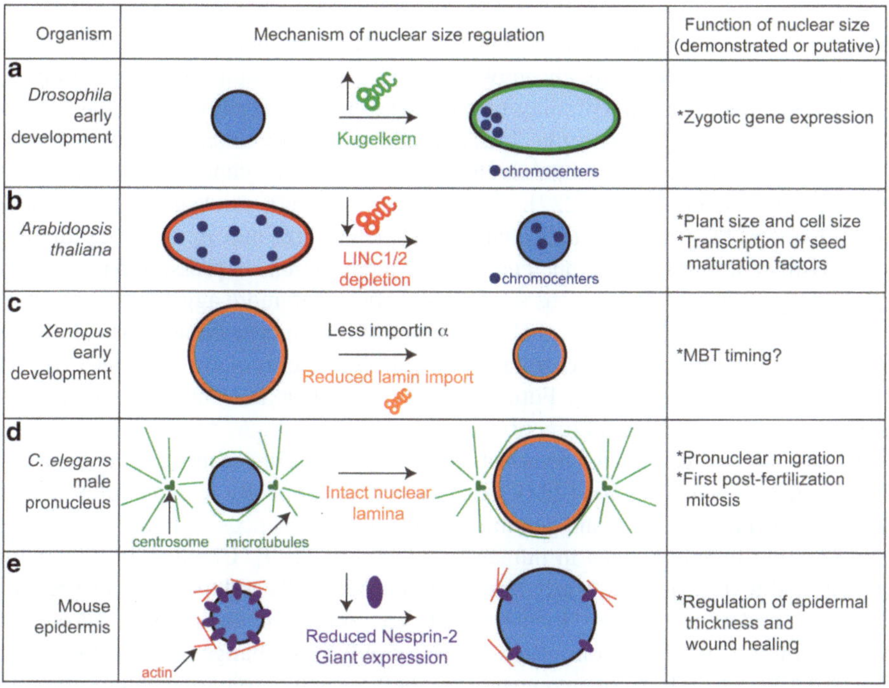

Fig. 4 Functions of nuclear size. In all diagrams, the NE is depicted as a *black circle* or *oval* and chromatin is shaded *blue*. (**a**) Expression of the lamin-like protein Kugelkern (*green*) during *Drosophila* cellularization increases nuclear volume and is important for proper zygotic gene expression. Chromocenters, clusters of centromeric chromatin, are depicted as *dark blue circles* that become apically localized during cellularization [131]. (**b**) Deletion of *Arabidopsis* lamin-like proteins Linc1 and Linc2 (*red*) leads to reductions in nuclear, cell, and plant size, as well as defects in gene expression during seed maturation. Small nuclei also have fewer chromocenters, depicted as *dark blue circles* [137, 138, 236]. (**c**) During early *Xenopus* embryonic development, nuclear size decreases concomitantly with reductions in importin α levels and lamin B3 (*orange*) import [51, 55]. We speculate that nuclear size changes may contribute to developmental timing events. (**d**) The *C. elegans* male pronucleus must have sufficient NE surface area for efficient attachment of microtubules from both centrosomes (*green*). Loss of integrity of the nuclear lamina (*orange*) results in small nuclei, centrosome attachment defects, and failed pronuclear migration and mitosis [124]. (**e**) Nesprin-2 Giant (*purple*) is an outer nuclear membrane protein that contacts the actin cytoskeleton (*red*). Reduced expression of Nesprin-2 Giant in the mouse leads to increased nuclear size and epidermal thickness, as well as defective wound healing [153]. Image adapted with permission from [84]

breast epithelium [219]. One explanation for this "field effect" is that genetic alterations leading to cancer occur in a stepwise fashion, and cells in the field around the tumor represent a clonal population arising from an early genetic change that was a precursor to carcinogenesis. Indeed, evidence supports the idea that many somatic mutations occur prior to tumor initiation [220]. By this model, precancerous mutations that alter nuclear size could be priming events in cancer development. Restructured nuclear morphology would disrupt chromatin positioning, thereby altering transcriptional profiles. This, in turn, could directly contribute to cancer progression or promote additional genetic alterations [66]. Critical questions are if enlarging nuclei in primary cells is sufficient to confer malignant growth, and conversely, if reducing nuclear size in cancer cells mitigates their proliferation.

Within the nuclear space, chromosomes and genes are non-randomly arranged. The three-dimensional positioning of chromatin and its attachments to the lamina influence gene expression profiles [221]. Spatial proximity of specific genes underlies the chromosomal translocations that give rise to many cancers, especially leukemias and lymphomas [222]. How might altered nuclear size promote tumorigenesis? One idea is that changes in nuclear size and morphology disrupt chromatin positioning in the nucleus, thereby altering gene expression patterns and converting a cell into a cancerous state or contributing to more aggressive growth of an already malignant cell [66, 213]. In support of this, increased nuclear volume in the G_1 phase of the cell cycle correlated with decreased chromatin condensation and enhanced proliferation [223]. Future work will address how nuclear size affects subnuclear organization and function.

A number of studies have demonstrated that chromatin organization is dynamic and impacts gene expression [224]. Chromatin is dynamically regulated during cell differentiation [225–227], and genome organization influences cell migration [228] and regional mutation rates in human somatic cells [229]. Changes in chromatin pattern and texture are frequently observed in cancer cells and there are a variety of potential underlying causes including altered chromatin positioning, chromatin remodeling and modifications, altered DNA ploidy, and changes in proteins of the NE and NPC [2, 23]. An open question is whether such chromatin changes directly modulate nuclear size or if altered genome organization influences expression of nuclear scaling factors, thus increasing nuclear size in cancer cells.

Embryo development and cancer progression exhibit many similarities with respect to cell proliferation, cell migration, epigenetic regulation, gene expression, and signaling. Studies suggest that cancer may arise from reactivation of developmental programs that are downregulated after embryogenesis [230–232]. Understanding mechanisms that control developmental nuclear scaling and how nuclear size contributes to developmental progression will inform how nuclear size and function become deregulated in cancer.

Acknowledgements We thank Lisa Edens and Karen White for critical reading of the manuscript. This work was supported by start-up funds from the University of Wyoming. P.J. is supported by a graduate assistantship from the University of Wyoming Agricultural Experiment Station.

References

1. Pienta KJ, Partin AW, Coffey DS (1989) Cancer as a disease of DNA organization and dynamic cell structure. Cancer Res 49(10):2525–2532
2. Dey P (2010) Cancer nucleus: morphology and beyond. Diagn Cytopathol 38(5):382–390
3. Schaff Z, Lapis K, Safrany L (1971) The ultrastructure of primary hepatocellular cancer in man. Virchows Arch A Pathol Pathol Anat 352(4):340–358
4. Rashid F, Ul Haque A (2011) Frequencies of different nuclear morphological features in prostate adenocarcinoma. Ann Diagn Pathol 15(6):414–421

5. Veltri RW, Miller MC, Partin AW, Coffey DS, Epstein JI (1996) Ability to predict biochemical progression using Gleason score and a computer-generated quantitative nuclear grade derived from cancer cell nuclei. Urology 48(5):685–691
6. Veltri RW, Partin AW, Epstein JE, Marley GM, Miller CM, Singer DS, Patton KP, Criley SR, Coffey DS (1994) Quantitative nuclear morphometry, Markovian texture descriptors, and DNA content captured on a CAS-200 Image analysis system, combined with PCNA and HER-2/neu immunohistochemistry for prediction of prostate cancer progression. J Cell Biochem Suppl 19:249–258
7. Khan MA, Walsh PC, Miller MC, Bales WD, Epstein JI, Mangold LA, Partin AW, Veltri RW (2003) Quantitative alterations in nuclear structure predict prostate carcinoma distant metastasis and death in men with biochemical recurrence after radical prostatectomy. Cancer 98(12):2583–2591
8. Blom JH, Ten Kate FJ, Schroeder FH, van der Heul RO (1990) Morphometrically estimated variation in nuclear size. A useful tool in grading prostatic cancer. Urol Res 18(2):93–99
9. Ottesen GL (2003) Carcinoma in situ of the female breast. A clinico-pathological, immunohistological, and DNA ploidy study. APMIS Suppl 108:1–67
10. Tan PH, Goh BB, Chiang G, Bay BH (2001) Correlation of nuclear morphometry with pathologic parameters in ductal carcinoma in situ of the breast. Mod Pathol 14(10):937–941
11. Abdalla F, Boder J, Markus R, Hashmi H, Buhmeida A, Collan Y (2009) Correlation of nuclear morphometry of breast cancer in histological sections with clinicopathological features and prognosis. Anticancer Res 29(5):1771–1776
12. Rajesh L, Dey P, Joshi K (2002) Automated image morphometry of lobular breast carcinoma. Anal Quant Cytol Histol 24(2):81–84
13. Thunnissen FB, Diegenbach PC, van Hattum AH, Tolboom J, van der Sluis DM, Schaafsma W, Houthoff HJ, Baak JP (1992) Further evaluation of quantitative nuclear image features for classification of lung carcinomas. Pathol Res Pract 188(4–5):531–535
14. Thunnissen FB, Diegenbach PC, Baak JP, Houthoff HJ (1988) The influence of variations in the measuring procedure on quantitative nuclear image features in histologic sections of lung tissue. Anal Quant Cytol Histol 10(5):349–354
15. Mossbacher U, Knollmayer S, Binder M, Steiner A, Wolff K, Pehamberger H (1996) Increased nuclear volume in metastasizing "thick" melanomas. J Invest Dermatol 106(3): 437–440
16. Zeimet AG, Fiegl H, Goebel G, Kopp F, Allasia C, Reimer D, Steppan I, Mueller-Holzner E, Ehrlich M, Marth C (2011) DNA ploidy, nuclear size, proliferation index and DNA-hypomethylation in ovarian cancer. Gynecol Oncol 121(1):24–31
17. Taira T, Kawahara A, Yamaguchi T, Abe H, Ishida Y, Okabe Y, Naito Y, Yano H, Kage M (2011) Morphometric image analysis of pancreatic disease by ThinPrep liquid-based cytology. Diagn Cytopathol 40(11):970–975
18. van Velthoven R, Petein M, Oosterlinck WJ, Zandona C, Zlotta A, Van der Meijden AP, Pasteels JL, Roels H, Schulman C, Kiss R (1995) Image cytometry determination of ploidy level, proliferative activity, and nuclear size in a series of 314 transitional bladder cell carcinomas. Hum Pathol 26(1):3–11
19. Helander KG, Tribukait B (1988) Modal DNA values of normal and malignant urothelial cells of the bladder in relation to nuclear size. Anal Quant Cytol Histol 10(2):127–133
20. Yan J, Zhuo S, Chen G, Wu X, Zhou D, Xie S, Jiang J, Ying M, Jia F, Chen J, Zhou J (2012) Preclinical study of using multiphoton microscopy to diagnose liver cancer and differentiate benign and malignant liver lesions. J Biomed Opt 17(2):026004
21. Camargo RY, Tomimori EK, Knobel M, Medeiros-Neto G (2007) Preoperative assessment of thyroid nodules: role of ultrasonography and fine needle aspiration biopsy followed by cytology. Clinics (Sao Paulo) 62(4):411–418
22. Samarnthai N, Elledge R, Prihoda TJ, Huang J, Massarweh S, Yeh IT (2012) Pathologic changes in breast cancer after anti-estrogen therapy. Breast J 18(4):362–366
23. Chow KH, Factor RE, Ullman KS (2012) The nuclear envelope environment and its cancer connections. Nat Rev Cancer 12(3):196–209

24. Wilson EB (1925) The karyoplasmic ratio. In: The cell in development and heredity, 3rd edn. The Macmillan Company, New York, pp 727–733
25. Gregory T (2005) Genome size evolution in animals. In: Gregory T (ed) The evolution of the genome, vol 1. Elsevier Academic Press, London, pp 4–87
26. Animal Genome Size Database (2011) http://www.genomesize.com.
27. Cell Size Database (2011) http://www.genomesize.com/cellsize.
28. Levy DL, Heald R (2012) Mechanisms of intracellular scaling. Annu Rev Cell Dev Biol 28:113–135
29. Goehring NW, Hyman AA (2012) Organelle growth control through limiting pools of cytoplasmic components. Curr Biol 22(9):R330–R339
30. Marshall WF (2008) Controlling size within cells. Semin Cell Dev Biol 19(6):479
31. Marshall WF (2008) Engineering design principles for organelle size control systems. Semin Cell Dev Biol 19(6):520–524
32. Buxboim A, Ivanovska IL, Discher DE (2010) Matrix elasticity, cytoskeletal forces and physics of the nucleus: how deeply do cells "feel" outside and in? J Cell Sci 123(Pt 3):297–308
33. Gregory TR (2001) Coincidence, coevolution, or causation? DNA content, cell size, and the C-value enigma. Biol Rev Camb Philos Soc 76(1):65–101
34. Horner HA, Macgregor HC (1983) C value and cell volume: their significance in the evolution and development of amphibians. J Cell Sci 63:135–146
35. Khatau SB, Hale CM, Stewart-Hutchinson PJ, Patel MS, Stewart CL, Searson PC, Hodzic D, Wirtz D (2009) A perinuclear actin cap regulates nuclear shape. Proc Natl Acad Sci U S A 106(45):19017–19022
36. Baetcke KP, Sparrow AH, Nauman CH, Schwemmer SS (1967) The relationship of DNA content to nuclear and chromosome volumes and to radiosensitivity (LD50). Proc Natl Acad Sci U S A 58(2):533–540
37. Cavalier-Smith T (1982) Skeletal DNA and the evolution of genome size. Annu Rev Biophys Bioeng 11:273–302
38. Fankhauser G (1945) The effects of changes in chromosome number on amphibian development. Q Rev Biol 20(1):20–78
39. Henery CC, Kaufman MH (1992) Relationship between cell size and nuclear volume in nucleated red blood cells of developmentally matched diploid and tetraploid mouse embryos. J Exp Zool 261(4):472–478
40. Altman PL, Katz DD (1976) Cell biology. Federation of American Societies for Experimental Biology, Bethesda, MD
41. Butler JT, Hall LL, Smith KP, Lawrence JB (2009) Changing nuclear landscape and unique PML structures during early epigenetic transitions of human embryonic stem cells. J Cell Biochem 107(4):609–621
42. Oh SK, Kim HS, Ahn HJ, Seol HW, Kim YY, Park YB, Yoon CJ, Kim DW, Kim SH, Moon SY (2005) Derivation and characterization of new human embryonic stem cell lines: SNUhES1, SNUhES2, and SNUhES3. Stem Cells 23(2):211–219
43. Faro-Trindade I, Cook PR (2006) A conserved organization of transcription during embryonic stem cell differentiation and in cells with high C value. Mol Biol Cell 17(7):2910–2920
44. Thomson JA, Itskovitz-Eldor J, Shapiro SS, Waknitz MA, Swiergiel JJ, Marshall VS, Jones JM (1998) Embryonic stem cell lines derived from human blastocysts. Science 282(5391):1145–1147
45. Gerdes AM, Morales MC, Handa V, Moore JA, Alvarez MR (1991) Nuclear size and DNA content in rat cardiac myocytes during growth, maturation and aging. J Mol Cell Cardiol 23(7):833–839
46. Mitchell JP, van der Ploeg M (1982) Nuclear changes accompanying cell differentiation in stems of Pisum sativum L. Histochemistry 75(3):327–340
47. Jorgensen P, Edgington NP, Schneider BL, Rupes I, Tyers M, Futcher B (2007) The size of the nucleus increases as yeast cells grow. Mol Biol Cell 18(9):3523–3532
48. Neumann FR, Nurse P (2007) Nuclear size control in fission yeast. J Cell Biol 179(4): 593–600

49. Maeshima K, Iino H, Hihara S, Funakoshi T, Watanabe A, Nishimura M, Nakatomi R, Yahata K, Imamoto F, Hashikawa T, Yokota H, Imamoto N (2010) Nuclear pore formation but not nuclear growth is governed by cyclin-dependent kinases (Cdks) during interphase. Nat Struct Mol Biol 17(9):1065–1071
50. Chan RC, Forbes DI (2006) In vitro study of nuclear assembly and nuclear import using Xenopus egg extracts. Methods Mol Biol 322:289–300
51. Levy DL, Heald R (2010) Nuclear size is regulated by importin alpha and Ntf2 in Xenopus. Cell 143(2):288–298
52. Nieuwkoop PD, Faber J (1967) Normal table of Xenopus laevis (Daudin), 2nd edn. North-Holland Publishing Company, Amsterdam
53. Montorzi M, Burgos MH, Falchuk KH (2000) Xenopus laevis embryo development: arrest of epidermal cell differentiation by the chelating agent 1,10-phenanthroline. Mol Reprod Dev 55(1):75–82
54. Yanai I, Peshkin L, Jorgensen P, Kirschner MW (2011) Mapping gene expression in two Xenopus species: evolutionary constraints and developmental flexibility. Dev Cell 20(4): 483–496
55. Gerhart JC (1980) Mechanisms regulating pattern formation in the amphibian egg and early embryo. In: Goldberger RF (ed) Biological regulation and development, vol 2. Plenum, New York, pp 133–316
56. Cook M, Tyers M (2007) Size control goes global. Curr Opin Biotechnol 18(4):341–350
57. Fankhauser G (1939) Polyploidy in the salamander, Eurycea bislineata. J Hered 30:379–388
58. Sugimoto-Shirasu K, Roberts K (2003) "Big it up": endoreduplication and cell-size control in plants. Curr Opin Plant Biol 6(6):544–553
59. Lee HO, Davidson JM, Duronio RJ (2009) Endoreplication: polyploidy with purpose. Genes Dev 23(21):2461–2477
60. Leiva-Neto JT, Grafi G, Sabelli PA, Dante RA, Woo YM, Maddock S, Gordon-Kamm WJ, Larkins BA (2004) A dominant negative mutant of cyclin-dependent kinase A reduces endoreduplication but not cell size or gene expression in maize endosperm. Plant Cell 16(7):1854–1869
61. Schnittger A, Weinl C, Bouyer D, Schobinger U, Hulskamp M (2003) Misexpression of the cyclin-dependent kinase inhibitor ICK1/KRP1 in single-celled Arabidopsis trichomes reduces endoreduplication and cell size and induces cell death. Plant Cell 15(2):303–315
62. Liautaud-Roger F, Teyssier JR, Ferre D, Dufer J, Coninx P (1992) Can chromatin texture predict structural karyotypic changes in diploid cells from thyroid cold nodules? Anal Cell Pathol 4(6):421–428
63. Teyssier JR, Liautaud-Roger F, Ferre D, Patey M, Dufer J (1990) Chromosomal changes in thyroid tumors. Relation with DNA content, karyotypic features, and clinical data. Cancer Genet Cytogenet 50(2):249–263
64. Brahmi U, Rajwanshi A, Joshi K, Ganguly NK, Vohra H, Gupta SK, Dey P (2001) Role of immunocytochemistry and DNA flow cytometry in the fine-needle aspiration diagnosis of malignant small round-cell tumors. Diagn Cytopathol 24(4):233–239
65. Sorensen FB (1996) Quantitative analysis of nuclear size for prognosis-related malignancy grading. In: Heppner G, Bittar EE (eds) Advances in oncobiology, vol 1. JAI Press Inc, Greenwich, CT, pp 221–255
66. Zink D, Fischer AH, Nickerson JA (2004) Nuclear structure in cancer cells. Nat Rev Cancer 4(9):677–687
67. Nielsen K, Petersen SE, Orntoft T (1989) A comparison between stereological estimates of mean nuclear volume and DNA flow cytometry in bladder tumours. APMIS 97(10): 949–956
68. Dangou JM, Kiss R, DePrez C, Jeannot MC, Fastrez R, Pasteels JL, Verhest A (1993) Heterogeneity of DNA ploidy, proliferation index and nuclear size in human colorectal carcinomas. Anal Quant Cytol Histol 15(1):23–31
69. Lindberg LG (1992) Nuclear DNA ploidy in mammary carcinomas; using nuclear size as co-parameter reveals more complex patterns. Anal Cell Pathol 4(5):389–394

70. Miyamoto H, Isobe H, Akita H, Ishiguro A, Endo T, Inoue S, Kawakami Y (1992) The flow cytometric nuclear-DNA content, tumor-origin, nuclear size and prognosis in squamous-cell lung-cancer. Int J Oncol 1(3):325–329
71. Sorensen FB, Erlandsen M (1990) Intralesional and metastatic heterogeneity in malignant melanomas demonstrated by stereologic estimates of nuclear volume. Lab Invest 62(5): 646–654
72. Sorensen FB, Bichel P, Jakobsen A (1992) DNA level and stereologic estimates of nuclear volume in squamous cell carcinomas of the uterine cervix. A comparative study with analysis of prognostic impact. Cancer 69(1):187–199
73. Wang N, Stenkvist BG, Tribukait B (1992) Morphometry of nuclei of the normal and malignant prostate in relation to DNA ploidy. Anal Quant Cytol Histol 14(3):210–216
74. Cremer M, Kupper K, Wagler B, Wizelman L, von Hase J, Weiland Y, Kreja L, Diebold J, Speicher MR, Cremer T (2003) Inheritance of gene density-related higher order chromatin arrangements in normal and tumor cell nuclei. J Cell Biol 162(5):809–820
75. Fischer AH, Bond JA, Taysavang P, Battles OE, Wynford-Thomas D (1998) Papillary thyroid carcinoma oncogene (RET/PTC) alters the nuclear envelope and chromatin structure. Am J Pathol 153(5):1443–1450
76. Debes JD, Sebo TJ, Heemers HV, Kipp BR, Haugen DL, Lohse CM, Tindall DJ (2005) p300 modulates nuclear morphology in prostate cancer. Cancer Res 65(3):708–712
77. Huber MD, Gerace L (2007) The size-wise nucleus: nuclear volume control in eukaryotes. J Cell Biol 179(4):583–584
78. Harris H (1967) The reactivation of the red cell nucleus. J Cell Sci 2(1):23–32
79. Merriam RW (1969) Movement of cytoplasmic proteins into nuclei induced to enlarge and initiate DNA or RNA synthesis. J Cell Sci 5:333–349
80. Gurdon JB (1976) Injected nuclei in frog oocytes: fate, enlargement, and chromatin dispersal. J Embryol Exp Morphol 36:523–540
81. Conklin E (1912) Cell size and nuclear size. J Exp Embryol 12:1–98
82. Sato S, Burgess SB, McIlwain DL (1994) Transcription and motoneuron size. J Neurochem 63(5):1609–1615
83. Schmidt EE, Schibler U (1995) Cell size regulation, a mechanism that controls cellular RNA accumulation: consequences on regulation of the ubiquitous transcription factors Oct1 and NF-Y and the liver-enriched transcription factor DBP. J Cell Biol 128(4):467–483
84. Edens LJ, White KH, Jevtic P, Li X, Levy DL (2013) Nuclear size regulation: from single cells to development and disease. Trends Cell Biol 23(4):151–159
85. Webster MT, McCaffery JM, Cohen-Fix O (2010) Vesicle trafficking maintains nuclear shape in Saccharomyces cerevisiae during membrane proliferation. J Cell Biol 191(6):1079–1088
86. Lum PY, Wright R (1995) Degradation of HMG-CoA reductase-induced membranes in the fission yeast, Schizosaccharomyces pombe. J Cell Biol 131(1):81–94
87. Tange Y, Hirata A, Niwa O (2002) An evolutionarily conserved fission yeast protein, Ned1, implicated in normal nuclear morphology and chromosome stability, interacts with Dis3, Pim1/RCC1 and an essential nucleoporin. J Cell Sci 115(Pt 22):4375–4385
88. Stewart M (2007) Molecular mechanism of the nuclear protein import cycle. Nat Rev Mol Cell Biol 8(3):195–208
89. Madrid AS, Weis K (2006) Nuclear transport is becoming crystal clear. Chromosoma 115(2):98–109
90. Newport JW, Wilson KL, Dunphy WG (1990) A lamin-independent pathway for nuclear envelope assembly. J Cell Biol 111(6 Pt 1):2247–2259
91. Misteli T, Spector DL (2011) The nucleus. Cold Spring Harbor Laboratory Press, Cold Spring Harbor, NY
92. Maul GG, Deaven LL, Freed JJ, Campbell GL, Becak W (1980) Investigation of the determinants of nuclear pore number. Cytogenet Cell Genet 26(2–4):175–190
93. Maul GG, Deaven L (1977) Quantitative determination of nuclear pore complexes in cycling cells with differing DNA content. J Cell Biol 73(3):748–760
94. Shen X, Yu L, Weir JW, Gorovsky MA (1995) Linker histones are not essential and affect chromatin condensation in vivo. Cell 82(1):47–56

95. Iwamoto M, Mori C, Kojidani T, Bunai F, Hori T, Fukagawa T, Hiraoka Y, Haraguchi T (2009) Two distinct repeat sequences of Nup98 nucleoporins characterize dual nuclei in the binucleated ciliate tetrahymena. Curr Biol 19(10):843–847
96. Malone CD, Falkowska KA, Li AY, Galanti SE, Kanuru RC, LaMont EG, Mazzarella KC, Micev AJ, Osman MM, Piotrowski NK, Suszko JW, Timm AC, Xu MM, Liu L, Chalker DL (2008) Nucleus-specific importin alpha proteins and nucleoporins regulate protein import and nuclear division in the binucleate Tetrahymena thermophila. Eukaryot Cell 7(9):1487–1499
97. Theerthagiri G, Eisenhardt N, Schwarz H, Antonin W (2010) The nucleoporin Nup188 controls passage of membrane proteins across the nuclear pore complex. J Cell Biol 189(7): 1129–1142
98. Tamura K, Hara-Nishimura I (2011) Involvement of the nuclear pore complex in morphology of the plant nucleus. Nucleus 2(3):168–172
99. Yasuhara N, Shibazaki N, Tanaka S, Nagai M, Kamikawa Y, Oe S, Asally M, Kamachi Y, Kondoh H, Yoneda Y (2007) Triggering neural differentiation of ES cells by subtype switching of importin-alpha. Nat Cell Biol 9(1):72–79
100. Yasuhara N, Oka M, Yoneda Y (2009) The role of the nuclear transport system in cell differentiation. Semin Cell Dev Biol 20(5):590–599
101. D'Angelo MA, Gomez-Cavazos JS, Mei A, Lackner DH, Hetzer MW (2012) A change in nuclear pore complex composition regulates cell differentiation. Dev Cell 22(2):446–458
102. Andrade R, Alonso R, Pena R, Arlucea J, Arechaga J (2003) Localization of importin alpha (Rch1) at the plasma membrane and subcellular redistribution during lymphocyte activation. Chromosoma 112(2):87–95
103. Dahl E, Kristiansen G, Gottlob K, Klaman I, Ebner E, Hinzmann B, Hermann K, Pilarsky C, Durst M, Klinkhammer-Schalke M, Blaszyk H, Knuechel R, Hartmann A, Rosenthal A, Wild PJ (2006) Molecular profiling of laser-microdissected matched tumor and normal breast tissue identifies karyopherin alpha2 as a potential novel prognostic marker in breast cancer. Clin Cancer Res 12(13):3950–3960
104. Gluz O, Wild P, Meiler R, Diallo-Danebrock R, Ting E, Mohrmann S, Schuett G, Dahl E, Fuchs T, Herr A, Gaumann A, Frick M, Poremba C, Nitz UA, Hartmann A (2008) Nuclear karyopherin alpha2 expression predicts poor survival in patients with advanced breast cancer irrespective of treatment intensity. Int J Cancer 123(6):1433–1438
105. Wang CI, Wang CL, Wang CW, Chen CD, Wu CC, Liang Y, Tsai YH, Chang YS, Yu JS, Yu CJ (2010) Importin subunit alpha-2 is identified as a potential biomarker for non-small cell lung cancer by integration of the cancer cell secretome and tissue transcriptome. Int J Cancer 128(10):2364–2372
106. Kau TR, Silver PA (2003) Nuclear transport as a target for cell growth. Drug Discov Today 8(2):78–85
107. Kau TR, Way JC, Silver PA (2004) Nuclear transport and cancer: from mechanism to intervention. Nat Rev Cancer 4(2):106–117
108. Poon IK, Jans DA (2005) Regulation of nuclear transport: central role in development and transformation? Traffic 6(3):173–186
109. Rensen WM, Mangiacasale R, Ciciarello M, Lavia P (2008) The GTPase Ran: regulation of cell life and potential roles in cell transformation. Front Biosci 13:4097–4121
110. Yashiroda Y, Yoshida M (2003) Nucleo-cytoplasmic transport of proteins as a target for therapeutic drugs. Curr Med Chem 10(9):741–748
111. Marshall WF (2012) Organelle size control systems: from cell geometry to organelle-directed medicine. Bioessays 34(9):721–724
112. Gruenbaum Y, Margalit A, Goldman RD, Shumaker DK, Wilson KL (2005) The nuclear lamina comes of age. Nat Rev Mol Cell Biol 6(1):21–31
113. Worman HJ, Courvalin JC (2005) Nuclear envelope, nuclear lamina, and inherited disease. Int Rev Cytol 246:231–279
114. Walters AD, Bommakanti A, Cohen-Fix O (2012) Shaping the nucleus: factors and forces. J Cell Biochem 113(9):2813–2821
115. Dechat T, Adam SA, Taimen P, Shimi T, Goldman RD (2010) Nuclear lamins. Cold Spring Harb Perspect Biol 2(11):a000547

116. Parnaik VK, Chaturvedi P, Muralikrishna B (2011) Lamins, laminopathies and disease mechanisms: possible role for proteasomal degradation of key regulatory proteins. J Biosci 36(3): 471–479
117. Jenkins H, Holman T, Lyon C, Lane B, Stick R, Hutchison C (1993) Nuclei that lack a lamina accumulate karyophilic proteins and assemble a nuclear matrix. J Cell Sci 106(Pt 1): 275–285
118. Stick R, Hausen P (1985) Changes in the nuclear lamina composition during early development of Xenopus laevis. Cell 41(1):191–200
119. Lehner CF, Stick R, Eppenberger HM, Nigg EA (1987) Differential expression of nuclear lamin proteins during chicken development. J Cell Biol 105(1):577–587
120. Rober RA, Weber K, Osborn M (1989) Differential timing of nuclear lamin A/C expression in the various organs of the mouse embryo and the young animal: a developmental study. Development 105(2):365–378
121. Benavente R, Krohne G, Franke WW (1985) Cell type-specific expression of nuclear lamina proteins during development of Xenopus laevis. Cell 41(1):177–190
122. Rober RA, Sauter H, Weber K, Osborn M (1990) Cells of the cellular immune and hemopoietic system of the mouse lack lamins A/C: distinction versus other somatic cells. J Cell Sci 95(Pt 4):587–598
123. Prufert K, Vogel A, Krohne G (2004) The lamin CxxM motif promotes nuclear membrane growth. J Cell Sci 117(Pt 25):6105–6116
124. Meyerzon M, Gao Z, Liu J, Wu JC, Malone CJ, Starr DA (2009) Centrosome attachment to the C. elegans male pronucleus is dependent on the surface area of the nuclear envelope. Dev Biol 327(2):433–446
125. Berger R, Theodor L, Shoham J, Gokkel E, Brok-Simoni F, Avraham KB, Copeland NG, Jenkins NA, Rechavi G, Simon AJ (1996) The characterization and localization of the mouse thymopoietin/lamina-associated polypeptide 2 gene and its alternatively spliced products. Genome Res 6(5):361–370
126. Dechat T, Vlcek S, Foisner R (2000) Review: lamina-associated polypeptide 2 isoforms and related proteins in cell cycle-dependent nuclear structure dynamics. J Struct Biol 129(2–3): 335–345
127. Yang L, Guan T, Gerace L (1997) Lamin-binding fragment of LAP2 inhibits increase in nuclear volume during the cell cycle and progression into S phase. J Cell Biol 139(5): 1077–1087
128. Gant TM, Harris CA, Wilson KL (1999) Roles of LAP2 proteins in nuclear assembly and DNA replication: truncated LAP2beta proteins alter lamina assembly, envelope formation, nuclear size, and DNA replication efficiency in Xenopus laevis extracts. J Cell Biol 144(6):1083–1096
129. O'Brien LL, Wiese C (2006) TPX2 is required for postmitotic nuclear assembly in cell-free Xenopus laevis egg extracts. J Cell Biol 173(5):685–694
130. Olins AL, Rhodes G, Welch DB, Zwerger M, Olins DE (2010) Lamin B receptor: multi-tasking at the nuclear envelope. Nucleus 1(1):53–70
131. Brandt A, Papagiannouli F, Wagner N, Wilsch-Brauninger M, Braun M, Furlong EE, Loserth S, Wenzl C, Pilot F, Vogt N, Lecuit T, Krohne G, Grosshans J (2006) Developmental control of nuclear size and shape by Kugelkern and Kurzkern. Curr Biol 16(6):543–552
132. Gonzalez Y, Saito A, Sazer S (2012) Fission yeast Lem2 and Man1 perform fundamental functions of the animal cell nuclear lamina. Nucleus 3(1):60–76
133. Hiraoka Y, Maekawa H, Asakawa H, Chikashige Y, Kojidani T, Osakada H, Matsuda A, Haraguchi T (2011) Inner nuclear membrane protein Ima1 is dispensable for intranuclear positioning of centromeres. Genes Cells 16(10):1000–1011
134. King MC, Lusk CP, Blobel G (2006) Karyopherin-mediated import of integral inner nuclear membrane proteins. Nature 442(7106):1003–1007
135. Kruger A, Batsios P, Baumann O, Luckert E, Schwarz H, Stick R, Meyer I, Graf R (2012) Characterization of NE81, the first lamin-like nucleoskeleton protein in a unicellular organism. Mol Biol Cell 23(2):360–370

136. DuBois KN, Alsford S, Holden JM, Buisson J, Swiderski M, Bart JM, Ratushny AV, Wan Y, Bastin P, Barry JD, Navarro M, Horn D, Aitchison JD, Rout MP, Field MC (2012) NUP-1 Is a large coiled-coil nucleoskeletal protein in trypanosomes with lamin-like functions. PLoS Biol 10(3):e1001287
137. Dittmer TA, Stacey NJ, Sugimoto-Shirasu K, Richards EJ (2007) LITTLE NUCLEI genes affecting nuclear morphology in Arabidopsis thaliana. Plant Cell 19(9):2793–2803
138. van Zanten M, Koini MA, Geyer R, Liu Y, Brambilla V, Bartels D, Koornneef M, Fransz P, Soppe WJ (2011) Seed maturation in Arabidopsis thaliana is characterized by nuclear size reduction and increased chromatin condensation. Proc Natl Acad Sci U S A 108(50):20219–20224
139. Foster CR, Przyborski SA, Wilson RG, Hutchison CJ (2010) Lamins as cancer biomarkers. Biochem Soc Trans 38(Pt 1):297–300
140. Tilli CM, Ramaekers FC, Broers JL, Hutchison CJ, Neumann HA (2003) Lamin expression in normal human skin, actinic keratosis, squamous cell carcinoma and basal cell carcinoma. Br J Dermatol 148(1):102–109
141. Moss SF, Krivosheyev V, de Souza A, Chin K, Gaetz HP, Chaudhary N, Worman HJ, Holt PR (1999) Decreased and aberrant nuclear lamin expression in gastrointestinal tract neoplasms. Gut 45(5):723–729
142. Broers JL, Raymond Y, Rot MK, Kuijpers H, Wagenaar SS, Ramaekers FC (1993) Nuclear A-type lamins are differentially expressed in human lung cancer subtypes. Am J Pathol 143(1):211–220
143. Oguchi M, Sagara J, Matsumoto K, Saida T, Taniguchi S (2002) Expression of lamins depends on epidermal differentiation and transformation. Br J Dermatol 147(5):853–858
144. Sun S, Xu MZ, Poon RT, Day PJ, Luk JM (2010) Circulating Lamin B1 (LMNB1) biomarker detects early stages of liver cancer in patients. J Proteome Res 9(1):70–78
145. Hudson ME, Pozdnyakova I, Haines K, Mor G, Snyder M (2007) Identification of differentially expressed proteins in ovarian cancer using high-density protein microarrays. Proc Natl Acad Sci U S A 104(44):17494–17499
146. Willis ND, Cox TR, Rahman-Casans SF, Smits K, Przyborski SA, van den Brandt P, van Engeland M, Weijenberg M, Wilson RG, de Bruine A, Hutchison CJ (2008) Lamin A/C is a risk biomarker in colorectal cancer. PLoS One 3(8):e2988
147. Kaufmann SH, Mabry M, Jasti R, Shaper JH (1991) Differential expression of nuclear envelope lamins A and C in human lung cancer cell lines. Cancer Res 51(2):581–586
148. de Las Heras JI, Batrakou DG, Schirmer EC (2013) Cancer biology and the nuclear envelope: a convoluted relationship. Semin Cancer Biol 23(2):125–137
149. Bussolati G, Marchio C, Gaetano L, Lupo R, Sapino A (2008) Pleomorphism of the nuclear envelope in breast cancer: a new approach to an old problem. J Cell Mol Med 12(1):209–218
150. Crisp M, Liu Q, Roux K, Rattner JB, Shanahan C, Burke B, Stahl PD, Hodzic D (2006) Coupling of the nucleus and cytoplasm: role of the LINC complex. J Cell Biol 172(1):41–53
151. Mierke CT (2013) Physical break-down of the classical view on cancer cell invasion and metastasis. Eur J Cell Biol 92(3):89–104
152. Mejat A, Misteli T (2010) LINC complexes in health and disease. Nucleus 1(1):40–52
153. Luke Y, Zaim H, Karakesisoglou I, Jaeger VM, Sellin L, Lu W, Schneider M, Neumann S, Beijer A, Munck M, Padmakumar VC, Gloy J, Walz G, Noegel AA (2008) Nesprin-2 Giant (NUANCE) maintains nuclear envelope architecture and composition in skin. J Cell Sci 121(Pt 11):1887–1898
154. Lu W, Schneider M, Neumann S, Jaeger VM, Taranum S, Munck M, Cartwright S, Richardson C, Carthew J, Noh K, Goldberg M, Noegel AA, Karakesisoglou I (2012) Nesprin interchain associations control nuclear size. Cell Mol Life Sci 69(20):3493–3509
155. Sjoblom T, Jones S, Wood LD, Parsons DW, Lin J, Barber TD, Mandelker D, Leary RJ, Ptak J, Silliman N, Szabo S, Buckhaults P, Farrell C, Meeh P, Markowitz SD, Willis J, Dawson D, Willson JK, Gazdar AF, Hartigan J, Wu L, Liu C, Parmigiani G, Park BH, Bachman KE, Papadopoulos N, Vogelstein B, Kinzler KW, Velculescu VE (2006) The consensus coding sequences of human breast and colorectal cancers. Science 314(5797):268–274

156. Mokarram P, Kumar K, Brim H, Naghibalhossaini F, Saberi-firoozi M, Nouraie M, Green R, Lee E, Smoot DT, Ashktorab H (2009) Distinct high-profile methylated genes in colorectal cancer. PLoS One 4(9):e7012
157. Doherty JA, Rossing MA, Cushing-Haugen KL, Chen C, Van Den Berg DJ, Wu AH, Pike MC, Ness RB, Moysich K, Chenevix-Trench G, Beesley J, Webb PM, Chang-Claude J, Wang-Gohrke S, Goodman MT, Lurie G, Thompson PJ, Carney ME, Hogdall E, Kjaer SK, Hogdall C, Goode EL, Cunningham JM, Fridley BL, Vierkant RA, Berchuck A, Moorman PG, Schildkraut JM, Palmieri RT, Cramer DW, Terry KL, Yang HP, Garcia-Closas M, Chanock S, Lissowska J, Song H, Pharoah PD, Shah M, Perkins B, McGuire V, Whittemore AS, Di Cioccio RA, Gentry-Maharaj A, Menon U, Gayther SA, Ramus SJ, Ziogas A, Brewster W, Anton-Culver H, Pearce CL (2010) ESR1/SYNE1 polymorphism and invasive epithelial ovarian cancer risk: an Ovarian Cancer Association Consortium study. Cancer Epidemiol Biomarkers Prev 19(1):245–250
158. Marme A, Zimmermann HP, Moldenhauer G, Schorpp-Kistner M, Muller C, Keberlein O, Giersch A, Kretschmer J, Seib B, Spiess E, Hunziker A, Merchan F, Moller P, Hahn U, Kurek R, Marme F, Bastert G, Wallwiener D, Ponstingl H (2008) Loss of Drop1 expression already at early tumor stages in a wide range of human carcinomas. Int J Cancer 123(9):2048–2056
159. West M, Zurek N, Hoenger A, Voeltz GK (2011) A 3D analysis of yeast ER structure reveals how ER domains are organized by membrane curvature. J Cell Biol 193(2):333–346
160. Voeltz GK, Prinz WA, Shibata Y, Rist JM, Rapoport TA (2006) A class of membrane proteins shaping the tubular endoplasmic reticulum. Cell 124(3):573–586
161. Hu J, Shibata Y, Voss C, Shemesh T, Li Z, Coughlin M, Kozlov MM, Rapoport TA, Prinz WA (2008) Membrane proteins of the endoplasmic reticulum induce high-curvature tubules. Science 319(5867):1247–1250
162. Shibata Y, Shemesh T, Prinz WA, Palazzo AF, Kozlov MM, Rapoport TA (2010) Mechanisms determining the morphology of the peripheral ER. Cell 143(5):774–788
163. Orso G, Pendin D, Liu S, Tosetto J, Moss TJ, Faust JE, Micaroni M, Egorova A, Martinuzzi A, McNew JA, Daga A (2009) Homotypic fusion of ER membranes requires the dynamin-like GTPase atlastin. Nature 460(7258):978–983
164. Hu J, Shibata Y, Zhu PP, Voss C, Rismanchi N, Prinz WA, Rapoport TA, Blackstone C (2009) A class of dynamin-like GTPases involved in the generation of the tubular ER network. Cell 138(3):549–561
165. Anderson DJ, Hetzer MW (2007) Nuclear envelope formation by chromatin-mediated reorganization of the endoplasmic reticulum. Nat Cell Biol 9(10):1160–1166
166. Anderson DJ, Hetzer MW (2008) Reshaping of the endoplasmic reticulum limits the rate for nuclear envelope formation. J Cell Biol 182(5):911–924
167. Anderson DJ, Vargas JD, Hsiao JP, Hetzer MW (2009) Recruitment of functionally distinct membrane proteins to chromatin mediates nuclear envelope formation in vivo. J Cell Biol 186(2):183–191
168. Razafsky D, Hodzic D (2009) Bringing KASH under the SUN: the many faces of nucleo-cytoskeletal connections. J Cell Biol 186(4):461–472
169. Bjorling E, Lindskog C, Oksvold P, Linne J, Kampf C, Hober S, Uhlen M, Ponten F (2008) A web-based tool for in silico biomarker discovery based on tissue-specific protein profiles in normal and cancer tissues. Mol Cell Proteomics 7(5):825–844
170. Rahbari R, Kitano M, Zhang L, Bommareddi S, Kebebew E (2013) RTN4IP1 is down-regulated in thyroid cancer and has tumor-suppressive function. J Clin Endocrinol Metab 98(3):E446–E454
171. Shirahata M, Oba S, Iwao-Koizumi K, Saito S, Ueno N, Oda M, Hashimoto N, Ishii S, Takahashi JA, Kato K (2009) Using gene expression profiling to identify a prognostic molecular spectrum in gliomas. Cancer Sci 100(1):165–172
172. van de Velde HJ, Senden NH, Roskams TA, Broers JL, Ramaekers FC, Roebroek AJ, Van de Ven WJ (1994) NSP-encoded reticulons are neuroendocrine markers of a novel category in human lung cancer diagnosis. Cancer Res 54(17):4769–4776

173. Senden N, Linnoila I, Timmer E, van de Velde H, Roebroek A, Van de Ven W, Broers J, Ramaekers F (1997) Neuroendocrine-specific protein (NSP)-reticulons as independent markers for non-small cell lung cancer with neuroendocrine differentiation. An in vitro histochemical study. Histochem Cell Biol 108(2):155–165
174. Yang YS, Strittmatter SM (2007) The reticulons: a family of proteins with diverse functions. Genome Biol 8(12):234
175. Morrone FB, Oliveira DL, Gamermann P, Stella J, Wofchuk S, Wink MR, Meurer L, Edelweiss MI, Lenz G, Battastini AM (2006) In vivo glioblastoma growth is reduced by apyrase activity in a rat glioma model. BMC Cancer 6:226
176. Petersen I, Kotb WF, Friedrich KH, Schluns K, Bocking A, Dietel M (2009) Core classification of lung cancer: correlating nuclear size and mitoses with ploidy and clinicopathological parameters. Lung Cancer 65(3):312–318
177. Zaidel-Bar R, Itzkovitz S, Ma'ayan A, Iyengar R, Geiger B (2007) Functional atlas of the integrin adhesome. Nat Cell Biol 9(8):858–867
178. Maniotis AJ, Chen CS, Ingber DE (1997) Demonstration of mechanical connections between integrins, cytoskeletal filaments, and nucleoplasm that stabilize nuclear structure. Proc Natl Acad Sci U S A 94(3):849–854
179. Bissell MJ, Weaver VM, Lelievre SA, Wang F, Petersen OW, Schmeichel KL (1999) Tissue structure, nuclear organization, and gene expression in normal and malignant breast. Cancer Res 59(Suppl 7):1757s–1763s, discussion 1763s–1764s
180. Dahl KN, Ribeiro AJ, Lammerding J (2008) Nuclear shape, mechanics, and mechanotransduction. Circ Res 102(11):1307–1318
181. Pajerowski JD, Dahl KN, Zhong FL, Sammak PJ, Discher DE (2007) Physical plasticity of the nucleus in stem cell differentiation. Proc Natl Acad Sci U S A 104(40): 15619–15624
182. Dahl KN, Kahn SM, Wilson KL, Discher DE (2004) The nuclear envelope lamina network has elasticity and a compressibility limit suggestive of a molecular shock absorber. J Cell Sci 117(Pt 20):4779–4786
183. Friedl P, Wolf K, Lammerding J (2011) Nuclear mechanics during cell migration. Curr Opin Cell Biol 23(1):55–64
184. Discher DE, Janmey P, Wang YL (2005) Tissue cells feel and respond to the stiffness of their substrate. Science 310(5751):1139–1143
185. Mammoto T, Ingber DE (2010) Mechanical control of tissue and organ development. Development 137(9):1407–1420
186. Engler AJ, Sen S, Sweeney HL, Discher DE (2006) Matrix elasticity directs stem cell lineage specification. Cell 126(4):677–689
187. Briand P, Nielsen KV, Madsen MW, Petersen OW (1996) Trisomy 7p and malignant transformation of human breast epithelial cells following epidermal growth factor withdrawal. Cancer Res 56(9):2039–2044
188. Danen EH, Yamada KM (2001) Fibronectin, integrins, and growth control. J Cell Physiol 189(1):1–13
189. Malhas A, Goulbourne C, Vaux DJ (2011) The nucleoplasmic reticulum: form and function. Trends Cell Biol 21(6):362–373
190. Titus LC, Dawson TR, Rexer DJ, Ryan KJ, Wente SR (2010) Members of the RSC chromatin-remodeling complex are required for maintaining proper nuclear envelope structure and pore complex localization. Mol Biol Cell 21(6):1072–1087
191. Hodge CA, Choudhary V, Wolyniak MJ, Scarcelli JJ, Schneiter R, Cole CN (2010) Integral membrane proteins Brr6 and Apq12 link assembly of the nuclear pore complex to lipid homeostasis in the endoplasmic reticulum. J Cell Sci 123(Pt 1):141–151
192. Friederichs JM, Ghosh S, Smoyer CJ, McCroskey S, Miller BD, Weaver KJ, Delventhal KM, Unruh J, Slaughter BD, Jaspersen SL (2011) The SUN protein Mps3 is required for spindle pole body insertion into the nuclear membrane and nuclear envelope homeostasis. PLoS Genet 7(11):e1002365

193. Zwerger M, Kolb T, Richter K, Karakesisoglou I, Herrmann H (2010) Induction of a massive endoplasmic reticulum and perinuclear space expansion by expression of lamin B receptor mutants and the related sterol reductases TM7SF2 and DHCR7. Mol Biol Cell 21(2):354–368
194. Webster M, Witkin KL, Cohen-Fix O (2009) Sizing up the nucleus: nuclear shape, size and nuclear-envelope assembly. J Cell Sci 122(Pt 10):1477–1486
195. Bidwell JP, Fey EG, van Wijnen AJ, Penman S, Stein JL, Lian JB, Stein GS (1994) Nuclear matrix proteins distinguish normal diploid osteoblasts from osteosarcoma cells. Cancer Res 54(1):28–32
196. Leman ES, Getzenberg RH (2002) Nuclear matrix proteins as biomarkers in prostate cancer. J Cell Biochem 86(2):213–223
197. Leman ES, Arlotti JA, Dhir R, Greenberg N, Getzenberg RH (2002) Characterization of the nuclear matrix proteins in a transgenic mouse model for prostate cancer. J Cell Biochem 86(2):203–212
198. Lakshmanan Y, Subong EN, Partin AW (1998) Differential nuclear matrix protein expression in prostate cancers: correlation with pathologic stage. J Urol 159(4):1354–1358
199. Keesee SK, Briggman JV, Thill G, Wu YJ (1996) Utilization of nuclear matrix proteins for cancer diagnosis. Crit Rev Eukaryot Gene Expr 6(2–3):189–214
200. Coffey DS (2002) Nuclear matrix proteins as proteomic markers of preneoplastic and cancer lesions. Clin Cancer Res 8(10):3031–3033
201. Chen YB, Guo LC, Yang L, Feng W, Zhang XQ, Ling CH, Ji C, Huang JA (2010) Angiosarcoma of the lung: 2 cases report and literature reviewed. Lung Cancer 70(3): 352–356
202. Zhuang PY, Zhang JB, Zhu XD, Zhang W, Wu WZ, Tan YS, Hou J, Tang ZY, Qin LX, Sun HC (2008) Two pathologic types of hepatocellular carcinoma with lymph node metastasis with distinct prognosis on the basis of CK19 expression in tumor. Cancer 112(12):2740–2748
203. Hokamp HG, Grundmann E (1983) Histological grading and clinical staging of plasmacytoma. J Cancer Res Clin Oncol 105(2):197–198
204. Nandakumar V, Kelbauskas L, Hernandez KF, Lintecum KM, Senechal P, Bussey KJ, Davies PC, Johnson RH, Meldrum DR (2012) Isotropic 3D nuclear morphometry of normal, fibrocystic and malignant breast epithelial cells reveals new structural alterations. PLoS One 7(1):e29230
205. Hanai S, Kishi K, Watanabe S, Shimosato Y (1985) Morphometry of bladder cancer cells in tissue sections according to histological grade of malignancy. Jpn J Clin Oncol 15(4): 671–678
206. Nakayama H, Kondo Y, Saito N, Sarashina H, Okui K (1988) Morphometric analysis of cytological atypia in colonic adenomas. Virchows Arch A Pathol Anat Histopathol 413(6):499–504
207. van den Einden LC, Grefte JM, van der Avoort IA, Vedder JE, van Kempen LC, Massuger LF, de Hullu JA (2012) Cytology of the vulva: feasibility and preliminary results of a new brush. Br J Cancer 106(2):269–273
208. Wahi PN, Kehar U, Mali S, Misra GD (1962) Histopathogenesis of carcinoma in situ of the uterine cervix. Bull World Health Organ 26:661–674
209. Srivastava S, Henson DE, Gazdar A (1999) Molecular pathology of early cancer. Ios Press, Amsterdam
210. Newport J, Kirschner M (1982) A major developmental transition in early Xenopus embryos: I. Characterization and timing of cellular changes at the midblastula stage. Cell 30(3):675–686
211. Newport J, Kirschner M (1982) A major developmental transition in early Xenopus embryos: II. Control of the onset of transcription. Cell 30(3):687–696
212. Clute P, Masui Y (1995) Regulation of the appearance of division asynchrony and microtubule-dependent chromosome cycles in Xenopus laevis embryos. Dev Biol 171(2):273–285
213. Gasser SM (2002) Visualizing chromatin dynamics in interphase nuclei. Science 296(5572):1412–1416
214. Chai H, Brown RE (2009) Field effect in cancer-an update. Ann Clin Lab Sci 39(4):331–337
215. Brunagel G, Schoen RE, Bauer AJ, Vietmeier BN, Getzenberg RH (2002) Nuclear matrix protein alterations associated with colon cancer metastasis to the liver. Clin Cancer Res 8(10):3039–3045

216. Konety BR, Nguyen TS, Dhir R, Day RS, Becich MJ, Stadler WM, Getzenberg RH (2000) Detection of bladder cancer using a novel nuclear matrix protein, BLCA-4. Clin Cancer Res 6(7):2618–2625
217. Franklin WA, Gazdar AF, Haney J, Wistuba II, La Rosa FG, Kennedy T, Ritchey DM, Miller YE (1997) Widely dispersed p53 mutation in respiratory epithelium. A novel mechanism for field carcinogenesis. J Clin Invest 100(8):2133–2137
218. Veltri RW, Khan MA, Miller MC, Epstein JI, Mangold LA, Walsh PC, Partin AW (2004) Ability to predict metastasis based on pathology findings and alterations in nuclear structure of normal-appearing and cancer peripheral zone epithelium in the prostate. Clin Cancer Res 10(10):3465–3473
219. Deng G, Lu Y, Zlotnikov G, Thor AD, Smith HS (1996) Loss of heterozygosity in normal tissue adjacent to breast carcinomas. Science 274(5295):2057–2059
220. Tomasetti C, Vogelstein B, Parmigiani G (2013) Half or more of the somatic mutations in cancers of self-renewing tissues originate prior to tumor initiation. Proc Natl Acad Sci U S A 110(6):1999–2004
221. Cope NF, Fraser P, Eskiw CH (2010) The yin and yang of chromatin spatial organization. Genome Biol 11(3):204
222. Meaburn KJ, Misteli T, Soutoglou E (2007) Spatial genome organization in the formation of chromosomal translocations. Semin Cancer Biol 17(1):80–90
223. Roca-Cusachs P, Alcaraz J, Sunyer R, Samitier J, Farre R, Navajas D (2008) Micropatterning of single endothelial cell shape reveals a tight coupling between nuclear volume in G1 and proliferation. Biophys J 94(12):4984–4995
224. Zuleger N, Robson MI, Schirmer EC (2011) The nuclear envelope as a chromatin organizer. Nucleus 2(5):339–349
225. Solovei I, Wang AS, Thanisch K, Schmidt CS, Krebs S, Zwerger M, Cohen TV, Devys D, Foisner R, Peichl L, Herrmann H, Blum H, Engelkamp D, Stewart CL, Leonhardt H, Joffe B (2013) LBR and lamin A/C sequentially tether peripheral heterochromatin and inversely regulate differentiation. Cell 152(3):584–598
226. Peric-Hupkes D, Meuleman W, Pagie L, Bruggeman SW, Solovei I, Brugman W, Graf S, Flicek P, Kerkhoven RM, van Lohuizen M, Reinders M, Wessels L, van Steensel B (2010) Molecular maps of the reorganization of genome-nuclear lamina interactions during differentiation. Mol Cell 38(4):603–613
227. Kohwi M, Lupton JR, Lai SL, Miller MR, Doe CQ (2013) Developmentally regulated subnuclear genome reorganization restricts neural progenitor competence in Drosophila. Cell 152(1–2):97–108
228. Gerlitz G, Bustin M (2011) The role of chromatin structure in cell migration. Trends Cell Biol 21(1):6–11
229. Schuster-Bockler B, Lehner B (2012) Chromatin organization is a major influence on regional mutation rates in human cancer cells. Nature 488(7412):504–507
230. Ma Y, Zhang P, Wang F, Yang J, Yang Z, Qin H (2010) The relationship between early embryo development and tumourigenesis. J Cell Mol Med 14(12):2697–2701
231. Heiser PW, Hebrok M (2004) Development and cancer: lessons learned in the pancreas. Cell Cycle 3(3):270–272
232. Calvo R, Drabkin HA (2000) Embryonic genes in cancer. Ann Oncol 11(Suppl 3):207–218
233. Tang Y, Kitisin K, Jogunoori W, Li C, Deng CX, Mueller SC, Ressom HW, Rashid A, He AR, Mendelson JS, Jessup JM, Shetty K, Zasloff M, Mishra B, Reddy EP, Johnson L, Mishra L (2008) Progenitor/stem cells give rise to liver cancer due to aberrant TGF-beta and IL-6 signaling. Proc Natl Acad Sci U S A 105(7):2445–2450
234. Donhuijsen K, Schulz S, Leder LD (1991) Nuclear grading of renal cell carcinomas–is morphometry necessary? J Cancer Res Clin Oncol 117(1):73–78
235. Gregory TR (2001) The bigger the C-value, the larger the cell: genome size and red blood cell size in vertebrates. Blood Cells Mol Dis 27(5):830–843
236. van Zanten M, Carles A, Li Y, Soppe W (2012) Control and consequences of chromatin compaction during seed maturation in Arabidopsis thaliana. Plant Signal Behav 7(3):338–341

Control of Nuclear Size by NPC Proteins

Masatoshi Takagi and Naoko Imamoto

Abstract The architecture of the cell nucleus in cancer cells is often altered in a manner associated with the tumor type and aggressiveness. Therefore, it has been the central criterion in the pathological diagnosis and prognosis of cancer. However, the molecular mechanism behind these observed changes in nuclear morphology, including size, remains completely unknown. Based on our current understanding of the physiology of the nuclear pore complex (NPC) and its constituents, which are collectively referred to as nucleoporins (Nups), we discuss how the structural and functional ablation of the NPC and Nups could directly or indirectly contribute to the changes in nuclear size observed in cancer cells.

Keywords Nuclear size • Nuclear pore complex (NPC) • Nucleoporin (Nup) • Nuclear envelope (NE) • Cancer

Abbreviations

AF	Atrial fibrillation
INM	Inner nuclear membrane
KTIP	Kap121p transport inhibitory pathway
LINC	Linker of nucleoskeleton and cytoskeleton
MAC	Macronucleus
MIC	Micronucleus
NE	Nuclear envelope
NPC	Nuclear pore complex

M. Takagi • N. Imamoto (✉)
Cellular Dynamics Laboratory, RIKEN, WAKO, Saitama 351-0198, Japan
e-mail: mtakagi@riken.jp; nimamoto@riken.jp

E.C. Schirmer and J.I. de las Heras (eds.), *Cancer Biology and the Nuclear Envelope*, Advances in Experimental Medicine and Biology 773, DOI 10.1007/978-1-4899-8032-8_26,

NTR	Nuclear transport receptor
Nup	Nucleoporin
ONM	Outer nuclear membrane

Introduction

The nucleus is the cellular compartment that encloses the nuclear genome with a double-lipid bilayer nuclear envelope (NE) [1]. The active and passive exchange of molecules between the nucleus and cytoplasm is mediated through nuclear pore complexes (NPCs) embedded in the NE. The NPC is a huge macromolecular assembly with a calculated molecular mass of 60 MDa in vertebrates [2]. It is formed from multiple copies of 30 different proteins that are collectively referred to as nucleoporins (Nups). The molecular architecture of the NPC was modeled through integrative approaches using diverse biophysical and proteomic data [3–5]. Importantly, many Nups have phenylalanine-glycine (FG) repeats that extend into and fill the central transport channel, thus determining the limitations and selectivity of pore permeability by creating a diffusion barrier. The molecular mechanisms regulating nucleocytoplasmic transport have also been extensively studied [6–8]. Briefly, the nuclear transport of most macromolecular cargo begins when it binds to soluble nuclear transport receptors (NTRs), such as importins/exportins/transportins/karyopherins, p10/NTF2, TAP/p15, or Hikeshi [7–9]. The complex then moves through the NPC diffusion barrier by interacting with the FG repeat-containing Nups (FG-Nups) until it reaches its destination compartment where the cargo is released. In the case of importins, the small GTPase Ran plays a critical role in this transport termination step.

The structural integrity of the nucleus is predominantly supported internally by the nuclear lamina, a proteinaceous layer composed of proteins encoded by the lamin A/C and lamin B genes and their associated proteins [10]. Cancer cells are often distinguished by a large nucleus with abnormal shapes [11]. Consistent with this observation, mutations and changes in the abundance of lamina proteins are often associated with different cancers [12, 13]. Because the nuclear import of lamina protein components through NPCs is a prerequisite for enlarging the nuclear volume, malfunction of NPCs and/or soluble nuclear transport factors may also play a critical role. Not only can the genetic mutation of the nuclear transport machinery (NPC, NTRs) play a determining role in this malfunction, but the genetic mutation in the NLS of their transport cargo could also be a primary cause of malfunction. In addition to the classic role of the NPC as a portal for nucleocytoplasmic transport, transport-independent roles of the NPC and Nups, such as their roles in nuclear organization, transcriptional regulation, and mitotic progression, have been extensively studied during the past decade [14–16]. Taking these findings into consideration, another scenario is also possible. The malfunction of the NPCs and Nups during these transport-independent roles may occur first and disrupt cellular homeostasis, which could then indirectly result in changes in NPC function and consequent increase in nuclear volume (Fig. 1).

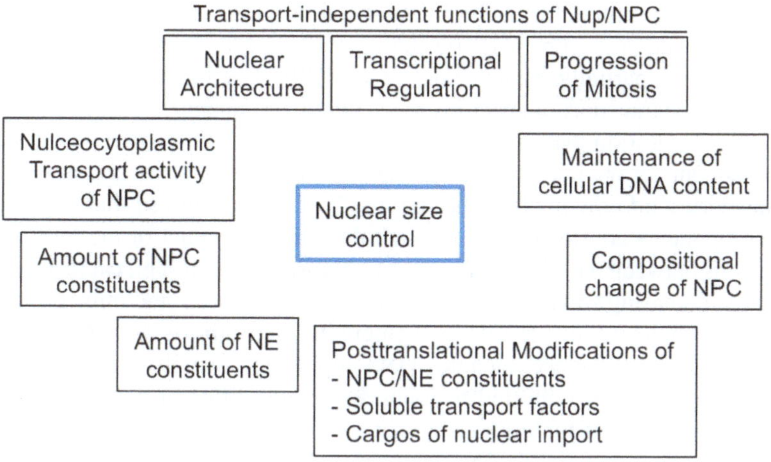

Fig. 1 Nup/NPCs may contribute to nuclear size control directly or indirectly. Possible determinants of the nuclear size are listed in the *boxes* drawn in *black*. Some of them are intimately correlated. Nup/NPCs are involved in nuclear size control directly through their functions as a portal for nucleocytoplasmic transport of macromolecules or indirectly through their functions as regulators of nuclear architecture, transcription, or progression of mitosis (see detail in the main text)

In this review, we first discuss how NPCs may contribute to controlling nuclear size. Next, we list the physical and functional changes that may occur to the NPC and Nups in cancer cells and discuss how these changes could account for the nuclear enlargement often observed in these cells. We propose that while Nup malfunction does not necessarily play a causative role, it does play a pivotal role in a cascade of events that promote cancer. We hope that this review will serve as a starting point for the development of novel cancer therapy strategies that target Nups.

How the NPC May Contribute to Controlling Nuclear Size

It is easy to assume that nuclear size is controlled by the amount of its constituents transported into the nucleus through the NPCs. In living mammalian cells, an absence in NPC assembly prevents nuclear growth [17, 18]. Studies using an in vitro reconstitution system with *Xenopus* egg extracts have demonstrated that the NPC and nuclear import are required for nuclear growth after the NE has assembled [19–21]. On the other hand, it is unlikely that an increase in the number of NPCs accounts for nuclear growth because inhibiting de novo NPC assembly does not impair nuclear growth [22]. It has been shown that individual NPCs possess an enormous capacity for transport, and therefore, the number of NPCs, unless drastically altered, is most likely not the rate-determining factor for the net flux of molecules that cross the NE through the NPC [23]. Consistent with this hypothesis, a mathematical model suggests that it is the amount of soluble transport factors, such as NTF2 and importin α, that are critical for determining the transport rate in importin α/β-mediated transport [24]. Furthermore, NTF2 and importin α have been shown to at least partially regulate nuclear size [20]. Whether it is the global nuclear transport or the transport of specific cargos that is responsible for nuclear growth is an intriguing question that should be addressed.

Components of the nucleus are synthesized in the cytoplasm, and most of them are transported into the nucleus through the NPC. It is believed that a majority of these components are transported through the central channel of the NPC through interactions between their NTRs and the FG-Nups. During the cell cycle, the nuclear size starts increasing soon after the formation of sister nuclei and immediately after mitosis, which continues until the cell enters the next mitosis [22]. Generally, this increase in nuclear size accompanies DNA replication, which is associated with a vast number of newly imported histones and many different chromatin proteins. Nevertheless, reported findings show that the inhibition of DNA replication does not inhibit nuclear growth [22, 25]. It appears that major constituents of the nucleus, such as the DNA and chromatin, are not primarily responsible for nuclear size determination, even though the concentrations of these constituents are associated with a large number of nuclear transport events.

NE Component Limitations

On the other hand, it has been shown that the expression levels and nuclear import of lamins, major components of the nuclear lamina, and lamin-binding proteins regulate nuclear size in *Xenopus*, *Drosophila*, and *Arabidopsis* [20, 26, 27]. These findings indicate that these transport substrates play an important role in defining nuclear size. A recent study suggests that the NPC regulates nuclear size by restricting the passage of integral membrane proteins from the outer nuclear membrane (ONM) to the inner nuclear membrane (INM) [25]. This function of the NPC is considered to be different from the transport that occurs through the central channel of the NPC because many INM proteins could potentially traverse the NPCs through the pore membrane, a membrane area facing the NPC where the INM and ONM are connected [28] (Fig. 2). Depletion of Nup188, an evolutionarily conserved scaffold Nup, specifically inhibits this function of NPCs. Nup188 is dispensable for NPC formation, and it is not critical for the nuclear transport that proceeds through the central channel of the NPC. The nuclei, however, will increase severalfold in size when they are reconstituted in a *Xenopus* egg extract depleted of Nup188, and this correlates with an increase in translocation of INM proteins, presumably as Nup188 can no longer regulate/sterically hinder translocation. Although the mechanism of Nup188-mediated transport is currently unknown, recent findings emphasize that the nuclear framework consisting of the INM proteins and lamins plays an important role in determining nuclear size and that the NPC acts as a gatekeeper in the nucleocytoplasmic transport of these framework constituents.

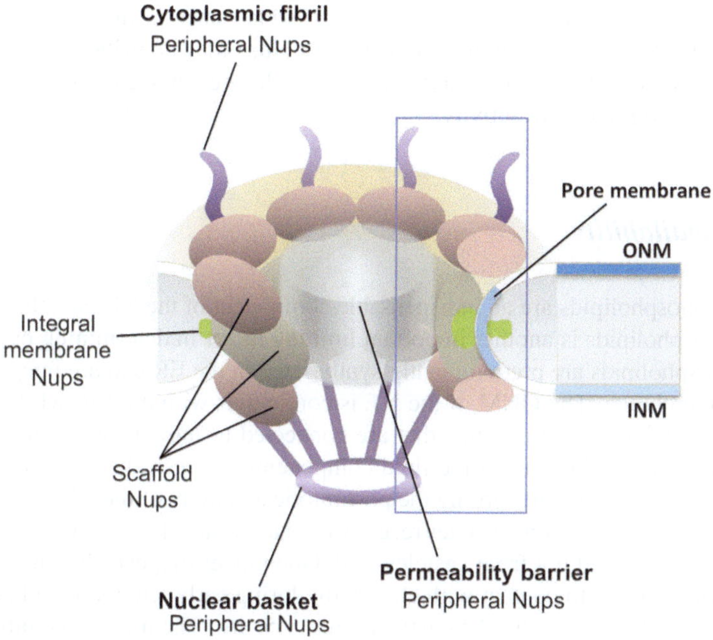

Fig. 2 Structural organization of an NPC embedded in the NE

NPCs are embedded in the NE and, therefore, may physically contribute to the integrity of nuclear architecture. NPCs associate with some INM proteins and the underlying lamins. Nup53, Nup88, and Nup153 have been shown to directly interact with lamins [29–31]. NPCs also associate with linker of nucleoskeleton and cytoskeleton (LINC) complexes, which play a role in nuclear membrane shaping. LINC complexes span the NE double membrane and physically connect nuclear structures to cytoskeletal elements [32]. The depletion of Nup53 or Nup153, which associate with lamins, or Pom121, which functionally associates with SUN1 [33], a component of the LINC complex, indeed results in nuclear structure abnormalities [29, 34, 35].

NPC–Gene Associations

The "gene-gating" hypothesis proposes that the NPC participates in gene regulation independently of its role as the gate for nucleocytoplasmic transport at the NE [36]. The NPC constituents, Nups, are indeed connected to chromatin at the NE and regulate the expression of certain genes [37–40]. Furthermore, recent findings show that some Nups are mobile and that a fraction of these Nups are within the nucleoplasm [41] where they directly participate in transcription activation away from the NPC. These findings have been reported in *Drosophila* cells [14, 42, 43]. Moreover, the expression of many developmental and cell cycle genes is directly activated by intranuclear Nups. These findings indicate that Nups may contribute to various cellular processes, including those associated with changes in nuclear size, through its transcriptional regulating activity.

Lipid Availability

Nuclear phospholipids are an indispensable component of the NE, and the availability of phospholipids is another important limiting factor in determining nuclear size [44]. Phospholipids are predominantly synthesized in the ER and are integrated into the ER membrane. The ONM of the NE is continuous with the ER, while the INM contains specific sets of proteins that are connected to intranuclear structures. The similarities and differences in the lipid components between these two membranes are currently unknown. Regarding the postmitotic assembly of the NE, though it was long believed to form from vesicles recent work has argued that the membrane comes from ER tubules and not from vesicles [45]. One model suggests that the tips of the ER tubules come into contact with chromatin during early telophase and are subsequently immobilized on the chromatin surface [45, 46]. The membrane tubules then flatten into sheets with the aid of INM proteins, which have DNA binding activities [46], spread across the chromatin, and reorganize into a sealed NE. Postmitotic NE assembly is spatially and temporally coordinated with NPC assembly [47]. Membrane recruitment, flattening, and sealing, however, can proceed in the absence of NPC assembly [18, 48, 49], which shows that NE and NPC assembly can be uncoupled.

Once the NE is assembled, it must expand to the proper size. To determine the proper nuclear size, cells must regulate the amount of ER membrane to incorporate into the NE (first into the ONM and then into the INM). It may be that the NPCs sterically limit the movement of phospholipids from the ONM to the INM because of their location at the junction of these different lipid domains. Alternatively, the NPC may regulate nuclear size at the level of phospholipid metabolism by controlling the subcellular localization of the responsible enzymes, such as lipin [50] and many lipid-generating proteins identified in the NE proteome [51], or by controlling the expression of these enzymes through the transport-independent functions previously discussed.

While the NPC (or FG-Nups) and components of the nuclear transport machinery (including NTRs, Ran, and Ran regulators) have established roles in the nucleocytoplasmic trafficking of macromolecules during interphase, they have distinct roles during mitosis, such as centrosome positioning (RanBP2/Nup358 [52], Nup133 [53]), normal spindle morphology (Rae1), kinetochore function (Nup107–160 complex) [54], and mitotic signal transduction (Rae1 and Nup98 [55], Nup107–160 complex [56], Nup153 [35, 57]). The functional ablation of such Nups often results in defects in the progression of mitosis that sometimes results in aneuploidy. One may assume that the different genomic size in aneuploid cells may account for the difference in nuclear size compared with normal cells. The findings from many studies, however, argue against this assumption and instead support the proposal that nuclear size is determined by cytoplasmic volume rather than DNA content [44, 58, 59]. It has also been reported that nuclear size does not strictly correlate with ploidy in some cancers [60].

Genetic Changes in Nups in Cancer Cells

Genetic changes in four Nups have been linked to cancers. Nup88 is overexpressed in a broad spectrum of neoplasia [61, 62]. Tpr, Nup98, and Nup214 are linked to cancers through their involvement in chromosomal translocations that encode chimeric fusion proteins. Xu and Powers provide an extensive review on the normal function of these Nups and their major translocation partners as well as what is currently known about their mechanistic contribution to carcinogenesis [63]. Interestingly Nup214–ABL1 must localize to the NPC for its transforming potential [64], and Nup98 fusions cause leukemia by inhibiting the CRM1-mediated nuclear export of certain transcription factors [65]. Although both of these examples show how altering NPC function can result in carcinogenesis, they do not explain why the nuclei are enlarged in these situations. Because of the complexity of this entire process, we do not currently know whether changes in nuclear size are caused by defects in the transport activity or the transport-independent activity of the NPC such as transcriptional regulating activity within the nucleoplasm. Other genetic changes to Nups (overexpression of Nup133 and downregulation of Nup214) were observed in breast tumors using an integrative analysis of a cancer somatic mutome [66], but the significance in nuclear size was not addressed.

Functional and Physical Changes in the NPCs in Cancer Cells

Changes in nuclear transport are frequently observed in transformed cells [67]. These occur at multiple levels because of changes that occur in a diverse range of mechanisms, such as altering the expression levels or posttranslational modification of each nuclear transport cargo and its partner protein, altering the expression levels or function of the NTRs, and possibly altering the constituents or the organization of the NPCs. For example, NF-κB, a transcriptional activator, predominantly localizes to the nucleus of many cancer cells because of a disruption in IκB function, which masks the nuclear localization signal in NF-κB in normal cells. The disruption of IκB function is caused by the hyperphosphorylation and subsequent degradation of IκB [68–70]. Another example is that an elevated importin α2 level is a known and reliable biomarker for aggressive breast and non-small-cell lung cancers [71], which is consistent with its ability to increase the dynamic range of transport [72].

While there are a number of reports describing transport regulation mechanisms at the level of cargo alterations, few reports exist describing mechanisms at the level of NPC alteration/reorganization. The NPC function is so central for cells that such changes would cause embryonic lethality, which may have hampered the study. But recently, many reports address cell type and cell context-specific changes to the NPC constituents, which suggests that individual Nups play specific roles. Because each NTR interacts with different FG-Nups, changes to an NPC constituent could affect the transport pathways mediated by specific NTRs without affecting the pathways mediated by the other NTRs present in eukaryotic cells. It is important to mention, however, that data on the cargo selectivity of each NTR is still limited and prevents us from making specific assumptions on how cell type-specific NPC changes affect cell physiology. If a comprehensive chart showing the cargo–NTR–FG–Nups combinations was available, then we would be able to better choose the appropriate cargos to focus on that could elucidate the functional differences among the NPCs from different cancer cell types. The transport efficiency of these cargos could be examined using live imaging or a reconstituted transport assay with purified soluble transport factors to ensure that only NPC function was being evaluated. Once a functional change in the NPCs specific to a cancer type has been identified, identifying the causative alterations in the Nups would be extremely valuable.

The NPC density is known to double from the G1 to G2 phase of the cell cycle [73]. In the initiation step of the de novo assembly of NPCs in human cells, Cdk activity is required [22]. Based on the observations in replicative senescent cells [74] and mechanically stretched smooth muscle cells [75], the positive involvement of MAPK–ERK signaling has also been implicated in this process. Importantly, nuclear growth requires MAPK–ERK signaling but not Cdk 1 or Cdk 2. MAPK–ERK signaling may increase the NPC density through its downstream CDK signaling independent of regulating nuclear growth. Although a higher NPC number is associated with more aggressive tumors and could be a good diagnostic marker [76], this increase may not be anything causal but a consequence of enhanced MAPK–ERK signaling.

While looking for Nups that have concentrations that correlate with cancer states and/or nuclear size using a comprehensive analysis without bias is important, analyses focusing on individual Nups in cancer cells are equally valuable for accuracy and reliability. Recent findings have shown that the NPC composition changes in various different cellular contexts (described below) and that these changes are associated with changes in specific NPC functions. The altered Nups identified in these studies may be essential for NPC function and, therefore, are important targets to study to determine whether similar changes also occur in cancer cells. Furthermore, changes in the absolute level of each Nup and the posttranslational modifications made to Nups may also regulate NPC function.

Compositional Alterations to the NPC

Below, we discuss the known NPC compositional changes that occur in a variety of cellular contexts and the posttranslational modifications made to Nups. We chose the cases in which the physiological outcomes were thoroughly explained. Even though these physical alterations made to the NPC/Nups are not necessarily currently associated with nuclear size control, this information may offer insight for future studies exploring this connection.

Differentiation and Disease

One prominent concept that has recently emerged concerning the NPC is that its composition is not invariant among cell types and tissues. Currently, at least six Nups, Nup210 (also known as gp210), Nup45, Nup50, Nup133, Nup155, and Aladin, have been shown to exhibit different expression levels in different cell types or tissues, which suggests that there is compositional heterogeneity in NPCs (reviewed in [77]). Among the Nups previously listed, the physiological significance of the tissue-specific expression has been most thoroughly described for Nup210, a vertebrate-specific transmembrane Nup [78, 79]. Nup210 expression is induced during myogenic and neuronal differentiation. By knocking down Nup210 expression with specific shRNA and then subsequently rescuing its expression in cell differentiation model systems, the requirement of Nup210 for differentiation has been clearly demonstrated. Importantly, depleting Nup210 does not affect the global nuclear transport rates or the targeting of INM proteins to the NE. Instead, Nup210 contributes to cell differentiation by regulating the expression pattern of the genes involved in this process. Even though the molecular mechanism in which Nup210 regulates gene expression has not been elucidated, this study clearly suggests that cells contain NPCs with various compositions to dictate distinct cellular functions.

Consistent with the tissue-specific function of specific Nups, several "nucleoporinopathies" with tissue-specific defects have been reported [80]. Achalasia–addisonianism–alacrima syndrome (triple-A syndrome) is a recessively inherited disorder that is characterized by adrenal insufficiency, dysfunction of the lower esophageal sphincter that interferes with normal swallowing (achalasia), and dry eyes (alacrima). This disease is caused by loss-of-function mutations in *Aladin* [81], which encodes a Nup with four WD repeats. A recessively inherited missense mutation in Nup155 leads to atrial fibrillation (AF), which causes sudden death [82]. Heterozygous *Nup155*$^{+/-}$ mice also exhibit the AF phenotype [82]. Cases where mutations in Nup genes cause CNS disorders have also been reported. A recessively inherited missense mutation in *Nup62* causes infantile bilateral striatal necrosis [83]. The last of these identified to date is a dominantly inherited missense mutation in *Nup358/RanBP2* that causes susceptibility to infection-triggered acute necrotizing encephalopathy [84].

Regulation of Nuclear Size and Permeability

Like most ciliates, the binucleate *Tetrahymena thermophila* has two different sized nuclei with distinct functions in a single cell. This organism offers us a unique system for exploring the mechanisms used by cells to define nuclear size. The smaller nucleus, called the micronucleus (MIC), contains a diploid genome that originated in the zygote. The bigger nucleus, called the macronucleus (MAC), is also generated in the zygote using programmed DNA rearrangements and amplification. A linker histone H1 encoded by the *HHO* gene is specifically localized in the MAC [85], whereas linker proteins (α, β, γ, and δ) collectively referred to as MicLH are specifically localized in the MIC [86]. Depletion of either histone H1 or MicLH causes the corresponding nucleus to enlarge without affecting the other [87], indicating the decisive role of the linker proteins in nuclear size control, possibly through their chromatin packaging activities. To specifically address whether there is an intimate relationship between nuclear size and chromatin compaction, however, further studies in different experimental systems are required. The nucleus-selective transport of linker proteins, and most likely other proteins as well, may be explained by the biased localization of all 13 importin α proteins; 9 of them only localize to the MIC [88]. Interestingly, the composition of the NPCs is also different between the MIC and MAC; two of the four Nup98 homologs are exclusively found in the MIC, and the other two are exclusively found in the MAC [89]. It has been shown using chimeric proteins of different Nup98 homologs that MAC-localizing Nup98, which harbors GLFG repeats, functions to block the nuclear import of MicLH. Furthermore, the MIC-localizing Nup98, which harbors unusual NIFN repeats instead of GLFG repeats, functions to block the nuclear import of histone H1. This difference is a clear example demonstrating how the NPC composition contributes to the selectivity in the nuclear transport substrate and the subsequent determination in nuclear size.

More dynamic regulation of the NPC composition in a single cell cycle has also been reported in mice [90]. The authors found that the protein level of Nup96, a constituent of the Nup107/160 subcomplex, is preferentially downregulated during mitosis in a manner dependent on the ubiquitin–proteasome pathway. The regulated decrease in the Nup96 level is not critical for the progression of mitosis but surprisingly for the progression from the G1 to S phase, which occurs later in the cell cycle. In cells from Nup96$^{+/-}$ mice, which express a low level of Nup96, the cell cycle rate was accelerated. Furthermore, the authors also showed that the export of specific mRNAs, including those encoding the G1 cell cycle regulators, such as cyclin D3 and CDK6, is inversely related to the concentration of Nup96. This finding indicates that the composition of the NPC has the ability to affect the cellular physiology by regulating the export efficiency of specific mRNAs.

Another example of NPC compositional alterations has been reported in aging cells in *C. elegans* [91]. The authors began by investigating the fate of the NPCs in postmitotic cells. First, they showed that scaffold Nups, such as the Nup93 complex, are extremely long-lived and remain incorporated in the NE during the entire lifespan of the cell. This same group demonstrated that there is age-related deterioration of the NPCs that results in an increase in NPC permeability. During the aging process, a subset of Nups, including Nup93, suffers oxidative damage and is completely eliminated from the NPC. Because the Nup93 complex acts as a linker between the Nup107/160 complex and the FG-Nups located in the central channel [4], a functional change in Nup93 could affect the organization of the FG-Nups, which could result in a deterioration of the permeability barrier. Consistent with this proposal, Nup93 has been shown to be required for maintaining a fully functional permeability barrier in *C. elegans* [92]. Furthermore, Nup93 has also been shown to be selectively degraded by caspases in apoptotic cells resulting in efficient disruption of NPC function [93]. Taken together, these observations strongly suggest that Nup93 is an essential component of the NPC.

Posttranslational Modifications Made to Nups

Regulation by Phosphorylation

In metazoans, new NPCs are assembled during two different cell cycle stages, the end of mitosis that coincides with NE formation and during interphase in an already closed NE. Postmitotic NPC assembly is accomplished in a stepwise manner [94]. Briefly, the Nup107/160 complex first associates with chromatin through its DNA-binding ELYS subunit. Next, membrane Nups and Pom121 associate with the newly forming pore, followed by recruitment of the Nup93/205 and Nup62 complexes. Finally, more peripheral Nups are inserted to complete the assembly of the NPC. These events are thought to be regulated spatiotemporally by the phosphorylation of several Nups [95, 96] and by many other cell cycle-regulated events. The process of NPC assembly that occurs during interphase is different from the postmitotic NPC assembly process in several aspects. For example, ELYS appears to be dispensable [97], Pom121 accumulates before Nup107 and Nup93 [97, 98], and the overall kinetics is much slower [98]. Furthermore, it has been shown that Cdk activity is required for the initial step of interphase NPC assembly, but not for postmitotic NPC assembly, which indicates that there is a difference between these two NPC assembly processes [22]. Currently, the responsible kinase substrate of Cdk in this process remains elusive. Moreover, the molecular mechanisms underlying NPC disassembly were investigated, and the phosphorylation of Nup98 was identified as a crucial initial step [99].

Two lines of observations have identified Nups that have a phosphorylation status that may direct the functionality of the NPC. Osmani and colleagues observed that the NPC is partially disassembled under the control of NIMA and CDK1 kinases in the semi-closed mitosis of *Aspergillus nidulans* and that this disassembly leads to changes in the NPC permeability [100]. The NIMA activity is correlated with the phosphorylation of $SonBn^{Nup98}$ and its dispersal from the NPC. Another study from this same group showed that not only $SonBn^{Nup98}$ but also virtually all peripheral Nups, including all FG repeat Nups, are dispersed during mitosis, while other Nups remain associated with the NE [101]. The molecular rearrangement of specific Nups within the NPC, which results from the phosphorylation of Nups and leads to functional changes in the NPC, has also been reported during closed mitosis in the yeast *Saccharomyces cerevisiae* [102]. During interphase, Nup53p is bound to Nup170p within the NPC. Before the kinetochore–microtubule interaction has been established in metaphase, Mad1p, a protein known to localize both on the NPC and kinetochores, directs a structural reorganization of the NPC to expose a high-affinity Kap121p-binding site on Nup53p. As a result, Kap121p is trapped by Nup53p, and Kap121p-mediated nuclear import is inhibited [103]. This mechanism, known as the Kap121p transport inhibitory pathway (KTIP), requires Mad1p to cycle between the kinetochores and the NPCs and Ipl1p, Aurora B-like kinase, activity at the kinetochores [103]. The authors speculate that phosphorylation of Mad1p by Ipl1 at the kinetochores and the subsequent interaction between phosphorylated Mad1p and Nup53p at the NPC may be the mechanistic basis for activating the KTIP.

Regulation by Glycosylation

NPCs maintain a permeability barrier between the nucleus and the cytoplasm through FG-Nups. Several models have been proposed to explain how NPCs maintain a barrier against inert molecules while simultaneously conducting the facilitated transport of NTRs. Among these models, the "selective phase model" [23] has provided conceptual features of the FG-Nups from experiments using reconstituted nuclear pores [6, 104] and follow-up experiments using NPCs reconstituted in *Xenopus* egg extracts [105]. According to this model, the FG repeats interact with one another to form a sievelike barrier that can be locally disrupted by the binding of NTRs. The multivalent cohesion between FG repeats, especially between the FG repeats in Nup98, is required. Interestingly, it has also been shown that the cohesiveness of FG repeats is modified by O-GlcNAc modification of Nup98 and Nup62 [106]. It has been proposed that this posttranslational sugar modification, which was discovered in early studies [107–109], reduces the cohesiveness of the FG repeat domains and thereby increases the transport rate of NTRs [106]. Because the O-GlcNAc modification is reversible, cells may exploit this modification to dynamically regulate the permeability of the NPCs.

In addition to the previously discussed sites, a large number of phosphorylation sites in Nups have been identified through several phosphoproteomic studies [110–115]. Moreover, other posttranslational modifications of Nups, such as sumoylation [116], ubiquitination [117, 118], and acetylation [119], have also been identified (Fig. 3). The significance of these modifications may be revealed by comparing modifications in various types of cancer cells. For example, this strategy successfully identified that lamin A phosphorylation is associated with metastatic propensity [120].

Future Perspective

Nup/NPCs clearly can play a role in nuclear size control through its primary function in mediating nuclear transport, though they could also affect size through other less direct functions. Accumulating evidence supports the proposal that the nuclear framework, consisting of INM proteins and lamins (and lamin-like proteins), is an important determinant in nuclear size. While lamins have classical NLS and are predicted to be transported into the nucleus by importin α and β through the central channel of NPCs [121], it has been proposed that membrane-associated B-type lamins may get into the nucleus also through the pore membrane of NPC as well as other INM proteins [122]. Molecular mechanism of how INM proteins cross through the NPC to reach the inner membrane is still an intriguing problem. Recent studies revealed the existence of several distinct mechanisms in terms of the usage of energy or Ran [123]. Based on early prominent studies in yeast [124, 125] and the existence of NLS-like sequences in many mammalian INM proteins, NLS receptor-mediated translocation of INM proteins has also been proposed. However, unlike transport that occurs through the central channel of the NPC, the venue of which was directly visualized by different microscopic approaches, where and how the reported INM

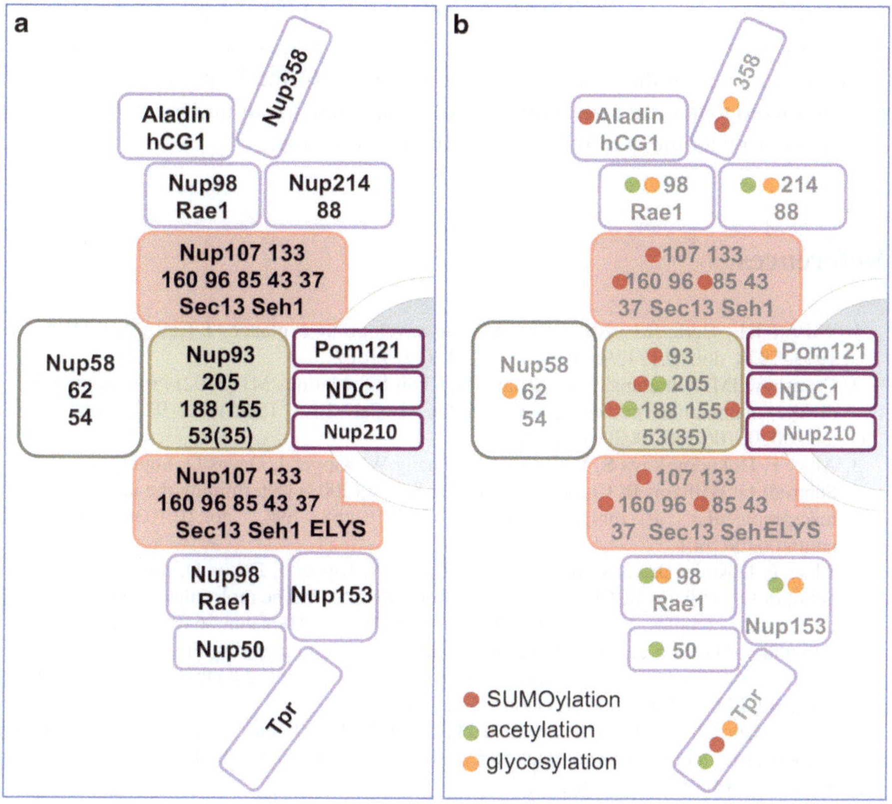

Fig. 3 Schematic diagram of vertebrate Nups. (**a**) The approximate positions of Nups in the NPC are shown. The diagram corresponds to the portion of the NPC that is *boxed* in Fig. 2. Substructures of the NPC are color coded according to Fig. 2. (**b**) Posttranslational modifications made to Nups. Nups that have been shown in proteome-based analyses to be subjected to SUMOylation [116], acetylation [119], and glycosylation (O-GlcNAc modification) [126] are marked with *circles* with different colors. Almost all Nups are subjected to phosphorylation [111–115] and ubiquitination [117, 118] in certain biological situations, and not included in this diagram are those modifications

proteins migrate within the NPC structure are still a large mystery. The transport might accompany some structural rearrangement of NPCs that we do not know yet. How and whether these different transport pathways collaborate to establish a nuclear framework are intriguing questions that should be addressed in the future.

While a number of studies show that the nuclear framework plays an important role in determining nuclear size, studies conducted in several model systems from yeast to mammals suggest that nuclear size is not determined by the DNA content but by the cytoplasmic volume around the nucleus. However, whether the chromatin state, like its compaction state, plays a role in controlling nuclear size needs to be addressed. Because it is widely accepted that the nuclear framework affects the chromatin organization and state, we also need to further investigate this relationship.

Changes in cell size and nuclear size within a single organism are often associated with developmental programs, and unwanted changes may induce diseases, including cancer. Studies focusing on the functions of the Nup/NPCs and the NE that are linked to nuclear size control should contribute to our understanding of such higher order phenomena in relation to cellular phenotype.

References

1. Wilson KL, Berk JM (2010) The nuclear envelope at a glance. J Cell Sci 123(Pt 12): 1973–1978. doi:10.1242/jcs.019042, 123/12/1973 [pii]
2. Cronshaw JM, Krutchinsky AN, Zhang W, Chait BT, Matunis MJ (2002) Proteomic analysis of the mammalian nuclear pore complex. J Cell Biol 158(5):915–927. doi:10.1083/jcb.200206106, jcb.200206106 [pii]
3. Alber F, Dokudovskaya S, Veenhoff LM, Zhang W, Kipper J, Devos D, Suprapto A, Karni-Schmidt O, Williams R, Chait BT, Rout MP, Sali A (2007) Determining the architectures of macromolecular assemblies. Nature 450(7170):683–694. doi:10.1038/nature06404, nature06404 [pii]
4. Alber F, Dokudovskaya S, Veenhoff LM, Zhang W, Kipper J, Devos D, Suprapto A, Karni-Schmidt O, Williams R, Chait BT, Sali A, Rout MP (2007) The molecular architecture of the nuclear pore complex. Nature 450(7170):695–701. doi:10.1038/nature06405, nature06405 [pii]
5. Fernandez-Martinez J, Rout MP (2012) A jumbo problem: mapping the structure and functions of the nuclear pore complex. Curr Opin Cell Biol 24(1):92–99. doi:10.1016/j.ceb.2011.12.013, S0955-0674(11)00175-X [pii]
6. Frey S, Gorlich D (2007) A saturated FG-repeat hydrogel can reproduce the permeability properties of nuclear pore complexes. Cell 130(3):512–523. doi:10.1016/j.cell.2007.06.024, S0092-8674(07)00791-X [pii]
7. Terry LJ, Shows EB, Wente SR (2007) Crossing the nuclear envelope: hierarchical regulation of nucleocytoplasmic transport. Science 318(5855):1412–1416. doi:10.1126/science.1142204, 318/5855/1412 [pii]
8. Wente SR, Rout MP (2010) The nuclear pore complex and nuclear transport. Cold Spring Harb Perspect Biol 2(10):a000562. doi:10.1101/cshperspect.a000562, cshperspect.a000562 [pii]
9. Kose S, Furuta M, Imamoto N (2012) Hikeshi, a nuclear import carrier for Hsp70s, protects cells from heat shock-induced nuclear damage. Cell 149(3):578–589. doi:10.1016/j.cell.2012.02.058, S0092-8674(12)00407-2 [pii]
10. Gerace L, Huber MD (2012) Nuclear lamina at the crossroads of the cytoplasm and nucleus. J Struct Biol 177(1):24–31. doi:10.1016/j.jsb.2011.11.007, S1047-8477(11)00322-4 [pii]
11. Zink D, Fischer AH, Nickerson JA (2004) Nuclear structure in cancer cells. Nat Rev Cancer 4(9):677–687. doi:10.1038/nrc1430, nrc1430 [pii]
12. Prokocimer M, Davidovich M, Nissim-Rafinia M, Wiesel-Motiuk N, Bar DZ, Barkan R, Meshorer E, Gruenbaum Y (2009) Nuclear lamins: key regulators of nuclear structure and activities. J Cell Mol Med 13(6):1059–1085. doi:10.1111/j.1582-4934.2008.00676.x, JCMM676 [pii]
13. Chow KH, Factor RE, Ullman KS (2012) The nuclear envelope environment and its cancer connections. Nat Rev Cancer 12(3):196–209. doi:10.1038/nrc3219, nrc3219 [pii]
14. Capelson M, Liang Y, Schulte R, Mair W, Wagner U, Hetzer MW (2010) Chromatin-bound nuclear pore components regulate gene expression in higher eukaryotes. Cell 140(3):372–383. doi:10.1016/j.cell.2009.12.054, S0092-8674(09)01681-X [pii]
15. Nakano H, Wang W, Hashizume C, Funasaka T, Sato H, Wong RW (2011) Unexpected role of nucleoporins in coordination of cell cycle progression. Cell Cycle 10(3):425–433, 14721 [pii]
16. Strambio-De-Castillia C, Niepel M, Rout MP (2010) The nuclear pore complex: bridging nuclear transport and gene regulation. Nat Rev Mol Cell Biol 11(7):490–501. doi:10.1038/nrm2928, nrm2928 [pii]

17. Walther TC, Alves A, Pickersgill H, Loiodice I, Hetzer M, Galy V, Hulsmann BB, Kocher T, Wilm M, Allen T, Mattaj IW, Doye V (2003) The conserved Nup107-160 complex is critical for nuclear pore complex assembly. Cell 113(2):195–206, S0092867403002356 [pii]
18. Franz C, Walczak R, Yavuz S, Santarella R, Gentzel M, Askjaer P, Galy V, Hetzer M, Mattaj IW, Antonin W (2007) MEL-28/ELYS is required for the recruitment of nucleoporins to chromatin and postmitotic nuclear pore complex assembly. EMBO Rep 8(2):165–172. doi:10.1038/sj.embor.7400889, 7400889 [pii]
19. D'Angelo MA, Anderson DJ, Richard E, Hetzer MW (2006) Nuclear pores form de novo from both sides of the nuclear envelope. Science 312(5772):440–443. doi:10.1126/science.1124196, 312/5772/440 [pii]
20. Levy DL, Heald R (2010) Nuclear size is regulated by importin alpha and Ntf2 in Xenopus. Cell 143(2):288–298. doi:10.1016/j.cell.2010.09.012, S0092-8674(10)01060-3 [pii]
21. Newport JW, Wilson KL, Dunphy WG (1990) A lamin-independent pathway for nuclear envelope assembly. J Cell Biol 111(6 Pt 1):2247–2259
22. Maeshima K, Iino H, Hihara S, Funakoshi T, Watanabe A, Nishimura M, Nakatomi R, Yahata K, Imamoto F, Hashikawa T, Yokota H, Imamoto N (2010) Nuclear pore formation but not nuclear growth is governed by cyclin-dependent kinases (Cdks) during interphase. Nat Struct Mol Biol 17(9):1065–1071. doi:10.1038/nsmb.1878, nsmb.1878 [pii]
23. Ribbeck K, Gorlich D (2001) Kinetic analysis of translocation through nuclear pore complexes. EMBO J 20(6):1320–1330. doi:10.1093/emboj/20.6.1320
24. Riddick G, Macara IG (2005) A systems analysis of importin-{alpha}-{beta} mediated nuclear protein import. J Cell Biol 168(7):1027–1038. doi:10.1083/jcb.200409024, jcb.200409024 [pii]
25. Theerthagiri G, Eisenhardt N, Schwarz H, Antonin W (2010) The nucleoporin Nup188 controls passage of membrane proteins across the nuclear pore complex. J Cell Biol 189(7): 1129–1142. doi:10.1083/jcb.200912045, jcb.200912045 [pii]
26. Brandt A, Papagiannouli F, Wagner N, Wilsch-Brauninger M, Braun M, Furlong EE, Loserth S, Wenzl C, Pilot F, Vogt N, Lecuit T, Krohne G, Grosshans J (2006) Developmental control of nuclear size and shape by Kugelkern and Kurzkern. Curr Biol 16(6):543–552. doi:10.1016/j.cub.2006.01.051, S0960-9822(06)01060-8 [pii]
27. Dittmer TA, Stacey NJ, Sugimoto-Shirasu K, Richards EJ (2007) LITTLE NUCLEI genes affecting nuclear morphology in Arabidopsis thaliana. Plant Cell 19(9):2793–2803. doi:10.1105/tpc.107.053231, tpc.107.053231 [pii]
28. Antonin W, Ungricht R, Kutay U (2011) Traversing the NPC along the pore membrane: targeting of membrane proteins to the INM. Nucleus 2(2):87–91. doi:10.4161/nucl.2.2.14637, 1949-1034-2-2-3 [pii]
29. Hawryluk-Gara LA, Shibuya EK, Wozniak RW (2005) Vertebrate Nup53 interacts with the nuclear lamina and is required for the assembly of a Nup93-containing complex. Mol Biol Cell 16(5):2382–2394. doi:10.1091/mbc.E04-10-0857, E04-10-0857 [pii]
30. Lussi YC, Hugi I, Laurell E, Kutay U, Fahrenkrog B (2011) The nucleoporin Nup88 is interacting with nuclear lamin A. Mol Biol Cell 22(7):1080–1090. doi:10.1091/mbc.E10-05-0463, mbc.E10-05-0463 [pii]
31. Smythe C, Jenkins HE, Hutchison CJ (2000) Incorporation of the nuclear pore basket protein nup153 into nuclear pore structures is dependent upon lamina assembly: evidence from cell-free extracts of Xenopus eggs. EMBO J 19(15):3918–3931. doi:10.1093/emboj/19.15.3918
32. Starr DA, Fridolfsson HN (2010) Interactions between nuclei and the cytoskeleton are mediated by SUN-KASH nuclear-envelope bridges. Annu Rev Cell Dev Biol 26:421–444. doi:10.1146/annurev-cellbio-100109-104037
33. Talamas JA, Hetzer MW (2011) POM121 and Sun1 play a role in early steps of interphase NPC assembly. J Cell Biol 194(1):27–37. doi:10.1083/jcb.201012154, jcb.201012154 [pii]
34. Funakoshi T, Maeshima K, Yahata K, Sugano S, Imamoto F, Imamoto N (2007) Two distinct human POM121 genes: requirement for the formation of nuclear pore complexes. FEBS Lett 581(25):4910–4916. doi:10.1016/j.febslet.2007.09.021, S0014-5793(07)01003-4 [pii]
35. Mackay DR, Elgort SW, Ullman KS (2009) The nucleoporin Nup153 has separable roles in both early mitotic progression and the resolution of mitosis. Mol Biol Cell 20(6):1652–1660. doi:10.1091/mbc.E08-08-0883, E08-08-0883 [pii]

36. Blobel G (1985) Gene gating: a hypothesis. Proc Natl Acad Sci U S A 82(24):8527–8529
37. Brickner JH, Walter P (2004) Gene recruitment of the activated INO1 locus to the nuclear membrane. PLoS Biol 2(11):e342. doi:10.1371/journal.pbio.0020342
38. Casolari JM, Brown CR, Drubin DA, Rando OJ, Silver PA (2005) Developmentally induced changes in transcriptional program alter spatial organization across chromosomes. Genes Dev 19(10):1188–1198. doi:10.1101/gad.1307205, 19/10/1188 [pii]
39. Casolari JM, Brown CR, Komili S, West J, Hieronymus H, Silver PA (2004) Genome-wide localization of the nuclear transport machinery couples transcriptional status and nuclear organization. Cell 117(4):427–439, S0092867404004489 [pii]
40. Taddei A, Van Houwe G, Hediger F, Kalck V, Cubizolles F, Schober H, Gasser SM (2006) Nuclear pore association confers optimal expression levels for an inducible yeast gene. Nature 441(7094):774–778. doi:10.1038/nature04845, nature04845 [pii]
41. Rabut G, Doye V, Ellenberg J (2004) Mapping the dynamic organization of the nuclear pore complex inside single living cells. Nat Cell Biol 6(11):1114–1121. doi:10.1038/ncb1184, ncb1184 [pii]
42. Kalverda B, Pickersgill H, Shloma VV, Fornerod M (2010) Nucleoporins directly stimulate expression of developmental and cell-cycle genes inside the nucleoplasm. Cell 140(3):360–371. doi:10.1016/j.cell.2010.01.011, S0092-8674(10)00012-7 [pii]
43. Vaquerizas JM, Suyama R, Kind J, Miura K, Luscombe NM, Akhtar A (2010) Nuclear pore proteins nup153 and megator define transcriptionally active regions in the Drosophila genome. PLoS Genet 6(2):e1000846. doi:10.1371/journal.pgen.1000846
44. Webster M, Witkin KL, Cohen-Fix O (2009) Sizing up the nucleus: nuclear shape, size and nuclear-envelope assembly. J Cell Sci 122(Pt 10):1477–1486. doi:10.1242/jcs.037333, 122/10/1477 [pii]
45. Anderson DJ, Hetzer MW (2007) Nuclear envelope formation by chromatin-mediated reorganization of the endoplasmic reticulum. Nat Cell Biol 9(10):1160–1166. doi:10.1038/ncb1636, ncb1636 [pii]
46. Anderson DJ, Vargas JD, Hsiao JP, Hetzer MW (2009) Recruitment of functionally distinct membrane proteins to chromatin mediates nuclear envelope formation in vivo. J Cell Biol 186(2):183–191. doi:10.1083/jcb.200901106, jcb.200901106 [pii]
47. Clever M, Mimura Y, Funakoshi T, Imamoto N (2013) Regulation and coordination of nuclear envelope and nuclear pore complex assembly. Nucleus 4(2):105–114. doi:10.4161/nucl.23796, 23796 [pii]
48. Harel A, Orjalo AV, Vincent T, Lachish-Zalait A, Vasu S, Shah S, Zimmerman E, Elbaum M, Forbes DJ (2003) Removal of a single pore subcomplex results in vertebrate nuclei devoid of nuclear pores. Mol Cell 11(4):853–864, S1097276503001163 [pii]
49. Rasala BA, Orjalo AV, Shen Z, Briggs S, Forbes DJ (2006) ELYS is a dual nucleoporin/kinetochore protein required for nuclear pore assembly and proper cell division. Proc Natl Acad Sci U S A 103(47):17801–17806. doi:10.1073/pnas.0608484103, 0608484103 [pii]
50. Siniossoglou S (2013) Phospholipid metabolism and nuclear function: roles of the lipin family of phosphatidic acid phosphatases. Biochim Biophys Acta 1831(3):575–581. doi:10.1016/j.bbalip.2012.09.014, S1388-1981(12)00211-9 [pii]
51. Korfali N, Wilkie GS, Swanson SK, Srsen V, de Las HJ, Batrakou DG, Malik P, Zuleger N, Kerr AR, Florens L, Schirmer EC (2012) The nuclear envelope proteome differs notably between tissues. Nucleus 3(6):552–564. doi:10.4161/nucl.22257, 22257 [pii]
52. Splinter D, Tanenbaum ME, Lindqvist A, Jaarsma D, Flotho A, Yu KL, Grigoriev I, Engelsma D, Haasdijk ED, Keijzer N, Demmers J, Fornerod M, Melchior F, Hoogenraad CC, Medema RH, Akhmanova A (2010) Bicaudal D2, dynein, and kinesin-1 associate with nuclear pore complexes and regulate centrosome and nuclear positioning during mitotic entry. PLoS Biol 8(4):e1000350. doi:10.1371/journal.pbio.1000350
53. Bolhy S, Bouhlel I, Dultz E, Nayak T, Zuccolo M, Gatti X, Vallee R, Ellenberg J, Doye V (2011) A Nup133-dependent NPC-anchored network tethers centrosomes to the nuclear envelope in prophase. J Cell Biol 192(5):855–871. doi:10.1083/jcb.201007118, jcb.201007118 [pii]

54. Wozniak R, Burke B, Doye V (2010) Nuclear transport and the mitotic apparatus: an evolving relationship. Cell Mol Life Sci 67(13):2215–2230. doi:10.1007/s00018-010-0325-7
55. Jeganathan KB, Malureanu L, van Deursen JM (2005) The Rae1-Nup98 complex prevents aneuploidy by inhibiting securin degradation. Nature 438(7070):1036–1039. doi:10.1038/nature04221, nature04221 [pii]
56. Platani M, Santarella-Mellwig R, Posch M, Walczak R, Swedlow JR, Mattaj IW (2009) The Nup107-160 nucleoporin complex promotes mitotic events via control of the localization state of the chromosome passenger complex. Mol Biol Cell 20(24):5260–5275. doi:10.1091/mbc.E09-05-0377, E09-05-0377 [pii]
57. Mackay DR, Makise M, Ullman KS (2010) Defects in nuclear pore assembly lead to activation of an Aurora B-mediated abscission checkpoint. J Cell Biol 191(5):923–931. doi:10.1083/jcb.201007124, jcb.201007124 [pii]
58. Edens LJ, White KH, Jevtic P, Li X, Levy DL (2012) Nuclear size regulation: from single cells to development and disease. Trends Cell Biol. doi:10.1016/j.tcb.2012.11.004, S0962-8924(12)00217-6 [pii]
59. Neumann FR, Nurse P (2007) Nuclear size control in fission yeast. J Cell Biol 179(4):593–600. doi:10.1083/jcb.200708054, jcb.200708054 [pii]
60. Cremer M, Kupper K, Wagler B, Wizelman L, von Hase J, Weiland Y, Kreja L, Diebold J, Speicher MR, Cremer T (2003) Inheritance of gene density-related higher order chromatin arrangements in normal and tumor cell nuclei. J Cell Biol 162(5):809–820. doi:10.1083/jcb.200304096, jcb.200304096 [pii]
61. Martinez N, Alonso A, Moragues MD, Ponton J, Schneider J (1999) The nuclear pore complex protein Nup88 is overexpressed in tumor cells. Cancer Res 59(21):5408–5411
62. Agudo D, Gomez-Esquer F, Martinez-Arribas F, Nunez-Villar MJ, Pollan M, Schneider J (2004) Nup88 mRNA overexpression is associated with high aggressiveness of breast cancer. Int J Cancer 109(5):717–720. doi:10.1002/ijc.20034
63. Xu S, Powers MA (2009) Nuclear pore proteins and cancer. Semin Cell Dev Biol 20(5):620–630. doi:10.1016/j.semcdb.2009.03.003, S1084-9521(09)00041-X [pii]
64. De Keersmaecker K, Rocnik JL, Bernad R, Lee BH, Leeman D, Gielen O, Verachtert H, Folens C, Munck S, Marynen P, Fornerod M, Gilliland DG, Cools J (2008) Kinase activation and transformation by NUP214-ABL1 is dependent on the context of the nuclear pore. Mol Cell 31(1):134–142. doi:10.1016/j.molcel.2008.05.005, S1097-2765(08)00336-5 [pii]
65. Takeda A, Sarma NJ, Abdul-Nabi AM, Yaseen NR (2010) Inhibition of CRM1-mediated nuclear export of transcription factors by leukemogenic NUP98 fusion proteins. J Biol Chem 285(21):16248–16257. doi:10.1074/jbc.M109.048785, M109.048785 [pii]
66. Hernandez P, Sole X, Valls J, Moreno V, Capella G, Urruticoechea A, Pujana MA (2007) Integrative analysis of a cancer somatic mutome. Mol Cancer 6:13. doi:10.1186/1476-4598-6-13, 1476-4598-6-13 [pii]
67. Kau TR, Way JC, Silver PA (2004) Nuclear transport and cancer: from mechanism to intervention. Nat Rev Cancer 4(2):106–117. doi:10.1038/nrc1274, nrc1274 [pii]
68. Beg AA, Ruben SM, Scheinman RI, Haskill S, Rosen CA, Baldwin AS Jr (1992) I kappa B interacts with the nuclear localization sequences of the subunits of NF-kappa B: a mechanism for cytoplasmic retention. Genes Dev 6(10):1899–1913
69. Ganchi PA, Sun SC, Greene WC, Ballard DW (1992) I kappa B/MAD-3 masks the nuclear localization signal of NF-kappa B p65 and requires the transactivation domain to inhibit NF-kappa B p65 DNA binding. Mol Biol Cell 3(12):1339–1352
70. Henkel T, Zabel U, van Zee K, Muller JM, Fanning E, Baeuerle PA (1992) Intramolecular masking of the nuclear location signal and dimerization domain in the precursor for the p50 NF-kappa B subunit. Cell 68(6):1121–1133, 0092-8674(92)90083-O [pii]
71. Gluz O, Wild P, Meiler R, Diallo-Danebrock R, Ting E, Mohrmann S, Schuett G, Dahl E, Fuchs T, Herr A, Gaumann A, Frick M, Poremba C, Nitz UA, Hartmann A (2008) Nuclear karyopherin alpha2 expression predicts poor survival in patients with advanced breast cancer irrespective of treatment intensity. Int J Cancer 123(6):1433–1438. doi:10.1002/ijc.23628

72. Riddick G, Macara IG (2007) The adapter importin-alpha provides flexible control of nuclear import at the expense of efficiency. Mol Syst Biol 3:118. doi:10.1038/msb4100160, msb4100160 [pii]
73. Maul GG, Maul HM, Scogna JE, Lieberman MW, Stein GS, Hsu BY, Borun TW (1972) Time sequence of nuclear pore formation in phytohemagglutinin-stimulated lymphocytes and in HeLa cells during the cell cycle. J Cell Biol 55(2):433–447
74. Maeshima K, Yahata K, Sasaki Y, Nakatomi R, Tachibana T, Hashikawa T, Imamoto F, Imamoto N (2006) Cell-cycle-dependent dynamics of nuclear pores: pore-free islands and lamins. J Cell Sci 119(Pt 21):4442–4451. doi:10.1242/jcs.03207, 119/21/4442 [pii]
75. Richard MN, Deniset JF, Kneesh AL, Blackwood D, Pierce GN (2007) Mechanical stretching stimulates smooth muscle cell growth, nuclear protein import, and nuclear pore expression through mitogen-activated protein kinase activation. J Biol Chem 282(32):23081–23088. doi:10.1074/jbc.M703602200, M703602200 [pii]
76. Lewin JM, Lwaleed BA, Cooper AJ, Birch BR (2007) The direct effect of nuclear pores on nuclear chemotherapeutic concentration in multidrug resistant bladder cancer: the nuclear sparing phenomenon. J Urol 177(4):1526–1530. doi:10.1016/j.juro.2006.11.048, S0022-5347(06)03131-4 [pii]
77. Raices M, D'Angelo MA (2012) Nuclear pore complex composition: a new regulator of tissue-specific and developmental functions. Nat Rev Mol Cell Biol 13(11):687–699. doi:10.1038/nrm3461, nrm3461 [pii]
78. D'Angelo MA, Gomez-Cavazos JS, Mei A, Lackner DH, Hetzer MW (2012) A change in nuclear pore complex composition regulates cell differentiation. Dev Cell 22(2):446–458. doi:10.1016/j.devcel.2011.11.021, S1534-5807(11)00531-4 [pii]
79. Olsson M, Scheele S, Ekblom P (2004) Limited expression of nuclear pore membrane glycoprotein 210 in cell lines and tissues suggests cell-type specific nuclear pores in metazoans. Exp Cell Res 292(2):359–370, S0014482703004889 [pii]
80. Dauer WT, Worman HJ (2009) The nuclear envelope as a signaling node in development and disease. Dev Cell 17(5):626–638. doi:10.1016/j.devcel.2009.10.016, S1534-5807(09)00438-9 [pii]
81. Tullio-Pelet A, Salomon R, Hadj-Rabia S, Mugnier C, de Laet MH, Chaouachi B, Bakiri F, Brottier P, Cattolico L, Penet C, Begeot M, Naville D, Nicolino M, Chaussain JL, Weissenbach J, Munnich A, Lyonnet S (2000) Mutant WD-repeat protein in triple-A syndrome. Nat Genet 26(3):332–335. doi:10.1038/81642
82. Zhang X, Chen S, Yoo S, Chakrabarti S, Zhang T, Ke T, Oberti C, Yong SL, Fang F, Li L, de la Fuente R, Wang L, Chen Q, Wang QK (2008) Mutation in nuclear pore component NUP155 leads to atrial fibrillation and early sudden cardiac death. Cell 135(6):1017–1027. doi:10.1016/j.cell.2008.10.022, S0092-8674(08)01314-7 [pii]
83. Basel-Vanagaite L, Muncher L, Straussberg R, Pasmanik-Chor M, Yahav M, Rainshtein L, Walsh CA, Magal N, Taub E, Drasinover V, Shalev H, Attia R, Rechavi G, Simon AJ, Shohat M (2006) Mutated nup62 causes autosomal recessive infantile bilateral striatal necrosis. Ann Neurol 60(2):214–222. doi:10.1002/ana.20902
84. Neilson DE, Adams MD, Orr CM, Schelling DK, Eiben RM, Kerr DS, Anderson J, Bassuk AG, Bye AM, Childs AM, Clarke A, Crow YJ, Di Rocco M, Dohna-Schwake C, Dueckers G, Fasano AE, Gika AD, Gionnis D, Gorman MP, Grattan-Smith PJ, Hackenberg A, Kuster A, Lentschig MG, Lopez-Laso E, Marco EJ, Mastroyianni S, Perrier J, Schmitt-Mechelke T, Servidei S, Skardoutsou A, Uldall P, van der Knaap MS, Goglin KC, Tefft DL, Aubin C, de Jager P, Hafler D, Warman ML (2009) Infection-triggered familial or recurrent cases of acute necrotizing encephalopathy caused by mutations in a component of the nuclear pore, RANBP2. Am J Hum Genet 84(1):44–51. doi:10.1016/j.ajhg.2008.12.009, S0002-9297(08)00630-7 [pii]
85. Wu M, Allis CD, Richman R, Cook RG, Gorovsky MA (1986) An intervening sequence in an unusual histone H1 gene of Tetrahymena thermophila. Proc Natl Acad Sci U S A 83(22):8674–8678
86. Wu M, Allis CD, Sweet MT, Cook RG, Thatcher TH, Gorovsky MA (1994) Four distinct and unusual linker proteins in a mitotically dividing nucleus are derived from a 71-kilodalton polyprotein, lack p34cdc2 sites, and contain protein kinase A sites. Mol Cell Biol 14(1):10–20

87. Shen X, Yu L, Weir JW, Gorovsky MA (1995) Linker histones are not essential and affect chromatin condensation in vivo. Cell 82(1):47–56, 0092-8674(95)90051-9 [pii]
88. Malone CD, Falkowska KA, Li AY, Galanti SE, Kanuru RC, LaMont EG, Mazzarella KC, Micev AJ, Osman MM, Piotrowski NK, Suszko JW, Timm AC, Xu MM, Liu L, Chalker DL (2008) Nucleus-specific importin alpha proteins and nucleoporins regulate protein import and nuclear division in the binucleate Tetrahymena thermophila. Eukaryot Cell 7(9): 1487–1499. doi:10.1128/EC.00193-08, EC.00193-08 [pii]
89. Iwamoto M, Mori C, Kojidani T, Bunai F, Hori T, Fukagawa T, Hiraoka Y, Haraguchi T (2009) Two distinct repeat sequences of Nup98 nucleoporins characterize dual nuclei in the binucleated ciliate tetrahymena. Curr Biol 19(10):843–847. doi:10.1016/j.cub.2009.03.055, S0960-9822(09)00909-9 [pii]
90. Chakraborty P, Wang Y, Wei JH, van Deursen J, Yu H, Malureanu L, Dasso M, Forbes DJ, Levy DE, Seemann J, Fontoura BM (2008) Nucleoporin levels regulate cell cycle progression and phase-specific gene expression. Dev Cell 15(5):657–667. doi:10.1016/j.devcel.2008.08.020, S1534-5807(08)00344-4 [pii]
91. D'Angelo MA, Raices M, Panowski SH, Hetzer MW (2009) Age-dependent deterioration of nuclear pore complexes causes a loss of nuclear integrity in postmitotic cells. Cell 136(2):284–295. doi:10.1016/j.cell.2008.11.037, S0092-8674(08)01512-2 [pii]
92. Galy V, Mattaj IW, Askjaer P (2003) Caenorhabditis elegans nucleoporins Nup93 and Nup205 determine the limit of nuclear pore complex size exclusion in vivo. Mol Biol Cell 14(12):5104–5115. doi:10.1091/mbc.E03-04-0237, E03-04-0237 [pii]
93. Patre M, Tabbert A, Hermann D, Walczak H, Rackwitz HR, Cordes VC, Ferrando-May E (2006) Caspases target only two architectural components within the core structure of the nuclear pore complex. J Biol Chem 281(2):1296–1304. doi:10.1074/jbc.M511717200, M511717200 [pii]
94. Antonin W, Ellenberg J, Dultz E (2008) Nuclear pore complex assembly through the cell cycle: regulation and membrane organization. FEBS Lett 582(14):2004–2016. doi:10.1016/j.febslet.2008.02.067, S0014-5793(08)00190-7 [pii]
95. Glavy JS, Krutchinsky AN, Cristea IM, Berke IC, Boehmer T, Blobel G, Chait BT (2007) Cell-cycle-dependent phosphorylation of the nuclear pore Nup107-160 subcomplex. Proc Natl Acad Sci U S A 104(10):3811–3816. doi:10.1073/pnas.0700058104, 0700058104 [pii]
96. Onischenko EA, Gubanova NV, Kiseleva EV, Hallberg E (2005) Cdk1 and okadaic acid-sensitive phosphatases control assembly of nuclear pore complexes in Drosophila embryos. Mol Biol Cell 16(11):5152–5162. doi:10.1091/mbc.E05-07-0642, E05-07-0642 [pii]
97. Doucet CM, Talamas JA, Hetzer MW (2010) Cell cycle-dependent differences in nuclear pore complex assembly in metazoa. Cell 141(6):1030–1041. doi:10.1016/j.cell.2010.04.036, S0092-8674(10)00490-3 [pii]
98. Dultz E, Ellenberg J (2010) Live imaging of single nuclear pores reveals unique assembly kinetics and mechanism in interphase. J Cell Biol 191(1):15–22. doi:10.1083/jcb.201007076, jcb.201007076 [pii]
99. Laurell E, Beck K, Krupina K, Theerthagiri G, Bodenmiller B, Horvath P, Aebersold R, Antonin W, Kutay U (2011) Phosphorylation of Nup98 by multiple kinases is crucial for NPC disassembly during mitotic entry. Cell 144(4):539–550. doi:10.1016/j.cell.2011.01.012, S0092-8674(11)00013-4 [pii]
100. De Souza CP, Osmani AH, Hashmi SB, Osmani SA (2004) Partial nuclear pore complex disassembly during closed mitosis in Aspergillus nidulans. Curr Biol 14(22):1973–1984. doi:10.1016/j.cub.2004.10.050, S096098220400853X [pii]
101. Osmani AH, Davies J, Liu HL, Nile A, Osmani SA (2006) Systematic deletion and mitotic localization of the nuclear pore complex proteins of Aspergillus nidulans. Mol Biol Cell 17(12):4946–4961. doi:10.1091/mbc.E06-07-0657, E06-07-0657 [pii]
102. Makhnevych T, Lusk CP, Anderson AM, Aitchison JD, Wozniak RW (2003) Cell cycle regulated transport controlled by alterations in the nuclear pore complex. Cell 115(7):813–823, S0092867403009863 [pii]
103. Cairo LV, Ptak C, Wozniak RW (2013) Mitosis-specific regulation of nuclear transport by the spindle assembly checkpoint protein Mad1p. Mol Cell 49(1):109–120. doi:10.1016/j.molcel.2012.10.017, S1097-2765(12)00894-5 [pii]

104. Frey S, Gorlich D (2009) FG/FxFG as well as GLFG repeats form a selective permeability barrier with self-healing properties. EMBO J 28(17):2554–2567. doi:10.1038/emboj.2009.199, emboj2009199 [pii]
105. Hulsmann BB, Labokha AA, Gorlich D (2012) The permeability of reconstituted nuclear pores provides direct evidence for the selective phase model. Cell 150(4):738–751. doi:10.1016/j.cell.2012.07.019, S0092-8674(12)00887-2 [pii]
106. Labokha AA, Gradmann S, Frey S, Hulsmann BB, Urlaub H, Baldus M, Gorlich D (2012) Systematic analysis of barrier-forming FG hydrogels from Xenopus nuclear pore complexes. EMBO J 32(2):204–218. doi:10.1038/emboj.2012.302, emboj2012302 [pii]
107. Finlay DR, Newmeyer DD, Price TM, Forbes DJ (1987) Inhibition of in vitro nuclear transport by a lectin that binds to nuclear pores. J Cell Biol 104(2):189–200
108. Hanover JA, Cohen CK, Willingham MC, Park MK (1987) O-linked N-acetylglucosamine is attached to proteins of the nuclear pore. Evidence for cytoplasmic and nucleoplasmic glycoproteins. J Biol Chem 262(20):9887–9894
109. Holt GD, Snow CM, Senior A, Haltiwanger RS, Gerace L, Hart GW (1987) Nuclear pore complex glycoproteins contain cytoplasmically disposed O-linked N-acetylglucosamine. J Cell Biol 104(5):1157–1164
110. Daub H, Olsen JV, Bairlein M, Gnad F, Oppermann FS, Korner R, Greff Z, Keri G, Stemmann O, Mann M (2008) Kinase-selective enrichment enables quantitative phosphoproteomics of the kinome across the cell cycle. Mol Cell 31(3):438–448. doi:10.1016/j.molcel.2008.07.007, S1097-2765(08)00493-0 [pii]
111. Dephoure N, Zhou C, Villen J, Beausoleil SA, Bakalarski CE, Elledge SJ, Gygi SP (2008) A quantitative atlas of mitotic phosphorylation. Proc Natl Acad Sci U S A 105(31): 10762–10767. doi:10.1073/pnas.0805139105, 0805139105 [pii]
112. Nousiainen M, Sillje HH, Sauer G, Nigg EA, Korner R (2006) Phosphoproteome analysis of the human mitotic spindle. Proc Natl Acad Sci U S A 103(14):5391–5396. doi:10.1073/pnas.0507066103, 0507066103 [pii]
113. Kosako H, Yamaguchi N, Aranami C, Ushiyama M, Kose S, Imamoto N, Taniguchi H, Nishida E, Hattori S (2009) Phosphoproteomics reveals new ERK MAP kinase targets and links ERK to nucleoporin-mediated nuclear transport. Nat Struct Mol Biol 16(10): 1026–1035. doi:10.1038/nsmb.1656, nsmb.1656 [pii]
114. Santamaria A, Wang B, Elowe S, Malik R, Zhang F, Bauer M, Schmidt A, Sillje HH, Korner R, Nigg EA (2011) The Plk1-dependent phosphoproteome of the early mitotic spindle. Mol Cell Proteomics 10(1):M110 004457. doi:10.1074/mcp.M110.004457, M110.004457 [pii]
115. Kettenbach AN, Schweppe DK, Faherty BK, Pechenick D, Pletnev AA, Gerber SA (2011) Quantitative phosphoproteomics identifies substrates and functional modules of Aurora and Polo-like kinase activities in mitotic cells. Sci Signal 4(179):rs5. doi:10.1126/scisignal.2001497, 4/179/rs5 [pii]
116. Galisson F, Mahrouche L, Courcelles M, Bonneil E, Meloche S, Chelbi-Alix MK, Thibault P (2011) A novel proteomics approach to identify SUMOylated proteins and their modification sites in human cells. Mol Cell Proteomics 10(2):M110 004796. doi:10.1074/mcp.M110.004796, M110.004796 [pii]
117. Oshikawa K, Matsumoto M, Oyamada K, Nakayama KI (2012) Proteome-wide identification of ubiquitylation sites by conjugation of engineered lysine-less ubiquitin. J Proteome Res 11(2):796–807. doi:10.1021/pr200668y
118. Wagner SA, Beli P, Weinert BT, Nielsen ML, Cox J, Mann M, Choudhary C (2011) A proteome-wide, quantitative survey of in vivo ubiquitylation sites reveals widespread regulatory roles. Mol Cell Proteomics 10(10):M111 013284. doi:10.1074/mcp.M111.013284, M111.013284 [pii]
119. Choudhary C, Kumar C, Gnad F, Nielsen ML, Rehman M, Walther TC, Olsen JV, Mann M (2009) Lysine acetylation targets protein complexes and co-regulates major cellular functions. Science 325(5942):834–840. doi:10.1126/science.1175371, 1175371 [pii]
120. Xie X, Feng S, Vuong H, Liu Y, Goodison S, Lubman DM (2010) A comparative phosphoproteomic analysis of a human tumor metastasis model using a label-free quantitative approach. Electrophoresis 31(11):1842–1852. doi:10.1002/elps.200900752

121. Loewinger L, McKeon F (1988) Mutations in the nuclear lamin proteins resulting in their aberrant assembly in the cytoplasm. EMBO J 7(8):2301–2309
122. Mical TI, Monteiro MJ (1998) The role of sequences unique to nuclear intermediate filaments in the targeting and assembly of human lamin B: evidence for lack of interaction of lamin B with its putative receptor. J Cell Sci 111(Pt 23):3471–3485
123. Zuleger N, Kelly DA, Richardson AC, Kerr AR, Goldberg MW, Goryachev AB, Schirmer EC (2011) System analysis shows distinct mechanisms and common principles of nuclear envelope protein dynamics. J Cell Biol 193(1):109–123. doi:10.1083/jcb.201009068, jcb.201009068 [pii]
124. King MC, Lusk CP, Blobel G (2006) Karyopherin-mediated import of integral inner nuclear membrane proteins. Nature 442(7106):1003–1007. doi:10.1038/nature05075, nature05075 [pii]
125. Meinema AC, Laba JK, Hapsari RA, Otten R, Mulder FA, Kralt A, van den Bogaart G, Lusk CP, Poolman B, Veenhoff LM (2011) Long unfolded linkers facilitate membrane protein import through the nuclear pore complex. Science 333(6038):90–93. doi:10.1126/science.1205741, science.1205741 [pii]
126. Wang Z, Udeshi ND, Slawson C, Compton PD, Sakabe K, Cheung WD, Shabanowitz J, Hunt DF, Hart GW (2010) Extensive Crosstalk between O-GlcNAcylation and phosphorylation regulates cytokinesis. Sci Signal 3(104):ra2. doi: 10.1126/scisignal.2000526

Do Lamins Influence Disease Progression in Cancer?

Christopher J. Hutchison

Abstract For nearly 60 years, diagnosis of cancer has been based on pathological tests that look for enlargement and distortion of nuclear shape. Because of their involvement in supporting nuclear architecture, it has been postulated that the basis for nuclear shape changes during cancer progression is altered expression of nuclear lamins and in particular lamins A and C. However, studies on lamin expression patterns in a range of different cancers have generated equivocal and apparently contradictory results. This might have been anticipated since cancers are diverse and complex diseases. Moreover, whilst altered epigenetic control over gene expression is a feature of many cancers, this level of control cannot be considered in isolation. Here I have reviewed those studies relating to altered expression of lamins in cancers and argue that consideration of changes in the expression of individual lamins cannot be considered in isolation but only in the context of an understanding of their functions in transformed cells.

Keywords Lamin • Nuclear lamina • Lung cancer • Breast cancer • Ovarian cancer • Prostate cancer • Colon cancer

Abbreviations

BCC	Basal cell carcinoma
BBD	Benign breast disease
EMT	Epithelial to mesenchymal transition

C.J. Hutchison (✉)
School of Biological and Biomedical Sciences, Durham University,
South Road, Durham DH1 3LE, UK
e-mail: c.j.hutchison@durham.ac.uk

E.C. Schirmer and J.I. de las Heras (eds.), *Cancer Biology and the Nuclear Envelope*, Advances in Experimental Medicine and Biology 773, DOI 10.1007/978-1-4899-8032-8_27,

LDMDs	Lamin B-deficient microdomains
SCLC	Small-cell lung carcinoma
SCC	Squamous cell carcinoma

Cancers and Cancer Progression

Cancers are diseases that arise through growth regulation defects but progress because cells lose cohesion with their tissue of origin, undergo architectural changes that allow them to invade surrounding tissues, and colonize distant organ systems. Broadly, cancers are classified as benign or malignant. Benign cancers grow locally without invading surrounding tissues and are only harmful if they press on and damage vital organs. In contrast, malignant cancers invade surrounding tissues and give rise to secondary tumors or metastases in organ systems that are removed from the primary tumor. Most human cancers arise in epithelial tissues, and these account for ~80 % of cancer-related deaths. In order to spread, these cancers must undergo very dramatic architectural changes in a process referred to as an epithelial to mesenchymal transition (EMT). The remaining tumors fall broadly into three groups: sarcomas arise in soft tissues of mesenchymal origin and represent only 1 % of human cancers; hematological malignancies arise from blood-forming tissues including cells of the immune system; finally neuroectodermal tumors include those cancers that arise in tissue derived from neuroectoderm and, whilst rare (~1.3 %), account for a disproportionate number of cancer-related deaths [1]. As alluded to above, cancers arise through a complex series of events, some of which occur through a sequence of somatic mutations and some through architectural remodeling that might arise through epigenetic processes. Of particular importance in architectural remodeling is altered expression of cytoskeleton proteins, cell adhesion molecules, and components of the extracellular matrix. Over the past 15 years the contribution of nuclear architectural proteins such as the nuclear lamins to this architectural remodeling has been investigated, but as yet no clear consensus on their contribution to cancer progression has emerged [2]. Historically, it makes every sense to assume that changes in nuclear architecture are central to cancer progression because historically the most common pathological feature of cancers is altered nuclear architecture. In this chapter I will summarize the historical context in which lamins and cancer collide and attempt to resolve the apparent contradictions in the underpinning literature.

Nuclear Morphology as Diagnostic and Prognostic Markers

For several decades now, nuclear morphology has been used to both diagnose malignancy and to predict patient outcomes [3]. Today the best known diagnostic test that uses nuclear morphology to detect malignancy is the cervical Papanicolaou (PAP) smear test [4]. In the test, enlarged and irregularly shaped nuclei are used for the initial diagnosis of cervical and uterine cancer. However, the same criteria can be applied to ovarian cancer, which, in combination with genetic changes, can be used to distinguish low- from high-grade cancers [5]. Clinical studies on breast cancer developed a semiquantitative scoring system based on three criteria: tubule formation within the tumor, nuclear pleomorphism, and mitotic counts. Of these three criteria nuclear pleomorphism included both quantitative and qualitative features, namely, nuclear size, regularity of shape, and uniformity of chromatin. In low-grade tumors (well differentiated) nuclei are small and round and have uniformly distributed chromatin. In contrast, in high-grade (poorly differentiated) tumors, nuclei vary in size, have "bizarre" shapes, and possess prominent and multiple nucleoli [6]. In breast cancer alteration in nuclear morphology might precede malignancy. Benign breast disease (BBD) is a condition that includes proliferative changes in breast tissue and is a risk factor associated with malignancy. In one control study it was reported that altered nuclear shape might identify those individuals with BBD that will go on to develop breast cancer [7]. Thus changes that lead to altered nuclear morphology might also promote malignancy. As a consequence, understanding which proteins support nuclear morphology may lead to better and more refined diagnostic and prognostic tools.

The Nuclear Matrix and Cancer

A number of studies have used sequential extraction of DNA, RNA, and soluble proteins to attempt to understand nuclear architecture and to identify proteins that support it. Early studies termed this structure the nuclear matrix or the nucleoskeleton, and at an ultrastructural level it appears as a network of filaments throughout the nucleus [8]. Whether or not this structure actually exists as opposed to being an artefact of the conditions of extraction has been questioned [9]. Nevertheless, comparisons of the biochemical composition of nuclear matrices between normal and malignant cells suggest that there are characteristic subsets of proteins in this compartment, which distinguish cancerous from normal cells and could be used as diagnostic or prognostic tools [10]. Most prominent amongst proteins that might form the nuclear matrix are non-myogenic nuclear actin, NuMA, nuclear lamins, LAP2a, and nuclear pore-associated proteins Nup153 topoisomerases I and II and Tpr [11]. Of these only the lamins have been systematically tested as potential diagnostic or prognostic tools, and so the rest of the chapter concentrates on studies relating to lamins.

Lamins and Lung Cancer

Lung cancers can be divided into two major disease types termed non-small-cell lung carcinoma (non-SCLC) and small-cell lung carcinoma (SCLC). Of these two types SCLC is poorly differentiated and has a markedly more rapid rate of progression. Since treating metastases of these two cancers involves completely different chemotherapy regimens, their correct diagnosis is critical [12]. In these cancers chromatin organization is the basis of diagnosis. At the level of light microscopy, chromatin is diffuse in SCLC whereas it is coarse and clumped in non-SCLC. At an ultrastructural level, this difference in appearance arises from a greater abundance of heterochromatin in the non-SCLC [13]. Two studies have attributed these chromatin structural differences to altered patterns of expression of nuclear lamins. By comparing levels of expression of lamins in non-SCLC and SCLC cell lines it was found that the expression of B-type lamins did not vary. In contrast, there was an 80 % reduction in the level of expression of lamins A/C in SCLC compared to non-SCLC lines. Furthermore, when SCLC cell lines were induced to differentiate into large (non-small)-cell carcinomas following transfection with the v-ras oncogene, expression of lamins A/C increased >10-fold whereas expression of a range of other nuclear proteins including B-type lamins, topoisomerases I and II, and B23/nucleophosmin did not change [14]. Different results were reported in a separate study. Similarly comparing lamin expression in lung carcinoma cell lines, B-type lamin expression was not noticeably different in those derived from SCLC and non-SCLC. In contrast, A-type lamins were expressed in all non-SCLC cell lines but were absent from 14 out of 16 SCLC cell lines—the opposite result from the first study. However, the patterns of expression were much more complex in cancer tissues. By staining of frozen sections of SCLC and non-SCLC, B-type lamins were expressed in all tumors, but in some non-SCLC (11 out of 23) a considerable proportion of tumor cells were negative for B-type lamins. In contrast, A-type lamins were absent or expressed at low levels in 14 out of 15 cases of SCLC but were expressed in all non-SCLC [15]. Neither of these studies attempted to understand whether these altered patterns of lamin expression contributed to the more rapid progression of SCLC. In retrospect, however, it is possible to understand the major diagnostic distinctions between the two cancer types (altered abundance of heterochromatin) in the light of lamin expression. A major feature of laminopathy disease is a loss of peripheral heterochromatin [16]. Thus in SCLC the dispersal and relative absence of heterochromatin may well arise from the lack of expression of A-type lamins.

Lamins and Breast and Ovarian Cancer

There is no particular reason to group breast and ovarian cancer (other than as examples of gender-specific cancers), and the grouping here is due to a paucity of studies as opposed to any other reason. Relating back to an earlier discussion, pathologists use nuclear morphology as a basis for diagnosis and in some cases prognosis in both diseases [5, 6]. In each diagnosis, nuclear shape is not considered in isolation but in combination with other disease features such as mitotic index and genetic defects such as aneuploidy [17]. Therefore it could be useful to understand if these two features are linked by a single determining factor. In a recent study, the expression of lamins was investigated in a very limited number of breast cancers and breast cancer cell lines. In breast cancer tissue, lamins A/C were either absent or aberrantly expressed in cancerous cells but were highly expressed in cells of surrounding normal tissue. In breast cancer cell lines, patterns of expression were more variable with some cell lines expressing normal levels of lamins A/C, whilst in other cell lines both A-type lamins were absent. In the same cell lines expression of lamin B1 and nuclear envelope transmembrane protein emerin were either unaffected or increased relative to normal breast epithelial cells [18]. In an attempt to link lamin expression to diagnostic features of the disease, lamin A/C expression was silenced in normal breast epithelial cells using shRNA. Silencing of lamin A/C expression led to irregularly shaped nuclei and aneuploidy, suggesting that loss of lamin A/C expression is linked to disease progression [18]. Using very similar approaches, lamin A/C expression and the consequences of their loss were investigated in ovarian cancer. In both ovarian cancer tissue and ovarian cancer cell lines lamin A/C expression was absent in 47 % of cases and heterogeneous in the remainder. Lamin A/C silencing in ovarian cancer cell lines led to an increase in nuclear volume and aneuploidy. Interestingly, in this instance those cells that became aneuploid appeared to undergo p53- and p21-induced growth arrest [19]. Thus, in this example loss of expression of lamin A/C is associated with diagnostic features of the disease but might retard rather than promote cancer progression.

Lamins and Hematological Malignancies

One of the problems in interpreting some of the studies referred to thus far is that findings were based on relatively small numbers of cell lines or cancer tissue samples and therefore lacked statistical power. More recent studies have looked at much larger numbers of patients and have attempted to link lamin expression to either disease-free survival or overall survival using Kaplan–Meier statistical tests. A very early study noted an absence of lamin A and C transcripts (but not B-type lamins) in neoplastic cells from patients with acute lymphoblastic leukaemia and non-Hodgkin's lymphoma. Interestingly, lamin A and C transcripts could be detected in normal peripheral blood lymphocytes but only after mitogenic stimulation indicating that repression of *LMNA*, which codes for both lamin A and lamin C, was in some way related to the differentiation of lymphoid cells [20]. To understand the mechanism underlying lamin A/C repression and to investigate its link to survival, Agrelo and co-workers analyzed the promoter methylation status in human cancer cell lines from 17 tumor types as well as in primary leukaemias and lymphomas [21]. Importantly, the patient numbers were sufficient to guarantee statistical power in the study.

Although the cell lines used (70 in total) represented 17 different tumor types, only hematological malignancies were hypermethylated at *LMNA* promoters. Importantly, when *LMNA* CpG island-promoter methylation was found, it was directly linked to silencing. In primary cancers *LMNA* hypermethylation was detected in a minority (34 %) of nodal diffuse large B-cell lymphoma and Burkitt's lymphoma

(17 %) but was absent from cutaneous T-cell lymphoma, follicular lymphoma, and extranodal diffuse large B-cell lymphoma. In the case of nodal diffuse large B-cell lymphoma, *LMNA* promoter hypermethylation was associated with a statistically significant decrease in disease-free survival and overall survival, implying that *LMNA* silencing is a predictor of poor outcome only in this hematological cancer [21].

Lamins and Cancers of the Gastrointestinal Tract

Perhaps the best-studied cancers with respect to lamin expression are those of the gastrointestinal tract including the oesophagus, stomach, colon, and rectum. In an early study, immunohistochemistry was used to investigate the expression of lamins A/C and B1 in a variety of gastrointestinal cancers. On the whole, the results were quite variable although several key findings emerged. First, lamin A/C expression was reduced or absent from cancer cells in the majority of gastrointestinal cancers compared to normal epithelium and the protein was sometimes redistributed to the cytoplasm. Similar results were also found for lamin B1, although complete absence was less common, except in gastric adenocarcinoma. Whilst important, this study was mainly descriptive and did not attempt to link expression to cancer progression or survival [22]. In a more extensive study, lamin A/C expression was investigated in a large cohort (656) of colorectal cancer patients for whom survival data was available. Patients were found to either strongly express these lamins in cancer tissue (70 % of patients) or to display a complete absence of expression in cancer tissue (30 % of patients). Unexpectedly, patients that expressed lamin A/C in their tumor were twice as likely to die from cancer-related causes compared to patients without lamin A/C expression in their tumors (Cox's hazard ratio = 1.85, $p \leq 0.005$—[23]). In stark contrast to this report, an investigation of cancer tissue collected from 370 patients with stage II and III colon cancer showed an association of lamin A/C absence with disease recurrence and poor survival. However, poor prognosis was only found in patients with stage III colon cancer who had not received adjuvant chemotherapy ($p \leq 0.01$—[24]). There were several important differences between the two studies that might account for these different results. Firstly, in the earlier study stage I adenocarcinomas were included as were rectal cancers and the study did not distinguish between patients who had and patients who had not received adjuvant chemotherapy [23]. If the differences are due to the inclusion of other regions of the intestinal tract, it implies that variation in lamin expression might reflect some aspect of gut architecture that differs depending upon location. This suggestion is reinforced by a third study on primary gastric carcinomas (GC). In this study 126 GC tissue samples were investigated for lamin A/C expression at both an mRNA and a protein level. Decreased levels of lamin A/C expression were observed in 56 % of patients, and this was correlated with poor differentiation ($p \leq 0.034$) and poorer prognosis ($p \leq 0.034$—[25]).

Lamins and Prostate Cancer

In a recent proteomic study, paired (benign and tumor) samples were collected from 23 low-grade and 26 high-grade tumors and subjected to two-dimensional differential gel electrophoresis (2D-DIGE). Of 19 abundant proteins that were identified by mass spectrometry as differentially expressed between tumor grades, lamin A was statistically highly discriminatory between low- and high-grade tumors ($p \leq 0.0003$—[26]). In a follow-up study, the same group found that lamin A/C expression was concentrated at the invasive front of prostate cancer tissue. They went on to show that silencing of lamin A/C in prostate cancer cell lines inhibited cell growth, colony formation, migration, and invasion, whilst over-expression of lamin A/C stimulated the same processes [27]. In a complementary study Helfand and co-workers showed that instead of its normal uniform perinuclear distribution, prostate cancer cell lines are enriched for lamin B-deficient microdomains (LDMDs), which in turn over-express lamins A/C. In human prostate cancer tissue, the frequency of occurrence of LDMDs increased with tumor grade, again suggesting that increased expression of lamin A/C, in this case associated with lamin B1 deficiency, is correlated with tumor progression [28].

Lamins and Skin Cancers

Skin cancers are very heterogeneous and include sunlight-induced cancers such as melanoma, which can progress very rapidly; squamous cell carcinomas (SCCs), which have intermediate progression rates; and basal cell carcinomas (BCCs), which grow very slowly and rarely metastasize [29]. As their names suggest, the different skin cancers reflect their origins within the epidermis. Melanomas arise from melanocytes, BCCs from basal keratinocytes, and SCCs from the squamous layers of the skin. Studies on the expression of lamins in skin cancers also highlight why it might be that it has been hard to associate lamin expression with a consistent outcome in different and even in similar patient groups. In normal skin lamin A is expressed in suprabasal keratinocytes but is absent from basal keratinocytes. Lamin C is mostly absent from basal keratinocytes but can be observed in some basal cells. In contrast, lamins B1 and B2 are expressed in all epidermal layers although curiously lamin B1 is depleted in dermal fibroblasts [30]. In BCC lamin B1 and B2 expression was relatively constant whilst lamin A and lamin C varied considerably. In BCCs showing a high proliferative index, lamin A was typically absent or expressed at low levels. In contrast lamin A expression was higher in BCCs with a lower proliferation index but lamin C was absent or displaced from the nuclear periphery [30, 31]. In contrast, in SCCs lamin A was expressed at relatively high levels but lamin C was absent or displaced [31]. In order to explain these observations the authors suggested that the origins and progression of the cancers mapped to their lamin expression patterns in the different layers of a particular tissue. Since BCCs arise from basal cells which lack lamin A at an early stage they are lamin A negative and lamin C positive and have a relatively high proliferative index. As these cancers differentiate and their growth rates slow down, lamin A expression is up-regulated but lamin C is downregulated. Finally, SCCs that arise from squamous (lamin A positive) cells express lamin A, but this does not appear to impair proliferation rates [31]. What these two studies illustrate is a need to understand how lamin expression is linked to normal tissue organization before any strong links with tumor progression can be made or understood.

Mechanisms and Conclusions

At face value it is hard to reconcile some of the apparent contradictions in the current literature relating to lamins and cancers. Even within clinical studies that have statistical power, some point to downregulation of lamins A/C being associated with poor patient outcome [21, 24, 25] whilst others point to up-regulated expression of lamins as a prelude to metastatic disease [23, 27]. To some extent this illustrates the complexity of cancer biology and the different types of cancer being investigated. For example in B-cell lymphoma it is likely that expression of lamins A/C promotes differentiation [21], and therefore it might be expected that epigenetic changes that prevent this expression are likely to maintain B-cells in a primitive and more aggressive state. Cohort studies are also inherently problematic, and this is illustrated in the two large cohort studies on colorectal cancer. The Willis study utilized a cohort collected throughout the Netherlands dating back to 1986 [32]. Patients within the study group are thus unlikely to have received adjuvant chemotherapy. In contrast, the cohort specimens used in the Belt study were collected between 1996 and 2005 from patients undergoing surgical resection in the same hospital, some of whom were treated with adjuvant chemotherapy [24]. In the latter study, the follow-up period was 57 months, whereas in the Willis study the follow-up period was 84 months. Finally, the Belt study was limited to patients with colonic cancer, whereas the Willis study included patients with rectal cancer. Thus it is hard to compare two studies, which at face value appear to ask exactly the same question and arrive at directly contradictory results. Finally, the use of single biomarkers can be problematic, and, indeed, the studies of Tilli and co-workers suggest that subtle changes in the balance of lamin expression are usually observed, at least in skin cancers. Therefore studies including the expression of multiple lamin subtypes are definitely preferable.

The idea that lamin expression patterns in cancers are linked to where they arise harks back to a notion of how cancers arise. In skin cancers as has been described the different cancers are classified according to the cells from which they arose (basal keratinocytes, squamous cells, or melanocytes). Whilst a detailed study on lamin expression in melanoma has not been performed it does appear that for BCC and SCC the patterns of lamin expression do map quite nicely to the patterns observed in the basal cells and squamous cells. This might also be true of colorectal

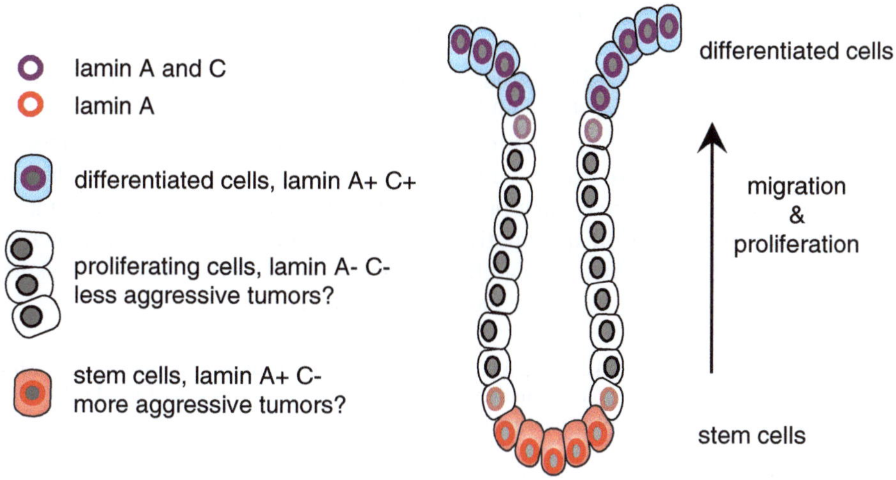

Fig. 1 Lamin A/C expression in the colonic crypt. The different cell layers/stages in the colonic crypt as well as those of several other types of epithelia have different levels of lamin A/C expression. Cells at the base of the crypt are the precursor stem cells and the least differentiated, yet they express high levels of lamin A and no lamin C. As the cells differentiate and migrate up the crypt, lamin A expression is greatly reduced, and then as they approach the top of the crypt both lamins A and C become more and more expressed until there are very high levels of both in the most differentiated cells at the top of the crypt

cancers. In the colon stem cells express lamin A but not lamin C. The transit amplifying cells do not express lamin A or lamin C, whilst the differentiated epithelial cells express both A-type lamins (Fig. 1—[23]). It is not absolutely clear where colonic adenocarcinomas arise, but there are two theories. The top-down theory supposes that the cancers arise in differentiated epithelial cells at the top of crypts and become more stem cell like. The bottom-up theory supposes that the cancers arise in the stem cell niche but somehow maintain a progenitor phenotype as they migrate upwards [1]. Either way it might be expected that the outcome of this switch to or maintenance of a progenitor phenotype is that lamin C expression is lost but lamin A expression is maintained. This again suggests that looking at individual lamin subtypes in the same study is important.

One area of convergence in the literature is the implication that expression of lamins in certain epithelial cancers might promote invasive behavior. In prostate cancer cell lines, over-expression of lamin A/C promotes cell survival, cell motility, and invasiveness, and this appears to be coordinated through the phosphaditylinositol-signaling pathway PI3K/AKT/PTEN [27]. In SW480 colorectal cancer cells expression of lamin A also promotes cell motility by remodeling the actin cytoskeleton as a result of up-regulation of actin-bundling proteins such as T- and F-plastin which leads to loss of expression of cell adhesion molecules such as E-cadherin, apparently promoting an epithelial transition [23, 33, 34]. Similar properties have been attributed to vimentin in breast cancer [35], which signals via the transcription factor slug and oncogenic Ras.

Thus future studies could be concentrated in two areas. Firstly, focusing on how lamins A/C as well as other intermediate filament proteins such as vimentin promote cell motility and invasiveness in several epithelial cancers might reveal convergent signaling pathways that could be influenced by an overall balance of the expression patterns of different lamin subtypes, as has been suggested in a recent study [36]. Secondly, a thorough understanding of how subtle changes in lamin expression influence pathways that promote either metastasis or survival of cancer cells or both, the design of quantitative methods to interrogate archival material from cancer patient cohorts, and careful consideration of factors that might skew data together could help to generate more certainty and consistency.

Acknowledgements I would like to thank all members of my laboratory, past and present, who have contributed to our current understanding of the function of nuclear lamins. I would also like to thank BBSRC, Wellcome Trust, and the Association of International Cancer Research for their financial support.

References

1. Weinberg RA (2007) The biology of cancer. Garland Science, New York, NY
2. Foster CR, Przyborski SA, Wilson RG, Hutchison CJ (2010) Lamins as cancer biomarkers. Biochem Soc Trans 38(Pt 1):297–300. doi:10.1042/BST0380297, BST0380297 [pii]
3. Zink D, Fischer AH, Nickerson JA (2004) Nuclear structure in cancer cells. Nat Rev Cancer 4(9):677–687. doi:10.1038/nrc1430, nrc1430 [pii]
4. Papanicolaou GN (1942) A new procedure for staining vaginal smears. Science 95(2469): 438–439. doi:10.1126/science.95.2469.438, 95/2469/438 [pii]
5. Hsu CY, Kurman RJ, Vang R, Wang TL, Baak J, Shih Ie M (2005) Nuclear size distinguishes low- from high-grade ovarian serous carcinoma and predicts outcome. Hum Pathol 36(10):1049–1054. doi:10.1016/j.humpath.2005.07.014, S0046-8177(05)00373-4 [pii]
6. Elston CW, Ellis IO (1991) Pathological prognostic factors in breast cancer. I. The value of histological grade in breast cancer: experience from a large study with long-term follow-up. Histopathology 19(5):403–410
7. Cui Y, Koop EA, van Diest PJ, Kandel RA, Rohan TE (2007) Nuclear morphometric features in benign breast tissue and risk of subsequent breast cancer. Breast Cancer Res Treat 104(1):103–107. doi:10.1007/s10549-006-9396-4
8. Penman S (1995) Rethinking cell structure. Proc Natl Acad Sci U S A 92(12):5251–5257
9. Pederson T (2000) Half a century of "the nuclear matrix". Mol Biol Cell 11(3):799–805
10. Nickerson JA (1998) Nuclear dreams: the malignant alteration of nuclear architecture. J Cell Biochem 70(2):172–180. doi:10.1002/(SICI)1097-4644(19980801)70:2<172::AID-JCB3>3.0.CO;2-L
11. Pederson T, Aebi U (2005) Nuclear actin extends, with no contraction in sight. Mol Biol Cell 16(11):5055–5060. doi:10.1091/mbc.E05-07-0656, E05-07-0656 [pii]
12. True LD, Jordan CD (2008) The cancer nuclear microenvironment: interface between light microscopic cytology and molecular phenotype. J Cell Biochem 104(6):1994–2003. doi:10.1002/jcb.21478
13. Thunnissen FB, Diegenbach PC, van Hattum AH, Tolboom J, van der Sluis DM, Schaafsma W, Houthoff HJ, Baak JP (1992) Further evaluation of quantitative nuclear image features for classification of lung carcinomas. Pathol Res Pract 188(4–5):531–535
14. Kaufmann SH, Mabry M, Jasti R, Shaper JH (1991) Differential expression of nuclear envelope lamins A and C in human lung cancer cell lines. Cancer Res 51(2):581–586

15. Broers JL, Raymond Y, Rot MK, Kuijpers H, Wagenaar SS, Ramaekers FC (1993) Nuclear A-type lamins are differentially expressed in human lung cancer subtypes. Am J Pathol 143(1):211–220
16. Maraldi NM, Lattanzi G, Capanni C, Columbaro M, Mattioli E, Sabatelli P, Squarzoni S, Manzoli FA (2006) Laminopathies: a chromatin affair. Adv Enzyme Regul 46:33–49. doi:10.1016/j.advenzreg.2006.01.001, S0065-2571(06)00002-1 [pii]
17. Rajagopalan H, Lengauer C (2004) Aneuploidy and cancer. Nature 432(7015):338–341. doi:10.1038/nature03099, nature03099 [pii]
18. Capo-chichi CD, Cai KQ, Simpkins F, Ganjei-Azar P, Godwin AK, Xu XX (2011) Nuclear envelope structural defects cause chromosomal numerical instability and aneuploidy in ovarian cancer. BMC Med 9:28. doi:10.1186/1741-7015-9-28, 1741-7015-9-28 [pii]
19. Capo-chichi CD, Cai KQ, Smedberg J, Ganjei-Azar P, Godwin AK, Xu XX (2011) Loss of A-type lamin expression compromises nuclear envelope integrity in breast cancer. Chin J Cancer 30(6):415–425, 1944-446X201106415 [pii]
20. Stadelmann B, Khandjian E, Hirt A, Luthy A, Weil R, Wagner HP (1990) Repression of nuclear lamin A and C gene expression in human acute lymphoblastic leukemia and non-Hodgkin's lymphoma cells. Leuk Res 14(9):815–821
21. Agrelo R, Setien F, Espada J, Artiga MJ, Rodriguez M, Perez-Rosado A, Sanchez-Aguilera A, Fraga MF, Piris MA, Esteller M (2005) Inactivation of the lamin A/C gene by CpG island promoter hypermethylation in hematologic malignancies, and its association with poor survival in nodal diffuse large B-cell lymphoma. J Clin Oncol 23(17):3940–3947. doi:10.1200/JCO.2005.11.650, JCO.2005.11.650 [pii]
22. Moss SF, Krivosheyev V, de Souza A, Chin K, Gaetz HP, Chaudhary N, Worman HJ, Holt PR (1999) Decreased and aberrant nuclear lamin expression in gastrointestinal tract neoplasms. Gut 45(5):723–729
23. Willis ND, Cox TR, Rahman-Casans SF, Smits K, Przyborski SA, van den Brandt P, van Engeland M, Weijenberg M, Wilson RG, de Bruine A, Hutchison CJ (2008) Lamin A/C is a risk biomarker in colorectal cancer. PLoS One 3(8):e2988. doi:10.1371/journal.pone.0002988
24. Belt EJ, Fijneman RJ, van den Berg EG, Bril H, Delis-van Diemen PM, Tijssen M, van Essen HF, de Lange-de Klerk ES, Belien JA, Stockmann HB, Meijer S, Meijer GA (2011) Loss of lamin A/C expression in stage II and III colon cancer is associated with disease recurrence. Eur J Cancer 47(12):1837–1845. doi:10.1016/j.ejca.2011.04.025, S0959-8049(11)00304-2 [pii]
25. Wu Z, Wu L, Weng D, Xu D, Geng J, Zhao F (2009) Reduced expression of lamin A/C correlates with poor histological differentiation and prognosis in primary gastric carcinoma. J Exp Clin Cancer Res 28:8. doi:10.1186/1756-9966-28-8, 1756-9966-28-8 [pii]
26. Skvortsov S, Schafer G, Stasyk T, Fuchsberger C, Bonn GK, Bartsch G, Klocker H, Huber LA (2011) Proteomics profiling of microdissected low- and high-grade prostate tumors identifies Lamin A as a discriminatory biomarker. J Proteome Res 10(1):259–268. doi:10.1021/pr100921j
27. Kong L, Schafer G, Bu H, Zhang Y, Klocker H (2012) Lamin A/C protein is overexpressed in tissue-invading prostate cancer and promotes prostate cancer cell growth, migration and invasion through the PI3K/AKT/PTEN pathway. Carcinogenesis 33(4):751–759. doi:10.1093/carcin/bgs022, bgs022 [pii]
28. Helfand BT, Wang Y, Pfleghaar K, Shimi T, Taimen P, Shumaker DK (2012) Chromosomal regions associated with prostate cancer risk localize to lamin B-deficient microdomains and exhibit reduced gene transcription. J Pathol 226(5):735–745. doi:10.1002/path.3033
29. Marks R (1995) An overview of skin cancers. Incidence and causation. Cancer 75(2 Suppl):607–612
30. Venables RS, McLean S, Luny D, Moteleb E, Morley S, Quinlan RA, Lane EB, Hutchison CJ (2001) Expression of individual lamins in basal cell carcinomas of the skin. Br J Cancer 84(4):512–519. doi:10.1054/bjoc.2000.1632, S000709200091632X [pii]
31. Tilli CM, Ramaekers FC, Broers JL, Hutchison CJ, Neumann HA (2003) Lamin expression in normal human skin, actinic keratosis, squamous cell carcinoma and basal cell carcinoma. Br J Dermatol 148(1):102–109, 5026 [pii]

32. van den Brandt PA, Goldbohm RA, van't Veer P, Volovics A, Hermus RJ, Sturmans F (1990) A large-scale prospective cohort study on diet and cancer in The Netherlands. J Clin Epidemiol 43(3):285–295, 0895-4356(90)90009-E [pii]
33. Foran E, McWilliam P, Kelleher D, Croke DT, Long A (2006) The leukocyte protein L-plastin induces proliferation, invasion and loss of E-cadherin expression in colon cancer cells. Int J Cancer 118(8):2098–2104. doi:10.1002/ijc.21593
34. Foster CR, Robson JL, Simon WJ, Twigg J, Cruikshank D, Wilson RG, Hutchison CJ (2011) The role of Lamin A in cytoskeleton organization in colorectal cancer cells: a proteomic investigation. Nucleus 2(5):434–443. doi:10.4161/nucl.2.5.17775, 17775 [pii]
35. Vuoriluoto K, Haugen H, Kiviluoto S, Mpindi JP, Nevo J, Gjerdrum C, Tiron C, Lorens JB, Ivaska J (2011) Vimentin regulates EMT induction by Slug and oncogenic H-Ras and migration by governing Axl expression in breast cancer. Oncogene 30(12):1436–1448. doi:10.1038/onc.2010.509, onc2010509 [pii]
36. Dreesen O, Chojnowski A, Ong PF, Zhao TY, Common JE, Lunny D, Lane EB, Lee SJ, Vardy LA, Stewart CL, Colman A (2013) Lamin B1 fluctuations have differential effects on cellular proliferation and senescence. J Cell Biol 200(5):605–617. doi:10.1083/jcb.201206121, jcb.201206121 [pii]

ERRATUM TO

The Non-random Repositioning of Whole Chromosome and Individual Gene Loci in Interphase Nuclei and Its Relevance in Disease, Infection, Aging, and Cancer

Joanna M. Bridger, Halime D. Arican-Gotkas, Helen A. Foster, Lauren S. Godwin, Amanda Harvey, Ian R. Kill, Matty Knight, Ishita S. Mehta, and Mai Hassan Ahmed

E.C. Schirmer and J.I. de las Heras (eds.), *Cancer Biology and the Nuclear Envelope*, Advances in Experimental Medicine and Biology 773, DOI 10.1007/978-1-4899-8032-8_1,

DOI 10.1007/978-1-4899-8032-8_28

The publisher regrets that in the table of contents and chapter 12 opening page of the print and online versions of this book, the name of the co-author Halime D. Arican-Goktas was misspelled as Halime D. Arican Gotkas.

The online version of the original chapter can be found at http://dx.doi.org/10.1007/978-1-4899-8032-8_12

E.C. Schirmer and J.I. de las Heras (eds.), *Cancer Biology and the Nuclear Envelope*, Advances in Experimental Medicine and Biology 773, DOI 10.1007/978-1-4899-8032-8_12,

Index

E.C. Schirmer and J.I. de las Heras (eds.), *Cancer Biology and the Nuclear Envelope*, Advances in Experimental Medicine and Biology 773, DOI 10.1007/978-1-4899-8032-8,

MIX
Papier aus verantwortungsvollen Quellen
Paper from responsible sources
FSC® C105338

If you have any concerns about our products,
you can contact us on
ProductSafety@springernature.com

In case Publisher is established outside the EU,
the EU authorized representative is:
Springer Nature Customer Service Center GmbH
Europaplatz 3, 69115 Heidelberg, Germany

Printed by Libri Plureos GmbH
in Hamburg, Germany